# Semiconductor Devices and Integrated Electronics

# Semiconductor Devices and Integrated Electronics

A.G. MILNES, D.Sc.
Carnegie-Mellon University
Pittsburgh, Pennsylvania

VNR VAN NOSTRAND REINHOLD COMPANY
NEW YORK    CINCINNATI    ATLANTA    DALLAS    SAN FRANCISCO
LONDON    TORONTO    MELBOURNE

Van Nostrand Reinhold Company Regional Offices:
New York    Cincinnati    Atlanta    Dallas    San Francisco

Van Nostrand Reinhold Company International Offices:
London    Toronto    Melbourne

Library of Congress Catalog Card Number: 79-11452
ISBN: 0-442-23660-3

Manufactured in the United States of America

Published by Van Nostrand Reinhold Company
135 West 50th Street, New York, N.Y. 10020

Published simultaneously in Canada by Van Nostrand Reinhold Ltd.

15 14 13 12 11 10 9 8 7 6 5 4 3 2

Library of Congress Cataloging in Publication Data

Milnes, Arthur George.
    Semiconductor devices and integrated electronics.

    Bibliography: p.
    Includes index.
    1. Semiconductors.    2. Integrated circuits.
I. Title.
TK7871.85,M5715        621.3815'2        79-11452
ISBN 0-442-23660-3

*To my students —*

*past, present and future*

# Preface

For some time there has been a need for a semiconductor device book that carries diode and transistor theory beyond an introductory level and yet has space to touch on a wider range of semiconductor device principles and applications. Such topics are covered in specialized monographs numbering many hundreds, but the voluminous nature of this literature limits access for students. This book is the outcome of attempts to develop a broad course on devices and integrated electronics for university students at about senior-year level. The educational prerequisites are an introductory course in semiconductor junction and transistor concepts, and a course on analog and digital circuits that has introduced the concepts of rectification, amplification, oscillators, modulation and logic and switching circuits. The book should also be of value to professional engineers and physicists because of both, the information included and the detailed guide to the literature given by the references. The aim has been to bring some measure of order into the subject area examined and to provide a basic structure from which teachers may develop themes that are of most interest to students and themselves.

Semiconductor devices and integrated circuits are reviewed and fundamental factors that control power levels, frequency, speed, size and cost are discussed. The text also briefly mentions how devices are used and presents circuits and comments on representative applications. Thus, the book seeks a balance between the extremes of device physics and circuit design.

Study questions and further reading references are offered at the end of each chapter. An instructor can maintain the integrated-electronics balance by his choice of questions and further reading matter. Where the needs of the curriculum require it, either device physics or circuit technology may be emphasized by appropriate selection of the assigned reading and study questions. Many of the questions are numerical and can be answered from treatment of the appropriate subject matter in the text; others require extensive use of a library. To avoid pressure on particular issues of journals it is recommended that classes be divided into groups to study five or six different questions. Such work may then form the basis for general class discussion each week.

The topics that are covered include junction *pn* diodes, Schottky barrier diodes, varactors, bipolar transistors, JFETs, MOSFETs, integrated circuit fundamentals and applications, charge-coupled devices, Impatt oscillators, Gunn oscillators, solar cells, light detecting devices, electron emission devices, light emitting devices, injection lasers, and semiconductor sensors. For a one-semester course the material covered may be limited to Chaps. 1-10. If two semesters are available, a suitable division is Chaps. 1-9 and 10-15.

No attempt has been made to cover topics such as thermoelectrics, cryogenics, bulk or surface wave acoustic devices, and gas or dye lasers. Neither has there been space for description of semiconductor processing, crystal growing, epitaxy methods and thin- and thick-film circuit fabrication technologies and display systems.

Some sections of the book have been reviewed for content and emphasis by authorities in various fields. Their comments have eliminated some errors and misunderstandings and I offer them my thanks. Carnegie-Mellon University provided a short leave of absence in 1975 in order that progress could be made and I am grateful for this, and to the University of California, Berkeley, for receiving me during this time. The library and editorial assistance of M. Shure and B. Smith, and the drafting assistance of E. Lipanovich, have contributed substantially to the outcome and I am greatly in their debt. I am also indebted to students for problems and comments.

A. G. MILNES
PITTSBURGH, PA

# List of Tables

# Curves and Tables Useful for Reference Purposes

Absorption coefficient for Ge, Si, GaAs, GaP, CdTe, Fig. 12.11.

Atmospheric transmission spectrum, Fig. 13.11.

Capacitance in abrupt junctions, Fig. 1.18.

Color of $SiO_2$ and $Si_3N_4$ films, Table 7.1.

Depletion layer thickness, Fig. 1.17, 1.18.

Diffusion coefficients in Si, Fig. 1.32.

Drift velocities in GaAs, Si and Ge, Fig. 11.16.

Electron affinity values, Table 4.2.

Energy bandgaps, Table 4.2.

Energy band-structure parameters of III-V ternary materials, Fig. 14.5.

Energy bandgap for quaternary compounds, Fig. 14.43, 14.44.

Fermi levels in Si, Ge and GaAs, Fig. 1.2.

Heterojunction semiconductor pairs, Table 4.2.

Ionization rates in Si, Fig. 1.13, 11.6.

Photoemission spectral response data, Fig. 13.22 and Table 13.2.

Resistivity of Si and GaAs, Fig. 1.18.

Schottky barrier heights on Si, Fig. 2.9, 2.14, Table 2.1.

$SiO_2$ growth on Si, Fig. 7.23.

Solar material resources, InP, GaAs, CdS, Fig. 12.29.

Solar spectra, Fig. 12.1.

Solid solubilities in Si, Fig. 1.31.

Voltage breakdown in junctions, Fig. 1.15-1.20.

Work function values, Fig. 2.9.

# Contents

# 14    Light Emitting Diodes and Injection Lasers    814

# 15    Semiconductor Sensors and Transducers    891

# Semiconductor Devices and Integrated Electronics

# 1

# Semiconductor Junctions and Diodes

## CONTENTS

A great deal of important and interesting advanced material is discussed in this book. Some introductory material is included in each chapter to set the stage but there is no space to consider in detail elementary concepts of semiconductors or circuits. Some familiarity is therefore assumed with analog and digital circuits, amplifiers, oscillators, modulators and gates; and with semiconductor concepts of bandgaps, mobilities, density of states, Fermi levels, doping, minority carrier diffusion and lifetime, the simple *pn* junction model and general ideas about bipolar and field-effect transistors. We begin, therefore, with a list of some standard equations of semiconductors. If the form and notation are not familiar some time should be spent with the introductory semiconductor books that are included in the reference list at the end of the volume.

## 1.1 INTRODUCTORY SEMICONDUCTOR EQUATIONS AND CONCEPTS

Three important semiconductors are silicon (Si), germanium (Ge), and gallium arsenide (GaAs).

The bandgaps of Si, Ge and GaAs are 1.1, 0.66 and 1.43 eV at 300°K and their respective electron mobilities are 1350, 3600 and about 5000 $cm^2V^{-1}sec^{-1}$. If perfectly pure they would have intrinsic resistivities of about $2.5 \times 10^3$, 50 and $10^8$ ohm-cm, respectively, corresponding to intrinsic carrier densities of $1.5 \times 10^{10}$, $2.5 \times 10^{13}$ and $1.7 \times 10^6$ $cm^{-3}$ at 300°K. Semiconductors of Si and Ge may be doped $n$ type by group V impurities such as P, As and Sb, and $p$ type by group III impurities such as B, Al, Ga and In. Gallium arsenide is doped $n$ type by group VI impurities such as S, Se and Te, and $p$ type by group II impurities such as Zn and Cd. Group IV impurities such as Si, Ge, and Sn have doping effects in GaAs that depend on the lattice site occupied by the impurity. For instance, Ge in liquid-phase epitaxially grown material tends to be on the As site and be a $p$ type dopant whereas Sn is on the Ga site and is an $n$ type dopant.

Nowadays, Si has almost completely replaced Ge for diodes and transistors because the larger energy gap allows higher temperature operation, the raw material is lower in cost, the oxide is favorable to masking operations and the dopant properties in Si are good (low segregation coefficients, good solubilities, shallow energy levels and easily controlled diffusion conditions).

An important semiconductor is GaAs because the electron mobility and saturation drift velocity is high so it is suited to high frequency operation for microwave transistors and diodes. Also, GaAs has a band structure that allows transferred-electron (Gunn) oscillations (see Chap. 11) and good solar cell performance (see Chap. 12) and efficient light emission (see Chap. 14). On the other hand, GaAs is much more expensive than Si because of higher raw material costs and extra difficulties of preparation.

Tables of the physical properties of Si, Ge and GaAs and curves of resistivity, mobility and other properties such as diffusion coefficients, energy levels and solubilities for impurities are found in various books. Many other semiconductors in the III-V group and the II-VI group are also of great value in special applications. Some data for these are given in Chaps. 12, 13 and 14.

Now let us review briefly some of the simple equations of semiconductor bulk and transport theory.

- Thermal energy $= kT/q = 0.026$ eV at 300°K where $q = 1.6 \times 10^{-19}$ coulombs per electron.
- Photon energy of light $E = hc/\lambda = 1.24 /\lambda$ eV. Wavelength range of visible light is about 0.40-0.72 μm corresponding to an energy range 3.0-1.7 eV.
- The resistivity, at low field strengths, is given by

$$\rho = 1/(qn\mu_n + qp\mu_p) \tag{1.1}$$

and

$$J = q(n\mu_n + p\mu_p)\,\mathscr{E}$$ (1.2)

- The intrinsic concentration in a semiconductor varies with temperature as

$$n_i^2 = AT^3\,e^{-E_g/kT}$$ (1.3)

- Mobility varies with temperature as $T^{-m}$ where $m$ is 2.5 for electrons and 2.7 for holes in Si. The mobility concept begins to be modified at drift velocities in excess of $10^6$ cm sec$^{-1}$ ($\mathscr{E} > 10^3$ V cm$^{-1}$).
- The concepts of doping and Fermi levels lead to the following expressions

$$n = N_c\,e^{-(E_c - E_F)/kT}$$ (1.4)

$$p = N_v\,e^{-(E_F - E_v)/kT}$$ (1.5)

or

$$n = n_i\,e^{(E_F - E_i)/kT}$$ (1.6)

$$p = n_i\,e^{(E_i - E_F)/kT}$$ (1.7)

where

$$N_{c,v} = 2\left(\frac{2\mu\,m^*kT}{h^2}\right)^{3/2}$$ (1.8)

For Si at 300°K, $N_c = 2.8 \times 10^{19}$ cm$^{-3}$ and $N_v = 1.04 \times 10^{19}$ cm$^{-3}$ (these are the effective density of energy states imagined to be concentrated at the band edges $E_c$ and $E_v$). For GaAs the values are $N_c = 4.7 \times 10^{17}$ cm$^{-3}$ and $N_v = 7.0 \times 10^{18}$ cm$^{-3}$.

If the Fermi level approaches within a few $kT$ of the band edge because the doping is heavy, these equations may not be used and account must be taken of the variation of the energy state density with distance from the band edge— typically this variation is proportional to $(E - E_c)^{1/2}$.

- The product of Eqs. (1.4) and (1.5) gives

$$np = N_cN_v\,e^{-(E_c - E_v)/kT}$$ (1.9)

or

$$np = n_i^2$$ (1.10)

Thus if $n_i^2$ for Si at 300°K is $2.25 \times 10^{20}$ cm$^{-3}$ and if the doping is $10^{15}$ atoms cm$^{-3}$ of As the value of $n$ is $10^{15}$ cm$^{-3}$ and the density $p$ is $2.25 \times 10^5$ cm$^{-3}$.

The product $np$ is a constant only in equilibrium. This condition is not valid if carriers are being injected into the semiconductor by light or from a junction.

- Minority carriers in a condition of disturbed equilibrium diffuse by a random walk process and so in density gradients the net flow is given by

$$\text{Flow cm}^{-2}\sec^{-1} = -D_{n,p}\frac{d(n,p)}{dx} \tag{1.11}$$

- The Einstein relationship shows that mobility and the diffusion coefficient are related by

$$D = \mu\frac{kT}{q} \tag{1.12}$$

Thus, if $\mu_n$ is 1350 cm$^2$ V$^{-1}$ sec$^{-1}$ for Si the value of $D_n$ is 35 cm$^2$ sec$^{-1}$ at 300°K.

- The transport equations for a bulk semiconductor containing both density gradients and electric fields are

$$J_n = q\mu_n n \mathcal{E} + qD_n\frac{dn}{dx} \tag{1.13}$$

and

$$J_p = q\mu_p p \mathcal{E} - qD_p\frac{dp}{dx} \tag{1.14}$$

- If the minority carrier average lifetime in a semiconductor is $\tau$, the effect of a pulse of light is to cause a conductivity increase that returns to the equilibrium conductivity with an exponential time constant expression

$$\Delta\sigma = \Delta\sigma_0 e^{-t/\tau} \tag{1.15}$$

- Consideration of the net balance of recombination and generation of minority carriers in a semiconductor leads to expressions of the form

$$\frac{dp}{dt} = -\frac{p - p_0}{\tau_p} \tag{1.16}$$

where $p_0$ is the equilibrium hole density.

- In a spatial element of an n-type semiconductor with a hole density gradient present and recombination present the expression becomes

$$\frac{\partial p}{\partial t} = -\frac{p - p_0}{\tau_p} + D_p\frac{d^2 p}{dx^2} \tag{1.17}$$

- In an $n$ type semiconductor with an injected carrier density of $p_{x=0}$ sustained at one face and in which these injected carriers diffuse away from this face into the bulk, the expression becomes

$$\frac{d^2p}{dx^2} = \frac{p - p_0}{D_p \tau_p} \tag{1.18}$$

where

$$p = p_{x=0} \, e^{-x/L_p} \tag{1.19}$$

where $L_p$ is known as the diffusion length and equals $(D_p \tau_p)^{1/2}$

• A simplified lifetime expression for recombination through a single recombination center $N_{rc}$ is

$$\tau = \left(\frac{n_0 + n_i}{n_0 + p_0}\right) \tau_{p0} + \left(\frac{p_0 + p_i}{n_0 + p_0}\right) \tau_{n0} \tag{1.20}$$

where

$$\tau_{p0} = (\sigma_p v_p N_{rc})^{-1}$$

and

$$\tau_{n0} = (\sigma_n v_n N_{rc})^{-1}$$

where $\sigma_p$ and $\sigma_n$ are capture cross sections.

• The expressions for the diffusion of impurities in a semiconductor are approximated by one of two forms. For a constant surface concentration $C_s$, as in a predeposition cycle, the form is a complementary-error-function distribution

$$C(x,t) = C_s \, \text{erfc} \, (x/2 \sqrt{Dt}) \tag{1.21}$$

The total number of dopant atoms $\text{cm}^{-2}$ is then

$$N = 2C_s \sqrt{Dt/\pi} \tag{1.22}$$

For a drive-in diffusion the fixed number of dopant atoms is redistributed according to a Gaussian expression

$$C(x,t) = \frac{N}{\sqrt{\pi Dt}} \exp{(-x^2/4Dt)} \tag{1.23}$$

• The built-in voltage $V_D$ between a $p$ and $n$ region in a semiconductor junction is given by

$$V_D = \frac{kT}{q} \ln{(p_{p0}/p_{n0})} \tag{1.24}$$

where $p_{po} = N_A$ and $p_{no} = n_i^2/N_D$ (for moderate doping densities).

Hence
$$V_D = \frac{kT}{q} \ln \left( \frac{N_A N_D}{n_i^2} \right)$$
(1.25)

- The application of a forward bias voltage $V_{ext}$ to a $p$-$n$ junction raises the hole density near the junction in the $n$ region to a value $p_{n_{x=0}}{}^+$ given by the Boltzmann relationship

$$\frac{p_{n_{x=0}}{}^+}{p_{no}} = \exp\left(q V_{ext}/kT\right)$$
(1.26)

Hence an external applied voltage of, say, 0.26 V raises the hole density by a factor of $e^{10}$ (namely $2 \times 10^4$). Conversely a reverse voltage of 0.26 V lowers the hole density by a factor of $e^{10}$. The corresponding expression for the electron density is

$$\frac{n_{p_{x=0}}{}^-}{n_{po}} = \exp\left(q V_{ext}/kT\right)$$
(1.27)

- If the bulk region of a semiconductor contains a gradient of donor doping impurities there is a graded built-in bulk field given by

$$\mathcal{E} = \frac{kT}{q} \frac{1}{N_D(x)} \frac{d N_D(x)}{dx}$$
(1.28)

Hence for an exponential variation of doping with a characteristic length of, say, 1 $\mu$m, $N = N_o \exp(x/10^{-4})$, the built-in field is $\frac{kT}{q} \times 10^4$ V cm$^{-1}$. Such drift fields may aid in the transport of minority carriers across the base regions of bipolar transistors.

- Poisson's equation

$$\frac{d^2 V}{dx^2} = \frac{-\rho}{\epsilon}$$
(1.29)

where $\rho$ is the charge density and $\epsilon$ the dielectric constant, may be used to determine the depletion region width and maximum field strength and capacitance of a junction under a reverse bias $|V_{rev}|$.

For a junction with an abrupt doping step the depletion width for the $n$ side of the junction is

$$x_n = \left[ \frac{2}{q} \frac{N_A \epsilon}{N_D(N_D + N_A)} (V_D + |V_{rev}|) \right]^{1/2} \tag{1.30}$$

and for the $p$ side

$$x_p = \left[ \frac{2}{q} \frac{N_D \epsilon}{N_A(N_D + N_A)} (V_D + |V_{rev}|) \right]^{1/2} \tag{1.31}$$

The maximum field strength occurs at the junction and is given by

$$|\mathcal{E}_m| = \frac{q N_D x_n}{\epsilon} = \frac{q N_A x_p}{\epsilon} \tag{1.32}$$

The relative dielectric constants for Si, Ge and GaAs are about 11.5, 16 and 12, respectively, and the dielectric constant of free space is $8.86 \times 10^{-14}$ F/cm. The maximum field strength for avalanche in Si is $2\text{-}4 \times 10^5$ V/cm for dopings in the $10^{14}\text{-}10^{16}$ cm$^{-3}$ range.

The voltages sustained across the depleted $n,p$ regions in this abrupt junction are about

$$V_{n,p} = \frac{|\mathcal{E}_m| x_{n,p}}{2} \tag{1.33}$$

Curves of voltage ratings of abrupt and graded junctions are given later in the chapter.
- The small signal capacitance of an abrupt junction is given by

$$C = A\epsilon/(x_n + x_p) \tag{1.34}$$

where $A$ is the junction area.

If $N_A \gg N_D$ then $x_p \ll x_n$ and from Eq. (1.30)

$$\frac{C}{A} = \frac{\epsilon}{x_n} = \left( \frac{q\epsilon N_D}{V_D + |V_{rev}|} \right)^{1/2} \tag{1.35}$$

A Si diode doped at $10^{14}$ cm$^{-3}$ with a total potential of 10 V applied has a depletion width of about 10 $\mu$m and a capacitance of about $10^3$ pF/cm$^2$.
- If there is departure from space-charge neutrality in a semiconductor the return to equilibrium is with a dielectric relaxation time

$$\tau_d = \frac{\epsilon}{\sigma} \tag{1.36}$$

Hence for 100 ohm-cm Si ($\sigma = 10^{-2}$) the dielectric relaxation time is $10^{-10}$ sec.

- For an abrupt $pn$ diode, the Shockley model for the relationship of current density to voltage is

$$J = J_0 \left( \exp\left( qV/kT \right) - 1 \right) \tag{1.37}$$

where (neglecting generation in the depletion region) $J_o$ is the current of minority carriers generated within one diffusion length on either side of the depletion region. Hence

$$J_0 = qL_p g_{(N_D)} + q\, L_n g_{(N_A)}$$

$$= qL_p \left( \frac{p_{no}}{\tau_p} \right) + qL_n \left( \frac{n_{po}}{\tau_n} \right)$$

$$= q\, \frac{D_p\, p_{no}}{L_p} + q\, \frac{D_n\, n_{po}}{L_n}$$

$$= q\, \frac{D_p\, n_i^2}{L_p N_D} + q\, \frac{D_n\, n_i^2}{L_n N_A} \tag{1.38}$$

For a narrow-base $p^+n$ diode of width $w$ ($\ll L_p$) the expression for $J_o$ becomes $q\, D_p p_{no}/w$.

These preceding equations are well known to those who have taken an introductory semiconductor course and represent our starting point.

## 1.2 PN JUNCTION FORWARD CHARACTERISTICS

Consider a $pn$ junction structure with an abrupt doping step as shown in Fig. 1.1. The Fermi levels equalize across the junction and across the metal contacts when no external voltage is applied and the current flow is zero, as in Fig. 1.1(c). With a forward voltage applied, as shown in Fig. 1.1(d), the Fermi levels separate and electrons are injected into the $p$ region from the tail of the electron energy distribution in the $n$ region. Similarly holes flow from the $p$ region into the $n$ region as minority carriers. If the $p$ region is more heavily doped than the $n$ region the main part of the current flow will be by hole injection and a lesser amount is carried by the electron injection. The nonequilibrium concentration of injected minority carriers may be represented by quasi-Fermi levels (imrefs) as indicated in Fig. 1.1(d). If reverse voltage is applied the depletion region widens and the reverse current is small and determined by thermal generation processes.

In Fig. 1.1(c) the contacts are represented by metal $-p^+$ and metal $-n^+$ interfaces at which thin energy spikes occur. In practical devices these energy barriers are only a few tens of angstroms thick and therefore tunneling of majority carriers through them can take place for a voltage drop of only a few millivolts at the desired current levels. The contacts are, therefore, not truly ohmic but in a well designed device cause no major voltage loss.

Fermi levels vary with respect to the conduction or valence band edges as in Eqs. (1.4) and (1.5) with doping and with temperature. Figure 1.2 shows this variation for Si, GaAs and Ge. The built-in electrostatic voltages across junctions, $V_D$ in Fig. 1.1(c), can be estimated from such diagrams.

In a few specialized applications $pn$ junctions are formed between two different semiconductors, such as $pAl_xGa_{1-x}As$ on $nGaAs$, and these may have energy steps at the junction interface that perform useful functions.[86] However, consideration of such heterojunctions is reserved to later chapters.

A comparison of the forward characteristics of Ge, Si and GaAs diodes is shown in Fig. 1.3. The voltage required for substantial current flow is 0.3 V for Ge but 0.6 V for Si and 1.2 V for GaAs. The current intercepts at zero voltage show the GaAs diode to have a much lower $J_0$ current than for Si or Ge. A reverse current of, say, 10 pA at 10 V reverse voltage and 25°C for a Si diode becomes more than $1\mu A$ at 10 V at 200°C. For a GaAs diode the corresponding current would be about 1 pA at 25°C and rise to about 0.1 $\mu A$ at 200°C and still be less than 1 $\mu A$ at 300°C. So the wider bandgap material allows a wider temperature range of use for the junction, although at the expense of an increased forward voltage. The voltage required for a given forward current in a Si junction tends to decrease at about 2.5 mV/°C around room temperature (see Fig. 1.4), and the reverse saturation current about doubles every 10°C rise in temperature.

### 1.2.1 The Exponential Ideality Factor, $n$

The simple Shockley diode model that predicts

$$J = J_0 \left( \exp \left( qV/kT \right) - 1 \right) \tag{1.39}$$

does not adequately represent the behavior of the Si and GaAs diodes in Fig. 1.3. A better representation is

$$J = J_0 \left( \exp \left( qV/nkT \right) - 1 \right) \tag{1.40}$$

where the factor $n$, called the ideality factor of the diode, varies according to the current and voltage range being considered. In the limit of high injection $n$ may be expected to approach two. If the injected carrier density tends to exceed the majority carrier density the diffusion model must be modified by a drift field which allows pile-up of the majority carrier density to match the

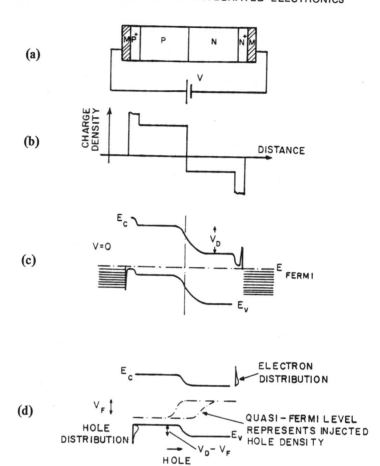

(a)

(b)

(c)

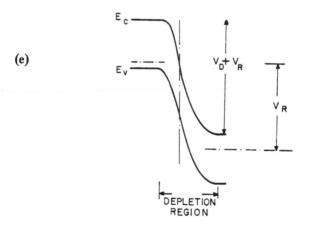

(d)

(e)

high injection level. The current-voltage expression for a $p^+n$ junction then becomes[98]

$$J_p = \frac{2qD_p n_i}{L_p} \left( \exp\left(qV/2kT\right) - \frac{p_{n_0}}{n_i} \right) \tag{1.41}$$

Figure 1.5 shows an example of a Si power rectifier which has the ideality factor $n$ changing from 1 to 2 for a substantial region until series resistance effects cause nonlinearity at high currents.

However, some diodes show regions of $n = 2$ at low current densities as suggested by the GaAs curve of Fig. 1.3. This is explained to a reasonable extent if the current at low injection levels is dominated by the effect of recombination at Shockley-Read-Hall traps uniformly distributed in the depletion layer. Analysis shows the results depend on a term

$$\epsilon_T = \left( \frac{E_T - E_i}{kT} - \frac{1}{2}\ln\left(\frac{\tau_n}{\tau_p}\right) \right) \tag{1.42}$$

where $E_T - E_i$ is the trap level with respect to the center of the bandgap and $\tau_n$ and $\tau_p$ are the minority lifetimes.[112]

The type of characteristic predicted is shown in Fig. 1.6(a) where the factor $n$ is seen to vary between 1 and 2 depending on the current and voltage level. Figure 1.6(b) shows the dependence of $n$ on the trap term $\epsilon_T$ and the voltage. If the trap is near the center of the bandgap $\epsilon_T$ tends to have a value near zero and the theoretical $n$ value is near 1.9 over a substantial voltage range which is the basis for the generally accepted view that Sah-Noyce-Shockley (SNS) theory predicts $n = 2$.

Although the SNS explanation is acceptable for some junctions, other junctions exhibit intermediate values of ideality factors ($n = 1.30$ or $1.60$) over substantial voltage ranges. Channel and surface recombination effects are sometimes considered to be contributing to the observed effect but bulk effects may also be involved. Two characteristics for diodes with long regions of intermediate values of $n$ are given in Fig. 1.7(a). One effect to be considered is that the Shockley-Read-Hall traps may be nonuniformly distributed in the depletion region, for instance, more heavily present in the $p$ side than in the $n$ side of the

Fig. 1.1. Energy band diagrams for a $PN$ junction:
(a)  $MP^+ PNN^+M$ structure,
(b)  Charge density assuming the $P$ region is about twice as heavily doped as the $N$ region,
(c)  Energy diagram with no external applied voltage; $V_D$ is the diode internal electrostatic voltage;
(d)  Junction energy diagram with an applied forward voltage $V_F$, and
(e)  Energy diagram with an applied reverse voltage $V_R$.

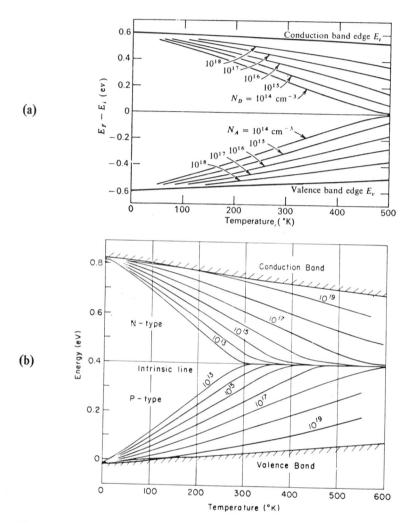

Fig. 1.2. Variation of Fermi level with doping and temperature in Si, Ge and GaAs. (After A.S. Grove, *Physics and Technology of Semiconductor Devices*, John Wiley & Sons, Inc., NY, 1967), (After A.K. Jonscher, *Principles of Semiconductor Device Operation*, John Wiley & Sons, Inc., NY, 1960.)

junction. With this kind of model, reasonable agreement can be obtained with experimental results as shown in Fig. 1.7(b).

Further examinations of this kind are to be found in the literature and are worth study to gain understanding of the role of recombination centers in junctions.[4] For characterization of diode junctions there is no single best method or expression, apart, perhaps, from the representation of complete *I-V* curves for several temperatures with effective *n* values for various sections roughly indicated by the slopes of dotted lines.

**(c)**

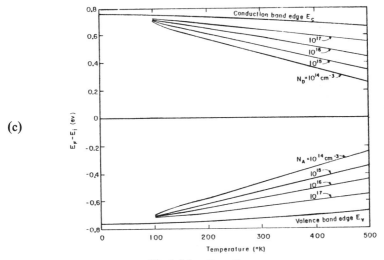

Fig. 1.2 (continued)

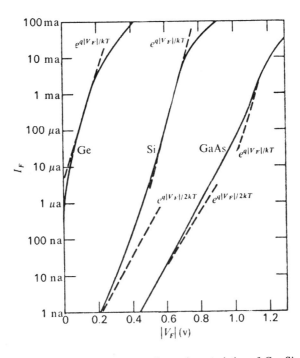

Fig. 1.3. Comparison of the forward current-voltage characteristics of Ge, Si and GaAs diodes at 25°C. Dashed lines indicate slopes $e^{q|V_F|/kT}$ and $e^{q|V_F|/2kT}$ dependences. (After A.S. Grove, *Physics and Technology of Semiconductor Devices*, John Wiley & Sons, Inc., NY, 1967.)

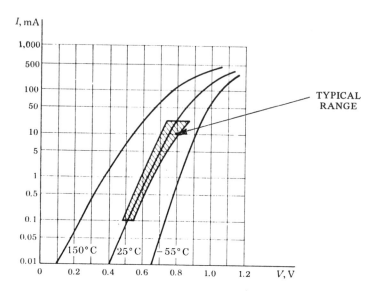

Fig. 1.4. $I/V$ for a typical Si diffused diode. The shaded rectangle indicates the tolerance range within which most of the characteristics lie at 25°C. (After Millman and Halkias,[55] courtesy General Electric Co.)

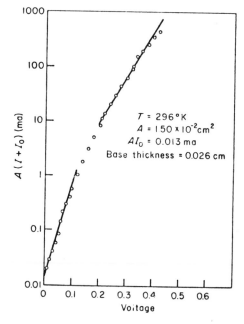

Fig. 1.5. Current-voltage relation for an actual alloyed rectifier showing separate regions of slope $q/kT$ and $q/2kT$ corresponding to low current and high current level regions. (After McKelvey,[75] courtesy H.F. John.)

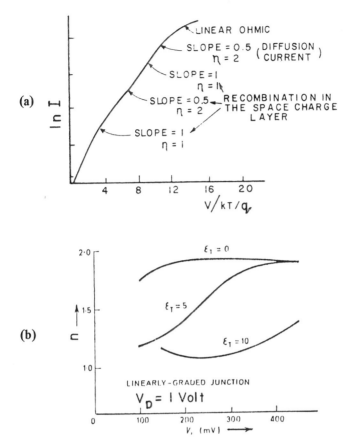

Fig. 1.6. Curves illustrating the role of recombination centers on the slope of log $I/V$ characteristics for *pn* junctions:
(a)  Characteristic calculated for a recombination-center near the middle of the bandgap, (After Sah, Noyce, and Shockley. Reprinted with permission from *Proc. IRE*, **45**, 9, 1957, p. 1235) and
(b)  Variation of slope factor with voltage calculated for recombination centers of various depths. (After Buckingham and Faulkner, from *Radio and Electronic Engineer*, **38**, 1969, p. 34.)

Some diodes can be found that have forward characteristics adequately represented by an expression of the form

$$J = J_{01} \exp \left[ q(V\text{-}IR_s)/n_1 kT - 1 \right]$$

$$+ J_{02} \exp \left[ q(V\text{-}IR_s)/n_2 kT - 1 \right] + \frac{V\text{-}IR_s}{R_{sh}} \qquad (1.43)$$

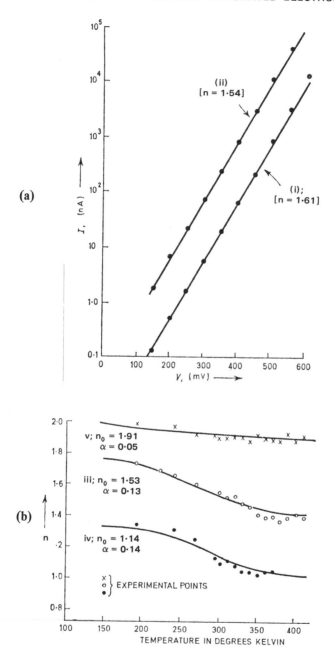

Fig. 1.7. Variation of the slope factor $\dfrac{q}{kT}\dfrac{dV}{d(\ln I)}$ for diode exponential characteristic:
(a) Diodes showing long regions of constant slope in the range 1.6-1.54. (i) step recovery diode, Hewlett Packard, (ii) diode 1S922 (Texas Instruments), and

where $n_1$ is taken as unity and $n_2$ usually has a value near 2. The symbol $R_s$ is the resistance in series with the junction and $R_{sh}$ is a shunt resistance term perhaps related to channel or surface resistance. The current $J_{01}$ in this model represents the bulk recombination-generation effects and $J_{02}$ the depletion region recombination-generation effects. However, such an approach often has serious limitations of fit to experimental data and of tracking over a temperature range.

### 1.2.2 Some Factors Influencing the Forward Characteristics

In Si power rectifier diodes the reverse voltage rating must be high for useful applications, hence the doping in the base region must be low. For instance, a rating of 1000 V for an $n^+p$ step-junction requires a doping level of about $10^{14}$ cm$^{-3}$ on the lightly doped side corresponding to a resistivity of about 100 ohm-cm and the depletion region width is 100 $\mu$m at full reverse voltage. The diode wafer may be about 200 $\mu$m (8 mils) thick for ease of handling. At forward current densities of several hundreds of amperes cm$^{-2}$, the bulk forward resistance of the diode would cause significant voltage drop if it were not for conductivity modulation effects associated with the injected carriers. The extent of the injected carrier effect depends on the effective carrier lifetime and this itself may be a function of the injection level and temperature.

Another factor that is involved is that the low-high junction needed to make majority carrier contact to the metal at the base contact in say an $n^+pp^+$ structure acts as a minority carrier reflecting boundary preventing injected electrons from freely reaching the contact. This barrier is shown in Fig. 1.8(b) and determines the junction minority carrier leakage current. This leakage current may be related to an effective surface recombination velocity $S_{pp^+}$ of the high-low junction. For uniform doping in the $p^+$ region and neglecting recombination within the high-low junction space charge region, the effective surface recombination velocity is

$$S_{pp^+} = \frac{D_{np^+}}{L_{np^+}} \left[ \coth \left( \frac{W_{p^+}}{L_{np^+}} \right) \right] \frac{n_{p^+}}{n_p} , \qquad (1.44)$$

where $D_{np^+}$ and $L_{np^+}$ are the electron diffusion constant and diffusion length in the $p^+$ region.

---

(b) Variation of the slope factor with temperature (iii) diode 1S922, (iv) Mullard diode OA202, (v) Hughes diode HS9008. The numbers given on the graph are the parameters used in the calculations by Buckingham and Faulkner (1969) for a model in which the recombination centers are nonuniformly distributed across the junction. (After Buckingham and Faulkner from *Radio and Electronic Engineer*, 38, 1969, p. 35.)

The ratio of minority carrier densities across the high-low junction is related to the junction voltage $V_{hl}$ by

$$n_{p^+}/n_p = \exp(-q\,V_{hl}/kT) \tag{1.45}$$

where, in equilibrium, $n_p{}^+/n_p \sim N_p/N_p{}^+$. However, at high injection levels

$$n_{p^+}/n_p = \left(1 + \frac{n_p}{N_p}\right)\frac{N_p}{N_{p^+}} \tag{1.46}$$

and the interface surface recombination velocity becomes

$$S_{pp^+} = \frac{D_{np^+}N_p}{L_{np^+}N_{p^+}}\left[\coth\left(\frac{W_{p^+}}{L_{np^+}}\right)\right]\left(1 + \frac{n_p}{N_p}\right) \tag{1.47}$$

At low injection this is a constant but the value increases linearly with $n_p$ at high injection. This causes the minority carrier current into the high-low junction to increase linearly with carrier density $n_p$ at low injection and to increase as $n_p{}^2$ at high injection. The potential around a $pp^+$ junction for various electron injection levels is shown in Fig. 1.8(c). The high-low junction electron leakage current given by

$$J_{nhl} = qn_pS_{pp^+} \tag{1.48}$$

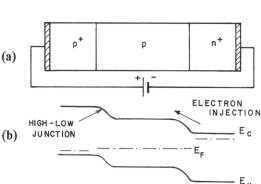

Fig. 1.8.  p$^+$p High-low junction behavior:
(a)  Junction structure,
(b)  Energy diagram under forward bias conditions,
(c)  Electrostatic potential around high-low junction at various injection levels (2000 ohm-cm $p$-region), and
(d)  High-low junction leakage current as a function of injection level on $p$-side of junction. (After Hauser and Dunbar.[46] Reprinted with permission from *Solid-State Electronics*, Pergamon Press Ltd.)

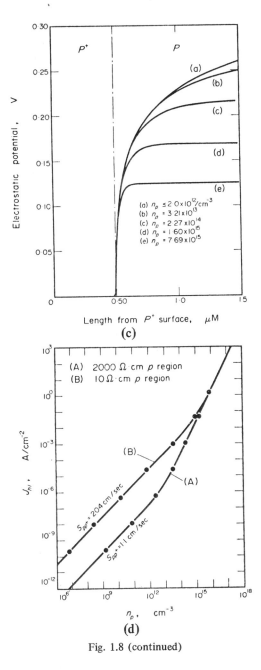

Fig. 1.8 (continued)

is shown in Fig. 1.8(d). The initial linear dependence on $n_p$ is seen to be followed by $n_p^2$ dependence at high electron injection levels.

In analyses of rectifier diodes it is, however, more usual to consider the complete structure rather than concentrate attention on one part of it such as the $pp^+$ region. Models are capable of showing that in the absence of recombination current in the heavily doped end regions, there exists an optimum base lifetime giving a minimum forward voltage. This minimum occurs because for increasing lifetime, the increase in junction voltage due to carrier buildup at the junction edge eventually overtakes the reduction in base voltage due to conductivity modulation. On the other hand, when the recombination currents in the end regions predominate over those in the base, their presence tends to inhibit carrier buildup and, for sufficiently large values of base lifetime,

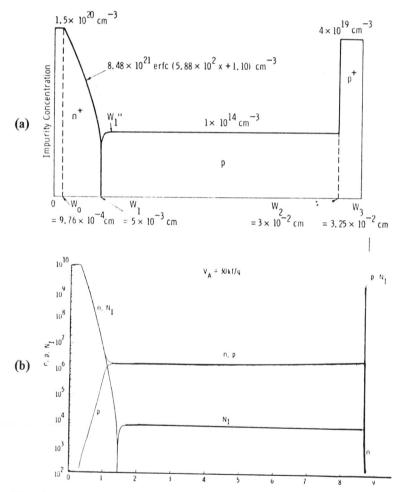

Fig. 1.9. Analytical solution of $n^+pp^+$ diode performance with $V_A = 30\,kT/q$:
(a)   Assumed doping distribution (one-dimensional model),
(b)   Calculated carrier concentrations,

the forward voltage falls to a limiting value. Little is to be gained by further increase in lifetime beyond a certain value which depends on both the properties of the base and those of the end regions. In a typical structure this lifetime may be a few tens of microseconds and correspond to a voltage drop of 1.1 V. The turn-off transient time of a rectifier is of course increased if the minority carrier lifetime is long and a design trade-off is sometimes necessary between forward voltage and speed.

A typical analytical treatment for a diffused $n^+pp^+$ structure is shown in Fig. 1.9. Quasineutrality extends across the base and reaches slightly into the diffused region at this high injection level. At the boundary between the effective base region and the diffused region, the hole current varies more or less directly as the carrier concentration, in contrast to the situation in a step-junction rectifier where the minority carrier current in the heavily doped regions varies as the square of the carrier concentration at the base edge. As a consequence of this linear relationship, the injection efficiency of the diffused junction is roughly independent of the injection or current levels. It may be

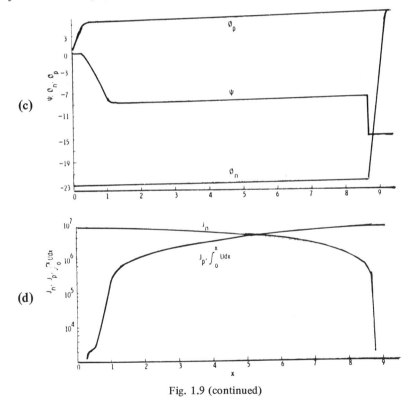

Fig. 1.9 (continued)

(c)  Electrostatic and quasi-Fermi potentials, and
(d)  Electron current and hole recombination rate integral. (After Choo. Reprinted with permission from *IEEE Trans. Electron Devices*, **ED-19**, 8, 1972, pp. 955, 959.)

high or low depending on whether the concentration of recombination centers in the diffused region is small or large compared to that in the base region. This is unlike the situation in a step junction where the injection efficiency is initially high and then degrades as the current increases.

### 1.2.3 PN Junction Capacitance

From Poisson's equation the reverse bias space-charge capacitance per unit area of a step junction is given by

$$C_S = \left( \frac{q\epsilon}{2} \left( \frac{N_A N_D}{N_A + N_D} \right) \left( \frac{1}{V_D + |V_{rev}|} \right) \right)^{1/2} \tag{1.49}$$

However, for a forward bias voltage condition the expression tends to become infinite as $V_D - V_F$ approaches zero, and the depletion edge approximation is invalidated by the injected charge. There is then a need to represent the injected carriers by a diffusion capacitance component.

For the step junction shown in Fig. 1.10(a) at zero forward voltage the net charge distribution and field distribution have spike-like forms [see Fig. 1.10(b)]

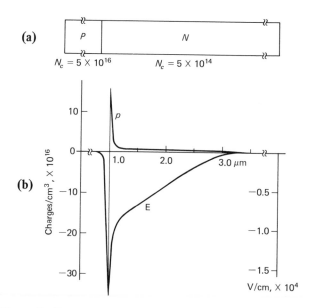

Fig. 1.10.  Charge distributions in a junction:
(a)  Dopings assumed,
(b)  Charge and voltage spikes caused by charge spillover at zero bias voltage, and
(c), (d), (e) Charge distributions with forward bias voltage applied ( —— $V$=0.5 V, - - - - $V$=0.45 V). (After Heald et al.[47] Reprinted with permission from *Solid-State Electronics*, Pergamon Press, Ltd.)

caused by charge spillover effects. When forward voltage is applied to the junction the carrier distributions become as shown in Fig. 1.10(c), (d) and the net charge density is that given in Fig. 1.10(e). The value of total capacitance may be computed from

$$C_1 = K_s \epsilon_0 \frac{\partial E(x_e, V)}{\partial V} + q \int_{x_e}^{L} \frac{\partial p(x, V)}{\partial V} dx + q \int_{0}^{x_e} \frac{\partial n(x, V)}{\partial V} dx \qquad (1.50)$$

The first term represents the space-charge limited capacitance and dominates at small forward bias (or at reverse bias). The second term represents the diffusion capacitance which becomes important as soon as a significant forward current flows.

Another approach to determining capacitance of a junction is via the stored energy and its variation with voltage across the junction. In this approach attention must be given to the electrochemical as well as electrostatically stored energy of the diode. The result obtained is

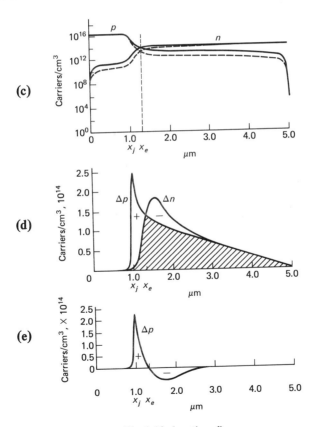

Fig. 1.10 (continued)

$$C_1 = \frac{1}{V}\frac{d}{dV}\left(\frac{K_s\epsilon_0}{2}\int_0^L E^2 dx\right) + \frac{1}{V}\frac{d}{dV}\left[kT\int_0^L \left(p\ln\left(\frac{p}{n_i}\right)\right.\right.$$

$$\left.\left. - p + n\ln\left(\frac{n}{n_i}\right) - n\right)dx\right] \tag{1.51}$$

Equations (1.50) and (1.51) give the same capacitance variation with voltage but the values are somewhat larger, particularly in the forward bias direction, than given by the first-order equation (1.49).

For most purposes however a small-signal analytical approach to diffusion capacitance is all that is needed. With an applied voltage given by $V_{DC} + V_1 e^{j\omega t}$ the simple diode equations (1.37) and (1.38) yield

$$J_1 = \frac{qV_1}{kT}\left[\frac{qD_p P_{no}}{L_p/\sqrt{1+j\omega\tau_p}} + \frac{qD_n n_{po}}{L_n/\sqrt{1+j\omega\tau_n}}\right]\exp\left(\frac{qV_{DC}}{kT}\right) \tag{1.52}$$

The junction ac admittance follows from $J_1/V_1$. This gives for the diode conductance at low frequencies (such that $j\omega\tau_{n,p}$ are much less than unity)

$$G_d = \frac{q}{kT}\left(\frac{qD_p P_{no}}{L_p} + \frac{qD_n n_{po}}{L_n}\right)\exp\left(\frac{qV_{DC}}{kT}\right) \text{ for } f\to 0$$

or

$$\frac{1}{r_d} = \frac{q}{kT}(I_{DC} + I_0) \tag{1.53}$$

Hence at a dc bias of about 10 mA this simple model predicts an ac resistance of 2.6 ohms for 300°K.

The small signal diffusion capacitance from Eq. (1.52) is

$$C_d = \frac{q}{kT}\left(\frac{qL_p P_{no} + qL_n n_{po}}{2}\right)\exp\left(\frac{qV_{DC}}{kT}\right) \tag{1.54}$$

The diffusion capacitance can be represented as

$$C_d = \frac{q}{kT}\left(\frac{Q_p + Q_n}{2}\right) \text{ for } f\to 0 \tag{1.55}$$

where $Q_p$ is the stored charge of holes injected in the $n$ region of the semiconductor and $Q_n$ is the stored charge of electrons in the $p$ region.

The equivalent circuit for the diode is then a conductance $r_d^{-1}$ in parallel with a capacitance $C_d$ and the variations of these quantities with frequency if $\tau_n = \tau_p = \tau$ are as shown in Fig. 1.11.

Consider a $p^+n$ diode so that only $\tau_p$ is significant—then

$$C_d = \frac{qQ_p}{2kT} \cong \frac{q}{kT} I_{DC} \tau_p \qquad (1.56)$$

and

$$r_d \simeq \frac{kT}{qI_{DC}} \qquad (1.57)$$

Hence, the time constant product $r_d C_d$ is determined by $\tau_p$ as might be expected.

The admittance of a diode may be measured as a function of frequency with a lock-in phase-sensitive amplifier arrangement.[65]

## 1.3 DIODE REVERSE CHARACTERISTICS

Some diode reverse voltage characteristics are shown in Fig. 1.12. Curve (a) is the form expected for the simple Shockley diode model and exhibits almost constant reverse current associated with generation a diffusion length on either side of a negligibly-wide depletion region. At high voltages the leakage current goes into a turn-over knee because of avalanche multiplication. In semiconductor junctions which have small diffusion lengths the generation in the depletion width may be a substantial part of the leakage current and since the depletion thickness increases with reverse voltage, as $V^{1/2}$, $V^{1/3}$, the total leakage current increases with voltage approximately as a straight line, as in Fig. 1.12(b), on a $\ln I / \ln V$ plot until the onset of avalanche conditions. If the temperature is

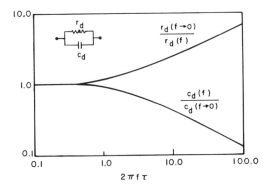

Fig. 1.11 Variation of the components of a small signal equivalent circuit model with frequency for a simple junction model at fixed bias current.

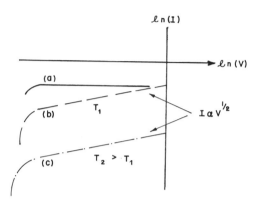

Fig. 1.12. Schematic variation of diode reverse characteristics:
(a)   Simple Shockley model,
(b)   Diode with substantial depletion region component of leakage current, and
(c)   Effect of temperature increase for curve (b).

increased the leakage current is substantially larger as suggested by Fig. 1.12(c) and the breakdown voltage may be a little larger.

In the depletion region the field strength is high and the free carrier electrons and holes may acquire enough energy in a mean free path to impact-ionize secondary hole-electron pairs and so an avalanche process ensues if the depletion region is thick enough. If $E_{op}$ is the energy of an optical phonon involved in the collision process and $l_{op}$ is the mean free path for scattering by optical phonons and $\mathcal{E}$ is the field strength, then in a simple approach that neglects the main complexities of the problem[92] the electron effective temperature is

$$kT_e = \sqrt{\frac{(q\mathcal{E}l_{op})^2}{3E_{op}}} \tag{1.58}$$

If $l_{op}$ is 100 Å this equation gives an electron energy of 0.21 eV for a field strength of $2 \times 10^5$ V cm$^{-1}$ in Si. The threshold energy for generating secondary carrier pairs in Si is greater than the indirect bandgap and is about 1.8 eV for electrons and 2.4 eV for holes. The crude calculation above suggests the onset of secondary ionization at field strengths in the mid $10^5$ V cm$^{-1}$.

The number of ionization events $\alpha$ per cm of travel is a measure of the rate of avalanche. The variations of $\alpha_n$ and $\alpha_p$ with field strength for Si are given in Fig. 1.13. We are usually interested in the multiplication factor that occurs for electrons and holes in passing through the junction depletion region. This obviously depends on the variation of field strength through the junction. If the ratio of $\alpha_p/\alpha_n$ is taken as a constant $\gamma$ to facilitate a quantitative treatment and if we take $\alpha = \alpha_\infty e^{-b/\mathcal{E}}$, then for a p-i-n junction where the field strength is constant for a thickness $W$ the multiplication factors are

$$1 - \frac{1}{M_p} = \frac{\gamma}{1-\gamma}\left(e^{-(\gamma-1)\alpha_n W} - 1\right)$$

$$1 - \frac{1}{M_n} = \frac{1}{\gamma-1}\left(e^{(\gamma-1)\alpha_n W} - 1\right) \tag{1.59}$$

and the junction breakdown voltage when the multiplication becomes infinite is

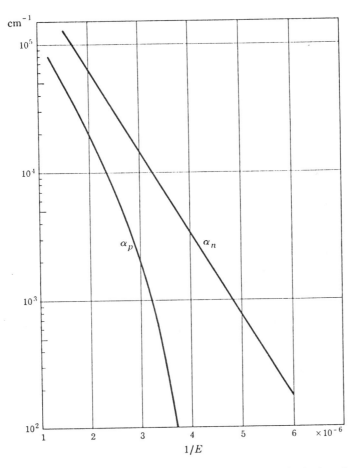

Fig. 1.13. Ionization rates for electrons and holes in Si. (The data obtained are from A.G. Chynoweth, *Phys. Rev.*, **109**, 1958 p. 1537 and J.L. Moll and R. Van Overstraeten, *Solid-State Electronics*, **6**, 1963, pp. 147-157 for electrons. The hole data are from Moll and Van Overstraeten for electric field greater than 4 x 10⁵ V/cm and from C.A. Lee, *Phys. Rev.*, **134**, 1964 p. A761 for electric field less than 4 x 10⁵ V/cm). For temperature dependence of electron ionization rate see Fig. 11.6. (After J. Moll, *Physics of Semiconductors*, McGraw-Hill, 1964, with permission of McGraw-Hill Book Co., NY.)

$$V_B = \frac{Wb}{\ln[\alpha_\infty W(\gamma - 1)/\ln\gamma]} \tag{1.60}$$

Since the denominator does not vary rapidly with $W$, the breakdown voltage is almost proportional to the thickness of the intrinsic region in a *p-i-n* diode.

The behavior of the junction at a voltage $V$ where the multiplication is small is given by

$$1 - \frac{1}{M_n} = \alpha_\infty \, W e^{-b\dot{W}/V}$$

$$\approx \frac{\ln\gamma}{\gamma - 1} \left(\frac{V}{V_B}\right)^m \tag{1.61}$$

where

$$m = \frac{Wb}{V_B} \tag{1.62}$$

This is the basis of the empirical law

$$1 - \frac{1}{M} = \left(\frac{V}{V_B}\right)^m \tag{1.63}$$

which is often used to characterize the shape of the reverse voltage knee of a diode or to estimate the voltage at which the current gain of a transistor, $a_{npn}M_n$, becomes unity. These approaches give rough estimates only since the ionization coefficient dependences have been oversimplified.

For step junctions and linearly graded junctions the same type of analysis may be developed, however, with differences in the exponents $m$ that arise because of the differences in the variation of field strength through the depletion regions. Figure 1.14 shows the charge, field strength and voltage variations for *PIN*, step and linearly graded junctions.

The observed breakdown voltages for Si and Ge step junctions as functions of the low side doping are shown in Fig. 1.15(a). For linearly graded junctions the results are those given in Fig. 1.15(b). For GaAs the breakdown voltages are about 10% higher than for Si and for GaP about 50% higher as seen from Fig. 1.16. Depletion thickness and field strengths at breakdown are given in Fig. 1.17. Depletion widths, capacitances and breakdown voltages for abrupt junctions of Si and GaAs may also be found conveniently from the nomograms of Fig. 1.18.

Most junctions are, of course, formed by diffusion, and the breakdown voltage is then conditioned by both the gradient and the base doping level. The resulting curves are shown in Fig. 1.19. However, it must also be recognized that diffusion through an oxide window gives a planar junction with roughly circular cylindrical edges and that electric field concentration occurs at these edges. The breakdown voltage is therefore lowered (see Fig. 1.20) by an extent

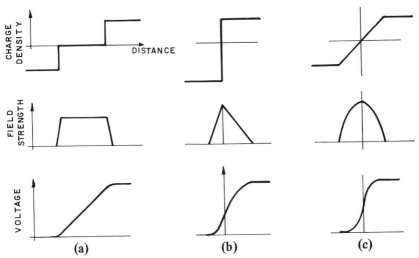

Fig. 1.14. Charge, field strength and voltage variations across *p-i-n*, abrupt and linearly graded junctions.

that depends on the radius of the edge which may be taken as approximately the junction depth.

For some junctions the breakdown voltage may be related to depletion layer thickness $X_T$ by an empirical relationship, such as

$$V_B = 5.8 \times 10^4 \, X_T{}^{0.84} \tag{1.64}$$

However, the results for planar $p^+n$ Gaussian diffused diodes only agree with this line for fabrication procedures corresponding to a particular degree of grading[136] but if the breakdown voltage is normalized as $V_B/x_j{}^2 C_b$ (where $x_j$ is the junction depth and $C_b$ is the background doping) and plotted against $x_j C_b$, all of the points lie on the same line. Such relationships, although empirical, or at best semi-theoretical, are occasionally useful.

In high voltage rectifiers or thyristors it is desirable to bevel the junction surface to reduce the electric field there so that body instead of surface breakdown occurs. Surface breakdown may be influenced by impurities at the surface and the leakage current and energy dissipation are less predictable than for bulk breakdown. Figure 1.21(a) shows the equipotential voltage lines for a $p^+n$ junction with a 6° bevel from which it is seen that the maximum field strength is considerably less at the surface than in the bulk. However if the structure is $p^+nn^+$ as in Fig. 1.21(b) the voltage distribution becomes crowded at the $nn^+$ boundary and a bevel angle of about 60° is preferable as seen from Fig. 1.21(c).

Another approach to achieving ideal breakdown voltages consists of a partial etch into the heavily doped side of either a plane or planar junction as shown in

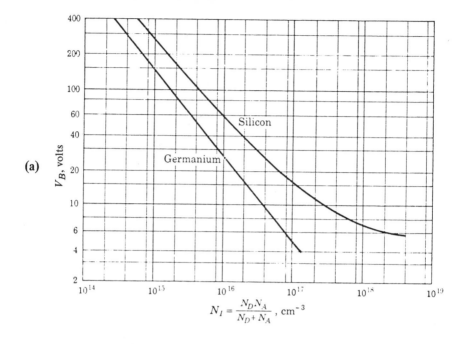

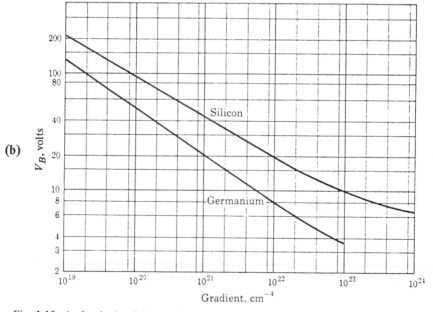

Fig. 1.15. Avalanche breakdown voltages predicted for step and linearly-graded Si and Ge junctions. (Observed results tend to be lower than predicted by avalanche theory at voltages below 8 V because of the onset of tunneling or Zener mechanisms.) (After J. Moll, *Physics of Semiconductors*, McGraw-Hill, 1964, with permission of McGraw-Hill Book Co., NY.)

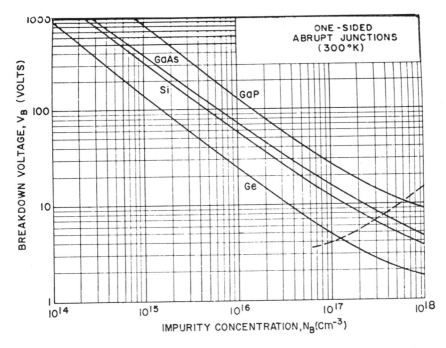

Fig. 1.16. Avalanche breakdown voltage vs impurity concentration for one-sided abrupt junctions in Ge, Si, GaAs, GaP. Tunnel mechanisms dominate to the right of the dashed line. (After S.M. Sze, *Physics of Semiconductors*, John Wiley & Sons, Inc., NY 1969.)

Fig. 1.22. This uses less area than a typical bevel and mechanical contouring is unnecessary. Good control of the etch depth, though, is required.

Local doping fluctuations, such as resistivity striations or swirls in semiconductor crystals may be possible causes of high field concentrations in junctions.[95] Numerical results, however, show that small spikes having tenfold doping concentration only increase the maximum field strength by 10%.[27] Some specimens of high-voltage junctions that are unexpectedly low in rating may have inclusions of metal precipitates, such as Fe, Cu and Au, in the high field region. Junctions may also suffer from inclusions of $SiO_x$.

However, variation of doping within a semiconductor is a matter of some concern in several classes of device applications such as Si vidicon display targets and Gunn oscillator diodes where epi-GaAs layers are used. Doping profiles are normally determined by small-signal capacitance studies as a function of reverse bias voltage that sets the depth at which the concentration is being determined. Schematics for two such profilers are shown in Fig. 1.23.

Bulk Si may have resistivity fluctuations of ±10% and the variation across a large diameter wafer may be considerably greater. If close control of doping

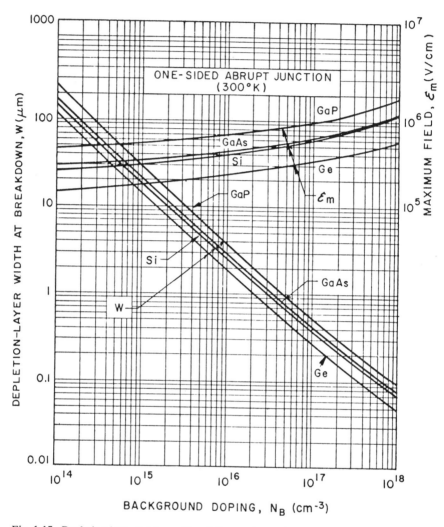

Fig. 1.17. Depletion layer thickness $W$ and the maximum field strength at breakdown voltage for Ge, Si, GaAs and GaP $p^+n$ or $pn^+$ abrupt junctions. (After S.M. Sze, *Physics of Semiconductors*, John Wiley & Sons, Inc., NY, 1969.)

is essential, consideration should be given to Si that has been $n$ type doped by P created by fast neutron transmutation of the Si itself.

Avalanche multiplication is of practical use in increasing the sensitivity of structures such as photodiodes. Very uniform field strength is desirable in these devices and attention must be given to edge effects. Frequently guard rings are used to reduce edge effects. Another approach is to use Ga as the diffusant through a $SiO_2-Si_3N_4$ mask since this gives an extraordinarily elongated diffused layer under the masked region. This diffusion broadening at the edge reduces and displaces the edge field and the photon active region of the diode is therefore more uniform. Such a diode of 500-$\mu$m diameter is capable of operating with a multiplication factor $M$ of 60 or greater and a rise time of a fraction of a nanosecond. The avalanche noise power is proportional to about $M^{2.5}$.

## 1.4 JUNCTION TRANSIENT CHARACTERISTICS

The charge of injected minority carriers in a forward biased $p^+n$ diode is given, per unit area, by

$$Q = \int_0^\infty qp(x=0^+) \exp^{(-x/L_p)} dx$$

and from the Boltzmann expression

$$p(x=0^+) = p_{no} \exp{(qV_{ext}/kT)} \tag{1.65}$$

Hence

$$Q = q L_p p_{no} \exp{(qV_{ext}/kT)}$$

and from Eqs. (1.37) and (1.38) for a simple model

$$Q \cong J\tau_p \tag{1.66}$$

Thus, the stored charge is proportional to the current flowing and to the minority carrier lifetime.

The minority carrier lifetime in Si can range from about $10^{-4}$ to $10^{-9}$ sec depending on the density of the recombination centers in the material. Recombination of electrons with holes directly across the bandgap is an unlikely process in a perfect Si crystal because the bandgap is indirect and phonon cooperation is needed. However, all Si crystals contain some degree of imperfection either as missing or displaced Si atoms or inclusion of impurity atoms such as oxygen, carbon in significant quantities ($10^{16}$-$10^{18}$ cm$^{-3}$) or metallic impurities such as Au, Cu, Fe, etc. that may be present in such minute densities as $10^{13}$ cm$^{-3}$ due to chance or may be specifically added in greater quantities to control the lifetime. For instance, the presence of Au in a concentration of $5 \times 10^{14}$ cm$^{-3}$ (about 1 part in $10^8$) and with a capture cross-section of $2 \times 10^{-16}$ cm$^{-2}$ will limit the lifetime to $10^{-6}$ sec (from $\tau = (v_{th} \sigma N_{Au})^{-1}$). Radiation damage in Si also has a considerable effect in lowering the lifetime.[137]

GaAs is a direct-gap semiconductor and the intrinsic lifetime tends to be much lower than in Si. In this material, too, the degree of crystal perfection and

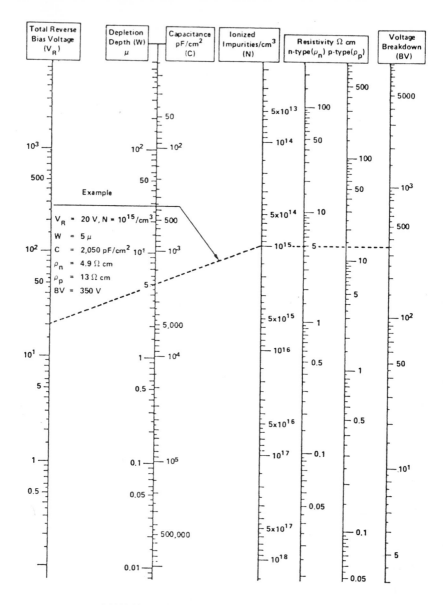

NOMOGRAPH FOR Si STEP JUNCTION AT 300°K

Fig. 1.18. Nomograms for depletion depth (thickness), capacitance and voltage breakdown for Si and GaAs abrupt junctions. (300°K). (After Miller, Lang and Kimerling.[78] Reproduced with permission from the *Annual Review of Materials Science*, Vol. 7, 1977, copyright 1977 by Annual Reviews, Inc. All rights reserved.)

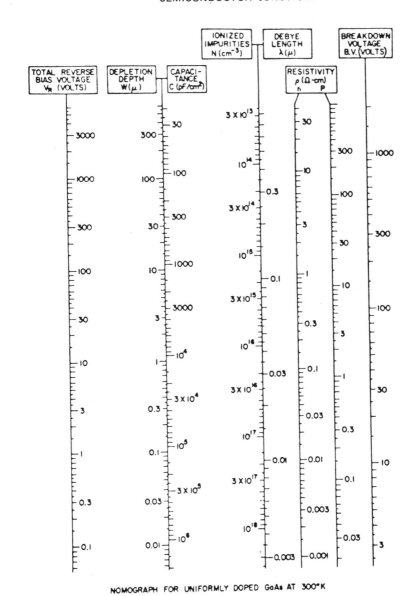

NOMOGRAPH FOR UNIFORMLY DOPED GaAs AT 300°K

Fig. 1.18 (continued)

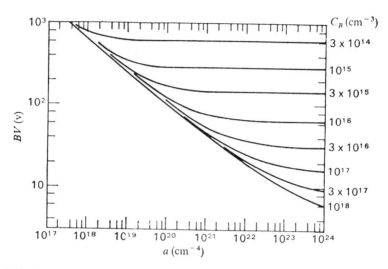

Fig. 1.19. Breakdown voltage of diffused Si junctions with a graded erfc type distribution of impurities. (After Kennedy & O'Brien, 1962. Reprinted from A.S. Grove, *Physics and Technology of Semiconductor Devices*, John Wiley & Sons, Inc., NY, 1967.)

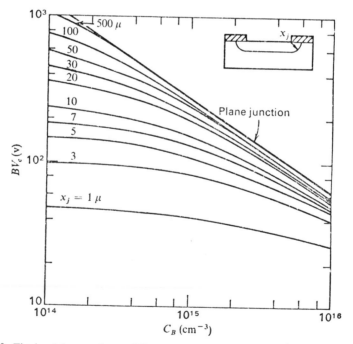

Fig. 1.20. The breakdown voltage of Si one-sided step junctions is reduced by field concentration at the curved junction edge. (After Armstrong. Reprinted with permission from *IRE Trans. Electron Devices*, **ED-4**, 1957, p. 16.)

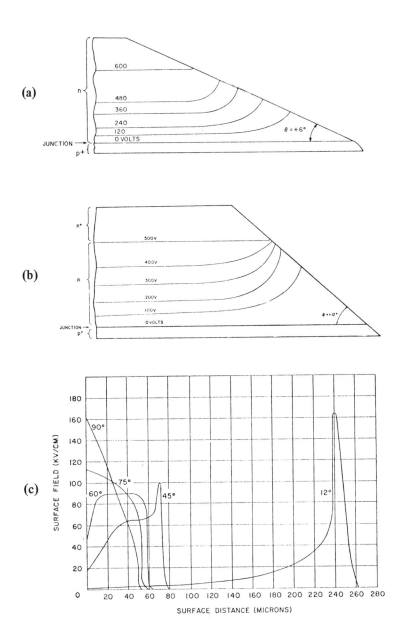

Fig. 1.21. Computed voltage and field distributions in Si junctions with bevelled surfaces:

(a) Voltage in $p^+n$ (3.5 x $10^{14}$ cm$^{-3}$) junction with 6° bevel,

(b) Voltage in a $p^+nn^+$ diffused junction with 12° bevel shows crowding at the $nn^+$ junction surface. (The diffused junctions are 60 $\mu$m deep with surface concentrations of 7 x $10^{18}$ atoms/cc. The lightly-doped $n$-region is 50-$\mu$m thick and the bias 500 V)

(c) Field distribution along the bevel of the structure shown in (b) for several angles. The distance is measured from the $p^+$-$n$ junction toward the $n$-$n^+$ junction. (After Davis and Gentry, Reprinted with permission from *IEEE Trans. Electron Devices*, **ED-11**, 1964, pp. 318, 319.)

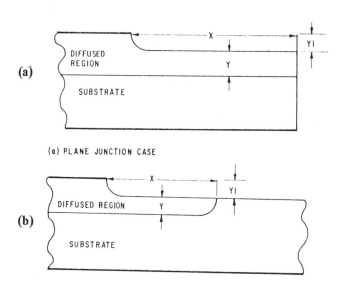

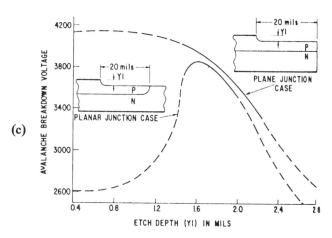

Fig. 1.22. Control of surface breakdown voltage by depletion junction etching:
(a), (b)  Plane and planar junctions, and
(c)   The avalanche breakdown voltage in the planar junction is very dependent on the etch
depth in relation to the doping profile. (After Temple and Adler. Reprinted with permission, from *International IEEE Electron Devices Meeting Technical Digest*, 1975,
pp. 171, 172.)

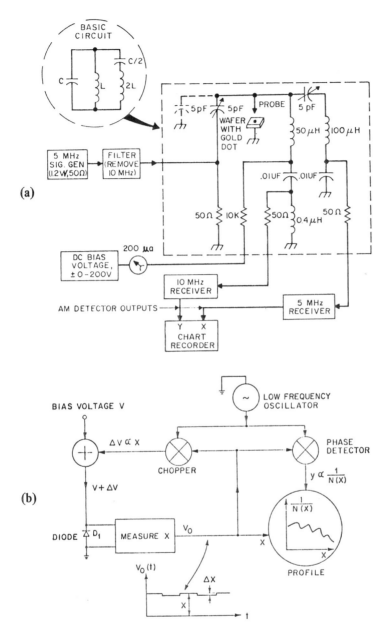

Fig. 1.23. Circuit schematics for profilers for study of doping concentration as a function of depth:

(a) The inverse doping profile $n^{-1}(x)$ is obtained by monitoring the voltage across the diode at the fundamental frequency, which is proportional to the depth $x$, and the second harmonic voltage, which is proportional to $n^{-1}$, (After Copeland, Reprinted with permission, from *IEEE Trans. Electron Devices*, **ED-16**, 1969, p. 447), and (b) (After Miller, Reprinted with permission, from *IEEE Trans. Electron Devices*, **ED-19**, 1972, p. 1105.)

purity has a significant effect on the lifetime. The hole diffusion lengths for numerous specimens in GaAs prepared in several ways are shown in Fig. 1.24. Most of the results are seen to lie substantially below the theoretical line for diffusion length variation with doping density.

There are many methods of measuring minority carrier lifetime in bulk semiconductors and devices[87] and a number of them are listed in Table 1.1.

**Table 1.1. Methods of Measuring Carrier Lifetime.**

| | |
|---|---|
| *Conductivity Decay Methods* | 11. Sweepout effects |
|   1. Photoconductive decay |     Pulse injection |
|     Direct observation of resistivity |     Photoinjection |
|     $Q$ changes | 12. Drift field |
|     Microwave reflection | 13. Pulse delay |
|     Microwave absorption | 14. Emitter point efficiency |
|     Spreading resistances | |
|     Eddy current losses | *Junction Methods* |
|   2. Pulse decay | 15. Open-circuit voltage decay |
|     Direct observation of resistivity | 16. Reverse recovery |
|     Microwave absorption | 17. Reverse current decay |
|   3. Bombardment decay | 18. Diffusion capacitance |
| | 19. Junction photocurrent |
| *Conductivity Modulation Methods* |     Steady state |
|   4. Photoconductivity |     Decay |
|     Steady state | 20. Junction photovoltage |
|     Modulated source | 21. Stored charges |
|     Infrared detection, steady state | 22. Current distortion effects |
|     $Q$ changes | 23. Current-voltage characteristics |
|     Microwave absorption | |
|     Eddy current losses | *Transistor Methods* |
|     Spreading resistance, | 24. Base transport |
|       modulated source | 25. Collector response |
|   5. Pulse injection spreading resistance | 26. Alpha cutoff frequency |
| | 27. Beta cutoff frequency |
| *Magnetic Field Methods* | |
|   6. Suhl effect (and related effects) | *Other Methods* |
|   7. Photoelectromagnetic effect | 28. MOS capacitance |
|     Steady state | 29. Charge collection efficiency |
|     Modulated source | 30. Noise |
|     Transient decay | 31. Surface photovoltage |
| |     Steady state |
| *Diffusion Length Methods* |     Decay |
|   8. Traveling spot | 32. Bulk photovoltage |
|     Steady state |     Steady state |
|     Modulated source |     Modulated source |
|   9. Flying spot | 33. Electroluminescence |
| 10. Dark spot | 34. Photoluminescence |
| | 35. Cathodoluminescence |

After Bullis, 1968, Measurement of Carrier Lifetime in Semiconductors: An annotated Bibliography covering the Period 1949-1967, Document AD 674 627 NTIS.[14]

In this chapter several junction transient techniques will be briefly described since they not only give lifetime information but may be of importance in actual circuit use.

### 1.4.1 Junction Recovery from Forward Bias

Consider a $p^+n$ junction that is forward biased and then subjected to an open-circuit and the junction voltage observed. From the Boltzmann relationship, see Eq. (1.26), if $\Delta p$ is the excess hole density at $x=0^+$ in the $n$ region

$$V = \frac{kT}{q} \ln \left(1 + \frac{\Delta p}{p_{no}}\right) \qquad (1.67)$$

So if $V_0$ is the junction voltage and $\Delta p_0$ the excess hole density immediately after removal of the forward current pulse, and if the excess carrier concentration decays with a time constant $\tau_p$ such that

$$\Delta_p = \Delta p_0 \, e^{-t/\tau_p} \qquad (1.68)$$

then

$$V = \frac{kT}{q} \ln \left[1 + (e^{q V_0 \; kT} - 1)e^{-t/\tau_p}\right] \qquad (1.69)$$

For $t/\tau_p$ very small, and $V_0 \gg kT/q$ this simplifies to

$$V \approx V_0 - \frac{kT}{q} \frac{t}{\tau_p} \qquad (1.70)$$

and the initial voltage variation with time is a linear decrease. Hence

$$\tau_p = -\frac{kT}{q} \left(\frac{\Delta t}{\Delta V}\right)$$

$$= -\frac{kT}{q} \left(\frac{1}{\text{slope of the linear decay}}\right) \qquad (1.71)$$

This open-circuit voltage decay technique for estimating lifetime is interesting because it may show a transition from a high-injection-level lifetime to a low-injection-level lifetime in a single transient.

For circuit reasons it may be easier to reverse bias a junction than open-circuit it and this leads to a second method of determining lifetime that depends on observing part of the reverse current transient. After a reverse step of bias voltage, as in Fig. 1.25(a), the current transient waveform (b) has a horizontal plateau period of time duration $t_1$. During this time the current is almost

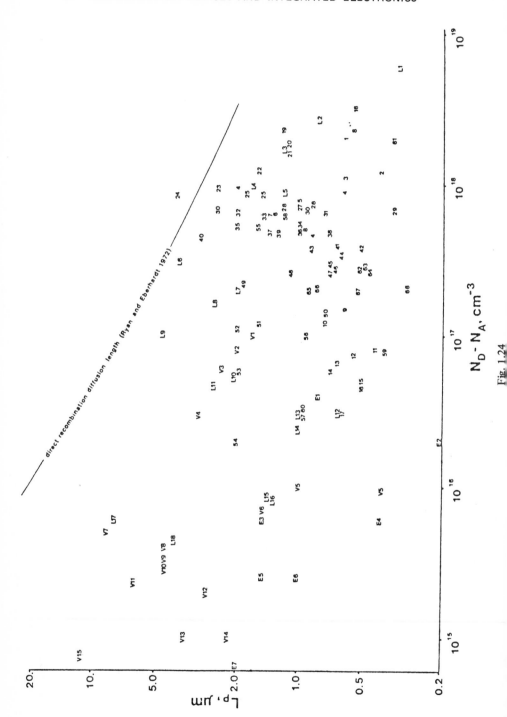

Fig. 1.24

constant and the hole density at the junction depletion edge in the $n$ region is falling to the equilibrium value $p_{no}$ as the reverse current flow extracts carriers and as they recombine within the diode, as shown in Fig. 1.25(d). Only when the hole density falls below $p_{no}$ can reverse voltage begin to build up across the junction, [see Fig. 1.25(c)], and the current transient period $t_2$ commence. The problem when analyzed yields the transcendental solutions

$$\text{erf}\sqrt{\frac{t_1}{\tau_p}} = \frac{1}{1 + \dfrac{I_r}{I_f}} \tag{1.72}$$

$$\text{erf}\sqrt{\frac{t_2}{\tau_p}} + \frac{\exp\left(-\dfrac{t_2}{\tau_p}\right)}{\sqrt{\pi\dfrac{t_2}{\tau_p}}} = 1 + 0.1\left(\frac{I_r}{I_f}\right) \tag{1.73}$$

Figure 1.26 allows easy graphical interpretation of these equations for an abrupt diode with an $n$ region many· diffusion lengths deep. To obtain a representative range of values for the lifetime it is advisable to make transient measurements for several injection levels and several ratios of reverse to forward current.

For a large ratio $I_r/I_f$ and $W \gg L_p$ the total transient time is given by

$$t_1 + t_2 = \frac{\tau_p}{2}\left/\frac{I_r}{I_f}\right. \tag{1.74}$$

More usually in actual circuit applications $I_r/I_f$ is unity. Then from Fig. 1.26 the plateau current time is seen to be $0.3\tau_p$ and the decay time $0.6\tau_p$ so the total transient time is about $0.9\tau_p$.

The presentation given so far relates only to a step junction. For a linearly graded junction the expression is more nearly of the form[91]

$$t_1 = \tau_p \ln\left(1 + \frac{I_f}{I_r}\right) \tag{1.75}$$

Fig. 1.24 Hole diffusion length as a function of net electron concentration for $n$ type GaAs samples. Sample numbers prefixed with a "V" are vapor epitaxial material, "L" samples are liquid epitaxial, "E" are unspecified epitaxial samples, and those with no prefix are from single crystal boules prepared either by the Czochralski or horizontal Bridgman process. (After Sekela, Feucht and Milnes, from *Gallium Arsenide and Related Compounds 1974*, Institute of Physics Conf. Series No. 24.)

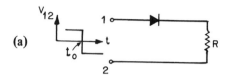

(a)

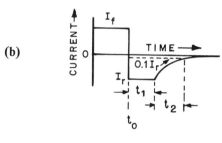

(b)

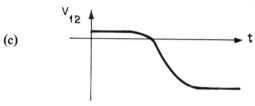

(c)

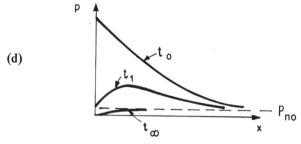

(d)

Fig. 1.25. Turn-off transient of a $p^+n$ diode:
(a)  Circuit,
(b)  Reverse current shows a plateau region $I_r$ before the reverse voltage build up across the junction as in (c), and
(d)  The hole density stored decreases as shown during the switching transient.

The difference between Eqs. (1.72) and (1.75) can be fairly substantial, as seen from Table 1.2.

If the lifetimes of the minority carriers in the diode are short, the switching circuit may use techniques similar to those in time domain reflectometry.[29] The circuit schematic and the observed response at the oscilloscope are shown in

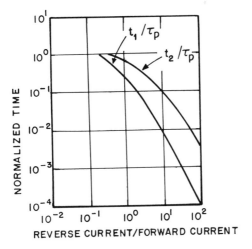

Fig. 1.26. Diode transient recovery times $t_1$ and $t_2$ as a function of the ratio of reverse-to-forward current for an abrupt junction. (After Kingston. Reprinted with permission from *Proc. IRE*, **42**, 1954, p. 834.)

Fig. 1.27 with the graph for interpretation of the results. To use this graph, one must know $T_s$, $(T_s + T_r)$, and the ratio $I_F/I_R$. The curves slanting left and up correspond to values of $T_s/(T_s + T_r)$. The solution occurs at the intersection of the appropriate pair of curves. The abscissa below gives the ratio $\tau/(T_s + T_r)$. Using $(T_s + T_r)$ one can immediately determine $\tau$. The ordinate to the left gives the grading parameter $L/L_D$. With $\tau$ and an independent knowledge of the diffusion constant, $L_D$ is known and then $L$, the average injected carrier penetration depth, can be determined. For an ideally abrupt junction $L = L_D$ but for a graded junction a retarding field crowds the injected carriers close to the junction and $L < L_D$. The technique is capable of measuring lifetimes down to a few $10^{-9}$ secs.

If there is a retrograded hyperabrupt impurity distribution in the junction this establishes a minority-carrier potential barrier that sweeps carriers away

Table 1.2. Normalized Storage Phase Time
$t_1/\tau_p$ vs $I_r/I_f$ for Abrupt and Graded
Junctions.

| $I_r/I_f$ | $t_1/\tau_p$ | |
|---|---|---|
| | Abrupt Junction | Graded Junction |
| 0.3 | 0.8 | 1.47 |
| 1.0 | 0.3 | 0.69 |
| 3.0 | 0.06 | 0.29 |

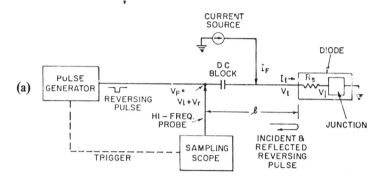

(a)

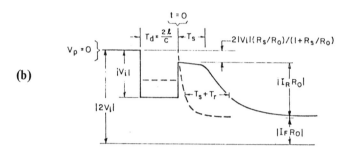

(b)

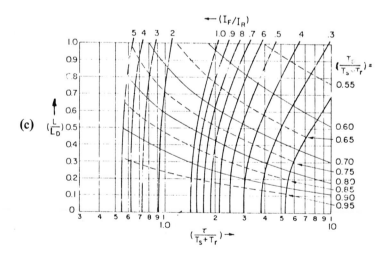

(c)

Fig. 1.27

from the junction during injection and prevents them from being retrieved by the reversing step. In this case, the step recovery measurement erroneously indicates a strongly graded profile with $L/L_D$ approaching zero. A hyperabrupt junction distribution would be indicated by a capacitance-voltage measurement which gives the form $C \sim V^{-1/n}$, where $n$ is significantly less than 2.0.

Another technique for measuring short lifetimes uses the response to a sinusoidal waveform. Sinusoidal excitation is applied to the diode and series resistor, as in Fig. 1.28 and the diode current (resistor voltage) is observed with a sampling oscilloscope. A ballast network may be necessary to maintain uniform loading of the signal generator during the cycle and thereby minimize distortion of the excitation waveform. For optimum matching, the ballast resistor $R_b$ should approximate the characteristic impedance of the generator and line, and the diode load resistor $R_s$ should be high. However, if $R_s$ is too high, discrimination will be lost between storage and capacitive conduction, and the test will lose its value. Insertion loss of the ballast network must usually be minimized in order to maintain an adequate operating signal level. Usually $R_s$ is the 50-ohm input impedance of the sampling oscilloscope.

If $\psi_0$ is the turn-on voltage of the diode (roughly related to the electrostatic junction potential), the peak value of forward conduction $i_{pf}$ is

$$i_{pf} = \frac{E_p - \psi_0}{R_c} = \frac{E_p}{R_c}\left(1 - \frac{\psi_0}{E_p}\right) \tag{1.75}$$

where, assuming that the generator and line are matched,

$$R_c = R_s + \frac{R_b R_g}{R_b + R_g} + R_f \tag{1.76}$$

and

$$E_p = \frac{E_g \sqrt{2R_b}}{R_b + R_g} \tag{1.77}$$

$E_g$ is the rms value of generator voltage and $R_g$ the generator impedance. These comprise the equivalent Thevenin generator for the signal source.

---

Fig. 1.27 Nanosecond lifetimes may be measured by reflection transient techniques:
(a) Apparatus using coaxial lines and 1 nsec pulse generator,
(b) Idealized oscilloscope traces ($V_p$) for a reversing step voltage. Zero time occurs when the reflected voltage step just reaches the probe. The solid curve indicates the response with a moderate forward biasing current $I_F$. The dotted curve indicates the response with small but finite $I_F$ and with $V_i$ adjusted for the same $V_p$ at $t=\infty$, and
(c) Chart for determining $\tau$ and $L/L_D$ from $T_s$, $T_s + T_r$, and $I_F/I_R$. (After Dean and Nuese. Reprinted with permission from *IEEE Trans. Electronic Devices*, ED-18, 1971, pp. 155, 156, 154.)

Then, if $i_{pr}/i_{pf}$ is measured the lifetime follows from

$$\tau \approx \frac{i_{pr}}{i_{pf}} \frac{\left(1 - \dfrac{\psi_0}{E_p}\right)}{\omega_1} \tag{1.78}$$

An excitation frequency of about 30 MHz with 3 V available into 50 ohms is needed to give convenient lifetime readings of a few $10^{-9}$ sec.

### 1.4.2 Measurement of Diffusion Length

Another useful technique for measuring the minority carrier lifetime in the region of a junction is to use the beam of a scanning electron microscope to inject minority carriers and to observe the junction current, under slight reverse bias, as a function of beam position. A straight line is obtained as in Fig. 1.29 from which the diffusion length may be determined and the lifetime inferred. The technique is limited to diffusion length measurements of greater than 0.5-1 μm because of the size of the pear-shaped region of carriers induced by the electron beam. The depth of this region below the surface depends on the electron beam voltage and the measured result depends to some extent on this voltage and on the surface recombination velocity.

Several lifetime measuring techniques are known that involve injection of carriers by light. For instance the amplitude or phase-shift dependence of the short-circuit photocurrent of a junction or Schottky barrier illuminated by sine wave modulated light may be used.[109]

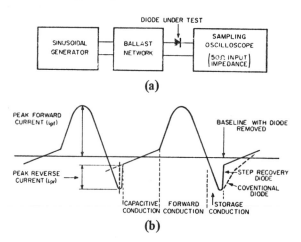

**(a)**

**(b)**

Fig. 1.28. Minority carrier lifetime in a diode may be measured by observing the response to a sinusoidal voltage. The difference in response of a conventional and a step-recovery diode is shown. (After Krakauer. Reprinted with permission from *Proc. IRE*, **50**, 1962, p. 1675.)

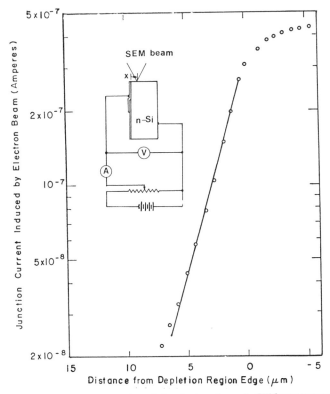

Fig. 1.29. Junction current vs position of the electron beam in SEM measurement of diffusion length for $n$-type Si containing Pt. The calculated value of $L_p$ from the plot is 2.6 $\mu$m. This corresponds to a lifetime of about 5 nsec. Dark current is less than 10 nA. The measurement temperature was 300°K. (After Lisiak and Milnes.[71])

### 1.4.3  Lifetime Control

High speed Si diodes may have switching recovery times of a few nanoseconds achieved by recombination centers in the material. Creation of disorder in the Si lattice is one method of achieving recombination centers and another is the addition of specific impurity centers to facilitate recombination.[137]

Quenching of Si from a high temperature creates donor defect centers at levels $E_c - 0.26$ and $E_c - 0.54$ eV but the density of such centers is not large ($< 10^{14}$ cm$^{-3}$).[134] Lifetime reduction effects achieved in this way are not promising since quenching introduces strain into Si wafers and distortion may occur. Irradiation of Si devices in an electron flux of 1-3 MeV from a Van de Graaff generator or in a neutron flux in the vicinity of atomic reactors is another way of creating recombination centers. The possible centers created include the $A$ center (oxygen interstitial paired with a Si vacancy giving an acceptor level at $E_c - 0.17$ eV); the $C$ center (a Si divacancy apparently associated with levels at $E_v + 0.25$ eV and $E_v + 0.54$ eV) and the $E$ center (an arsenic or P-Si-vacancy pair which has energy levels at $E_c - .16$ eV and at $E_v + .27$ eV). The electron and hole capture cross sections of the $A$ and $E$ centers are of the order of $10^{-15}$-$10^{-16}$ cm$^2$ and densities of $10^{16}$-$10^{17}$ cm$^{-3}$ are readily achieved so considerable lifetime reduction is possible. However, neutron irradiation tends to create P transmutation doping of the Si. Furthermore, certain elements must not be present in the structures irradiated because they would become radioactive with a long half-life, and thus would present a health hazard. Lifetime control by electron irradiation is more usual, but deep-impurity doping is also used.

The ideal properties of a deep impurity dopant for lifetime control include the following:

- The impurity must be easy to apply and it must not be gettered by certain regions of the device at the expense of other regions, or be incompatible with subsequent device processing.
- The lifetime achieved must be adequately low and its variation with temperature must not be inconveniently large.
- The effects of the deep impurity on the forward voltage drop of the junction, on the reverse leakage current and on the maximum reverse voltage rating of the junction must be small.

A Si junction of high reverse voltage rating (about 500 V) requires a base region doping of $4 \times 10^{14}$ cm$^{-3}$ (see Fig. 1.16). The concentration of the deep impurity therefore must not exceed $3 \times 10^{14}$ cm$^{-3}$ since otherwise its compensating effect would raise the resistivity of the base and have the effect of increasing the junction forward voltage drop. If the required lifetime is about $3 \times 10^{-8}$ sec and if the simple lifetime expression $\tau = (\sigma v_{th} N_T)^{-1}$ is valid, then the capture cross section required for the impurity must be on the order of $10^{-14}$ cm$^{-2}$.

The reverse leakage current of a junction with a deep impurity present is related to the trap concentration and the emission coefficient that is generation rate limiting. If the impurity energy level is somewhat removed from the center of the bandgap the emission coefficient to the remote band edge may be small and this may favor a low generation current density. On the other hand, it may make the lifetime rather more temperature sensitive than for an impurity level at the center of the bandgap.

The most commonly used lifetime control impurity in Si is Au.[87] In $n$ type Si this has an acceptor level at $E_c - 0.54$ eV, slightly above mid-bandgap, and in $p$ type Si the Au level is a donor at $E_v + 0.35$ eV. The capture cross sections of these levels at 300°K are in the range $10^{-14}$-$10^{-16}$ as seen from Table 1.3 and at 77°K are about an order of magnitude larger.

The observed variation of lifetime with Au concentration, and diffusion temperature of the Au for $p^+n$ Si diodes is shown in Fig. 1.30. An Au concen-

**Table 1.3. Capture Cross Sections of Gold in Silicon**[a].

| Temper- ature (°K) | Acceptor Level | | Donor Level | |
|---|---|---|---|---|
| | $\sigma_n{}^0$ (cm$^2$) | $\sigma_p{}^-$ (cm$^2$) | $\sigma_n{}^+$ (cm$^2$) | $\sigma_p{}^0$ (cm$^2$) |
| 300 | $5 \times 10^{-16}$ | $1 \times 10^{-15}$ | $3.5 \times 10^{-15}$ | $\geq 10^{-16}$ |
| 300 | $1.7 \times 10^{-16}$ | $1.1 \times 10^{-14}$ | $6.3 \times 10^{-15}$ | $2.4 \times 10^{-15}$ |
| 77 | $3 \times 10^{-15}$ | $1 \times 10^{-13}$ | $6 \times 10^{-14}$ | $3 \times 10^{-15}$ |
| 77 | $5 \times 10^{-16}$ | $2.3 \times 10^{-13}$ | $1 \times 10^{-13}$ | |

[a]The convention adopted is that $\sigma_n$ represents the capture of electrons and $\sigma_p$ that of holes. The superscript 0, −, or + indicates the charge state of capturing center before capture. Thus $\sigma_n{}^0$ indicates the capture of an electron by a neutral center. The thermal velocity assumed was $10^7$ cm sec$^{-1}$ for the values in the table. (Reprinted with permission from Academic Press Inc.[87].)

tration of $10^{15}$ cm$^{-3}$ corresponds to lifetimes of a few tens of nanoseconds. The maximum solubility of Au in Si is about $10^{17}$ cm$^{-3}$ (see Fig. 1.31) but high concentrations of interstitial impurities sometimes result in metal precipitates that can have adverse effects on junction breakdown voltages. The diffusion coefficient of Au in Si is quite high (see Fig. 1.32) so flooding a Si wafer with Au does not involve exceptionally high temperatures and may sometimes be done at 850°C or less after the main diffusions of the device structure have been made at temperatures such as 1100°C or 1200°C. However, Au does tend to accumulate in $n$ type P diffused regions and on surfaces and at oxide-semiconductor interfaces.[71] Such effects are generally not desirable. However, there are instances where a Si slice may be gettered of metal impurities by a P diffusion that is then removed to leave the lifetime of the rest of the wafer increased by this treatment.

Another lifetime reducing impurity is Pt. Its main levels are an acceptor at $E_c - 0.26$ eV and a donor at $E_v + 0.32$ eV.[84] From the Shockley-Read-Hall model the lifetimes are

$$\tau_n = \tau_{n0}(1 + n_1/N_D)  \tag{1.79}$$

and

$$\tau_p = \tau_{p0}(1 + p_1/N_A)  \tag{1.78}$$

where

$$\tau_{n0,p0} = \frac{1}{\sigma_{n,p}v_{n,p}N_T}  \tag{1.79}$$

and $n_1$ and $p_1$ are the electron and hole concentrations that would be in the conduction and valence bands if the Fermi level were at the trap level $E_T$. For a trap level somewhere near the center of the bandgap, such as the Au acceptor level, the $n_1$ value is small. The minority carrier lifetime then depends

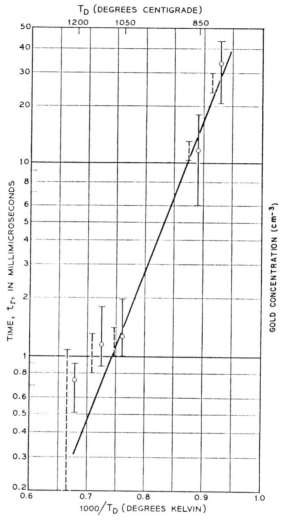

Fig. 1.30. Si diode reverse recovery time as a function of the inverse of the Au diffusion temperature; Au concentrations shown on the right-hand side were calculated from solubility data. (After Bakanowski and Forster.[7] Reprinted with permission from *The Bell System Tech. J.*, copyright 1960, The American Telephone and Telegraph Company.)

primarily on the capture cross-section and varies slowly with temperature and depends hardly at all on background doping. However, for an impurity such as Pt there is a substantial variation of lifetime with doping and ambient temperature as seen in Fig. 1.33. Recombination levels far from the center of the bandgap have this disadvantage. On the other hand, they usually have the advantage of a smaller generation current since the rate-determining emission

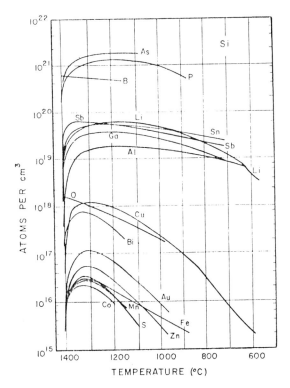

Fig. 1.31. Solid solubilities of some impurity elements in Si vs temperature. (After Trumbore.[128] Reprinted with permission from *The Bell System Tech. J.*, copyright 1960, The American Telephone and Telegraph Company.)

coefficient is less than for a more centrally located generation-recombination level. For Pt in Si there is also an advantage that the gettering effect of a P diffused layer or phosphorus oxide glass is less for Pt than for Au.

### 1.4.4 Snap-action Step-recovery Diodes

If a junction is doped to have a finite lifetime but is provided with a high junction retarding field, the injected minority carriers are constrained to the vicinity of the junction. On reversal of a forward bias voltage, reverse current flows for the storage period but when the stored minority carriers are depleted a very abrupt turn-off of current occurs. The cut-off occurs in a time approximately $x_0^2/D$ where $x_0$ is the distance from the junction to the center of the injected charge distribution and $D$ is the average diffusion coefficient for the injected carriers. The high junction retarding field may be achieved from the doping gradient obtained by diffusing B and P into opposite sides of a thin slice (25-50 $\mu$m) of Si of moderate or high minority carrier lifetime.

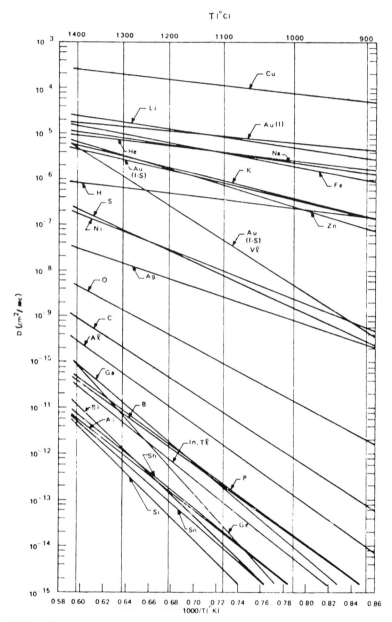

Fig. 1.32. Diffusion coefficients in Si at the low impurity concentration limit. For Au diffusion, the three curves denote the interstitial ($I$), the interstitial-substitutional ($IS$) with unlimited vacancy supply through dislocations, and the vacancy-limited interstitial-substitutional process ($IS$, $V\ell$). (After Burger and Donovan, *Fundamentals of Silicon Integrated Device Technology:* Vol. 1, "Oxidation, Diffusion, and Epitaxy," copyright 1967. Adapted by permission of Prentice-Hall, Inc. Englewood Cliffs, NJ.)

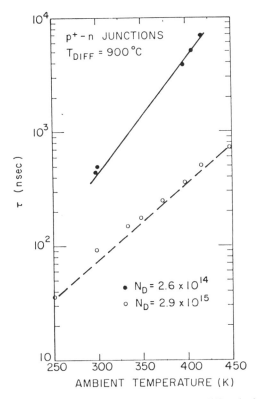

Fig. 1.33. Minority carrier (hole) lifetime in $p^+n$ Si junctions diffused with Pt at 900°C and slow cooled. Since the Pt energy level ($E_c$ - 0.22 eV) is not in the center of the bandgap, the hole lifetime is a function of the doping and of the temperature, as the Fermi level affects the occupancy of the recombination center. (After Miller, Schade and Neuse. Reprinted with permission from *International IEEE Electron Devices Meeting Technical Digest*, 1975, p. 182.)

The step turn-off characteristic that results is illustrated in Fig. 1.28. Such a step is rich in harmonics and so may be used in harmonic generation circuits of the kind discussed in Chap. 3.

## 1.5 RECTIFIER CIRCUITS

Diodes used for frequency-mixing and demodulation purposes or for harmonic generation or for switches in microwave circuits are discussed in Chap. 3. In a few applications diodes may be used for their logarithmic characteristics. The voltage and current, neglecting series resistance, are related by

$$I = I_0 \left[ \exp\left( q V_a / nkT \right) - 1 \right] \qquad (1.80)$$

Hence

$$V_a \cong \frac{nkT}{q}(\ln I - \ln I_0) \tag{1.81}$$

Therefore circuits are known, based on this logarithmic relation, that perform multiplication and division by adding or subtracting the logarithms of individual signals and then taking the antilog.[53] The accuracy of such an approach is not high and is not sustained over a wide temperature range, and so in these days of low-cost computers most functions of this kind are performed by electronic computation. Let us turn, therefore, to a discussion of power applications of rectifiers.

Consider first the series connection of two diodes to increase the reverse voltage rating. The equivalent circuit is shown in Fig. 1.34(a). There are three factors that may cause unequal distribution of reverse voltage between the two diodes. One is the steady-state effect of the reverse leakage resistances $R_R$ of the diodes not being equal, therefore producing unequal voltage division as shown in Fig. 1.34(b). The second factor is that unequal charge storage times in the diodes may cause unequal voltage sharing during circuit turn-off transients as illustrated in Fig. 1.34(c). The solution is to shunt each unit with a capacitor that slows the rise time of voltage and allows more equal distribution. The size of capacitance $C$ needed for each of $n$ rectifiers is such that $R_L C/n$ equals the storage time of an average diode. Hence, if the storage time is $10^{-6}$ sec and $R_L$ is 100 ohms, the required capacitance for a chain of three diodes is 0.03 $\mu$F. These added capacitors swamp any effect of unequal voltage division that might be caused by unequal reverse bias capacitances $C_s$ of the rectifiers. The diode-to-ground capacitances in Fig. 1.34(a) give rise to displacement currents that can disturb the high-frequency voltage division and it is best to mount the units so that $C_G$ is minimized.

Parallel connection of rectifier units to increase current rating also presents problems of unequal sharing. A diode that carries too much current becomes hot and the process is not self-limiting because the diode characteristic displaces with increased temperature in the direction of even more current flow for the same forward voltage drop.* A little series resistance added to the Si junction can help the current sharing at the expense of both the forward voltage drop and the efficiency. Preferably, however, the current rating required should be obtained from a single large-area diode so that the current density vs voltage characteristic is uniform and tracking with temperature is the same. Failing that, the diodes that are to be paralleled should be carefully selected for matching performance, and special circuits used to handle current surge problems.

---

*The surge absorbing capability of a semiconductor such as Se is greater than that of Si because local heating causes a crystalline form to develop at the hot spot that is of high resistance and so the overheated area tends to seal off. Unfortunately the polycrystalline nature of Cd/$p$Se rectifiers limits the reverse voltage rating to a few tens of volts/cell.

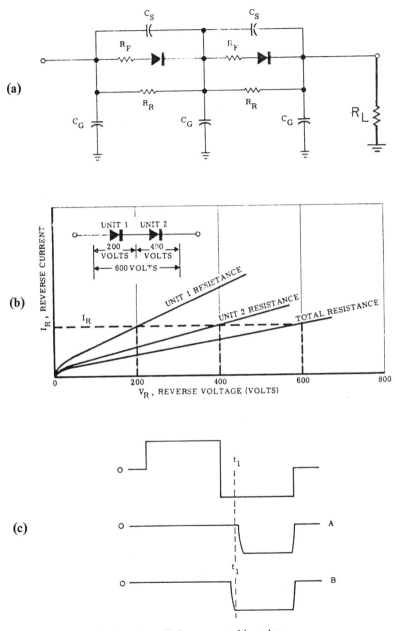

Fig. 1.34. Voltage sharing of two diodes connected in series:
(a)  Equivalent circuit for two rectifiers in series,
(b)  Reverse voltage distribution may be unequal because of different leakage currents, and
(c)  If rectifier 1 recovers from the switching slowly (*A*), the second rectifier may be forced to sustain all the reverse voltage (*B*). (Courtesy Motorola Corp.)

Now consider ac to dc rectifiers using diodes. Half-wave rectification with capacitance smoothing of the load voltage, or inductance smoothing of the load current, is sometimes used. However, this loads the ac supply on only one half cycle, which is usually not desirable. Full-wave rectification by either the center-tap transformer circuit of Fig. 1.35(a), (c), or the full-wave bridge circuit (d) is preferable. The bridge arrangement has two rectifiers in series in each path and so has a higher forward voltage drop, but the transformer rating is smaller.

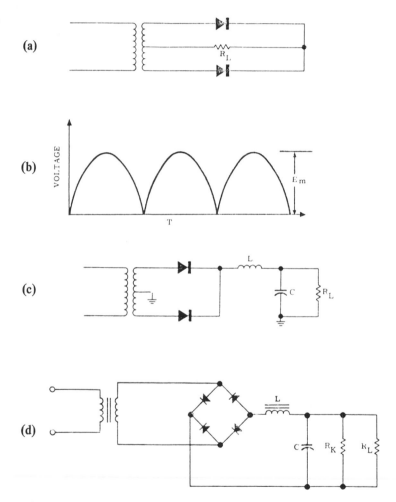

Fig. 1.35. Some single-phase rectifier circuits:
(a)  Center-tap full wave rectifier,
(b)  Rectified waveform,
(c)  Full-wave rectifier with an L-section filter, and
(d)  Full-wave bridge rectifier with an L-section filter. (Courtesy Motorola Corp.)

Some three-phase rectifier circuits are shown in Fig. 1.36. They have less intrinsic ripple than single-phase circuits and therefore smoothing presents less of a problem; also, for handling large powers the cost is less. Table 1.4 summarizes the voltage and current ratings of various rectifier configurations and the required transformer ratings, intrinsic ripple levels and conversion efficiencies. The advantages of three-phase circuits for large power rectification are very apparent.

In electrical apparatus where very high voltage is required use may be made of cascaded rectifiers in a Cockroft-Walton circuit to obtain voltage multiplication. The schematic of a three-stage circuit that produces six-fold voltage multiplication is shown in Fig. 1.37. Analysis of the circuit behavior is an interesting study problem and reveals that preferably $C_3 > C_2 > C_1$.

Heat dissipation from rectifiers is usually with the aid of vertical metal fins in air. If the fin surface temperature (°C) has some average value $T_S$ and the free air temperature is $T_A$, then a fin of area $A$ cm$^2$ has a heat dissipation by free-convection and radiation of

$$P_{HD} = (h_c + h_r) A \text{ (2 sides)} (T_S - T_A) \text{ watts} \qquad (1.82)$$

where $h_c$ is a convection coefficient and $h_r$ is a radiation coefficient.

For a 7.5 × 7.5 cm$^2$ square fin mounted vertically $h_c$ is 0.73 mW cm$^{-2}$ °C$^{-1}$. The value of $h_c$ decreases slightly with increase in fin height because of the changed temperature distribution. The radiant heat transfer effect depends on

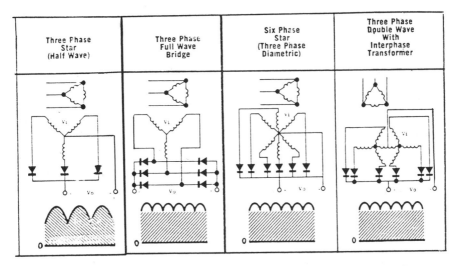

Fig. 1.36. Some three-phase rectifier circuits and wave forms. (Courtesy Motorola Corporation.)

the third power of absolute temperature with an expression such as

$$h_r = 2.3 \times 10^{-11}\, \epsilon(1-F)\left(\frac{T_S + T_A}{2} + 273\right)^3 \text{ W cm}^{-2}\,{}^{\circ}\text{C}^{-1} \qquad (1.82)$$

where $\epsilon$ is the surface emissivity factor, typically about 0.9 and $F$ is a shielding factor associated with the proximity of neighboring fins. This factor is zero for a single fin and about 0.5 for fins spaced a distance apart corresponding to a

### Table 1.4.  Rectifier Circuit Characteristics.

LOAD CURRENT

$I_0 = \dfrac{V_0}{R_0}$

POWER OUTPUT

$P_0 = I_0 V_0$

| Characteristic | Load | Single Phase Half Wave | Single Phase Full Wave Center-Tap | Single Phase Full Wave Bridge | Three Phase Star (Half Wave) | Three Phase Full Wave Bridge | Six Phase Star (Three Phase Diametric) | Three Phase Double Wave With Interphase Transformer |
|---|---|---|---|---|---|---|---|---|
| R.M.S. Input Voltage Per Transformer Leg ($V_1$) | Resistive & Inductive | $2.22\,V_0$ | $1.11\,V_0$ | $1.11\,V_0$ | $0.855\,V_0$ | $0.428\,V_0$ | $0.741\,V_0$ | $0.855\,V_0$ |
| | Capacitive | $0.707\,V_0$ | $0.707\,V_0$ | $0.707\,V_0$ | $0.707\,V_0$ | $0.408\,V_0$ | $0.707\,V_0$ | $0.707\,V_0$ |
| Peak Inverse Voltage Per Rectifier (P.I.V.) | R & L | $3.14\,V_0$ | $3.14\,V_0$ | $1.57\,V_0$ | $2.09\,V_0$ | $1.05\,V_0$ | $2.09\,V_0$ | $2.09\,V_0$ |
| | C | $2.00\,V_0$ | $2.00\,V_0$ | $1.00\,V_0$ | $2.00\,V_0$ | $1.00\,V_0$ | $2.00\,V_0$ | $2.00\,V_0$ |
| Average Current Through Rectifier $I_F$ | R, L & C | $1.00\,I_0$ | $0.50\,I_0$ | $0.50\,I_0$ | $0.333\,I_0$ | $0.333\,I_u$ | $0.167\,I_0$ | $0.167\,I_0$ |
| Peak Current Through Rectifier $I_F$ | R | $3.14\,I_0$ | $1.57\,I_0$ | $1.57\,I_0$ | $1.21\,I_0$ | $1.05\,I_0$ | $1.05\,I_0$ | $0.525\,I_0$ |
| | L | | $1.00\,I_0$ | $1.00\,I_0$ | $1.00\,I_0$ | $1.00\,I_0$ | $1.00\,I_0$ | $0.500\,I_0$ |
| | C | Depends on Size of Capacitor | | | | | | |
| Transformer Total Secondary VA | Sine Wave | $3.49\,P_0$ | $1.75\,P_0$ | $1.23\,P_0$ | $1.50\,P_0$ | $1.05\,P_0$ | $1.81\,P_0$ | $1.49\,P_0$ |
| | Sq. Wave | $3.14\,P_0$ | $1.57\,P_0$ | $1.11\,P_0$ | $1.48\,P_0$ | $1.05\,P_0$ | $1.81\,P_0$ | $1.48\,P_0$ |
| Transformer Total Primary VA | Sine Wave | $3.49\,P_0$ | $1.23\,P_0$ | $1.23\,P_0$ | $1.23\,P_0$ | $1.05\,P_0$ | $1.28\,P_0$ | $1.06\,P_0$ |
| | Sq. Wave | $3.14\,P_0$ | $1.11\,P_0$ | $1.11\,P_0$ | $1.21\,P_0$ | $1.05\,P_0$ | $1.28\,P_0$ | $1.05\,P_0$ |
| % Ripple | Sine Wave Resistive Load | 121% | 47% | 47% | 17% | 4% | 4% | 4% |
| Lowest Ripple Frequency | – | $1\,F_I$ | $2\,F_I$ | $2\,F_I$ | $3\,F_I$ | $6\,F_I$ | $6\,F_I$ | $6\,F_I$ |
| Conversion Efficiency | – | 40.6% | 81.2% | 81.2% | 97% | 99.5% | 99.5% | 99.5% |

(Courtesy Motorola Corporation)

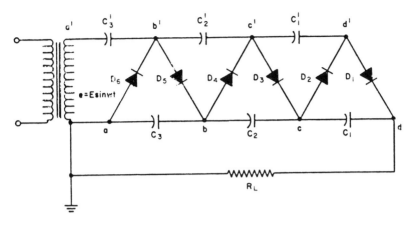

Fig. 1.37. Schematic of a three-stage diode capacitor multiplier.

third of their height. For a fin with a surface temperature of $100°C$ in an ambient temperature of $40°C$, $h_r$ is about 0.84 mW cm$^{-2}$ $°C^{-1}$ or half this if $F$ is 0.5.

A convenient rule for approximate first-order design is that $(h_c + h_r)$ is roughly 1 mW cm$^{-2}$ $°C^{-1}$ for a vertical fin in free air.

## 1.6 ZENER REFERENCE DIODES

If diodes are made on various slices of Si with increasing doping the avalanche reverse breakdown voltage decreases as expected from Fig. 1.16. At doping levels on the order of $5 \times 10^{17}$ cm$^{-3}$ corresponding to a diode breakdown voltage of about 5 V the breakdown knee shape is rather rounded as shown in Fig. 1.38(a). This is because the depletion region is very narrow and a modified form of avalanching occurs, or even some tunneling which tends to be rather dependent on the buildup of field strength with reverse voltage. This field emission tunneling becomes greater as the temperature increases, so the breakdown voltage in this condition has a negative temperature coefficient. However, at lower doping levels the avalanche effect has a positive temperature coefficient because the mobility of carriers decreases with increased temperature and more field is required for the same avalanche current. This is shown in Figs. 1.38(b) and 1.39. By designing reference diodes for 5 or 6 V, where the breakdown mechanism is a mixture of field emission and avalanche, there is the possibility of obtaining breakdown voltages with low temperature coefficients.

Another important factor in the design of reference diodes is the dynamic impedance which should be as low as possible. This is illustrated in Fig. 1.40 with a circuit example for a Zener diode with a dynamic impedance of 10 ohms. If the diode current and voltage are 7.5 mA and 6 V, respectively, with no load

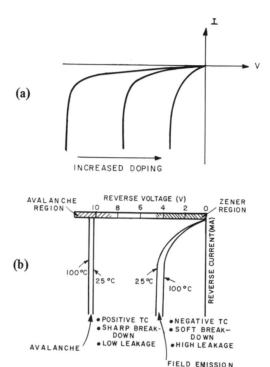

Fig. 1.38.  Zener reference diode characteristics:
(a)   Effect of increased doping, and
(b)   Effects of   temperature on reference and avalanche diodes. (After Henderson and
      Glasser[48], Courtesy *Electronic Products Magazine.*)

current drawn, the application of a 6 kilohm load draws 1 mA of load current
and the diode current drops to 6.5 mA and the Zener voltage is reduced by
10 mV (namely $R_Z \Delta I$).

Obviously the smaller $R_Z$ and the smaller the load current, the better the
voltage regulation. The use of an operational amplifier as in Fig. 1.41(a) or (b)
reduces the current variation through the Zener diode and thus gives better
voltage regulation.

Zener diodes are manufactured for a wide range of rated voltage, for example
from 3 to 100 V and with power ratings from 0.1 to 50 W. A 5-W unit at 6 V
may have a dynamic impedance of 1 ohm and a 70-V unit of the same rating
may have 30 ohms impedance. The self heating effects of reference diodes have
considerable influence on performance and must be included in any modeling
of the diode circuit for computer aided circuit design. Figure 1.42 shows some
linear and nonlinear models of reference diodes.

Since high-voltage reference diodes have positive temperature coefficients,
one or more standard forward-biased Si diodes may be connected in series so

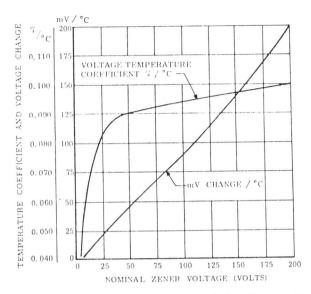

Fig. 1.39. Voltage-temperature coefficients of Si Zener diodes. (Courtesy Motorola Corporation.)

that their negative temperature coefficients of voltage may cancel to some extent the positive coefficient. A 10-V Zener diode may have a temperature coefficient of +6 mV °C$^{-1}$ and hence is compensated by three forward-biased diodes each of which have a variation of 2 mV °C$^{-1}$. At 0.6 V for 10 mA of forward current, the overall assembly has a reference voltage of 11.8 V. However, the improved temperature performance involves a larger dynamic resistance. Dynamic resistance effects can sometimes be minimized by the use of a bridge arrangement, shown in Fig. 1.43(a), or by the use of transistors as in Fig. 1.43(b) or (c). The temperature coefficient of voltage of a reference diode is a function of the diode current and manufacturer's data should always be consulted on its value.

High performance reference diodes are now usually made by expitaxial processes since these give better manufacturing control than bulk diffusion processes. Excessive microplasma noise in Zener diodes has been associated with diffusion-induced imperfections.[66,133]

Zener diodes are also used to provide protection against voltage transients. Figure 1.44 shows the clipping effect of a Zener diode on an oscillatory transient on a dc busbar. For a transient on an ac system two Zener diodes may be used connected back-to-back.

## 1.7 DIODES WITH NEGATIVE RESISTANCE

In general, devices that exhibit differential negative resistance with two terminals instead of three are not very convenient for circuit use. A semi-insulating bar of

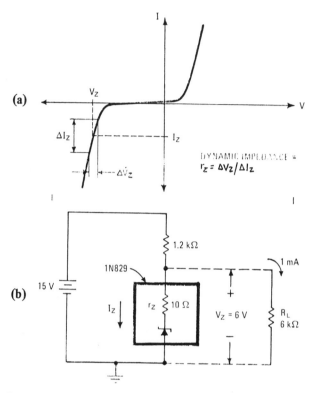

Fig.1.40. Effect of Zener diode internal resistance on the reference voltage accuracy. Because of the finite slope of the Zener diode's transfer characteristic (a) in the reverse breakdown region, Zener current is affected by both supply and load variations. In circuit (b), for instance, without the load the Zener current is 7.5 mA, producing a Zener voltage of 6 V. When the 1-mA load is connected, however, the Zener voltage drops by 10 mV, corresponding to a load regulation of 0.16%. (After Miller and DeFreitas.[83])

material with two injecting contacts may give the characteristic form shown in Fig. 1.45(a) if there are traps to be filled and conducting filaments develop.[87] A four-layer diode *pnpn* also has such a characteristic but with a contact added to one of the center regions becomes much easier to use as a three-terminal device, the thyristor.

The N shaped characteristic shown in Fig. 1.45(b) is typical of a tunnel diode. These have some circuit uses, as discussed in Chap. 3.

One useful negative resistance device is the double-base diode that depends on conductivity modulation to give an S shaped negative resistance in a three-terminal structure as in Fig. 1.45(a). Because of the three terminals and the single junction, it is generally termed the unijunction transistor. In its simplest form it is a bar of *n* Si of long minority carrier lifetime with contacts at each end and a $p^+$ junction alloyed partway along the bar as in Fig. 1.46. With a voltage $V$

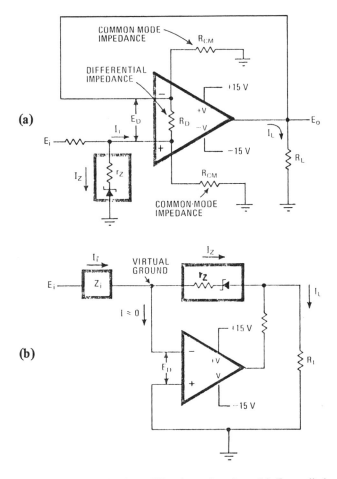

Fig. 1.41. The use of operational amplifiers in conjunction with Zener diodes to obtain a more constant reference voltage. If a Zener diode is buffered by an op amp, it will be isolated from load fluctuations. For both circuits shown here, the Zener current is independent of the load current, which is supplied by the op amp. In (a), the op amp acts as a high-impedance buffer for the Zener diode. In (b), because of the virtual ground, the input current, however derived, determines the Zener current. If the input current for either of these circuits is obtained from an op amp, then the Zener diode will also be isolated from supply variations. (After Miller and DeFreitas.[83])

applied, the potential in the bar at the junction is $R_{x_1} V/R_{x_2}$. If this exceeds the emitter voltage $V_E$ no injection occurs. However, if $V_E$ is raised by an external voltage source, as for instance by charging of the capacitor $C$ through the resistor $R$, injection begins. The injected holes (with their long lifetimes) conductivity modulate the resistance of the bar in region $x_1$ and so $R_{x_1}$ decreases and the voltage in the bar under the emitter $E$ decreases and the emission rises until the

**(a)**

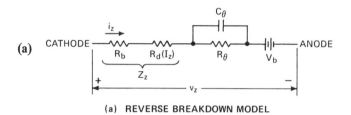

**(a)  REVERSE BREAKDOWN MODEL**

**(b)**

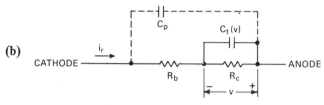

**(b)  REVERSE-BIASED MODEL**

**(c)**

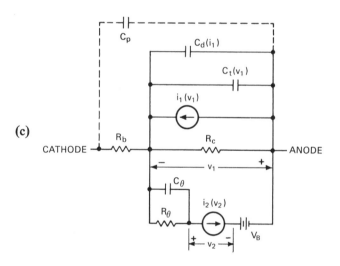

Fig. 1.42. Zener diode models for computer aided circuit design. Circuit (a) models the Zener in reverse breakdown. It consists of an $I_z$ dependent dynamic resistance $R_d$, bulk resistance $r_b$, breakdown back-voltage $V_b$, and thermal-effect components $R_\theta$ and $C_\theta$. Reverse-biased model (b) includes leakage resistance $R_c$, package capacitance $C_p$, and voltage-dependent transition capacitance. Circuit (c) is a nonlinear model for the full (forward and reverse) voltage range, addition of $C_p$, $R_\theta$, $C_\theta$, back-voltage $V_B$, and voltage-dependent current source $i_2$ to an Ebers-Moll diode model accounts for Zener self-heating effects and nonlinear voltage regulation. (Courtesy Emanuel Schnall.[116] Reprinted with permission from *Electronics*, Oct. 11, 1971; copyright McGraw-Hill, Inc., NY, 1971.)

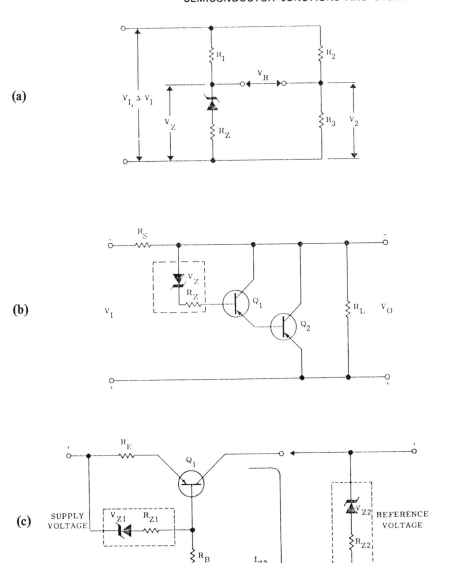

Fig. 1.43.  Circuits that give improved Zener action:
(a)  Impedance cancellation bridge,
(b)  Transistor circuit to reduce current demand on the reference diode, and
(c)  The reference diode $Z2$ is driven by a constant current supply determined by $Z1$ in the circuit to the left. (Courtesy Motorola Corp.)

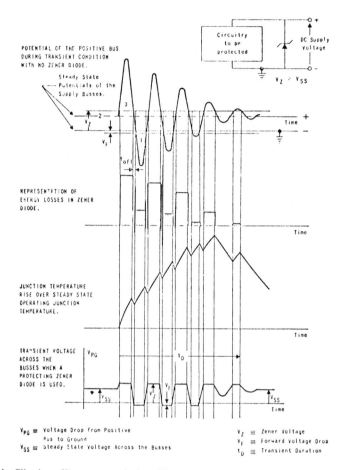

POTENTIAL OF THE POSITIVE BUS
DURING TRANSIENT CONDITION
WITH NO ZENER DIODE.

Steady State
Potentials of the
Supply Busses.

REPRESENTATION OF
ENERGY LOSSES IN ZENER
DIODE.

JUNCTION TEMPERATURE
RISE OVER STEADY STATE
OPERATING JUNCTION
TEMPERATURE.

TRANSIENT VOLTAGE
ACROSS THE
BUSSES WHEN A
PROTECTING ZENER
DIODE IS USED.

$V_{PG}$ ≡ Voltage Drop from Positive
Bus to Ground
$V_{SS}$ ≡ Steady State Voltage Across the Busses

$V_Z$ ≡ Zener Voltage
$V_F$ ≡ Forward Voltage Drop
$t_D$ ≡ Transient Duration

Fig. 1.44. Clipping effect on a typical oscillatory voltage transient achieved by a protective Zener diode. (After Acosta. Reprinted, with permission, from *IEEE Trans. on Industry and General Applications*, **IGA-5**, 1969, p. 483.)

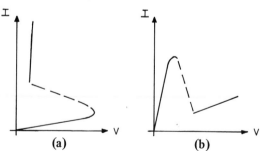

(a)                    (b)

Fig. 1.45. Negative resistance *I-V* characteristics:
(a)   *S* type, and
(b)   *N* type tunnel diode form.

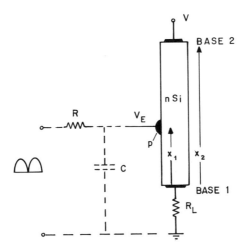

Fig. 1.46. The unijunction transistor or double-base diode depends on injection conductivity modulation, in the region $x$, for its negative resistance action.

capacitance is fully discharged. The circuit therefore produces spikes of current in the load, a small resistor $R_L$, and these may be used to trigger thyristor gate circuits.

The convenience of the device has led in recent years to the development of both alloyed and diffused unijunction transistor structures, some of which are shown in Fig. 1.47. The switching speed of the device is not high but nevertheless it fulfills a circuit need in conjunction with thyristors.

## 1.8 ELECTRON-BEAM BOMBARDED SEMICONDUCTOR DIODES AS AMPLIFIERS

EBS structures are hybrid vacuum tube-semiconductor devices with high voltage, high gain and high frequency performance. As shown in Fig. 1.48(a) a grid-controlled pulsed electron beam injects carriers into a reverse-biased Si semiconductor diode. The pair creation energy in Si is about 3.6 eV so a 10-20 keV incident beam can create thousands of carrier pairs in the diode and give current amplifications of 2000 or more. Figure 1.48(b) shows the predicted pulsed RF output power capability of a class B amplifier. This shows that 200 W of pulsed RF power with a bandwidth from dc to 1 GHz can be obtained in a load of 50 ohms. With a gain of 27 dB, the gain-bandwidth product is 500 GHz and the power bandwidth product is $2 \times 10^{11}$ W.sec$^{-1}$. A pulse amplifier has shown a rise time of 0.75 nsec for a pulse of 800 V which represents a voltage rise of $10^{12}$ V.sec$^{-1}$. These figures of merit are two orders of magnitude better than for comparable individual vaccum tube or semiconductor devices.

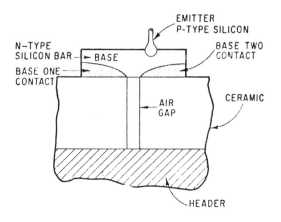

BAR UNIJUNCTION TRANSISTOR

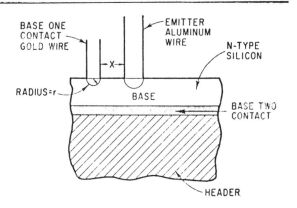

CUBE UNIJUNCTION TRANSISTOR

Fig. 1.47. Various unijunction transistor structures and geometries. (After Clark. Reprinted with permission, from *Electronics*, June 14, 38, 1965, p. 94.)

Two basic modulation techniques are used:
- Density modulation such as occurs in a close-spaced planar cathode-grid structure which varies the total current bombarding the diode.
- Deflection modulation where a relatively long circular or rectangular beam is deflected in either one or two orthogonal directions which controls the fraction of total beam current illuminating the semiconductor target.

Velocity modulation where the longitudinal velocity of electrons within a linear electron beam is varied, by interaction with a longitudinal electric field traveling on a slow-wave structure, producing density variations in the electron beam striking the target is less suitable.

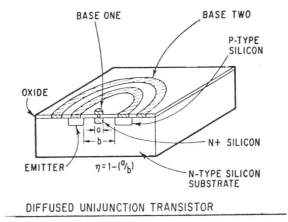

DIFFUSED UNIJUNCTION TRANSISTOR

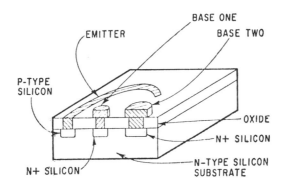

DIFFUSED (DOUBLE) UNIJUNCTION TRANSISTOR

Figure 1.47 (continued)

A deflection modulation EBS amplifier is shown in Fig. 1.49 with a traveling wave deflection structure. A typical diode target is a Si passivated planar diode of area cm² or more but mesa diodes are also used. The ideal parameters for an EBS diode material are: 1) high breakdown voltage, (2) high saturation velocity, 3) high low-field mobility, 4) high thermal conductivity, 5) low pair creation energy, and 6) adequate minority carrier diffusion length.

In some applications a high operating temperature and a low electron penetration depth are additional desirable features. Although most work has been with Si, other materials such as SiC, GaAs, and InP show promise.

Several EBS devices have demonstrated operating lifetimes well in excess of

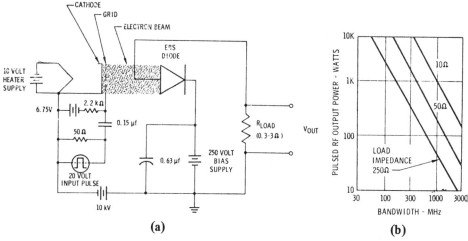

**(a)**                                                                **(b)**

Fig. 1.48.  Electron-beam bombarded semiconductor diode amplifier:
(a)  Test circuit used in generating current pulses, (After Bell, Knight and Silzars. Reprinted with permission from *IEEE Trans. Electron Devices*, ED-22, 1975, p. 361.)
(b)  Predicted pulsed RF output power capability of a class-B high-efficiency EBS amplifier as a function of the bandwidth for various load impedances. (After Silzars, Bates, and Ballonoff. Reprinted with permission from *Proc. IEEE*, **62**, 1974, pp. 1120, 1121.)

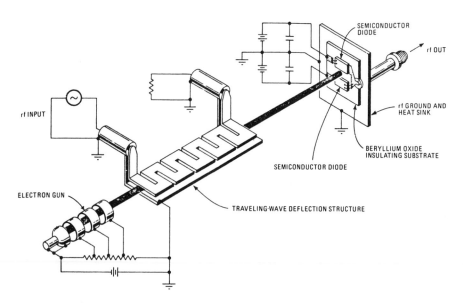

Fig. 1.49.  Schematic of a deflection-modulated EBS amplifier using a  cylindrical beam electron gun, distributed deflection structure, and a  two diode class-B connected semiconductor target. (After Bates. Reprinted with permission from *Electronics*, July 25, 1974; copyright McGraw-Hill, Inc., NY, 1974.)

**Table 1.5. Summary List of Advanced Lumped Element EBS Devices Resulting from the Combination of Electron Beam and Semiconductor Technologies.**

## EBS DEVICES

### ELECTRON BEAM

1 – TYPICALLY IN THE mA RANGE; VERY LOW CATHODE LOADING; LONG LIFE
2 – NO FOCUSING REQUIRED BEYOND THE ELECTRON GUN
3 – MODULATION EASILY APPLIED TO BEAM
4 – SIMPLE TUBE CONSTRUCTION

### SEMICONDUCTOR TARGET

1 – SIMPLE DIODE STRUCTURE MOUNTED DIRECTLY ON HEAT SINK
2 – CURRENT INJECTION CONTROLLED BY ELECTRON BEAM
3 – DIODE STRUCTURE RELATIVELY SIMPLE TO SHIELD FROM MONOENERGETIC BEAM RADIATION

### RF POWER AMPLIFIERS

1 BROADBAND CLASS B OPERATION
2 HIGH POWER – KILOWATTS OR MORE
3 HIGH EFFICIENCY – 50 TO 60%
4 HIGH GAIN – 25 TO 35 dB
5 BANDPASS CHARACTERISTIC POSSIBLE WITH OUTPUT TUNING
6 MULTIFUNCTION OUTPUT CAPABILITY

### PULSE AMPLIFIERS

1 LINEAR RESPONSE WITH RISE-TIMES LESS THAN 1 ns
2 HIGH POWER – KILOWATTS OR MORE
3 HIGH EFFICIENCY – 80 TO 90%
4 PULSE SHAPING POSSIBLE WITH REACTIVE ELEMENTS

### SWITCHES AND MODULATORS

1 HIGH CURRENT CAPABILITY – 200 AMPERES OR MORE
2 HIGH VOLTAGE CAPABILITY – 500 VOLTS OR MORE
3 HIGH POWER – TENS OF KILOWATTS OR MORE
4 FAST SWITCHING CAPABILITY WITH SHORT SIGNAL DELAY

### SIGNAL PROCESSORS

1 SINGLE EVENT SAMPLERS
2 A TO D AND D TO A CONVERTERS
3 PHASE CONTROL DEVICES
4 LIMITERS
5 MULTIFUNCTION SIGNAL CONTROL AND PROCESSING DEVICES

After Silzars, Bates and Ballonoff (Reprinted with permission from *Proc. IEEE*, Vol. 62, 1974, p. 1120.)

5000 hr and have met severe operating conditions comparable to those for other high power vacuum tube equipment.

Table 1.5 summarizes some of the characteristics being demonstrated by EBS devices in the fields of amplification, switching, modulation and signal processing, for special applications.

## 1.9 QUESTIONS

1. A Si *pn* diode structure is assumed to have the following doping profile and dimensions: $N_A = 10^{17}$ cm$^{-3}$, 0.01 cm; linear transition from $N_A\text{-}N_D = 10^{17}$ cm$^{-3}$ to $N_A\text{-}N_D = 10^{14}$ cm$^{-3}$ over a distance 0.01 cm; uniform doping $N_D = 10^{14}$ cm$^{-3}$ over a distance 0.01 cm; abrupt doping rise to $N_D = 10^{17}$ cm$^{-3}$ in a base substrate of thickness 0.02 cm.

   (a) At what reverse voltage will the depletion region extend across the $10^{14}$ cm$^{-3}$ doped *n* region? Sketch and number the voltage and field strength diagrams for this condition.

   (b) At what higher voltage value would avalanche occur in this one-dimensional structure?

   (c) If the $10^{14}$ cm$^{-3}$ region lifetime is dominated by a recombination center at the center of the bandgap of concentration $5 \times 10^{13}$ cm$^{-3}$ with $\sigma_p = \sigma_n = 10^{-15}$ cm$^2$, estimate the forward $I$ vs $V$ characteristic of the diode at 300°K.

2. An $n^+p$ diode has a uniform $p$ base region of doping $10^{14}$ cm$^{-3}$ and thickness 50 $\mu$m and infinite recombination at the ohmic contact.

   (a) What must the electron minority carrier lifetime be so that 90% of the electrons injected into the base diffuse across to the contact?

   (b) Assume the lifetime in the base region is not constant but varies linearly from $10^{-8}$ to $10^{-6}$ sec as a function of electron concentration from $10^{13}$-$10^{15}$ cm$^{-3}$. At what current density will conductivity modulation become significant in this diode?

3. For an $n^+p$ Si diode of area $10^{-2}$ cm$^2$ with $N_A = 10^{14}$ cm$^{-3}$ and thickness 50 $\mu$m and electron lifetime $10^{-6}$ sec, what will be the variation of diode open circuit voltage when a forward current of density 100 A cm$^{-2}$ is abruptly terminated? If the diode is in series with a resistance of 50 ohms and the voltage is swung abruptly from 50 V forward to 100 V reverse, sketch and dimension the current transient that ensues.

4. Consider a Si $p^+nn^+$ junction diode where $N_D = 3.10^{14}$ cm$^{-3}$ and $n$ region is 10 $\mu$m thick and contains $1.10^{14}$ cm$^{-3}$ Au as a recombination level. Determine the forward current variation with voltage that best characterizes the diode and the effective minority carrier lifetime at 300°K and at 400°K. If the diode is $n^+pp^+$ in form with $N_A = 3.10^{14}$ cm$^{-3}$ and $N_{Au} = 1.10^{14}$ cm$^{-3}$ rework the problem to determine the difference in the recombination center behavior.

5. An $n$ type bar (2 cm $\times$ 1 cm $\times$ 0.01 cm) of Si semiconductor doped $10^{15}$ cm$^{-3}$ has a bulk lifetime of $10^{-5}$ sec (300°K) and a surface recombination velocity of $10^5$ cm/sec.

   (a) Does the surface recombination velocity have an influence on the photoconductive effective lifetime?

   (b) For a hole recombination center of density $10^{14}$ cm$^{-3}$ at $E_V + 0.25$ eV and capture cross section $10^{-16}$ cm$^2$, what will be the effective photoconductive lifetime at 300°K? At 400°K?

   (c) What would be the lifetime at those two temperatures if the recombination center was at mid-bandgap?

6.  A Si $p^+nn^+$ junction diode is of area $10^{-3}$ cm$^2$ and the $n$ region (doping $N_D = 3.10^{14}$ cm$^{-3}$ and Au $1.10^{14}$ cm$^{-3}$) is 10 $\mu$m thick. Estimate the ac admittance at 0.26 V forward bias (300°K) for frequencies of $10^3$ Hz and $10^6$ Hz.

7.  An $n^+p$ GaAs diode of area $10^{-3}$ cm$^{-2}$ has a $10^{16}$ cm$^{-3}$ uniformly-doped $p$ region and is illuminated with bandgap light that creates a reverse current of $10^{-9}$ A at zero bias voltage. The reverse voltage across the diode is now increased and the capacitance suggests that the depletion thickness has become 2 $\mu$m.
    (a)  What will be the capacitance and the required voltage?
    (b)  If the photo-induced reverse current of the device becomes $3 \times 10^{-9}$ A with this voltage applied, what can be inferred about the electron minority carrier lifetime in the $p$ region?

8.  Diode circuits may often be understood by various simplified representations of the diode (see Fig. P1.8), such as a linearization of zero forward voltage drop and zero reverse current.
    (a)  Apply this approximation to the circuit shown that contains a current source of 10 mA, and sketch $V_2$ versus $V_1$, and $I_1$ versus $V_1$, for $V_1$ between $-10$ and $+10$ V.
    (b)  Repeat the analysis for the diodes represented as having a constant forward voltage drop of 1 V.

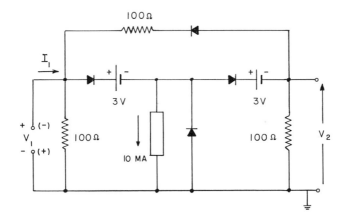

9.  A one-dimensional diffusion equation of form $\dfrac{\partial N}{\partial t} = D \dfrac{\partial^2 N}{\partial x^2}$, gives for a source of undepleted concentration a complementary error function expression $N(x,t) = N_s$ erfc $(x/2\sqrt{Dt})$. For a source of initial surface concentration $C_s$ atoms per unit area, a Gaussian expression is obtained

$$N(x,t) = \frac{C_s}{\sqrt{\pi Dt}} \exp(-x^2/4Dt).$$

    (a)  Consider an impurity such as P diffused at 1100°C into $p$ type $10^{15}$ cm$^{-3}$ doped Si. Assume an infinite source surface concentration of $10^{20}$ cm$^{-3}$ and a diffusion coefficient of $2 \times 10^{-13}$ cm$^2$/sec. How long must the diffusion be continued for a junction depth of 4 $\mu$m? What would be the skin resistance of this diffusion?

(b) If the source was an oxide of P concentration $10^{16}$ cm$^{-2}$ what must be the diffusion time (1100°C) for a 4 $\mu$m junction depth?

(c) What are the relative merits of P, As and Sb for $n$ type diffusions in Si?

(d) What are the relative merits of B, Al, Ga and In for $p$ type diffusions in Si?

10. The maximum electric field in an abrupt junction diode is given by $\mathscr{E}_m = 2V/W_d$ where $V$ is the total voltage across the junction and $W_d$ is the depletion region thickness. Does it make physical sense to use such an expression to design Zener diodes of heavy doping? Calculate the design of a 5 V Si Zener diode of capacitance less than 100 pF and dynamic resistance less than 10 ohms for a current level of 10 mA. Estimate the effect of a rise in temperature from 300°K to 400°K. Could some degree of temperature compensation be achieved by the use of a series forward-biased Si diode?

11. The electron beam of a scanning electron microscope is used to make measurements on the cleaved surface of a $p^+n$, GaAs junction for which $N_D$ is $10^{16}$ cm$^{-3}$ and $\tau_p$ is $10^{-8}$ sec. If the beam voltage is 25 kV, the beam current $10^{-10}$ A and the carriers are assumed all created in a sphere of radius 1 $\mu$m, is it likely that a low-level or a high-level injection lifetime is being measured?

12. For the expression $I_0 [\exp(qV/nkT-1)]$, the value of $n$ can be determined from the ratio of two average currents produced by two different voltage sine waves.[25] Develop this idea for automatic instrumentation of the $n$ values of diodes in production testing.

13. Discuss the variation of the diode ideality factor $n$ with the density, energy level and spatial distribution of recombination centers. Then consider Pt in $n$ type Si as the recombination center and discuss the effects of temperature for chosen doping levels corresponding to a 200 V rated $p^+n$ diode.

14. Explain in detail the energetic concepts of calculating $pn$ junction capacitance involving electrostatic and chemical energies. Compare the result obtained with the capacitance vs voltage relationship obtained by other methods.

15. Describe small-signal phase-shift and amplitude variation with frequency in a Si abrupt junction diode with a dc bias. What differences would be expected for phase-shift variation (a) in $pin$ or retrograded junctions, (b) in linearly-diffused junctions and (c) in snap-action diodes?

16. Discuss the current-voltage characteristics of $nn^+$ low-high junctions (a) with ohmic contacts applied and (b) as part of a $pnn^+$ rectifier.

17. Discuss the models for high level injection $I$-$V$ characteristics in Si junctions.

18. Describe the charge, electric-field and voltage distribution for a Si $pn^-n^+$ diode. If the $n^-$ region is almost intrinsic would any time delays in transmitting current when voltage pulses are applied be expected?[68]

19. Discuss bevel angle effects on the breakdown voltage of Si step or diffused junctions.

20. Describe guard-ring technologies for Si junctions that must have very uniform avalanche characteristics.

21. How are ionization rates measured for electrons and holes on Si?

22. Several semiempirical expressions have been proposed for the relationship between breakdown voltage and depletion layer thickness in Si junctions—for instance see Eq. (1.64) et seq. Explore any justification and define their limitations.

23. To what extent does carrier generation within the space-charge region of a Si $p$-$n$ junction affect the breakdown voltages?[15] What experiment might be made to verify the effect?

24. Review ohmic contact technologies for Si discrete devices and integrated circuits.

25. Review ohmic contact technologies for the III-V semiconductors, GaAs and GaP.

26. Decaying exponential signals are characteristic of semiconductors relaxing back to equilibrium following an abrupt change in the populations of carriers occupying the available states. The magnitudes of such signals indicate the density of states involved, while the decay time constants (as a function of temperature) provide information

on the energy levels of the states in question. As such, both the amplitudes and the time constants of the signals are of physical interest. Discuss how a processor might be constructed to perform a continuous real-time cross correlation between the experimental signals and an appropriately synchronized locally-generated exponential waveform, for use with transient space-charge measurements on reverse-biased Au-doped Si junction diodes, providing both correlation spectra that yield energy levels and also trap profiles that expose the spatial distribution of the Au inside the junctions.[80]

27. Several methods of measuring the switching time and minority carrier lifetime of a diode are known. Discuss their relative merits for a 0.1 A, 100 V Si diode in which the lifetime is expected to be about 50 nsec.

28. Discuss the relative merits of Au and Pt as lifetime control impurities in Si.[71,84]

29. In studies of Au in Si lifetime maps have been constructed.[87] Construct part of a similar map for Pt in Si at $300°K$ and compare. Could a useful nomogram be constructed for Pt doping effects? Describe methods of processing (or gettering) Si to achieve high minority carrier lifetimes.

30. Describe some thermally-stimulated capacitance studies in Si junctions for studying impurities in the depletion region.

31. Review optical-induced carrier methods for determining minority carrier lifetime in Si.

32. Review capacitance and current transients in Si junction depletion regions to determine emission coefficients and lifetimes.

33. Discuss Auger recombination processes in semiconductors.

34. Discuss the variation of minority carrier lifetimes with temperature in indirect gap and direct gap semiconductors.

35. Discuss the role of surface effects in the measurement of minority carrier lifetime.

36. Discuss the design compromises necessary in a snap-action diode.

37. Examine the relative merits of the doping profilers of Copeland, Miller and other commercial units such as spreading-resistance probes for diffusion profiling.

38. Consider the relative merits of P, As and Sb diffusion in $p$ type Si. Strain effects and strain compensation should be included in the study.[67]

39. Review the technology of oxide growth on Si.

40. Review the technology of native oxide growth on GaAs and $SiO_2$ and $Al_2O_3$ deposition.

41. $SiO_2$ oxide layers are used as diffusion masks on Si. Discuss how thick such masks must be for $p$ and $n$ type diffusions.

42. Discuss failure mechanisms in semiconductor junctions and the role of accelerated testing to reveal them.[45]

43. Devise a circuit that uses the logarithmic characteristics of diodes to multiply two small voltages. What degree of accuracy may be expected and what is the voltage range of operation.

44. If $n$ diodes are connected in series to increase the reverse voltage that can be withstood, it is advisable to connect a capacitance $C$ across each diode to improve voltage sharing during transients. Show that $C$ should be chosen so that $R_L C/n$ approximately equals the storage time of an average diode.

45. Obtain data sheets on the rated heat dissipation of some commercial finned heat sinks having areas of at least 100 $cm^2$. Then compare the stated performance with that expected from (a) the dissipation approximation of 1 mW $cm^{-2}$ $°C^{-1}$ and (b) from the convection and radiation coefficients given in handbooks or elsewhere.

46. Explain the action of a three-stage Cockroft-Walton voltage multiplier circuit (see Fig. 1.37), and discuss the relative size needed for the capacitors.

47. What are the design factors controlling the dynamic impedance in Zener diodes? How is the impedance related to the reference voltage and the current level? How can dynamic impedance effects be minimized in circuit applications?

48. How can temperature effects be minimized in Zener diodes? Would a GaAs Zener diode be potentially better than a Si diode in temperature stability, because of the larger bandgap?

49. Discuss circuit models for Zener diodes that include thermal effects. Select a simple regulator circuit involving the Zener model and apply a computer-aided-design program to demonstrate the usefulness of the model.[116]

50. Use of an operational amplifier does not fully overcome Zener imperfections but it can create nearly ideal circuit conditions for the diode by isolating and buffering it against both supply and load variations. Hence a voltage temperature stability of 1 ppm $°C^{-1}$ is obtainable. Discuss such circuit combinations of Zener diodes and operational amplifiers.[83]

51. Zener diodes are not good in applications involving fast pulses because of high junction capacitance. Analyze the problem for a 7-V Zener diode rated at 0.1 A. Can this circuit limitation be readily overcome?

52. Discuss voltage transient protection in ac and dc circuits by the use of Zener diodes. Illustrate by waveforms for specific circuits.

53. Discuss general design considerations for Si unijunction transistors. Then discuss specifically the high frequency design limitations.

54. In a Si diode target for an electron beam bombarded semiconductor amplifier, the diode may be reverse-biased so that the depletion region extends throughout the semiconductor. In general, the rise time of the target will be determined by a combination of the transit time and capacitance effects. Since these two effects vary inversely with respect to each other as a function of $w$, it is possible to minimize the overall rise time by properly choosing the drift-region width. Consider this optimization problem and also the output capability in current, voltage and power.

55. Review recent performance data for electron-bombarded semiconductor amplifiers, modulators, pulsers, etc.

56. Derive the relationship between output power and bandwidth shown in Fig. 1.48(b) for an electron beam bombarded semiconductor diode amplifier.

## 1.10 REFERENCES AND FURTHER READING SUGGESTIONS

1. Acosta, O.N., "Zener diode—A protecting device against voltage transients," *IEEE Trans. Industry and General Applications*, IGA-5, July/August 1969, p. 481.

2. Aldridge, R.V., "On the behaviour of forward biased silicon diodes at low temperatures," *Solid-State Electronics*, 17, 1974, p. 617.

3. Armstrong, H.L., "A theory of voltage breakdown of cylindrical *p-n* junctions, with applications," *IRE Trans. Electron Devices*, ED-4, 1957, p.15.

4. Ashburn, P., D.V. Morgan, and M.J. Howes, "A theoretical and experimental study of recombination in silicon *p-n* junctions," *Solid-State Electronics*, 18, 1975, p. 569.

5. Ashley, K.L., and J.R. Biard, "Optical microprobe response of GaAs diodes," *IEEE Trans. Electron Devices*, ED-14, 1967, p. 429.

6. Authier, A., D. Simon, and A. Senes, "Correlation between lattice defects and electrical properties of silicon Zener diodes," *Phys. Stat. Sol.*, 10, 1972, p. 233.

7. Bakanowski, A.E., and J.H. Forster, "Electrical properties of gold-doped diffused silicon computer diodes," *Bell System Tech. J.*, 39, 1960, p. 87.

8. Bakowski, M., and B. Hansson, "Influence of bevel angle and surface charge on the breakdown voltage of negatively beveled diffused *p-n* junctions," *Solid-State Electronics*, 18, 1975, p. 651.

9. Baliga, B.J., and E. Sun, "Comparison of gold, platinum and electron irradiation for controlling lifetime in power rectifiers and thyristors," *IEEE Trans. Electron Devices*, ED-24, 1977, p. 685.

10. Bates, D.J., "Semiconductors inside tubes make high-performance rf amplifiers," *Electronics*, July 25, 1974, p. 85.
11. Bloem, J., L.J. Giling, and M.W.M. Graef, "The incorporation of phosphorus in silicon epitaxial layer growth," *J. Electrochem. Soc.*, 121, 1974, p. 1354.
12. Bora, J.S., and A.K. Babar, "Temperature dependence of transistors and Zener diodes and their life testing procedures," *J. of the Institution of Telecommunications Engineers*, New Delhi, 16, 1970, p. 6777.
13. Brown, M., C.L. Jones, and A.F.W. Willoughby, "Solubility of gold in *n*-type silicon," *Solid-State Electronics*, 18, 1975, p. 763.
14. Bullis, W.M., "Measurement of carrier lifetime in semiconductors: an annotated bibliography covering the period 1949-1967," Document AD 674 627 National Technical Information Service, Springfield, Virginia (1968).
15. Bulucea, C.D., and D.C. Prisecaru, "The calculation of the avalanche multiplication factor in silicon *p-n* junctions taking into account the carrier generation (thermal or optical) in the space-charge region," *IEEE Trans. Electron Devices*, ED-20, 1973, p. 692.
16. Burger, R.M., and R.P. Donovan, eds., "Fundamentals of silicon integrated device technology," Vol. 1, *Oxidation Diffusion and Epitaxy*, Prentice-Hall, Englewood Cliffs, NJ, 1967.
17. Choo, S.C., "Analytical approximations for an abrupt *p-n* junction under high-level conditions," *Solid-State Electronics*, 16, 1973, p.793.
18. Choo, S.C., "Carrier generation-recombination in the space-charge region of an asymmetrical *p-n* junction," *Solid-State Electronics*, 11, 1968, p. 1069.
19. Choo, S.C., "Theory of a forward-biased diffused-junction *p-i-n* rectifier—II. analytical approximations," *Solid-State Electronics*, 16, 1973, p. 197.
20. Choo, S.C., "Theory of a forward-biased diffused-junction *p-i-n* rectifier—part III: further analytical approximations," *IEEE Trans. Electron Devices*, ED-20, 1973, p. 418.
21. Choo, S.C., and R.G. Mazur, "Open circuit voltage decay behavior of junction devices," *Solid-State Electronics*, 13, 1970, p. 553.
22. Choo, S.C. and A.C. Sanderson, "Bulk trapping effect on carrier diffusion length as determined by the surface photovoltage method: theory," *Solid-State Electronics*, 13, 1970, p. 609.
23. Clark, L., "Now new unijunction geometries," *Electronics*, 38, 1965, p. 93.
24. Clement, R.D., and R.L. Starliper, "Four-layer diode circuit outregulates Zener by 5:1," *Electronics*, June 7, 1971, p. 81.
25. Coerver, L.E., "New method of measurement of diode junction parameters," *IEEE Trans. Electron Devices*, ED-16, 1969, p. 1082.
26. Copeland, J.A., "A technique for directly plotting the inverse doping profile of semiconductor wafers," *IEEE Trans. Electron Devices*, ED-16, 1969, p. 445.
27. Cornu, J., and R. Sittig, "The influence of doping inhomogeneities on the reverse characteristics of semiconductor power devices," *IEEE Trans. Electron Devices*, ED-22, 1975, p. 108.
28. Davies, R.L., and F.E. Gentry, "Control of electric field at the surface of *p-n* junctions," *IEEE Trans. Electron Devices*, ED-11, 1964, p. 313.
29. Dean, R.H., and C.J. Neuse, "A refined step-recovery technique for measuring minority carrier lifetimes and related parameters in asymmetric *p-n* junction diodes," *IEEE Trans. Electron Devices*, ED-18, 1971, p. 151.
30. DeSmet, L., and R. van Overstraeten, "Calculation of the switching time in junction diodes," *Solid-State Electronics*, 18, 1975, p. 557.
31. Dewan, S.B., and A. Straughem, *Power Semiconductor Circuits*, Wiley-Interscience, NY 1975.

32. Drew, R.C., "Special report on Zener diodes," *Electronic Products*, August 1969, p. 74.
33. Dumin, D.J., "Emission of visible radiation from extended plasmas in silicon diodes during second breakdown," *IEEE Trans. Electron Devices*, ED-16, 1969, p. 479.
34. Edwards, W.D., W.A. Hartman, and A.B. Torrens, "Specific contact resistance of ohmic contacts to gallium arsenide," *Solid-State Electronics*, 15, 1972, p. 387.
35. EerNisse, E.P., "Accurate capacitance calculations for *p-n* junctions containing traps," *Appl. Phys. Lett.*, 18, 1971, p. 183.
36. Engstrom, O., and H.G. Grimmeiss, "Thermal activation energy of the gold-acceptor level in silicon," *J. Appl. Phys.*, 46, 1975, p. 831.
37. Esaki, L., and L.L. Chang, "New phenomenon in semiconductor junctions—GaAs duplex diodes," *Phys. Rev. Lett.*, 25, 1970, p. 653.
38. Evwaraye, A.O., and B.J. Baliga, "Identification of the dominant recombination center in electron irradiated *n*-silicon using DLTS and lifetime measurement," *Electrochem. Soc. Mtg. Abstracts*, Oct., 1976.
39. Farrell, R., "A SiC backward diode," *IEEE Trans. Electron Devices*, ED-16, 1969, p. 253.
40. Fivecoate, J.E., S.E. Miller, and J.E. Reynolds, "Characterization of high power Zener diode," *Conference Record of 1970, Fifth Annual Meeting of the IEEE Industry and Applications Group*, Chicago, Ill., USA, October 5-8, 1970, (NY, USA, 1970) p. 437.
41. Friis, H.T., "Analysis of harmonic generator circuits for step recovery diodes," *Proc. IEEE*, 55, 1967, 1192.
42. Ghoshtagore, R.N., "Silicon dioxide masking of phosphorus diffusion in silicon," *Solid-State Electronics*, 18, 1975, p. 399.
43. Groezinger, O. and W. Haecker, "Influence of tunneling processes on avalanche breakdown in Ge and Si," *J. Appl. Phys.*, 44, 1973, p. 1307.
44. Grove, A.S., *Physics and Technology of Semiconductor Devices*, John Wiley and Sons, NY, 1967.
45. Guzski, D.P., and A. Fox, "Reliability technology in accelerated testing," *Proc. 1968 Annual Symposium Reliability*, January, 1968, IEEE, p. 91.
46. Hauser, J.R., and P.M. Dunbar, "Minority carrier reflecting properties of semiconductor high-low junctions," *Solid-State Electronics*, 18, 1975, p. 715.
47. Heald, D.I., et al., "Thermodynamic considerations of *p-n* junction capacitance," *Solid-State Electronics*, 16, 1973, p. 1055.
48. Henderson, B., and S. Glasser, "Constant voltage diodes are getting sharper," *Electronic Products Magazine*, 16, March 18, 1974, p. 43.
49. Hess, D.W., and B.E. Deal, "Kinetics of the thermal oxidation of silicon in $O_2/N_2$ mixtures at 1200°C," *J. Electrochem. Soc.*, 122, 1975, p. 579.
50. Higuchi, H., and S. Nakamura, "Gettering of electrically active gold in silicon," *J. Electrochem. Soc.*, 122, 1975, p. 85C, Abstract No. 175, Toronto, Canada.
51. Howes, M.J., D.V. Morgan, and P. Ashburn, "The small signal admittance of carbon implanted *p-n* diodes," *Solid-State Electronics*, 18, 1975, p. 491.
52. Huff, H.R., and R.R. Burgess, (eds.), *Semiconductor Silicon 1973*, Electochemical Soc. Symposium Series, 1973.
53. Hull, R.W., and K. Simonyan, "Log diodes can be counted on," *Electronics*, March 31, 1969, p. 104.
54. Jain, R.K., and R.J. Van Overstraeten, "Accurate theoretical arsenic diffusion profiles in silicon from processing data," *J. Electrochem. Soc.*, 122, 1975, p. 552.
55. Jonscher, A.K., *Principles of Semiconductor Device Operation*, John Wiley and Sons, NY, 1960.
56. Kamins, T.I., and B.E. Deal, "Characteristics of $Si-SiO_2$ interfaces beneath thin silicon films defined by electrochemical etching," *J. Electrochem. Soc.*, 122, 1975, p. 557.

57. Kennedy, D.P., "Reverse transient characteristics of a *p-n* junction diode due to minority carrier storage," *IRE Trans. Electron Devices*, ED-9, 1962, p. 174.
58. Kingston, R.H., "Switching time in junction diodes and junction transistors," *Proc. IRE*, 42, 1954, p. 829.
59. Kircher, C.J., "Comparison of leakage currents in ion-implanted and diffused *p-n* junctions," *J. Appl. Phys.*, 46, 1975, p. 2167.
60. Kobayashi, M., A. Shimizu, and T. Ishida, "Dependence of Zener breakdown voltage on impurity distribution in hyper-abrupt junction," *Electronics and Comm. in Japan*, 51, 1968, p. 168.
61. Kokkas, A.G., "Empirical relationships between thermal conductivity and temperature for silicon and germanium," *RCA Review*, 35, 1974, p. 579.
62. Kotzebue, K.L., "A circuit model of the step-recovery diode," *Proc. IEEE*, 53, 1965, p. 2119.
63. Krakauer, S.M., "Harmonic generation, rectification, and lifetime evaluation with step recovery diode," *Proc. IRE*, 50, 1962, p. 1665.
64. Kroll, W.J., Jr., and R.L. Titus, "Formation of silica films on silicon using silane and carbon dioxide," *J. Electrochem. Soc.*, 122, 1975, p. 573.
65. Lanyon, H.P.D., and A.E. Sapega, "Measurement of semiconductor junction parameters using lock-in amplifiers," *IEEE Trans. Electron Devices*, ED-2, 1973, p. 487.
66. Lecoy, G., R. Alabedra, and B. Barban, "Noise studies in internal field emission diodes," *Solid-State Electronics*, 15, 1972, p. 1273.
67. Lee, Y.T., N. Miyamoto, and J. Nishizawa, "The lattice misfit and its compensation in the Si-epitaxial layer by doping with germanium and carbon," *J. Electrochem. Soc.*, 122, 1975, p. 530.
68. Leenov, D., "Unusual transient properties of a wide silicon *p-i-n* diode," *Proc. IEEE*, 58, 1970, p. 1156.
69. Lehovec, K., "C-V analysis of a partially depleted semiconducting channel," *Appl. Phys. Lett.*, 26, 1975, p. 82.
70. Lewis, D.C., "On the determination of the minority carrier lifetime from the reverse recovery transient of *pnR* diodes," *Solid-State Electronics*, 18, 1975, p. 87.
71. Lisiak, K.P., and A.G. Milnes, "Platinum as a lifetime control deep impurity in silicon," *J. Appl. Phys.*, 46, 1975, p. 5229.
72. Lukaszek, W.A., A. van der Ziel, and E.R. Chenette, "Investigation of the transition from tunneling to impact ionization multiplication in silicon *p-n* junctions," *Solid-State Electronics*, 19, 1976, p. 57.
73. Maerfeld, C., Ph. Defranoud, and P. Tournois, "Acoustic storage and processing device using *p-n* diodes,", *Appl. Phys. Lett.*, 27, 1975, p. 577.
74. McGreivy, D.J., and C.R. Viswanathan, "The effect of arsenic in improving lifetime in silicon-on-sapphire films," *Appl. Phys. Lett.*, 25, 1974, p. 505.
75. McKelvey, J.P., *Solid State and Semiconductor Physics,* Harper and Row, NY, 1966.
76. Meek, R.L., T.E. Seidel, and A.G. Cullia, "Diffusion gettering of Au and Cu in silicon," *J. Electrochem. Soc.*, 122, 175, p. 786.
77. Mielke, W., "Penetration of gold and platinum through phosphorus-doped *n+* layers in silicon," *J. Electrochem. Soc.*, 122, 175, p. 965.
78. Miller, G.L., D.V. Lang, and Kimerling, "Capacitance transient spectroscopy," *Ann. Rev. Mater. Sci.*, 1977, p. 377.
79. Miller, G.L., "A feedback method for investigating carrier distributions in semiconductors," *IEEE Trans. Electron Devices*, ED-19, 1972, p. 1103.
80. Miller, G.L., J.V. Ramirez, and D.A.H. Robinson, "A correlation method for semiconductor transient signal measurements," *J. Appl. Phys.*, 46, 1975, p. 2638.
81. Miller, S.E., J.E. Reynolds, and J.R. Washburn, "Absorption of transient energy by a

high power Zener diode," *IEEE Conference Record*, 4th Annual Meeting, Ind. & Gen. Appl. Group, Oct. 12-16, 1969, Detroit, Michigan.

82. Miller, S.E., J.E. Reynolds, and J.R. Washburn, "Design of a high-power Zener diode and its energy absorption capability," *IEEE Trans. on Industry and General Applications*, IGA-7, 1971, p. 208.

83. Miller, W.D., and R.E. DeFreitas, "Op amp stabilizes Zener diode in reference-voltage source," *Electronics*, Feb. 20, 1975, p. 101.

84. Miller, M.D., H. Schade, and C.J. Neuse, "Use of platinum for lifetime control in power devices," *IEEE International Electron Devices Meeting Technical Digest*, New York, 1975, p. 180; 1976, p. 491.

85. Millman, J., and C.C. Halkias, *Integrated Electronics: Analog and Digital Circuits and Systems*, McGraw-Hill, NY, 1972.

86. Milnes, A.G., and D.L. Feucht, *Heterojunctions and Metal Semiconductor Junctions*, Academic Press, NY, 1972.

87. Milnes, A.G., *Deep Impurities in Semiconductors*, John Wiley & Sons, NY, 1973.

88. Mimura, T., "Voltage-controlled DNR in unijunction transistor structure," *IEEE Trans. Electron Devices*, ED-21, 1974, p. 604.

89. Mlynar, P. (ed.), *Westinghouse High-Voltage Silicon Rectifier Designer's Handbook*, Westinghouse Semiconductor Division, Youngwood, PA, 1963.

90. Moline, R.A., and G.F. Foxhall, "Ion-implanted hyperabrupt junction voltage variable capacitors," *IEEE Trans. Electron Devices*, ED-19, 1972, p. 267.

91. Moll, J.L., "*p-n* Junction charge-storage diodes," *Proc. IRE*, 50, 1962, p. 43.

92. Moll, J., *Physics of Semiconductors*, McGraw-Hill, NY, 1964.

93. Mortenson, K.E., *Variable Capacitance Diodes*, Artech House, Dedham, MA, 1975.

94. Mortenson, K.E., and J.M. Borrego, *Design Performance and Applications of Microwave Semiconductor Control Components*, Artech House, Dedham, MA, 1972.

95. Muhlbauer, A., F. Sedlak, and P. Voss, "Investigation of breakdown and resistivity striations in high voltage silicon diodes," *J. Electrochem. Soc.*, 122, 1975, p. 1113.

96. Muñoz, E., J.P. Monico, and I. Padilla, "Breakover and negative resistance in polycrystalline silicon *n-π-n* DIAC's," *Proc. IEEE*, 62, Oct. 1974, p. 1394.

97. Namordi, M.R., and H.W. Thompson, Jr., "Aluminum-silicon Schottky barriers as semiconductor targets for EBS devices," *Solid-State Electronics*, 18, 1975, p. 499.

98. Nanavati, R.P., *An Introduction to Semiconductor Electronics*, McGraw-Hill, NY, 1963.

99. Nishida, K., T. Takakawa, and M. Nakajima, "New Si planar junction diodes with uniform avalanche multiplication," *Appl. Phys. Lett.*, 25, 1974, p. 669.

100. Norris, C.B., Jr., "Optimum design of electron beam-semiconductor linear low-pass amplifiers—Part II: output capabilities," *IEEE Trans. Electron Devices*, ED-20, 1973, p. 827.

101. Nuttall, K.I., "An investigation into the behaviour of trapping centres in microplasmas," *Solid-State Electronics*, 18, 1975, p. 13.

102. Okuto, Y., and C.R. Crowell, "Threshold energy effect on avalanche breakdown voltage in semiconductor junctions," *Solid-State Electronics*, 18, 1975, p. 161.

103. Olson, H.M., "Design calculations of reverse bias characteristics for microwave *p-i-n* diodes," *IEEE Trans. Electron Devices*, ED-14, 1967, p. 418.

104. O'Shaughnessy, T.A., and H.D. Barber, "The solid solubility of gold in doped silicon by oxide encapulation," *J. Electrochem. Soc.*, 121, 1974, p. 1350.

105. Otani, Y., K. Matsubara, and Y. Nishida, "New photomagneto diode (PMD)," *Proc. of the First Conf. on Solid-State Devices*, Tokyo, 1969, Suppl. of the J. of the Japan Society of Appl. Phys., 39, 1970, p. 248.

106. Pal, B.B., and S.K. Roy, "Small-signal analysis of *p-n* junction avalanche diode having a

uniform avalanche zone and a drift zone for unequal ionization rates and drift velocities of electrons and holes," *Phys. D. Appl. Phys.*, **4**, 1971, p. 2041.

107. Parrott, J.E., "Carrier temperature effects and energy transfers in non-radiative recombination," *Solid-State Electronics*, **19**, 1976, p. 229.

108. Pronin, B.V., and I.V. Ryzhikov, "Investigation of tunnel breakdown and photoelectric properties of reverse-biased *p-n* junctions in $Al_xGa_{1-x}As$ and $GaAs_{1-x}P_x$," *Soviet Physics–Semiconductors*, **6**, 1973, p. 1247.

109. Reichl, H., and H. Bernt, "Lifetime measurements in silicon epitaxial materials," *Solid-State Electronics*, **18**, 1975, p. 453.

110. Ringo, J.A., and P.O. Lauritzen, "1/f Noise in uniform avalanche diodes," *Solid-State Electronics*, **16**, 1973, p. 327.

111. Ryan, R.D., and Eberhardt, J.E., "Hole Diffusion Length in High Purity *n*-GaAs," *Solid-State Electronics*, **15**, 865, (1972).

112. Sah, C.T., R.N. Noyce, and W. Shockley, "Carrier generation and recombination in *p-n* junction and *p-n* junction characteristics," *Proc. IRE*, **45**, 1957, p. 1228.

113. Sah, C.T., C.T. Wang, and S.H. Lee, "Junction edge region thermally stimulated capacitance (TSCAP) of Ni-Si doped with phosphorus and bismuth," *Appl. Phys. Lett.*, **25**, 1974, p. 524.

114. Sankur, H., and J.O. McCaldin, "Interface effects in the dissolution of silicon into thin gold films," *J. Electrochem. Soc.*, **122**, 1975, p. 565.

115. Schnable, G.L., W. Kern, and R.B. Comizzoli, "Passivation coatings on silicon devices," *J. Electrochem. Soc.*, **122**, 1975, p. 1092.

116. Schnall, E., "A convenient way to model the handy Zener diode," *Electronics*, Oct. 11, 1971, p. 67.

117. Seidman, T.I., and S.C. Choo, "Iterative scheme for computer simulation of semiconductor devices," *Solid-State Electronics*, **15**, 1972, p. 1229.

118. Siekanowicz, W.W., et al., "Current-gain characteristics of Schottky-barrier and *p-n* junction electron-beam semiconductor diodes," *IEEE Trans. Electron Devices*, **ED-21**, 1974, p. 691.

119. Silzars, A., D.J. Bates, and A. Ballonoff, "Electron bombarded semiconductor devices," *Proc. IEEE*, **62**, 1974, p. 1119.

120. Smiljanic, M., D. Tjapkin, and Z. Djuric, "Influence of carriers on the capacitance of *p-n* junctions with deep donors," *Solid-State Electronics*, **17**, 1974, p. 931.

121. Spear, W.E., et al., "Amorphous silicon *p-n* junction," *Appl. Phys. Lett.*, **28**, 1976, p. 105.

122. Sze, S.M., *Physics of Semiconductor Devices*, Wiley-Interscience, NY, 1969, p. 117.

123. Sze, S.M., and G. Gibbons, "Avalanche breakdown voltages of abrupt and linearly graded *p-n* junctions in Ge, Si, GaAs, and GaP," *Appl. Phys. Lett.*, **8**, 1966, p. 111.

124. Takeshima, M., "Effect of electron-hole interaction on the Auger recombination process in a semiconductor," *J. Appl. Phys.*, **46**, 1975, p. 3082.

125. Temple, V.A.K., and M.S. Adler, "Calculation of the diffusion curvature related avalanche breakdown in high-voltage planar *p-n* junctions," *IEEE Trans. Electron Devices*, **ED-22**, 1975, p. 910.

126. Temple, V.A.K., and M.S. Adler, "A simple etch contour for near ideal breakdown voltage in plane and planar *p-n* junctions," *IEEE International Electron Devices Meeting Technical Digest*, December, 1975, Washington, D.C. p. 171.

127. Toshiaki, I., and B. Jeppsson, "Determination of hole and electron traps from capacitance measurements," *Jpn. J. Appl. Phys.*, **12**, 1973, p. 1011.

128. Trumbore, F.A., "Solid solubilities of impurity elements in germanium and silicon," *Bell System Tech. J.*, **39**, 1960, p. 205.

129. Tyagi, M.S., "Zener and avalanche breakdown in silicon alloyed *p-n* junctions–I," *Solid-State Electronics*, **11**, 1968, p. 99.

130. Valdman, H., "Diodes regulatrices de tension: diodes Epi-Z ou nouvelles conceptions des diodes Zener," *L'onde Electrique*, **51**, April 1971, p. 320.
131. van Vliet, K.M., et al., "Noise in single injection diodes, I. A survey of methods," *J. Appl. Phys.*, **46**, 1975, p. 1804.
132. van Vliet, K.M., et al., "Noise in single injection diodes. II. Applications," *J. Appl. Phys.*, **46**, 1975, p. 1814.
133. Varker, C.J., "An investigation of microplasma noise in Zener diodes with the SEM," *IEEE Ninth Ann. Proc. Reliability Phys.*, Las Vegas, NV, March 31-Apr. 2, 1971.
134. Yau, L.D., and Sah, C.T., "Quenched-in centers in silicon $p^+n$ junctions," *Solid-State Electronics*, **17**, 1974, p. 193.
135. Wolf, M. *Silicon Semiconductor Data*, Pergamon, NY, 1969.
136. Wilson, P.R., "Avalanche breakdown voltage of diffused junctions in silicon," *Solid-State Electronics*, **16**, 1973, p. 991.
137. "Electron irradiation speeds switching," *Electronics*, June 12, 1975, p. 34.

# 2

# Metal-Semiconductor Schottky-Barrier Diodes

## CONTENTS

## 2.1 ELEMENTARY METAL-SEMICONDUCTOR JUNCTION CONCEPTS

In a metal-semiconductor junction the difference between the work function of the metal $\phi_m$ and the electron affinity $\chi_s$ of an $n$ type semiconductor, Fig. 2.1(a), determines the barrier height $(\phi_m - \chi_s)$ for the simple model shown on Fig. 2.1(b). This barrier forms by equalization of the Fermi levels across the junction due to the movement of electrons from the semiconductor to the metal interface. The barrier in the semiconductor itself is $qV_D = \phi_m - \phi_s$ and the junction conducts in the forward direction by movement of electrons over this barrier when the $n$ type semiconductor is made negative with respect to the metal. If the work function of the metal is less than that of the $n$ type semiconductor the barrier does not exist in this model and the junction is ohmic. However, if the semiconductor is $p$ type and $\phi_m < \phi_s$ then, as in Fig. 2.1(c) and (d), a rectifying junction is expected, and forward current flows when the semiconductor is positively biased with respect to the metal.

Thermionic conduction conditions in a metal-$n$ semiconductor junction are illustrated in Fig. 2.2 for forward, zero and reverse bias. The leakage current under reverse bias conditions therefore is $J_r$ and is determined by thermionic emission over the barrier $\phi_{bn}$. This emission current is similar in form to the emission of electrons from a hot metal into a vacuum, so

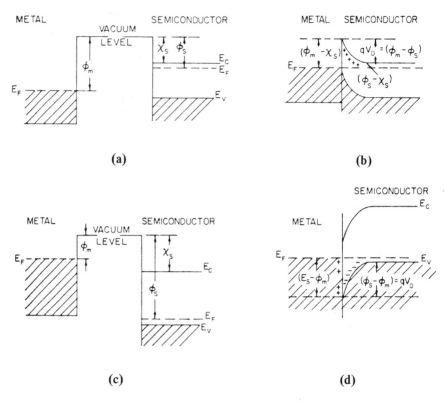

Fig. 2.1. Simple model of metal-semiconductor Schottky barrier heights: (a), (b) Metal-$n$-semiconductor barrier to electron flow is $\phi_m - \chi_s$, and (c), (d) Metal-$p$-semiconductor, barrier to hole flow is $E_s - \phi_m$, where $E_s$ is $\chi_s - E_g$.

$$J_r = A^* T^2 \exp\left(\frac{-\phi_{bn}}{kT}\right) \qquad (2.1)$$

where $A^*$ is the effective Richardson constant, which in vacuum theory should have a value of 120 A cm$^{-2}$ °K$^{-2}$. The emission current as a function of barrier height for this $A^*$ value is given in Fig. 2.3. Useful barrier heights corresponding to low reverse currents tend to be in the range 0.5 eV and up. The values of $A^*$ observed for semiconductors tend to be a few tens of A cm$^{-2}$ °K$^{-2}$ and are affected by band structure, crystal orientation, strain and sample preparation technique. However values of A$^*$ as much as twice the vacuum value for the Richardson constant may be possible for n-type Si.[21]

The barrier height does not remain entirely unaffected by the field strength applied. This is seen most clearly by considering again a metal-vacuum interface. An electron at a distance $x$ from the metal surface experiences an attractive

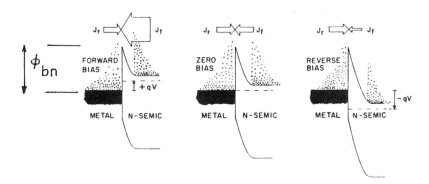

Fig. 2.2 Schematic representation of thermionic rectification in a Schottky barrier for forward, zero, and reverse applied bias conditions. (After Rideout.[106] Reprinted with permission from *Solid-State Electronics*, Pergamon Press, Ltd.)

force to the metal. At the interface the lines of field must be perpendicular to the metal surface since it is assumed the surface is a perfectly conducting sheet. Therefore, the field lines are as though the electron of charge $-q$ induces an image $+q$ at a distance $-x$ inside the metal [see Fig. 2.4(a)]. By Coulomb's equation the force attracting the electron to the metal is therefore

$$F = q^2/4\pi\epsilon_0(2x)^2 \qquad (2.2)$$

where $\epsilon_0$ is the dielectric constant of free space, $8.85 \times 10^{-14}$ F cm$^{-1}$. Integration of electric field $F/q$ from $x = \infty$ to a finite $x$ provides the expression

$$\phi(x) = -q/(16\pi\epsilon_0 x) \qquad (2.3)$$

for the electron energy (in units of electron-volts) near the metal. [In Eq. (2.3) at $x = 0$, $\phi(x)$ goes to minus infinity instead of to $\phi_{WF}$. This, however, is not physically significant since substituting $x$ equal to say 3 Å shows that $\phi(x)$ is then still only about $-1$ eV, and Eq. (2.3) is mostly applied at distances of some hundreds of angstroms or more.]

If an electric field $\mathcal{E}$ V cm$^{-1}$ is now applied in the vicinity of the metal-vacuum interface, the energy of an electron at a distance $x$ becomes

$$\phi(x) = -(q/16\pi\epsilon_0 x) - \mathcal{E}x \qquad (2.4)$$

This has a maximum value at a value of $x$ given by $(q/16\pi\epsilon_0\mathcal{E})^{1/2}$. Hence, the barrier lowering produced is $\Delta\phi$ eV [see Fig. 2.4(c)] where

$$\Delta\phi = (q\mathcal{E}/4\pi\epsilon_0)^{1/2} \qquad (2.5)$$

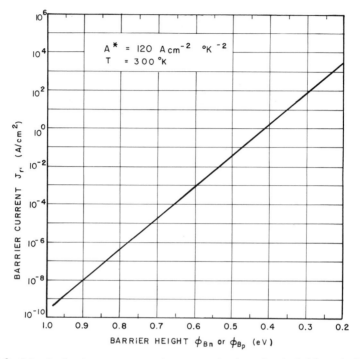

Fig. 2.3. Schottky barrier reverse saturation current density vs barrier height, calculated for a Richardson constant of 120 A cm$^{-2}$ °K$^{-2}$ and 300°K.

If $\&$ is $10^4$ V cm$^{-1}$, then $\Delta\phi$ is about 0.039 eV and occurs at a distance 235 Å from the metal surface.

Extending this concept to a metal-semiconductor junction the $J_r$ expression becomes

$$J_r = AT^2 \exp\left[-(\phi_B - \Delta\phi)/kT\right]$$

$$= AT^2 \exp\left(-\phi_B/kT\right)\exp\left[(q\&/4\pi\epsilon\epsilon_0)^{1/2}/kT\right] \qquad (2.6)$$

where the barrier lowering $\Delta\phi$ is related to the maximum field strength $\&$ at the junction. Since this field strength for a forward applied voltage $V_a$ is given by

$$\& = [2qN_D(V_D - V_a)/\epsilon\epsilon_0]^{1/2} \qquad (2.7)$$

the reverse current $I_r$ should plot as $1n(I_r)$ proportional to $(V_D - V_a)^{1/4}$.

The basic Schottky model then gives for the current-voltage relationship

$$J = J_r\left[\exp(qV_a/nkT) - 1\right] \qquad (2.8)$$

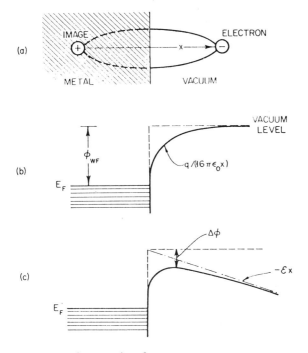

Fig. 2.4.  Barrier at a metal-vacuum interface:
(a)  Electron in vacuum with image charge in the metal,
(b)  Electron energy barrier in the absence of applied field, and
(c)  External applied field, $\mathcal{E}$, reduces the barrier height $\Delta\phi$. (After Milnes and Feucht, *Heterojunctions and Metal-Semiconductor Junctions*, Academic Press, Inc., NY, 1972, p. 174.)

with an ideality factor $n$ of unity, or very slightly larger (about 1.05) if the analysis is refined.

Study of practical Si and GaAs metal-semiconductor junctions shows that such $n$ values are readily obtained (see Fig. 2.5 for an Al-$n$Si junction), but that the barrier heights are not at all those expected from the $\phi_m - \chi_s$ concept. The reason is that interface states between the semiconductor and the metal control the barrier height.

## 2.2 BARRIER HEIGHT MEASUREMENTS

Several electrical and optical techniques may be used for determination of Schottky barrier heights.

The basic equation is of the form

$$J = A^*T^2 \exp(-\phi_B/kT) \, [\exp(qV_a/kT) - 1] \qquad (2.9)$$

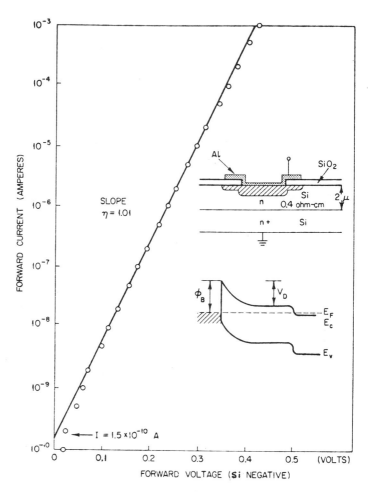

Fig. 2.5. Forward $I$-$V$ characteristic of an Al on $n$-Si Schottky diode. Insert shows the Al overlapping the SiO$_2$ for control of the depletion at the diode perimeter. Area of the diode is $0.85 \times 10^{-5}$ cm$^2$. (After Yu and Mead.[144] Reprinted with permission from *Solid-State Electronics*, Pergamon Press, Ltd.)

and the current intercept Fig. 2.6(a) for $V_a = 0$, cannot be used directly since it contains two unknowns $A^*$ and $\phi_B$. However, if the characteristics are taken over a temperature range and $\ln(J_0/T^2)$ is plotted vs $1/T$, a line is obtained from which $\phi_B$ and $A^*$ can be determined. The $\phi_B$ values so obtained are reasonably consistent for a given metal over a range of semiconductor resistivity. The $A^*$ values, however, may be somewhat scattered depending on the processing of the diode since a residual thin oxide layer may be present at the metal-semiconductor interface; $A^*$ values in the range 80-100 A cm$^{-2}$ °K$^{-2}$ are not unusual.

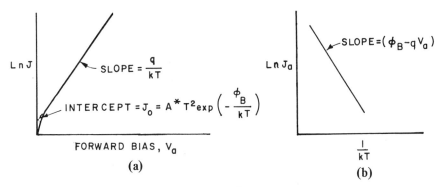

Fig. 2.6. Metal-semiconductor junction characteristics.
(a)   I-V characteristic of forward-biased metal-semiconductor contact
(b)   Activation energy plot of forward-biased metal-semiconductor contact. (After Mead.[81] Reprinted with permission from *Solid-State Electronics*, Pergamon Press, Ltd.)

Another technique is to apply a fixed forward-bias voltage $V_a$ and observe the forward current $J_a$ for various temperatures, then if $1nJ_a$ is plotted vs $1/kT$ as in Fig. 2.6(b) the slope of the line is $(\phi_B - qV_a)$ and hence gives the barrier height.

The barrier height may also be determined from capacitance vs reverse voltage measurements. For a uniformly doped semiconductor, $C^{-2}$ vs voltage is a straight line and its voltage intercept provides the diffusion barrier height, $V_D$, although usually not with high accuracy.

Spectral response measurements of photoexcitation in metal-semiconductor junctions provide another method of determining barrier heights. As shown in Fig. 2.7, light may be applied to the front of the photocell, if the metal is very thin, or through the semiconductor to the junction. If the photon energy exceeds the barrier height but is less than the semiconductor bandgap, photo-emission of electrons from the metal into the semiconductor is observed [see process (a) in Fig. 2.7(b)]. If the photon energy exceeds the bandgap of the semiconductor, direct band-to-band excitation occurs, as shown in Fig. 2.7(c) by the sharp increase in response.

The short-circuit photocurrent expected for excitation over a barrier is proportional[119] to $(h\nu - \phi_B)^2$ provided $(h\nu - \phi_B)$ is more than a few $kT$. There-fore, the square root of the photocurrent response, $R^{1/2}$, plotted against $h\nu$ should give a straight line, and extrapolation of this line to intercept the energy axis gives the barrier height $\phi_B$. Although $R^{1/2}$ is suggested by the Fowler model, better linearity for certain ranges of semiconductor doping may be obtained by plotting $R^{1/3}$ or $(Rh\nu)^{1/2}$ vs $h\nu$. Models taking account of the density of states function have been suggested for both of these approaches.

Simple plotting of $R^{1/2}$ vs $h\nu$ works well in Fig. 2.8(a), (b) for Al on $n$- and $p$-GaAs. The barrier heights $\phi_{Bn}$ and $\phi_{Bp}$ sum to 1.35 eV which is very nearly

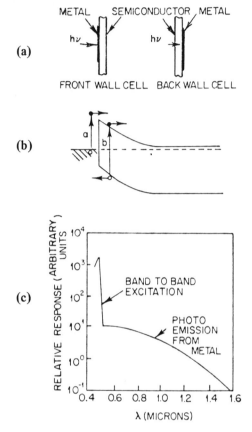

Fig. 2.7. Photoresponse of metal-semiconductor junctions:
(a)  Photocells illuminated from front or rear,
(b)  Photoexcitation processes depending on photon energy, and
(c)  Photoemission of electrons from the metal into the semiconductor develops into band-to-band excitation as the photon energy is increased. (After Mead.[81] Reprinted with permission from *Solid-State Electronics*, Pergamon Press, Ltd.)

the bandgap value of GaAs. This might be expected from the model of Fig. 2.1 or from a model in which interface states, highly localized in the bandgap, tend to pin the Fermi level of the junction.

Consider now observed barrier heights of various metals, of a wide work-function range, on $n$ type Si. The values plotted in Fig. 2.9 show an increase of barrier height with increasing work-function, but the dependence is not of the form $\phi_B = \phi_m - \chi_s$ where $\chi_s$ is the electron affinity of Si (4.01 eV). Instead, the variation seems to be more of the form $\phi_B = \phi_m/3 - \phi_C$ where $\phi_C$ is between 0.6 and 0.7 eV although this relationship is without physical justification. If the same barrier heights are plotted against the electronegativities of the metals

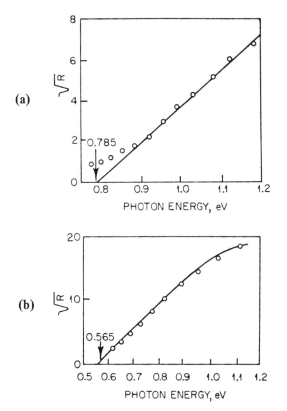

Fig. 2.8. Barrier height determinations from the photoresponse of Al-GaAs metal-semiconductor photojunctions (The vertical scale is arbitrary):
(a)  n-type GaAs, and
(b)  p-type GaAs. (After Spitzer and Mead, *J. Appl. Phys.*, **34**, 1963, pp. 3066, 3067.)

the results are more scattered. All that can be safely stated is that for Si the interface states are exerting a considerable influence on the barrier heights observed.

However, in studying barrier heights on other semiconductors a pattern emerges. Barrier heights for metals on $n$ GaAs show very little dependence on the work function, or electronegativity, of the metal (see Fig. 2.10). Apparently the Fermi level is being pinned by a high density of interface states at about $E_g/3$ above the valence band edge so the barrier height is between 0.8 and 0.9 eV for all metals. However, for ZnS the barrier heights are seen to be almost directly proportional to the electronegativities. The difference is probably related to the covalent-bonding nature of GaAs and the ionic-bonding nature of ZnS.

Many covalent-bonded semiconductors have a large density of surface states that form a narrow band centered about one-third of the way up the forbidden

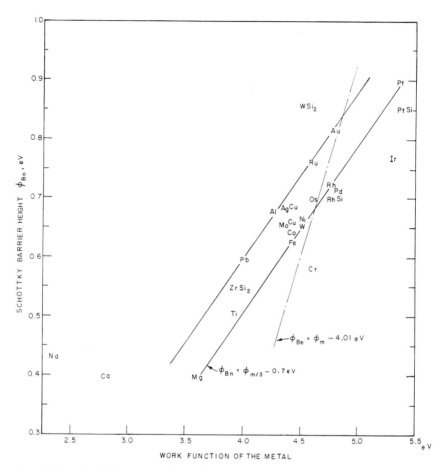

Fig. 2.9. Barrier heights vs. work function for metals on *n*-type Si. (The work function values were taken from Formenko; see Sze, *Physics of Semiconductor Devices*, John Wiley & Sons, NY, copyright 1969, p. 366.)

gap from the valence band edge. Since this large density of interface states can supply or accept charge with relatively little shift of the state Fermi level, the Fermi level of the metal (irrespective of which metal is used) tends to lock on at this position above the valence band. This suggests that the barrier heights for various semiconductors might be proportional to the energy gap of the semiconductor and Fig. 2.11 shows this to be roughly so.

At this stage a reasonable question is whether intrinsic surface states on a semiconductor surface persist in the presence of a metal overlayer, or whether the metal atoms induce new surface states in the bandgap. The matter is under examination still, but theoretical modeling and experiments at present suggest

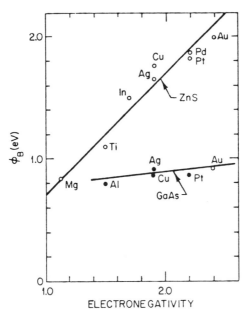

Fig. 2.10. Barrier heights vs electronegativity for various metals on ZnS and GaAs (surface-state-controlled). (After Mead.[81] Reprinted with permission from *Solid-State Electronics*, Pergamon Press, Ltd.)

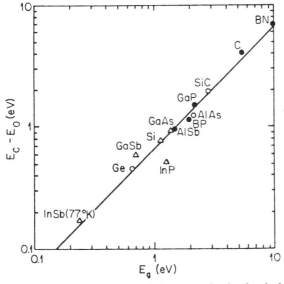

Fig. 2.11. Location of interface Fermi-level relative to conduction band edge for gold contacts on various surface state controlled materials. The line is $\phi_{Bn} = E_c - E_O = \frac{2}{3}E_g$: (○) *n*-type, (●) *p*-type, and (△) both. (After Mead.[81] Reprinted with permission from *Solid-State Electronics*, Pergamon Press, Ltd.)

that intrinsic surface states are replaced by metal-induced surface bandgap states for most crystal orientations of Si, Ge or GaAs (the (110) orientation being possibly an exception).

Furthermore, the rule that $\phi_{Bn} \approx \frac{2}{3} E_g$, however, does not survive the closer scrutiny that is possible by studying the barrier height of a given metal on ternary semiconductor compounds such as $Ga_x In_{1-x} P$ or $Ga_x In_{1-x} As$. For such compounds, the bandgap changes as $x$ changes. Study of the barrier height variation then shows that the metal Fermi level tends to maintain a constant energy above the valence band edge. This energy depends essentially on the anion and becomes larger as one progresses from Sb to As to P. This is shown in Fig. 2.12 with Au as the reference metal. Results for Te, Se and S compounds are also included.

The intrinsic surface states on a clean semiconductor surface can not be expected to persist in the presence of a metal overlayer. If the metal is reactive it forms interface compounds and if it is less reactive it induces new interface

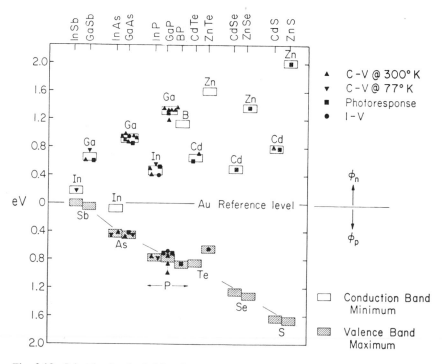

Fig. 2.12. Schottky barrier heights for Au on $n$-type III-V and II-VI semiconductors. The anion is positioned in the diagram from $E_g - \phi_{Bn}$. For instance, the barrier height of Au with respect to the conduction band of GaAs is about 0.9 eV, therefore the valence band edge is about 0.55 eV below the Au level and this is labeled As. (After McCaldin, McGill, Mead.[79])

states within the bandgap. The barrier height that is observed depends on the dipole layer that forms and the chemical reactivities of elements at the interface play a role in this.[147]

Chemical reactivity may be expressed in terms of heat of formation, or in terms of electronegativity, a somewhat related concept. The difference in electronegativity $\Delta E_N$ of the two constituents of a binary compound is a measure of the degree of ionicity. Figure 2.13 shows the results of plotting $d\phi_B/d\chi_M$ versus $\Delta E_N$ for a large number of semiconductors. There is evidently a transition between the interface properties corresponding to "ionic" materials and those corresponding to "covalent" materials. Apparently in ionic semiconductors the wavefunctions of electrons associated with the cations and anions overlap insufficiently to create interface states with energies near the center of the band gap that if present would control the barrier heights by Fermi-level stabilization at the metal-semiconductor interface. If in Fig. 2.13 the difference of electronegativity axis is replaced by a "difference of heat of formation" axis, $\Delta h_f$, the shape of the curve is almost unchanged. Covalent semiconductors are more likely to interact with metals on the surface because they have lower $\Delta h_f$ values than ionic semiconductors. Compound formation at the interface may be expected to control the local charge redistribution and the atomic positions of the elements

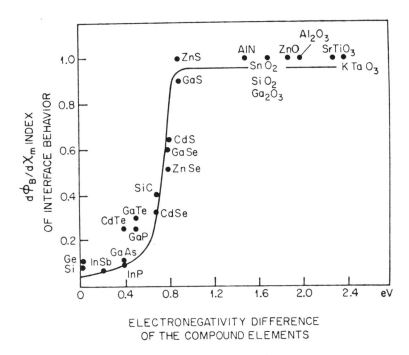

Fig. 2.13. Variation of $d\phi_B/d\chi_m$ vs the electronegativity difference which is large for a high degree of ionic bonding. (After Kurtin, McGill, Mead.[65])

at the interface and so be a strong factor in determining the barrier height.

The data plotted in Fig. 2.13 appear to show an abrupt rise at a 0.8 eV electronegativity difference, and a tending to saturation at $d\phi_B/d\chi_m$ equal to unity for compounds of high electronegativity differences. However recent examination of data suggests that the scatter is large and that $d\phi_B/d\chi_m$ values (symbolized by S) well in excess of unity are found for some compounds.[148] Indeed recent theoretical studies have suggested that values of S between 2.0 and 3.0 may be possible.

Another approach to the determination of barrier heights considers a metal monolayer and a simple representation of the (110) semiconductor surface with the cation sites occupied by metal atoms. Application of a semi-empirical tight-binding calculation then gives quite reasonable barrier heights for metals of various work functions on GaAs and GaP.[150] The results, however, are less satisfactory for barriers on ZnSe and ZnS.

Barriers for many metals on $n$ type semiconductors are given in Table 2.1. The structure of Table 2.1 provides space for the addition of new-found values.

Barrier heights $\phi_{Bp}$ on $p$ type Si are given in Table 2.2 together with $\phi_{Bn}$ values. The sums $\phi_{Bp} + \phi_{Bn}$ are seen to approximate the bandgap of Si. The $\phi_{Bp}$ values are shown plotted against metal work-functions[35] in Fig. 2.14. The dashed line $\phi_{Bp} = 1.9 - \phi_m/3$ was chosen to complement the expression $\phi_{Bn} = (\phi_m/3 - 0.7)$ eV from Fig. 2.9, so that the sum $\phi_{Bn} + \phi_{Bp}$ is approximately the same as the bandgap of Si.

For some metal-semiconductor junctions the barrier heights observed are affected by the details of the processing. This may be attributed to existence, or development, of a thin foreign layer at the interface between the metal and the semiconductor. The layer may be an oxide, from 10 to 30-Å thick, having locked-in charge that alters the barrier height. Current flow takes place by tunneling through the oxide and the $I/V$ characteristics of the junction reflect this action.[1, 20, 95]

In some junctions, compounds between the metal and the semiconductor elements may form and change the barrier heights—PtSi, $WSi_2$ and other silicides are examples shown in Table 2.1.

The metal itself may dope the semiconductor if a moderate amount of heat treatment is applied and so create a $pn$ junction that alters the junction behavior. For example, Al on $n$ type Si has a barrier height of 0.68 eV. However, heat treatments of 1 hr at 400°C and 550°C result in barrier heights of 0.79 and 0.89 eV because of formation of a thin $p$ layer.[10] The ideality factors for the junction may still have values such as $n = 1.06$ but the $J_0$ values are lower as seen in Fig. 2.15.

Many metal-semiconductor junctions need short-duration postmetal anneals to develop good Schottky barrier characteristics, perhaps by absorbing impurities, such as oxides, at the interface. However, such processes must be well controlled to prevent barrier changes.[87, 117, 118]

**Table 2.1. Barrier Heights for Metals on n-Type Semiconductors.**

| Metal | Observed Barrier Heights (eV) 300°K | | | | | | | | | | | | | |
| --- | --- | --- | --- | --- | --- | --- | --- | --- | --- | --- | --- | --- | --- | --- |
| | Si | Ge | SiC | GaP | GaAs | GaSb | InP | InAs | InSb | ZnS | ZnSe | CdS | CdSe | CdTe |
| Al | 0.68-0.74 | 0.48 | 2.0 | 1.05<br>1.20 | 0.73-0.80<br>0.88 | | | Ohmic | 0.18<br>(77°) | 0.8<br>1.65 | 0.74<br>1.22 | Ohmic<br>0.35-0.56 | 0.43 | 0.76<br>0.66-0.78 |
| Ag | 0.56-0.79<br>0.69 | | | | | | 0.54 | Ohmic | 0.17<br>(77°K) | | | | | |
| Au | 0.81-0.83 | 0.45 | 1.95 | 1.18 | 0.90 | 0.61 | 0.40-0.49 | Ohmic | | 2.0 | 1.35-1.51 | 0.68-0.78 | 0.70 | 0.86 |
| Au/Ti | | | | | | | | | | | | | | |
| Bi | | | | | 0.89-0.92<br>0.56 | | 0.53 | | | | 1.14 | 0.84 | | 0.78 |
| Ca | 0.40 | 0.5 | | | | | | | | | | | | |
| Co | 0.64 | | | | | | | | | | | | | |
| CoSi | 0.68 | | | | | | | | | | | | | |
| Cr | 0.57-0.59 | 0.48 | | 1.18<br>1.20 | 0.82 | | | | | 1.75 | 1.10<br>1.11<br>0.86 | 0.36-0.50 | 0.33 | 0.82<br>0.78<br>0.69 |
| Cu | 0.66-0.79 | 0.42 | | | | | | | | | | | | |
| Fe | 0.65 | | | | 0.83 | | | | | | | | | |
| In | | | | | | | | | | | | | | |
| Ir | 0.77 | | | | | | | | | | | | | |
| Mg | 0.4 | 0.42 | | 1.04 | 0.65 | | | | | 0.82 | 0.49 | | | |
| Mo | 0.56-0.68 | | | | | | | | | | | | | |
| Na | 0.43 | | | | | | | | | | | 0.45 | | |
| Ni | 0.66-0.70 | | | | 0.78-0.83 | | | | | | 1.14 | | | 0.83<br>0.68 |
| Pb | 0.6 | | | | | | | | | | | | | |
| Os | 0.70 | 0.40 | | | | | | | | | | | | |
| Pd | 0.71 | | | | | | | | | 1.87 | | 0.53<br>0.62 | | 0.86 |
| Pd₂Si | 0.74 | | | 1.45 | | | | | | | | | | |
| Pt | 0.90 | | | | 0.86 | | | | | 1.84 | 1.40 | 0.85-1.1 | 0.37 | 0.89 |
| PtSi | 0.85 | | | | | | | | | | | | | |
| Rh | 0.72 | 0.40 | | | | | | | | | | | | |
| RhSi | 0.70 | | | | | | | | | | | | | |
| Ru | 0.76 | 0.38 | | | | | | | | | | | | |
| (SN)ₓ polymer | | | | | 1.0-1.2<br>0.86<br>0.75 | | 0.8-0.9 | | | 2.7-3.0 | 1.7<br>1.34 | 1.1 | 0.6-0.7 | 0.76 |
| Sb | 0.50 | | | | | | | | | | | | | |
| Sn | 0.66 | | | | | | | | | | | | | |
| Ta | 0.86 | | | | | | | | | | | | | |
| Ti | 0.50 | 0.48 | | | 0.75-0.84<br>0.66-0.71 | | | | | | | | | |
| W | 0.66 | 0.48 | | | | | | | | | | | | |
| WSi₂ | 0.86 | | | | | | | | | | | | | |
| Zn | 0.75 | | | | | | | | | | | | | |
| ZrSi₂ | 0.55 | | | | | | | | | | | | | |

**Table 2.2. Some Barrier Height Measurements on $p$ and $n$ Silicon.**

| Metal | $\phi_m$ eV | Barrier Height, eV | | $\phi_{Bp} + \phi_{Bn}$ |
|---|---|---|---|---|
| | | $\phi_{Bp}{}^a$ | $\phi_{Bn}$ | |
| Au | 4.8 | 0.34 | 0.81 | 1.15 |
| Ni | 4.5 | 0.50 | 0.66 | 1.16 |
| Cu | 4.4 | 0.46 | 0.69 | 1.14 |
| Ag | 4.3 | 0.53 | 0.69 | 1.22 |
| Al | 4.25 | 0.57 | 0.68 | 1.25 |
| Pb | 4.0 | 0.54 | 0.6 | 1.14 |
| Hf | 3.5 | $0.63^b$ | — | — |
| CoSi | | 0.38 | 0.68 | 1.06 |

[a]From Smith and Rhoderick[120] (Reprinted with permission from *Solid-State Electronics*, Pergamin Press Ltd.)
[b]From Beguwala and Crowell[11] (Reprinted with permission from *Journal of Applied Physics*.)

## 2.3 SCHOTTKY BARRIER CURRENT-VOLTAGE CHARACTERISTICS

The first-order approach to the modeling of current in a metal-semiconductor junction is to assume that the carrier flow is caused by thermionic emission over the barrier and to neglect all tunnel effects and image-force barrier lowering effects. Figure 2.2 illustrates the effect of forward and reverse voltage bias on the barrier between a metal and an $n$-type semiconductor. At zero bias the

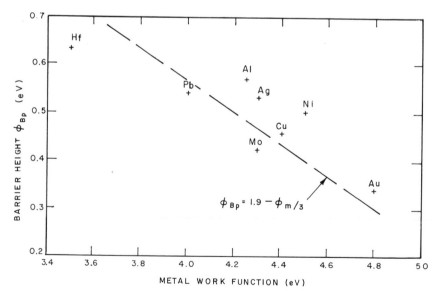

Fig. 2.14. Schottky barrier heights on $p$-type Si.

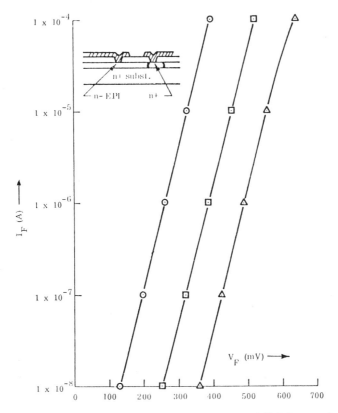

Fig. 2.15. Forward current-voltage characteristics of Al and Al-Si SBDs as a function of post-metal heat treatment that results in development of a *pn* junction.:
(○)   Al, 400°C for 1 hr. $\phi_B$ = 0.68 V, $n$ = 1.06,
(□)   Al-Si, 400°C for 1 hr. $\phi_B$ = 0.79 V, $n$ = 1.06, and
(△)   Al-Si, 550°C for 1 hr. $\phi_B$ = 0.89 V, $n$ = 1.07. (After Reith and Schick, *Journal of Appl. Physics,* 25, 1974, p. 525.)

electron flux from the semiconductor into the metal is given by the number of electrons having an energy equal to or greater than $qV_D$, directed per second towards unit area of the interface. For a Maxwellian distribution the result is

$$J_{f0} = qN_D(kT/2\pi m^*)^{1/2}\exp(-qV_D/kT) \qquad (2.10)$$

where $m^*$ is the effective electron mass.

At zero voltage the net flow over the barrier must of course be zero, and $J_r$ therefore may also be written in terms of the electron flux from the metal into the semiconductor over the barrier $\phi_B$. The barrier height is the diffusion barrier plus the doping step $E_C - E_F$,

$$\phi_B = qV_D + \delta \tag{2.11}$$

From simple semiconductor theory, with $N_c$ the effective density of states at $E_c$,

$$\exp(-\delta/kT) = N_D/N_C = N_D/2(2\pi m^* kT/h^2)^{3/2} \tag{2.12}$$

With the aid of Eq. (2.12), the expression for $J_{f0}$ from Eq. (2.10) may be rewritten as

$$J_{f0} = \frac{4\pi}{h^3} qm^* k^2 T^2 \exp(-\phi_B/kT) = AT^2 \exp(-\phi_B/kT) \tag{2.13}$$

This is similar to the expression obtained for the emission of electrons from a metal into vacuum over a barrier $\phi_B$, and $A$ would be the Richardson constant if the free electron mass could be used.[132]

As the slopes of experimental $\ln J$ vs forward voltage $V_a$ curves are usually slightly greater than unity, an empirical factor $n$ is introduced so that

$$J = J_{f0} [\exp(qV_a/nkT) - 1] \tag{2.14}$$

This form, however, is not consistent with thermodynamics ($J_f = J_r$ at $V_a = 0$) or time reversal invariance and a better form would be[26]

$$J = J_0\{\exp(qV_a/nkT) - \exp[(-1 + 1/n)qV_a/kT]\} \tag{2.15}$$

The difficulty with this approach is that the ideality factor $n$ is not generally a constant over a large temperature range (although a few diodes are exceptions). The $I$-$V$ forward characteristics of an Au-GaAs diode are shown in Fig. 2.16(a) for the temperature range $373°K$ to $4.2°K$. The $n$ values vary as shown in Fig. 2.16(b): hence, $n$ is not an ideal parameter to use for characterization although convenient at room temperature.

Further examination shows that current-voltage relationships can be better represented by the expression

$$J = J_0 \exp(V_a/V_0)$$

where

$$V_0 = k(T + T_0)/q \tag{2.16}$$

and $T_0$ is an empirical factor that is temperature independent. The expression then is

$$J = AT^2 \exp[-\phi_B/k(T + T_0)] \exp[qV_a/k(T + T_0)) - 1] \tag{2.17}$$

For the Au-$n$GaAs junction of Fig. 2.16 a $T_0$ value of about $45°$K provides a straight line that fits the data at constant current from $373°$K to $4.2°$K, as shown in Fig. 2.17. Not all Schottky barriers fit this model with $T_0$ independent of temperature; this may be seen from Fig. 2.18 which is for a Ni-$n$GaAs diode.

Several attempts have been made to explain the factor $T_0$ in an acceptable physical way, such as by an assumed temperature dependence of the work function.[111] However, one of the most interesting[69, 70, 105] shows that a term such as $T_0$ would be a consequence of an interface state energy distribution

$$\phi_{ss} + \phi_m = Q_C \exp(-\phi_B/E_0) \tag{2.18}$$

where $E_0$ is a constant of the distribution that is related to $T_0$ and $\phi_m$ is the charge in the metal. The analysis is too unwieldly to include here but an attempt will be made to convey the idea.

The interface charge distribution assumed is shown in Fig. 2.19 for forward-, zero- and reverse-bias conditions. Analysis shows that

$$T_0 = -T \left(\frac{\partial Q_{SC}}{\partial(\phi_b - eV_a)}\right) \Big/ \left(\frac{\partial Q_{ss}}{\partial \phi_B}\right) \tag{2.19}$$

where $Q_{SC}$ is the space charge (cm$^{-2}$) in the depletion layer of the Schottky barrier. In general, $T_0$ is seen to be a function of temperature and of forward voltage. The temperature dependence disappears if the interface state distribution is taken as exponential.

A development of this approach can also be used to predict that the reverse current $J^-$ of a Schottky junction should vary as a power law of the reverse voltage according to

$$\frac{J^-}{A^*T^2} = \left(\frac{F}{F_C}\right)^{E_0/kT} \tag{2.20}$$

where $F$ is the electric field at the interface and $F_C$ is a normalizing constant. For Si and GaAs with various metals $E_0/kT$ tends to be of the order 0.5 to 2.0 and the test of Eq. (2.20) is not conclusive. However, for Pt/ZnS, $E_0/kT$ is large ($\sim 28$) since $E_0$ from the forward characteristics is determined to be 0.72 eV. For Ag/GaP, $E_0/kT$ is also large (22) with $E_0$ determined to be 0.58 eV. Reverse characteristics for these two junctions [see Fig. 2.19(d)] do exhibit the expected steep power-law dependence on voltage.

In all reverse-current studies edge effects must be eliminated. This may be done by the use of guard rings—a few forms of which are illustrated in Fig. 2.20. The image-force barrier-lowering effect discussed earlier is often found to be insufficient to explain the increase of reverse current with voltage, as can be seen in Fig. 2.21(a). The effect can be taken care of by postulating an intrinsic barrier

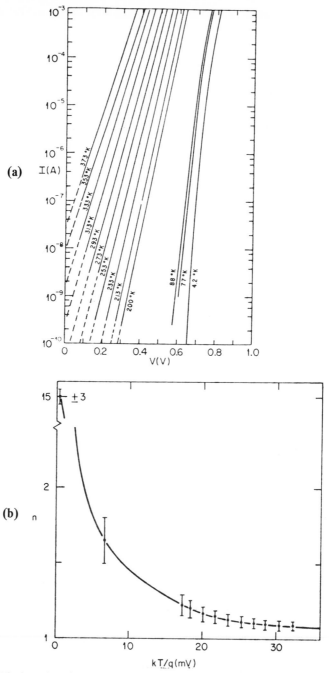

Fig. 2.16. The behavior of Au-GaAs Schottky junction with temperature:

(a) Variation of *V-I* forward characteristics., (After Padovani and Sumner, 1965, *Journal of Applied Physics*, **36**, 1965, p. 3745)

(b) Temperature dependence of the parameter *n*. $I = I_s \exp(qV/nkT)$. (After Padovani, in *Semiconductors and Semimetals*, Academic Press, NY, 1970, Vol. 7a)

104

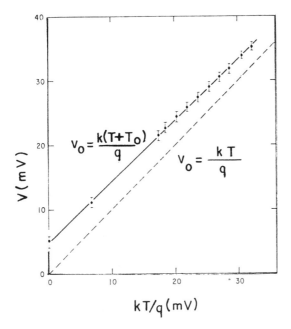

Fig. 2.17. Temperature dependence of the voltage $V_0$ for a Au-GaAs Schottky barrier. (After Padovani and Sumner, *J. of Appl. Phys.*, 36, 1965, p. 3746.)

height lowering with field strength $\partial\phi/\partial\mathscr{E}$, perhaps associated with a dipole-layer effect. The maximum reverse voltage that a metal-$n$ semiconductor can sustain is limited basically by avalanche in the depletion region as for a $p^+n$ junction structure.

The effects of interfacial layers have been studied by several workers.[18, 19, 20, 31] Energy diagrams and characteristics for Au-$n$Si junctions with oxide interface layers from 10-40 Å are shown in Fig. 2.22 where there is seen to be a voltage increase. In Chap. 12 such layers are shown to have beneficial effect in Schottky barrier solar cells.

Return for a moment to a consideration of more ideal metal-semiconductor interfaces. Figure 2.23(a) shows the quasi-Fermi levels in junctions limited by thermionic emission and by diffusion theory. Junctions in which the semiconductors are fully depleted at zero bias voltage or at low reverse voltage are termed Mott junctions and the sweep-out action in such units is sometimes advantageous in high frequency applications. More often the reverse voltage band diagram is that of Fig. 2.23(b) where full depletion has not occurred.

For a few metal-semiconductor junctions (such as Pt on $n$GaP) the energy barrier may be greater than $E_g/2$. Forward current flow then takes place by electrons from the semiconductor gaining sufficient energy to enter a recombination zone where the Fermi level is near the center of the bandgap

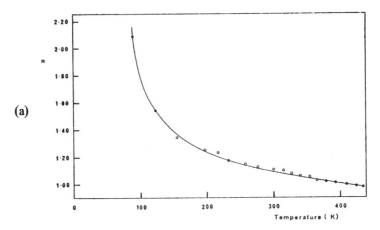

Variation of the ideality factor with temperature.

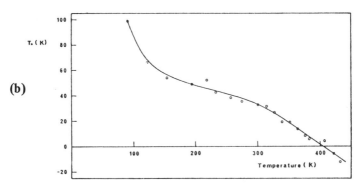

$T_0$ versus temperature.

Fig. 2.18. Variation of $n$ and $T_0$ with temperature for a Ni-$n$GaAs diode. (After Hackam and Harrop. Reprinted with permission from *IEEE Trans. Electron Devices*, **ED-19**, 1972, p. 1233.)

[see Fig. 2.24(a)] so that recombination centers are active there. The forward current characteristic may then be dominated by the recombination-current path rather than by the over-the-barrier current as seen in Fig. 2.24(b) and the slope of the characteristic corresponds to $\phi_B/0.5E_g$.

The energy band diagram for a Schottky barrier with deep donors in the semiconductor is shown in Fig. 2.25. Since the Fermi level cuts through the deep donor level the occupancy of the deep level states is a function of the applied reverse voltage. A step-change of this voltage therefore produces capacitance and current transients that may be used to gain information on the emission and capture coefficients of the impurity.

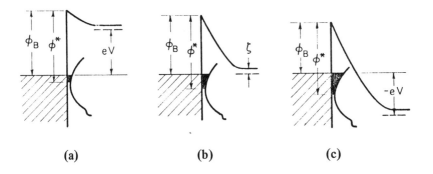

(a)                    (b)                    (c)

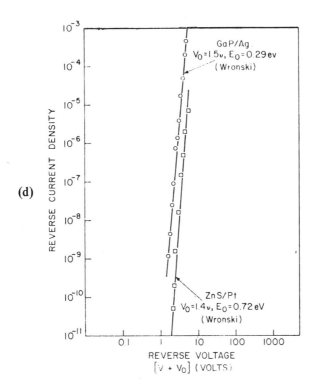

Fig. 2.19. The Levine model of interface state effects in a Schottky barrier:

(a), (b), (c): forward, zero and reverse bias conditions. The amount of interface charge is proportional to the dark area. The surface-state energy distribution is assumed fixed at a characteristic energy $\phi^*$ with respect to the conduction band edge, and

(d)    The reverse current-voltage relationships for GaP/Ag and ZnS/Pt show large power law dependences as expected for the model. (After Levine.[69] Reprinted with permission from *Solid-State Electronics*, Pergamon Press, Ltd.)

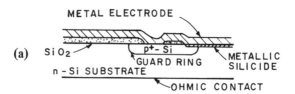

(a)

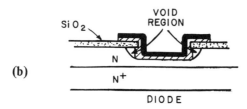

(b)

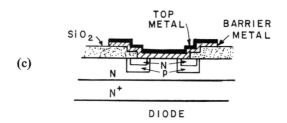

(c)

Fig. 2.20. Some examples of guard rings in Schottky barrier structures:
(a)   $p^+$ guard ring reduces edge leakage of metal-silicide junction, (After Andrews and Lepselter.[6] Reprinted with permission from *Solid-State Electronics*, Pergamon Press, Ltd.)
(b), (c) Moat etched and double guard-ring structures. (After Rhee.[102] Reprinted with permission from *Solid-State Electronics*, Pergamon Press, Ltd.)

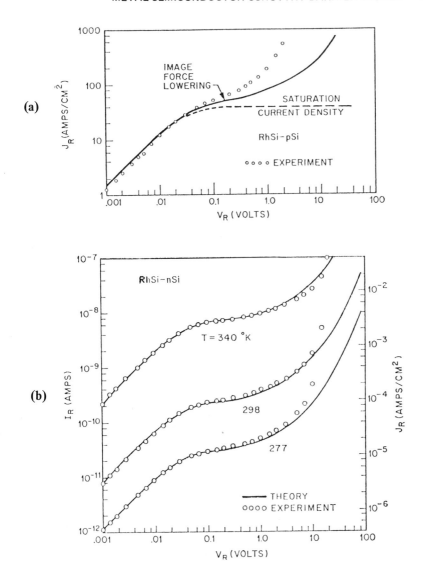

Fig. 2. 21.  Reverse characteristics of RhSi barriers on $p$ and $n$Si:
(a)  Image-force concept fails to fully account for soft reverse characteristics of a RhSi-$p$Si barrier, and
(b)  Postulation of an intrinsic barrier height lowering effect with field strength (perhaps a dipole layer effect) produces good agreement between theory and experiment for a RhSi-$n$Si barrier. (After Andrews and Lepselter.[6] Reprinted with permission from *Solid-State Electronics*, Pergamon Press, Ltd.)

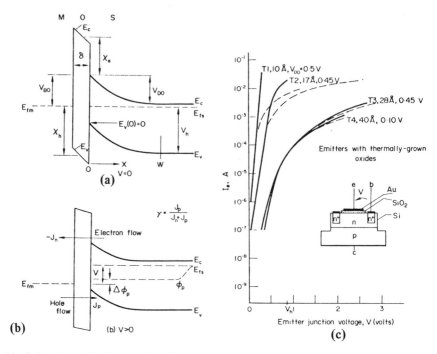

Fig. 2. 22. Interfacial layer oxide effects in Au-$n$Si barrier:
(a)   Zero bias energy diagram postulated,
(b)   Diagram with forward bias, and
(c)   Forward current-voltage characteristics showing the slope change and increase in voltage for oxide layers of 10-40 Å. (After Card and Rhoderick.[19] Reprinted with permission from *Solid-State Electronics*, Pergamon Press Ltd.)

Interface states also affect the capacitance-frequency dependence of metal-semiconductor junctions[8] and state density vs energy level profiles can be obtained from forward bias admittance studies.

Consider now a metal-semiconductor junction where the semiconductor is heavily doped so that the barrier depletion region is thin enough to allow significant tunneling. The electron flow may take place either from the Fermi level of the semiconductor or over some range of energy levels partway up the barrier. The latter effect has been termed thermionic field emission.[92] For a forward bias the two effects are as shown in Fig. 2.26(a). For reverse bias the emission is from the metal into the semiconductor, and this is shown in Fig. 2.26(b).

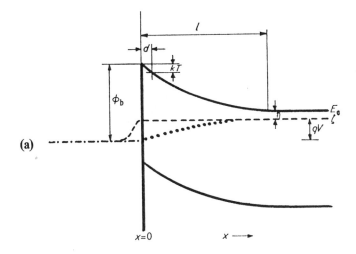

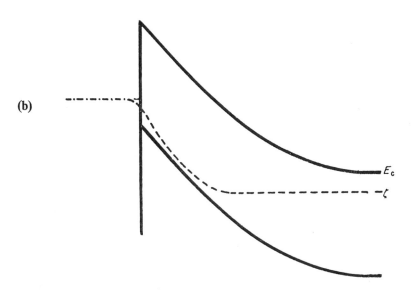

Fig. 2.23. Band diagrams in a AuSi Schottky barrier under forward and reverse characteristics:
(a)  Under forward bias, showing electron quasi-Fermi level: ---- according to thermionic
     emission theory;. . . . . . according to diffusion theory; ---- Fermi level in metal,
(b)  Under reverse bias showing electron quasi-Fermi level obtained by analysis of experi-
     mental data. (After Rhoderick.[104] Reprinted with permission from the Institute
     of Physics, UK)

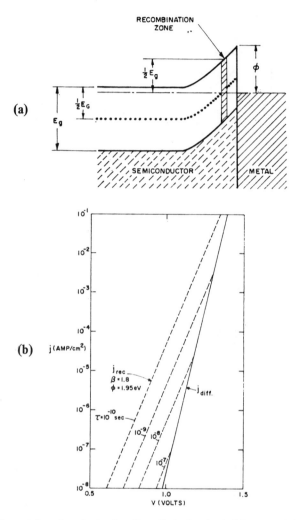

Fig. 2.24. In metal-semiconductor junctions if the barrier height exceeds $E_g/2$, the characteristic does not conform to the usual model:

(a) Forward conduction at low current flow takes place by electrons entering a recombination zone and recombining with holes there,

(b) The recombination current then dominates the over-the-barrier current and the $J$ vs $V$ characteristic has an $n$ value of larger than unity ($n \simeq \phi_B/0.5\,E_g$). The results shown are calculated for Pt on $n$GaP with several values of lifetime $\tau$ and $\phi_B/0.5\,E_g = 1.8$. (After R. Williams, *RCA Review*, 30, 1969, p. 306.)

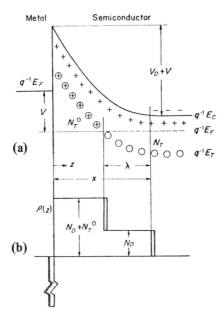

Fig. 2.25. Energy band diagram of a reverse-biased Schottky barrier with deep donors. The corresponding charge distribution based on the depletion approximation is shown in (b). (After Zohta.[146] Reprinted with permission from *Solid-State Electronics*, Pergamon Press, Ltd.)

The expression obtained for the current density is

$$J = \frac{AT^2 \pi E_{00} \exp[-2E_B^{3/2}/3E_{00}(E_B - E)^{1/2}]}{kT[E_B/(E_B - E)]^{1/2} \sin\{\pi kT[E_B/(E_B - E)]^{1/2}/E_{00}\}} \qquad (2.21)$$

for low temperature conditions in which $F$-emission dominates. The term $E_{00}$ is

$$E_{00} = q[N_D/2\epsilon]^{1/2} \Big/ [(2m^*)^{1/2}/\hbar] \qquad (2.22)$$

These equations are not tractable enough for further development here. Such equations have been able to explain the forward and reverse characteristics of Schottky barriers on GaAs doped $10^{17}$ cm$^{-3}$ and higher.[91] Figure 2.27 illustrates the effect of progressively increased doping on the characteristics, and shows the range of doping for $n$GaAs over which thermionic field emission is important.

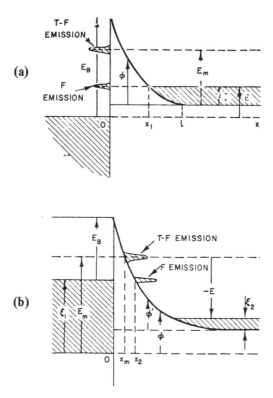

Fig. 2.26. Field emission and thermionic-field emission currents in heavily-doped metal-semiconductor junctions:
(a) Energy diagram for forward bias, and
(b) Energy diagram for reverse bias. (After Padovani and Stratton.[92] Reprinted with permission from *Solid-State Electronics*, Pergamon Press, Ltd.)

The thermionic-field emission approach has also been successful in explaining various discrepancies between barrier heights deduced from photothreshold, capacitance-voltage and current-voltage characteristics.[26] Quantum mechanical transmission effects must be considered.[3]

## 2.4 MINORITY CHARGE IN SCHOTTKY JUNCTIONS

In a metal-$nn^+$ semiconductor junction the forward current flow is by the passage of electrons over the barrier $V_D$-$V_a$ from the $n$ region into the metal. Minority carrier flow therefore does not play the role it does in $pn$ junctions and the switching speed of the majority carrier device is limited primarily by the circuit RC time constant. However, a small minority carrier effect is present that can possibly affect the switching speed. The holes involved are generated

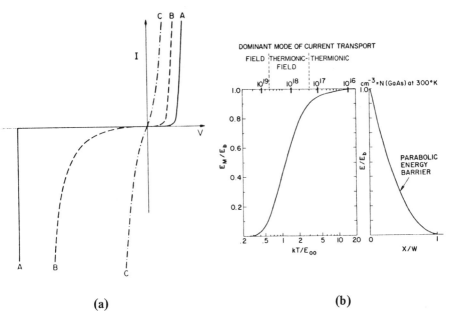

(a)                                              (b)

Fig. 2.27. Curves showing progression from thermionic emission to field-emission tunneling in a Schottky barrier such as Au on *n*GaAs:
(a)  Curve A, doping $< 10^{17}$ cm$^{-3}$, thermionic over-the-barrier emission dominates,
     Curve B, $10^{18}$-$10^{19}$ cm$^{-3}$, thermionic-field tunneling dominates,
     Curve C, doping $> 10^{19}$ cm$^{-3}$, thermionic emission tunneling dominates, and
(b)  Relative position of maximum transmission through or over a Schottky barrier vs the parameter $kT/E_{oo}$. Note the narrow range of doping in *N*-type GaAs over which thermionic field emission dominates conduction. (After Rideout.[106] Reprinted with permission from *Solid-State Electronics*, Pergamon Press, Ltd.)

in the neutral region of the *n* epi-layer and some hole pile-up occurs at the $nn^+$ barrier, as shown in Fig. 2.28, while on their way to the ohmic contact of the $n^+$ region. The effect is assisted by the electric field $\mathcal{E}_n$ resulting from the majority carrier current flow.

This field produces a minority carrier current many orders of magnitude above that predicted by diffusion theory. The ratio of the minority carrier current to the total current is defined as the ratio ($\gamma$). This term increases with forward current with the relationship

$$\gamma = n_i^2 J / b N_D^2 J_{ns}$$

where $n_i$ and $N_D$ are the intrinsic and doping concentrations, $b$ the mobility ratio, $J_{ns}$ the Schottky diode saturation current density, and $J$ the diode forward current density. By way of example, a 5 ohm-cm *n*-type Si-Au diode will have a current ratio of 5% at a current density of 350 A cm$^{-2}$.[113]

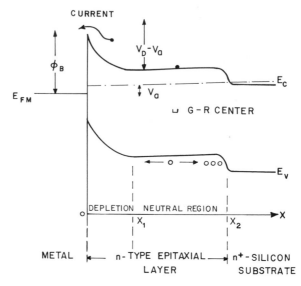

Fig. 2.28. Energy diagram showing holes accumulating at the low-high $nn^+$ barrier in a Schottky diode.

The minority carrier stored charge per unit area ($Q$), for Schottky diodes made on thin epitaxial layers, depends upon the characteristics of the epitaxy-substrate interface and can become very significant when this interface is highly reflecting (i.e., has a low value of surface recombination velocity). For large applied bias and negligible bulk recombination the stored charge is given by

$$Q = qn_i^2 D_p J \Big/ N_D J_{ns} S \qquad (2.24)$$

where $D_p$ is the hole diffusion constant, and $S$ the surface recombination velocity. In measurements on experimental Si epitaxial diodes the interface was found to be characterized by a recombination velocity of about 2000 cm sec.$^{-1}$. This value applied to the 5 ohm-cm Si-Au diode yields a storage time ($Q/J$) of about 1/3 nsec.

Edge effects in Schottky diodes are avoided by the use of guard-ring structures[109, 129] but care must be taken that minority carrier effects are not introduced by such protection. In Fig. 2.29(a) a $p$-type diffused guard ring extends in normal planar fashion under the oxide. The Schottky barrier, which is formed in the interior of the ring, is electrically contacting the $pn$ junction so that the depletion region of the total reverse-biased structure has the general shape indicated in Fig. 2.29(b). The regions of highest electric field strength will depend upon the depth and profile of the diffused junction. For an abrupt diffused junction of small curvature radius, the highest field strength will be

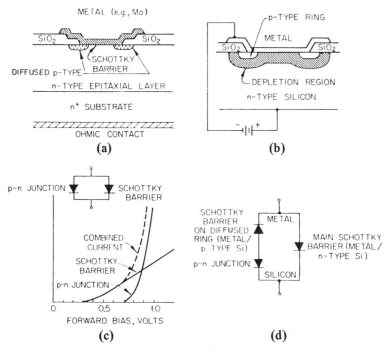

Fig. 2.29.  Schottky diode with *p-n* junction guard ring:
(a)  Structure for "hybrid" diode,
(b)  Reverse-biased hybrid diode, showing depletion region shape,
(c)  Hybrid-diode equivalent circuit, assuming contact between metal and diffused ring is ohmic. Qualitative I-V characteristics expected from the structure, and
(d)  Equivalent circuit of hybrid diode, assuming rectifying contact between metal and diffused ring. (After Zettler and Cowley, Reprinted with permission from *IEEE Trans. Electron Devices*, **ED-16**, 1969, p. 59.)

at the periphery of the junction and the breakdown will be lower than that of a plane parallel junction. On the other hand, for an ideal linearly graded junction, the breakdown voltage is higher than that for a plane junction, so that a hybrid diode with a linearly graded *pn* junction would be expected to have its breakdown limited first by the abrupt junction Schottky barrier. With this approach, control of the avalanche breakdown voltage and location by tailoring the diffused junction profile has been demonstrated. Hybrid diodes may have voltage ratings in excess of 100 V, compared with a few tens of volts for conventional Schottky diodes.

The equivalent circuit of a hybrid structure is shown in Fig. 2.29(c) together with a sketch of the anticipated *I-V* characteristics of the two diode components alone and the total *I-V* characteristic. It is seen that the composite characteristic will be dominated by the *pn* junction at voltages at which the *pn* junction injects appreciable minority carrier charge and hence modulates the conductivity of the epitaxial layer. Injection by the *pn* junction will limit the high-speed switching operation at high-current levels due to minority carrier charge storage. At low levels, the *I-V* characteristic is essentially that of the Schottky barrier and the switching speed will not be limited by storage effects.

The model shown in Fig. 2.29(c) assumes that the metal used to form the Schottky barrier to the *n*-type Si forms an ohmic contact to the diffused *p* region. This is effectively the case for a $p^+$ diffusion with virtually any metal since the current flow can proceed by a tunneling mechanism. The "ohmic" contact can also exist independently of the diffused region surface concentration with the use of a metal (or silicide) which has a high barrier height on *n*-type Si and therefore a low barrier on *p*-type Si. For instance, the barrier for PtSi is 0.85 eV on *n*-Si and 0.25 eV on *p*-Si. For such conditions there is a very high saturation current for the (reverse polarity) Schottky barrier diode formed on the diffused *p* region. However, it is possible to form opposite polarity rectifying Schottky barriers over a diffused *pn* junction provided that the concentration at the surface of the diffused region is not so high as to permit large tunneling currents. The equivalent circuit of Fig. 2.29(c) is then modified to the three-diode model shown in Fig. 2.29(d). The essential feature of this modification is that the metal *p*-type Schottky barrier in series with the *pn* junction is reverse biased so that most of the applied forward voltage for this branch appears across the reverse-biased Schottky barrier. Hence, the current which the *pn* junction can inject is limited by the reverse current of the metal *p*-type Schottky barrier which, in turn, is determined by the effective barrier height of the metal on the diffused ring. Thus, the rectifying contact to the *p* region can be used to eliminate almost completely the charge storage under heavy forward bias.

Consider a Mo on *n*-Si Schottky diode with a guard-ring construction as in Fig. 2.29(d). The barrier of the Mo on the *n*-Si is 0.68 eV and the barrier for the Mo on the *p*-Si ($N_A = 5 \times 10^{17}$ cm$^{-3}$) guard ring region is 0.42 eV. With this barrier of 0.42 eV and an area of $10^{-5}$ cm$^2$ the leakage current of the guard-ring Schottky barrier is only 160 $\mu$A. Thus, the Mo on *p*-Si guard-ring junction prevents the *pn* junction, Fig. 2.29(d), from contributing significantly to charge storage under forward bias conditions for the main Mo on *n*-Si Schottky junction.

Since the barrier for Au on *p*-type Si is about 0.30 eV compared with 0.42 eV for Mo, guard-ring hybrid diodes based on Au are found to exhibit more charge storage than those based on Mo. In a comparative study, Au-based diodes were found to be storing more than 100 pC of charge for 15 mA forward-current, whereas Mo-based diodes at the same current had only 1.4 pC of stored charge. The hybrid Mo-Si diodes are found to have low $f^{-1}$ noise with corner frequencies (at which the $f^{-1}$ noise value falls below the value for thermal and shot noise) of a few hundred hertz.

Noise studies of Schottky barriers have been made by workers in several laboratories.[48, 85, 88, 138]

Since a Schottky-barrier junction is a majority carrier device, a metal layer cannot be used as the emitter of a transistor. However, the semiconductor might form the emitter and the metal side a very thin base region, as shown in Fig. 2.30(a). Hot-electrons injected into the base, say, with an energy 0.85 eV above the Fermi level, may have a mean free path of 230 Å in a Au film. If the metal base is 100-Å thick, a fraction of the injected hot electrons may be collected at the metal-semiconductor collector. However, in experiments the collected fraction $I_c/I_e$ has tended to be 0.3 at most. This transfer inefficiency and the impedance levels involved suggest that the power gains for SMS transistor structures are too low to be of interest.[123]

Of greater interest is the use of a semiconductor $p^+n$ emitter junction with a metal Schottky barrier collector junction on the $n$ base region, as shown in Fig. 2.30(b), (c). For saturated switching operations this eliminates the injection of minority-carriers from the collector into the base region, as well as minority charge storage in the collector region. Theoretically, the storage time is approximately $\beta$ times less than that of the best possible Au-doped transistor with similar geometry and $\beta$.

## 2.5 SCHOTTKY BARRIERS IN INTEGRATED CIRCUITS

Schottky junctions are extensively used as clamping diodes between the base and collector of bipolar transistors to prevent excess minority carrier charge storage that would limit switching speed. An $npn$ transistor and a characteristic line are shown in Fig. 2.31(a), (b). If the base current $I_B$ is made large enough that $I_B > V_{CC}/\beta R_L$ the transistor is at the operating point $Q$ on the load line in Fig. 2.31(b). Hence, $V_{CB}$ is negative and the transistor base-collector junction is positively-biased instead of reverse-biased as in normal operation. The effect is an excess charge buildup everywhere within the transistor as suggested by Fig. 2.31(c). This excess charge considerably slows the turn-off of the transistor along the load line later in the cycle. The problem is solved by connection of a Schottky diode between the base and collector as shown in Fig. 2.31(d). The tendency of the base-collector junction to become forward biased is limited because the metal-semiconductor junction turns on at a low forward drop and carries the excess base current $\Delta I_B$.

Prior to the development of Schottky-barrier technology, Au doping was used to reduce storage time in transistors. For example in a particular logic transistor a storage time of 34 nsec was reduced to 7 nsec by the use of Au doping. However, the lifetime in the Au-doped unit was temperature dependent and became 15 nsec at 125°C. By the use of a Schottky clamp the storage time became 1 nsec and was temperature independent. The use of Au doping has

(a)

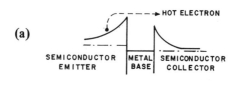

(b)

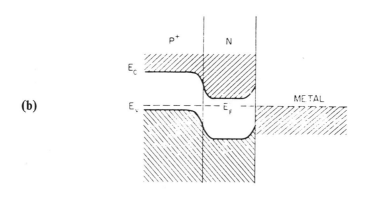

(c)

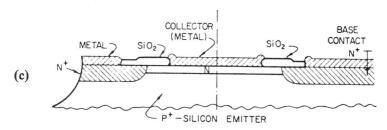

Fig. 2.30. Transistors involving Schottky barriers:
(a) SMS transistor must have very thin metal base since the mean free path of the injected hot electron is very low,
(b) This SSM transistor has a Schottky barrier collector for the holes injected in the $n$ base region, and
(c) Structure of the SSM transistor. (After May.[78] Reprinted with permission from *Solid-State Electronics*, Pergamon Press, Ltd.)

other limiting features: it reduces the $h_{FE}$ of the transistor because of reduced base-lifetime, it influences all the devices on the chip and it raises device leakage currents. In comparison, the provision of metal-semiconductor clamps presents almost no problem in integrated circuit processing. Fig. 2.32(a) shows the integrated circuit (IC) structure for a conventional *NPN* transistor and (b) shows the variation needed to include an Al-Schottky diode between the base and the collector. If a guard ring form is needed for the Schottky junction it is readily provided as shown in Fig. 2.32(c). Many integrated-circuit planar components,

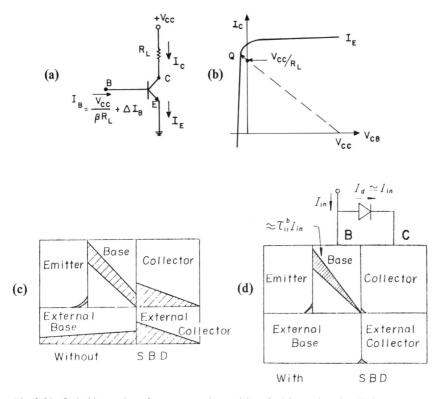

Fig. 2.31. Switching action of an *npn* transistor with and without clamping diode:
(a)   *npn* transistor with base overdrive $\Delta I_B$,
(b)   In saturation at point $Q$ because of overdrive,
(c)   Excess minority carrier density caused by saturation, and
(d)   Addition of the clamping Schottky diode diverts the overdrive and reduces the stored charge. [c,d. (After Tarui. Reprinted with permission from *IEEE Journal of Solid-State Circuits*, SC-4, 1969, 1, p. 4.)]

fast and slow, are compatible with the Schottky technology as indicated in Fig. 2.33.

Metal semiconductor junctions also have applications as modulators and detectors at high frequencies and as light-sensitive diodes as discussed in later chapters.

## 2.6 HIGH POWER SCHOTTKY BARRIER RECTIFIERS

Schottky junctions tend to be confined to small-signal low power applications, such as the diode and integrated circuit uses already mentioned, where high speed is a desired feature. However, the low forward-voltage drop associated

(a)

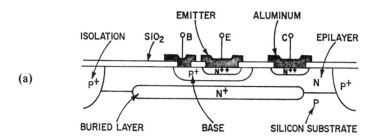

CONVENTIONAL NPN TRANSISTOR

(b)

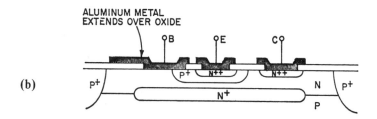

SCHOTTKY EXTENDED-METAL TRANSISTOR

(c)

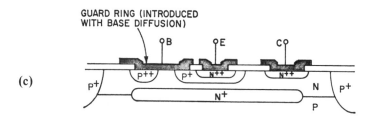

SCHOTTKY GUARD-RING TRANSISTOR

Fig. 2.32. A conventional transistor may have a Schottky diode clamp added by extension of the base metalization as shown in (a) and (b). A guard ring on the Schottky diode may be introduced as shown in (c). (After Noyce, Bohn, Chua.[90] Reprinted from *Electronics*, July 21, 1969; copyright McGraw-Hill, Inc. NY 1969.)

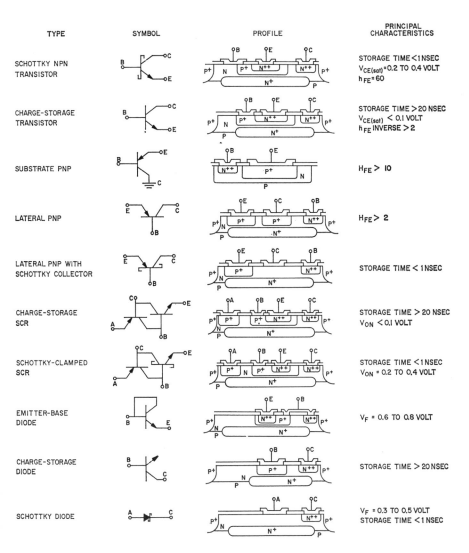

| TYPE | SYMBOL | PROFILE | PRINCIPAL CHARACTERISTICS |
|------|--------|---------|---------------------------|
| SCHOTTKY NPN TRANSISTOR | | | STORAGE TIME < 1 NSEC<br>$V_{CE(sat)}$ = 0.2 TO 0.4 VOLT<br>$h_{FE}$ = 60 |
| CHARGE-STORAGE TRANSISTOR | | | STORAGE TIME > 20 NSEC<br>$V_{CE(sat)}$ < 0.1 VOLT<br>$h_{FE}$ INVERSE > 2 |
| SUBSTRATE PNP | | | $H_{FE}$ > 10 |
| LATERAL PNP | | | $H_{FE}$ > 2 |
| LATERAL PNP WITH SCHOTTKY COLLECTOR | | | STORAGE TIME < 1 NSEC |
| CHARGE-STORAGE SCR | | | STORAGE TIME > 20 NSEC<br>$V_{ON}$ < 0.1 VOLT |
| SCHOTTKY-CLAMPED SCR | | | STORAGE TIME < 1 NSEC<br>$V_{ON}$ = 0.2 TO 0.4 VOLT |
| EMITTER-BASE DIODE | | | $V_F$ = 0.6 TO 0.8 VOLT |
| CHARGE-STORAGE DIODE | | | STORAGE TIME > 20 NSEC |
| SCHOTTKY DIODE | | | $V_F$ = 0.3 TO 0.5 VOLT<br>STORAGE TIME < 1 NSEC |

Fig. 2.33. Some integrated circuit components that are readily compatible with the addition of Schottky diodes. (After Noyce, Boyn, Chua.[90] Reprinted from *Electronics*, July 21, 1969; copyright McGraw-Hill, Inc. NY 1969.)

with ideality factors of only slightly greater than unity is another feature of interest, particularly if rectification of a low-voltage supply is needed. In this context high-current (10-50 A) Schottky rectifiers have been studied.

The *I-V* characteristics of a Cr barrier of area 0.11 cm$^2$ on a 0.5 ohm-cm 2-$\mu$m-thick $nn^+$Si epilayer are shown in Fig. 2.34 and 2.35. The forward voltage drop is less than 0.5 V at 50 A and the series resistance is about 0.003 ohms.

The reverse voltage rating is in excess of 20 V. The zero bias capacitance is 3500 pF which at 500 kHz represents an impedance of only 90 ohms. However, if the load resistance is low the rectification efficiency of a halfwave circuit remains almost constant from dc to 500 kHz. The rectifier has reasonable surge resistance capability, for instance withstanding 10 surges of 400 A at 60 Hz.[97]

The epitaxial layer introduces a series parasitic resistance that depends on

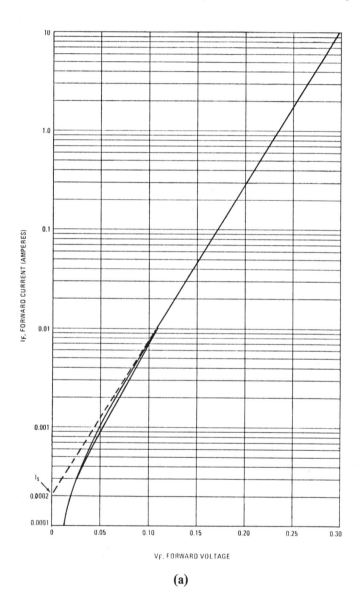

(a)

Fig. 2.34

the layer thickness and doping. Since the device does not function by minority carrier flow, this resistance cannot be reduced by conductivity modulation in forward conduction as in a *pn* junction. The parasitic voltage drop is proportional to the forward current density and is

$$V_F' = J_F \rho d \qquad (2.25)$$

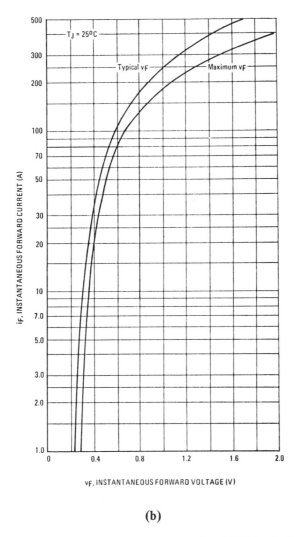

**(b)**

Fig. 2.34. Forward current-voltage characteristics of Cr *nn*⁺ Si Schottky barrier power rectifier. (After Polgar, Mouyard, Shiner. Reprinted with permisssion from *IEEE Trans. Electron Devices*, **ED-17**, 1970, pp. 728.)

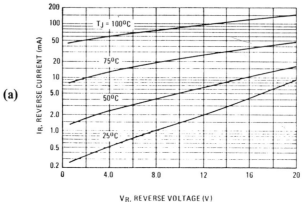

(a)

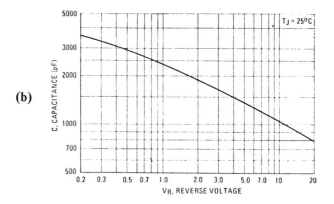

(b)

Fig. 2.35. Reverse characteristics of a Cr $nn^+$ Si Schottky power rectifier:
(a) Leakage current as a function of temperature, and
(b) Typical capacitance variation with reverse voltage. (After Polgar, Mouyard, Shiner. Reprinted with permission from *IEEE Trans. Electron Devices*, ED-17, 1970, pp. 729, 728.)

where $\rho$ is the resistivity of the epitaxial layer and $d$ is the thickness.

The minimum thickness of the depletion (epitaxial) region is given by

$$d = \frac{2V_R}{\mathscr{E}_{max}} \qquad (2.26)$$

$\mathscr{E}_{max}$ can be obtained for a given value of $V_R$ and $\rho$ can be determined from the corresponding value of $N_d$. Hence, a relationship may be obtained between the reverse blocking voltage $V_B$ and the parasitic forward drop, $V'_F$. This is shown in Fig. 2.36 for densities of forward conduction of 100 and 500 A cm$^{-2}$. The parasitic voltage drop is negligible for small blocking voltages but as the blocking voltage is raised, the parasitic forward drop increases as about $V_B{}^{2.5}$.

Since a parasitic voltage drop of 0.6 V is the maximum that can be tolerated if the total rectifier forward voltage is to be 1.0 V or less, it follows from Fig. 2.36 that a reverse voltage of 100 V is the maximum that can be obtained in a 500 A cm$^{-2}$ design. Hence, in a Schottky junction, high values of blocking voltage may be obtained only at the expense of increased forward voltage drop. The reverse blocking voltage achievable by moat etching may be 80%-90% of the ideal bulk breakdown voltage for the Si doping.[102]

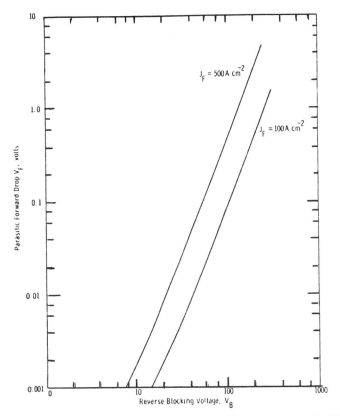

Fig. 2.36. Variation of the parasitic voltage drop in the depletion region of a Schottky barrier rectifier with reverse blocking voltage. (After Page.[94] Reprinted with permission from *Solid-State Electronics*, Pergamon Press, Ltd.)

## 2.7 QUESTIONS

1. A Schottky barrier diode is made by evaporating a dot of Au of area $10^{-3}$ cm$^2$ onto $10^{14}$ cm$^{-3}$ doped Si. What is the $I_O$ likely to be in the expression $I = I_O \left[ \exp \dfrac{qV}{kT} - 1 \right]$ if the Si is $n$ type? If $p$ type?

2. A Schottky diode has the following forward biased characteristic (approximate values) at 300°K

| $I_f$ | $5 \times 10^{-9}$ | $2 \times 10^{-8}$ | $2 \times 10^{-7}$ | $10^{-6}$ | $10^{-5}$ | $7 \times 10^{-5}$ | A |
|---|---|---|---|---|---|---|---|
| $V_f$ | 0.05 | 0.10 | 0.15 | 0.20 | 0.25 | 0.30 | V |

   What may be concluded about the diode if the area is $10^{-4}$ cm$^2$?

3. An Al Schottky barrier diode on $Al_{0.5}Ga_{0.5}As$ is forward biased at 0.25 V and the current measured at various temperatures with the following results:

| Temperature, C | 27 | 34 | 44 | 56 | 68 | 79 | 92 |
|---|---|---|---|---|---|---|---|
| Current, nA | 2.1 | 6.0 | 17.6 | 55.2 | 162 | 440 | 1170 |

   What information can be determined from the experiment? If it is uncertain whether the diode is on Si or on $Al_{0.5}Ga_{0.5}As$ is there a non-destructive test that could determine this? What test could be made to establish the parasitic resistance of the diode?

4. A Schottky diode on Si by microscopic inspection appears to have an area of $10^{-4}$ cm$^2$. By experiment it is found to have a capacitance $C$ in Farads that varies as a function of diode voltage $V$ according to $C^{-2} = 0.70 \times 10^{24} - 1.2 \times 10^{24}$ V where V is between 0 and $-10$ volts. The $I_O$ current of the diode is about $10^{-12}$ A at 300°K. What is the diffusion potential of the barrier, what is the doping of the semiconductor and what is the depletion region thickness at zero voltage? What else can be said about the diode?

5. Consider a Au Schottky barrier of area $10^{-2}$ cm$^2$ on $nn^+$ Si where the 10 $\mu$m thick epilayer is doped $10^{14}$ cm$^{-3}$ and has $\tau_p = 10^{-5}$ sec and the $n^+$ substrate is doped $10^{18}$ cm$^{-3}$, is 0.010 cm thick and has $\tau_p = 10^{-8}$ sec. For a current of 200 mA estimate the stored hole charge (pC) in the epilayer at 300°K. If the current is increased by a factor of 10 does the stored hole charge increase in proportion?

6. A Si hybrid diode structure is shown in Fig. 2.29(d). Design a similar guard ring structure for a Au/GaAs Schottky barrier hybrid rated at 100 mA, 100 V. Sketch the $I$ vs $V$ forward and reverse characteristic expected. Estimate the charge storage expected for 100 mA at 300°K and 400°K. What will be the small signal equivalent circuit of the diode at 1 MHz? (The diffusion length of $n$ GaAs doped $\leqslant 10^{15}$ cm$^{-3}$ may be taken as 5 $\mu$m and this may be considered to decrease to 1 $\mu$m for a doping $\geqslant 10^{16}$ cm$^{-3}$).

7. Attempt the design of a GaAs power Schottky diode for 100 A, 200 V, 300°K ambient operation.
   (a) Specify the design principles and select dimensions and dopings. Assume cooling is to be with a 10 × 10 cm vertical fin which dissipates 1 mW/cm$^2$/°C.
   (b) Estimate the $I$ vs $V$ characteristics for a junction temperature of 450°K and compare with the characteristics expected of an alternative $p^+nn^+$ junction design.

8. A semiconductor (emitter area $10^{-4}$ cm$^2$)/metal-base/semiconductor (collector area $10^{-3}$ cm$^2$) transistor test structure (see Fig. 2.30) is fabricated with the semiconductor $10^{17}$ cm$^{-3}$ $n$ type Si. The metal base is 125 Å of Au. Hot electrons with an energy 0.85 eV above the Fermi level in the Au are expected to have a mean free path of about 250 Å.

   (a)  What is the expected ratio of the dc collector and emitter currents, $I_c/I_e$, under typical bias conditions?

   (b)  If connected as a small signal 100 MHz common emitter amplifier what would be a suitable load resistance, or impedance, for good power gain? Estimate this power gain.

9.  Barrier heights of Schottky diodes may be determined in several different ways, electrically and electro-optically. Show with specific examples the extent to which these methods give consistent answers.

10.  Gather evidence to show if the rule $\phi_{Bn} + \phi_{Bp} = E_g$ is well obeyed.

11.  Discuss the reverse voltage-current characteristics of Schottky diodes before breakdown. To what extent does the image-force barrier-height reduction explain the observed results if edge-controlled specimens are used?

12.  Discuss the temperature coefficients of Schottky barrier heights for several semiconductors.[44,58]

13.  Heat treatment applied to an Al-$n$Si Schottky junction results in a change of apparent barrier height because of development of a $pn$ junction. Discuss this effect.[10]

14.  Discuss the interaction of various metals with Si to form conducting silicides and their role in Schottky barriers.

15.  In GaAs Schottky junctions changes have been associated with out-migration of Ga or As or in-migration of metals. Discuss the problem.[62,75]

16.  Discuss the effects of an interface layer of oxide between the metal and the semiconductor in Si or GaAs Schottky diodes.[1,19,24,25,103,105]

17.  Some Schottky diodes obey the law $I = I_0 \exp(V/V_0)$ where $V_0$ is independent of temperature. Discuss this with respect to ZnSe and present evidence of its validity for other semiconductors.[142]

18.  Discuss the expression[93]

$$I = A^*ST^2 \exp\left[qV_B/k\,(T + T_o)\right]\left\{\exp\left[qV/k(T + T_o)\right] - 1\right\}$$

Gather and summarize the evidence for such a relationship for Si, GaAs and other semiconductors. To what extent does the evidence support Levine's model involving an exponential distribution of interface states?

19.  How do the density of states at a metal-coated semiconductor interface compare with the density of states at an oxide-coated interface and at a cleaved surface?

20.  Discuss Schottky barrier height variations on ternary III-V compounds such as $Al_xGa_{1-x}As$ or $GaAs_yP_{1-y}$.[79]

21.  Discuss the relative merits of guard-ring structures, mesa etching and chamfering as means of obtaining good reverse voltage characteristics and eliminating edge effects in metal-semiconductor junctions.[102,109,129]

22.  Discuss tunneling, field emission and thermionic field emission in Schottky barriers.[3,92,107]

23.  Discuss minority carrier charge storage in a Schottky $mnn^+$ structure in a quantitative way. How is the effective recombination velocity $s$ at the $nn^+$ interface derived?[40,113]

24.  Discuss the circumstances in which the small signal impedance of a Schottky diode becomes inductive.[40]

25.  Discuss the effect of deep impurities in Schottky barrier junctions (a) on frequency response, (b) on capacitance transients, (c) on reverse characteristics and (d) on forward characteristics.[37,146]

26.  An inversion layer may be present at the metal-semiconductor interface of Schottky diodes if the barrier is high and the semiconductor lightly doped. Discuss its effects.[30]

27.  Discuss the Williams model for the effect of recombination on the ideality factor of a Schottky junction that results in the form $n = \phi_{Bn}/0.5E_g$.

28. Characterize the current, voltage and capacitance behavior of a two terminal structure: $n$ semiconductor 1/metal/$n$ semiconductor 2 where $\phi_{ms_1}$ is about twice $\phi_{ms_2}$. Consider some specific examples.

29. Discuss noise in Schottky barrier diodes. Is noise related to the ideality factor of the diode?[25,48,85,137,138]

30. Discuss the practical realization and use of Mott barriers instead of Schottky barriers.[80]

31. In the use of a Schottky diode clamp between the base and collector of a transistor it is stated that because the Schottky diode has a lower temperature coefficient than the emitter-base diode, $V_{CE(sat)}$ decreases with increasing temperature. This behavior of the Schottky diode amounts to partial temperature compensation of the emitter-base diode and hence of $V_{CE(sat)}$. It helps to maintain a high worst-case logic 0 noise margin at high temperature when the Schottky transistor drives fan-out circuits that operate at a $2V_{BE}$ logic threshold as standard DTL and TTL ICs do.[90] Discuss this effect.

32. In a TTL logic circuit with and without an SB-clamp the low-level voltage margin is related to storage charge reduction. For a low-level voltage margin $M$ given by $V_T - V_{to}$ where $V_T$ is the input threshold voltage and $V_{to}$ is the low-level output voltage, develop relationships of $M$ to the rise-time charge $Q_r$ and to the decay-time charge $Q_d$ and sketch the results for $M$ between 0.3 and 0.6 V.[125]

33. Discuss the possible use of Schottky diode clamps in conjunction with thyristor $pnpn$ structures.[112]

34. In a power Si Schottky diode the frequency of operation may be limited by an $RC$ time constant. Discuss this effect. Compare this limitation with the recovered charge limitation of an alloy $p^+n$Si junction of comparable rating.[94,97]

35. Discuss thermal instability problems as functions of barrier heights in Si Schottky diodes.[94]

36. A relationship has been demonstrated between the series resistance of the epi-layer of a Si Schottky power rectifier and the thickness and doping and breakdown voltage. Consider how this relationship would be changed if either: (a) the epi-layer were graded in doping, or (b) the epi-layer were uniformly doped but of GaAs.[94]

37. Describe five circuit applications where Schottky power rectifiers of current rating over 1 A would represent an improvement over the use of ordinary diodes.

38. Consider "ohmic" contacts to semiconductors in terms of Schottky barrier action.[83,124]

39. A Schottky diode acousto-electric memory correlator has been proposed by Ingebrigtsen.[52] Discuss the principles and review any recent developments.

## 2.8 REFERENCES AND FURTHER READING SUGGESTIONS

1. Adams, A.C., and B.R. Pruniaux, "Gallium arsenide film evaluation by ellipsometry and its effect on Schottky barriers," *J. Electrochem. Soc.*, 120, 1973, p. 408.

2. Altman, L., "Shottky-TTL controller put on a chip," *Electronics*, March 7, 1974, p. 159.

3. Anderson, C.L., C.R. Crowell, and T.W. Kao, "Effects of thermal excitation and quantum-mechanical transmission on photothreshold determination of Schottky barrier height," *Solid-State Electronics*, 18, 1975, p. 705.

4. Anderson, W.A., A.E. Delahoy, and R.A. Milano, "An 8% efficient layered Schottky-barrier solar cell," *J. Appl. Phys.*, 45, 1974, p. 3913.

5. Anderson, R.M, and T.M. Reith, "Microstructural and electrical properties of thin PtSi films and their relationships to deposition parameters," *J. Electrochem. Soc.*, 122, 1975, p. 1337.

6. Andrews, J.M., and M.P. Lepselter, "Reverse current-voltage characteristics of metal-silicide Schottky diodes," *Solid-State Electronics*, 13, 1970, p. 1011.

7. Banbury, P.C., "Theory of the forward characteristic of injecting point contacts," *Proc. Phys. Soc.* (London), **B, 66**, 1953, p. 833.
8. Barret, C., and A. Vapaille, "Determination of the density and the relaxation time of silicon-metal interfacial states," *Solid-State Electronics,* **18**, 1975, p 25.
9. Barret, C., and A. Vapaille, "Interfacial states spectrum of a metal-silicon junction," *Solid-State Electronics,* **19**, 1976, p. 73.
10. Basterfield, J., J.M. Sharron, and A. Gill, "The nature of barrier height variations in alloyed Al-Si Schottky barrier diodes," *Solid-State Electronics,* **18**, 1975, p. 290.
11. Beguwala, M., and C.R. Crowell, "Determination of hafnium-$p$-type silicon Schottky barrier height," *J. Appl. Phys.,* **45**, 1974, p. 2793.
12. Best, J.S., et al., "HgSe, a highly electronegative stable metallic contact for semiconductor devices," *Appl. Phys. Lett.,* **29**, 1976, p. 433.
13. Blattner, R.J., et al., "Effect of oxidizing ambients on platinum silicide formation. II. Auger and Backscattering Analyses," *J. Electrochem. Soc.,* **122**, 1975, p. 1732.
14. Borrego, J.M., R.J. Gutmann, and S. Ashok. "Interface state density in Au and Al*n*GaAs Schottky diodes," *IEEE International Electron Devices Meeting Technical Digest*, December 1976, Washington, D.C., p. 365.
15. Borrego, J.M., and R.J. Gutmann, "Changes in Au-GaAs Schottky barrier diodes with low neutron fluence," *Appl. Phys. Lett.,* **28**, 1976, p. 280.
16. Burgess, R.E., "The influence of mobility variation in high fields on the diffusion theory of rectifier barriers," *Proc. Phys. Soc.,* (London), **B, 66**, 1953, p. 430.
17. Canali, C., S.U. Campisano, S.S. Law, Z.L. Liau, and J.W. Mayer, "Solid-phase epitaxial growth of Si through palladium silicide layers," *J. Appl. Phys.,* **46**, 1975, p. 2831.
18. Card, H.C., "On the direct currents through interface states in metal-semiconductor contacts," *Solid-State Electronics,* **18**, 1975, p. 881.
19. Card, H.C., and E.H. Rhoderick, "The effect of an interfacial layer on minority carrier injection in forward-biased silicon Schottky diodes," *Solid-State Electronics,* **16**, 1973, p. 365.
20. Card, H.C., and E.H. Rhoderick, "Studies of tunnel MOS diodes I. Interface effects in silicon Schottky diodes," *J. Phys. D: Appl. Phys.,* **4**, 1971, p. 1589.
21. Carver, L., "High-power Schottky diodes—no barrier to system efficiency," *Electronic Products Magazine*, March 19, 1973, p. 229.
22. Carver, L., "Use Schottky diodes to solve your power-switching problems," *Electron Device News*, October 20, 1973, p. 46.
23. Chaika, G.E., "Negative resistance of metal-semiconductor contacts," *Soviet Physics—Semiconductors*, **6**, 1972, p. 115.
24. Cowley, A.M., "Depletion capacitance and diffusion potential of gallium phosphide Schottky-barrier diodes," *J. Appl. Phys.,* **37**, 1966, p. 3024.
25. Cowley, A.M., "Titanium-silicon Schottky barrier diodes," *Solid-State Electronics,* **13**, 1970, p. 403.
26. Crowell, C.R., and V.L. Rideout, "Normalized thermionic-field emission in metal-semiconductor (Schottky) barriers," *Solid-State Electronics,* **12**, 1969, p. 89.
27. Danyluk, S., and G.E. McGuire, "Platinum silicide formation: Electron spectroscopy of the platinum-platinum silicide interface," *J. Appl. Phys.,* **45**, 1974, p. 5141.
28. Day, H.M., A. Christou, W.H. Weisenberger, and J.K. Hirrionen, "Reactions between the Ta-Pt-La-Au metallization and PtSi ohmic contacts," *J. Electrochem. Soc.,* **122**, 1975, p. 769.
29. deKock, A.J.R., S.D. Ferris, L.C. Kimerling, and H.J. Leamy, "Investigation of defects and striations in as-grown Si crystals by SEM using Schottky diodes," *Appl. Phys. Lett.,* **27**, 1975, p. 313.
30. Demoulia, E., and F. van DeWick, "Inversion layer at the interface of Schottky diodes," *Solid-State Electronics,* **17**, 1974, p. 825.

31. Eimers, G.W., and E.H. Stevens, "Composite model for Schottky diodes barrier height," *IEEE Trans. Electron Devices*, **ED-18**, 1971, p. 1185.
32. Engemann, J., and J. Naumann, "Temperaturverhalten von Au-$n$-GaAs-Schottky-kontahten," *Solid-State Electronics*, **15**, 1972, p. 899.
33. Flietner, H., "The $E(h)$ relation for a two-band scheme of semiconductors and the application to the metal-semiconductor contact," *Phys. Stat. Sol.*, (b) **54**, 1972, p. 201.
34. Fonash, J., "Current transport in metal semiconductor contacts—a unified approach," *Solid-State Electronics*, **15**, 1972, p. 783.
35. Formenko, V.S., *Handbook of Thermionic Properties*, (G.V. Samsonov, ed.), Plenum Press, New York 1966.
36. Franson, P., "Schottky TTL blunts ECL growth," *Electronics*, March 1, 1971, p. 69.
37. Glover, G.H., "Determination of deep levels in semiconductors from $C$-$V$ measurements," *IEEE Trans Electron Devices*, **ED-19**, 1972, p. 138.
38. Glover, G.H., "Ionization rate in GaAs determined from photomultiplication in a Schottky barrier," *J. Appl. Phys.*, **44**, 1973, p. 3253.
39. Gossick, B.R., "Electrical characteristics of a metal-semiconductor contact. II," *Surface Science*, **25**, 1971, p. 465.
40. Green, M.A., and J. Shewchun, "Minority carrier effects upon the small-signal and steady-state properties of Schottky diodes," *Solid-State Electronics*, **16**, 1973, p. 1141.
41. Guétin, P., and G. Schréder, "Quantitative aspects of the tunneling resistance in $n$-GaAs Schottky barriers," *J. Appl. Phys.*, **42**, 1971, p. 5689.
42. Gupta, H.M., and R.J. VanOverstraeten, "Role of trap states in the insulator region for MIM characteristics," *J. Appl. Phys.*, **46**, 1975, p. 2675.
43. Gutknecht, P., "Aluminum Schottky barriers on sputter-etched silicon, " *Electronics Lett.*, **7**, 1971, p. 298.
44. Hackam, R., and P. Harrop, "Electrical properties of nickel-low-doped $n$-type gallium arsenide Schottky-barrier diodes," *IEEE Trans. Electron Devices*, **ED-19**, 1972, p. 1231.
45. Handu, V.K., and M.S. Tyagi, "In-GaAs Schottky barriers," *J. Institution of Telecommunication Engineers*, New Delhi, **18**, 1972, p. 527.
46. Hariu, T., and Y. Shibata, "Optimum Schottky-barrier height for high-efficiency microwave transferred-electron diodes," *Proc. IEEE*, **63**, 1975, p. 823.
47. Helman, J.S., and F.S. Sinencio, "Resonant tunneling through Schottky barriers," *Appl. Phys. Lett.*, **28**, 1976, p. 34.
48. Hsu, S.T., "Flicker noise in metal semiconductor Schottky barrier diodes due to multistep tunneling processes," *IEEE Trans. Electron Devices*, **ED-18**, 1971, p. 882.
49. Huang, C.I., and S.S. Li, "Reverse $I$-$V$ characteristics in Au-GaAs Schottky diode in the presence of interfacial layer," *Proc. IEEE*, **61**, 1973, p. 477.
50. Hughes, G.W. and R.M. White, "Microwave properties of nonlinear MIS and Schottky-barrier microstrip," *IEEE Trans. Electron Devices*, **ED-22**, 1975, p. 945.
51. Ingebrigtsen, K.A., "Simultaneous storage of spatially orthogonal acoustic beams in a Schottky-diode memory correlator," *Electronics Lett.*, **11**, 1975, p. 585.
52. Ingebrigtsen, K.A., R.A. Cohen, and R.W. Mountain, "A Schottky-diode acoustic memory and correlator," *Appl. Phys. Lett.*, **26**, 1975, p. 596.
53. Jäger, H., and W. Kosak, "Modulation effect by intense hole injection in epitaxial silicon Schottky-barrier-diodes," *Solid-State Electronics*, **16**, 1973, p. 357.
54. Kajiyama, K., Y. Mizushima, and S. Sakata, "Schottky barrier height of $n$-In$_x$Ga$_{1-x}$As diodes," *Appl. Phys. Lett.*, **23**, 1973, p. 458.
55. Kajiyama, K., S. Sakata, and O. Ochi, "Barrier height of Hf/GaAs diode," *J. Appl. Phys.*, **46**, 1975, p. 3221.
56. Kajiyama, K., S. Sakata, and Y. Mizushima, "Schottky-barrier devices with low barrier height," *Proc. IEEE*, **62**, 1974, p. 1287.

57. Kingzett, T.J., and C.A. Ladas, "Effect of oxidizing ambients on platinum silicide formation, 1. Electron microprobe analysis," *J. Electrochem. Soc.*, 122, 1975, p. 1729.
58. Klose, H., "On the potential barrier of Schottky junctions," *Phys. Stat. Sol.* (a) 14, 1972, p. 457.
59. Korol, A.N., M. Ye Kitsai, V.I. Strikka, and D.I. Sheba, "Effect of deep impurity levels on Schottky barrier diode characteristics," *Solid-State Electronics*, 18, 1975, p. 375.
60. Kassing, R. and E. Kähler, "Low-frequency current oscillations in high-resistivity, Au-doped silicon junctions with two Schottky contacts," *Phys. Stat. Sol.* (a) 12, 1972, p. 209.
61. Keen, N.J., "Evidence for coherent noise in pumped Schottky-diode mixers," *Electronics Lett.*, 13, 1977, p. 282.
62. Kim, H.B., G.G. Sweeney, and T.M.S. Heng, "Analysis of metal-GaAs Schottky barrier diodes by secondary ion mass spectrometry," Gallium Arsenide and Related Compounds 1974, Inst. Phys. Conf. Ser. No. 24, London.
63. Korwin-Pawlowski, M.L., and E.L. Heasell, "The properties of some metal-InSb surface barrier diodes," *Solid-State Electronics*, 18, 1975, p. 849.
64. Kräutle, H., M.A. Nicolet, and J.W. Mayer, "Kinetics of silicide formation by thin films of V on Si and SiO$_2$ substrates," *J. Appl. Phys.*, 45, 1974, p. 3304.
65. Kurtin, S., T.C. McGill, and C.A. Mead, "Fundamental transition in electronic nature of solids," *Phys. Rev. Lett.*, 22, 1969, p. 1433.
66. Landbrooke, P.H., "Reverse current characteristics of some imperfect Schottky barriers," *Solid-State Electronics*, 16, 1973, p. 743.
67. Losee, D.L., "Admittance spectroscopy of impurity levels in Schottky barriers," *J. Appl. Phys.*, 46, 1975, p. 2204.
68. Lau, S.S., and D. Sigurd, "An investigation of the structure of Pd$_2$Si formed on Si," *J. Electrochem. Soc.*, 121, 1974, p. 1538.
69. Levine, J.D., "Power law reverse current-voltage characteristics in Schottky barriers," *Solid-State Electronics*, 17, 1974, p. 1083.
70. Levine, J.D., "Schottky-barrier anomalies and interface states," *J. Appl. Phys.*, 42, 1971, p. 3991.
71. Liau, Z.L., S.U. Campisano, C. Canali, S.S. Lau, and J.W. Mayer, "Kinetics of the initial stage of Si transport through Pd-silicide for epitaxial growth," *J. Electrochem. Soc.*, 122, 1975, p. 1696.
72. Livingstone, A.W., K. Lurvey, and J.W. Allen, "Electroluminescence in forward-biased zinc selenide Schottky diodes," *Solid-State Electronics*, 16, 1973, p. 351.
73. Lubberts, G., "Barrier modulation in Ag-diffused Au-CdS diodes," *J. Appl. Phys.*, 47, 1976, p. 365.
74. Lubberts, G., and B.C. Burkey, "A method for calculating band profile and capacitance of nonuniformly doped Schottky barriers," *Solid-State Electronics*, 18, 1975, p. 805.
75. Madams, C.J., D.V. Morgan, M.J. Hawes, "Outmigration of gallium from Au-GaAs interfaces," *Electronics Lett.*, 11, 1975, p. 574.
76. Mahoney, G.E., "Retardation of Impatt diode aging by use of tungsten in the electrodes," *Appl. Phys. Lett.*, 27, 1975, p. 613.
77. Mark, P., and W.F. Creighton, "Chemical trends in oxygen uptake rates at the ordered surfaces of tetrahedrally coordinated compound semi-conductors," *Appl. Phys. Lett.*, 27, 1975, p. 400.
78. May, G.A., "The Schottky-barrier-collector transistor," *Solid-State Electronics*, 11, 1968, p. 613.
79. McCaldin, J.O., T.C. McGill, and C.A. Mead, "Correlation for III-V and II-VI semiconductors of the Au Schottky barrier energy with anion electronegativity," *Phys. Rev. Lett.*, 36, 1976, p. 56.

80. McColl, M. and M.F. Millea, "Advantages of Mott barrier mixer diodes," *Proc. IEEE*, April 1973, p. 499.
81. Mead, C.A., "Metal-semiconductor surface barriers," *Solid-State Electronics*, 9, 1966, p. 1023.
82. Mead, C.A., and W.G. Spitzer, "Fermi level position at semiconductor surfaces," *Phys. Rev. Lett.*, 10, 1963, p. 471.
83. Milnes, A.G., and D.L. Feucht, *Heterojunctions and Metal Semiconductor Junctions*, Academic Press, NY, 1972.
84. Minden, H.T., "Gallium arsenide dual Schottky barrier diodes," *Solid-State Electronics*, 16, 1973, p. 1185.
85. Minniti, R.J., Jr., G.W. Neudeck, and R.M. Anderson, "Shot noise in platinum-gallium arsenide Schottky barrier diodes," *J. Appl. Phys.*, 42, 1971, p. 1886.
86. Marino, A and T. Sugano, "Forward characteristics of Si Schottky diodes," *Jpn. J. Appl. Phys.*, 9, 1970, p. 1484.
87. Murarka, S.P., "Forward *I-V* characteristics of Pt/$n$-GaAs Schottky barrier contacts," *Solid-State Electronics*, 17, 1974, p. 985.
88. Neudeck, G.W., "High-frequency shot noise in Schottky barrier diodes," *Solid-State Electronics*, 13, 1970, p. 1249.
89. Neudeck, G.W., R.J. Minniti, Jr., and R.M. Anderson, "Ideality factor and the high-frequency noise of Schottky-barrier-type diodes," *IEEE J. Solid-State Circuits*, SC7, 1972, p. 89.
90. Noyce, R.N., R.E. Bohn, and H.L. Chua, "Schottky diodes make IC scene," *Electronics*, July 21, 1969, p. 74.
91. Padovani, F.A., and G.G. Sumner, "Experimental study of gold-gallium arsenide Schottky barriers," *Solid-State Electronics*, 9, 1966, p. 3744.
92. Padovani, F.A. and R. Stratton, "Field and thermionic-field emmission in Schottky barriers," *Solid-State Electronics*, 9, 1966, p. 695.
93. Padovani, F.A., "Metal-semiconductor barrier devices," *Semiconductor and Semimetals*, Academic Press, New York, Vol. 7a, 1970, p. 75.
94. Page, D.J., "Theoretical performance of the Schottky barrier power rectifier," *Solid-State Electronics*, 15, 1972, p. 505.
95. Peckerar, M., "On the origin of the increase in Schottky barrier height with interfacial oxide thickness," *J. Appl. Phys.*, 45, 1974, p. 4652.
96. Pellegrini, B., "Current-voltage characteristics of silicon metallic-silicide interfaces," *Solid-State Electronics*, 18, 1975, p. 417.
97. Polgar, P., A. Mouyard, and B. Shiner,"A high-current metal-semiconductor rectifier," *IEEE Trans. Electron Devices*, ED-17, 1970, p. 725.
98. Rand, M.J., "*I-V* characteristics of PtSi-Si contacts made from CVD platinum," *J. Electrochem. Soc.*, 122, 1975, p. 811.
99. Reith, T.M., "Aging effects in Si-doped Al Schottky barrier diodes," *Appl. Phys. Lett.*, 28, 1976, p. 152.
100. Reith, T.M., and J.D. Schick, "The electrical effect on Schottky barrier diodes of Si crystallization from Al-Si metal films," *Appl. Phys. Lett.*, 25, 1974, p. 524.
101. Rusu, A., and C. Bulucca, "Enhanced breakdown voltage in planar metal-overlap laterally diffused (MOLD) Schottky diodes," *Appl. Phys. Lett.*, 27, 1975, p. 620.
102. Rhee, C., J. Saltich, and R. Zwernemann, "Moat-etched Schottky barrier diode displaying near ideal *I-V* characteristics," *Solid-State Electronics*, 15, 1972, p. 1181.
103. Rhoderick, E.H., "The physics of Schottky barriers," *J. Phys. D.*, 3, 1970, p. 1153.
104. Rhoderick, E.H., "Comments on the conduction mechanism in Schottky diodes," *J. Phys. D: Appl. Phys.*, 5, 1972, p. 1920.
105. Rhoderick, E.H., "A note on Levine's model of Schottky barriers," *J. Appl. Phys.*, 46, 1975, p. 2809.

106. Rideout, V.L., "A review of the theory and  technology for ohmic contacts to group III-V compound semiconductors," *Solid-State Electronics*, 18, 1975, p. 541.
107. Rideout, V.L., and C.R. Crowell, "Effects of image force and tunneling on current transport in metal-semiconductor (Schottky barrier) contacts," *Solid-State Electronics*, 13, 1970, p. 993.
108. Riezeniman, M.J., "Schottky components are byte-sized," *Electronics*, Nov. 28, 1974, p. 131.
109. Saltich, J.L., and L.E. Clark, "Use of a double diffused guard ring to obtain near ideal *I-V* characteristics in Schottky barrier diodes," *Solid-State Electronics*, 13, 1970, p. 857.
110. Sato, Y., M. Ida, M. Uchida, and K. Shimada, "GaAs Schottky-barrier diodes for ultrahigh frequency communication systems," *Electronics and Comm. in Japan*, 55-C, 1972, p. 82.
111. Saxena, A.N., "Forward current-voltage characteristics of Schottky barriers on *n* type silicon," *Surface Science*, 13, 1969, p. 151.
112. Schade, P.A., and H.C. Lin, "Characteristics of a *P-N-P-N* switch with an aluminumsilicon Schottky diode clamp," *IEEE Trans Electron Devices*, ED-20, 1973, p. 1159.
113. Scharfetter, D.L., "Minority carrier injection and charge storage in epitaxial Schottky barrier diodes," *Solid-State Electronics*, 8, 1965, p. 299.
114. Scranton, R.A., et al., "Highly electronegative metallic contacts to semiconductors using polymeric sulfur nitride," *Appl. Phys. Lett.*, 29, 1976, p. 47.
115. Shannon, J.M., "*I-V* impurity profiling with a Schottky barrier," *Appl. Phys. Lett.*, 31, 1977, p. 707.
116. Shepherd, F.D., Jr., A.C. Yang, R.W. Taylor, "A 1 to 2 $\mu$m silicon avalanche photodiode," *Proc. IEEE*, 58, 1970, p. 1160.
117. Sinha, A.K., "Metallization scheme for *n*-GaAs Schottky diodes incorporating sintered contacts and a W diffusion barrier," *Appl. Phys. Lett.*, 26, 1975, p. 171.
118. Sinha, A.K., S.E. Haszko, and T.T. Sheng, "Thermal stability of PtSi films on polysilicon layers," *J. Electrochem. Soc.*, 122, 1975, p. 1714.
119. Smith, R.A., *Semiconductors*, Cambridge University Press, London-New York, 1959.
120. Smith, B.L., and E.H. Rhoderick, "Schottky barriers on *p*-type silicon," *Solid-State Electronics*, 14, 1971, p. 71.
121. Spicer, W.E., P.E. Gregory, P.W. Chye, I.A. Babalola, and T. Sukegawa, "Photoemission study of the formation of Schottky barriers," *Appl. Phys. Lett.*, 27, 1975, p. 617.
122. Spitzer, W.G., and C.A. Mead, "Barrier height studies on metal-semiconductor systems," *J. Appl. Phys.*, 34, 1963, p. 3061.
123. Sze, S.M., and H.K. Gummel, "Appraisal of semiconductor-metal-semiconductor transistor," *Solid-State Electronics*, 9, 1966, p. 751.
124. Tantraporn, W., "Determination of low barrier heights in metal-semiconductor contacts," *J. Appl. Phys.*, 41, 1970, p. 4669.
125. Tarui, Y., Y. Hayashi, H. Teshima, and T. Sekigawa, "Transistor Schottky-barrierdiode integrated logic circuit," *IEEE J. Solid-State Circuits*, SC-4, 1969, p. 3.
126. Terry, L.E., and J. Saltich, "Schottky barrier heights of nickel-platinum silicide contacts on *n*-type Si," *Appl. Phys. Lett.*, 28, 1976, p. 229.
127. Thomas, S., and L.E. Terry, "Auger spectroscopy analysis of palladium silicide films," *Appl. Phys. Lett.*, 26, 1975, p. 433.
128. Thomas, S., and L.E. Terry, "Composition profiles and Schottky barrier heights of silicides formed in NiPt alloy films," *J. Appl. Phys.*, 47, 1976, p. 301.
129. Tove, P.A., S.A. Hyder and G. Susila, "Diode characteristics and edge effects of metalsemiconductor diodes," *Solid-State Electronics*, 16, 1973, p. 513.
130. Tyagi, M.S., "Surface states and barrier height at metal-GaAs interface," *Surface Science*, 64, 1977, p. 323.

131. van den Dries, J.G.A.M., and A.G. Post, "Reverse characteristics of Ag-GaS Schottky barriers," *Solid State Communications*, **12**, 1973, p. 709.
132. van der Ziel, A., *Solid-State Electronics*, Prentice Hall, Englewood Cliffs, NJ 1968.
133. van Gurp., G.J., "Cobalt silicide layers on Si. II. Schottky barrier height and contact resistivity," *J. Appl. Phys.*, **46**, 1975, p. 4308.
134. van Gurp, G.J., and C. Langereis, "Cobalt silicide layers on Si I. Structure and growth," *J. Appl. Phys.*, **46**, 1975, p. 4301.
135. Vernon, S.M., and W.A. Anderson, "Temperature effects in Schottky-barrier silicon solar cells," *Appl. Phys. Lett.*, **26**, 1975, p. 707.
136. Vincent, G., D. Bois, and P. Pinard,"Conductance and capacitance studies in GaP Schottky barriers," *J. Appl. Phys.*, **46**, 1975, p. 5173.
137. Viola, L.J., Jr., and R.J. Mattauch, "High-frequency noise in Schottky barrier diodes," *Proc. IEEE*, **61**, 1973, p. 393.
138. Viola, L.J., Jr., and R.J. Mattauch, "Unified theory of high-frequency noise in Schottky barriers," *J. Appl. Phys.*, **44**, 1973, p. 2805.
139. Wang, C.T., and S.S. Li, "A new grating-type gold-$n$-type silicon Schottky-barrier photodiode," *IEEE Trans. Electron Devices*, **ED-20**, 1973, p. 522.
140. Wilkinson, J.M., "Current diffusion effects in titanium-$n$ silicon Schottky diodes," *Solid-State Electronics*, **17**, 1974, p. 583.
141. Williams, R., "The effect of barrier recombination on production of hot electrons in a metal by forward bias injection in a Schottky diode," *RCA Rev.*, **30**, June 1969, p. 306.
142. Wilson, J.I.B., and J.W. Allen, "A compensation law for reverse-biased ZnSe Schottky diodes," *Solid-State Electronics*, **18**, 1975, p. 759.
143. Wortmann, A., K. Heime, and H. Beneking, "A new concept for microstrip integrated GaAs Schottky-diodes," *IEEE Trans. Electron Devices*, **ED-22**, 175, p. 198.
144. Yu, A.Y.C., and C.A. Mead, "Characteristics of aluminum-silicon Schottky barrier diode," *Solid-State Electronics*, **13**, 1970, p. 97.
145. Zettler, R.A., and A.M. Cowley, "p-n junction-Schottky barrier hybrid diode," *IEEE Trans. Electron Devices*, **ED-16**, 1969, p. 58.
146. Zohta, Y., "Frequency dependence of C and $\Delta V/\Delta(C^{-2})$ of Schottky barriers containing deep impurities," *Solid-State Electronics*, **16**, 1973, p. 1029.
147. Brillson, L.J., "Transition in Schottky barrier formation with chemical reactivity," *Phys. Rev. Letts.*, **40**, 1978, p. 260.
148. Schlüter, M., "Chemical Trends in metal-semiconductor barrier heights," *Phys. Rev. B*, **17**, 1978, p. 5044.
149. Wronski, C.R., D.E. Carlson and R.E. Daniel, "Schottky-barrier characteristics of metal-amorphous-silicon diodes," *App. Phys. Letts.*, **29**, 1976, p. 602.
150. Ždánský and Z. Šroubek, "Semiempirical tight-binding calculation of the Schottky barrier heights," *Solid State Communications*, **27**, 1978 p. 1459.

# 3

# Microwave Applications of Diodes, Varactors and Tunnel Diodes

## CONTENTS

Diodes provide many circuit functions (e.g., detection of signals, modulation of carrier frequencies with signal frequencies and subsequent demodulation) at high frequencies. Diodes are also used as RF switches and as harmonic generators for efficient frequency multiplication. The variable capacitance feature of varactor diodes is used in voltage-controlled oscillators for tuning purposes and in parametric amplifier circuits. Tunnel diodes also have similar uses.

Such areas of activity are important at frequencies above a few hundred megahertz and involve special techniques, styles and language.

This sub-discipline therefore merits separate examination. The concepts introduced here will also be useful later in considering transistors at GHz frequencies. Avalanche diode oscillators and transferred-electron (Gunn) oscillators are considered in a later chapter.

Microwave systems cover the frequency band from 900 MHz to 100 GHz. Some of the use allocations in the USA are:

- 470-890   MHz   UHF Television Channels 14-83
- 890-942   MHz   Instruments, scientific and medical
- 960-1215  MHz   Ground beacon/airborne interrogator
- 1.30-1.35  GHz   Air route surveillance radar
- 1.435-1.54 GHz   Aeronautical telemetering
- 1.60-1.66  GHz   Radio altimeter
- 2.70-2.90  GHz   Airport surveillance radar

- 3.70-4.20 GHz    Satellite to earth
- 5.92-6.42 GHz    Earth to satellite
- 8.50-9.50 GHz    Airborne doppler, precision approach and weather radars
- 11.7-12.20 GHz    Common carrier, land and mobile
- 14.4-15.25 GHz    Radar TV remoting
- 22-22.25 GHz    Instruments, scientific and medical
- 33.4-36.0 GHz    Radar
- 40.0-100 GHz    Experimental uses

Some of the commonly used frequency bands are designated by letters, e.g., L band, 1.0-2.6 GHz; S band, 2.60-3.95 GHz; C band, 4.90-7.05 GHz; X band 8.20-12.40 GHz; Ku band, 15.3-18.0 GHz and K band, 18.0-26.50 GHz. At 10 GHz the wavelength is 3 cm and above 30 GHz the wavelengths enter the millimeter range.

## 3.1 DETECTORS, MIXER DIODES AND RELATED DEVICES

Consider a diode used as a detector with a voltage applied $A \cos \omega_s t + V_0$, where $A$ is the amplitude to be detected, and $\omega_s$ is the signal frequency, and $V_0$ is a dc bias voltage applied to control the impedance level and the general performance. If the current-voltage characteristic of the diode is $i = f(v)$, then the applied voltage $A \cos \omega_s t + V_0$ produces (based on a power series expression and neglecting small terms) the following signal related components:

$$i = \left[ \frac{A^2}{4} f^{(2)} + \frac{A^4}{64} f^{(4)} \right] + I_0 + \left[ A f^{(1)} + \frac{A^3}{8} f^{(3)} \right] \cos \omega_s t \quad (3.1)$$

where $f^{(1)}, \ldots, f^{(4)}$ are derivatives of $f(v)$ with respect to $v$, evaluated at $V_0$, and $I_0$ is $f(V_0)$.

The dc term depending on $A^2$ and $A^4$ is the detected current $\Delta i$. The cosine term may be multiplied by $A \cos \omega_s t$ and integrated over a complete cycle to give the average microwave power absorbed by the device,

$$P = \frac{A^2}{2} \left[ f^{(1)} + \frac{A^2}{8} f^{(3)} \right] \quad (3.2)$$

The ratio of $\Delta i$ to $P$ is called the current sensitivity $\beta$, and can be written in the form

$$\frac{\Delta i}{P} = \beta = \beta_0 \left[ \frac{1 + \Delta_1}{1 + \Delta_2} \right] \quad (3.3)$$

where the low-current-level sensitivity is

$$\beta_0 = \frac{1}{2}\frac{f^{(2)}}{f^{(1)}} \tag{3.4}$$

and

$$\Delta_1 = \frac{A^2}{16}\frac{f^{(4)}}{f^{(2)}} \tag{3.5}$$

and

$$\Delta_2 = \frac{A^2}{8}\frac{f^{(3)}}{f^{(1)}} \tag{3.6}$$

The output of a square-law detector is therefore proportional to the input RF power, Eq. (3.3), and hence to the square of the input voltage to a first-order approximation. The sensitivity depends on

• the rectification efficiency as given by the current sensitivity factor $\beta$ which is larger the greater the nonlinearity of the device and depends on the bias point. In unbiased detector operation, the self-bias current is extremely small resulting in low current sensitivity. Generally detector diodes are externally forward biased (0.25-30 $\mu$A) in order to improve their performance.

• the output impedance and noise properties of the diode.

• the input impedance, bandwidth and noise properties of the amplifier that follows the diode, and

• the effectiveness of the RF antenna matching structure.

Voltage sensitivity $\Delta v/P$ is sometimes used instead of short-circuit current sensitivity.

The equivalent circuit of a detector diode consists of the small signal resistance $R_j$ at the bias point, the junction barrier capacitance $C_j$, a parasitic series resistance $R_s$ and package capacitance and inductance $C_p, L_p$, as shown in Fig. 3.1(a). The series resistance represents the spreading resistance of the undepleted epi-layer or bulk region of the diode. The small-signal junction resistance for a diode characterized by

$$I = I_s \left[ \exp\left(\frac{qV}{nkT}\right) - 1 \right] \tag{3.7}$$

is

$$R_j = \frac{nkT}{q(I_0 + I_s)} \tag{3.8}$$

where $I_0$ is the bias current. If $n$ is unity, $T = 300°$K, $I_s = 10^{-13}$ A and $I_0 = 10^{-5}$ A, then $R_j$ is 2600 ohms. If the signal level is large, i.e., comparable to $10^{-5}$ A, then $R_j$ varies throughout the cycle and a representative average value $R_v$ must be assumed and this is known as the video impedance.

The barrier area is made small (typically $10^{-6}$ cm$^{-2}$ corresponding to a diameter of $10^{-3}$ cm) so that the barrier capacitance $C_j$ is 0.5 pF or less. At

(a)

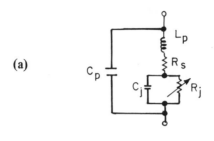

(b)

Fig. 3.1. Detector diode concepts:
(a)  Equivalent circuit, and
(b)  Oscilloscope display used in determining tangential signal sensitivity (TSS).

2 GHz a capacitance of 0.5 pF represents a reactance of 160 ohms. The package inductance may be 1 nH which at 2 GHz is a reactance of 12 ohms. (Thin lead wires must be avoided because a thin wire of length a small fraction of a wavelength can have significant inductance.)

Signal sensitivity measurements are used to characterize detection in the presence of unavoidable noise. Tangential signal sensitivity (TSS) is a widely used characterizing parameter for a detector receiver. The measurement is made with a pulse signal, the level of which is adjusted so that the highest noise peaks observed on an oscilloscope in the absence of signal are the same level as the lowest noise peaks in the presence of signal, as shown in Fig. 3.1(b). The signal power level thus determined gives the TSS value which corresponds to a signal-to-noise ratio of approximately 2.5. Typical TSS values for Schottky barrier detectors are in the range −46 to −56 dBm for 1-20 GHz applications.

Another measure of sensitivity is the nominal detectable signal (NDS) which is defined as the microwave power required to produce an output power equal to the noise power (a signal-to-noise ratio of unity) and is represented as

$$NDS = \frac{2}{\gamma}\sqrt{kTR_v \left[\left(t_w + \frac{T_0}{T}[F_v - 1]\right) B + B_x \ln\frac{f_h}{f_l}\right]} \qquad (3.9)$$

where $\gamma$ is a performance factor for the diode, $t_w$ is the white noise temperature ratio, $F_v$ is the noise figure of the video amplifier, $B$ is the video bandwidth, and $B_x \ln(f_h/f_l)$ represents flicker noise of the diode.

If $F_v = 1 + (R_A/R_V)$, where $R_A$ is the amplifier resistance, and if certain noise terms are negligible, then

$$NDS = \frac{2}{M_c} \sqrt{kTB}.$$  (3.10)

where $M_c$ is a circuit performance factor.

The TSS is found empirically to be 4 dB above NDS under ordinary conditions. Since TSS and NDS vary as the square root of the bandwidth, the value at which measurements are made must be quoted in specifying the detector; a usual value is a 1- or 2-MHz video bandwidth.

Other characterizing parameters that are sometimes used are minimum detectable signal (MDS), noise-equivalent-power (NEP) and a figure of merit $\beta R_V / \sqrt{R_V + R_A}$.[5]

Mixing is the conversion of a low power signal $\omega_s$ to sideband frequencies $\omega_c \pm \omega_s$ by combining the signal with a carrier (local oscillator) frequency $\omega_c$. The expression for an amplitude modulated wave is

$$e = E_c \left(1 + m \cos \omega_s t\right) \cos \omega_c t$$  (3.11)

where $m$ is the extent of modulation (modulation factor). The expression may be rewritten as

$$e = E_c \cos \omega_c t + \frac{mE_c}{2} \cos (\omega_c + \omega_s) \, t + \frac{mE_c}{2} \cos (\omega_c - \omega_s) \, t$$  (3.12)

If a voltage that is the sum of a carrier voltage and a signal voltage is applied to a nonlinear device represented by

$$i = a_1 v + a_2 v^2$$  (3.13)

then amplitude modulation mixing is obtained.

When square-law demodulation is required the process is reversed and the amplitude modulated waveform is applied to a nonlinear device represented by Eq. (3.13). The signal frequency term obtained is then $a_2 E_c^2 m \cos \omega_s t$, but also present are dc components, the carrier frequency and its second harmonic, the original sideband frequencies and their second harmonics, and a second harmonic of the signal. These undesired components are removed by filtering.

An approximate high frequency limit for a mixer diode is $1/C_j R_v$.[143, 146] For a detector diode an approximate figure of merit (neglecting certain noise contributions) is

$$M = \beta \sqrt{R_v}$$  (3.14)

The current sensitivity factor $\beta$ is influenced by the series resistance $R_s$ and by the $C_j R_v$ time constant and M may be rewritten as[84]

$$M = \frac{q\sqrt{R_v}}{2nkT} \left[ \frac{1}{[1 + (R_s/R_v)]^2} \left( \frac{1}{1 + (\omega^2 C_j^2 R_s R_v^2/R_s + R_v)} \right) \right] \qquad (3.15)$$

Hence, if $R_v \gg R_s$ and $n = 1$

$$M \simeq \frac{q\sqrt{R_v}}{2kT} \left( \frac{1}{1 + \omega^2 C_j^2 R_s R_v} \right) \qquad (3.16)$$

Thus, $C_j^2 R_s$ is a term that it is desirable to minimize. The value of $R_v$ must usually be adjusted by the bias current to less than a few thousand ohms otherwise the RF matching problems become severe. On the other hand, $R_v$ must be greater than $R_s$ if the tangential signal sensitivity is not to be adversely affected. From Eq. (3.8) an $I_0$ bias value of $10^{-5}$ A corresponds to an $R_v$ value of about 2600 ohms at $300°K$, if $I_0 \gg I_s$.

In a detector diode of normal barrier height, $\phi_B$ between 0.5 and 0.9 eV, the $I_s$ value is small and a bias current $I_0$ must be used. This is no great circuit problem but does, perhaps, add to the flicker noise of the detector. To increase $I_s$, Schottky barrier detector diodes may be fabricated with low barrier heights, such as the 0.34 eV obtainable for Au on $p$ Si. The current density over the barrier is then on the order of 10 A cm$^{-2}$ and so for diode areas of $10^{-5}$-$10^{-6}$ cm$^2$ the $I_s$ term is about $10^{-5}$ A. Hence, the $R_v$ value is a few thousand ohms or lower[85, 122, 129] with no externally provided bias currents. Such detectors are termed zero-bias Schottky diodes. Since $I_s$ increases very rapidly with temperature, the $R_v$ term at the origin decreases greatly with temperature. For a good compromise between thermal stability and sensitivity, diodes are used with origin resistances that are from one to three orders of magnitude higher than the source impedance, and shunted by a constant matching resistor, as in Fig. 3.2. For a 50-ohm system, this condition places the room-temperature origin resistance in the kilohm range which corresponds to a saturation current of $10^{-4}$-$10^{-5}$ A. The relative performances of biased and zero-biased-Schottky detectors and tunnel diode detectors (see Sec. 3.5) are summarized in Table 3.1.

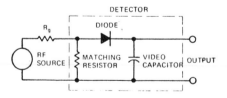

Fig. 3.2. Circuit for a Schottky barrier detector with zero-bias current. The detector match is practically determined by the matching resistor if the diode is high-impedance. This minimizes temperature effects on the match. (After Szente, Adam, Riley, *Microwave Journal*, Feb. 1976, 19, 2, p. 42.)

Table 3.1.  Typical Performance Comparison of Microwave Detectors
(over octave bands to 12.4 GHz)

| DETECTOR TYPE ►<br><br>PERFORMANCE ▼ | BIASED<br>SCHOTTKY | ZERO-BIAS<br>SCHOTTKY | GERMANIUM<br>TUNNEL | GaAs<br>TUNNEL |
|---|---|---|---|---|
| CIRCUIT<br>CONFIGURATION | RF⟶ | | | ⟶ VIDEO |
| BIAS | 100<br>to 300µA | 0 | 0 | 0 |
| TSS in dBm<br>(2-MHz VIDEO BW,<br>NF = 3dB) | −52 to<br>−50 | −52 | −51 to<br>−49 | −49 |
| VOLTAGE SENSITIVITY<br>K (mV/mW) | 1200 to<br>2000 | 2000 | 400 to<br>1200 | 300 to<br>700 |
| VIEDO RESISTANCE Ω<br>(SL, SQUARE-LAW<br>RANGE) | 200<br>to<br>400 | 400<br>to<br>600 | 60 to<br>120 | 100 to<br>150 |
| INPUT SWR<br>(SQUARE-LAW RANGE) | 2:1 to<br>4:1 | 6:1 | 1.5:1 to<br>2:1 | 3:1 |
| FREQUENCY RESPONSE<br>(FLATNESS in dB) | ±0.5 | ±0.5 | ±0.3 | ±0.5 |
| TEMPERATURE<br>STABILITY in dB<br>(−55 to +85°C) | ±1 | ±2 | ±0.5 | ±0.5 |
| POWER RATING dBm | +20 | +20 | +17 | +17 |
| RELATIVE RISE AND<br>DECAY TIME | MODERATE | MODERATE | SHORT | SHORT |

After Siegal and Pendleton, 1975. Courtesy Aertech Industries.

Zero-bias-Schottky diodes have $f^{-1}$ noise corner frequencies between 1 KHz and 10 KHz and match the noise performance of tunnel- or back-diodes. Conventional Schottky diode noise performance is highly dependent on operating bias necessitating a trade-off between sensitivity or TSS and $f^{-1}$ noise performance. Point contact diodes, although offering comparable sensitivity without external bias, exhibit considerably higher values of $f^{-1}$ noise (see Fig. 3.3). Thus Schottky barrier zero-bias diode detectors are excellent for low-level RF power monitoring, square-law detectors for Doppler radar, zero IF mixers and Doppler radar mixer applications.

From Eq. (3.16), for high frequency performance in a detector $C_j^2 R_s$ is a term that should be made small. The diode capacitance is given by

$$C = \pi a^2 \left[ \frac{eqN}{2(\phi - V)} \right]^{1/2} \tag{3.17}$$

(a)

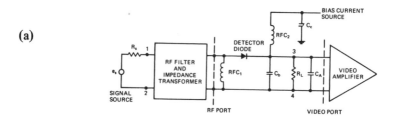

(b)

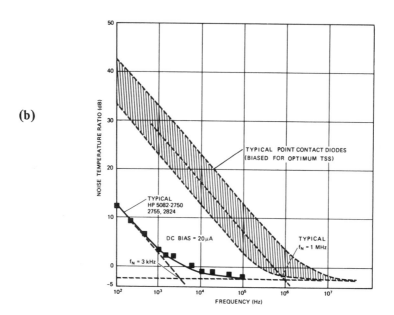

Fig. 3.3. Detector diode performance:
(a) Typical video receiver circuit, and
(b) Noise performance of Schottky detector diodes compared with previously used point-contact diodes. (Courtesy Hewlett Packard Company, Application Note 923.)

where $a$ is the diode radius, $\epsilon$ is the semiconductor dielectric constant, $N$ is the epitaxial doping and $\phi$ is the built-in junction voltage.

$R_s$ is the sum of the unswept epitaxial resistance, the substrate resistance and the resistance of the ohmic contact, $R_o$. Thus,

$$R_s = \frac{L - W}{N\mu(N)q\pi a^2} + \frac{\rho_s}{4a} + R_0 \qquad (3.18)$$

where $L$ is the epitaxial thickness, $\mu$ is the doping-dependent mobility, $\rho_s$ is the substrate resistivity. The thickness of the junction depletion region $W$ is

$$W = \left[ \frac{2\epsilon(\phi - V)}{Nq} \right]^{1/2} \tag{3.19}$$

With some manipulation it may be shown that

$$\phi - V = \psi_{ms} (1 - n) - E_g/2$$

$$-\frac{kT}{q} [1 - \log (N/N_i) + n(\log I - \log \pi a^2 A_R T^2)] \tag{3.20}$$

where $A_R$ is the Richardson constant of the diode. Hence if $n = 1$, $\psi_{ms}$ drops out of the theory. Then using Eqs. (3.17) through (3.20) and referring back to Eq. (3.16) it may be concluded that the frequency dependence of the current sensitivity $\beta$ is not particularly dependent on the metal-semiconductor barrier height.[85] However, the barrier height at a given bias affects the zero frequency sensitivity $\beta_0$, through Eq. (3.4). The higher electron mobilities of GaAs and Ge tend to make diodes of those materials lower in parasitic resistance than for Si diodes and therefore less frequency sensitive.

Consider now mixer diode conversion loss, $L$. This loss is the ratio of signal power available from the source to IF power available from the mixer. It may be represented approximately as the product of three losses:

$$L(\text{dB}) = L_1 + L_2 + L_3 \tag{3.21}$$

Then following closely a discussion by Anand and Moroney[5]

"mismatch loss $L_1$ depends on the degree of match obtained at the RF and IF ports. The mixer diode is often matched to the local-oscillator power in a fixed or tunable mount. However, since only the RF signal power is converted to the intermediate frequency, it is more important to match the diode at the signal frequency. Signal and local-oscillator optimum match settings do not generally occur at the same tuner positions. The IF impedance $R_{IF}$ appears at the mixer output terminals when the diode is biased by a local oscillator. This impedance is a function of the local-oscillator power level, the diode properties, and the circuit.

$L_2$ is a loss of signal power due to the series resistance $R_s$ and the junction capacitance $C_j$ in the diode. This loss is the ratio of the input RF signal power to the power delivered to the barrier resistance $R_j$:

$$L_2 (\text{dB}) = 10 \log \frac{P_{in}}{P_{out}} .$$

In terms of diode parasitic parameters $R_s C_j$ and the diode's nonlinear resistance $R_j$

$$L_2(\text{dB}) = 10 \log \left(1 + \frac{R_s}{R_j} + [\omega C_j]^2 R_s R_j\right) \qquad (3.22)$$

where $R_j$ is the time average value, established by the local-oscillator drive."
$L_2$ is found to have a minimum value

$$(L_2)_{\text{min}} = 1 + 2\omega C_j R_s \qquad (3.24)$$

which occurs when

$$R_j = 1/\omega C_j. \qquad (3.25)$$

Further increase in the drive results in increased $L_2$ due to dissipation in $R_s$, while decreasing drive also gives insertion-loss increase due to the shunting effect of the junction capacitance.

$L_3$ is the actual conversion loss at the diode junction. This depends mainly on the voltage vs current characteristics of the diode and the circuit conditions at the RF and IF ports. The nonlinear behavior of the diode is represented by a time-varying conductance $g = 1/R_j$ which is dependent on the dc characteristics of the diode and local-oscillator voltage waveform across the diode. Torrey and Whitmer[133] discuss the mixer as a linear passive three-port network having terminals at signal, image, and intermediate frequencies and describe the network parameters in terms of Fourier coefficients of the time-dependent conductance. Conversion loss and impedance values can then be calculated for the various image terminations by means of linear network theory. The theory predicts, for an ideal mixer, 3-dB conversion loss for a broadband case (same source resistance at both signal and image frequencies) and 0-dB conversion loss for open- or short-circuited image terminations.

Noise at the receiver output is the sum of the noise arising from the input termination (source), noise generated within the mixer diode, and noise from the following IF amplifier. The noise figure (NF) is the ratio of the available output noise power of the receiver to the noise power which would be available if the receiver were noise free. Alternatively:

$$\text{NF} = \frac{S_i/N_i}{S_0/N_0} \qquad (3.26)$$

where $S_i(S_o)$ is the available signal power at the receiver input/output and $N_i(N_o)$ the available noise power at the receiver input/output. The receiver noise figure

$$\text{NF} = L(t_m + F_{\text{IF}} - 1) \qquad (3.27)$$

depends on mixer diode conversion loss $L$, noise temperature ratio $t_m$, and IF amplifier noise figure $F_{IF}$.

Hence, $L$ and $t_m$ are the important diode parameters which must be low to obtain a small overall NF.

The conversion loss and noise figures for $n$ type Si and GaAs Schottky barrier mixer diodes are shown in Fig. 3.4(a) as a function of frequency. The

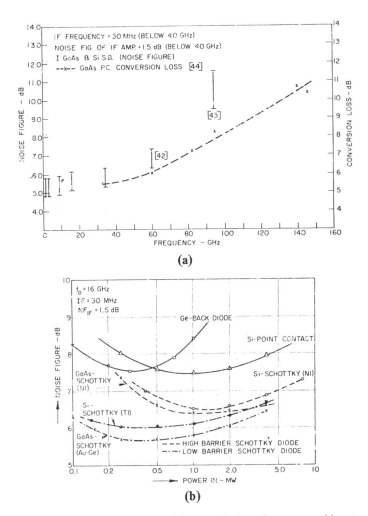

**(a)**

**(b)**

Fig. 3.4. Noise figure and conversion loss of detector diodes vs frequency and input power: (a) Noise and conversion loss for $n$-type Si and GaAs Schottky-barrier diodes, and (b) Noise figure vs local-oscillator power for point-contact and $n$-type Schottky-barrier diodes. (After Anand and Moroney. Reprinted, with permission, from *Proc. IEEE*, 59, 1971, p. 1187.)

noise figures for several kinds of diodes as a function of local oscillator power are given in Fig. 3.4(b).

Cryogenically cooling a Schottky-barrier mixer only slightly increases the conversion loss while giving a considerable reduction in mixer noise. The dc bias and local oscillator drive must be appropriately scaled. In conjunction with a cooled paramp IF amplifier, single-sideband (SSB) receiver noise temperatures of $\sim 350°K$ at 85 GHz, and $\sim 260°K$ at 33 GHz are obtainable—an improvement by a factor of 6 at 85 GHz and 4 at 33 GHz over room-temperature mixer receivers.[144]

The burn-out resistance of detector and mixer diodes is also important in practice and depends on the fabrication techniques and heat-sinking provided.[4,47]

In a mixer circuit various output frequencies are obtained by the squaring and interaction of the local oscillator and RF signal. The intermediate frequency (LO – RF) is the one of direct interest, but the sum frequency (LO + RF) and the image frequency (2LO – RF) are also important. In principle it is possible to minimize the conversion loss by choosing the proper reactive termination for the sum and image frequencies.

The schematic for a simple image termination mixer circuit is given in Fig. 3.5(a). However, operation is restricted to a narrow frequency range because of the fixed-frequency tuned filter. Figure 3.5(b) shows a circuit in which image power generated by the mixer diodes adds at the difference port of the input hybrid and cancels at the RF input terminals. With the proper reactance placed at the input hybrid difference terminals, the conversion loss and noise figure are minimized. However, this circuit is also limited in frequency because of the long equivalent-line length between each mixer diode and the terminating image reactance. Also, the insertion loss of the input hybrid restricts the maximum reflection obtainable from the image termination and so a smaller conversion loss improvement is obtained.

Better circuits result if the mixer diode-quads are in direct RF connection: good bandwidths, for instance 1.0 to 2.6 GHz, are then obtainable. Such a circuit is shown in Fig. 3.5(c) where the opposing image voltages generated by each diode-quad act to give a shorted image condition.

A monolithic X band (10 GHz) balanced two-diode mixer in stripline form is shown in Fig. 3.6(a). The two-branch directional coupler that precedes the diodes has the following characteristics:

- with a signal applied to terminal 1, outputs appear at terminals 2 and 4 that are equal in amplitude and differ in phase by 90°.
- a standing-wave ratio that is related to the amount of power, assumed to be coming from a matched 50-ohm source, that is reflected by the coupler when matched loads are connected to the coupler outputs. A VSWR of 1.9 corresponds to 10% reflected power.
- an isolation factor that characterizes the amount of power coupled into

(a)

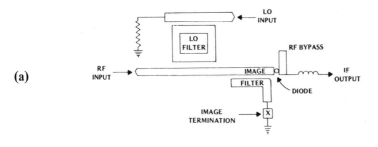

(b)

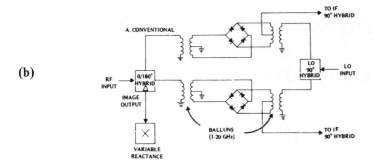

(c)

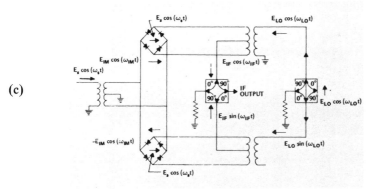

Fig. 3.5. Some mixer diode circuits:
(a)   Conventional single-diode image-recovery circuit,
(b)   Modified image-rejection mixer with narrow-band tunable image reactance, and
(c)   Circuit with shorted-image condition. (After Neuf, courtesy of RHG Electronics Labs, Deer Park, NY. Reprinted with permission from *Microwave Journal*, May, 1973, **16**, p. 29, 30.)

the theoretically decoupled arm. An isolation factor of 10 dB means that 10% of the incident power is coupled into the isolated arm.

• an insertion loss: in X-band microstrip attenuation is caused primarily

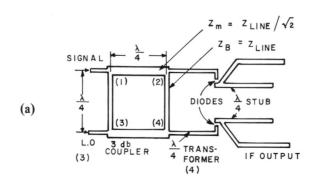

(a)

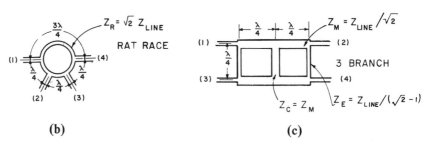

(b)                                    (c)

Fig. 3.6.  Balanced stripline coupled mixer circuits:
(a)   X-band two-branch monolithic hybrid coupler balanced mixer,
(b)   Rat-race coupler, and
(c)   Three-branch coupler.

by resistive loss in the microstrip conductors. At higher frequencies, radiation loss may also be significant. The attenuation is a function of line impedance and standing-wave ratio in the line.

Two other directional coupler configurations are shown in Fig. 3.6(b) and (c). The rat-race configuration is, in theory, less influenced by junction reactances and dimensional tolerances and has a greater bandwidth than the two-branch coupler. However, the rat-race coupler has the disadvantage that the output arms are not adjacent, so a cross-over connection is needed. Three-branch couplers have bandwidths comparable to the rat-race coupler but are more sensitive to junction reactances and dimensional tolerances.[80]

The structures of two Schottky barrier mixer diodes on semiinsulating GaAs that could be part of a stripline are shown in Fig. 3.7. The Schottky junction diameter in Fig. 3.7(a) may be 10 $\mu$m or less.

## 3.2 *PIN* DIODES AS ATTENUATORS AND SWITCHES

A *PIN* diode for RF use has a low capacitance because of the lightly-doped almost intrinsic region between the $P$ and $N$ layers. When forward bias is applied minority carrier injection increases the conductivity of the intrinsic region. In reverse bias the $I$ region is fully depleted of carriers and the field strength across the region is constant. The maximum voltage rating is determined by the critical field strength for avalanche ($\sim 2 \times 10^5$ Vcm$^{-1}$ for Si) and the thickness of the $I$ region.

A $PIN$ diode has the ability to act as a bias-current-controlled resistor at RF frequencies with good linearity and low distortion as shown in Fig. 3.8(a). With the control current varied continuously the $PIN$ diode is used for amplitude modulating functions and for attenuating and leveling as in Fig. 3.8(b). With the control current switched between on and off conditions, the diode is useful for pulse modulating and switching actions.

The conductance of the diode is proportional to the stored charge $Q_d$ and this is related to the bias current by

$$I_d = \frac{Q_d}{\tau} + \frac{dQ_d}{dt} \tag{3.28}$$

where

$$Q_d(\omega) = \frac{I_d \tau}{1 + j\omega/\omega_0} \tag{3.29}$$

where $\omega_0 = 2\pi f_0 = 1/\tau$. At frequencies below $f_0$ the $PIN$ diode acts as an ordinary $PN$ junction diode and the RF signal undergoes distortion associated with partial rectification. At frequencies above $f_0$ the diode behaves more as a linear resistor although the conductivity modulation effect decreases by 6 dB/octave with frequency. The minority carrier lifetime $\tau$ in $PIN$ diodes is chosen to give the desired frequency capability and switching speed. Typically $\tau$ is in the range 0.03 $\mu$sec to 3 $\mu$sec: for a value of 0.1 $\mu$sec, $f_0$ is 1.6 MHz.

Figure 3.8(b) shows the use of a 3 diode $\pi$ attenuator between the mixer and the IF amplifier stages of a TV video strip for automatic gain control. Figure 3.9 shows attenuation performance of a single shunt diode chip in several package forms, in a 50-ohm 500 MHz system. The solid lines are contours of constant attenuation calculated for $R_p$ and $jX_p$ in parallel across a 50-ohm load $Z_0$. The diode lines show the role of package parasitics in limiting performance. The stripline package structure gives the best performance since it has the equivalent circuit of Fig. 3.10 with a cutoff frequency of about 30 GHz.

Series and shunt attenuator circuits of simple form are shown in Fig. 3.11 together with curves of insertion loss (the residual attenuation that exists in the ON state) and isolation loss which is the attenuation in the OFF state. The series configuration has good isolation but may have higher insertion loss at high frequencies than the shunt arrangement.

Since $PIN$ diodes usually reflect power by the change of circuit impedance produced, they can control considerably more power than is dissipated. For

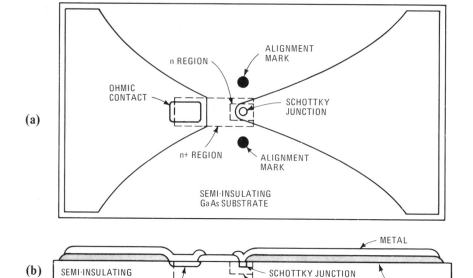

Fig. 3.7 Schottky-barrier mixer diode structures:
(a), (b) Typical $nn^+$ GaAs structure . (Reprinted from *Electronics*, Jan. 17, 1972; copyright McGraw-Hill, Inc., NY, 1972), and
(c), (d) Microstrip line integrated semi-planar GaAs structure and performance. (After Wortman, Heime, Beneking. Reprinted with permission from *IEEE Trans. Electron Devices*, **ED-22**, 1975, p. 198.)

instance, in a 50-ohm system about 2% of input power will be absorbed by a series diode of impedance range from 1 to 10 kilohms in either the ON or the OFF state. Thus, if the diode can dissipate 3 W it can control 150 W of signal power. In a shunt circuit the absorbed power is higher and the same diode can control 40 W of power.[57]

Single-pole single-throw two-port switches or attenuators meet many needs but multipole arrangements are possible as shown in Fig. 3.12. In Fig. 3.12(a) the RF power flows from port 3 to port 2, and port 1 is isolated when diode $D_1$ is forward biased and diode $D_2$ is zero or reverse-biased. With the bias conditions changed the flow is from port 3 to port 1. The $\lambda/4$ spacing is necessary to minimize the reactive loading of the open port by the closed port. The capacitors provide RF grounds for the diodes and the RF choke (RFC) provides the return path for the bias currents.

In some RF systems the return power of reflective switch circuits cannot be tolerated and the need is for switch arrangements that remain constant in impedance. This means that the *PIN* diodes must absorb the switched power

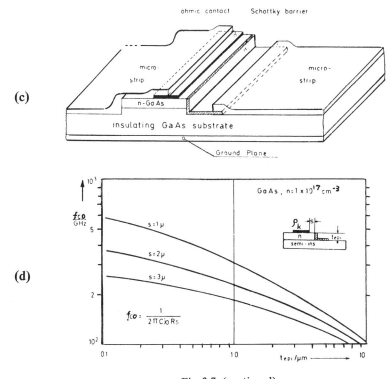

Fig. 3.7 (continued)

as in Fig. 3.13(a), (b), (c) or the power must be reflected to a different RF port for absorption as in Fig. 3.13(d), (e), (f). For instance, in Fig. 3.13(d) the power input at $A$ in the 3 dB coupler divides equally between ports $B$ and $C$. The reflected powers from $C$ and $B$ produced by the diodes then combine and emerge at $D$. Hence, the impedance of port $A$ matches the supply impedance whatever the state of the attenuator.

$PIN$ diodes also find applications in RF phase shifters.[57]

RF stripline may be symmetrical with the conductor spaced equally between two ground planes or single-sided as shown in the sketch in Fig. 3.14. The conductor line impedance then depends on the ratio $w/h$ as in Fig. 3.14 for a dielectric constant of 12 corresponding to semi-insulating GaAs. The wavelength in microstrip $\lambda_M$ normalized against $\lambda_{TEM}(= \lambda_{AIR}/\sqrt{\epsilon_r})$ is also a function of $w/h$ (for instance $\dfrac{\lambda_M}{\lambda_{TEM}}$ being 1.23 for $w/h = 1.0$) and must be allowed for in the design of $\lambda/4$ stubs on stripline.

Schematics for shunt and series diode transmit-receive single-pole double-throw switches are given in Fig. 3.15. The function of a transmit-receive switch

**(a)**

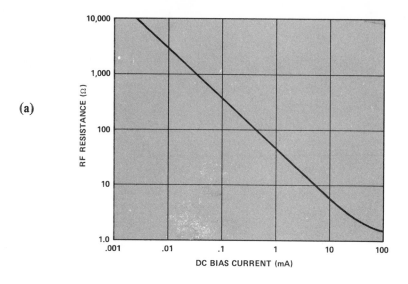

**(b)**

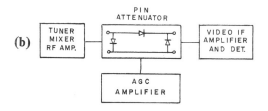

Fig. 3.8. *PIN* diode attenuators:
(a)  Typical *PIN* diode resistance vs control current (Courtesy Hewlett Packard Company, Application Note 936), and
(b)  Use of a *PIN* diode attenuator as an interstage for providing automatic gain control.

is made clear in the block-diagram of Fig. 3.16 which is for a 10 W, L band (1-2.6 GHz) module. An X band (9 GHz) stripline version of such a switch on Si is shown in Fig. 3.17. The substrate is 1500 ohm-cm $p$ type Si and the stripline width for 50 ohms is 153 $\mu$m. The series switching diodes are surface-oriented *PIN* diodes with a stripline gap forming the $I$ region in the substrate material as shown in the inset sketch.[41]

The switching of a *PIN* diode from low impedance to high impedance is accomplished by switching the bias current from forward to reverse, as shown for the diode under test in Fig. 3.18. The reverse current flows for a time such that $\int_{t_0}^{\infty} i_r dt$ equals the charge stored under the forward bias condition minus any charge that disappears by recombination (which may be small if $(t_2 - t_0)$ is much less than $\tau$ the minority carrier lifetime). The delay time $t_d$ is inversely proportional to the reverse current. The subsequent time $t_r$ is the transition

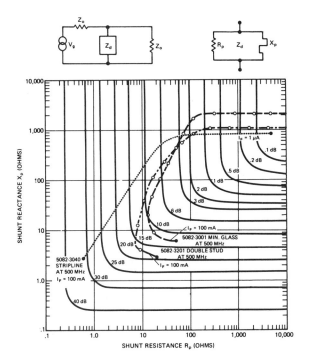

Fig. 3.9. Attenuation produced by a shunt *PIN* diode as its impedance varies with changes of bias current in several package forms to show package parasitic effects. (Courtesy Hewlett Packard Company, Application Note 923.)

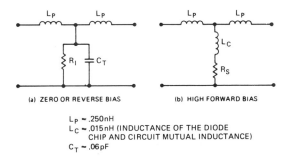

Fig. 3.10. Equivalent circuits of a stripline *PIN* diode (HP 5082-3040). (Courtesy Hewlett Packard Company, Application Note 922.)

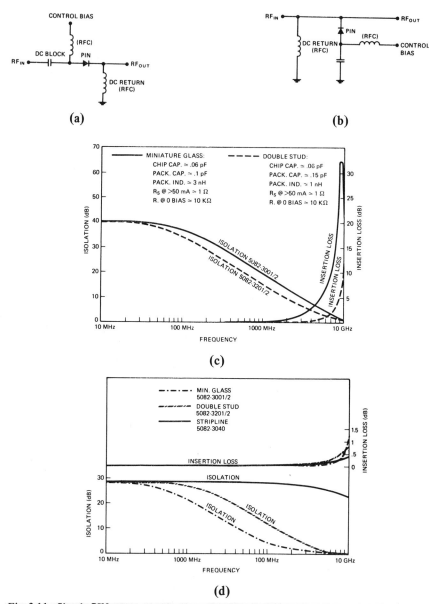

Fig. 3.11. Simple *PIN* attenuator circuits and performance:
(a)  Series attenuator circuit,
(b)  Shunt attenuator circuit,
(c)  Curves of insertion loss (ON state) and isolation loss (OFF state) for series diodes used as a switch in a 50-ohm system, and
(d)  Curves of insertion and isolation loss for shunt diodes in a 50-ohm system (Courtesy Hewlett Packard Company, Application Note 922.)

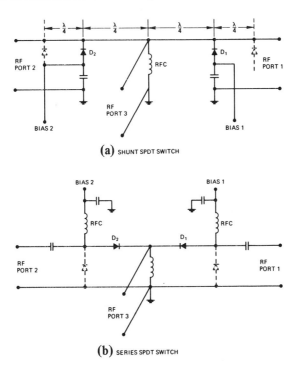

(a) SHUNT SPDT SWITCH

(b) SERIES SPDT SWITCH

Fig. 3.12. Shunt and series single-pole double-throw *PIN* switch circuits. (Courtesy Hewlett Packard Company, Application Note 922.)

time during which the impedance of the diode is changing from a low to a high value. This depends on doping profiles and diode geometry and not too strongly on the forward or reverse bias currents. In a snap-action diode the doping profiles can be chosen to minimize this transition time but in an RF switching diode this must not be at the sacrifice of the *PIN* form of the diode since this influences the diode distortion behavior in some applications.

The reverse recovery time is the sum of the delay time and the transition time. Provision of a large reverse current from the driver stage helps shorten the delay time. If the reverse current $I_R$ is constant during the delay time

$$t_d \simeq \tau \ln(1 + I_F/I_R) \qquad (3.30)$$

Hence, if $t_d$ is measured for several values of $I_F/I_R$ and fitted to Eq. (3.30) the value of minority carrier lifetime $\tau$ can be determined. Diodes with low values of $\tau$ have short delay times. However, they also have higher values of resistance in the turn-on condition and so increased insertion losses. An

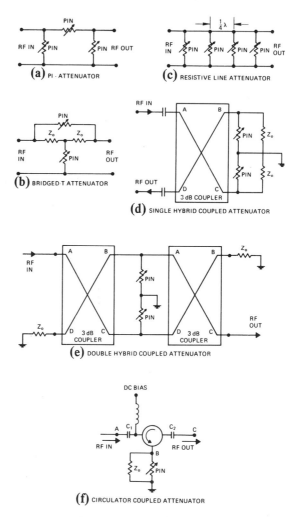

Fig. 3.13. Constant impedance *PIN* diode attenuator circuits. (Courtesy Hewlett Packard Company, Application Note 922.)

approximate expression for the forward resistance of a Si *PIN* diode is

$$R = \frac{2\,kT/q}{I_F} \sinh\left(\frac{W}{2\sqrt{D\tau}}\right) \tan^{-1}\left[\sinh\left(\frac{W}{2\sqrt{D\tau}}\right)\right] \qquad (3.31)$$

where $W$ is the thickness of the $I$-layer and $D$ is a diffusion constant of about 18 cm$^2$ sec$^{-1}$. This expression has been used to generate the curves of series resistance vs lifetime in Fig. 3.19 for a 25 $\mu$m $I$-layer. Reduction of $R$ by increase

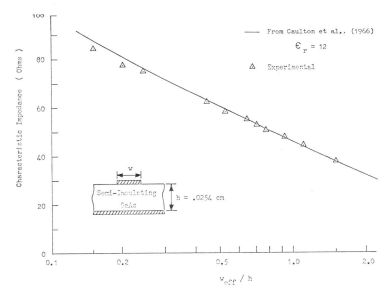

Fig. 3.14. Characteristic impedance vs $w/h$ for a single-sided stripline on semi-insulating GaAs at 10 GHz. (Courtesy T.M. Eppinger.)

of the bias current is not always practical since the increased current demand on the driver stage tends to slow the rise time of the bias current.

From Fig. 3.19 if the insertion loss specification requires a series resistance of not more than 1 ohm and if the permitted forward bias current is 20 mA and if a 25 $\mu$m base thickness allows a suitable RF peak power rating, then the required lifetime is 200 nsec. From Eq. (3.30) if the switching time needed is about 50 nsec, the required ratio of $I_F/I_R$ is about 0.25, hence the driver must have a reverse current capability of 80 to 100 mA.

The total time for switching a shunt diode in a 3 GHz circuit from forward bias (the isolation state of the circuit) to reverse bias (the insertion loss state) is shown in Fig. 3.20 for several forward currents and reverse voltage pulses of 2 5 and 10 V. The times are about 3 to 7 nsec. The times for switching from reverse to forward bias are generally less than these, namely $<$ 2 nsec.

## 3.3 VARACTOR DIODES AND PARAMETRIC FREQUENCY MULTIPLICA- TION AND AMPLIFICATION

A diode whose circuit function is primarily determined by the variation of capacitance with voltage is known as a varactor. For an abrupt *pn* junction the depletion capacitance varies as $V_R^{0.5}$ and the diode may be used as a voltage controlled capacitance. However, in many circuit applications a capacitance

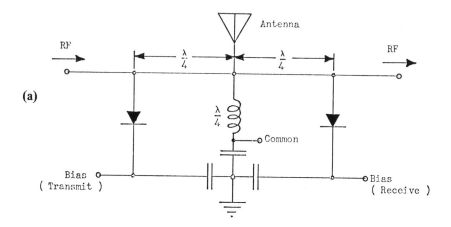

**(a)**

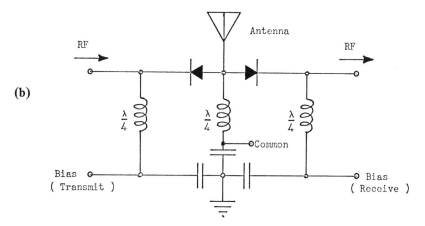

**(b)**

Fig. 3.15. Schematic diagrams of *PIN* diode transmit-receive switches (single-pole double-throw):
(a)  Using shunt diodes, and
(b)  Using series diodes.

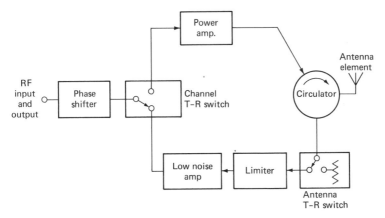

Fig. 3.16. An L band (1.0-2.6 GHz) transmit-receive module. (After Vincent and Wallace.[140] Reprinted from *Microwave Journal*, **12**, 1969, p. 53.)

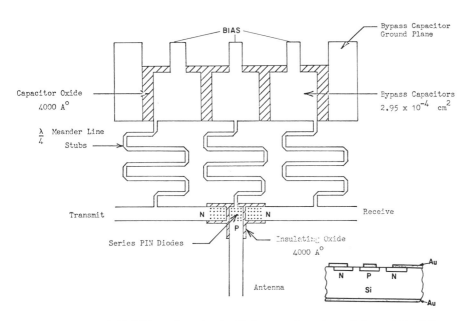

Fig. 3.17. X-band (9 GHz) version of series diode transmit-receive stripline switch. (After Ertel, courtesy Texas Instruments, Inc. Reprinted from *Electronics*, Jan. 23, 1967; copyright McGraw-Hill, Inc., NY, 1967.)

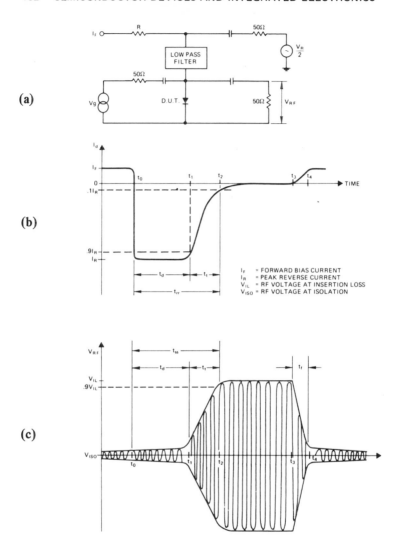

Fig. 3.18. Switching time test circuit and waveforms for an RF switching diode:
(a)  Circuit, with D.U.T. being the diode under test,
(b)  Current waveform drive, and
(c)  Switched RF voltage. (Courtesy Hewlett Packard Company, Application Note 929.)

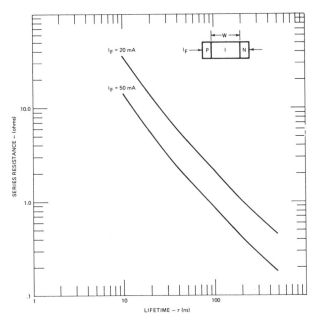

Fig. 3.19. Increase of minority carrier lifetime decreases the series ON resistance of a *PIN* diode. Given above are calculated results for an intrinsic region width of 25 $\mu$m. (After McDade and Schiavone. Reprinted with permission from *Microwave Journal,* 17, Dec. 1974, p. 65.)

voltage dependence of $V_R{}^n$ where $n$ is greater than unity is preferred. This may be achieved by making the doping profile hyperabrupt (retrograded) so that as the voltage is increased the depletion edge enters regions of lower and lower doping and the capacitance therefore decreases rapidly.

In Si junctions, retrograded doping profiles may be obtained by careful control of diffusion conditions or by ion-implantation followed by a PtSi Schottky barrier junction.[92] A typical hyperabrupt UHF tuning varactor may have a capacitance of 9 pF at 3 V reverse bias and 2 pF at 25 V reverse bias. The series inductance is 0.8 nH and the package capacitance 0.15 pF. At 4 V and 50 MHz the $Q$ of the diode is 600 or more.

For microwave frequency operation the series resistance of the undepleted section of the diode must be low and there is an advantage in using Schottky barrier $nn^+$GaAs structures because of the high electron mobility. Good control of epi-layer thickness and doping profile in GaAs can be achieved by molecular beam epitaxial growth. MBE growth involves evaporation of elements under high vacuum conditions.[25] For a junction doping profile $(N_D - N_A)$ proportional to $x^{-1.5}$, as in Fig. 3.21, the variation of capacitance goes as $(\phi - V)^2$. By varying the profile the power law $n$ may be varied from 0.5 to 4.5 as seen in Fig. 3.22.

High performance GaAs varactors have been made with zero bias junction capacitance as low as 0.12 pF and cutoff frequency in excess of 700 GHz as inferred from 70 GHz measurements. Used as frequency doublers (35 to 70 GHz)

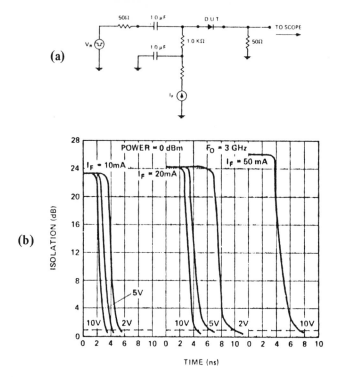

**(a)**

**(b)**

Fig. 3.20. Switching transient of a shunt *PIN* diode in a 2-3 GHz circuit:
(a)   Test circuit for switching from forward bias to reverse bias, and
(b)   Variation of isolation with time, for forward currents of 10, 20 and 50 mA and reverse voltage pulses of 2,5 and 10 V (H.P. diode 5082-3041). (Courtesy Hewlett Packard Company.)

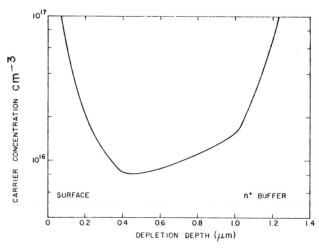

Fig. 3.21. Doping profile of a hyperabrupt GaAs Schottky barrier varactor giving a $C^{-1/2} \sim \varphi - V$ relationship for $0.3 \gtrsim V \gtrsim -1.7$ V (After Cho and Reinhart. Reprinted from *Journal of Applied Physics*, 45, 1974, p. 1815.)

164

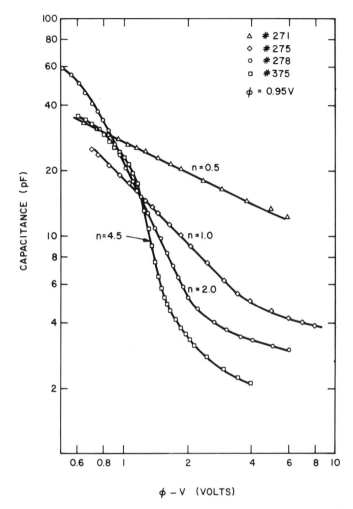

Fig. 3.22. Capacitance variation as a function of biasing voltage for GaAs molecular-beam-grown epitaxy layers with various doping profiles. (After Cho and Reinhart. Reprinted from *Journal of Applied Physics*, 45, 1974, p. 1815.)

and triplers (35 to 105 GHz), efficiencies greater than 40% and 25%, respectively, have been obtained. In a parametric amplifier pumped at 105 GHz and operated in the 55- to 65-GHz signal frequency band, the gain was 14 dB, for a 1-dB bandwidth of 670 MHz, and a single sideband noise factor of 5.9 dB.[18]

The lumped circuit diagram of a varactor frequency tripler is shown in Fig. 3.23. In addition to the input frequency $f_{in}$ and the output $3f_{in}$, an idler frequency circuit tuned to $2f_{in}$ is needed. Parametric up-converter circuit principles are defined by the Manley-Rowe power relations.[143] If the circuit can

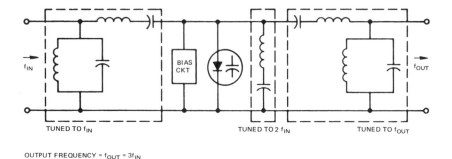

OUTPUT FREQUENCY = $f_{OUT}$ = 3$f_{IN}$
IDLER FREQUENCY = $f_{IDLER}$ = 2$f_{IN}$

Fig. 3.23. A simple varactor frequency tripler circuit. (After Siegel. Reprinted from *Microwave Systems News*, June/July, 1972, p. 64.)

be adjusted in a nonlinear-reactor harmonic generator so that only the desired harmonic output power is delivered to the load, and other harmonic frequencies are reactively terminated, the conversion efficiency for the desired harmonic approaches 100%. A sectional drawing of a 4 to 12 GHz tripler in the sidewall of a waveguide is shown in Fig. 3.24(a).

Since the waveguide cutoff frequency is 9.486 GHz, neither the fundamental nor the second harmonic can appear at the output port, and the waveguide constitutes a high-pass filter. A radial choke (5) is placed in the input coaxial line; this choke presents an open circuit at 12 GHz, thus preventing 12-GHz power from coming out of the input port. The position of this choke with respect to the diode is chosen to tune the reactance of the diode at the output frequency. The varactor diode capacitance and inductance are chosen so that the diode is series-resonant at the second harmonic 8 GHz. An idler stub (2), together with a tuning screw (3), forms a series-resonant path to ground at 8 GHz, thus completing the idler current path.

This tripler configuration with GaAs and Si varactors of about 0.5 pF zero bias junction capacitance, gives efficiencies of 83 and 65%, respectively, as shown in Fig. 3.24(b).

Frequency multiplication may also be achieved with snap-action RF Si diodes. Such a diode in a waveguide is capable of giving multiplication from 2 GHz to 16 GHz with an efficiency of about 10%.[58]

Returning to our consideration of varactor diodes, these may also be used as modulators. A waveguide configuration with a 70-MHz signal modulated onto a 6-GHz carrier is shown in Fig. 3.25. Because of the high up-conversion gain the power levels are favorable: For instance, a pump power of 320 mW and a signal power of 70 mW give an output power of 200 mW, so the microwave conversion efficiency is 62%.

Parametric amplifiers using varactor diodes are of importance in the field of low-noise nonrefrigerated and refrigerated amplifiers. At liquid-helium

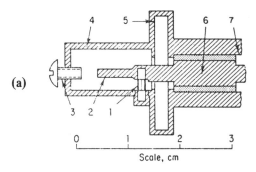

(a)

Scale, cm

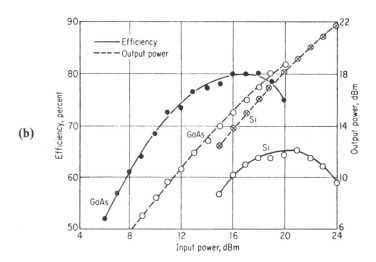

(b)

Fig. 3.24.  A 4 to 12 GHz varactor diode tripler circuit in a 50-ohm coaxial line:
(a)   Sectional view:
      (1)  varactor diode,
      (2)  transverse stub resonant at 8 GHz,
      (3)  idler turning screw,
      (4)  12-GHz output wave guide, 1.58 x 0.79 mm,
      (5)  12-GHz radial rejection choke on input coaxial line,
      (6)  4-GHz matching transformer, supported by dielectric sleeve,
      (7)  50-ohm input coaxial line.
(b)   Conversion efficiency and outpower of the tripler for GaAs and Si varactor diodes as
      a function of drive level. Varactor bias was adjusted at each point for maximum out-
      put. (After Uenohara and Gewartowski in *Microwave Semiconductor Devices and their
      Circuit Applications*, edited by H.A. Watson, McGraw-Hill, NY, 1969. Used with per-
      mission of McGraw-Hill Book Co.)

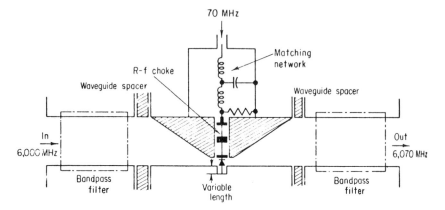

Fig. 3.25. A 70 MHz to 6 GHz varactor diode up-converter. (After Hefni and Spiwak in *Microwave Semiconductor Devices and their Circuit Applications*, edited by H.A. Watson, McGraw-Hill, NY, 1969. Used with permission of McGraw-Hill Book Co.)

temperature parametric amplifiers are comparable in noise performance with maser amplifiers.

Most parametric amplifiers are of the circulator type since this conveniently separates the input and output powers which are at the same frequency. In the equivalent circuit of Fig. 3.26 the varactor is the elastance $S(t)$ and series resistance $R_s$. The idler circuit is $C_2 L_2 R_2$ and $\boxed{\omega_2}$ represents a filter that passes only the idler frequency. The components $C_1 L_1 R_1$ and $\boxed{\omega_1}$ correspond to the signal frequency.

The voltage and current relationships are of the form

$$
\begin{bmatrix} v_1 \\ v_2 \end{bmatrix} = \begin{bmatrix} \dfrac{S_0}{j\omega_1} & -\dfrac{S_1}{j\omega_2} \\ \dfrac{S_1}{j\omega_1} & -\dfrac{S_0}{j\omega_2} \end{bmatrix} \begin{bmatrix} i_1 \\ i_2 \end{bmatrix} \tag{3.32}
$$

where $S_0$ and $S_1$ are Fourier coefficients in the expression $S(t) = S_0 + 2S_1 \cos \omega_p t$. With some manipulation the input impedance of the amplifier at $\omega_1$ can be shown to be

$$
Z_{in} = R_s + R_1 + Re \left( -\frac{Q_1 Q_2 R_s^2}{R_s + Z_{22}^*} \right) + j \left( \omega_1 L_1 - \frac{S_0}{\omega_1} \right)
$$

$$
+ j \cdot Im \left( -\frac{Q_1 Q_2 R_s^2}{R_s + Z_{22}^*} \right) \tag{3.33}
$$

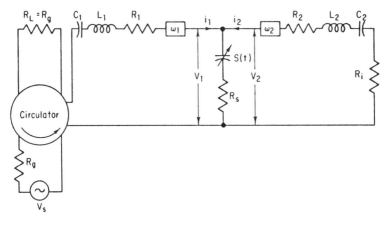

Fig. 3.26. Simplified equivalent circuit of a circulator-type parametric amplifier. (After Uenohara and Gewartowski in *Microwave Semiconductor Devices and Their Circuit Applications*, edited by H.A. Watson, 1969, McGraw-Hill, NY, 1969. Used with permission of McGraw-Hill Book Co.)

where $Z_{22}^* = R_i + R_2 - j(\omega_2 L_2 - S_0/\omega_2)$, and the dynamic quality factors are $Q_1 = |S_1|/\omega_1 R_s$ and $Q_2 = |S_1|/\omega_2 R_s$. If $R_1$ and $Z_{22}^*$ are small, the input resistance is $R_s(1 - Q_1 Q_2)$ and power gain is obtained if this is negative.

The reflection power gain for a one-port amplifier is equal to the power reflection coefficient and is given by

$$G_{11} = \frac{\left| (Z_{11}^* - R_1) - \left(R_s + R_1 - \dfrac{Q_1 Q_2 R_s{}^2}{R_s + Z_{22}^*}\right) \right|^2}{\left| (Z_{11} - R_1) + \left(R_s + R_1 - \dfrac{Q_1 Q_2 R_s{}^2}{R_s + Z_{22}^*}\right) \right|^2} \qquad (3.34)$$

where $Z_{11} = R_g + R_1 + j(\omega_1 L_1 - S_0/\omega_1)$. The power gain approaches infinity when $R_s + Z_{11} = Q_1 Q_2 R_s{}^2/(R_s + Z_{22}^*)$. Typical gains of practical amplifiers are 20 dB and bandwidths are of the order of a few percent for single-tuned signal and idler circuits. By multiple tuning, bandwidths of 10% or more are possible at frequencies of a few GHz.

Thermal noise is the dominant source of disturbance in parametric varactor amplifiers at normal temperatures. However, at low temperatures the thermal noise is less and shot noise of a few $°K/\mu A$ may become significant. Some calculated and experimental noise temperature levels for parametric amplifiers are shown in Fig. 3.27 for several ambient temperatures and show the low levels that are possible.

Multistage parametric amplifiers are sometimes used: Fig. 3.28 shows a two-

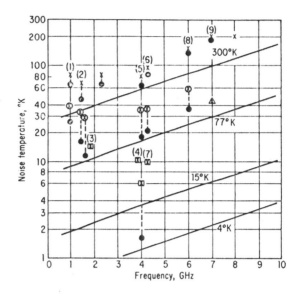

Fig. 3.27. Theoretical maximum noise temperatures of parametric amplifiers and experimental results of amplifiers ranging from 1 to 10 GHz. The points x show the measured overall noise temperatures at 300°K ambient temperature; ⊗ at 77°K; ⊠ at 30°K; ◬ at 4°K. The points indicate the computed noise temperature contribution of varactor and microwave circuits. On the curves the figures in parentheses refer to the results of different investigations. (After Uenohara and Gewartowski in *Microwave Semiconductor Devices and Their Circuit Applications*, edited by H.A. Watson, McGraw-Hill, NY, 1969. Used with permission of McGraw-Hill Book Co.)

stage 4 GHz receiver amplifier with the first stage at liquid nitrogen temperature. The first-stage noise temperature was less than 43°K and the contribution from the second stage and the mixer was less than 2°K, for an effective input noise temperature of less than 45°K. The first-stage gain was 20 dB and the second-stage gain was 18 dB. The bandwidth was 60 MHz.

More recent designs of parametric amplifiers for 4-30 GHz frequency range may have bandwidths of 500 MHz for a 30-dB gain and noise temperatures between 30° and 300°K depending on the cooling.

## 3.4 TUNNEL DIODES AND APPLICATIONS

The observation of tunneling in very-abrupt heavily-doped $p^+n^+$ diodes of Ge by Esaki (1958) led to very active study and great hopes for such devices because of the high speed potential. However, tunneling devices have not been as widely used in practical applications as initially expected for several reasons. One problem is the usual circuit difficulty in effectively using a two-terminal highly-nonlinear negative resistance device. There are also problems that are

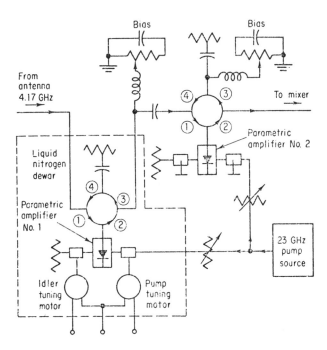

Fig. 3.28. Schematic circuit diagram of a 4 GHz two-stage parametric receiver amplifier. (After Uenohara and Gewartowski in *Microwave Semiconductor Devices and Their Circuit Applications*, edited by H.A. Watson, McGraw-Hill, NY, 1969. Used with permission of McGraw-Hill Book Co.)

inherent in tunnel diodes and there is competition from attractive alternative devices. The inherent problems are:

- Tunnel diodes are heavily doped devices and the capacitance per unit area is therefore high, so very small area structures are needed to obtain low *RC* time-constants. The structures used are usually alloyed abrupt junctions and good control of the short-time alloying and cooling cycle is needed. Furthermore this must be followed by etching of a small-area neck region to obtain the desired peak current and capacitance, and this is not a very controllable process. The rather fragile structure must then be packaged in a fashion that confers robustness, reliability and adequate heat sinking. Typical structures are shown in Fig. 3.29.

- The negative resistance region of a tunnel diode is developed over the voltage range 50 mV to 250 mV and this limits circuit uses with higher voltage semiconductor devices.[19]

- The peak-to-valley current ratio of Si tunnel diodes is not as great as for other semiconductors such as Ge, GaAs or GaSb for reasons of band structure. Typical characteristics are shown in Fig. 3.30. This tends to

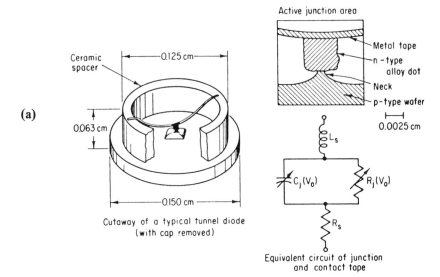

(a)

Cutaway of a typical tunnel diode
(with cap removed)

Equivalent circuit of junction
and contact tape

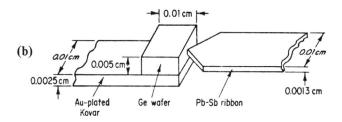

(b)

Fig. 3.29. Typical tunnel diode construction:
(a) Packaged construction and equivalent circuit, and
(b) Electronically pulsed diode structure. (After Dunn in *Microwave Semiconductor Devices and Their Circuit Applications*, edited by H.A. Watson, McGraw-Hill, NY, 1969. Used with permission of McGraw-Hill Book Co.)

limit Si tunnel diode performance and integrated circuit technology is less applicable to tunnel diode fabrication in other materials.

• Although GaAs tunnel diodes have excellent characteristics, these degrade in peak-to-valley ratio if the circuit action allows large forward current swings through the device. This is associated with impurity or defect redistributions involving Zn or vacancy movements.[86, 115, 116] This may be regarded as a potential risk inherent in a structure that depends for its action on abrupt impurity doping changes over distances of 50 Å or less.

• The competing devices in applications such as high-speed switching and signal detection are Schottky diodes. These are excellent in performance,

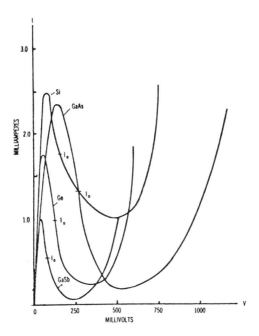

Fig. 3.30. Typical characteristic curves of tunnel diodes made from four materials—gallium antimonide (GaSb), germanium (Ge), silicon (Si) and gallium arsenide (GaAs). Curves are normalized to equal negative resistivity $R_d$ at $I_o$. (After Baasch, King and Sharpe. Adapted with permission from *IEEE Trans. Electron Devices*, **ED-11**, 1964, p. 263 and *Electronic Products*, November 1, 1969, p. 22.)

easy to fabricate, and very compatible with Si integrated circuit technology.

● The competition in amplifiers comes from low-noise parametric varactor amplifiers, and from GaAs FET performance in the 1-10 GHz range.

● The high frequency oscillator possibilities of tunnel diodes do not match these of Impatt avalanche multiplication diodes in efficiency and power output.

However, even with these tunnel diode limitations, a brief examination of tunnel principles is desirable because tunneling is present at many device interfaces including ohmic contacts and heavily doped Schottky barriers.

### 3.4.1 Tunnel Principles

If an electron is represented by the Schrödinger wave equation and the energy barrier is a step function of small width $a$ the transmission probability given by the quantum mechanical treatment is shown in Fig. 3.31. For barriers with $a > 1.5\hbar/p$ where $p$ is momentum, the transmission probability depends on $\exp(-2\ pa/\hbar)$, hence the exponential form means that barrier widths of not more than a few tens of angstrom units are needed for significant tunneling. For a triangular barrier, defined as shown in Fig. 3.32, the transmission probability is given by

$$T = \exp \left( -\frac{4\sqrt{2m^*}\, E_B^{3/2}}{3q\hbar\mathcal{E}} \right) \tag{3.35}$$

Tunnel diode action is obtained in junctions where the $n$ and $p$ sides are so heavily doped that the Fermi levels are inside the band edges. In Si dopings of a few $\times$ $10^{19}$ cm$^{-3}$ are required. At zero bias voltage the Fermi levels then line up as shown in Fig. 3.33, sketch 2. Application of a forward bias voltage of 20 or 30 mV (see Fig. 3.33, sketch 3) then causes a Fermi level displacement so that the conduction band electrons in the $n^+$ side line up with the empty band states (holes) in the $p^+$ side and current flow takes place by tunneling provided the energy barrier is thin enough. As the voltage is further increased (see Fig. 3.33, sketch 4) the overlap of electrons and available states becomes

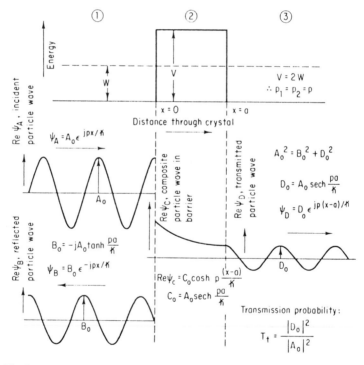

Fig. 3.31. Quantum mechanical transmission of a particle-wave incident on a rectangular energy barrier. (After Dunn in *Microwave Semiconductor Devices and Their Circuit Applications*, edited by H.A. Watson, McGraw-Hill, NY, 1969. Used with permission of McGraw-Hill Book Co.)

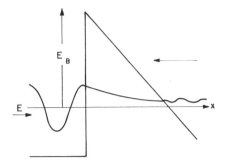

Fig. 3.32. Tunneling of a particle wave of energy $E$ through a triangular energy barrier of height $E_B$ and field strength $\mathcal{E}$ (Fowler-Nordheim tunneling.)

less and the current declines from its peak. Finally injection over the barrier dominates as in Fig. 3.33, sketch 5. In the reverse-bias direction, the valence band electrons of the $p$ region are in line with empty states in the conduction band of the $n$ region and easy tunneling conduction occurs. The shape of the forward tunnel current in the range from 0 to 100 mV, or so, is in general agreement with tunnel calculations[34] as suggested by Fig. 3.34(a), but at higher voltages the valley shape is usually affected by fringing of the band edges in the degenerately doped semiconductors. If the semiconductor contains deep impurity levels, tunneling to and from these may be possible across the junction

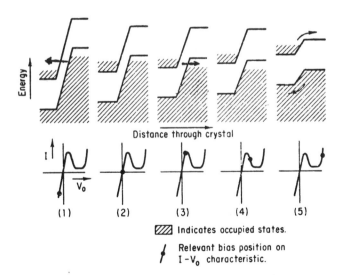

Fig. 3.33. Tunneling current is related to the energy band overlap which is determined by the applied voltage. (After Hall, reprinted with permission from *IRE Trans. Electron Devices*, ED-7, 1960, p. 4.)

and then the valley region shape is partly filled-in by this excess tunneling component.

Switching of a tunnel diode from a high current $P$ to a low current state $Q$ [see Fig. 3.34(b)] is accomplished by having a resistance in series and pulsing the voltage $V$ of the series combination to $V + \Delta V$. This raises the load line above the peak current, and the diode switches to $Q'$ and relaxes to $Q$ when $\Delta V$ is removed.

The peak-to-valley current ratio is one figure of merit for a tunnel diode since it represents the dynamic range of the negative resistance. However, for a measure of the switching speed the $RC$ time constant of the circuit is important. In this context $R$ may be taken as the negative resistance, which itself tends to approximately determine the series resistance value selected for the circuit.

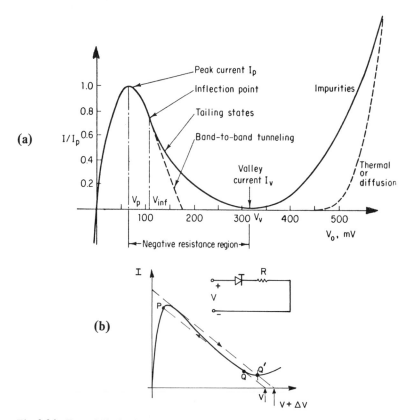

Fig. 3.34. Tunnel diode characteristics:
(a) Shows "excess" current from tailing states and impurities, (After Dunn in *Microwave Semiconductor Devices and Their Circuit Applications*, edited by H.A. Watson, Mc-Graw-Hill, NY, 1969. Used with permission of McGraw-Hill Book Co.)
(b) Switching action from $P$ to $Q'$ (or $Q$) is accomplished by an increment of voltage $\Delta V$.

The capacitance charging or discharging through this resistance may be taken as that at the valley voltage. Since the peak and valley voltages are roughly constant values for a given semiconductor (although depending somewhat on the doping levels), the negative resistance tends to be inversely proportional to the peak tunnel current. Hence, $I_p/C_v$, which is independent of the junction area, represents a speed factor for the tunnel diode. This speed index is increased if the tunnel junction region is narrower, as in Fig. 3.35(a), and if the doping is increased, as in Fig. 3.35(b).

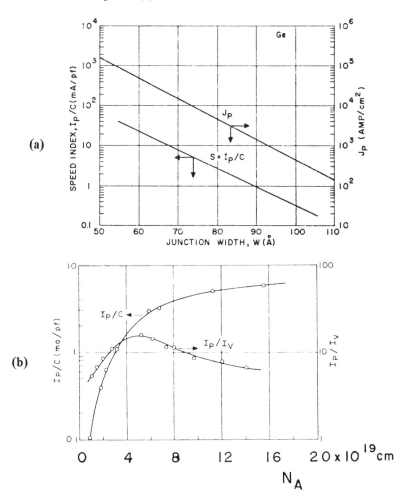

Fig. 3.35. Peak current and speed index variations calculated for Ge tunnel diodes at 300°K:
(a)   Variation of $I_p$ and $I_p/C$ with junction width, (After Davis and Gibbons[33])
(b)   Variation of $I_p/I_v$ and $I_p/C$ with doping concentration. (After Glicksman and Minton.[48] Reprinted with permission from *Solid-State Electronics*, Pergamon Press Ltd.)

The effects of temperature on Ge tunnel diode characteristics are shown in Fig. 3.36. The peak current may either decrease with temperature (lightly doped units) because of changes of density of states occupancy or increase because of the decrease of the bandgap with increasing temperature (more heavily doped units). The valley current always increases with a rise in temperature. If compressive stress is applied to a tunnel diode the peak current and the valley current increase. Consideration has been given to this as a stress measuring effect but the fragile nature of the tunnel diode and the large temperature sensitivity do not favor such device uses.[142]

If the doping densities in a tunnel diode are reduced below the degeneracy values, the Fermi levels are no longer within the band edges. Then the tunnel actions shown in sketches (3) and (4) of Fig. 3.33 cannot exist and the forward-bias tunnel current disappears. However, the reverse current as shown in sketch (1) continues to exist since it does not depend on degeneracy of the doping but rather on the abrupt narrow junction structure. The result is a so-called "backward diode" that conducts well with 50 or 100 mV (see Fig. 3.37) in the "reverse bias" direction, but has a rather small current in the forward direction up to several hundred millivolts at which forward conduction over the barrier begins to take place. Backward diodes may be used for the rectification of small signals in microwave signal mixing and detection. This role however is now being taken over increasingly by zero-bias Schottky diodes.

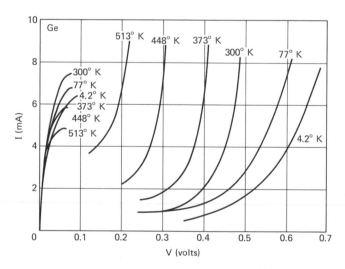

Fig. 3.36.  Typical current-voltage characteristics of a Ge tunnel diode as a function of temperature. (After Sze, *Physics of Semiconductors*, John Wiley and Sons, NY, 1969, p. 183.)

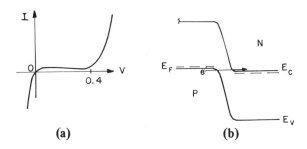

Fig. 3.37. Backward diode characteristic and energy diagram
(a)  The characteristic shows high conduction in the reverse bias condition but almost no tunnel hump in the forward bias direction, and
(b)  The energy band diagram shows how tunnel current flows in the reverse bias condition.

### 3.4.2 Tunnel Diode Applications

Schottky diode detectors have advantages over tunnel (or backward) diode detectors in terms of high tangential signal sensitivity, wide dynamic range, good square-law sensitivity, and reduced tendency to burn out.

The video-impedance of a zero-bias Schottky diodes may be 1000 ohms (assuming it is not further reduced by bias) and so looking into a high input impedance amplifier of about 5 pF capacitance, the rise-time at the detector amplifier junction may be about 2 $RC$ where $C$ is 20 pF to include an RF bypass capacitor. Hence, the rise time is the order of 40 nsec. This would allow processing of signal pulse widths down to 200 nsec. On the other hand, for a tunnel diode detector the video impedance may be 75 ohms or so. The video-impedance of a Schottky detector can, of course, be shunted by a parallel resistor, or reduced by bias current or by selection of a lower barrier height junction, but these remedies tend to be at the expense of lower sensitivity.

In comparing tunnel diode and Schottky detectors consideration must also be given to the effective dynamic range after compensation in the amplifier, to the tracking achievable over a temperature range, to the voltage standing wave ratio (VSWR) and to the burn-out resistance. All of these matters must be considered before settling for the most effective approach although nowadays this is likely to turn out to be the use of low-barrier-height Schottky diodes.

A tunnel diode relaxation oscillator circuit is shown in Fig. 3.38 with typical performance characteristics. Related circuits have been used for frequency, amplitude and digital phase modulation, for instance in binary repeaters with carrier frequencies of 1.3 GHz and bit rates of 306 Mbits/sec. Figure 3.39 shows the use of a tunnel diode in conjunction with a transistor and its employment in an over-current protection circuit.

Tunnel diode amplifiers have been used as simple moderately-low-noise pre-amplifiers for RF microwave receivers where the complication of a parametric amplifier is not justified. Such amplifiers may have a 17-dB gain with a 20-MHz

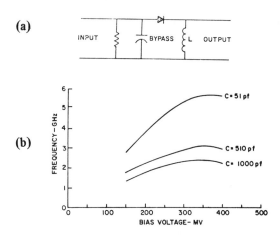

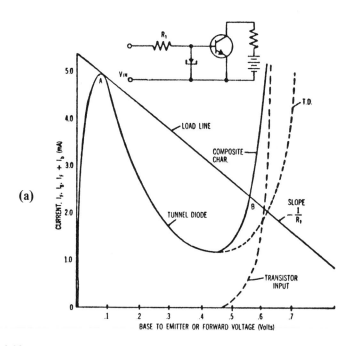

Fig. 3.38. Tunnel diode oscillator circuit and frequency:
(a) Typical relaxation oscillator, and
(b) Computed output frequency vs input voltage for several values of diode capacitance. ($R_L$ = 150 ohms) (Courtesy J. E. Goell.)

Fig. 3.39. Tunnel diode circuits:
(a) Composite input characteristic of tunnel diode-transistor hybrid circuit, and
(b) "Crowbar" power supply overcurrent protection circuit. Tunnel diode across current sense resistor initiates SCR trigger within nanoseconds of overload, limited only by transistor and SCR speed. (After Palmer.[103] Reprinted with permission from *Electronics Industry and Electronic Products*, Nov. 1, 1969.)

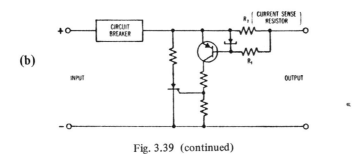

**(b)**

Fig. 3.39 (continued)

bandwidth and 60-dB dynamic signal range. However, with the advent of micro-wave GaAs FET transistors the tunnel diode amplifier is becoming obsolete.[32] The lumped circuit and integrated circuit form of a 2.8-4.1 GHz tunnel diode amplifier is shown in Fig. 3.40. The gain peaked from 9 dB to 12 dB in this range and the measured noise figure was 6.2 dB at 300°K. The transforming

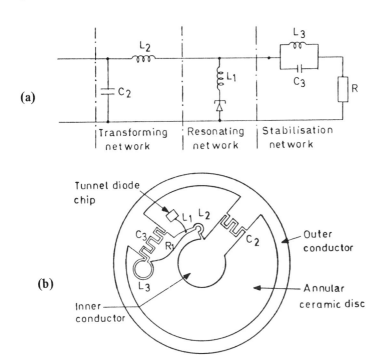

**(a)**

**(b)**

Fig. 3.40. Tunnel diode amplifier, 2.8-4.1 GHz:
(a)  Lumped element circuit form, and
(b)  Physical realization to fit in a coaxial line (After Aitchison, et. al., courtesy Philips Research Laboratories, reprinted with permission from *IEEE Trans. on Microwave Theory and Technique*, **MTT-19**, 1971, p. 934.)

network matches the amplifier impedance to the circulator that precedes it. The stabilizing network allows the negative conductance to be developed only at the frequency that must be amplified. The annular ceramic disk structure allows the amplifier to be inserted in a 50-ohm coaxial line.

The noise temperature levels of various microwave amplifiers including tunnel diode amplifiers, are illustrated in Fig. 3.41.

Tunnel diodes have potential as photodetectors where very high speed of response is needed. For instance, conventional tunnel diodes follow subnanosecond light pulses from mode-locked He-Ne lasers.[1] Metal-barrier-metal tunnel diodes (MBM) may have speeds of response of less than $10^{-12}$ sec and are being examined for submillimeter infrared and optical devices.

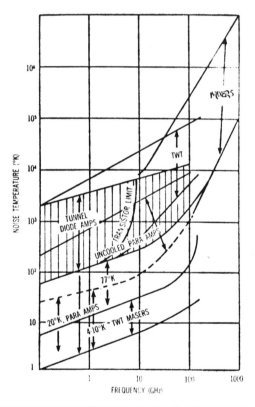

Fig. 3.41. Tunnel diode amplifier performance in comparison with competing devices. This chart shows that tunnel diode amplifiers are usable over the same noise temperature band and frequency range, except that transistor amplifier noise exceeds that of TD amplifiers above the 4 to 6 GHz band. Recent developments of low noise GaAs MESFET amplifiers, however, have closed the noise gap since the chart was prepared, and lessened the interest in tunnel diode amplifiers. (After Baasch.[9] Reprinted with permission from *Electronics Industry and Electronics Products*, Nov. 1, 1969.)

## 3.6 QUESTIONS

1. Design a Si detector diode for a 1-GHz signal of 10-mV amplitude to feed into an amplifier of 600-ohms impedance and 2-MHz video bandwidth. What is the expected current sensitivity at 300°K? At what signal level does serious departure from linearity occur? Estimate either the tangential signal sensitivity (TSS) or the nominal-detectable signal (NDS) of your design. Would the performance have been improved by the choice of GaAs instead of Si?

2. A small-signal Si detector diode is represented in the complex frequency plane of resistance $R_s$ and series reactances $X_s$ by a semicircle with its origin at 250 ohms and having a radius of 240 ohms. The value $R_s$ is 250 ohms at $\omega = 2\pi \times 10^9$ rad/sec. What is the diode resistance $R_d$, the series resistance $r_d$, the diode capacitance $C_d$ and the cutoff frequency in a low-level detector circuit? What would be the source and load impedance needed for maximum power gain?

3. A 100-kHz amplitude modulated voltage on a 1-GHz carrier is given by the expression

$$v(t) = 5 \; [1 + 1/2 \sin(2\pi \, 10^5 t)] \; \cos(2\pi \, 10^9 t)$$

The signal modulation may be recovered by an $RC$ envelope detector circuit if the voltage across the filter capacitance can follow the modulation envelope. Find the condition this imposes on the $RC$ product and proportion the circuit. What is the frequency spectrum involved? What would happen if the modulating signal amplitude became twice that of the carrier signal? For the waveform expression given would it be possible to make an integrated circuit form of the $RC$ envelope detector circuit?

4. The bridge circuit shown, Fig. P3.4, uses two Si diodes as varactors to chop a dc signal, $f_m \rightarrow 0$, into an ac form at the driver frequency $f_c$ for subsequent amplification. Alternatively the input may be a low frequency signal $f_m$ that appears as a modulation on the carrier frequency $f_c$. Assume the diodes to be $p^+n$ junctions of area $10^{-4}$ cm² and $N_d = 2 \times 10^{14}$ cm⁻³. The bridge is balanced for the condition

$$\frac{R_1}{R_2} = \frac{C_2}{C_1}$$

The application of a small dc voltage (or a slowly varying ac signal) at the modulation terminals (input terminals) causes the capacitance of the forward-biased diode to increase and of the reverse-biased diode to decrease. The bridge is therefore unbalanced and a signal at the carrier frequency (generated from the transformer) is developed across the output terminals. The phase of the output signal varies by 180° as a function of the polarity of the modulation input.

a) The bridge (excluding the transformer) is just a series and parallel combination of $R$ and $C$. Show that the equivalent output impedance of the bridge at the carrier frequency is given by

$$Z_{out} = \frac{R_1 \, R_2}{R_1 + R_2} + \frac{1}{j\omega(C_1 + C_2)}$$

For maximum power gain the load impedance should be conjugate-matched to this output impedance.

b) The modulation power required to vary the capacitance of the diodes is extremely small since the leakage resistance is very high. The sideband power developed across the load resistance can be correspondingly large. Attempt with suitable approximations, and assuming $C \propto V^{-1/2}$, to show that the power gain and operating gain of the bridge under small signal operation and with a conjugate match at the output are

$$\text{Power gain} = \left(\frac{v_c}{4V_{bi}}\right)^2 \frac{R_2 R_{in}}{(R_L + R_p)^2}$$

$$\text{Operating gain} = \left(\frac{v_c}{4V_{bi}}\right)^2 \left(\frac{R_2 R_g}{(R_L + R_p)^2}\right) \left(\frac{R_{in}}{R_{in} + R_g}\right)$$

where $v_c$ is the rms voltage across the secondary of the coupling transformer and $V_{bi}$ is the built-in voltage of the Si diodes. $R_{in}$ is the input resistance to the modulating signal, not shown in the circuit, $R_g$ is the generator resistance, $R_L$ is the load resistance and $R_p$ is the real part of the output resistance of the bridge at the carrier frequency.

c) If $R_1$, $R_2$, $R_{in}$, $R_g$ and $R_L$ are all 600 ohms, calculate the input/output characteristics for a dc signal and indicate the onset of nonlinearity.

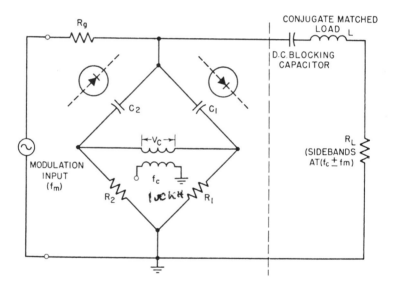

Fig. P 3.4

5. The circuit shown in Fig. P3.5 is known as a ring modulator since it may also be drawn with the diodes forming a series ring or Wheatstone bridge configuration. Show that it acts as a phase sensitive rectifier with the large voltage $e_p$ switching or polarizing the diodes so that when a small signal $e_s$ of the same frequency is applied, the output on the right-hand side of the circuit changes polarity according to the phase of the signal. Draw the circuit waveforms for 0, 90, 180° phase differences between the signal and the reference frequency. Then with $p^+n$ Si diodes (area $10^{-4}$ cm$^2$, $N_d = 2 \times 10^{14}$ cm$^{-3}$) design such a circuit to operate as a phase sensitive rectifier at 1 MHz, and compute the input/output characteristic and the linearity limit.

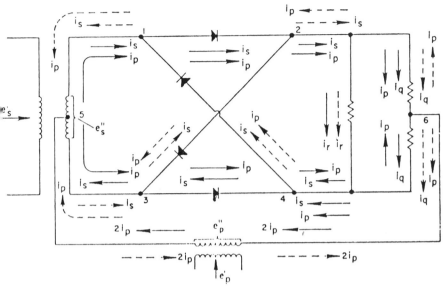

Fig. P 3.5

6. Commencing with the *PIN* diode characteristic of Fig. 3.8, design a three diode $\pi$ attenuator to interpose between a mixer and an IF amplifier both of 600-ohms resistive impedance at 450 MHz. Calculate the attenuation vs the dc bias signal as from an AGC amplifier at $300°$ K. If the attenuator temperature becomes $400°$ K what changes may be expected? How does the input impedance of the attenuator vary with the attenuator ratio? What is the effect of increasing input signal strength on the attenuator input/output linearity? Would there be any advantage in going to the bridged T attenuator circuit of Fig. 3.13?

7. Discuss power handling capability of a *PIN* diode in reflective-mode series and shunt connections. A series diode of 3 W dissipation with a resistance range of 1-10 kilohms can handle 150 W. In a shunt connection the same diode is stated to handle only 40 W of signal power. Is this statement correct?

8. Design a varactor frequency tripler circuit using Si diodes to generate 10 mW of power in a 50-ohm load at 2.4 GHz (Instrument, Scientific and Medical band), from an 800-MHz transistor oscillator of output impedance 50 ohms. Select from commercially available diodes and specify all other circuit components. Estimate the power conversion efficiency of your circuit.

9. Consider a circulator-type parametric amplifier with an input frequency $f_1$ of 500 MHz and an idler frequency $f_2$ of 2.5 GHz. The dynamic quality factors $Q_1$ and $Q_2$ are both equal to 10. What else needs to be known in order to calculate the input impedance of the amplifier at $\omega_1$ and the reflection power gain? Select a Si varactor diode and proportion the remaining circuit components so that the impedance and gain may be calculated. What is the expected bandwidth of your amplifier?

10. Design a very low frequency parametric amplifier to amplify a 20-Hz signal with a 120-Hz idler frequency. What are the design difficulties and what performance do you expect?

11. For a Schottky barrier characterized by an ideal exponential equation

$$I = I_0 \exp(\phi_B/kT) \exp(qV/kT - 1)$$

calculate the mixer current sensitivity $\beta$ as a function of bias current density and barrier height at $300°\text{K}$. Assume reasonable values for any other parameters needed.

12. Discuss scattering parameters as applied to mixer circuits.

13. Discuss the $\beta\sqrt{R_V}$ figure of merit for diode detectors.

14. In a typical Schottky barrier detector circuit why is there an optimum bias current which yields maximum signal sensitivity? How does this vary with the RF signal frequency?

15. How would you characterize and measure the RF burn-out ratio of $X$ band mixer diodes? How would you design for increased burn-out rating?

16. Some discussion has taken place (*Proc. IEEE*, June 1975 p. 980) on the merits of biasing Schottky barrier detectors instead of the use of low barrier height units. Can you add anything to this discussion?

17. It has been stated that the barrier height of a Schottky diode has an optimum value of about 0.3 eV for detector, harmonic generator and frequency converter uses at high frequencies. This has been given as reason for the use of $In_xGa_{1-x}As$ Schottky detectors. Discuss the merits of this approach.

18. What determines the zero-bias cut-off frequency of a Schottky diode? How does the impedance-plane plot of a Schottky diode depend on the packaging? How can one derive the equivalent circuit components from $S$ parameter curves?

19. Discuss free-carrier absorption in a low temperature semiconductor as a mechanism for detection in the 50-100 GHz frequency range.

20. Review the merits of cryogenic cooling of mixer diodes and varactors.

21. Discuss the merits of Mott-Barrier mixer diodes.

22. Balanced mixer circuits involve 3-dB branch-line and rat-race couplers. Discuss the merits of balanced mixers and any problems.

23. Discuss *PIN* diode attenuators in automatic-gain-control circuits, and comment on the dynamic range.

24. Turn-off of a *PIN* diode involves sweep-out of minority carriers from the $I$ region. Discuss this process and consider in particular the trade-off between insertion loss and switching speed.

25. Describe in detail some constant impedance *PIN* attenuator or switch circuits, and *PIN* diode phase-shift circuits.

26. Discuss the concepts of single pole multithrow *PIN* switches and driver circuit requirements for a *PIN* switch with TTL control.

27. Discuss the fabrication of hyperabrupt varactor diodes. Describe varactor diode tuning circuits for radio and TV and varactor phase-shift circuits and their applications.

28. Review varactor frequency multiplying circuits and some applications in the GHz range.

29. Discuss step-recovery diodes for harmonic generation and compare their performance briefly with that of varactor multipliers.

30. Discuss the burn-out resistance of *PIN*, Schottky-barrier and tunnel diodes.

31. Discuss microwave demodulators and down-converters in the GHz range.

32. Discuss excess currents in tunnel diodes. What effect would $10^{16}$ $cm^{-3}$ Au have on the excess current of a Ge tunnel diode?

33. Discuss proposed mechanisms of tunneling via multiple steps involving states at an interface in a junction.

34. Review switching time analyses of tunnel diodes that include parasitics. Must diffusion capacitance be considered? In particular consider the relative merits of Ge and GaAs tunnel diodes in switching circuits.

35. Discuss temperature dependences in tunnel diodes. Which of the three semiconductors (Ge, Si and GaAs) has the least temperature sensitivity of the negative resistance?

36. Discuss the speed index of a tunnel diode $I_p/C_v$ as figure of merit for amplifier applications. Can you devise a more appropriate figure of merit for this or other applications?

37. How would you measure the series resistance of a tunnel diode? Is there significance in defining the series resistance as that under reverse bias conditions?

38. The load line for switching of a tunnel diode into a pure resistance load is a straight line. However, discuss the switching locus if the load is $R_L + jX_L$ where $R_L = X_L$ and if the load is $R_L - jX_C$ where $R_L = X_C$.

39. How is the power output and efficiency determined for a tunnel diode oscillator or amplifier? Briefly compare with the output and efficiency of other devices.

40. Discuss the degradation of GaAs tunnel diode characteristics. Is degradation observed in any other GaAs devices for similar reasons?

41. Discuss tunnel diode frequency dividers and multipliers.

42. Review proposals that have been made for the use of tunnel diodes in logic and memory circuits. Consider tunnel diode bistable trigger circuits. With two tunnel diodes how can a quasi three-terminal trigger circuit be obtained and does this have any circuit merit?

43. Compare backward diodes and zero-bias Schottky diodes (Si and GaAs) as microwave detection devices.

44. Review switching time analysis of backward diodes, and compare Si and GaAs backward diode characteristics.

45. Describe metal-barrier-metal (MBM) tunnel structures and consider their usefulness as submillimeter and infrared optical detectors.

## 3.6 REFERENCES AND READING SUGGESTIONS

1. Adams, R.F., and R.L. Rosenberg, "Tunnel-diode photoresponse," *Appl. Phys. Lett.*, 15, 1969, p. 414.

2. Aitchison, C.S., "Lumped components for microwave frequencies," *Philips Technical Review*, 32, 1971, p. 305.

3. Aitchison, C.S., R. Davies, I.D. Higgins, S.R. Longley, B.H. Newton, J.F. Wells, and J.C. Williams, "Lumped-circuit elements at microwave frequencies," *IEEE Trans. on Microwave Theory and Techniques*, MIT-19, 1971, p. 928.

4. Anand, Y., "RF burnout dependence on variation in barrier capacitance of mixer diodes," *Proc. IEEE*, 61, 1973, p. 247.

5. Anand, Y., and W.J. Moroney, "Microwave mixer and detector diodes," *Proc. IEEE*, 59, 1971, p. 1182.

6. Andersen, S.G., and J.K. Newheim, "Phase noise of various frequency doublers," *IEEE Trans. on Instrumentation and Measurement*, IM-22, 1973, p. 185.

7. Andreyev, M.I., "The use of a tunnel diode connected in parallel with a backward diode in pulsed systems," *Telecommunications and Radio Engineering*, Pt.2 21, 1966, p. 109.

8. Asai, S., S. Ishioka, H. Kurono, S. Takahashi, and H. Kodera, "Effects of deep centers on microwave frequency characteristics of GaAs Schottky barrier gate FET, " *Proceedings of the Fourth Conference on Solid State Devices*, Tokyo, 1972. Supplement to the *Journal of the Japan Society of Applied Physics*, 42, 1973, p. 71.

9. Baasch, T.L., "Special report on tunnel diodes," *Electronic Products*, November 1, 1969, p. 22.

10. Banerjee, A.R., and J.G. Gardiner, "Unbalance and temperature effects in Schottky barrier diode mixers," *Int. J. Electronics*, 34, 1973, p. 619.

11. Bessho, T., Y. Hiyama, and H. Niiyama, "Memory circuit using Esaki diode," *Review of the Electrical Communication Laboratory, Tokyo*, 17, 1969, p. 89.
12. Bhattacharyya, A.B., "Switching time analysis of backward diodes," *Proc. IEEE*, 58, 1970, p. 513.
13. Bhattacharyya, A.B., and S.L. Sarnot, "Effect of the junction capacitance variation and a small inductance on the switching time of a tunnel diode," *Proc. IEEE*, 58, 1970, p. 1957.
14. Bhattacharyya, A.B., and S.L. Sarnot, "Static design of a single tunnel diode bistable trigger," *Int. J. Electronics*, 27, 1969, p. 527.
15. Bhattacharyya, A.B., and S.L. Sarnot, "Step response of a bistable tunnel diode trigger," *Int. J. Electronics*, 27, 1969, p. 401.
16. Biard, J.R., "Low-frequency reactance amplifier," *International Solid-State Circuits Conference*, 1960 Digest of Technical Papers, N.Y., 1960, p. 88.
17. Bowers, R., and J. Frey, "A preliminary technology assessment of solid state microwave devices," *Third Biennial Cornell Electrical Engineering Conference*, Ithaca, NY, 1971, p. 29.
18. Calviello, J.A., J.L. Wallace, and P.R. Bie, "High performance GaAs quasi-planar varactors for millimeter waves," *IEEE Trans. Electron Devices*, ED-21, 1974, p. 624.
19. Canero, C.A., "Tunnel diode output voltage converter," *Review of Scientific Instruments*, 42, 1971, p. 264.
20. Carr, W.N., and A.G. Milnes, "Bias-controlled tunnel-pair logic circuits," *IRE Trans. Electron. Computers*, EC-11, 1962, p. 773.
21. Carruthers, T., "Bias-dependent structure in excess noise in GaAs Schottky tunnel junctions," *Appl. Phys. Lett.*, 18, 1971, p. 35.
22. Caulton, M., "Thin-film microwave components," *Nerem Record*, 10, 1968, p. 136.
23. Chang, L.L., "Conductance extrema in metal-insulator-semiconductor tunnel junctions," *J. Appl. Phys.*, 39, 1968, p. 1455.
24. Chang, L.L., and J.S. Moore, "Generalized current and conductance extrema in metal-insulator-semiconductor tunnel junctions," *J. Appl. Phys.*, 40, 1969, p. 5315.
25. Cho, A.Y., and F.K. Reinhart, "Interface and doping profile characteristics with molecular-beam epitaxy of GaAs:GaAs voltage varactor," *J. Appl. Phys.*, 45, p. 1812.
26. Chorney, P., "Broadband solid-state control devices," *Microwave Journal*, 16, July, 1973, p. 17.
27. Chorney, P., Multi-octave, multi-throw, *PIN*-diode switches, *Microwave Journal*, 17, September, 1974, p. 39.
28. Clifton, B.J., W.T. Lindley, R.W. Chick, and R.A. Cohen, "Materials and processing techniques for the fabrication of high quality millimeter wave diodes," *Third Biennial Cornell Electrical Engineering Conference*, Ithaca, NY, 1971, p. 463.
29. Cohen, L.D., "Microwave characterization of the properties and performance of GaAs Schottky barrier mixer diodes," *Proc. IEEE*, 59, 1971, p. 288.
30. Cooperman, M., "Gigahertz tunnel-diode logic," *RCA Review*, 28, 1967, p. 421.
31. Cuccia, L.C., and J.J. Spilker, Jr., "Varactor diodes key v-band data communications system," *Microwave Systems News*, June/July 1972, p. 17.
32. Daglish, H.N., J.G. Armstrong, J.C. Walling, C.A.P. Foxell, *Low-Noise Microwave Amplifiers*, Cambridge University Press, Cambridge, UK, 1968.
33. Davis, R.E., and G. Gibbon, "Design principles and construction of planar Ge Esaki diodes," *Solid-State Electronics*, 10, 1967, p. 461.
34. Demassa, T.A., and D.P. Knott, "The prediction of tunnel diode voltage-current chacteristics," *Solid-State Electronics*, 13, 1970, p. 131.
35. Dem' Yanchenko, A.G., and E.A. Khurtin, "A tunnel-diode harmonic-relaxation frequency divider," *Radio Engineering and Electronic Physics*, 16, 1971, p. 1397.
36. Denlinger, E.J., et al., "High performance mixer diode," *IEEE International Electron Devices Meeting Technical Digest*, Washington, D.C., December 1976, p. 87.

37. Dragone, C., "Performance and stability of Schottky barrier mixers," *Bell System Tech. J.*, **51**, 1972, p. 2169.
38. Dubrovosky, I.A., "Application of networks in the measurement of some tunnel diode parameters," *Radio Engineering and Electronic Physics*, **12**, 1967, p. 1486.
39. Dunn, C.N., "Tunnel diodes," in:*Microwave Semiconductor Devices and their Circuit Applications* (ed. by H.A. Watson), McGraw-Hill, NY, 1969 p. 396.
40. Eldumiati, I.I., and G.I. Haddad, "A microwave-biased millimeter-and-submillimeter-wave detector using InSb," *IEEE Trans. Electron Devices*, **ED-19**, 1972, p. 257.
41. Ertel, A., "Monolithic IC techniques produce first all silicon X-band switch," *Electronics*, Jan 23, 1967, p. 76.
42. Esaki, L., "Long journey into tunneling," *Proc. IEEE*, **62**, 1974, p. 825.
43. Etkin, V.S., "Solid-state microwave devices," *Telecommunications and Radio Engineering*, **26**, December, 1971, p. 52.
44. Farrell, R., "A SiC backward diode, *Proc. IEEE*, **57**, 1969, p. 221.
45. Fleri, D.A., and L.D. Cohen, "Nonlinear analysis of the Schottky-barrier mixer diode," *IEEE Trans. on Microwave Theory and Techniques*, **MIT-21**, 1973, p. 39.
46. Feldman, N.E., "Syllabus on low-noise microwave devices," *Microwave Journal*, July 1969.
47. Gerzon, P.H., J.W. Barnes, D.W. Waite, and D.C. Northrop, "The mechanism of r.f. spike burnout in Schottky barrier microwave mixers," *Solid-State Electronics* **18**, 1975, p. 343.
48. Glicksman, R., and R.M. Minton, "The effect of *p*-region carrier concentration on the electric characteristics of germanium epitaxial tunnel diodes," *Solid-State Electronics*, **8**, 1965, p. 517.
49. Goddard, N.E., "Solid-state microwave electronics," *Philips Technical Rev*, **32**, 1971, p. 282.
50. Green, M.A., F.D. King, and J. Shewchun, "Minority carrier MIS tunnel diodes and their application to electron-and photo-voltaic energy conversion—I. Theory, *Solid-State Electronics*, **17**, 1974, p. 551.
51. Green, M.A., and J. Shewchun, "Capacitance properties of MIS tunnel diodes," *J. Appl. Phys.*, **46**, 1975, p. 5185.
52. Guétin, P. and G. Schréder, "Tunnelling in metal-semiconductor contacts under pressure," *Philips Technical Review*, **32**, 1971, p. 211.
53. Gupta, K.C., and A. Singh, *Microwave Integrated Circuits*, John Wiley & Sons, Inc., NY, 1974.
54. Hall, R.N., "Tunnel diodes," *IRE Trans. Electron Devices*, **ED-7**, 1960, p. 1.
55. Hefni, I.E., and R.R. Spiwak, "High-efficiency ultraflat low-noise varactor frequency converter using low-frequency pumping," *IEEE International Solid State Circuits Conference*, *Dig. Tech. Papers*, **9**, 1966, p. 46.
56. Hewlett-Packard Co., "Improved frequency response for M/W detector," *Microwave Journal*, **18**, April 1975, p. 16.
57. Hewlett-Packard Co., "Applications of *PIN* diodes," Application Note 922, Palo Alto, CA, USA.
58. Hewlett-Packard Co., "Ku-band step recovery multipliers," Application Note 928, Palo Alto, CA, USA.
59. Hewlett-Packard Co., "Hot carrier diode video detectors," Application Note 923, Palo Alto, CA, USA.
60. Hewlett-Packard Co., "High performance *PIN* attenuator for low cost AGC applications," Application Note 936, Palo Alto, CA, USA.
61. Hopkins, J.B., "Microwave backward diodes in InAs," *Solid-State Electronics*, **13**, 1970, p. 697.
62. Hopkins, J.B., "Optimum design of germanium microwave backward diodes," *Proc. IEEE*, **57**, 1969, p. 1458.

63. Hughes, G.W., and R.M. White, "MIS and Schottky barrier microstrip devices," *Proc. IEEE*, **60**, 1972, p. 1460.
64. Hughes, G.W., and R.M. White, "Microwave properties of non-linear MIS and Schottky-barrier microstrip," *IEEE Trans. Electron Devices*, **ED-22**, 1975, p. 945.
65. Iizuka, H., and S. Kitaoka, "Low-noise GaAs Schottky-barrier beam-lead mixer diodes," *Proc. IEEE*, **58**, 1970, p. 1372.
66. Imai, T., H. Satoh, and T. Miyajima, "Drift of the electrical characteristics of the Esaki-diodes during life tests," *Review of the Electrical Communication Laboratory*, Tokyo, **15**, 1967, p. 459.
67. Immorlica, A.A., "Cross the barrier of varactor drift," *Microwaves*, August 1976, p. 36.
68. Jäger, D., and W. Rabus, "Bias-dependent phase delay of Schottky contact microstrip line," *Electronics Lett.*, **9**, 1973, p. 201.
69. Jäger, D., W. Rabus and W. Eickhoff, "Bias-dependent small-signal parameters of Schottky contact micro-strip lines," *Solid-State Electronics*, **17**, 1974, p. 777.
70. Jonkuhn, G., and C.H. Lembke, "Matched-varactor chip brings electronic tuning to a-m radios," *Electronics*, July 19, 1971, p. 60.
71. Kajiyama, K., S. Sakata and Y. Mizushima, Schottky-barrier devices with low barrier height, *Proc. IEEE*, **62**, 1974, p. 1287.
72. Khan, P.J., "Optimum design of varactor diode parametric amplifiers," *Proc. First Cornell Biennial Conference on Engineering Applications of Electronic Phenomena*, Ithaca, NY, 1967, p. 421.
73. King, B.G., and G.E. Sharpe, "Measurement of the spot noise of germanium, gallium antimonide, gallium arsenide and silicon Esaki diodes," *IEEE Trans. Electron Devices*, **ED-11**, 1964, p. 273.
74. Korablev, I.V., V.U. Potemkin, and F. Yunosov, "Low-frequency noise in backward diodes," *Radio Engineering and Electronic Physics*, **13**, 1968, p. 605.
75. Lambe, J., and R.C. Jaklevic, "Charge quantization studies using a tunnel capacitor," *Phys. Rev. Lett.*, **22**, 1969, p. 1371.
76. Lawley, K.L., J.A. Heilig, and D.L. Klein, "Preparation of restricted area GaAs tunnel diodes," *Electrochemical Technology*, **5**, 1967, p. 376.
77. Lecoy, G., and J. Carmassel, "Relation between $I/V$ characteristics and noise in metal-oxide-metal tunnel diodes," *Electronics Lett.*, **3**, 1967, p. 566.
78. Lee, T.P. and C.A. Burrus, "A millimeter-wave quadrupler and an up-converter using planar diffused gallium varactor diodes," *Proc. First Biennial Conference on Engineering Applications of Electronic Phenomena*, Ithaca, NY, 1967, p. 362.
79. Leedy, H.M., H.L. Stover, H.G. Morehead, R.P. Bryan, and H.L. Garvin, "Advanced millimeter-wave diodes, GaAs and silicon, and a broadband low-noise mixer," *Proc. Third Biennial Cornell Electrical Engineering Conference*, Ithaca, NY, 1971, p. 451.
80. Leighton, W.H., and A.G. Milnes, "Junction reactance and dimensional tolerance effects on X band 3 db directional couplers," *IEEE Trans. on Microwave Theory and Techniques*, **MIT-19**, 1971, p. 81.
81. Liberman, L.S., B.V. Sestroretskey, V.A. Shpirt, and L.M. Yakuben, "Semiconductor diodes for controlling microwave power," *Telecommunications and Radio Engineering*, **26-27**, 1972, p. 63.
82. Liechti, C.A., "Down-converters using Schottky-barrier diodes," *IEEE Trans. Electron Devices*, **ED-17**, 1970, p. 975.
83. Ma, T.P. and R.C. Barker, "Surface-state spectra from thick-oxide MOS tunnel junctions," *Solid-State Electronics*, **17**, 1974, p. 913.
84. MacPherson, A.C., and H.M. Day, "Design and fabrication of high burn-out Schottky barrier crystal video diodes," *Solid-State Electronics*, **15**, 1972, p. 409.
85. MacPherson, A.C., and H.M. Day, "Comments on "Schottky-barrier devices with low barrier height," *Proc. IEEE*, **63**, 1975, p. 980.

86. Marconato, R., and R.L. Anderson, "Partial recovery of GaAs tunnel diodes after stress-induced degradation," *J. Appl. Phys.*, **42**, 1971, p. 3209.
87. Margalit, S., J. Shappir, and I. Kidron, "Field-induced tunnel diode in indium antimonide," *J. Appl. Phys.*, **46**, 1975, p. 3999.
88. McColl, M., and M.F. Millea, "Advantages of Mott barrier mixer diodes," *Proc. IEEE*, **61**, 1973, p. 499.
89. McColl, M., R.J. Pedersen, M.F. Battjer, M.F. Millea, A.H. Silver, and F.L. Vernon, Jr., "The super-Schottky diode microwave mixer," *Appl. Phys. Lett.*, **28**, 1976, p. 159.
90. McDade, J.C., and F. Schiavone, "Switching time performance of microwave *PIN* diodes," *Microwave Journal*, **17**, December 1974, p. 65.
91. Milnes, A.G., and D.L. Feucht, *Heterojunctions and Metal Semiconductor Junctions*, Academic Press NY, 1972.
92. Moline, R.A., and G.F. Foxhall, "Ion-implanted hyperabrupt junction voltage variable capacitors," *IEEE Trans. Electron Devices*, **ED-19**, 1972, p. 267.
93. Morino, A., and T. Sugano, "Conversion loss of Schottky diode microwave downconverters," *Electronics and Comm. in Japan*, **54-C**, 1971, p. 68.
94. Mizuno, K., R. Kuwahara, and S. Ono, "Submillimeter detection using a Schottky diode with a long-wire antenna," *Appl. Phys. Lett.*, **26**, 1975, p. 605.
95. Nanavati, R.P., "On thermal and excess currents in GaSb tunnel diodes," *IEEE Trans. Electron Devices*, **ED-15**, 1968, p. 796.
96. Neuf, D., "A quiet mixer—a quadrature image enhancement technique for image recovery," *Microwave Journal*, May 1973, p. 29.
97. Neuf, D., "Varactor pair in new stripline circuit improves modulation," *Electronics*, July 31, 1972, p. 86.
98. Niiyama, H., S. Ogawa, T. Yagasaki, T. Araki, Y. Isii, and T. Imai, "Waveguide-type millimeter-wave diode," ECl-2173, *Electronics and Comm. in Japan*, **54-B**, 1971, p. 94.
99. Niiyama, H., and T. Imai, "Measurement of Esaki-diode series resistance," *Review of the Electrical Communication Laboratory*, Tokyo, **15**, 1967, p. 326.
100. Ohta, T., and H. Nakano, "Characteristics of reflected-type Schottky-barrier diode mixer," *Electronics and Comm. in Japan*, **54-B**, 1971, p. 101.
101. Orman, C., "Method for plotting frequency cutoff measurements for GaAs varactor diodes," *IEEE Trans. Electron Devices*, **ED-20**, 1973, p. 653.
102. Paffard, A.J., "O-band fast *p-i-n* diode switch," *Electronic Lett.*, **12**, May 27, 1976, p. 272.
103. Palmer, L.M., "The case for silicon tunnel diodes," *Electronic Products*, November 1, 1969, p. 33.
104. Pellegrini, B., "Negative diffusion capacitance in tunnel diodes," *Alta Frequenza*, **30**, 1970, p. 429.
105. Perkins, T.O., "Hot-switching high-power CW," *Microwave Journal*, May 1975, p. 59.
106. Pollard, R.D., M.J. Howes, and D.V. Morgan, "Consideration of accuracy in the determination of the capacitance-voltage law parameters of microwave diodes," *Proc. IEEE*, **63**, 1975, p. 323.
107. Polyakov, I.V., "Shaping of abrupt voltage steps in multistage circuits with tunnel diodes," *Radio Engineering and Electronic Physics*, **16**, 1971, p. 693.
108. Poulton, G., "Noise measure of metal-semiconductor-metal Schottky-barrier microwave diodes," *Electronics Lett.*, **7**, 1971, p. 667.
109. Rafuse, R.P., "Low noise and dynamic range in symmetric mixer circuits," *Proc. First Cornell Biennial Conference on Engineering Applications of Electronic Phenomena*, Ithaca, NY, 1967, p. 147.
110. Riccius, H.D., "High-frequency limitation of metal-insulator-metal point-contact diodes," *Appl. Phys. Lett.*, **27**, 1975, p. 232.

111. Roy, D.K., "Fabrication of GaAs tunnel diodes and detection of its recombination radiation," *Indian Journal of Pure and Applied Physics*, 6, 1968, p. 420.
112. Roy, D.K., "Technology of Esaki tunnel diodes," *J. Scientific and Industrial Research (India)*, 29, 1970, p. 310.
113. Sarnot, S.L., and P.K. Dubey, "Effect of band structure on the voltage current characteristics of metal-insulator-metal tunnel junctions," *Solid-State Electronics*, 15, 1972, p. 745.
114. Sato, Y., M. Uchida, Y. Ishibashi, and T. Araki, "Chip-type planar Schottky barrier diodes made from selectively grown GaAs," *Electronics and Comm. in Japan*, 55-C, 1972, p. 93.
115. Satoh, H., and T. Imai, "Degradation of Ge-doped and Zn-doped GaAs tunnel diode," *Japanese Journal of Applied Physics*, 7, 1968, p. 875.
116. Satoh, H., and T. Imai, "Reliability of GaAs junction diodes," *Review of the Electrical Communication Laboratory, Tokyo*, 18, 1970, p. 645.
117. Scanlan, J.O., and V.P. Kodali, "Characterisation of wave-guide-mounted tunnel diodes," *Proc. IEEE* (London), 114, 1967, p. 1844.
118. Shah, M.J., "Internal parasitics of the tunnel diode: theoretical explanation," *Proc. IEEE*, 58, 1970, p. 850.
119. Shen, L.Y.L., and J.M. Rowell, "Magnetic field and temperature dependence of the 'zero bias tunneling anomaly'," *Solid State Communications*, 5, 1967, p. 189.
120. Shevchenko, S.I., "Oscillations of the current in normal tunnel structures," *Soviet Physics–Semiconductors*, 7, 1973, p. 275.
121. Shewchun, J., M.A. Green, and F.D. King, "Minority carrier MIS tunnel diodes and their application to electron- and photo-voltaic energy conversion–II. Experiment." *Solid-State Electronics*, 17, 1974, p. 563.
122. Siegal, B., and E. Pendleton, "Zero-bias Schottky diodes as microwave detectors," *Microwave Journal*, September 1975, p. 40.
123. Siegal, B., "A practical guide to microwave semiconductors: Part I," *Microwave Systems News*, June/July 1972, p. 63.
124. Siegal, B., "A practical guide to microwave semiconductors: Part II," *Microwave Systems News*, August/September, 1972, p. 3.
125. Siegal, B., "A practical guide to microwave semiconductors: Part III," *Microwave Systems News*, October/November, 1972, p. 54.
126. Simonov, Yu. L., "Power and efficiency of a tunnel-diode oscillator," *Telecommunications and Radio Engineering, Pt. 2*, 21, 1966, p. 99.
127. Syrkin, L.N., and N.N. Feoktistova, "Effect of uniform pressure on the tunnel and excess currents in tunnel diodes," *Soviet Physics–Semiconductors*, 6, 1973, p. 1553.
128. Sze, S.M., *Physics of Semiconductor Devices*, Wiley-Interscience, NY, 1969.
129. Szente, P.A., S. Adam, and R.B. Riley, "Low barrier Schottky-diode detectors," *Microwave Journal*, 19, 2, February, 1976, p. 42.
130. Tabalski, J., E.S. Chambers, and J.D. Kinder, "Tunnel diode switching induced by transient ionizing radiation," *IEEE Trans. Nuclear Science*, NS-14, 1967, p. 221.
131. Taylor, R.J., and C.R. Westgate, "An evaluation of point contact tunnel diodes as microwave circuit elements," *1968 G-MTT International Microwave Symposium Digest and Technical Program*, Detroit, 1968, p. 179.
132. Tenenholtz, R., "Broadband MIC multithrow *PIN* diode switches," *Microwave Journal*, 16, July, 1973, p. 25.
133. Torrey, H.C. and C.A. Whitmer, *Crystal Rectifiers*, McGraw-Hill, NY, 1948.
134. Turner, R.J., "Broadband *p-i-n* attenuator has wide input dynamic range," *Electronics*, August 8, 1974, p. 106.

135. Uenohara, M., and J.S. Gewartowski, "Varactor applications," in: *Microwave Semiconductor Devices and Their Circuit Applications*, (ed. by H.A. Watson), McGraw-Hill, NY, 1969.

136. Ushirokawa, A., and M. Warashina, "Capacitance and field ionization in Esaki junction," *Electronics and Comm. in Japan*, 50, 1967, p. 111.

137. Ushirokawa, A., and M. Warashina, "Internal mechanisms and characteristics of Esaki diodes," *Electronics and Comm. in Japan*, 51-C, 1968, p. 89.

138. Veil, B.M., E.I. Zavaritskaya, and I.I. Ivanchik, "Dependence of the current-voltage characteristic of a tunnel diode on the Fermi levels in the $n$- and $p$-type regions," *Soviet Physics—Semiconductors*, 4, 1970, p. 268.

139. Vincent, B.T., Jr., "Integration techniques for microwave circuitry," *Proc. First Cornell Biennial Conference on Engineering Applications of Electronic Phenomena*, Ithaca, NY, 1967, p. 176.

140. Vincent, B.T., and M.E. Wallace, "Microwave integrated circuits for phased array applications," *Microwave Journal*, 12, 1969, p. 53.

141. Virk, S., "The state of tunnel diode technology," *Electronic Products*, November 1, 1969, p. 30.

142. Voznesenskii, N.N., B.M. Kolomytsev and L.N. Syrkin, "Tunnel diode in the generator condition as a piezosensitive element," *Instruments and Experimental Techniques*, 1, 1968, p. 137.

143. Watson, H.A., editor, *Microwave Semiconductor Devices and Their Circuit Applications*, McGraw-Hill, NY, 1969.

144. Weinreb, S., and A.R. Kerr, "Cryogenic cooling of mixers for millimeter and centimeter wavelengths," *IEEE J. Solid-State Circuits*, SC-8, 1973, p. 58.

145. White, C.E., "Microwave market forecast for 1978," *Microwave Journal*, 18, June 1975, p. 21.

146. Wortmann, A.K., and E.E. Kohn, "A monolithic integrated Schottky diode for microwave mixers," *Solid-State Electronics*, 18, 1975, p. 1095.

147. Wortmann, A., K. Heime, H. Beneking, "A new concept for microstrip-integrated GaAs Schottky diodes," *IEEE Trans. Electron Devices*, ED-22, 1975, p. 198.

# 4
# Bipolar Junction Transistors

## CONTENTS

## 4.1 GENERAL CHARACTERISTICS

Bipolar transistors are *npn* or *pnp* structures where the current flow across the base region is by minority carriers injected from an emitter region. This minority carrier flow is collected at a reverse-biased collector junction. The energy diagram for an *npn* transistor without bias is given in Fig. 4.1(a). With forward bias on the emitter-base junction, electrons are injected into the base and by diffusion make their way to the base-collector region which is reverse-biased so that the electric field in the depletion region aids the collection, as in Fig. 4.1(b). If the base region is graded in doping, as in Fig. 4.1(c), the transport

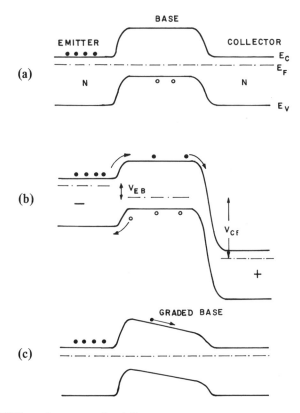

Fig. 4.1. *NPN* transistor energy-band diagrams:
(a)  With uniform base doping and zero bias voltage,
(b)  With emitter and collector bias voltages, and
(c)  With graded base doping (shown for the zero bias condition).

of electrons across the base is aided by the electric field there and the frequency response of the transistor is increased.

Transistor structures may be fabricated in various ways. Alloying of the emitter and collector regions was an early process much used for Ge transistors. This did not allow thin base thicknesses and small well-controlled dimensions, so base junctions grown during pulling of crystals were then used, particularly for Si transistors. Diffusion processes were then developed and diffusions from both sides of a wafer gave the kind of profile shown in Fig. 4.2(a). Successive $p$ and $n$ diffusions from the same side of the wafer, as in Fig. 4.2(b), were then found to give better control of thin base thicknesses and the emitter area could be defined by mesa etching, Fig. 4.2(c). Then $SiO_2$ was found to be easily grown and patterned using photoresist techniques, and used as a diffusion mask. This allowed a planar technology to be developed with surface metallization over

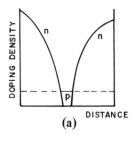 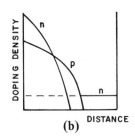

(a)            (b)

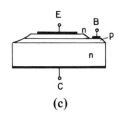 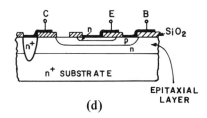

(c)            (d)

Fig. 4.2. *NPN* transistor construction:
(a) By diffusion from both faces of a wafer,
(b) By *p* and *n* diffusions of different profiles from the same face of the wafer,
(c) Mesa-type structure, and
(d) Planar-type structure with SiO$_2$ masking.

the oxide. The version shown in Fig. 4.2(d) shows the collector region to be a thin epitaxially-grown Si layer on an $n^+$ Si substrate. The lightly-doped epi-layer allows a relatively wide collector depletion region but is thin enough not to add a significant series resistance. More recently ion-implantation techniques have led to excellent doping control and tenth-micron base thicknesses.

The emitter region of a transistor is more heavily doped than the base and collector regions. The reason for the emitter-to-base doping ratio is to insure that most of the emitter current is carried by injection of electrons into the base region and that relatively few holes from the base are injected back into the emitter. Since the barrier heights are equal for both processes [see Fig. 4.1(b)] the ratio of the two injection processes is determined primarily by the free carrier concentrations available for injection. However, as shown in Fig. 4.3 the injection profiles are different: In the base region the collector gives a nearly zero density of electrons at the depletion edge $(x=W)$ and so the profile is nearly linear with distance across the base. In a simple model it is the sum of two exponentials $[A \exp \dfrac{-x}{L_n} + B \exp \dfrac{+x}{L_n}]$ and this sum appears nearly linear since the maximum value of $x$ is $W$ and $W \ll L_n$. The small number of holes injected from the base into the emitter has more nearly a single

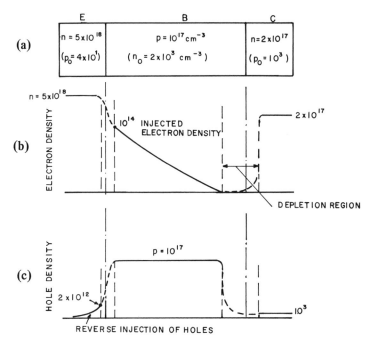

Fig. 4.3. Typical carrier densities in an *npn* transistor:
(a)  Majority and minority carrier levels in emitter, base and collector for Si,
(b)  Electron densities under injection conditions, and
(c)  Hole densities under the same base conditions.

exponential form since it has a greater distance to develop in before natural recombination reduces the excess hole density to zero. The emitter injection efficiency $\gamma$ [defined as $I_n/(I_n + I_p)$] may then be shown to be

$$\gamma = \frac{1}{1 + \dfrac{D_p P_{n0}}{D_n n_{p0}} \left(\dfrac{W}{L_p}\right)} \tag{4.1}$$

Then using the Einstein relationship between diffusion and mobility

$$\gamma = \frac{1}{1 + \left(\dfrac{\rho_{\text{Emitter}}}{\rho_{\text{Base}}}\right) \left(\dfrac{W}{L_p}\right)} \tag{4.2}$$

which shows the importance of the resistivity ratio.

The transport of the injected electrons by diffusion across the base region involves a further loss term $\beta_{rec}$ because of minority carrier recombination. Hence the dc current transfer ratio of the transistor, $\alpha$, is

$$\frac{I_C}{I_E} = \gamma\beta_{rec} \tag{4.3}$$

and simple modeling shows that

$$\beta_{rec} = 2/\left(e^{W/L_n} + e^{-W/L_n}\right)$$

$$= \left[\cosh\left(\frac{W}{L_n}\right)\right]^{-1} \tag{4.4}$$

$$\simeq 1/\left(1 + \frac{1}{2}\left(\frac{W}{L_n}\right)^2\right) \tag{4.5}$$

By way of an example, consider a Si transistor with the $5 \times 10^{18}$ cm$^{-3}$ and $10^{17}$ cm$^{-3}$ doping ratio of Fig. 4.3 and $W/L_p$ equal to unity and $W/L_n$ equal to 0.1. Then the injection efficiency $\gamma$ is 0.993, $\beta_{rec}$ is 0.995 and $\alpha$ is 0.988.

The characteristics and circuit configurations for an *npn* transistor are shown in Fig. 4.4. The equation relating dc collector current and emitter current for the common-base circuit configuration is

$$I_C = \alpha I_E + I_{C0} \tag{4.6}$$

where $I_{C0}$ is the reverse leakage current of the collector-base junction. If the transistor is arranged in the common-emitter configuration, as in Fig. 4.4(b), then $I_B$ is the independent variable and use of the expression $I_B = I_E - I_C$ in Eq. (4.6) gives

$$I_C = \left(\frac{\alpha}{1-\alpha}\right)I_B + \frac{I_{C0}}{1-\alpha} \tag{4.7}$$

The term $\alpha/(1-\alpha)$ is given the symbol $\beta_{DC}$ or $h_{FE}$ for the common-emitter forward dc current gain. An $\alpha$ of 0.988, as in the earlier numerical example, corresponds to a common-emitter current gain of 82. Hence, the input power for the common-emitter circuit configuration is much less than for common-base arrangement and the power gain is correspondingly higher. The common-collector (emitter-follower) configuration has a voltage gain of slightly less than unity as may be seen by inspection. The input impedance is therefore high and the circuit is of interest primarily for impedance step-down purposes.

From Eq. (4.7) it is seen that if the base is open-circuited ($I_B = 0$) the collector current is $I_{C0}$ $(1-\alpha)$. Hence, there is a multiplication of $I_{C0}$ by the term $1/(1-\alpha)$. Physically this comes about because in an *npn* structure

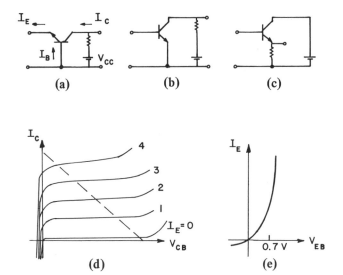

Fig. 4.4. Circuit configurations and characteristics of an *npn* transistor:
(a)  Common-base connection,
(b)  Common-emitter connection,
(c)  Common-collector (emitter-follower) connection,
(d)  Common-base $I_C$-$V_{CB}$ curves, and
(e)  Emitter-base curve.

the thermally generated holes associated with $I_{C0}$ cannot exit at the open-circuited base lead so they bias the emitter-base junction slightly in the forward direction; this allows holes to exit through the emitter but also causes injection of electrons in much greater number from the emitter into the base and thus the current rises to $I_{C0}/(1 - \alpha)$. This multiplication effect may be used in phototransistors to obtain gain.

The relationship between $I_E$ and $V_{BE}$ is of the form $I_E = I_{E0} [\exp(q V_{BE}/nkT) - 1]$ where $n$ is 2 or somewhat less. The temperature dependence of $V_{BE}$ for a given $I_E$ is a factor that must be watched in the arrangement of dc bias circuits for a transistor. If $I_C$ is measured instead of $I_E$ the relationship observed tends to be of the form

$$I_C = I_{C0} [\exp(q V_{BE}/n'kT) - 1] \tag{4.8}$$

where $n'$ is close to unity since the base current takes care of the recombination current component that is associated with $n$ values larger than unity. However, in diffused transistors the $n'$ value may be 10 or 20% above unity and tends to increase with the current level for reasons connected with voltage built into the base region.

Turning now to small-signal characteristics, equivalent circuits of bipolar transistors need not be developed in detail since they will already be familiar. Three such circuits are shown in Fig. 4.5. The common-emitter hybrid parameter equations are

$$v_{be} = h_{ie}i_b + h_{re}v_{ce} \qquad (4.9)$$

$$i_c = h_{fe}i_b + h_{oe}v_{ce} \qquad (4.9a)$$

The current gain for a resistance load $R_L$ is

$$\frac{i_c}{i_b} = \frac{h_{fe}}{1 + h_{oe}R_L} \qquad (4.10)$$

The voltage gain is

$$A_V = -\frac{h_{fe}R_L}{h_{ie} + \Delta^h R_L} \qquad (4.11)$$

where $\Delta^h = h_{ie}h_{oe} - h_{re}h_{fe}$.

The input impedance is

$$R_{in} = h_{ie} - \frac{h_{re}h_{fe}R_L}{1 + h_{oe}R_L} \qquad (4.12)$$

and the output impedance is

$$R_{out} = \frac{h_{ie} + R_s}{\Delta^h + h_{oe}R_s} \qquad (4.13)$$

where $R_s$ is the signal source resistance.

If the common-emitter hybrid parameters are known in value, they may be converted to the common-base or common-collector parameters by algebraic conversion and used in equivalent expressions for gain and impedance. For instance, for $h_{ie}$ = 2000 ohms, $h_{re}$ = 16 X 10$^{-4}$, $h_{fe}$ = 49 and $h_{oe}$ = 50 X 10$^{-6}$ mho, the results are $h_{ib}$ = 40 ohms, $h_{rb}$ = 4 X 10$^{-4}$, $h_{fe}$ = −0.98 and $h_{ob}$ = 10$^{-6}$ mho. The transistor gains as a function of $R_L$ for the three circuit configurations are shown in Fig. 4.6. The input and output impedances vary with $R_L$ and $R_S$ as shown in Fig. 4.7(a) and (b).

The frequency variation of the transistor behavior can approximately be taken care of by using $h_{fe}$ in the form

$$h_{fe}(\omega) = h_{fe}(0)\frac{1}{1 + j\dfrac{\omega}{\omega_c}} \qquad (4.14)$$

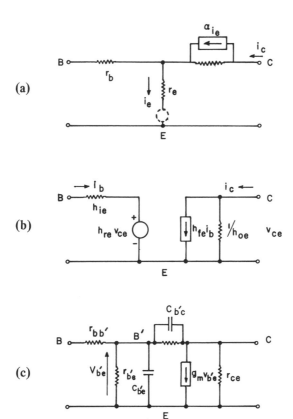

Fig. 4.5. Small-signal equivalent circuits for the common-emitter transistor configuration:
(a)  Tee equivalent circuit,
(b)  H parameter circuit, and
(c)  Hybrid $\pi$ equivalent circuit.

where $\omega_c$ is a cut-off frequency at which the short-circuit current gain is 3 dB down from the midband value.

However, for calculation of frequency response it is more usual to use the hybrid $\pi$ equivalent circuit as in Fig. 4.5(c). For a low-power bipolar transistor typical parameter values may be $r_{bb'} = 100$ ohms, $r_{b'e} = 1$ kilohms, $r_{b'c} = 4$ megohms, $r_{ce} = 80$ kilohms, $g_m = 50$ mA/V, $C_c = 3$ pF and $C_e = 100$ pF. The relationships to the $h$ parameters are $r_{bb'} = h_{ie} - r_{b'e}$, $r_{b'e} = h_{fe}/g_m$ where $g_m = I_c/(kT/q)$, $r_{b'c} = r_{b'e}/h_{re}$ and $r_{ce} = \left(h_{oe} - \dfrac{1 + h_{fe}}{r_{b'c}}\right)^{-1}$. The capacitance $C_c$ is the measured small-signal CB output capacitance with the emitter lead open, namely the collector-base reverse-biased depletion capacitance. The capacitance $C_e$ is essentially the forward-biased emitter-base junction diffusion capacitance.

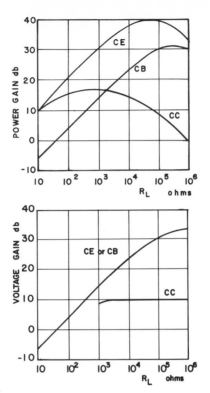

Fig. 4.6. Variation of power and voltage gain with load resistance for the three circuit configurations (calculated for $h_{ie} = 2000$ ohms, $h_{re} = 16 \times 10^{-4}$, $h_{fe} = 49$ and $h_{oe} = 50 \times 10^{-6}$mho).

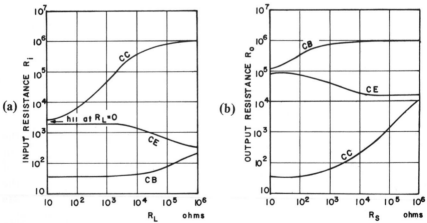

Fig. 4.7. Variation of the input and output resistance with resistance $R_L$ or $R_S$ for the three transistor circuit configurations:
(a) Input resistance dependence on load resistance, and
(b) Output resistance dependence on signal source resistance. (Calculated for the same transistor parameters as Fig. 4.6.)

For the hybrid $\pi$ parameters values given above, frequency analysis of a common-emitter circuit with a load of 2 kilohms gives a $-3$ dB frequency of about 3 MHz.

## 4.2 VOLTAGE RATING AND SECOND BREAKDOWN

As the collector-base voltage of a transistor is increased the field strength in the depletion region increases and the injected current passing through this region begins to undergo avalanche multiplication. Therefore,

$$I_c = \alpha \beta_{DC} M\, I_E \qquad (4.15)$$

where $M$ is a multiplication factor. Hence, the $h_{FE}$ of the transistor, $I_C/I_B$, approaches infinity and changes sign as $I_C$ equals and exceeds $I_E$. The voltage rating of a transistor is therefore usually chosen to be well below the value of $V_{CB}$ at which multiplication occurs. However, fast switching circuits were devised during the early uses of transistors that operated in this breakdown region. The collector-emitter avalanche breakdown voltage with $I_B = 0$ is rather less than the collector-base circuit voltage rating, as shown in Fig. 4.8(a), for

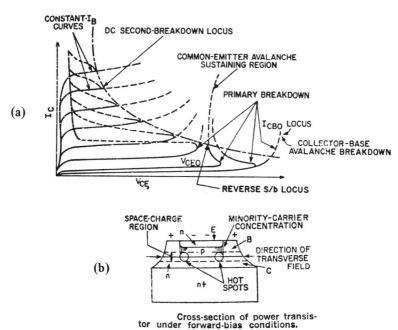

Fig. 4.8. Voltage limitations of transistors:
(a)   Avalanche and second-breakdown characteristics, and
(b)   Hot spots formed by minority carrier concentration. (Courtesy RCA Corporation.)

reasons connected with injection at the emitter. Also, a form of second-breakdown is seen to occur that considerably restricts the collector-emitter voltage rating under conditions of base drive.[141] The effect is thermal and is caused by hot spots developing in the base in regions of concentration of minority carrier injection as shown in Fig. 4.8(b). These regions of concentration under the emitter near its perimeter occur because the lateral current flow of recombination current from the center of the emitter to the base-lead causes a voltage drop that makes the perimeter more heavily biased than the center region. The local heating then produces the breakdown effect. The problem is reduced by designing the emitter and base contacts so they are interdigitated to maximize the perimeter-to-area ratio and to minimize the resistances of the lateral flow paths.

Another form of second-breakdown may occur under reverse bias conditions if the lateral voltage-drop effect changes sign because of reversal of base current flow and causes a hot spot to develop in the base under the center of the emitter.[130]

Other factors that can cause voltage limitations in transistors include fabrication defects such as diffusion spikes or inclusions at junctions that concentrate the field strength. The sketch in Fig. 4.9 gives an impression of diffusion spikes from the emitter that reduce the effective base thickness. As the depletion region, produced by the reverse bias across the collector-base junction, spreads into the base the voltage permitted is limited to $V_1$ due to the presence of the spikes. Anomolously high collector leakage currents and $h_{FE}$ degradation may also occur.

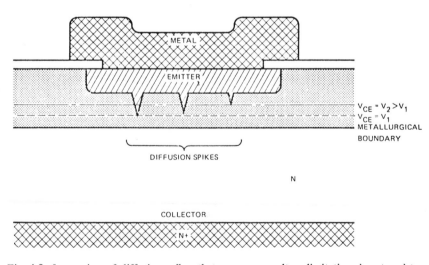

Fig. 4.9. Impression of diffusion spikes that may cause voltage limitations in a transistor. (After Wang and Kakihana. Reprinted with permission from *IEEE Trans. Electron Devices*, **ED-21**, 11, 1974, p. 669.)

Another possibility is that the $SiO_2$ in an integrated circuit may have charge locked in it, for instance, produced by $Na^+$ ions in the insulating $SiO_2$ layer, and that this may induce an accumulation layer in the underlying $n$-epitaxial Si layer. This, then, can introduce a premature breakdown path as suggested by the line $A'$-$A$ in Fig. 4.10.

The interaction between the effective base thickness of a transistor and the collector-base voltage can be minimized by having the collector more lightly doped than the base so that the depletion region spreads more in the collector than in the base. Base thickness reduction, however, can still limit the rating by producing depletion reach-through (punch-through).

Several other effects of base-thickness modulation have been discussed by Early.[48] One such effect is that the collected current increases as $V_{CB}$ increases, as shown in Fig. 4.11(a), because the reduced base thickness increases the emitter injection efficiency. Projections of these lines backward intercept the voltage axis at a value $V_E$ known as the Early intercept voltage. The collector impedance is defined as

$$r_c = \frac{dV_{CB}}{dI_C} \Big|_{I_E}$$

$$= \frac{1}{I_E} \frac{dV_{CB}}{d\gamma} \Big|_{I_E} \qquad (4.16)$$

From Eq. (4.1), $dW/d\gamma$ can be obtained and from the depletion thickness variation with voltage, a $dW/dV_{CB}$ expression may be derived. For an abrupt junction with the depletion region entirely in the base the collector impedance expression finally obtained from Eq. (4.16) is

$$r_C = \frac{1}{I_E} \left(\frac{\rho_B L_P}{\rho_E}\right) \frac{V_{CB}^{1/2}}{(\epsilon/2\,N_B q)^{1/2}} \qquad (4.17)$$

The collector impedance, therefore, is inversely proportional to $I_E$ and proportional to $(V_{CB})^{1/2}$. The recombination term $\beta_{DC}$ [see Eq. (4.5)] also depends on $W$ and if significant should also have been considered in arriving at $r_C$.

A further effect described by Early (1952) is that if the output current is increased by $\Delta I_C$ in a transistor which has a load resistance $R_L$, the voltage $V_{CB}$ decreases by $R_L \Delta I_C$. Hence, the effective base thickness increases by $\Delta W$ and the injected electron density at the emitter side of the base region must be increased by $\Delta n$, as in Fig. 4.11(b). This requires an increment $\Delta V_{EB}$ of emitter-base voltage (to satisfy the Boltzmann relationship) that is larger than would be needed in the absence of the voltage change $\Delta V_{CB}$. Hence, the effect is as though there is a feedback voltage from the load to the input side of the

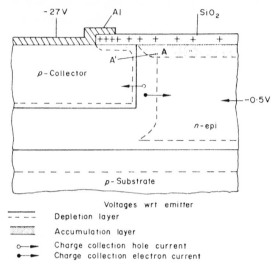

Fig. 4.10. Cross section of the collector junction of a planar transistor showing a premature voltage breakdown path AA′ formed by Na⁺ ions in the SiO₂. (After Last and Lucas.[100] Reprinted with permission from *Solid-State Electronics*, Pergamon Press Ltd.)

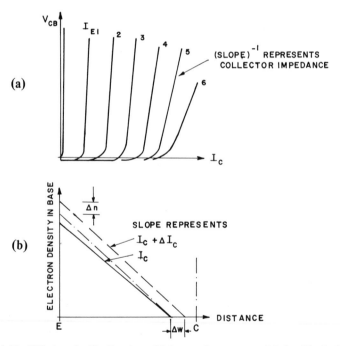

Fig. 4.11. Effects of effective base-thickness changes associated with depletion width changes:
(a) Characteristic slope changes with $I_E$, and
(b) The base-thickness increase $\Delta W$ produced by $R_L \Delta I_c$ requires an injected carrier increment $\Delta n$ and therefore an increase in the voltage $V_{EB}$.

transistor. The effect must be provided for in the small-signal equivalent circuit by a reverse feedback factor such as $h_{re}$ [see Fig. 4.5(b)].

An $n^{+}pn$ structure is shown in Fig. 4.12 that contains an added $p$ type region in the collector that is known as a lock layer. Although not generally used, this has some attractive features such as buffering the effective base thickness against change of voltage and increasing the usable collector-emitter breakdown voltage.[170]

## 4.3 FACTORS CONTROLLING THE CURRENT GAIN

The current gain of a transistor is low at very small collector currents because of dominance of recombination currents, but becomes approximately constant at moderate currents, presumably being limited by emitter injection efficiency $\gamma$, and declines at very high currents because of high-level injection effects.

For modern planar transistors the gain at moderate currents

$$\beta_\gamma = \gamma/(1 - \gamma) \qquad (4.18)$$

is not satisfactorily fitted by the injection efficiency expression [see Eq. (4.1)]. To secure a tentative fit the diffusion length $L_P$ in the emitter (assumed $n$ type) must be chosen as very small and then the recombination at low current levels predicts a $\beta$ fall-off with decreasing current that is faster than actually observed. Furthermore the temperature dependencies of $\beta$ are not right.

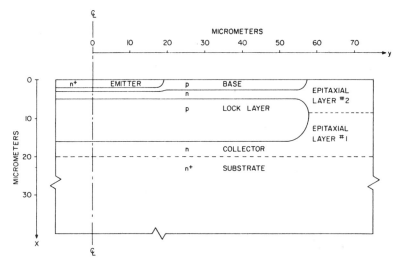

Fig. 4.12. Concept of a lock layer to buffer the effective base-thickness against change with collector-base voltage. (After Warner and Grung.[170] Reprinted with permission from *Solid-State Electronics*, Pergamon Press Ltd.)

The factors that have been neglected are those associated with the very heavy doping of the emitter region. This produces a bandgap decrease and band tailing or impurity band formation. A parameter that helps characterize this is the effective intrinsic carrier concentration $n_{ie}$ that increases with increasing doping level and with increasing doping compensation. The equation for the hole current in the emitter becomes

$$I_p = A \left[ \frac{kT}{q} \mu_p p q \left( \frac{1}{n_{ie}^2} \frac{dn_{ie}^2}{dx} - \frac{1}{N} \frac{dN}{dx} \right) - qD_p \frac{dp}{dx} \right] \qquad (4.19)$$

or

$$I_p = A \left( q\mu_p p \& - qD_p \frac{dp}{dx} \right) \qquad (4.20)$$

where $N = N_D - N_A$ and $\& = (kT/q)[1/n_{ie}^2)(dn_{ie}^2/dx) - (1/N)(dN/dx)]$. From Eqs. (4.19) and (4.20) it can be seen that the heavy doping effect modifies the electric field $\&$ acting on the minority carriers and this influences the transport in the emitter. Further development results in an expression for the current gain

$$\beta_\gamma = \frac{\int_0^{x_{eb}} \dfrac{N_D - N_A}{D_p(x)} \left( \dfrac{n_i}{n_{ie}} \right)^2 dx}{\int_{x_{eb}}^{x_{bc}} \dfrac{N_A - N_D}{D_n(x)} dx} \qquad (4.21)$$

For the doping profiles of Fig. 4.13(a), the calculated emitter-efficiency limited current gains are the solid lines of 4.13(b). The dashed lines represent the gains calculated in a conventional way. Obviously the heavy doping effect decreases the current gain but makes it much less dependent on the minority carrier lifetime in the emitter.

The mechanisms of bandgap narrowing, Shockley-Read-Hall (SRH) recombination, Auger recombination and carrier scattering have been considered in a recent study of a one-dimensional Si transistor model. The doping assumed was that of Fig. 4.14(a) and the theoretical and experimental results for current gain variation are shown in Fig. 4.14(b). The relative effects of the physical mechanisms involved can be seen from Fig. 4.15.

Consider now large collector currents so that the injection level in the base is high. The conductivity modulation that results gives increased base-defect recombination current and decreased emitter efficiency because the reverse injection increases with the buildup of base charge. Another effect is that the drift fields established by the doping gradients are swamped out and there is an apparent increase (approximately doubling) of the injected carrier diffusion coefficient. Emitter current-crowding of injection also tends to occur at the emitter perimeter because of lateral voltage drop effects in the base. Another

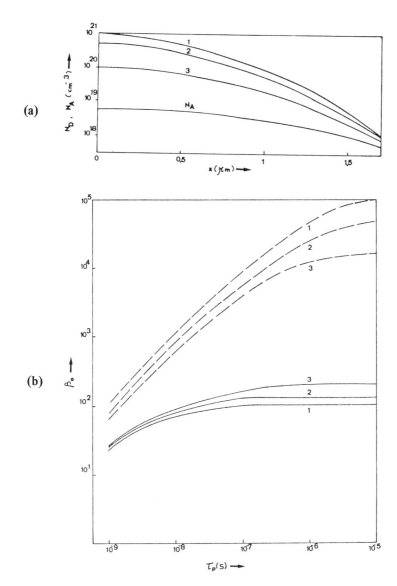

Fig. 4.13. The current gain of a transistor is affected by the emitter doping and minority carrier lifetime since these influence the emitter injection efficiency:
(a)  Emitter doping profiles $N_D$ used in the calculations, and
(b)  Emitter-efficiency-limited current gain calculated for various minority carrier hole lifetimes in the emitter. The dashed lines are the gain variations from more elementary theory that neglects heavy doping effects. (After Mertens, Deman, Van Overstraeten. Reprinted with permission from *IEEE Trans. Electron Devices*, **ED-20**, 1973, pp. 773, 774.)

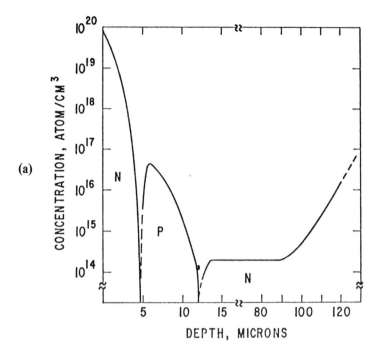

(a)

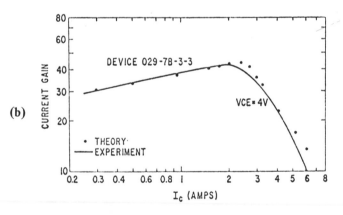

(b)

Fig. 4.14. Variation of current gain with collector current for a Si *npn* transistor:
(a)  Doping profiles assumed, and
(b)  Calculated current gain variation including emitter SRH traps and band-gaps narrowing
     effects. (After Adler et al. Reprinted with permission from *International IEEE Elec-
     tron Devices Meeting Technical Digest*, 1975, p. 178.)

WITH EMITTER TRAPS

**(a)**

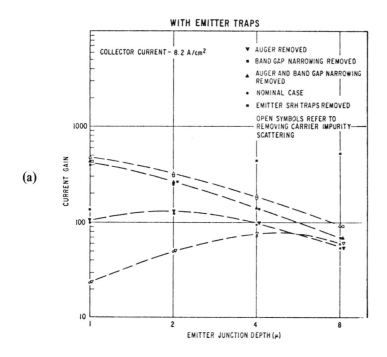

**(b)**

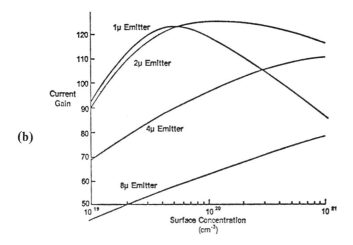

Fig. 4. 15. Various effects that control the current gain in the transistor of Fig. 4.14:

(a)  Shows the relative importance of effects such as Auger recombination, band-gap narrowing, emitter SRH traps and carrier impurity scattering, and

(b)  Effect of emitter surface concentration on the gain. (After Adler et al. Reprinted with permission from *International IEEE Electron Devices Meeting Technical Digest*, 1975, p. 179.)

important action is base-thickness increase caused by the mobile carriers in the base-collector depletion region. The undepleted collector region has increased voltage drop at high current levels and this may also modify the depletion width at low collector-base voltages.

Consider initially an *npn* transistor that is free of current crowding and of base-thickening effects. For such a transistor it is possible to demonstrate that $\beta(=h_{FE})$ should fall off approximately inversely with collector current. Begin with

$$h_{FE} = \beta = \frac{\beta_{DC}\,\gamma}{1 - \beta_{DC}\,\gamma} \tag{4.22}$$

where the base transport factor $\beta_{DC} = 1 - \delta$ and $\gamma = 1/(1 + \xi)$. Hence,

$$h_{FE} = \frac{1}{\delta + \xi} - \frac{\delta}{\delta + \xi}$$

$$\simeq \frac{1}{\delta + \xi} \tag{4.23}$$

The base defect term from Eq. (4.5) is

$$\delta = \frac{W_B{}^2}{2L_{nB}{}^2} \tag{4.24}$$

and the base injection term is

$$\xi = \frac{(qp_E D_{pE}')/(L_{pE}')}{(qn_B D_{nB}')/W_B} \tag{4.25}$$

Many assumptions are implicit in the use of Eqs. (4.24) and (4.25). Nevertheless let us proceed.

The base injection term may be rearranged as

$$\xi = \left(\frac{D_{pE'}}{D_{nB}}\right)\left(\frac{Q_{BO}}{Q_{EO}}\right) \tag{4.26}$$

where $Q_{BO}$ and $Q_{EO}$ are the majority (ionized impurity) charges in the base and emitter. When conductivity modulation occurs the majority charge in the base increases and the expression becomes

$$\xi = \left(\frac{D_{pE}}{D_{nB}}\right)\left(\frac{Q_{BO} + \Delta Q_B}{Q_{EO}}\right) \tag{4.27}$$

where the total charge in the base is related to the collector current by the equation

$$I_C = \frac{\Delta Q_B}{\tau_B} = \frac{2D_{nB}{'}\,\Delta Q_B}{W_B^{\,2}} \,, \tag{4.28}$$

where $\tau_B$ is the transit time in the base. Hence, the approximate relationship between $h_{FE}$ and $I_C$ is

$$h_{FE}I_C = \frac{2D_{nB}^{\,2}\,Q_{EO}}{D_{pE}\,W_B^{\,2}} \tag{4.29}$$

The current gains for $n^+pn^-n^+$ and $n^+pp^-n^+$ double-diffused transistors shown in Fig. 4.16 do exhibit variations approximately as $1/I_C$.

However if the effects of emitter crowding are included a two-dimensional analysis is needed, and such studies show

$$\beta \propto I_C^{-2}$$

The loss of current gain in the quasi-saturation region of a Si $n^+pn\,n^+$ power transistor may be seen from Fig. 4.17(a). Figure 4.17(b) illustrates the base-thickness increase that occurs as the current is increased. With base-thickness change the current gain in the quasi-saturated region $h_{FE}{'}$ is given by

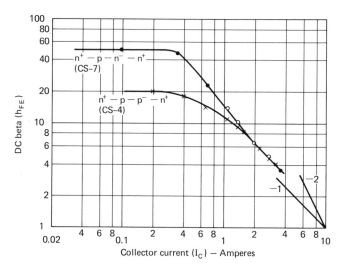

Fig. 4.16. Comparison of the performance of $n^+pp^-n^+$ and $n^+pn^-n^+$ transistors with identical geometry and structural dimensions. (After Olmstead et al., *RCA Review*, **32**, 1971, p. 235.)

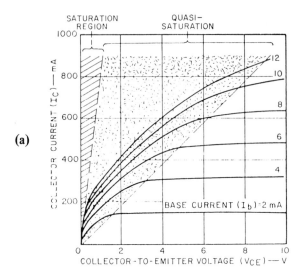

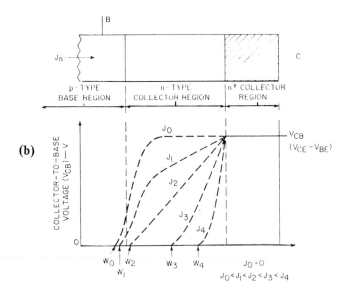

Fig. 4.17.  Some effects of increased current in a transistor:
(a)  Quasi-saturation region in a Si $n^+pn^-n^+$ transistor, and
(b)  Illustrates the base-collector voltage distribution for various collector current densities.
(After Olmstead et al., *RCA Review*, **32**, 1971, p. 227.)

$$h_{FE}' = \frac{W_B{}^2 h_{FE}}{\left(W_{BM} - \dfrac{V_{CB}}{J_C \rho_n}\right)^2} \qquad (4.30)$$

where $W_{BM} = W_B + W_n$.

Some results of an analytical study of base-thickening effects, and of the quasi-saturation region, are given in Fig. 4.18 to show the influence of $V_{CB}$ on the fall-off of $h_{FE}$ with $I_C$.

In a planar transistor the base current has three main components, as shown in Fig. 4.19(a) for a *pnp* structure. The intrinsic base current is $I_B{}^*$, the lateral current $I_L$ arises from the injection of the emitter into the side-wall region of the base and $I_R$ is the surface recombination current.

These quantities are expressed by

$$I_B^* = (1 - \alpha_N^*)I_1 = \frac{I_C}{\beta_N^*} \qquad (4.31)$$

$$I_L \simeq I_{SL} \exp\left(\frac{V_{EB}}{V_T}\right) \qquad (4.32)$$

$$I_R \simeq I_{SR} \exp\left(\frac{V_{EB}}{N_S V_T}\right) \qquad (4.33)$$

where $\beta_N^* = \dfrac{a_N^*}{1 - \alpha_N^*}$ is the intrinsic current gain, $I_1$ is the current crossing the

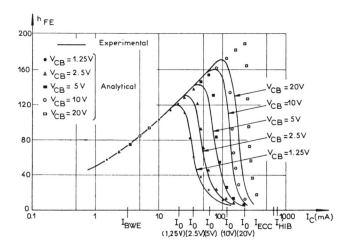

Fig. 4.18. The experimental and theoretical dependence of $h_{FE}$ on $I_C$ in the case of a quasi-saturation effect. (After Rey and Bailbe.[134] Reprinted with permission from *Solid-State Electronics*, Pergamon Press Ltd.)

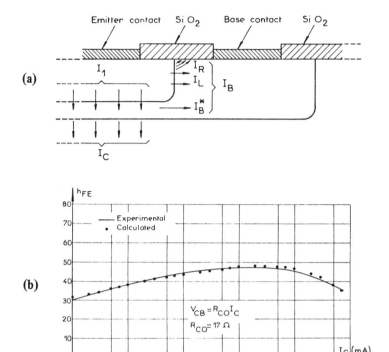

Fig. 4.19. The dependence of current gain on collector current for a Si transistor:
(a) Shows the main components of base current, and
(b) Dependence of $h_{FE}$ on $I_C$. (After Rey and Bailbe.[134] Reprinted with permission from *Solid-State Electronics*, Pergamon Press Ltd.)

intrinsic emitter base diode, $I_C$ is the collector current, $I_{SL}$ is the lateral emitter-base diode saturation current, $I_{SR}$ and $N_S$ are the parameters characteristic of surface current and $V_T$ is the thermal voltage.

At low bias conditions the surface recombination current may be dominant and then

$$\frac{1}{h_{FE}} \simeq \frac{I_R}{I_C} = \frac{I_{SR}}{(\alpha_{N0}^* I_{S10})^{1/N_S}} \cdot (I_C)^{1/N_S - 1} \qquad (4.34)$$

where $N_S$ is typically in the range 0.5-0.7. The correlation between $I_{SR}$ and the emitter-base junction perimeter $P_E$ and $N_S$ is of the form

$$I_{SR} \simeq P_E \gamma \cdot \exp\left(\frac{\xi}{N_S}\right) \qquad (4.35)$$

where $\gamma$ and $\xi$ are characteristics of the fabrication process. Typically $\gamma$ is $10^{-7}$-$10^{-8}$ A/$\mu$m and $\xi \simeq -30$.

A more general expression for $h_{FE}$ is

$$\frac{1}{h_{FE}}\bigg|_{V_{CB} = R_{C0}I_C} = \frac{1}{\beta_{N0}^*} + \frac{I_{SR}(1 + a/4)^{1/N_S}}{(\alpha_{N0}^* I_{S10})^{1/N_S}} (I_C)^{1/N_S - 1}$$

$$+ \frac{I_{SL}(1 + a/4)}{\alpha_{N0}^* I_{S10}} \tag{4.36}$$

where

$$a = \rho_B \frac{l I_B^*}{4nhW_B V_T} \tag{4.37}$$

and $n$, $l$ and $h$ represent the number, thickness and length, respectively, of the inter-digitated emitter fingers. Figure 4.19(b) shows the agreement obtained between theory and experiment with $\beta_{N0}^* = 52.3$ and $I_{SL} = 661 \times 10^{-20}$ A.

Although the analyses discussed have concentrated on $h_{FE}$ the dc current short-circuit gain, the dependence of the small-signal gain $h_{fe}$ on collector current is similar although not identical, as seen from Fig. 4.20(a). The variation of $h_{fe}$ with temperature for an npn double diffused Si mesa transistor is shown in Fig. 4.20(b). The decrease in $h_{fe}$ at low temperatures is partly explainable in terms of increased recombination current through traps in the emitter-base space charge region and the neutral base region. In shallow planar transistors tunneling currents through the emitter-base junction can cause $h_{FE}$ to decrease at low temperatures.[61]

Although for high injection-efficiency the emitter doping must be higher than the base doping this is not necessary for a structure $n^+npn$ where the emitter has an $n^+n$ form. Although carriers are injected from the $p$ type base into the $n$ region of the emitter, there is little hole current because the carriers are mostly reflected at the $n^+n$ barrier. Such a transistor with $n_E = 3 \times 10^{15}$ cm$^{-3}$ has been reported to have an $h_{FE}$ of 920 at $I_C = 1$ mA, $V_{CE} = 3$ V and an $f_T = 25$ MHz for a base thickness of 4 $\mu$m. One feature of an $n^+n$ emitter is that the base doping and thickness can be well controlled because emitter diffusion effects do not encroach into the base.[177] The n$^+$n emitter, however, does result in some loss of high frequency performance.

Modeling of bipolar transistors in one-dimensional forms has provided most of our understanding of device behavior but two-dimensional treatments have become common in the last few years.[112]

## 4.4 FREQUENCY PERFORMANCE AND MICROWAVE TRANSISTORS

### 4.4.1 First-Order Model of Frequency Response

Consider a *pnp* transistor with a uniformly-doped base region of thickness W. If the device is dc biased and a small ac signal of frequency $\omega$ is riding on the dc, the excess hole density in the base may be represented by

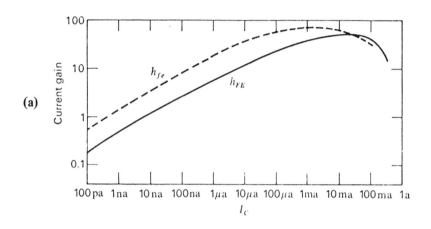

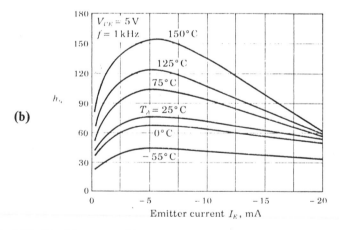

Fig. 4.20. Transistor curves of $h_{fe}$:
(a)  Curves showing the difference between $h_{fe}$ and $h_{FE}$ vs $I_c$ for a Si *pnp* transistor. (After Grove, *Physics and Technology of Semiconductor Devices*, John Wiley and sons, NY, 1967, p. 222.)
(b)  Variation of $h_{fe}$ with emitter current for the type 2N1573 Si transistor. (Courtesy of Texas Instruments, Inc.)

$$p = p_{DC}(x) + p_{AC}(x,\omega)e^{j\omega t} \tag{4.38}$$

The base transport equation, neglecting recombination, is

$$\frac{\partial p}{\partial t} = D_p \frac{\partial^2 p}{\partial x^2} \tag{4.39}$$

Substitution in this of Eq. (4.38) gives

$$P_{AC} = A \exp\left(\sqrt{\frac{j\omega}{D_p}} x\right) + B \exp\left(-\sqrt{\frac{j\omega}{D_p}} x\right) \tag{4.40}$$

and the boundary condition that $P_{AC} = 0$ at the depletion edge $W$ gives

$$\frac{A}{B} = -\exp\left(-2\sqrt{\frac{j\omega}{D_p}} W\right) \tag{4.41}$$

Then the ratio of the ac flows at the collector and the emitter is

$$\frac{i_c}{i_e} = \text{sech}\left(\sqrt{\frac{j\omega}{D_p}} W\right) \tag{4.42}$$

$$= \text{sech}\left[(1+j)\sqrt{\frac{\omega}{2D_p}} W\right]$$

$$\simeq \frac{1}{1 + \dfrac{j\omega}{2D_p} W^2}$$

$$= \frac{1}{1 + j\dfrac{\omega}{\omega_{co}}} \tag{4.43}$$

where $\omega_{co}$ is $2D_p/W^2$. Hence, if $D_p$ for Si is 12.5 cm$^2$s$^{-1}$ and if the base thickness is 5 $\mu$m the angular cutoff frequency, at which the current ratio $i_c/i_e$ is 3 dB down, is $10^8$ rad/sec and the phase shift is then $-45°$. If a common-emitter circuit is under consideration, $h_{fe} = \beta = \dfrac{\alpha}{1 - \alpha}$, and the result is

$$h_{fe}(\omega) = h_{fe}(\omega \to 0) \left(\frac{1}{1 + j\dfrac{\omega}{\omega_{co}(1 - \alpha)}}\right) \tag{4.44}$$

The $-3$ dB frequency is then lower by a factor $(1 - \alpha)$ as shown in Fig. 4.21.

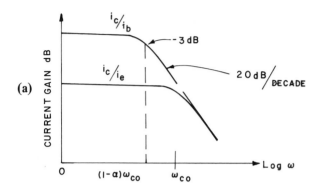

(a)

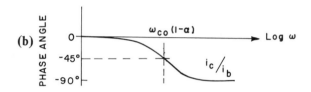

(b)

Fig. 4.21. Variation of current gains with signal frequency for the simple one-time-constant model of a transistor:
(a)  Gain variations with log $\omega$, and
(b)  Phase variation.

This elementary approach is of limited value in the calculation of actual device performance since few transistors have uniformly-doped base regions where the injected carrier flow is determined only by diffusion.

For a drift transistor the angular cutoff frequency expression is of the form

$$\omega_{co_\beta} = k(\epsilon)\frac{D_p}{W^2} \qquad (4.45)$$

where $k(\epsilon)$ is a function of the drift field $\&$ and where the parameter $\epsilon$ is

$$\epsilon = \frac{\mu_p \& W}{2D_p} \qquad (4.46)$$

The drift field $\&$ is arrived at from the doping gradient in the base region. The value of $k(\epsilon)$ depends upon the assumptions and simplifications made. Kroemer[94] in a simple treatment, for $\epsilon > 5$, obtains

$$k(\epsilon) = 2\epsilon^{3/2} \qquad (4.47)$$

Hence, a drift field can have a pronounced effect in increasing the transistor cutoff frequency.

Other studies result in more elaborate relationships that must be expressed graphically. The current transport factor $\beta$ of a drift transistor may be shown to be[39]

$$\beta = \sqrt{\left[\epsilon^2 + \frac{W^2}{L_p^2}\left(1 + j\omega\tau_p\right)\right]} \exp \epsilon \left/ \left(\epsilon \sinh \sqrt{\left[\epsilon^2 + \frac{W^2}{L_p^2}\left(1 + j\omega\tau_p\right)\right]}\right.\right.$$

$$\left.\left. + \sqrt{\left[\epsilon^2 + \frac{W^2}{L_p^2}\left(1 + j\omega\tau_p\right)\right]} \cosh \sqrt{\left[\epsilon^2 + \frac{W^2}{L_p^2}\left(1 + j\omega\tau_p\right)\right]}\right]\right) \quad (4.48)$$

Expanding the hyperbolic sine and cosine terms in the above expression, retaining five terms, and determining the frequency at which $|\beta|$ falls to 0.707 times its low frequency value, gives

$$k(\epsilon) = \left\{\frac{8\theta\epsilon^4}{\phi}\left[\sqrt{\left(1 + \frac{4\phi}{\theta^2}\right)} - 1\right]\left[1 + \gamma\frac{W^2}{L_p^2}\right]\right\}^{1/2} \quad (4.49)$$

where $\theta$, $\phi$ and $\gamma$ are functions of $\epsilon$. However, $\gamma W^2/L_p^2$ tends to be small compared with unity. Then Eq. (4.49) gives $k(\epsilon)$ values of 2.43, 8.35, 17.76, 29.69 and 43.7 for $\epsilon$ values of 0, 2, 4, 6 and 8, respectively. These are in good agreement with the values given by Eq. (4.47) if $\epsilon$ is large.

In the limit when $\epsilon \to 0$ Eq. (4.49) reduces to

$$k(0) = 2.43\left(1 + 0.40\frac{W^2}{L_p^2}\right) \quad (4.50)$$

which agrees reasonably well with studies for diffusion-limited base transport.

### 4.4.2 Frequency Response Dependence on $I_C$ Including the Base-Thickening Effect

The frequency response of a transistor is often characterized in terms of the common emitter $-3$ dB current gain point. However, another approach is to relate it to the frequency $f_T$ at which the common-emitter short-circuit current-gain $h_{fe}$ is unity. The frequency $f_T$ from Eq. (4.44) (which is an approximation) may be expected to be $h_{fe}(\omega \to 0)$ times $f_{co}(1-\alpha)$. However, for drift transistors $f_T$ depends on the collector current and on the collector voltage. The expression for the increase with collector current is

$$\frac{1}{f_T(V_c, J_c)} = 2\pi\alpha_0 \frac{kT}{qI_c} C_{TE} + \frac{1}{f_{T_{\max}}(V_c)} \qquad (4.51)$$

where the term $\alpha_0 kT/qI_c C_{TE}$ represents the phase delay associated with the emitter-junction transition region. Above a certain collector current, $f_T$ begins to decrease because of a buildup of minority-carrier space-charge in the collector transition layer of the transistor.

The current flow in the transition base-collector region of a *pnp* transistor is given by

$$\frac{d\mathscr{E}}{dx} = \frac{1}{\epsilon} \left[ q(N_D^+(x) - N_A^-(x)) + \frac{J_c}{v_p} \right] \qquad (4.52)$$

where $\epsilon$ is the dielectric constant and $J_c = qv_p p + qv_n n$.

In simple modeling of transistor action the collector transition layer problem is disposed of by assuming that the mobile charge carrier density is negligible compared with the fixed impurity-density in the layer. However, for a current density of, say, $10^3$ A cm$^{-2}$, $J = qv_{sat}p$ gives a minority carrier density of about $10^{15}$ cm$^{-3}$ which is comparable with the $10^{15}$-$10^{16}$ cm$^{-3}$ of the impurity doping space charge and so $J_c/v_p$ must be taken into account.

To simplify consideration of the problem it is usual to assume that the mobile carrier velocity through the transition region is a constant, $v_{sat}$, and then the minority carrier density throughout the region is $J_c/qv_{sat}$.

Solution of the transition region equation to determine the depletion edge position, as defined in Fig. 4.22, gives, according to Kirk,[91]

$$-x_1(V_{cb}, J_c) = x_m(V_{cb}, J_c) = x_{m0} \left\{ 1 + \frac{J_c}{J_1} \right\}^{-1/2} \qquad (4.53)$$

where

$$x_{m0} = \left\{ \frac{2\epsilon(-V_{cb} + V_0)}{qN_{db}} \right\}^{1/2} \qquad (4.54)$$

and

$$J_1 = qv_x N_{db} \qquad (4.55)$$

where $N_{db}$ is the base doping taken as a constant and the junction is assumed abrupt. At low current densities the depletion edge is at $-x_{m0}$ but at higher current densities the effective base thickness increases to $W_b + x_1$, as in Fig. 4.22(b).

The effect of this variation of the transition layer with $J_c$ on the cutoff frequency $f_T$ can be obtained by introducing the current-density-dependent behavior of $x_1$ into the expression for the total time delay across the transistor. The result is

$$\frac{1}{2\pi f_T} = \tau_{ec} = \tau_e + \tau_b + \tau_x + \tau_{cb} \qquad (4.56)$$

where

$$\tau_e = \alpha_0 \frac{kT}{qJ_c} C_{eb} \qquad (4.57)$$

$$\tau_b = \frac{W_b{}^2}{n D_p} \left\{ 1 - \frac{x_{m0}/W_b}{\sqrt{1 + J_c/J_1}} \right\}^2 + \frac{W_b}{v_x} \left\{ 1 - \frac{x_{m0}/W_b}{\sqrt{1 + J_c/J_1}} \right\} \qquad (4.58)$$

($n$ is between 2 and 5).

$$\tau_x = \frac{x_{m0}}{2v_{sc\ \lim}} \left\{ 1 + J_c/J_1 \right\}^{-1/2} \qquad (4.59)$$

and

$$\tau_{cb} = 0 \text{ (assuming } r'_c = 0) \qquad (4.60)$$

The theoretical change of $f_T$ with $I_c$ for an alloy transistor is shown in Fig. 4.22(c) together with the $f_T$ variation observed.

Since the analysis by Kirk,[91] several other studies have been made to further develop the concept of base push-out. In one such study[129] the structure is $n^+pn\ n^+$ where $n = 10^{15}$ cm$^{-3}$ and the electric field is determined as a function of

**(a)**

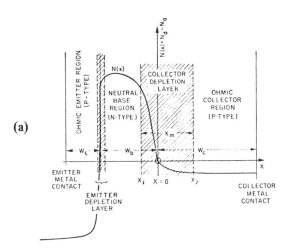

Fig. 4.22. Frequency response dependence on $I_c$ including the Kirk base-thickening effect:
(a)  One-dimensional $pnp$ transistor geometry considered,
(b)  Depletion region edge position depends on $V_{cb}$ and $J_c$, and
(c)  Variation of $f_T$ with $I_c$ for a 2N769 electrochemical-alloy $pnp$ transistor. (After Kirk. Reprinted with permission from *IEEE Trans. Electron Devices*, **ED-9**, 1962, pp. 165, 167, 169.)

distance for various collector currents. At low injection levels, a high-field region exists near the transition between base and epitaxial layer. At a current density of between $0.7 \times 10^3$ and $1 \times 10^3$ A cm$^{-2}$ the high field region moves to the interface between the epitaxial layer and the substrate. When this high-field relocation has occurred, the epitaxial layer acts as an extension of the base with an attendant increase in the delay time.

Some treatments involve two-dimensional models and introduce the concept of lateral base spreading. The base-thickening model is found to be successful in general, although base-spreading is significant in some transistors.[96]

The effects of emitter doping gradients on $f_T$ have been studied and related

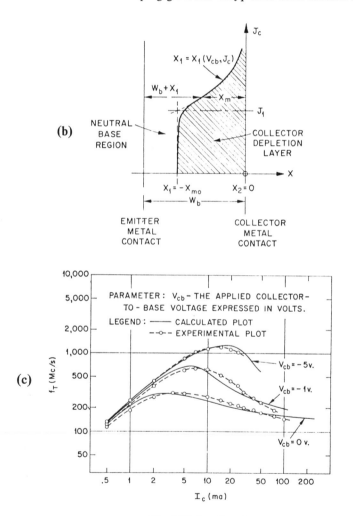

Fig. 4.22 (continued)

to the free-carrier storage in the emitter space-charge layer.[90] The doping gradient for a typical P diffused microwave emitter is $<3 \times 10^{23}$ cm$^{-4}$ whereas As diffusions give gradients of $10^{24}$ cm$^{-4}$. The general finding is that As-diffused $npn$ Si transistors have better values of $f_T$ and $\beta$ than P-diffused transistors.

### 4.4.3 Microwave Bipolar Transistors

In transistor technology the struggle for increased frequency response and power has been unremitting. The limitations in the early 1960s were those set by the minimum base thicknesses given by diffusion processes, by the minimum emitter strip widths possible with optical lithography and by the device parasitics. With the advent of ion-implantation doping and electron-beam lithography the base-thickness and strip-width limitations became less important and microwave transistor designs during the last decade have moved closer and closer in performance to the fundamental limits set by the material properties.

(a) *Fundamental Limits*    Semiconductor performance limits may be esti-mated by considering the unity-current-gain frequency $f_T$ to be $1/(2\pi\tau)$, where $\tau$ is the transit time across the emitter-collector distance with the minority carriers moving at the saturated drift velocity ($\sim6 \times 10^6$ cm/sec for Si). The drift velocity becomes saturated for field strengths above about $10^4$ V/cm. Reduction of the base thickness is advantageous in decreasing $\tau$, but the field strength for a given collector-emitter voltage is approximately $V_{ce}/W$ and must not exceed the breakdown field $\mathscr{E}_{crit}$ of $2 \times 10^5$ V/cm for Si. Hence,

$$V_{ce_{max}} f_T = \frac{\mathscr{E}_{crit} v_s}{2\pi} \qquad (4.61)$$

For Si this has a value of $2 \times 10^{11}$ V/sec, for Ge $1 \times 10^{11}$ V/sec and for GaAs $\sim5 \times 10^{11}$ V/sec.

In an actual transistor, the saturation drift velocity is not reached in all parts of the charge carrier path and the electric field is nonuniform and there must be a safety margin on $V_{ce}$. Hence, the limit of $2 \times 10^{11}$ V/sec is not likely to be reached in practice but at least serves as a measuring stick. Some results for $V_{CE}$ vs $f_T$ for Si and Ge transistors circa 1965 are shown in Fig. 4.23(a).

A relationship between current $I_{c_{max}}$ and frequency can be obtained by considering the limiting current density to be that determined by the onset of base-thickening, since this is a frequency-reducing effect. Thickening sets in when the mobile charge in the collector depletion region begins to be comparable with the fixed charge. Hence, if $\tau_c$ is the transit time through the collector depletion region

$$I_{c_{max}} \tau_c = C_c V_{ce} \qquad (4.62)$$

Substitution from Eq. (4.61) then gives

$$\frac{I_{c_{\max}}}{C_c} = \frac{\mathcal{E}_{\mathrm{crit}} \, v_s}{2\pi f_T \tau_c} \tag{4.63}$$

If $\tau_c$ can be taken as approximately the total transit time $\tau$ between emitter and collector (notice that rather sweeping assumptions are being made) then

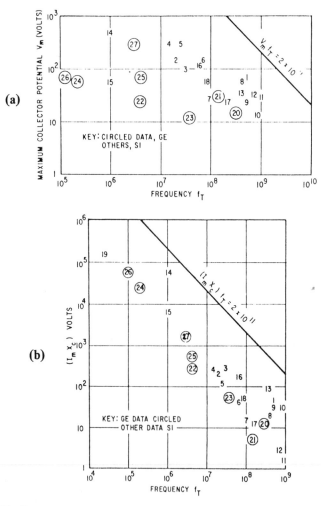

Fig. 4.23. Transistor voltage and current performance (about 1965) vs frequency $f_T$ compared with $2 \times 10^{11}$ ideal limit line:
(a) Voltage rating, and
(b) $I_{c_{\max}} X_c$ vs $f_T$. (After Johnson, *RCA Review*, 26, 1965, p. 163.)

$2\pi f_T \tau_c$ is unity and

$$\frac{I_{c_{max}}}{C_c} = \mathcal{E}_{crit}\, v_s \tag{4.64}$$

an expression that is independent of device area and other details of design. The ratio $I_{c_{max}}/C_c$ has the units of $v_s^{-1}$ and is a measure of the device's speed in switching.

Conversion of $C_c$ to reactance $X_c$ at the frequency $f_T$ gives

$$(I_{c_{max}} X_c)f_T = \frac{\mathcal{E}_{crit}\, v_s}{2\pi} \tag{4.65}$$

A plot of $I_{c_{max}} X_c$ vs $f_T$ for numerous transistors is shown in Fig. 4.23(b). The voltage-ampere rating or power handling potential of the transistor is

$$P_{max} = V_{ce_{max}} I_{c_{max}} \tag{4.66}$$

From Eq. (4.61) this may be written

$$P_{max} = I^2_{c_{max}} X_c \tag{4.67}$$

From Eq. (4.65)

$$(P_{max} X_c)^{1/2} f_T = \frac{\mathcal{E}_{crit}\, v_s}{2\pi} \tag{4.68}$$

So for transistors of a given impedance, such as 50 ohms for microwave uses, the power handling potential should decrease as $f_T^{-2}$. A power vs frequency graph for RF bipolar transistors (manufactured between 1965 and 1970) is shown in Fig. 4.24(a). The dashed line has a slope $f^{-2}$ and is placed at a position that has been estimated by one company as a conceivable limit for 3-dB power gain. The limit set by Eq. (4.68) is far from being reached, as seen from Fig. 4.24(b), and this might be expected because of the idealizations made in its derivation.

For increased power handling, devices may be connected in parallel and reasonably spaced to give good heat dissipation. Connection in parallel introduces problems of device uniformity and load-sharing, and lowers the impedance of the composite device.

Amplification properties may also be discussed in simplified fundamental terms. The current gain $G_I$ is greater than unity at a frequency less than $f_T$ and the simple form from Eq. (4.44) is

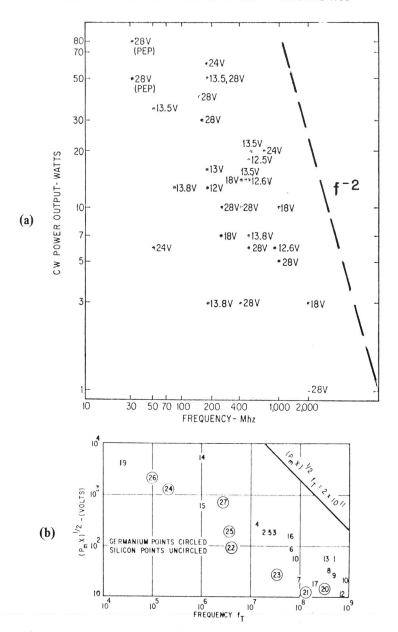

Fig. 4.24. Relationship of power output to the unity-current-gain frequency, $f_T$:
(a) For various European and Japanese transistors about 1970. The line $f^{-2}$ is for estimated practical 3-dB power gain, (After Magill. Reprinted with permission from *Electronics*, April 27, 1970, copyright, McGraw-Hill, Inc., NY 1970)
(b) $(P_m X_c)^{1/2}$ vs $f_T$ for transistors (about 1965) compared with the fundamental $2 \times 10^{11}$ limit. (After Johnson, *RCA Review*, **26**, 1965, p. 163.)

$$G_I = \frac{1}{2\pi f \tau} = \frac{f_T}{f} \qquad (4.67)$$

The power gain is therefore

$$G_p = G_I{}^2 \frac{Z_0}{Z_i} \qquad (4.68)$$

where $Z_o$ and $Z_i$ are the output and input impedances and can be represented as the output and input capacitances if resistive terms are neglected ($G_p$ is then of course not a real power expression). The input diffusion capacitance is given by

$$C_{in} = \frac{Q_{max}}{V_T} = \frac{I_{max}\tau_b}{V_T} \qquad (4.69)$$

where $\tau_b$ is the carrier transit time in the base and $V_T$ is the thermal voltage, $kT/q$. The output capacitance is

$$C_0 = C_c = \frac{I_{max}\tau_c}{V_{ce_{max}}} \qquad (4.70)$$

Hence from Eq. (4.68)

$$G_p = \left(\frac{f_T}{f}\right)^2 \frac{\tau_b}{\tau_c} \frac{V_{ce_{max}}}{V_T} \qquad (4.71)$$

$$\simeq \left(\frac{f_T}{f}\right)^2 \frac{V_{ce_{max}}}{V_T} \qquad (4.72)$$

This can be rearranged with the aid of Eq. (4.61) to become

$$(G_p V_{ce_{max}} V_T)^{1/2} f = \frac{\mathcal{E}_{crit} v_s}{2\pi} \qquad (4.73)$$

This again shows the importance of the volts/second rating of the material. The dynamic range of the transistor is determined by $V_{ce_{max}}$ and so for a given frequency there is a potential trade-off between power gain and dynamic range.

(b) *Some Microwave Transistor Designs*   All microwave transistors are planar in form and the bipolar units to be discussed here are almost all Si *npn* structures. As we have already seen, Ge does not have the same volts/second rating as Si and also does not have the natural passive oxide that is desirable in the fabrication of small structures. At present GaAs is not as well controlled in purity as Si, and processing bipolar GaAs transistors is even more difficult than processing Ge transistors.

The technology of Si *npn* microwave transistors now allows $f_T$ oscillation frequency limits of 15 GHz or higher and noise figures of 1 dB at 500 MHz and less than 6 dB at 6 GHz. The power output of power RF designs may be 100 W at 1 GHz, and 5 W CW (continuous wave) at 4 GHz with 4-dB gain. The technology of GaAs is still not advanced enough to compete with Si in bipolar applications although effective in field-effect transistor designs.

The transistor geometry illustrated in Fig. 4.25(a) shows emitter and base interdigitated to provide a large ratio of perimeter to area to reduce injection crowding and lateral resistance. For first-order design purposes the emitter current is often taken as 5-6 mA/100 $\mu$m of perimeter. For a 1-2 GHz transistor the emitter finger width may be 2.5 $\mu$m and the length 25-50 $\mu$m.

For higher power output the device may be divided into a number of emitter and base sites connected by overlay emitter and base metallizations, as in Fig. 4.25(b), or by mesh (also called emitter-grid or matrix) geometries, as in Fig. 4.25(c).

The maximum available gain $G_{MAG}$ from a transistor is obtained when the input and output are conjugately matched in impedance. The unilateral gain $G_U$ is that obtained with matched input and outputs and with a network that cancels feedback from output to input. Typically $G_U$ is 1-3 dB higher than $G_{MAG}$ and both go to unity at a frequency $f_{max}$.

The effective input resistance $r_b'$ and the collector capacitance $C_c$ are the principal factors in the simplified equivalent circuit of Fig. 4.26(a), and it may be shown that

$$f_{max} \simeq \sqrt{\frac{f_T}{8 \pi r_b' C_c}} \tag{4.74}$$

From Fig. 4.26(b) it should be noted that there are regions of potential instability in the gain vs frequency plots. Some of the distributed resistances and capacitances involved in a typical planar transistor geometry are shown in Fig. 4.27.

The total transit time is the sum of four terms, namely

$$\tau_t = \tau_e + \tau_b + \tau_c + \tau_{cb} \tag{4.75}$$

where the emitter-base junction time is

$$\tau_e = C_{eb} \frac{kT}{qI_c} \tag{4.76}$$

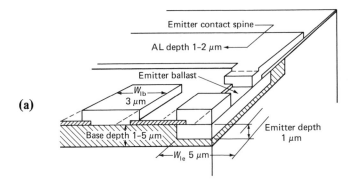

**(a)**

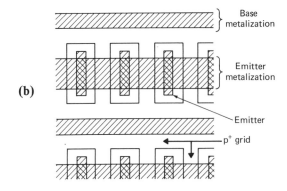

**(b)**

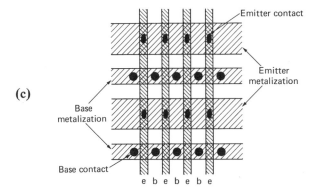

**(c)**

Fig. 4.25.  Typical Si power transistor geometries:
(a)   Interdigited emitter and base fingers, with emitter ballast resistor for load sharing,
(b)   Overlay emitter metallization, and
(c)   Mesh or matrix construction. (After Den Brinker.[45])

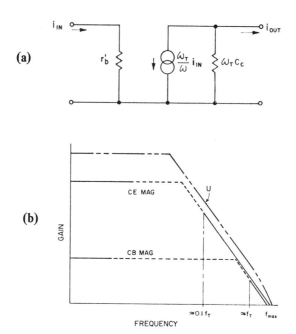

Fig. 4.26.  Gain variation with frequency for a microwave transistor:
(a)  Simplified equivalent circuit of a transistor for gain calculations, and
(b)  Maximum available gain (MAG) and unilateral gain (U) vs frequency. Broken line represents regions of potential instability. (After Cooke. Reprinted with permission from *Proc. IEEE*, 59, 1971, p. 1165.)

The base transport time is

$$\tau_b = \frac{Q_b}{I_c} \simeq \frac{W_b{}^2}{nD} \tag{4.77}$$

where $n$ is between 2 and 5.

The collector junction transport time is

$$\tau_c \simeq \frac{W_c}{2v_s} \tag{4.78}$$

where $W_c$ is the collector transition thickness and $v_s$ is the saturation drift velocity of the electrons.

The final term $\tau_{cb}$ is the collector bulk effect

$$\tau_{cb} = C_c r_c$$

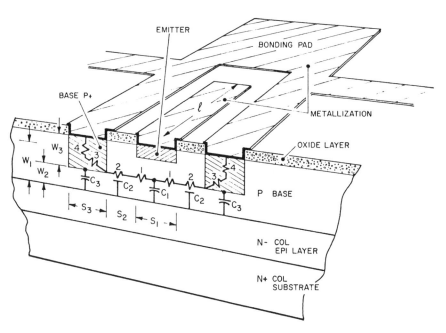

Fig. 4.27. Distributed resistances and capacitances in a typical planar transistor geometry; 1, 2, and 3 are parts of $r_b{}'$, and 4 is $R_c$ (contact resistance). (After Cooke. Reprinted with permission from *Proc. IEEE*, 59, 1971, p. 11.)

where $C_c$ is collector-base capacitance and $r_c$ is the bulk collector resistance through which the capacitance changes. In a typically proportioned microwave transistor the relative sizes of the terms are $\tau_e = 40\%$, $\tau_b = 10\%$, $\tau_c = 45\%$, and $\tau_{cb} = 5\%$.

The geometry of an overlay *npn* Si transistor with 24 P-diffused emitter strips, each $2.5 \times 41.5 \ \mu m$, is shown in Fig. 4.28(a) and the maximum available gain and equivalent circuit are given in (b) and (c).

Bipolar microwave transistors, although important for high power applications, are not able to match the very high frequency performance of GaAs field-effect transistors.

(c) *S Parameter Characterization* Microwave transistors, and other high frequency devices, are conveniently characterized by traveling wave transmission and reflection coefficients. The two port network of Fig. 4.29 can be characterized by the voltage equations

$$E_{1r} = S_{11} \, E_{1i} + S_{12} \, E_{2i} \qquad (4.79)$$

$$E_{2r} = S_{21} \, E_{1i} + S_{22} \, E_{2i} \qquad (4.80)$$

$S_{11}$ is therefore the reflection coefficient with the output matched and $S_{21}$ is the forward transmission coefficient under this condition. The ratio $S_{21}/S_{11}$ is the gain of the circuit. $S_{22}$ and $S_{12}$ are similar coefficients for the reverse flow condition. $S$ parameters are convenient to measure because (a) they involve incident and reflected voltage measurements for matched terminations usually

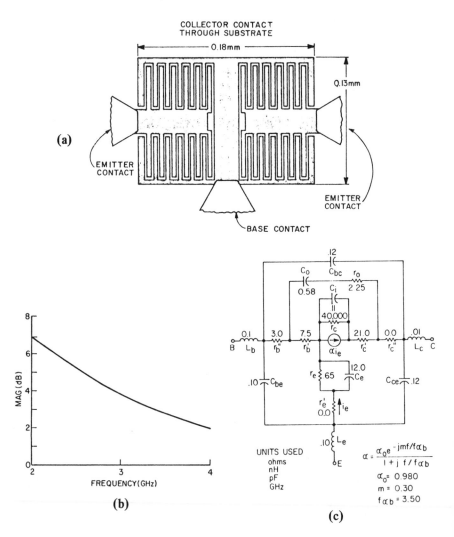

Fig. 4.28. Microwave *npn* Si transistor, frequency response and equivalent circuit:
(a) Geometry with 24 emitter stripes,
(b) Maximum available power gain vs frequency, and
(c) Equivalent circuit (at $I_E$ = 40 mA and $V_{CE}$ = 10 V). (After Shackle. Reprinted with permission from *IEEE Trans. Electron Devices*, **ED-21**, 1974, p. 33.)

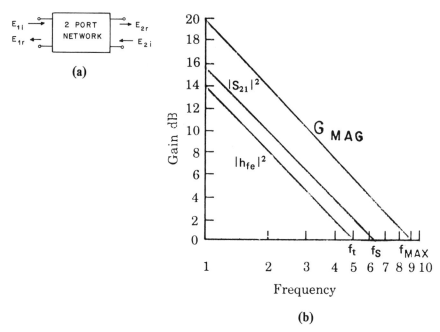

(a)

(b)

Fig. 4.29. $S$ parameter representation:
(a)  Forward and reflected waves for a 2 port network, and
(b)  $|S_{21}|^2$ for a transistor shown as a function of frequency and compared with $|h_{fe}|^2$ and $G_{MAG}$.

resistive, (b) they avoid the difficulties of open-circuit and short-circuit terminations at high frequencies and (c) parasitic oscillations that might disturb the
measurements are minimized.

A frequency $f_s$ can be defined at which the power gain $|S_{21}|^2$ is unity (zero
dB). With the transistor base driven by a 50-ohm voltage source and the collector
terminated in a 50-ohm load, $|S_{21}|$ may be measured and to obtain $f_s$ the gain
$|S_{21}|^2$ is plotted vs frequency and extrapolated to 0 dB (the fall-off should be
about 6 dB/octave).

This measurement is easier than attempting a short-circuit current gain
measurement to obtain $f_T$, which typically is a little smaller than $f_s$. With
conjugate matching at the input and output the line $G_{MAG}$ is obtained. In
actual circuit operation with proper matching most transistors can have useful
gain up to a frequency of $f_s$.

The $S$ parameters are conveniently displayed on a Smith chart. Fig. 4.30(a)
shows the plot of $S_{11}$ for a common-emitter Si RF transistor both in chip form
and as a packaged device. At low frequencies the reactance of the packaged
device is capacitive and as the frequency is increased the line moves into the
inductive reactance region, the equivalent circuit being as shown in Fig. 4.30(b).

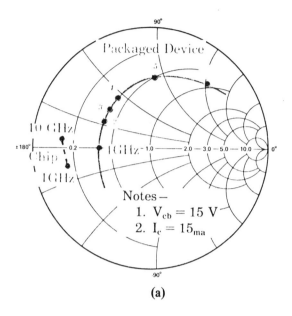

(a)

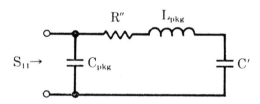

Input Equivalent Circuit (Package)

(b)

Fig. 4.30. Transistor $S_{11}$ and $S_{22}$ on a Smith chart display:
(a)  $S_{11}$ for chip and packaged device,
(b)  Equivalent circuit for $S_{11}$,
(c)  $S_{22}$ for chip and packaged device, and
(d)  Equivalent circuit for $S_{22}$. (Courtesy Hewlett Packard Company, Application Note 154.)

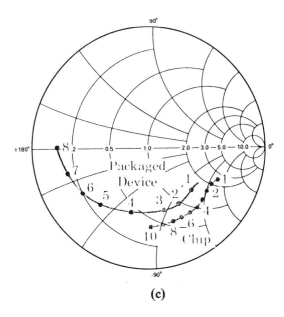

(c)

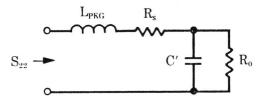

Output Equivalent Circuit (Package)

(d)

Fig. 4.30 (continued)

For $S_{22}$ the variation is that shown in Fig. 4.30(c) and the corresponding equivalent circuit is Fig. 4.30(d). $S_{21}$ [see Fig. 4.31(a)] is seen to cross the unity-gain circle at a frequency of about 4 GHz. Bode diagram plots of the parameters are as shown in Fig. 4.31(b).

From $S$ parameter measurements and Smith-chart plotting a combined equivalent circuit may be obtained for the microwave device under study.

## 4.5 POWER TRANSISTORS

Si transistors are readily available with voltage ratings of 350-450 V (a convenient rating for operating directly from a bridge-rectified 220 $V_{RMS}$ line), but higher voltage transistors are harder to obtain. The current rating of such units may be 10 A continuous, or 30 A peak pulsed. The operating region in the $I_c$-$V_c$ chart for such a transistor of 175 W dissipation rating is shown in Fig. 4.32. The line $I_{s/b}$ represents the limit set by second breakdown effects. Typical common-emitter characteristics are given in Fig. 4.33: the $h_{FE}$ gain is seen to be greatly influenced by the current density. The rise and fall switching times for 6 A are about 0.5 $\mu$sec, and the collector-to-emitter saturation voltage is 0.5-0.8 V. The maximum junction temperature allowed is 200°C. Figure 4.34 shows a dissipation derating chart of the fall-off that results from increase of case temperature or decrease of permitted junction temperature.

Multiple-emitter structures are used in most power transistors to improve the heat dissipation capability and to give a large emitter perimeter-to-area ratio and low resistance in the base. This limits emitter crowding and second-breakdown is then not a problem. Current sharing between the various emitter regions can be aided by including small stabilizing resistors in each emitter lead as shown in Fig. 4.35(a) and (b). The stabilizing resistors may be of Ni-Cr and are created during the metallization.

The structure shown in Fig. 4.35(b) is part of an integrated circuit block that includes protective circuitry to give thermal and current overload limiting. The device is basically a Darlington connection with protective circuitry added as shown in Fig. 4.36(a). Details of the protective circuits are given in Fig. 4.36(b). The double-emitter transistor $Q_{21}$ allows the power limiter to decrease the output current as the input voltage is increased so creating an approximate power hyperbola. Transistor $Q_{13}$ thermally limits the device by removing the base drive at high temperature. The actual temperature sensing circuit $Q_{11}, Q_{12}$ and $Q_{10}$, regulates the voltage across the temperature sensors so that the thermal-limiting temperature is independent of supply voltage. As temperature increases, the collector current of $Q_{11}$ increases, while the base-emitter voltage ($V_{BE}$) of $Q_{12}$ decreases. $C_1$, $Q_2$, and $Q_3$ boost operating currents during switching to obtain faster response time, and $Q_{17}$ and $Q_{18}$ compensate for variations in current gain ($h_{fe}$) of the power devices.

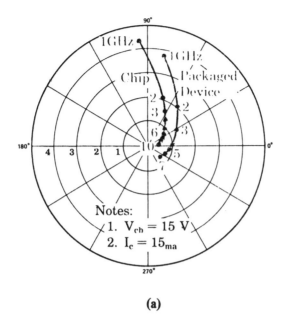

**(a)**

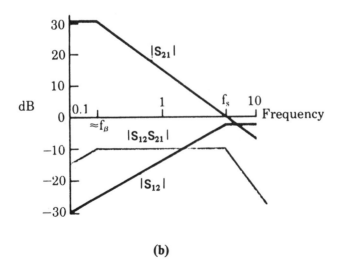

**(b)**

Fig. 4.31 Transistor $S$ parameters:
(a)   $S_{21}$ on a polar chart display, and
(b)   $S$ parameters on a Bode plot. (Courtesy Hewlett Packard Company, Application Note
154.)

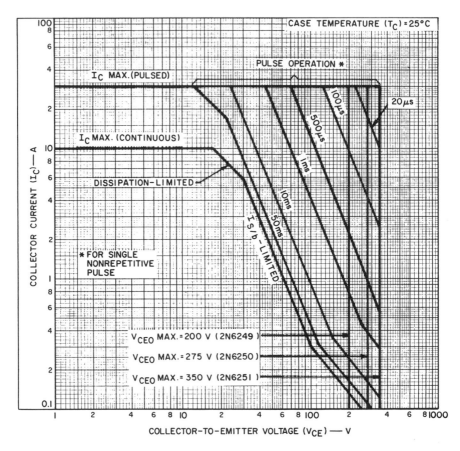

Fig. 4.32. Maximum operating regions for a Si transistor type 2N6249 for continuous and pulsed operation. (Courtesy RCA Corporation, *Solid State Databook-Power Transistors*, 2nd edition, 1973.)

The collector-emitter voltage rating is 40 V and the block is not protected against overvoltage damage. The peak collector current is internally limited to 2 A and the power dissipation is internally limited to 40 W. The thermal limit temperature is 165°C. The base input current requirement is 3 $\mu$A and the device readily interfaces with $T^2L$ and CMOS logic and has a switching speed of 0.5 $\mu$sec. The power transistor occupies about half of the total chip area. The combination of power and integrated circuit technology is an interesting development but thermal feedback effects must be guarded against.

Multiple connection of power transistors in parallel is practical, but only if measures are taken to assure equal RF and thermal load sharing. This may involve splitting the signal in an input circuit and adjusting the division to compensate for differences in performance of the power transistors.

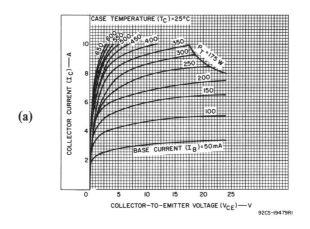

**(a)**

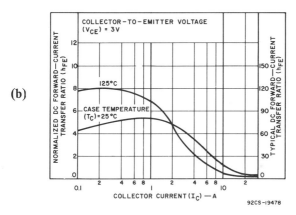

**(b)**

Fig. 4.33. Characteristics for the 2N6249, 175 W power transistor:
(a) $I_C$ vs $V_{CE}$ curves with the 175 W hyperbola, and
(b) Current gain as a function of collector current. (Courtesy RCA Corporation, *Solid State Databook-Power Transistors.*)

## 4.6 SWITCHING OF BIPOLAR TRANSISTORS

### 4.6.1 Ebers-Moll Model

Consider the common-base circuit configuration of Fig. 4.37(a). In the off-condition the collector current is small and the device is at point $P_0$ on the load line. If a current $I_{E1}$ is applied corresponding to $I_{B1}$, the device operating point moves to $P_1$ in the normal operating region of the transistor and the collector-base junction is still reverse-biased. However, if a current $I_{E2}$ is applied corresponding to $I_{B2}$ the operating point moves to $P_2$ and the collector-base

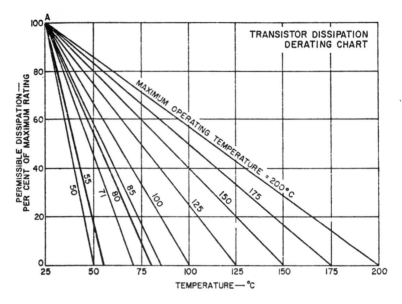

Fig. 4.34. Transistor dissipation derating chart. (Courtesy RCA Corporation.)

junction becomes slightly forward-biased. Figure 4.37(c) shows the same points $P_1$ and $P_2$ on the plot of common-emitter characteristics. Physically the reason for the collector-base junction becoming forward biased under the high-drive conditions is that the load is limiting the collector output current and so injected holes build up in density in the base until the recombination processes can supply the base current $I_{B2}$ necessary to satisfy the equation $I_{B2} = I_{E2} - I_{C2}$. Thus the base is flooded with holes and the depletion region at the base-collector junction is saturated. $V_{BC}$ becomes positive by the Boltzmann relationship,

$$\frac{p_c - p_{no}}{p_{no}} = (\exp)(qV_{BC}/kT) - 1 \qquad (4.81)$$

The hole density across the base region then has the form given by the line $P_2$ in Fig. 4.38(a). This can be considered as the result of adding two components $P_{2E}$ and $P_{2C}$. The $P_{2E}$ component may be considered the result of normal forward injection from the emitter and the $P_{2C}$ component as the result of reverse injection from the collector. Thus, one can write by superposition

$$I_E = a_{11}\left[\exp\left(\frac{qV_{EB}}{kT}\right) - 1\right] + a_{12}\left[\exp\left(\frac{qV_{CB}}{kT}\right) - 1\right] \qquad (4.82)$$

$$I_C = a_{21}\left[\exp\left(\frac{qV_{EB}}{kT}\right) - 1\right] + a_{22}\left[\exp\left(\frac{qV_{CB}}{kT}\right) - 1\right] \qquad (4.83)$$

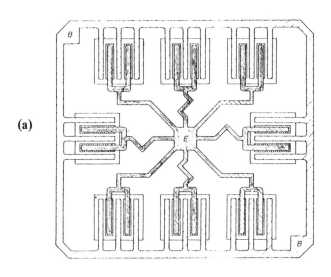

(a)

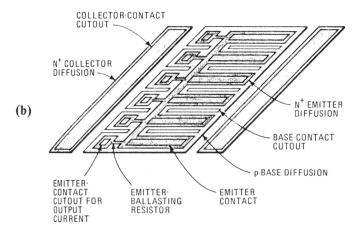

(b)

Fig. 4.35.  Power transistor structures with emitter ballast resistors

(a)  Second-breakdown proof power transistor. The large-area device is divided into eight cells, each having a stabilizing emitter resistor, (After Bergmann, Gerstner. Reprinted with permission from *IEEE Trans. Electron Devices*, **ED-13**, 1966, p. 634)

(b)  The transistor is a triple-diffused interdigitated structure with an emitter-ballasting resistor in each finger of the section of the device. Metallization is designed to eliminate any unbiasing caused by small voltage drops along the emitter fingers. (After Dobkin. Reprinted from *Electronics*, Feb. 7, 1974; copyright McGraw-Hill, Inc., NY, 1974.)

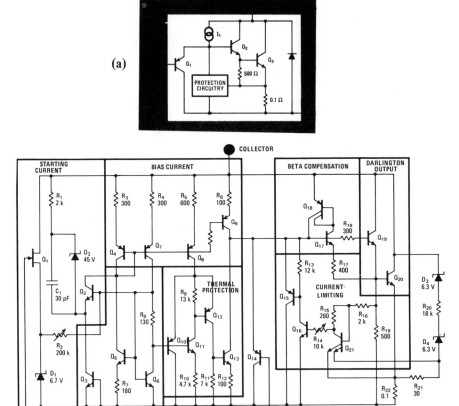

Fig. 4.36. Integrated circuit block that functions as an overload-protected power transistor:
(a)  The protective circuit provides thermal limiting and current limiting, and
(b)  The monolithic power transistor consists of seven sections, with the output transistor
     filling half of the chip area. (After Dobkin. Reprinted from *Electronics*, Feb. 7, 1974;
     copyright McGraw-Hill, Inc., NY, 1974.)

In the normal mode of operation

$$I_C = -\alpha_N I_E + I_{C0} \qquad (4.84)$$

and in the inverted mode of operation

$$I_E = -\alpha_I I_C + I_{E0} \qquad (4.85)$$

Hence the constants $a_{11}$-$a_{22}$ may be evaluated to obtain

$$I_E = \frac{-I_{E0}}{1 - \alpha_N \alpha_I} \left[ \exp \left( \frac{q V_{EB}}{kT} \right) - 1 \right] + \frac{\alpha_I I_{C0}}{1 - \alpha_N \alpha_I} \left[ \exp \left( \frac{q V_{CB}}{kT} \right) - 1 \right] \quad (4.86)$$

and

$$I_C = \frac{\alpha_N I_{E0}}{1 - \alpha_N \alpha_I} \left[ \exp \left( \frac{q V_{EB}}{kT} \right) - 1 \right] - \frac{I_{C0}}{1 - \alpha_N \alpha_I} \left[ \exp \left( \frac{q V_{CB}}{kT} \right) - 1 \right] \quad (4.87)$$

These are known as the Ebers-Moll equations for a bipolar transistor and are valuable in characterizing the operating curves from the action region into the saturated region. The equations may also be rearranged as

$$V_{EB} = \frac{kT}{q} \ell n \left( \frac{-I_E - \alpha_I I_C + I_{E0}}{I_{E0}} \right) \quad (4.88)$$

and

$$V_{CB} = \frac{kT}{q} \ell n \left( \frac{-\alpha_N I_E - I_C + I_{C0}}{I_{C0}} \right) \quad (4.89)$$

Hence

$$V_{CE} = -\frac{kT}{q} \ell n \left( \frac{I_{E0}}{I_{C0}} \left( \frac{\alpha_N I_E + I_C - I_{C0}}{I_E + \alpha_I I_C - I_{E0}} \right) \right) \quad (4.90)$$

Since analytical consideration suggests that $\alpha_{12} = \alpha_{21}$ and therefore $I_{E0}/I_{C0} = \alpha_I/\alpha_N$, simplification of Eq. (4.90 for $I_B \gg I_{E0}$, $I_{C0}$, gives the somewhat more convenient form

$$V_{CE} = -\frac{kT}{q} \ell n \left[ \frac{\dfrac{1}{\beta_I} \dfrac{I_C}{I_B} + \dfrac{1}{\alpha_I}}{1 - \dfrac{1}{\beta_N} \dfrac{I_C}{I_B}} \right] \quad (4.91)$$

Figure 4.38 shows the equivalent circuit of the Ebers-Moll model for a *pnp* transistor and a theoretical curve of $I_C/I_B$ vs $V_{CE}$ calculated from Eq. (4.90). The curve does not show a finite collector impedance because the model does not include base thickness modulation effects that are important in practice.

Although the Ebers-Moll equations are conceptually elegant, they are based on certain simplifications of modeling, such as diffusion-controlled current flow across the base, and neglect of all parasitic terms such as bulk series resistances. Additions to the equivalent circuit may be made to take care of some of those effects, as shown in Fig. 4.39. However, considerable complication results and computer modeling is needed.

### 4.6.2  Classical Switching Analysis

The large signal transient response of bipolar transistors has been treated by classic analysis[119] and by charge control concepts.[152,153] Consider first a treatment of the Moll type for the equivalent circuit of Fig. 4.40(a), with $r_e$ assumed negligibly small and $\alpha$ taken as $\alpha_N/(1 + j\omega/\omega_N)$. For a step input of base current $I_b'$ the collector current variation with time is given by

$$i_c = \frac{\alpha_N}{1 - \alpha_N} I_b' \left[ 1 - \exp^{-t} / \left( \frac{1}{\omega_N} + \frac{1}{\omega_C} \right) \left( \frac{1}{1 - \alpha_N} \right) \right] \qquad (4.92)$$

where $\omega_C = (R_L C_C)^{-1}$. For $t \to \infty$ this expression gives a collector current

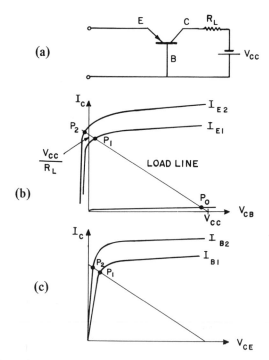

Fig. 4.37.  Switching of a common base transistor circuit:
(a)  Circuit,
(b)  $I_C$-$V_{CB}$ characteristics with switching from $P_0$ to $P_1$, and
(c)  $I_C$-$V_{CE}$ characteristics.

(a)

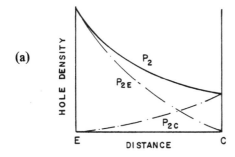

(b)

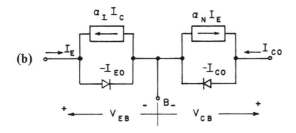

(c)

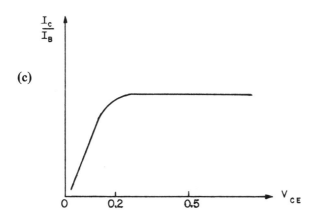

Fig. 4.38. Ebers-Moll model of a saturated transistor.
(a)   The hole density $P_2$ in the base region of a saturated *pnp* transistor may be considered
      as two components $P_{2E}$ and $P_{2C}$,
(b)   A two diode model results, and
(c)   A transistor curve calculated from the Ebers-Moll equations.

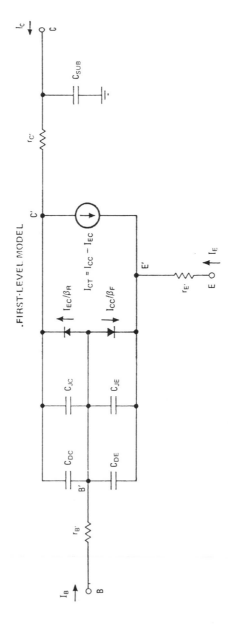

Fig. 4.39. For further development of the nonlinear model of a bipolar transistor, three resistances and five capacitances are added to the first-level model, which is drawn here in its nonlinear hybrid-$\pi$ form. To describe these additional elements, 12 more parameters are needed. (After Getreu. Reprinted from *Electronics*, Oct. 31, 1974; copyright, McGraw-Hill, Inc., NY, 1974.)

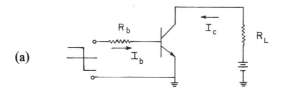

**(a)**

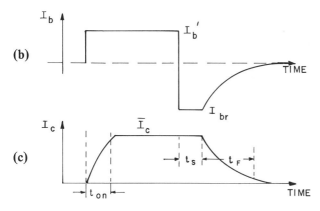

**(b)**

**(c)**

Fig. 4.40. Switching response of a bipolar transistor:
(a) Circuit,
(b) Base drive current, and
(c) Collector current transient showing turn on, storage and fall times.

$I_c'\left(= \dfrac{\alpha_N}{1-\alpha_N}\, I_b'\right)$ but this is not achieved because the external circuit limits the current to a lower value $\overline{I}_C(\sim V_{CC}/R_L)$. The switch-on time $t_{on}$ is defined as that for the current to rise to $0.9\,\overline{I}_C$. Thus, from Eq. (4.92)

$$t_{on} = \frac{1 + \omega_N R_L C_C}{(1-\alpha_N)\,\omega_N}\, \ell n \left[ \frac{1}{1 - \dfrac{0.9\overline{I}_C}{\left(\dfrac{\alpha_N}{1-\alpha_N}\right)I_b'}} \right] \tag{4.93}$$

In most switching applications ($\omega_N R_L C_C = \omega_N/\omega_C$) is small compared with unity so the equation reduces to

$$t_{on} = \frac{1}{(1-\alpha_N)\omega_N}\, \ell n \left[ \frac{1}{1 - \dfrac{0.9\overline{I}_C}{\left(\dfrac{\alpha_N}{1-\alpha_N}\right)I_b'}} \right] \tag{4.94}$$

If the base is driven just to the edge of saturation ($I_C' = \bar{I}_C$) the switch-on time is $2.3(1 - \alpha_N)\omega_N$. If $\omega_N$ is $10^7$ Hz and $\alpha_N$ is 0.9 then $\tau_{ON}$ is 2.3 $\mu$sec. Increasing the base drive shortens the switching time as would be expected. If $0.9 \, \bar{I}_C / \left( \dfrac{\alpha_N}{1 - \alpha_N} I_b' \right)$ is 0.5, the switch-on time is $0.69/(1 - \alpha_N)\omega_N$.

The reduced switching time achieved by base drive to saturation is at the expense of a large carrier buildup in the base region. This increases the storage time $t_s$ of the transistor observed when an attempt is made to turn it off by reversal of the base drive as shown in Fig. 4.40(b). If $\bar{I}_b$ is the base drive corresponding to the current $\bar{I}_c$, the magnitude of overdrive applied on closing the transistor switch is $\Delta_1 = I_b'/\bar{I}_b$ and this is a measure of the degree of carrier flooding produced in the base region. To "open" the switch a backdrive $\Delta_2 = |I_{br}|/\bar{I}_b$ is applied. The collector-emitter voltage cannot rise until the carrier density at the base-collector junction falls to zero and a time $t_s$ for the removal of stored holes ensues.

Moll's analysis of this process for the common emitter connection gives

$$t_s = \left( \frac{1}{\omega_N} + \frac{1}{\omega_I} \right) \frac{1}{(1 - \alpha_N \alpha_I)} \ell n \left[ \frac{\Delta_1 + \Delta_2}{1 + \Delta_2} \right] \qquad (4.95)$$

Thus, if the overdrive $\Delta_1$ is unity the storage time $t_s$ is zero, as would be expected. For a given overdrive the application of increased backdrive shortens the storage time. Once the collector-base voltage begins to rise the collector current falls to 0.1 of its saturation value in a time $t_F$ where

$$t_F = \frac{1}{(1 - \alpha_N)\omega_N} \ell n \left[ \frac{\bar{I}_c - \left( \dfrac{\alpha_N}{1 - \alpha_N} \right) I_{br}}{\dfrac{\bar{I}_c}{10} - \left( \dfrac{\alpha_N}{1 - \alpha_N} \right) I_{br}} \right] \qquad (4.96)$$

The general order of decay times is between $3\omega_N^{-1}$ and $[(1 - \alpha_N)\omega_N]^{-1}$.

### 4.6.4 Charge Control Analysis

The basic concept of bipolar transistor charge control theory is that the collector current equals the controlling base charge divided by the transit time.[152, 153] However, Gummel (1970) has shown that it is also possible to link the base charge and the collector current with the junction voltages. This considerably enhances the usefulness of the model.

Consider a one-dimensional *pnp* transistor where the dimension $x$ is zero at the emitter contact and $x_e$ and $x_c$ represent the outside edges of the emitter and collector transition regions as shown in Fig. 4.41(a). Under normal bias conditions ($V_{EB}$ positive and $V_{CB}$ negative) the electrostatic potential (EP)

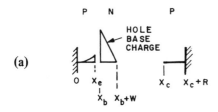

(a)

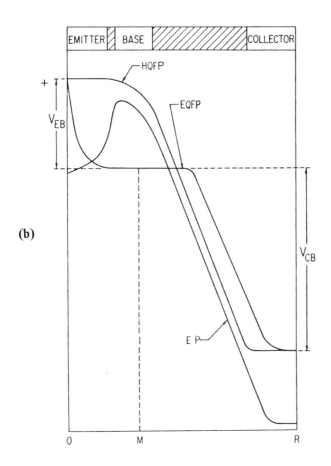

(b)

Fig. 4.41. One dimensional model of a bipolar transistor:
(a)  Labeling of distances,
(b)  Variation of electrostatic potential (EP) and hole and electron quasi-Fermi potentials
      (HQFP and EQFP) across the transistor under operating bias conditions. (After Gum-
      mel. Reprinted with permission from *IEEE Trans. Electron Devices*, **ED-11**, 1964,
      p. 457.)

and the hole and electron quasi-Fermi levels vary across the transistor as shown in Fig. 4.41(b). Let $\psi$ be the electrostatic potential in units of $kT/q$, and $\phi_p$ and $\phi_n$ be the hole and electron quasi-Fermi levels in the same units. Then

$$p(x) = n_i e^{\phi_p(x) - \psi(x)} \tag{4.97}$$

$$n(x) = n_i e^{\psi(x) - \phi_n(x)} \tag{4.98}$$

and

$$p(x) = p(x_1) e^{\psi(x_1) - \psi(x)} - \frac{j_c}{qD_0} \int_{x_1}^{x} a(t) e^{\psi(t) - \psi(x)} dt$$

$$- \frac{j_c}{qv_s} \int_{x_1}^{x} a(t) |\psi(t)'| e^{\psi(t) - \psi(x)} dt \tag{4.99}$$

where

$$a(x) = \frac{j_p(x)}{j_c}$$

Some manipulation of Eq. (4.99), with the aid of Eqs. (4.97) and (4.98), gives

$$j_c = \frac{qD_0 n_i [e^{\phi_p(x_e)} - e^{\phi_p(x_c)}]}{\int_{x_e}^{x_c} a(t) n_i e^{\psi(t)} dt + \frac{n_i D_0}{v_s} \int_{x_e}^{x_c} a(t) |\psi'(t)| e^{\psi(t)} dt} \tag{4.100}$$

The second integral in the denominator may be shown to be approximately $\frac{2 a_m D_0 n_i}{v_s} e^{\psi_{max}}$ where $\psi_{max}$ is the maximum value of $\psi$. The second integral therefore may be neglected in comparison with the first integral which has a value of about $wn_i e^{\psi_{max}}$, since $2D_0/v_s$ is of the order of 200 Å and is much less than a typical base thickness, $w$.

In Fig. 4.41(b) the electron quasi-Fermi level in the base region is seen to be almost constant. This is because a gradient in the electron quasi-Fermi level in the region where electrons are majority carriers would cause appreciable electron current to flow and for transistors of reasonable current gain, such currents are negligible. If we denote this value by $\phi_{nb}$, the emitter-base and collector-base voltages may be written as

$$V_{eb} = \frac{kT}{q} [\phi_p(x_e) - \phi_{nb}] \tag{4.101}$$

$$V_{cb} = \frac{kT}{q} [\phi_p(x_c) - \phi_{nb}]. \tag{4.102}$$

The next step is to define an average value $<a>_{av}$ of $a$:

$$<a>_{av} = \frac{\int_{x_e}^{x_c} a(t)n(t)dt}{\int_{x_e}^{x_c} n(t)dt} \tag{4.103}$$

Then using the approximation

$$\int_{x_e}^{x_c} n_i a(t) \left(e^{\psi(t)-\phi nb}\right) dt = \int_{x_e}^{x_c} a(t)n(t)dt = \frac{<a>_{av} q_b}{q} \tag{4.104}$$

Eq. (4.100) may be written

$$j_c = -\frac{(q^2 D_0 n_i^2 / <a>_{av}) \left[e^{(q V_{eb}/kT)} - e^{(q V_{cb}/kT)}\right]}{q_b} \tag{4.105}$$

where $q_b$ is the charge/unit area of the mobile electrons in the base region of the transistor. Conversion from current and charge densities to current and charge, where $I_c = -j_c A$ and $Q_b = q_b A$, gives

$$I_c = C \frac{e^{(q V_{eb}/kT)} - e^{(q V_{cb}/kT)}}{Q_b} \tag{4.106}$$

where

$$C = \frac{(q n_i A)^2 D_0}{<a>_{av}}$$

Equation (4.106) is true for any doping profile, within the limits of the assumptions and approximations made, but it is important to note that $Q_b$ depends on the bias. The form of this dependence is governed by the doping profile.

Equation (4.106) may also be written as

$$I_c = -a_{21} \left[e^{(q V_{eb}/kT)} -1\right] + a_{22} \left[e^{(q V_{cb}/kT)} -1\right] \tag{4.107}$$

with

$$a_{22} = -a_{21} = \frac{C}{Q_b}$$

which is the form of one of the Ebers-Moll equations. However, in the Ebers-Moll treatment the $'a'$ coefficients are regarded as constants, whereas the new treatment allows for their variation through the change of $Q_b$ with bias. It is this bias dependence of $Q_b$ that allows high-injection effects to be modeled.

### 4.6.4 Gummel-Poon Characterization of Bipolar Transistors

One of the most effective models available for the characterization of transistors in large signal operation is that developed by Gummel and Poon.[72] It provides the form by which bipolar transistors are represented in the computer-aided-design program known as SPICE (Simulation program with integrated circuit emphasis). This program has become widely used in the USA since its conception at the University of California, Berkeley.

The basic Ebers-Moll model of the large signal behavior of transistors fails to allow for many important effects in transistors of small base thicknesses and graded dopings. Among these are a finite collector-current-dependent output conductance due to base-thickness modulation (Early effect), space-charge-layer generation and recombination, conductivity modulation in the base and in the collector (Kirk effect), and emitter crowding. The Gummel-Poon model accommodates these and produces very exact transistor static and switching characteristics. It achieves this by modifying the Ebers-Moll approach and the charge control concepts first proposed by Sparkes.[153] The price that must be paid is the use of many model parameters. These are listed in Table 4.1, together with some representative values for a double-diffused Si *npn* transistor. Some of the characteristics that these parameters may be used to generate are shown in Figs. 4.42 and 4.43.

The approach begins by recognizing that the Ebers-Moll equations may be rewritten as

$$
\begin{pmatrix} I_e \\ I_c \end{pmatrix} = \begin{pmatrix} (1 + 1/\beta_N)I_s & -I_s \\ -I_s & (1 + 1/\beta_I)I_s \end{pmatrix} \begin{pmatrix} \exp \dfrac{qV_{eb}}{kT} - 1 \\ \exp \dfrac{qV_{cb}}{kT} - 1 \end{pmatrix} \tag{4.108}
$$

where $I_s$ is the intercept current obtained from a semilog plot of $I_c$ vs $V_{eb}$ where the collector current is extrapolated to zero voltage. From Eq. (4.108) the emitter and collector current are seen to have a common dominant component.

$$
I_{cc} = -I_s [\exp (qV_{eb}/kT) - \exp (qV_{cb}/kT)] \tag{4.109}
$$

In addition the emitter and collector currents each have a separate component $I_{be}$ and $I_{bc}$, which is proportional to the reciprocal forward and reverse common-emitter current gain:

$$
I_{be} = (I_s/\beta_f)[\exp (qV_{eb}/kT) - 1] \tag{4.110}
$$

$$
I_{bc} = (I_s/\beta_r)[\exp (qV_{cb}/kT) - 1] \tag{4.111}
$$

# Table 4.1. Gummel-Poon Model Parameters for a Bipolar Transistor

| Parameter Names | | Typical Values (300°K) |
|---|---|---|
| **Group 1:** Knee parameters and transit times | | |
| $I_k$ | Knee current* | $I_k = -1.875 \times 10^{-2}$ A |
| $v_k$ | Abs. value of knee voltage, in units of $kT/q$ | $v_k = 28.7$ |
| $\tau_f$ | Forward tau (forward transit time) | $\tau_f = 3.2 \times 10^{-10}$ secs. |
| $r_t$ | Tau ratio (ratio of reverse to forward transit time) | $r_t = 10.0$ |
| **Group 2:** Base Current | | |
| $i_1$ | Ideal base current coefficient | $i_1 = 2.35 \times 10^{-4}$ |
| $i_2$ | Nonideal base current coefficient | $i_2 = 2.19 \times 10^{-3}$ |
| $n_e$ | Forward base current emission coefficient | $n_e = 1.5$ |
| $i_3$ | Reverse base current coefficient | $i_3 = 7.65 \times 10^{-3}$ |
| $n_c$ | Reverse base current emission coefficient | $n_c = 1.5$ |
| **Group 3:** Emitter Capacitance | | |
| $v_{oe}$ | Abs. value of emitter offset voltage in units of $kT/q$ | $v_{oc} = 27.1$ |
| $m_e$ | Emitter grading coefficient | $m_e = 0.24$ |
| $a_{e_1}$ | Emitter zero-bias capacitance coefficient | $a_{e_1} = 0.337$ |
| $a_{e_2}$ | Emitter peak capacitance coefficient | $a_{e_2} = 1.03 \times 10^{-2}$ |
| **Group 4:** Collector Capacitance | | |
| $v_{oc}$ | Abs. value of collector offset voltage, in units of $kT/q$ | $v_{oc} = 27.1$ |
| $m_c$ | Collector grading coefficient | $m_c = 0.1265$ |
| $a_{c_1}$ | Collector zero-bias capacitance coefficient | $a_{c_1} = 0.187$ |
| $a_{c_2}$ | Collector peak capacitance coefficient | $a_{c_2} = 7.17 \times 10^{-3}$ |
| **Group 5:** Base Push-out | | |
| $v_{rp}$ | Abs. value of base push-out reference voltage, in units of $kT/q$ | $v_{rp} = 18.2$ |
| $r_w$ | Effective base thickness ratio | $r_w = 10.0$ |
| $r_p$ | Base push-out transition coefficient | $r_p = 4.55$ |
| $n_p$ | Base push-out exponent | $n_p = 3.0$ |

Auxiliary Quantities

$e_k = \exp(-v_k)$

$e_{ke} = \exp(-v_k/n_e)$

$e_{kc} = \exp(-v_k/n_c)$

A limited set of parameters for an npn 2N4124 may be the following: $\beta_f = 175$, $\beta_r = 2.4$, $I_s = 3.6 \times 10^{-5}$ A, $R_b = 25\Omega$, $I_k = -20$mA, $C_2$ (forward low-current nonideal base current coefficient) = 1400, $N_e$ (nonideal low-current base-emitter emission coefficient = 2, $I_{kr}$ (reverse high-current knee current) = 2mA, $C_4$ (reverse low-current nonideal base current coefficient) = 1, $C_{je}$ (zero-bias emitter-base junction capacitance) = 10 pF, $C_{jc}$ (zero-bias collector-base junction capacitance) = 5 pF.

*Quantity is negative for *npn* transistor.
(After Gummel and Poon[72]. Reprinted with permission from *The Bell System Technical J.*, copyright 1970, The American Telephone and Telegraph Company.)

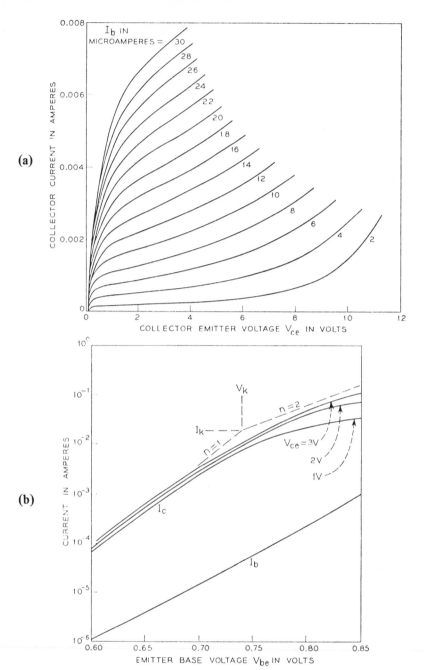

Fig. 4.42. Characteristics of an *npn* transistor calculated from the parameters of Table 4.1:
(a) Collector current vs collector-emitter voltage for various values of base current, and
(b) Collector and base currents for three values of collector-emitter voltage vs base-emitter voltage. Also shown are "knee" point ($V_k$, $I_k$), and slopes corresponding to values of 1 and 2 for the emission coefficient *n*. (After Gummel and Poon. Reprinted with permission from *The Bell System Tech. J.*, copyright 1970, The American Telephone and Telegraph Company.)

**(a)**

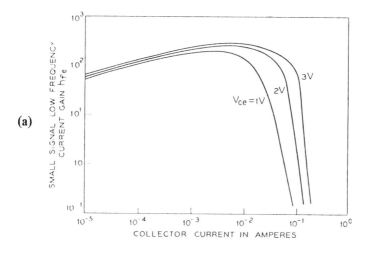

**(b)**

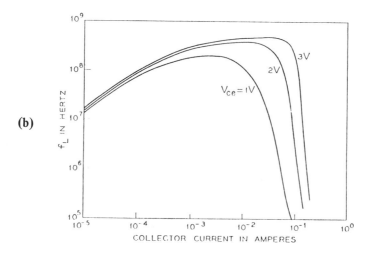

Fig. 4.43. Further characteristics calculated from the parameters of Table 4.1:
(a) Small-signal low-frequency current gain $h_{fe}$ vs collector current for three values of collector-emitter voltage, and
(b) Low-frequency approximation to unity-gain frequency $f_L$ vs collector current for three values of collector-emitter voltage. (After Gummel and Poon.[72] Reprinted with permission from *Bell System Tech. J.*, copyright 1970, American Telephone and Telegraph Co.)

The terminal currents are then given by

$$I_e = -I_{cc} + I_{be} \tag{4.112}$$

$$I_c = I_{cc} + I_{bc} \tag{4.113}$$

$$I_b = -I_{be} - I_{bc} \tag{4.114}$$

One of the features of the Gummel-Poon model is that it develops expressions for $I_{cc}, I_{be}$ and $I_{bc}$ that are representative of physical processes in the transistor.

The expression $I_{cc}$ is replaced by a charge control relationship that is valid for a wide range of bias conditions including those leading to high injection and base pushout. The expression derived from Eq. (4.106) is

$$I_{cc} = -I_s Q_{b0} \left( \frac{\exp(qV_{eb}/kT) - \exp(qV_{cb}/kT)}{Q_b} \right) \tag{4.115}$$

where $Q_b$ is the "base charge," that is the charge of all carriers of the type that communicate with the base terminal: electrons in a *pnp* transistor and holes in an *npn* transistor. The base charge is a function of bias; $Q_{b0}$ is the zero-bias value. The bias dependence of $Q_b$ is the subject of the conventional charge control theory which approximates the excess (or stored) base charge $Q_b - Q_{b0}$ as the sum of capacitive contributions and products of collector current components and transit times.

As in Ebers-Moll theory, this equation may be separated into emitter and collector components, or forward and reverse components,

$$I_{cc} = -I_s Q_{b0} \left( \frac{\exp(qV_{eb}/kT) - 1}{Q_b} \right)$$

$$+ I_s Q_{b0} \left( \frac{\exp(qV_{cb}/kT) - 1}{Q_b} \right) \equiv -I_f + I_r \tag{4.116}$$

The excess base charge may then be modeled as consisting of emitter and collector capacitive contributions $Q_e$ and $Q_c$ and of forward and reverse current-controlled contributions

$$Q_b = Q_{b0} + Q_e + Q_c - \tau_f B I_f - \tau_r I_r. \tag{4.117}$$

Here $\tau_f$ and $\tau_r$ are forward and reverse transit times. The coefficient $B$ is included to model the increase of the transit time when base push-out occurs; it has a value of unity in the absence of base push-out.

Normalizing with respect to the zero bias charge $Q_{b0}$ gives

$$q_b = 1 + q_e + q_c + \frac{1}{q_b}\left[\frac{I_s}{-Q_{bo}}\left\{\tau_f B[\exp(qV_{eb}/kT) - 1] + \tau_r[\exp(qV_{cb}/kT) - 1]\right\}\right]$$

$$\text{or } q_b = q_1 + \frac{1}{q_b}[q_2] \tag{4.118}$$

For high forward bias the charge $q_2$ is the dominant component of the base charge $q_b$. Except for the base push-out term $B$, it is characterized by four parameters: $I_s$, $Q_{bo}$, $\tau_f$, and $\tau_r$. It will be convenient to normalize these parameters. For this we define the knee voltage $V_k$ as the emitter voltage for which $q_2$ equals unity (for zero collector voltage). The knee occurs because the high level injection condition is setting in as in Fig. 4.42(b). Then

$$V_k = \frac{kT}{q}\ln\left(\frac{-Q_{bo}}{I_s\tau_f}\right) \tag{4.119}$$

and the low-injection-extrapolated collector current for $V_{eb} = V_k$ is

$$I_k = \frac{-Q_{bo}}{\tau_f} \tag{4.120}$$

It is convenient to normalize current terms with respect to $I_k$ and to express voltages in units of $kT/q$. This gives expressions

$$v_k = \frac{qV_k}{kT}$$

$$v_e = \frac{qV_{eb}}{kT} - v_k$$

$$v_c = \frac{qV_{cb}}{kT} - v_k \tag{4.121}$$

$$e_k = \exp(-v_k)$$

and

$$I_c = I_k i_c = I_k\left(-\frac{e^{v_e} - e^{v_c}}{q_b} + i_{bc}\right) \tag{4.122}$$

$$I_b = I_k i_b = I_k(-i_{be} - i_{bc}) \tag{4.123}$$

For the base charge

$$q_2 = B(e^{v_e} - e_k) + (\tau_r/\tau_f)(e^{v_c} - e_k)$$

$$q_b = q_1/2 + [(q_1/2)^2 + q_2]^{1/2} \tag{4.124}$$

$$Q_b = -I_k\tau_f q_b$$

With these normalizations, we can use the set $I_k$, $v_k$, $\tau_f$, $r_i$ ($\equiv \tau_f/\tau_r$) as the four model parameters describing $q_2$ for the case $B = 1$. These four parameters constitute Group 1 of the model parameters listed in Table 4.1.

The forward transit time $\tau_f \equiv \left(\dfrac{1}{2\pi f_L}\right)$ in terms of structural parameters is given approximately by

$$\tau_f = \frac{W_b^2}{2\eta D} \tag{4.125}$$

where $W_b$ is the base thickness and where $\eta$ represents the drift effect in the base. For uniform base doping $\eta$ is unity and for diffused-base transistors its typical values lie between two and ten.

The zero-bias base charge is approximately given by

$$Q_{b0} = -A_e q N_b \tag{4.126}$$

where $A_e$ is the emitter area, $q$ the electronic charge and $N_b$ the number of impurities per unit area in the base. Typical values for $N_b$ are a few times $10^{12}/\text{cm}^2$ (lower values would cause premature punch through and high values cause low injection efficiency and/or low $f_L$). The intercept current is given by

$$I_s = A_e \frac{q n_i^2 D}{N_b} = -\frac{A_e^2 q^2 n_i^2 D}{Q_{b0}} \tag{4.127}$$

where $n_i$ is the intrinsic carrier concentration and $D$ is the effective diffusivity of carriers in the base.

The Group 3 parameters model the emitter junction capacitance. The offset voltage $V_{oe}$ is approximately the conventional "built-in" voltage, which has a typical value near 0.7 V for Si at room temperature or $v_{oe} = V_{oe}/(kT/q) = 27$. The grading coefficient $m_e$ depends on the type of doping transition: It is one-fourth and one-sixth for ideal step and linearly graded junctions, respectively. Typical values for emitter junctions are approximately 0.2. Parameter $a_{e1}$ is related to the zero-bias capacitance $C_{oe}$ by

$$a_{e1} = \frac{C_{oe} V_{oe}}{Q_{b0}\,(1-2m_e)} \tag{4.128}$$

The term $a_{e2}$ is related to the forward bias capacitance $C_{ef}$ and may be deduced from the slope of the delay time vs reciprocal emitter current.

The parameters in Group 4 are similar to those in Group 3 but relate to the collector.

The Group 5 parameters model the base push-out effects. The resistive voltage drop across the collector $V_{rp} = (kT/q)v_{rp}$ is caused by a current of magnitude $I_k$. The ratio of the thickness of the collector epitaxial region to the thickness of the metallurgical base is designated $r_W$. The parameter $r_p$ determines the steepness of the variation of the forward delay time as a function of a collector current in the current range where base push-out is incipient. The base push-out exponent $n_p$ determines the fall-off of $f_L$ for high currents.

In summary, the Gummel-Poon approach is capable of modeling the large signal steady state and transient characteristics of bipolar transistors with parameters that are readily determined for the process technology under consideration. In conjunction with computer-aided-design programs such as SPICE it provides a significant approach to the simulation of integrated circuits prior to fabrication.

## 4.7 LATERAL TRANSISTORS

Lateral *pnp* transistors are readily fabricated in planar technology but have relatively low current gains because of the tendency of carriers to be injected towards the substrate and recombine instead of being collected at the laterally displaced collector. However, lateral transistors have some uses in integrated circuits for functions such as active loads, current sources and level shifters.

Simple one-dimensional modeling of lateral transistors is possible by considering independently lateral and vertical components of current. The results obtained in this way do not closely represent the behavior of actual transistors, but are worthy of brief examination. The geometry considered is shown in Fig. 4.44 and the current components are given by

$$I_L = -2qldD_p \left. \frac{dp}{dx} \right|_{\text{in base region}} \tag{4.129}$$

$$I_V = -qW_e lD_p \left. \frac{dp}{dy} \right|_{\substack{\text{at the bulk side} \\ \text{of emitter junction}}} \tag{4.130}$$

The lateral hole gradient $dp/dx$ is $p_0/W_b$ where $p$ is the excess hole density at the base side of the emitter depletion region with forward bias applied. The vertical hole gradient $dp/dy$ is determined by recombination and is $p_0/L_p$ as in a diode. Hence, the common-emitter current gain for this simple model is

$$\beta = \frac{I_L}{I_V} = \frac{2L_p d}{W_e W_b} \tag{4.131}$$

If $\tau_p$ is $10^{-6}$ sec and $D_p$ is 7 cm$^2$/sec, $L_p$ is 26 $\mu$m, so if $W_e$ is 3 $\mu$m, $W_b$ is 2 $\mu$m and $d$ is 2 $\mu$m, the resulting value of $\beta$ is 17.

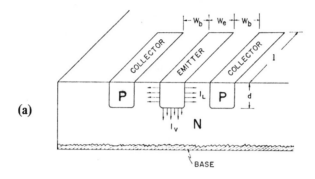

(a)

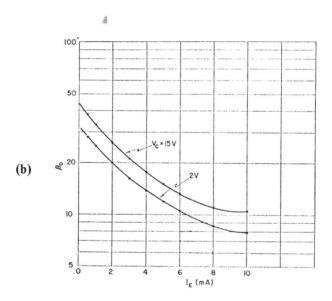

(b)

Fig. 4.44.  Lateral bipolar transistor model:
(a)  Geometry considered, and
(b)  Current gain dependence on emitter current for a simple model. (After Lindmayer and Schneider.[103] Reprinted with permission from *Solid-State Electronics*, Pergamon Press Ltd.)

The observed current gain for such a structure is shown in Fig. 4.44(b) from which it is seen that $\beta$ is a strong function of the injected current level, a dependency not provided for in the simple model. Various explanations have been presented for the effect, including the need for two-dimensional modeling and allowance for the injection levels.[29, 30]

The frequency performance of the simple one-dimensional model predicts a $-3$ dB cutoff frequency

$$f_{c\beta} = \frac{\sqrt{3}}{2\pi\tau_p} \tag{4.132}$$

The fall-off of $\beta$ with frequency is

$$|\beta| = \frac{\beta_0}{\sqrt[4]{(1 + \omega^2\tau_p{}^2}} \tag{4.133}$$

which is unusual since it predicts a frequency response that drops 3 dB/octave instead of the 6 dB/octave of normal transistors. The maximum frequency $f_T$ obtained for unity current gain is

$$f_T = \frac{\beta_0{}^2}{2\pi\tau_p} \tag{4.134}$$

and substitution from Eq. (4.131) gives

$$f_T = \frac{2D_p}{\pi} \left(\frac{d}{W_e W_b}\right)^2 \tag{4.135}$$

which is independent of base lifetime.

Although two-dimensional computer modeling of lateral transistors gives considerably different results, the simple model has been worth study since it emphasizes that the vertical component of current is undesirable. This component may be minimized by providing in the structure an $nn^+$ barrier to holes: Therefore, lateral transistors are usually fabricated on thin-epi $nn^+$ substrates or with a buried $n^+$ layer as shown in Fig. 4.45. Modeling of a buried layer structure in two-dimensional form[52] results in calculated current gains of 6.1 and 3.1, with and without the layer, and observed gains of 4 and 0.7, respectively. These may be compared with a value of 22 calculated from the one-dimensional model. A vertical *pnp* transistor of similar emitter- and base-thickness gave a calculated current gain of 35 and an observed gain of 20.

Curves of normalized injected hole density without and with a subdiffused field are shown in Fig. 4.46(a), (b). The current gain $\beta$ in this model changed from 2.8 to 5.0 with the addition of the field. The calculated and measured fall-off of $\beta$ with frequency is shown in Fig. 4.46(c), and is seen to approximate 20 dB/decade in slope.

In integrated circuit design, parasitic lateral transistors are liable to be formed by oversight and this must be guarded against.

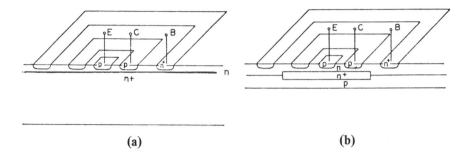

(a)                                    (b)

Fig. 4.45.  Lateral transistor geometries with improved lateral flow:
(a)  With $nn^+$ barrier as a result of an epitaxial layer on an $n^+$ substrate, and
(b)  With $n^+$ buried layer. (After Fulkerson.[52] Reprinted with permission from *Solid-State Electronics*, Pergamon Press, Ltd.)

## 4.8  HETEROJUNCTION TRANSISTORS

Heterojunctions involve junctions between two different semiconductors, and bandgap and electron-affinity differences allow unusual barrier conditions to occur.

The band diagram for an abrupt junction between $n$GaAs and $p$Ge may be constructed from the individual energy diagrams as in Fig. 4.47. In the absence of interface-state effects the difference of electron affinities is seen to produce an energy discontinuity $\Delta E_c$ in the conduction band where

$$\Delta E_c = \chi_{Ge} - \chi_{GaAs} \qquad (4.136)$$

In the valence band there is a step $\Delta E_v$ given by

$$\Delta E_v = \left( E_{g(GaAs)} - E_{g(Ge)} \right) - \Delta E_c \qquad (4.137)$$

Further details of the construction of such energy diagrams are given in books and review articles concerned with heterojunctions.[116, 145]

The reason for interest in heterojunctions for transistors is shown in Fig. 4.48 where the energy diagram for a Ge $npn$ transistor is compared with that for an $n$GaAs-$p$Ge-$n$Ge structure. The heterojunction transistor is seen to have a large barrier to the reverse injection of holes from the base region into the emitter. The ratio of the electron injected current $j_n$ to the hole injected current $j_p$ may be determined by the application of the Boltzmann relation.

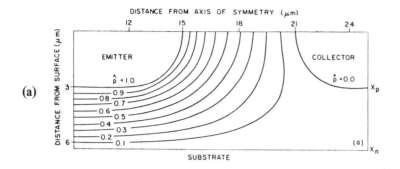

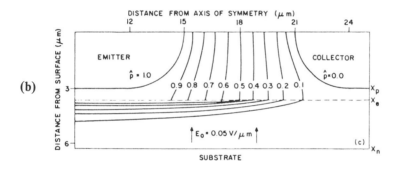

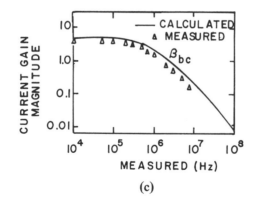

Fig. 4.46. Computer analysis of a lateral *pnp* transistor in which the subdiffused layer is modeled as an electric field $\mathcal{E}_0$:

(a) Normalized injected hole density with $\mathcal{E}_0 = 0$, i.e., no subdiffused layer, ($L_p$, $\tau_p$ taken as 33 $\mu$m, 1.2 $\mu$sec)

(b) Normalized hole density with $\mathcal{E}_0 = 0.05$ V/cm, and

(c) Current gain vs frequency ($\mathcal{E}_0 = 0.05$ V/cm, $\tau_p = 1.2$ $\mu$s, $\gamma = 0.9$). (After Seltz and Kidron. Reprinted with permission from *IEEE Trans. Electron Devices*, **ED-21**, 1974, pp. 589, 591.)

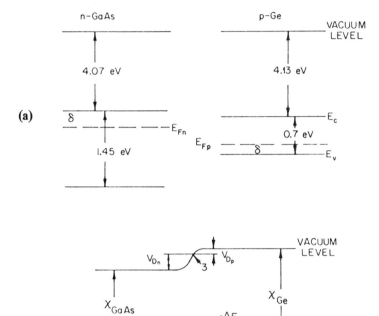

Fig. 4.47. Construction (not drawn to scale) of the energy band diagram for *n-p* GaAs-Ge. (After Milnes and Feucht, *Heterojunctions and Metal-semiconductor Junctions*, Academic Press, NY, 1972, p. 4.)

The injected electron density $n_b$ just inside the base region at the base-emitter junction is

$$n_b/n_e = \exp\left(-qV_n/kT\right) \tag{4.138}$$

Similarly the injected hole density $p_e$ just inside the emitter region at the base-emitter junction is

$$p_e/p_b = \exp\left(-qV_p/kT\right) \tag{4.139}$$

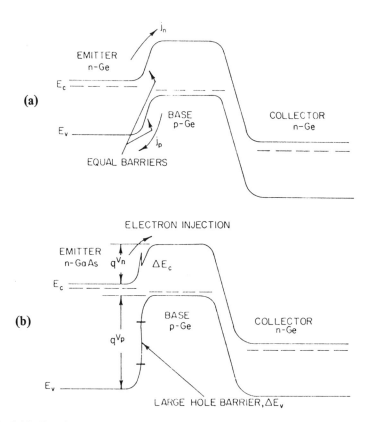

Fig. 4.48. Transistor energy band diagrams for homojunction and heterojunction emitters. (After Milnes and Feucht, *Heterojunctions and Metal-Semiconductor Junctions*, Academic Press, NY, 1972, p. 14.)

Following the simple Shockley diffusion model, the injected electron current is given by

$$j_n = qD_n dn/dx = qD_n n_n/W_b \qquad (4.140)$$

where $W_b$ is the base thickness and $D_n$ is the diffusion constant for the electrons. The reverse injected hole current in the emitter is given by

$$j_p = qD_p dp/dx = qD_p p_e/L_p \qquad (4.141)$$

where $L_p$ is the diffusion length for holes in the emitter (assuming the emitter thickness is several times the diffusion length). The ratio of the currents is then

$$j_n/j_p = L_p D_n n_b / W_b D_p p_e$$

$$= (L_p D_n n_e / W_b D_p p_b) \exp\left[(qV_p - qV_n)/kT\right] \qquad (4.142)$$

$$= (L_p D_n N_D / W_b D_p N_A) \exp(\Delta E_g/kT)$$

where $N_D/N_A$ is the emitter/base doping ratio.

For a homojunction $\Delta E_g$ is zero, hence the improvement factor created by the heterojunction is seen to be $\exp(\Delta E_g/kT)$. This factor is normally quite large in size. For example, if $\Delta E_g$ is 0.2 eV and $kT$ is 0.026 eV at room temperature, the factor $\exp(\Delta E_g/kT)$ is about 3000. In a homojunction transistor the emitter is normally doped at least a hundred times larger than the base for good injection efficiency. In a heterojunction transistor this ratio may be inverted allowing base dopings on the order of $10^{19}$ atoms cm$^{-3}$.

In homojunction transistors with lightly doped base regions, the alpha of the transistor falls off as the injected current is increased. This is partly attributable to the fact that at high injection the electrons cause increased hole density in the base (for space-charge limitation) and therefore increased reverse injection occurs. This conductivity modulation of the base would not be significant in the heavily doped bases of heterojunction transistors.

The gain-bandwidth figure of merit of a transistor is given, approximately by

$$f_{\max} = \frac{1}{4} \pi (r_b{}' C_c \tau_{ec})^{1/2} \qquad (4.143)$$

where $r_b{}'$ is the base resistance, $C_c$ is the collector capacitance and $\tau_{ec}$ equals $\tau_e + \tau_b + \tau_{csl} + \tau_c$, in which $\tau_e$ is the emitter diode charging time, $\tau_b$ is the base transit time, $\tau_{csl}$ is the collector depletion layer transit time (saturated-drift-velocity limited), and $\tau_c$ is the collector diode charging time.

Consideration of the various terms in $\tau_{ec}$ suggests that the overall $f_{\max}$ may in principle be increased for heterojunction transistors relative to present high-frequency bipolar transistors. Good switching characteristics may also be expected from heterojunction transistors stemming from a lowered $C_c$ and absence of hole storage effects in the emitter.

Some advantages may also be expected with respect to emitter crowding, base reach-through and second breakdown because of the reduced base resistance.

Heterojunction transistors, in spite of the potentially attractive features that have just been cataloged, have unfortunately never delivered performances comparable to those envisaged. They are difficult and expensive to make because they involve epitaxy procedures with semiconductors that are not under good technological control and the processes do not have the convenience of Si planar technology. Realization of the lifetimes, doping gradients and device parasitics needed to give the desired properties is not easy.

If semiconductor heterojunctions are to have good interfaces, relatively free of recombination centers, the lattice constants must match very closely, to perhaps better than 0.1%-0.5%. Some lattice matched pairs are listed in Table 4.2.

A Si-Ge heterojunction with 4% lattice mismatch has a density of dangling

**Table 4.2. Semiconductor Heterojunction Pairs with Good Lattice Match Conditions**

| Semi-conductor | Energy Gap (eV) | Lattice Con-stants (Å) | Energy Gap Struc-ture | Expansion Coefficient at 300°K ($\times 10^{-6}$ °$C^{-1}$) | Hetero-junction Preferred Doping | Typical Dopants | Electron Affinity (eV) |
|---|---|---|---|---|---|---|---|
| $Ge_{0.9}Si_{0.1}$ | 0.77 | (5.63) | Indirect | – | $n$ | P, As, Sb | (4.1) |
| Ge | 0.66 | 5.658 | Indirect | 5.7 | $p$ | Al, Ga, In | 4.13 |
| GaAs | 1.43 | 5.654 | Direct | 5.8 | $n$ | Se, Te | 4.07 |
| Ge | 0.66 | 5.658 | Indirect | 5.7 | $p$ | Al, Ga, In | 4.13 |
| ZnSe | 2.67 | 5.667 | Direct | 7.0 | $n$ | Al, Ga, In | 4.09 |
| Ge | 0.66 | 5.658 | Indirect | 5.7 | $p$ | Al, Ga, In | 4.13 |
| ZnSe | 2.67 | 5.667 | Direct | 7.0 | $n$ | Al, Ga, In | 4.09 |
| GaAs | 1.43 | 5.654 | Direct | 5.8 | $p$ | Zn, Cd | 4.07 |
| AlAs | 2.15 | 5.661 | Indirect | 5.2 | $p$ | Zn | 3.5 |
| GaAs | 1.43 | 5.654 | Direct | 5.8 | $n$ | Se, Te | 4.07 |
| GaP | 2.25 | 5.451 | Indirect | 5.3 | $n$ | Se, Te | 4.3 |
| Si | 1.11 | 5.431 | Indirect | 2.33 | $p$ | Al, Ga, In | 4.01 |
| AlSb | 1.6 | 6.136 | Indirect | 3.7 | $n/p$ | Se, Te/Zn, Cd | 3.65 |
| GaSb | 0.68 | 6.095 | Direct | 6.9 | $p/n$ | Zn, Cd/Se, Te | 4.06 |
| GaSb | 0.68 | 6.095 | Direct | 6.9 | $n$ | Se, Te | 4.06 |
| InAs | 0.36 | 6.058 | Direct | 4.5(5.3) | $p$ | Zn, Cd | 4.9 |
| ZnTe | 2.26 | 6.103 | Direct | 8.2 | $p$ | Cu | 3.5 |
| GaSb | 0.68 | 6.095 | Direct | 6.9 | $n$ | Se, Te | 4.06 |
| ZnTe | 2.26 | 6.103 | Direct | 8.2 | $p$ | Cu | 3.5 |
| InAs | 0.36 | 6.058 | Direct | 4.5(5.3) | $n$ | Se, Te | 4.9 |
| ZnTe | 2.26 | 6.103 | Direct | 8.2 | $p$ | Cu | 3.5 |
| AlSb | 1.6 | 6.136 | Indirect | 3.7 | $n$ | Se, Te | 3.65 |
| CdTe | 1.44 | 6.477 | Direct | 5 | $p/n$ | Li, Sb, P/I | 4.28 |
| PbTe | 0.29 | 6.52 | Indirect | 19.8 | $n/p$ | Cl, Br/Na, K | 4.6 |
| CdTe | 1.44 | 6.477 | Direct | 5 | $p$ | Li, Sb | 4.28 |
| InSb | 0.17 | 6.479 | Direct | 4.9 | $n$ | Se, Te | 4.59 |
| ZnTe | 2.26 | 6.103 | Direct | 8.2 | $p$ | Cu | 3.5 |
| CdSe(hex) | 1.7 | 4.3($\sqrt{2}$) (6.05) | Direct | 2.45-4.4 | $n$ | Cl, Br, I | 4.95 |

After Milnes & Feucht, 116. Reprinted with permission from *Heterojunctions and Metal-Semiconductor Junctions*, Academic Press, N.Y., 1972.

bonds of $10^{14}$ cm$^{-2}$ and minority carrier injection for transistor action is not observed. GaP-Si with a lattice mismatch of 0.4% has about $10^{13}$ cm$^{-2}$ interface states and some injection is seen. However, Si has an unusually small thermal lattice-expansion coefficient ($\sim$2.3 $\times$ $10^{-6}$ $^{\circ}$C$^{-1}$) compared with most semi-conductors (GaP $\sim$ 5.3 $\times$ $10^{-6}$ $^{\circ}$C$^{-1}$). This adversely affects the performance. Other problems that may arise in heterojunctions are associated with electron affinity differences that produce energy spikes in the band edge that may impede good injection characteristics. Such spikes may be reduced by grading the junction over several hundreds of angstroms, sometimes with improvements in performance. On the other hand grading may introduce fresh problems if cross-doping occurs. For instance, in $n$GaAs-$p$Ge junctions the As from the GaAs may enter the Ge and create a hidden $np$-Ge homojunction.

In a study of heterojunction transistor action, $n$Al$_x$Ga$_{1-x}$As-$p$GaAs-$n$GaAs structures made by liquid-phase epitaxy have shown common-emitter current gains of more than 100 at high injection levels but very low gains at low injection levels. This is because interface recombination becomes significant at low current levels.[93]

However, AlGaAs/GaAs junctions as we will see in later chapters are extremely valuable for carrier confinement and waveguide action in hetero-junction lasers and as window-effect junctions to reduce series resistance and surface recombination in GaAs solar cells.

## 4.10 QUESTIONS

1. Outline the design of a 2 GHz ($f_T$) $npn$ Si transistor intended to operate at a quiescent bias point of $I_c$ = 10 mA and $V_{ce}$ equal to 10 V. Estimate the probable performance of your design.

2. Outline the design of a Si $npn$ power transistor rated at 10 A, 300 V ($V_{CE0}$) and 150 W power dissipation (25$^{\circ}$C case temperature).

3. A $p^+np^+$ Si transistor has a base doping of 2 $\times$ $10^{14}$ cm$^{-3}$ of thickness 100 $\mu$m and emitter and collector doping of $10^{18}$ cm$^{-3}$ and $10^{15}$ cm$^{-3}$, respectively. When biased with $V_{CB}$ = 300 V it has a common emitter current gain of 30 at low current levels and this is assumed to be limited by recombination in the base. What may be inferred about the recombination lifetime? Estimate the device collector capacitance (cm$^{-2}$). If the current density is raised to 100 A/cm$^2$ what changes may be expected?

4. A Si $npn$ transistor with a uniformly doped base region has a unity gain frequency $f_T$ = 500 MHz at 300$^{\circ}$K. Estimate the base thickness. What will $f_T$ become at 400$^{\circ}$K? If a more realistic design is considered with the base doping decreasing exponentially from $10^{17}$ to $10^{14}$ over the same base thickness, what will be the unity gain frequency?

5. An $npn$ bipolar transistor has $\alpha_N$ = 0.98 and $\alpha_I$ = 0.8. $I_{E0}$ is 0.1 mA and $I_{C0}$ is 0.12 mA. The load is 100 ohms in a common base circuit and a collector current of 0.5 A is flowing. The resistance in series with the emitter lead is 10 ohms. The electron minority carrier lifetime in the base is $10^{-7}$ sec. The emitter-base voltage is reversed at time $t$ = 0. Sketch the circuit voltage and current waveforms.

6. Consider a conventional circuit of two common-emitter bipolar transistors linked by a coupling capacitance $C_b$ of negligible impedance. The emitter resistors $R_e$ are shorted

by capacitors of negligible impedance at the signal frequency and the base bias resistors are large enough to be neglected. The hybrid $\pi$ model parameters for the transistor are $r_{bb'} = 100$ ohms, $C_e = 100$ pF, $C_c = 3$ pF, $r_{b'e} = 1$ kilohm and $g_m = 50$ mA/V. The signal source resistance is 50 ohms in series with a signal voltage $V_s$. The effective load resistance between the two stages is 4 kilohms and the second-stage load $R_L$ is also 4 kilohms. Show that the transfer function has the form

$$A_v = \frac{V_{R_L}}{V_s} = \frac{C\,(s - s_{01})\,(s - s_{02})}{(s - s_{p1})\,(s - s_{p2})\,(s - s_{p3})\,(s - s_{p4})}$$

Determine the values of the poles and zeros $s_p$ and $s_0$ and plot the curve of $A_v$ vs frequency. Use any computer-aided-design program that is available and convenient. Compare your answer with that obtained by simplifying the problem by application of the Miller approximation that reduces the circuit response to a two-pole expression with one pole dominant.

7. An *npn* transistor is represented by the equivalent circuit Gummel-Poon SPICE parameters of Table 4.1. Design a three-stage $RC$ coupled amplifier with three of these transistors to have input and output impedances of about 600 ohms. Calculate the gain and phase response of your design (300°K) as a function of frequency. Calculate also the response to a step function input.

8. Design a Darlington emitter follower circuit with two transistors characterized by the SPICE parameters of Table 4.1. The supply voltage is to be 5 V and the load resistor in the emitter lead of the output transistor is to be 100 ohms. Fully characterize the circuit in terms of gain, input and output impedance (assume 600-ohms signal source impedance) dynamic range, frequency response and transient response.

9. Design a differential amplifier stage with two transistors characterized by the SPICE parameters of Table 4.1. The supply voltage $V_{cc}$ is to be +10 V and the voltage on the $R_E$ coupling resistor is to be $-10$ V. Assume the signal sources are of 600-ohms resistance and that the output loading is negligible. Calculate the transfer characteristic of the amplifier, the input and output impedances and the common-mode rejection ratio.

10. A Si *npn* transistor with a $10^{15}$ cm$^{-3}$ uniformly doped base has an alpha cutoff frequency of 500 MHz and $h_{FE} = 50$. If the collector junction capacitance is 1 pF and the collector series resistance is $10^3$ ohms, what parameters can be inferred for the transistor? If the base distance was unchanged but nonuniformly doped to produce a built-in field of $10^3$ V/cm with the doping decreasing to $10^{15}$ at the collector junction edge, estimate what then might be the changes in performance parameters.

11. Obtain the data sheets of a well characterized small signal transistor such as the 2N 5088 (or any convenient type that has many typical characteristic curves supplied). Extract from the given curves as many of the SPICE model parameters as possible. Explain what tests you would make to determine the remaining SPICE parameters. If convenient, obtain such a transistor and actually extract the SPICE parameters.

12. A Si *npn* transistor has the equivalent circuit of Fig. 4.39 and the SPICE parameters of Table 4.1. Determine the hybrid $\pi$ small signal equivalent circuit parameters that approximate the transistor performance. Knowing the hybrid $\pi$ equivalent circuit parameters how many SPICE parameters can one infer without measurements?

13. The transistor characterized by the SPICE parameters of Table 4.1 feeds a collector load of 5 kilohms and is switched on with a voltage of +5 V and then switched off by reversal of this voltage. Calculate the current turn-on and turn-off transients. Identify the turn-on delay, the turn-on time, turn-off delay and the turn-off time. To what extent does the calculation enable you to infer the minority carrier lifetime in the transistor base region?

14. Assume that the transistor characterized by Table 4.1, is available in both *npn* and *pnp* forms with similar SPICE parameters. The complementary pair then allows the design of a class B push-pull circuit without a transformer, provided balanced supply voltages are available, about ±15 V. Design such a circuit for maximum reasonable power output to a 150-ohm load. Calculate the transfer characteristic for a 1 kHz signal, the maximum power output and the collector circuit efficiency for an harmonic distortion of less than 1% in voltage.

15. Investigate the possibility of constructing a sinusoidal Colpitts oscillator of 10 MHz frequency with the transistor of Table 4.1, and the use of an external inductance of 1 $\mu$H. What values of capacitance $C_1$ and $C_2$ would be needed? Would it be conceivable to achieve these in integrated form?

16. Design a simple free-running cross-connected two transistor flip-flop with transistors having the SPICE parameters of Table 4.1. The frequency desired is 10 MHz. Assume an integrated circuit layout form and allowance for parasitics if necessary. Calculate the expected waveforms.

17. Derive Eq. (4.1) for emitter injection efficiency in a uniformly doped $p$ base region of an *npn* transistor.

18. Derive the common-emitter base-recombination transport factor $\beta_{rec}$ [(see Eq. (4.4)] for a uniformly doped base region.

19. Discuss the $I_E$ vs $V_{EB}$ relationship for an *npn* transistor and compare it to the $I_C$ vs $V_{EB}$ curve. What difference does base doping, whether uniform or graded, make in these relationships?

20. Discuss unilateralizing circuits that may be used to cancel feedback from the output to the input of a transistor through the term $h_{re}$. What effect does unilateralization have on the overall gain, and why is it not more frequently used?

21. Second breakdown has been attributed to hot spot formation in a transistor. Discuss physically how the hot spot produces the characteristics seen.

22. Discuss the collector impedance of a uniformly-doped-base transistor associated with base-thickness change [the Early effect, see Eq. (4.17)]. Consider how this must be modified if recombination in the base is not negligible. Finally consider how it must be modified if the base contains a uniform electric field produced by graduation of doping.

23. Discuss the probable variation of electric field across the base region of a diffused or double-diffused Si transistor. Illustrate with some typical values and discuss the effect on the base transit time.

24. The current gain of a Si transistor initially increases with increase of collector current. Describe this effect as quantitatively as possible. What experiment would you conduct to confirm your model?

25. In heavily-doped transistors at high current densities the decrease of transistor gain with $I_c$ is attributed to effects other than the simple emitter injection model of Eq. (4.1). Describe these effects and discuss how satisfactorily they explain the behavior actually observed.

26. A factor termed the Gummel number has been shown to be important in transistor design. Explain.

27. Discuss side-wall effects in transistors. Under what circumstances have tunneling currents at emitter-base junctions been found significant in Si planar transistors?

28. Discuss in detail the merits of $n^+n$ emitter structures in $n^+npn$ Si transistors.

29. Review noise effects in diffused Si transistors. What sort of design would you use to minimize noise?

30. Discuss the effect of irradiation of the kind experienced in outer space on the performance and degradation of Si bipolar transistors. How are the problems minimized for space probes and satellites?

31. Discuss the differences between the large-signal common-emitter current gain $h_{FE}$ and the small signal current gain $h_{fe}$.

32. The angular cutoff frequency of a *pnp* transistor with uniformly doped base is related to $2D_p/W^2$. Derive this relationship.

33. For a drift transistor the angular cutoff frequency has been shown to be $k(\epsilon)D_p/W^2$ where $\epsilon$ is $\mu_p W/2D_p$ and $k(\epsilon) \simeq 2\epsilon^{3/2}$. Derive this relationship.

34. Discuss the Kirk base-thickening effect and the more recent studies along this line.

35. Discuss the relative merits of P and As diffusions for *n* regions in Si transistors and the merits of B and Ga diffusions for *p* regions.

36. Discuss the relative merits of Si and Ge transistors. Consider metallurgical and other practical factors as well as fundamental limits.

37. Add to Figs. 4.23 and 4.24 the results for some new high-performance bipolar transistors to determine the further progress that has been achieved in recent years.

38. Discuss the differences between unilateral- and maximum-available-gains and frequency responses of a transistor. Explain the causes of regions of potential instability as shown in Fig. 4.26.

39. Explain what equipment is needed to measure the $S$ parameters of a transistor. In what way can $S$ parameters be used to create a useful equivalent circuit and give an example of such a circuit. Could it be made equivalent to a hybrid $\pi$ circuit? Could you derive $S$ parameters from the SPICE parameters of Table 4.1?

40. Discuss the design of emitter ballast resistances for load sharing in high-frequency power transistors.

41. Discuss the temperature dependences of all the parameters in the hybrid $\pi$ model of a Si *npn* transistor.

42. Discuss the thermal resistance of power transistors and the role it plays in limiting performance.

43. A statement has been made that "for bipolar transistor power applications, thin-film techniques and module construction offer the greatest chance of achieving further improvements." Explain and discuss this statement.

44. Discuss accelerated life tests as a way of predicting the long-term performance of transistors.

45. Does the idea of constructing power transistors with built-in protective integrated circuitry seem a good concept, and if so, why has it not been more generally adopted?

46. Describe or devise some circuits that could be used to insure proper load sharing between three Si power transistors that have to be connected in parallel to provide 300 W of CW output power at 1 kHz for an ambient temperature range of $-20$ to $+80°$C.

47. Find in the literature a technical paper in which the Ebers-Moll model has been used in a highly successful way to characterize the performance of a transistor and explain the merits of the application.

48. How do the Ebers-Moll equations have to be modified if applied to a transistor with a constant built-in electric field in the base region?

49. Discuss the relative merits of classical (Moll) switching analysis and charge-control analysis in the design and study of Si switching transistors.

50. The charge-control analysis of a transistor requires values for quantities such as $Q_{BF}$, $\tau_F$, $\tau_{BF}$, etc. Explain how these quantities are measured.

51. Linvill and Gibbons in an elegant book[104] have described a lumped model for a transistor that uses parameters that have been termed *mobilance, diffusance, combinance* and *storance*. Attempt to explain why these concepts have not been more generally adopted.

52. Give some examples of the uses of lateral transistors in integrated circuits. Would it have been equally practical to use conventional transistors?

53. Review the performances reported so far for GaAs bipolar transistors and predict the future potential of such transistors.
54. Discuss some of the experimental evidence for the presence of energy spikes in the conduction or valence bands of a heterojunction. How does grading of the two semiconductors affect the band diagram?

## 4.10 REFERENCES AND FURTHER READING SUGGESTIONS

1. Adler, M.S., et al., "Limitations on injection efficiency in power devices," *IEEE International Electron Devices Meeting Technical Digest,* Washington, D.C., December 1975, p. 175.
2. Ajdler, J., "Transistor circuits," *Electronic Engineering,* 37, 1965, p. 757.
3. Altman, L., "Bipolar LSI: 10,000 gates in sight," *Electronics,* March 6, 1975, p. 57.
4. Altman, L., "Hybrid technology solves tough design problems," *Electronics,* June 7, 1973, p. 89.
5. Altman, L., "ISSCC special report: bipolar moves up to LSI," *Electronics,* March 6, 1975, p. 100.
6. Altman, L., "The new LSI bipolar chips are best buy for designers of fast systems," *Electronics,* July 10, 1975, p. 81.
7. Antoniazzi, P., "CATV transistors function as low-distortion vhf preamplifiers," *Electronics,* October 11, 1973, p. 105.
8. Archer, J., "Low-noise implanted-base microwave transistors," *Solid-State Electronics,* 17, p. 387, 1974.
9. Arnold, R.P., and D.S. Zoroglu, "A quantitative study of emitter ballasting," *IEEE Trans. Electron Devices,* ED-21, 1974, p. 385.
10. Asaoka, T., "Minority carrier storage device for scanners," *IEEE International Electron Devices Meeting, Technical Digest,* Washington, D.C., December 1974, p. 63.
11. Ashley, K.L., and F.H. Doerbec, "Correlation of emission and electrical properties of gallium arsenide transistor structures, *IEEE Trans. Electron Devices,* ED-17, 1970, p. 633.
12. Auitisyan, G.K.L., et al., "Semiconductor transistors with $n$-GaAs-$p^+$-Ge heterojunctions," *Soviet Physics–Semiconductors,* 5, 1972, p. 1601.
13. Bakowski, M., "Experimental verification of inhomogeneous field distribution in negatively bevelled high-voltage $p$-$n$ junctions by means of photo-multiplication," *Electronics Lett.,* 10, 1974, p. 292.
14. Bakowski, M., and K.I. Lundström, "Depletion layer characteristics at the surface of beveled high-voltage $P$-$N$ junctions," *IEEE Trans. Electron Devices,* ED-20, 1973, p. 550.
15. Bardeen, J., and W.H. Brattain, "The transistor, a semi-conductor triode," *Physical Review,* 74, *Series 2, 1948, p. 230.*
16. Barson, F., "Emitter-collector shorts in bipolar devices," *IEEE International Electron Devices Meeting Technical Digest,* Washington, D.C., December 1975, p. 447.
17. Beardsley, C.W., "The future of discretes," *IEEE Spectrum,* 10, 1973, p. 36.
18. Beeke, H., D. Flatley, and D. Stolnitz, "Double diffused gallium arsenide transistors," *Solid-State Electronics,* 8, 1965, p. 255.
19. Bergmann, F., and D. Gerstner, "Some new aspects of thermal instability of the current distribution in power transistors," *IEEE Trans. Electron Devices,* ED-13, 1966, p. 630.
20. Bhattacharyya, A.B., A. Srivastava, and R. Kumar, "Switching properties of epitaxial planar transistors operating in saturation," *Solid-State Electronics,* 18, 1975, p. 277.

21. Bowler, D.L., and F.A. Lindholm, "High-current regimes in transistor collector region," *IEEE Trans. Electron Devices*, **ED-20**, 1973, p. 257.
22. Burger, R.M., and R.P. Donovan, eds., *Fundamentals of Silicon Integrated Device Technology, Vol. 2, Bipolar and Unipolar Transistors*, Prentice-Hall, Englewood Cliffs, NJ 1968.
23. Butler, E.M., "On-line transistor modeling in a manufacturing environment," *IEEE Trans. Circuit Theory*, **CT-20**, 1973, p. 683.
24. Carley, D.R., "A worthy challenger for r-f power honors," *Electronics*, February 19, 1968, p. 98.
25. Caulton, M., H. Sabol, and R.L. Ern, "Generation of microwave power parametric frequency multiplication in a single transistor," *RCA Review*, 26, 1965, p. 286.
26. Chang, F.Y., and A.W. Chang, "Characterization and modeling of biopolar transistors fabricated with combined oxide and diffused isolation structures," *IEEE International Electron Devices Meeting Technical Digest*, Washington, D.C., December 1975, p. 577.
27. Chen, J.T.C., and K. Verma, "A 6 GHz silicon bipolar power transistor, *IEEE International Electron Devices Meeting Technical Digest*, Washington, D.C., December 1974, p. 299.
28. Cho, A.Y., et al., "Preparation of GaAs microwave devices by molecular beam epitaxy," *IEEE International Electron Devices Meeting Technical Digest*, Washington, D.C., December 1974, p. 579.
29. Chou, S., "An investigation of lateral transistors—D.C. characteristics," *Solid-State Electronics*, 14, 1971, p. 811.
30. Chou, S., "Small-signal characteristics of lateral transistors," *Solid-State Electronics*, 15, 1972, p. 27.
31. Clark, L.E., "On the measurement of the specific emitter efficiency factor in bipolar transistors," *Solid-State Electronics*, 15, 1972, p. 1293.
32. Collins, T.W., "Collector capacitance versus collector current for a double-diffused transistor," *Proc. IEEE*, **57**, 1969, p. 840.
33. Conn, D.R., and R.H. Mitchell, "Small-signal modelling and characterization of microwave transistors," *Int. J. Electronics*, 34, 1973, p. 655.
34. Conti, M., "Surface and bulk effects in low frequency noise in *NPN* planar transistors," *Solid-State Electronics*, 13, 1970, p. 1461.
35. Cooke, H.F., "Microwave transistors: theory and design," *Proc. IEEE*, **59**, 1971, p. 1163.
36. Cornu, J., R. Sittig, and W. Zimmerman, "Analysis and measurement of carrier lifetimes in the various operating modes of power devices," *Solid-State Electronics*, 17, 1974, p. 1099.
37. Curran, L., "Electron beams shine on IC layouts," *Electronics*, June 21, 1971, p. 83.
38. Das, M.B., and J.M. Humenick, "Transient limitations of bipolar transistors under a step-function emitter-base voltage excitation," *Proc. IEEE*, **61**, 1973, p. 1163.
39. Daw, A.N., N.K.D. Choudhury, and T. Sinha, "On the variation of cut-off frequency at high injection level with emitter end concentration of a diffused base transistor," *Solid-State Electronics*, 17, 1974, p. 1108.
40. Daw, A.N., R.N. Mitra, and N.K.D. Choudhury, "Cut-off frequency of a drift transistor," *Solid-State Electronics*, 10, 1967, p. 359.
41. DeFalco, J.A., "Coming up fast from behind—denser bipolar devices," *Electronics*, July 19, 1971, p. 76.
42. Deger, E., and T.C. Jobe, "For the real cost of a design, factor in reliability," *Electronics*, August 30, 1973, p. 83.
43. DeMassa, T.A., and L. Rispin, "Analysis of modified lateral *PNP* transistors," *Solid-State Electronics*, 18, 1975, p. 481.

44. Demizu, K., and Y. Yamamoto, "Second breakdown of IC structured power transistors," *IEEE Trans. Electron Devices,* **ED-22,** 1975, p. 352.
45. Den Brinker, C.S. "The struggle for power, frequency and bandwidth," *Radio and Electronic Engineer,* 43, 1973, p. 49.
46. Digiondomenico, O.J., and T.M. Foster, "Hybrids for microwave gear get boost from thick-film technology," *Electronics,* August 16, 1973, p. 104.
47. Dobkin, R.C., "IC with load protection simulates power transistor," *Electronics,* February 7, 1974, p. 121.
48. Early, J.M., "Effects of space-charge layer widening in junction transistors," *Proc. IRE,* 40, 1952, p. 1401.
49. Engelbrecht, R., and K. Kurakawa, "A wide-band low noise L-band balanced transistor amplifier," *Proc. IEEE,* 53, 1965, p. 237.
50. Fair, R.B., "Optimum low-level injection efficiency of silicon transistors with shallow arsenic emitters," *IEEE Trans. Electron Devices,* **ED-20,** 1973, p. 642.
51. Feldmanis, C.J., "Network analog maps heat flow," *Electronics,* May 16, 1974, p. 116.
52. Fulkerson, D.E., "A two-dimensional model for the calculation of common-emitter current gains of lateral *p-n-p* transistors," *Solid-State Electronics,* 11, p. 821 (1968).
53. Gallace, L.J., and J.S. Vara, "Evaluating reliability of plastic-packaged power transistors in consumer applications," *IEEE Trans. Broadcast and Television Receivers,* **BTR-19,** 1973, p. 194.
54. Getreu, I., "A new series, Modeling the bipolar transistor, Part 1" *Electronics,* September 19, 1974, p. 114.
55. Getreu, I., "Modeling the bipolar transistor, Part 2," *Electronics,* October 31, 1974, p. 71.
56. Getreu, I., "Part 3, Modeling the bipolar transistor," *Electronics,* November 14, 1974, p. 137. (See also *Modeling the Bipolar Transistor* (Pub.) Tektronik, Inc., Beaverton, Oregon, 1976.)
57. Ghandhi, S.K., "Darlington's compound connection for transistors," *IRE Trans. Circuit Theory,* CT-4, 1957, p. 291.
58. Ghandi, S.K., *Semiconductor Power Devices,* Wiley-Interscience, NY, 1977.
59. Gokhale, B.V., "Numerical solutions for a one-dimensional silicon *n-p-n* transistor," *IEEE Trans. Electron Devices,* **ED-17,** 1970, p. 594.
60. Golden, F.B., "Analysis can take the heat off power semiconductors," *Electronics,* December 6, 1973, p. 103.
61. Gopen, H.J., and A.Y.C. Yu, "$h_{FE}$ Fallout at low temperatures," *IEEE Trans. Electron Devices,* **ED-18,** 1971, p. 1146.
62. Gordner, J.A., "Liquid cooling safe-guards high power semiconductors," *Electronics,* February 21, 1974, p. 103.
63. Gover, A., and A. Gaash, "Experimental model aid for planar design of transistor characteristics in integrated circuits," *Solid-State Electronics,* 19, 1976, p. 125.
64. deGraaff, H.C., and J.W. Slotboom, "Some aspects of low emitter concentration transistor behaviour," *Solid-State Electronics,* 19, 1976, p. 125.
65. Graham, Jr., E.D., C.W. Gwyn, and R.J. Chaffin, "Device physics simulation," *Microwave Journal,* 18, February, 1975, p. 37.
66. Graham, Jr., E.D., and C.W. Gwyn, *Microwave Transistors,* Artech House, Dedham, Massachusetts 1975.
67. Grossman, S.E., "The thermal demands of electronic design," *Electronics,* November 8, 1973, p. 98.
68. Grove, A.S., *Physics and Technology of Semiconductor Devices,* John Wiley and Sons, NY, 1967.

69. Gummel, H.K., "Measurement of the number of impurities in the base layer of a transistor," *Proc. IRE*, **49**, 1961, p. 834.
70. Gummel, H.K., "A self-consistent iterative scheme for one-dimensional steady state transistor calculations," *IEEE Trans. Electron Devices*, **ED-11**, 1964, p. 455.
71. Gummel, H.K., "A charge control relation for bipolar transistors," *Bell System Tech. J.*, **49**, 1970, p. 115.
72. Gummel, H.K., and H.C. Poon, "An integral charge control model of bipolar transistors," *Bell System Tech. J.*, **49**, 1970, p. 826.
73. Harrison, R.G., "Computer simulation of a microwave power transistor," *IEEE Solid-State Circuits*, **SC-6**, 1971, p. 226.
74. Hart, B.L., "The transistor charge model: a tutorial development," *Electronic Components*, October 29, 1971, p. 1113.
75. Hartman, K., W. Kotyczka, and M.J.O. Strutt, "Experimental gain parameters of three microwave-bipolar transistors in the 2- to 8-GHz range," *Proc. IEEE*, **59**, 1971, p. 1720.
76. Heimeier, H.H., "A two-dimensional numerical analysis of a silicon *n-p-n* transistor," *IEEE Trans. Electron Devices*, **ED-20**, 1973, p. 708.
77. Helms, W.J., "Designing class A amplifiers to meet specified tolerances," *Electronics*, August 8, 1974, p. 115.
78. Herskowitz, G.J., and R.B. Schilling, eds., *Semiconductor Device Modeling for Computer-Aided Design*, McGraw-Hill, NY, 1972.
79. Hewlett Packard Co., "*S*-Parameters . . . Circuit analysis and design," Application Note 95, 1968, Palo Alto, CA, USA.
80  Hewlett Packard Co., "*S*-Parameter Design," Application Note 154, 1972, Palo Alto, CA, USA.
81. Hodowanec, G., "High-power transistor m/w oscillators," *Microwave Journal*, **15**, October, 1972, p. 47.
82. Hodowanec, G., "Microwave transistor oscillators," *Microwave Journal*, **17** June, 1974, p. 39.
83. Horton, R.L., J. Englade, and G. McGee," $I^2L$ takes bipolar integration a significant step forward," *Electronics*, February 6, 1975, p. 83.
84. Hower, P.L., "Optimum design of power transistor switches," *IEEE Trans. Electron Devices*, **ED-20**, 1973, p. 426.
85. Johnson, E.O., "Physical limitations on frequency and power parameters of transistors," *RCA Review*, **26**, 1965, p. 163.
86. Johnson, E.O., "Simple general analysis of amplifier devices with emitter, control, and collector functions," *Proc. IRE*, **47**, 1959, p. 407.
87. Johnsson, J.H., and M.J. Mallinger, "You can depend on today's rf power transistors," *Electronics*, September 13, 1971, p. 90.
88. Kamioka, H., H. Takeda, and M. Takagi, "A new sub-micron emitter formation with reduced base resistance for ultra high speed devices, *IEEE International Electron Devices Meeting Technical Digest*, Washington, D.C., December 1974, p. 279.
89. Kellner, W., and A. Goetzberger, "Current gain recovery in silicon nitride passivated planar transistors by hydrogen implantation," *IEEE Trans. Electron Devices*, **ED-22**, 1975, p. 531.
90. Kerr, J.A., and F. Berz, "The effect of emitter doping gradient on $f_r$ in microwave bipolar transistors," *IEEE Trans. Electron Devices*, **ED-22**, 1975, p. 15.
91. Kirk, Jr., C.T., "A theory of transistor cutoff frequency ($f_T$) fallout at high current densities," *IRE Trans. Electron Devices*, **ED-9** 1962, p. 164.
92. Koji, T., "The effect of emitter-current density on popcorn noise in transistors," *IEEE Trans. Electron Devices*, **ED-22**, 1975, p. 24.

93. Konagai, M., and K. Takahashi, "(GaAl) As-GaAs Heterojunction Transistors with high injection efficiency," *J. Appl. Phys.*, 46, 1975, p. 2120.

94. Kroemer, H., *Transistor-I*, RCA Laboratories, 1956, p. 202.

95. Kronquist, R.L., et al., "Determination of a microwave transistor model based on an experimental study of its internal structure," *Solid-State Electronics*, 18, 1975, p. 949.

96. Kumar, R., and L.P. Hunter, "Collector capacitance and high-level injection effects in bipolar transistors," *IEEE Trans. Electron Devices*, ED-22, 1975, p. 51.

97. Kumar, R., and L.P. Hunter, "Prediction of $f_T$ and $h_{fe}$ at high collector currents," *IEEE Trans. Electron Devices*, ED-22, 1975, p. 1031.

98. Kuno, H.J., "Rise and fall time calculations of junction transistors," *IEEE Trans. Electron Devices*, ED-11, 1964, p. 151.

99. Last, J.D., D.W. Lucas, and G.W. Sumerling, "A numerical analysis of the d.c. performance of small geometry lateral transistors," *Solid-State Electronics*, 17, 1974, p. 1111.

100. Last, J.D., and D.W. Lucas, "An unusual surface breakdown phenomena in lateral transistors," *Solid-State Electronics*, 16, 1973, p. 1084.

101. Lee, H.C., "Microwave power generation using overlay transistors," *RCA Review*, 27, 1966, p. 199.

102. Leturcq, Ph., and C. Cavalier, "A thermal model for high power devices design," *IEEE International Electron Devices Meeting Technical Digest*, Washington, D.C., December 1974, p. 422.

103. Lindmayer, J., and W. Schneider, "Theory of lateral transistors," *Solid-State Electronics*, 10, 1967, p. 225.

104. Linvill, J.G., and J.F. Gibbons, *Transistors and Active Circuits*, McGraw-Hill, NY, 1961.

105. Lo, T.C., "Base transport factor calculation for transistors with complementary error function and Gaussian base doping profiles," *IEEE Trans. Electron Devices*, ED-18, 1971, p. 243.

106. Long, E.L., and T.M. Frederiksen, "High-gain 15-W monolithic power amplifier with internal fault protection," *IEEE J. Solid-State Circuits*, SC-6, 1971, p. 35.

107. Magdo, I.E., and S. Magdo, "A high-performance, low-power 2.5 × 2.5 $\mu$m emitter transistor," *IEEE International Electron Devices Meeting Technical Digest*, Washington, D.C., December 1974, p. 276.

108. Magdo, S., "Intrinsic junctions and their role in SCL transistors," *IEEE International Electron Devices Meeting Technical Digest*, Washington, D.C., December 1974, p. 426.

109. Magdo, S., "Theory and operation of space-charge-limited transistors with transverse injection," *IBM Journal of Research and Development*, 17, 1973, p. 443.

110. Magdo, S., and I. Magdo, "High speed transistor with double base diffusion," *IBM Journal of Research and Development*, 19, 1975, p. 146.

111. Magill, L.M., "Watt-megahertz ratings run second to high reliability in foreign rf power transistors," *Electronics*, April 26, 1970, p. 80.

112. Manck, O., H.H. Heimeier, and W.L. Engl, "High injection in a two-dimensional transistor," *IEEE Trans. Electron Devices*, ED-21, 1974, p. 403.

113. Matsushita, T., et al., "Highly reliable high-voltage transistors by use of the SIPOS Process," *IEEE International Electron Devices Meeting Technical Digest*, Washington, D.C., December 1975, p. 167.

114. Mertens, R.P., H.J. DeMan and R.J. Van Overstraeten, "Calculation of the emitter efficiency of bipolar transistors," *IEEE Trans. Electron Devices*, ED-20, 1973, p. 772.

115. Mertens, R.P., et al., "Influence of high doping on the design of bipolar transistors," *IEEE International Electron Devices Meeting Technical Digest*, Washington, D.C., December 1974, p. 411.

116. Milnes, A.G., and D.L. Feucht, *Heterojunctions and Metal-Semiconductor Junctions,* Academic Press, NY, 1972.
117. Mimura, R., "Planar analog transistor," *Proc. IEEE,* **62**, 1974, p. 1285.
118. Mock, M.S., "On heavy doping effects and the injection efficiency of silicon transistors," *Solid-State Electronics,* **17**, 1974, p. 819.
119. Moll, J.L., "Large-signal transient response of junction transistors," *Proc. IRE,* **42**, 1954, p. 1773.
120. Nagel, L.W., and D.O. Pederson, "Simulation program with integrated circuit emphasis," *Proc. Sixteenth Midwest Symposium on Circuit Theory,* Waterloo, Canada, April 12, 1973.
121. Nakamura, H., S. Ohyama, and C. Tadachi, "Boron diffusion coefficient increased by phosphorus diffusion," *J. Electrochem. Soc.,* **121**, 1974, p. 1377.
122. Naster, R.J., and W.H. Perkins, "Solid-state power amplifiers for L-band phased arrays," *Microwave Journal,* **18**, July, 1975, p. 56.
123. Nicholas, K.H., et al., "Reduced gain of ion-implanted transistors," *Appl. Phys. Lett.,* **26**, 1975, p. 320.
124. Olmstead, J., et al., "High-level current gain in bipolar power transistors," *RCA Review,* **32**, 1971, p. 221.
125. Peltzer, D., and B. Herndon, "Isolation method shrinks bipolar cells for fast, dense memories," *Electronics,* March 1, 1971, p. 52.
126. Perner, F.A., "Quasi-saturation region model of an $npn^-n$ transistor," *IEEE International Electron Devices Meeting Technical Digest,* Washington, D.C., December 1974, p. 418.
127. Petersen, K.E., D. Adler, and M.P. Shaw, "Operation of an amorphous-emitter transistor, *Appl. Phys. Lett.,* **25**, 1974, p. 586.
128. Petrov, B.K., A.I. Kochetkov, and V.F. Synorov, "A calculation of the equilibrium temperature distributions in multiple-emitter microwave transistors," *Radio Engineering and Electronic Physics,* **17**, 1972, p. 1738.
129. Poon, H.C., H.K. Grummel, and D.L. Scharfetter, "High injection in epitaxial transistors," *IEEE Trans. Electron Devices,* ED-16, 1969, p. 455.
130. Popescu, C., "The second breakdown in reverse biased transistor as an electrothermal switching," *IEEE Trans. Electron Devices,* ED-21, 1974, p. 428.
131. Presser, A., and E.F. Belohaubek, "1-2 GHz high-power linear transistor amplifier," *RCA Review,* **33**, 1972, p. 737.
132. Pritchard, R.L., et al., "Transistor internal parameters for small-signal representation," *Proc. IRE,* **49**, 1961, p. 725.
133. Reisman, A., "Germanium IC's point the way towards picosecond computers," *Electronics,* March 3, 1960, p. 88.
134. Rey, G., and J.P. Bailbe, "Some aspects of current gain variations in bipolar transistors," *Solid-State Electronics,* **17**, 1974, p. 1045.
135. Rey, G., F. Dupuy, and J.P. Bailbe, "A unified approach to the base widening mechanisms in bipolar transistors," *Solid-State Electronics,* **18**, 1975, p. 863.
136. Rosenberg, C., I. Arimura, and A.M. Unwin, "Statistical analysis of neutron-induced gain degradation of power transistors," *IEEE Trans. Nuclear Science,* NS-17, 1970, p. 160.
137. Rüegg, H.W., and W. Thommen, "Bipolar micropower circuits for crystal-controlled timepieces," *IEEE J. Solid-State Circuits,* SC7, 1972, p. 105.
138. Saltick, J.L., C.E. Volk, and L.E. Clark, "Use of specific emitter efficiency in evaluation and design of bipolar transistors," *Proc. IEEE,* **61**, 1973, p. 680.
139. Sayed, M.M., J.T.C. Chen, and S. Kakohana, "A new concept of ballasting mechanisms of microwave power transistors in class C operation," *IEEE International Electron Devices Meeting Technical Digest,* Washington, D.C., December 1974, p. 302.

140. Schafft, H.A., "Second breakdown—a comprehensive review," *Proc. IEEE*, **55**, 1967, p. 1272.
141. Schafft, H.A., and J.C. French, "Second breakdown in transistors," *IRE Trans. Electron Devices*, ED-9, 1962, p. 129.
142. Seidel, T.E., et al., "Transistors with boron bases predeposited by ion implantation and annealed in various oxygen ambients," *IEEE International Electron Devices Meeting Technical Digest*, Washington, D.C., December 1975, p. 581.
143. Seltz, D.S., and I. Kidron, "A two-dimensional model for the lateral *p-n-p* transistor," *IEEE Trans. Electron Devices*, ED-21, 1974, p. 587.
144. Shackle, P.W., "An experimental study of distributed effects in a microwave bipolar transistor," *IEEE Trans. Electron Devices*, ED-21, 1974, p. 32.
145. Sharma, B.L., and R.K. Purohit, *Semiconductor Heterojunctions*, Pergamon Press, Oxford, UK, 1974.
146. Sheng, W.W., "The effect of Auger recombination on the emitter injection efficiency of bipolar transistors," *IEEE Trans. Electron Devices*, ED-22, 1975, p. 25.
147. Shepherd, A.A., "Semiconductor device developments in the 1960's," *Radio and Electronic Engineer*, **43**, 1973, p. 11.
148. Shyne, N.A., "Bipolar transistor pair simulates unijunction," *Electronics*, January 24, 1974, p. 113.
149. Slotboom, J.W., "Computer-aided two-dimensional analysis of bipolar transistors," *IEEE Trans. Electron Devices*, ED-20, 1973, p. 669.
150. Sobol, H., and F. Sterzer, "Microwave power sources," *IEEE Spectrum*, April 1972, p. 20.
151. Sparkes, J.J., "The first decade of transistor development: a personal view," *Radio and Electronic Engineer*, **43**, 1973, p. 3.
152. Sparkes, J.J., "A study of charge control parameters of transistors," *Proc. IRE*, **48**, 1960, p. 1693.
153. Sparkes, J.J., and R. Beaufoy, "The junction transistor as a charge controlled device," *Proc. IRE*, **45**, 1957, p. 1740.
154. Statz, H., "Double diffused *pnp* GaAs transistor," *Solid-State Electronics*, **8**, 1965, p. 827.
155. Su, S.C., and J.D. Meindl, "A new complementary monolithic transistor structure," *IEEE J. Solid-State Circuits*, SC-7, 1972, p. 170.
156. Sutherland, A.D., and D.P. Kennedy, "A computer model for lateral thermal instabilities in power transistors," *IEEE International Electron Devices Meeting Technical Digest*, Washington, D.C., December 1975, p. 569.
157. Sze, S.M., *Physics of Semiconductor Devices*, Wiley-Interscience, NY, 1969.
158. Tada, K., and H. Yanai, "Basic considerations in charge control analysis of the transistor cutoff frequency," *Electronics and Comm. in Japan*, **48**, February 1965, p. 48.
159. Takagaki, T., and M. Mukogawa, "Gold diffusion transistor logic: a new LSI gate family," *IEEE International Electron Devices Meeting, Technical Digest*, Washington, D.C., December 1975, p. 559.
160. Tillman, J.R., "The expanding role of semiconductor devices in telecommunications," *Radio and Electronic Engineer*, **43**, 1973, p. 82.
161. Tokagi, M., et al., "Improvement of shallow base transistor technology by using a doped poly-silicon diffusion source, *Proc. of the Fourth Conference on Solid State Devices*, Tokyo 1972, *Supplement to the J. of the Japan Society of Applied Physics*, **42**, 1973, p. 101.
162. Uhler, P.T., "Doubling breakdown voltage with cascoded transistors," *Electronics*, January 4, 1973, p. 102.
163. Veloric, H.S., C. Fuselier, and D. Rauscher, "Base-layer design for high frequency transistors," *RCA Review*, **23**, 1962, p. 112.

164. Veloric, H.S., A. Presser, and F.J. Wozniak, "Ultra-thin RF silicon transistors with a copper-plated heat sink," *RCA Review*, **36**, 1975, p. 731.
165. Verma, K.B., and R.T. Westlake, "A microwave high voltage bipolar power transistor," *IEEE International Electron Devices Meeting Technical Digest*, Washington, D.C., December 1975, p. 155.
166. Wahl, A.J., "Distributed theory for microwave bipolar transistors," *IEEE Trans. Electron Devices*, **ED-21**, 1974, p. 40.
167. Wait, J.T., J.R. Hauser, "On the beta falloff in junction transistors," *Proc. IEEE*, **57**, 1969, p. 1293.
168. Wallace, Jr., R.L., and W.J. Pietenpol, "Some circuit properties and applications of *n-p-n* transistors," *Bell System Tech. J.*, **30**, 1951, p. 530.
169. Wang, A.S., and S. Kakihana, "Leakage and $h_{FE}$ degradation in microwave bipolar transistors," *IEEE Trans. Electron Devices*, **ED-21**, 1974, p. 667.
170. Warner, Jr., R.M., and B.L. Grung, "A bipolar lock-layer transistor," *Solid-State Electronics*, **18**, 1975, p. 323.
171. Whittier, R.J., and D.A. Tremere, "Current gain and cutoff frequency fall off at high current," *IEEE Trans. Electron Devices*, **ED-16**, 1969, p. 39.
172. Wieder, A.W., O. Manek, and W.L. Engl, "Two-dimensional analysis of a monolithic PNP transistor," *IEEE International Electron Devices Meeting Technical Digest*, Washington, D.C., December 1974, p. 414.
173. Wiedmann, S.K., "Pinch load resistors shrink bipolar memory cells," *Electronics*, March 7, 1974, p. 130.
174. Wiedmann, S.K., and H.H. Berger, "Superintegrated memory shares functions on diffused islands," *Electronics*, February 14, 1972, p. 83.
175. Te Winkel, J., "Extended charge-control model for bipolar transistors," *IEEE Trans. Electron Devices*, **ED-20**, 1973, p. 389.
176. Xylander, M.P., "Low-power bipolar technique begets low-power LSI logic," *Electronics*, July 31, 1972, p. 80.
177. Yagi, H., et al., "A novel bipolar device with low emitter impurity concentration structure," *IEEE International Electron Devices Meeting Technical Digest*, Washington, D.C., December 1974, p. 262.
178. Yanai, H., T. Sugano, K. Tada, "The minority carrier storage effect in the collector region and the storage time of transistors," *Proc. IEEE* **52**, 1964, p. 312.
179. "High-voltage power Darlingtons," *Electronics*, January 4, 1973, p. 42.

# 5

# Thyristors – Controlled *PNPN* and Related Switch Devices

## CONTENTS

## 5.1 BASIC CONCEPTS

Thyristors are triggered switching devices involving *pnpn* or more-layered structures. Usually three terminals are provided, namely anode, cathode and gate as shown in Fig. 5.1(a). The device is connected to a voltage supply (sometimes dc but usually ac) with the cathode generally grounded to a substantial heat sink. In the absence of gate voltage the $I$ vs $V$ curve of the device has the characteristic shown by the full line in Fig. 5.1(b). For the polarity shown in Fig. 5.1(a) and the upper right-hand quadrant of Fig. 5.1(b), the center junction $J_2$ sustains the applied voltage with very little current flow until the forward breakdown avalanche voltage $V_{FB}$ is reached. Then the center $n$ and $p$ regions become flooded with minority carriers. The depletion region voltage at $J_2$ collapses and the device turns on and the total voltage drop becomes less than 2 V and the current is primarily limited by the resistor. The dashed line in Fig. 5.1(b) is the load line for the resistance $R_L$. With a supply voltage $V_{DC}$ the current before device breakdown is $I_0$. A pulse of positive voltage of 1 or 2 V applied to the gate terminal then causes current to flow in junction $J_3$. Some of the carriers involved reach the junction $J_2$ and start the avalanche carrier flooding effect and the effective breakdown voltage of the device drops below $V_{FB}$ as suggested by the chain-dot line in Fig. 5.1(b). Since the peak of this line is below the load

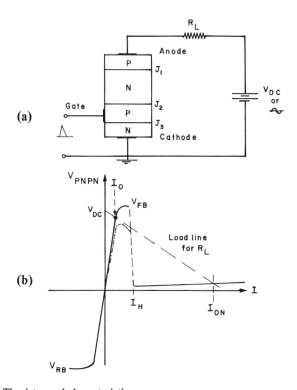

Fig. 5.1. Thyristor and characteristic:
(a)  Four layer *pnpn* structure, and
(b)  *V-I* characteristic with load line: The chain-dot line shows the effect of a positive voltage pulse on the gate.

line, switching occurs and the circuit current becomes $I_{ON}$. Notice that there is a holding current $I_H$ below which the device will not remain on. If the circuit has an ac supply voltage the device will turn off as the alternating voltage drops to zero and the reverse half-cycle voltage will be sustained by reverse bias voltages on junctions $J_1$ and $J_3$.

Another way of turning on the device is to add a voltage pulse to $V_{DC}$ that momentarily raises the load line above the breakdown value $V_{FB}$. In the early days of semiconductor switching devices, rectifiers or transistors occasionally exhibited such a switching action and the cause was probably the inadvertent formation of four layer structures because of ohmic contact or other processing problems. For a few years two-terminal switching rectifiers were commercially fabricated but the convenience of gate control soon led to three-terminal structures. The term silicon controlled rectifiers (SCR) was used in this period but has been universally replaced by the term thyristor. An important factor in the rapid growth of the thyristor industry was the ability to handle large amounts of power in a switching mode of operation. Depletion regions in thyristors can be made much thicker than in transistors (which must have narrow base regions for high current gain) so large voltage ratings are readily obtained. Since the gate in a thyristor is not involved in the current flow after turn-on, the fraction of wafer surface devoted to the gate can be smaller than that given to a transistor base so the current rating can be large. Air cooled single thyristor units are readily available today with voltage ratings of 3000 V or more and surge current ratings of 20,000 A. Thyristors have tended to replace mercury-arc and other gas discharge rectifiers for ac to dc conversion, electric motor drives and the control of electric mechanical servo systems. But thyristors are also convenient and economical circuit elements for operation at the level of a few amperes and 100 or 200 V. Hence the world-wide market for thyristors is substantial—in 1974 the USA market alone was \$204 million.

An initial understanding of the turn-on action of a thyristor can be obtained by considering the *pnpn* structure as a combination of *pnp* and *npn* transistors connected as shown in Fig. 5.2. The total current crossing the center junction is the sum of flux of holes across the base region of the *pnp* transistor, the flux of electrons coming from the *npn* transistor and the generation current $I_{C0}$ of the junction $J_2$ depletion region. This total current is the device current $I_A$. In the absence of gate current the cathode current is also $I_A$, hence at the center junction

$$I_A = \alpha_1 I_A + \alpha_3 I_A + I_{C0} \tag{5.1}$$

or

$$I_A = \frac{I_{C0}}{1 - (\alpha_1 + \alpha_3)} \tag{5.2}$$

where $\alpha_1$ and $\alpha_3$ are the current gains of the *pnp* and *npn* sections. If the field strength in the center region is high there may even be avalanche multiplication of holes, $M_h$, and of electrons, $M_e$, in this region, so that Eq. (5.2) becomes

$$I_D = \frac{I_{C0}}{1 - (M_h \alpha_1 + M_e \alpha_3)} \tag{5.3}$$

However, for simplicity assume that the circuit stand-off voltage is low enough that the $M_h$ and $M_e$ multiplication factors are unity. Inspection of Eq. (5.2) shows that the thyristor will turn on and conduct a large current if $(\alpha_1 + \alpha_3)$ equals unity. The device is therefore designed with the center layers thick enough that the minority carrier transport losses result in $\alpha_1$ and $\alpha_3$ values that do not quite sum to unity in the absence of gate current. These thick center layers, as previously explained, allow the structure to have a thick depletion region and therefore a large voltage rating for the center junction. If the center *n* region is the most lightly doped base, so that the depletion region is mainly

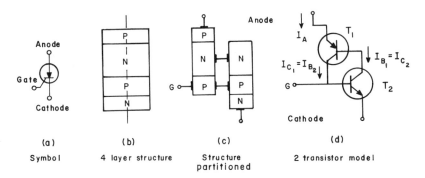

(a)                  (b)                    (c)                     (d)

Symbol      4 layer structure       Structure          2 transistor model
                                    partitioned

Fig. 5.2. Thyristor symbol and structure with representation as a two transistor model.

there, the voltage capability of the device is as shown in Fig. 5.3 (for uniform doping). Therefore, if the doping concentration $N_D$ is $10^{14}$ cm$^{-3}$ the voltage for avalanche will be about 1400 V and the depletion thickness is 140 $\mu$m. Hence, the thickness of the $n$ base region must be at least as large if reach-through is not to occur before avalanche. If a forward voltage bias is being considered the depletion region spreads from the center junction $J_2$ and reduces the effective base thickness of the *pnp* transistor section so $\alpha_1$ tends to increase. Therefore, the forward voltage may have to be limited to a lower value than the

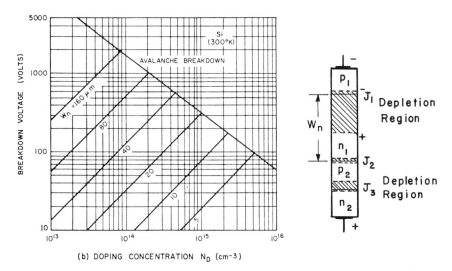

(b) DOPING CONCENTRATION $N_D$ (cm$^{-3}$)

Fig. 5.3. Reverse voltage limitations of a thyristor determined by avalanche or by depletion reach-through of the $n_1$ layer. (After Herlet.[39] Reprinted with permission from *Solid-State Electronics*, Pergamon Press, Ltd.)

reverse voltage rating in order that $(\alpha_1 + \alpha_3)$ does not exceed unity except under gate control.

Instead of supposing turn-on to occur when $I_A$ of Eq. (5.2) becomes infinite, a more appropriate turn-on point to consider is the peak of the voltage-current curve at which $dV_F/dI_A$ is zero. This occurs when the sum of the small signal alphas $(\alpha_{1s} + \alpha_{3s})$ equals unity rather than when the sum of the large signal alphas equals unity. Computed variations of small and large signal alphas with emitter current are shown in Fig. 5.4 for a typical transistor. The small signal alpha is seen to be significantly greater than the large signal (dc) alpha over most of the range of the curves.

The analysis that involves the sum of the small signal alphas follows. The gate current is considered to be increased by a small current $\Delta I_g$. The increment of cathode current is then

$$\Delta I_K = \Delta I_g + \Delta I_A \qquad (5.4)$$

The increment of hole current reaching the center junction $J_2$ is then $\alpha_{1s}\Delta I_A$ and the increment of electron current is $\alpha_{3s}\Delta I_K$. But the total increment of current across the center junction is $\Delta I_A$, hence

$$\Delta I_A = \alpha_{1s}\Delta I_A + \alpha_{3s}\Delta I_K \qquad (5.5)$$

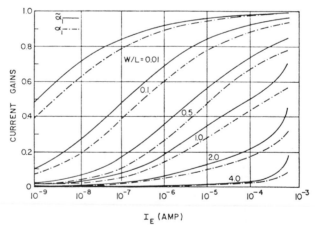

$$I_E \ (\text{AMP})$$

Fig. 5.4. Curves showing the dependence of small signal and dc alphas on current and base thickness in a *pnp* structure. These curves were calculated for $n_{no} = 3 \times 10^{14} \text{cm}^{-3}$, $p_{no} = 7.5 \times 10^5 \text{cm}^{-3}$, $A_s = 0.16$ mm$^2$, $\mu_n = 1400$ cm$^2$/V-sec, $\mu_p = 500$ cm$^2$/V-sec, $D = 13$ cm$^2$/sec, $\tau_p = 0.5$ $\mu$sec, $L_p = 25.5$ $\mu$m, $I_R = 2.5 \times 10^{-10}$ A, and $\eta = 1.5$. (After Yang and Voulgaris.[107] Reprinted with permission from *Solid-State Electronics*, Pergamon Press, Ltd.)

Combination of Eqs. (5.4) and (5.5) gives

$$\frac{\Delta I_A}{\Delta I_g} = \frac{\alpha_{3s}}{1 - (\alpha_{1s} + \alpha_{3s})} \tag{5.6}$$

Hence, if $\alpha_{1s} + \alpha_{3s}$ is unity, a slight change in $\Delta I_g$ causes $\Delta I_A$ to be large and $\Delta V_F/\Delta I_A$ becomes zero and the thyristor turns on.

Triggering of the thyristor must be achieved by the application of a gate signal that causes $\alpha_{3s}$ to increase so that the sum $(\alpha_{1s} + \alpha_{3s})$ exceeds unity. There are several ways in which $\alpha_{3s}$ may be made an increasing function of $I_g$. For instance the gate voltage causes a small drift voltage to occur in the $p$ base layer which assists the transport of injected minority carriers across this region in order that more arrive at the center junction. Another factor is that the $p$ base region may contain recombination or trapping centers that act to limit the injected electron transport efficiency. At a larger current flow, induced by the gate signal, the recombination actions may be relatively less significant so $\alpha_{3s}$ rises. Sometimes light Au doping or electron irradiation of the $p$ base region may be used to provide the recombination action needed.[32, 108] The variation of small signal alpha with emitter current for a constant recombination condition can be substantial as seen from Fig. 5.4.

However, a simpler method for achieving an increase in $\alpha_{3s}$ with increase of gate signal is generally used. This involves provision of a resistive path from the cathode contact to the $p$ base region that shunts the $np$ junction $J_3$. At very low gate-cathode voltages the current is carried mainly by the resistive shunt path, but as the gate-cathode drive voltage increases the $np$ junction injects more and more so the effective $\alpha_3$ increases and the device turns on. The shunt path may be modeled by the analogous transistor circuit of Fig. 5.5. A dc effective alpha may be defined for the $n_2 p_2 n_1$ structure as

$$\alpha_{eff} = \left(\frac{I_1}{I_1 + I_s}\right) \alpha_3 \tag{5.7}$$

The relationship between $I_1$ and $V_{BE}$ is of the form

$$I_1 \simeq I_0 \exp(q V_{BE}/nkT) \tag{5.8}$$

which for numerical illustration is assumed to be as in Fig. 5.5(b). Then

$$\alpha_{eff} = \alpha_3 \left[ \frac{I_0 \exp(q V_{BE}/nkT)}{I_0 \exp(q V_{BE}/nkT) + V_{BE}/R_s} \right] \tag{5.9}$$

If $\alpha_3$ is postulated to change with $I_1$ as shown in Fig. 5.5(c), the variation of $\alpha_{eff}$ becomes a pronounced function of $I_1$. Hence, the desired rapid dependence

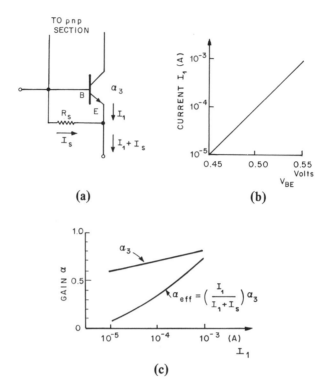

(a)

(b)

(c)

Fig. 5.5. Model of the *pn* section of a thyristor with a cathode shunt resistance $R_S$.

(a)   Circuit to show $a_{eff} = \left(\dfrac{I_1}{I_1 + I_s}\right) a_S$,

(b)   Assumed relationship between $I_1$ and $V_{BE}$, and

(c)   Assumed variation of $a_3$ with $I_1$ compared to the calculated $a_{eff}$ for a shunt resistance of $5 \times 10^3$ ohms.

of $(\alpha_1 + \alpha_{eff})$ on turn-on gate-cathode voltage is achieved by the shunt-resistance-path technique.

In practice the shunt path is usually provided by allowing the cathode metal contact to slightly overlap the perimeter of the $n_2$ region so that it provides a conducting path to the $p_2$ base region. A further feature of this cathode-short approach is that as current flows through the device it may divide as shown in Fig. 5.6 to produce a transverse voltage drop across the $n_2$ emitter that tends to drive the center of this region into increased forward bias and so further increases the effective $\alpha_3$. Multiple shorts are provided in large area thyristors.[32]

So far the doping profiles within a thyristor and the electron and hole distributions have not been discussed. Most thyristors are made by $p$ and $n$ diffusions into a uniformly doped $n$ type wafer and a profile for a low voltage unit might be as shown in Fig. 5.7(a). A one dimensional calculation of the terminal characteristic gives the result shown in Fig. 5.7(b) for an SRH

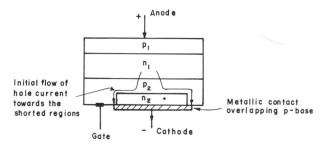

Fig. 5.6. Thyristor with cathode metallization shorted to the $p_2$ base at the perimeter to give alpha dependence on current. The flow lines show enchancement of the action by a transverse voltage drop across the $n_2$ emitter.

recombination model. The electron and hole distributions corresponding to various points on the terminal characteristic are given in Fig. 5.8. Large increases in electron and hole density in the region of the center junction are seen as the voltage across this region collapses, between points 3 and 5 in Fig. 5.7(b).

In more recent modeling of thyristors, the carrier lifetime has been taken as varying inversely as the 0.3 to 0.4 power of the doping in the diffused region. Consideration has also been given to carrier-carrier scattering and to Auger effects.[110]

Some thyristor structures are provided with an anode short in addition to the cathode short. The feature of the added anode short is that both $J_1$ and $J_3$ are then shunted in the forward blocking mode so the initial injection is low and the center junction sustains a higher voltage before turn-on and the junction temperatures may be as high as 150°C instead of a limit of about 125°C. The double-shorted structure is termed a reverse conducting thyristor (RCT) since it has no reverse voltage blocking rating because neither $J_1$ nor $J_3$ can sustain reverse voltage. There are some circuit applications where the higher forward blocking voltage and temperature capability of the RCT are advantageous and the loss of the reverse rating is not a serious disadvantage.

Important parameters in the performances of conventional thyristor switches include not only the voltage, current and power ratings but also the ability to turn on fast by handling large $dI/dt$ ratings, and to turn off fast and sustain large $dV/dt$ ratings without false triggering. These are matters that involve three dimensional considerations of the design so we must now consider partial turn-on conditions and problems of gate geometry.

## 5.2 THYRISTOR TURN-ON, TURN-OFF AND POWER CONSIDERATIONS
### TURN-ON

Complete turn-on of a thyristor cannot take place instantaneously and there is a period when the current through the device is rising and the voltage is dropping

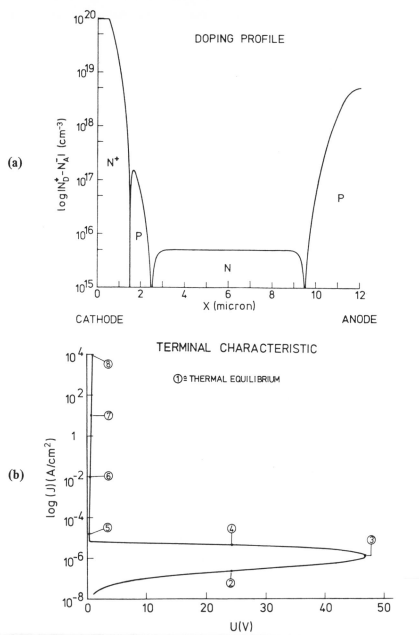

Fig. 5.7. Doping profile assumed for a low voltage thyristor and the resulting computed $V$ vs log $I$ characteristic if the $n$ base lifetime is about 20 μsec.

(a)  Doping profile of the one-dimensional structure, and

(b)  Calculated forward $I$-$V$ characteristics ($\tau_{no}$ = 20 μsec). The encircled numbers in this and the following figures refer to the same bias conditions. (After Anheier, et al. Reprinted with permission from *International IEEE Electron Devices Meeting*, 1975, p. 364.)

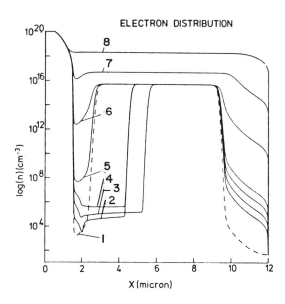

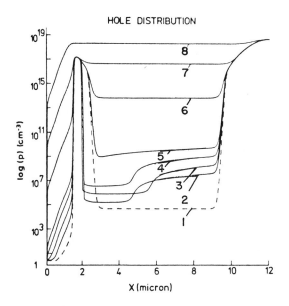

Fig. 5.8.  Electron and hole distributions calculated for thyristor of Fig. 5.7. (After Anheier, et al. Reprinted with permission from *International IEEE Electron Devices Meeting*, 1975, p. 365.)

from the stand-off value to the low "on" value. During this time the power dissipation is large, and the problem is compounded by the fact that initial turn-on is concentrated in the cathode region near the gate and takes some time to spread throughout the region. Hence, the power dissipation is high in a localized region and hot spots may occur. This is illustrated in Fig. 5.9(a) for a center-gate thyristor. The power dissipation in a thyristor peaks during the turn-on process when the current is rising but the device voltage has not fully declined as shown in Fig. 5.9(c). In Fig. 5.9(d) measured turn-on powers are shown vs time for a thyristor of 600 V initial voltage with various rates of current rise to a load current of 1000 A. This power is not uniformly distributed over the area of the wafer.

The current spreading during turn-on is mainly by a process of lateral diffusion of the initial plasma as shown in Fig. 5.10. The factors that influence the turn-on time of a thyristor include:

- The geometry of the gate. A large perimeter interlaced with that of the cathode is preferred so that no section of cathode area is remote from the gate.
- The amplitude and rise time of the gate voltage and current.
- The magnitude of the load current that is to be switched on and the resistive or resistive-inductive nature of the load.
- The stand-off voltage that has been sustained by the anode just before switching on.
- The temperature of the device.

The turn-on and turn-off times for a thyristor with a resistive load are illustrated in Fig. 5.11. The initial delay time $t_d$ after a step of gate voltage is associated with processes (a) and (b) of Fig. 5.10. Once the initial turn-on has been accomplished, which is taken (arbitrarily) as 10% of the final load current, a rise time $t_r$ ensues during which the lateral spread of the turn-on region proceeds as in Fig. 5.10(c). The rise time for convenience of measurement is taken as complete when 90% of the final load current is flowing.

Consider now a conducting thyristor which is turned off by a swing of anode voltage to a negative value as shown in Fig. 5.11(a). There is a storage time $t_s$ during which the voltage across the transistor changes very little because the stored charge in the two base layers initially prevents any depletion region forming at the junctions $J_1$ and $J_3$. As this charge is extracted or is decreased by recombination, voltage builds up across the thyristor and the reverse load current decreases from 90% to 10% during the fall time, $t_f$. The storage and fall times are decreased if $V_R$ is increased as the stored charge is extracted more rapidly by the flow of a larger reverse current.

The recovery time may be shortened by Au doping or by electron irradiation to reduce the lifetime of carriers in the center regions of the thyristor. However, most thyristors are high voltage devices and the $n$ type shallow doping in the $n$ base region is made low (mid $10^{14}$ cm$^{-3}$) to allow a thick depletion region to

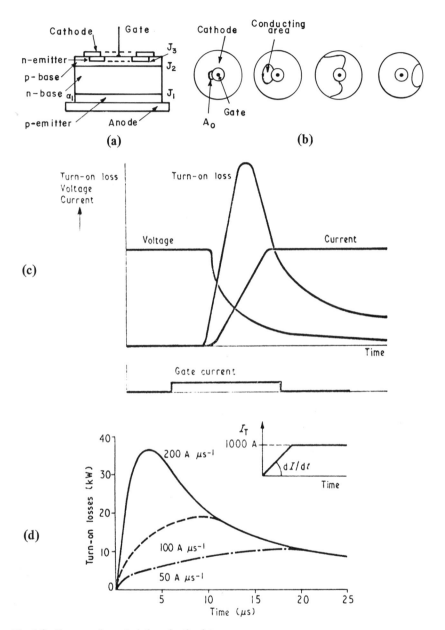

Fig. 5.9. Turn-on characteristics of a thyristor:

(a) Center gate structure,

(b) Turned-on gate developing, (After Cordingley.[19] Reprinted with permission from *J. of Science and Technology*)

(c) Turn-on losses. (After Herlet.[40] *Proc. of the European Semiconductor Device Research Conference*, copyright 1971, Inst. of Physics, London.)

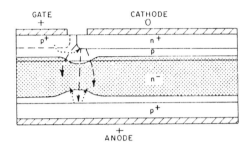

(i) Electrons cross depletion layer and forward bias *p* emitter.

(ii) Holes are injected by *p* emitter and enter depletion layer.

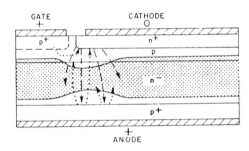

(iii) Holes cross depletion layer and add to forward bias on *n* emitter.

(iv) Additional electrons are injected, cross depletion layer, and add to forward bias on *p* emitter.

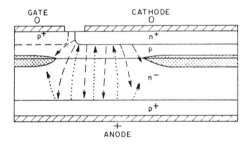

(v) Carrier injection continues to increase from both emitters.

(vi) Plasma develops, containing high concentration of holes and electrons.

(vii) Conduction continues to spread even if gate bias removed.

Fig. 5.10. Impression of the turn-on plasma spreading under the cathode of a thyristor. (After Becke, et al.[13], Courtesy of RCA Corp.)

form that will sustain the voltage. Also, the *n* base region must be quite thick to allow the *pnp* structure to have a low gain ($\alpha_1$). The Au doping should not be larger than the shallow doping since otherwise the *n* base develops high resistance, and therefore the lifetime must not be reduced to very low values. If the *n* base region develops high resistance the forward voltage drop of the

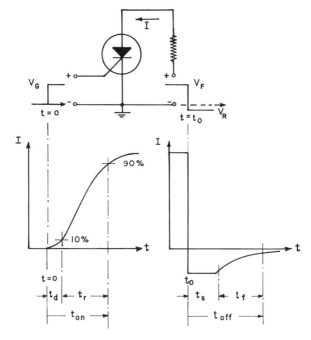

Fig. 5.11. Turn-on and turn-off transients of a thyristor. The sketch shows delay, rise, storage and fall times.

thyristor in the on-state is increased, even though this is mitigated by the conductivity modulation due to the high injected carrier densities when the device is conducting. Typical fall times for low current thyristors (about 20-A rating) may be 1 $\mu$sec. For larger thyristors the fall time is greater since injected carrier densities may be higher. However, 3-10 $\mu$sec recovery times for 500 A, 1000 V ratings are possible.[65]

The effect of increased junction temperature is usually to increase the fall time of a thyristor during turn-off mainly because the effect on recombination mechanisms is to lengthen the injected carrier lifetimes. This is a further reason to avoid hot-spot formation during turn-on, since this also increases turn-off time if the conduction period is short compared with the temperature redistribution time.

Thyristor gate geometry is often a simple center gate as in Fig. 5.9, but an outer ring gate may also be provided to encourage more uniform turn-on. The gate voltage and gate current must conform to the manufacturer's allowable maximum and minimum values. A chart as in Fig. 5.12(a) may be provided for each family of devices showing the preferred firing area and the maximum allowable instantaneous gate power. Figure 5.12(b) shows an enlarged section of Fig. 5.12(a) with some effects of temperature indicated.

**(a)**

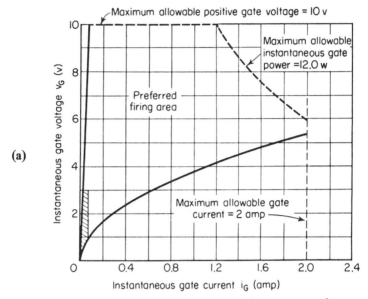

Notes: (1) Junction temperature − 65°C to + 150°C.

(2) Shaded areas represent locus of
possible firing points from
−65°C to +150°C.

**(b)**

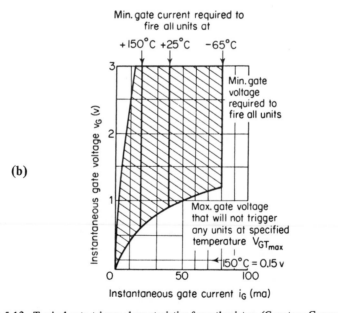

Fig. 5.12. Typical gate trigger characteristics for a thyristor. (Courtesy General Electric Co.)

A more elaborate gate arrangement is shown in Fig. 5.13. The gate with the spiral involute fingers[37, 97] is an approach which aims at keeping the distance from the gate to the cathode $n$ region constant for more uniform turn-on and improved $dI/dt$ performance. Table 5.1 shows the superior switching frequency performance of an involute gate thyristor compared with a more conventional gate unit.

The more complicated the gate structure, the more difficult it may be to achieve good heat sinking of the cathode. Also, the cathode area may be lessened by the increased area needs of the gate and so there may be some lowering effect on the current rating of a given wafer size, although a more effective turn-on performance may balance this effect. One measure of turn-on performance is the $dI_A/dt$ rating.[46] In typical design this may range from 100 A/$\mu$sec to 500 A/$\mu$sec.

Another measure of thyristor performance is the ability of the device to withstand, in a specified test circuit, a rapid application of forward voltage without false triggering caused by the capacitance current that flows in the device as the depletion region builds up around the center junction. If the depletion thickness in the $n_1$ base region is $\ell$ then the charge removed from this region is

$$Q = qAN_D\ell$$

$$= \left[\frac{2\epsilon\epsilon_0 A^2 q N_D(V_0 + V_A)}{1 + (N_D/N_A)}\right]^{1/2} \tag{5.10}$$

where $V_0$ is the built-in voltage of the junction $J_2$, $V_A$ is the applied voltage and $A$ is the junction area.

Table 5.1.  **Comparison of Current Ratings between an Involute-Type and a Noninvolute Type Thryistor**

| Switching Frequency (Hz) | Load Current Sinusoidal, Peak Value, 50% Duty, $T_c = 65°C$ | |
|---|---|---|
| | Involute Type C509 (A) | Noninvolute Type (A) |
| 60 | 775 | 700 |
| 1200 | 650 | 550 |
| 2500 | 600 | 275 |
| 5000 | 550 | 150 |

(After Storm and St. Clair. Reprinted with permission from *IEEE Trans. Electron Devices*, ED-21, 1974, p. 521.)

(a)

(b)

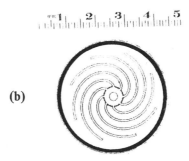

Fig. 5.13. Involute gate geometry for power thyristors:
Section of an involute gate for a 550-A, 1800-V thyristor. (Courtesy General Electric Co.)

The displacement current is then obtained from $dQ/dt$ and, neglecting $V_0$, is given by

$$I_{dis} = \frac{dV_A}{dt} \left[ \frac{\epsilon\epsilon_0 A^2 q N_D}{2 V_A (1 + N_D/N_A)} \right]^{1/2} \tag{5.11}$$

The capacitive current therefore is proportional to the rate of voltage rise. If the current is small it may be bypassed by the cathode shunt and have little effect in turning on the *npn* transistor region. But if the rate of rise of voltage is large the capacitive current acts rather like a gate current in raising $\alpha_3$ and turning on the device. Typical $dV_A/dt$ ratings of power thyristors are in the range 200-400 V/μsec up to the rated maximum forward voltage ($V_{DRM}$) for junction temperatures up to 125°C.

### 5.2.1 Gate Turn-off

The usual method of turn-off in a power thyristor is to allow the anode voltage to fall to zero; this drops the current below the sustaining current and allows voltage to build up across the device. This technique of commutative turn-off is convenient and practical in most instances since the ac voltage that is being regulated reverses every half-cycle.

However, in direct voltage circuits and in some alternating voltage circuits it may be desirable to have gate controlled turn-off. The fact that the load current may be flowing in regions of the thyristor that are laterally displaced from the gate presents some difficulties in achieving such control. Therefore it is desirable to use large gate perimeters and small lateral gate-to-cathode distances in the design of gate turn-off thyristors. The ratio of the local current to that which must be extracted by the gate ($I_{gf}$) for turn-off is the turn-off gain and this can be shown to be of the order of[31,56]

$$\beta_{off} = \frac{I_A}{I_{gf}} = \frac{\alpha_3}{(\alpha_1 + \alpha_3 - 1)} \qquad (5.12)$$

Hence, for good turn-off characteristics the thyristor should be designed to have $\alpha_3$ of the *npn* region large, say, 0.85 and the value of $(\alpha_1 + \alpha_3)$ in the on condition not too much larger than unity.

An impression is given in Fig. 5.14 of the processes in the turn-off of a laterally displaced region. Extinction of remote current filaments must be rapid if hot spots are not to form. Hence, turn-off designs tend to be limited to low current thyristors (20 A or less).

Turn-off action is also limited if a cathode shunt is present since it is difficult to achieve any substantial gate-cathode reverse bias.

### 5.2.2 Power Capability

The power capability of a thyristor may be defined as the product of the circuit peak-inverse voltage and the maximum load current amplitude.

High reverse voltage ratings can be achieved by the use of thick lightly-doped base regions. Values of 5000 V or more are achievable. The peak value of the input voltage $V_{DRM}$ is usually taken as 0.4-0.5 of peak reverse voltage to provide a safety factor. The maximum current amplitude $I_T$ in the forward direction is usually taken as about an eighth of the maximum peak surge current $I_{TSM}$ that can be withstood for a period of 10 msec. The power capability $V_{DRM} \times I_T$ is shown in Fig. 5.15 for a Si wafer of diameter 33 mm ($\sim$1.3 in.) using these safety factors. This power capability also roughly equals the dc power obtainable from a three-phase bridge consisting of six thyristors. The power capability is seen to increase with increasing voltage rating up to 2000 V but declines thereafter because of the decrease of current rating with increase of device voltage. The surge current rating declines because for high voltage ratings the thyristor must have a thick lightly-doped base region and the edge must be chamfered to reduce surface field strength as shown in Fig. 5.15(b). These effects raise the bulk resistance and reduce the device anode area with the result

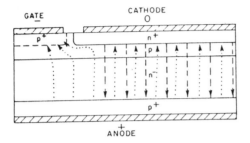

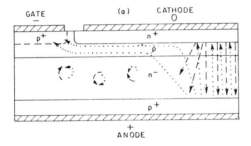

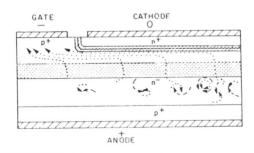

1. Negative bias applied to gate.
2. Holes from anode are removed through gate.
3. Electron injection stops in area adjacent to gate.
4. Gate draws holes from more remote areas as area closer to gate is depleted of carriers.
5. Carrier recombination reduces plasma density in non-injecting areas.
6. Electron injection stops over larger and larger area of *n* emitter.
7. Holes from conducting area must travel larger and larger lateral distance through *p* base to reach gate contact.
8. Anode current is crowded into high density filaments in area most remote from gate contact. This is the most critical portion of the gate-turn-off process because rapidly rising localized high temperature can damage crystal, unless filaments are extinguished rapidly. Filaments are extinguished more rapidly by higher gate voltage but maximum possible voltage is reverse gate breakdown voltage.
9. As final filaments are extinguished, electron injection stops and depletion layer builds on gate-cathode junction.
10. Depletion layer begins to build on forward blocking junction.
11. Cathode current stops, but anode-to-gate current decays exponentially as plasma concentration is reduced by recombination.

Fig. 5.14. Impression of turn-off the conducting plasma in a gate-turn-off thyristor when the gate is negatively pulsed. (After Becke, et al.[13] Courtesy of RCA Corp.)

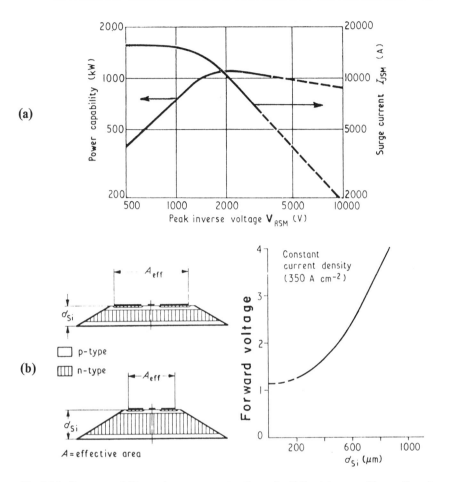

Fig. 5.15. Power capability and surge current ratings of a Si thyristor on a 33-mm-diameter wafer:
(a) Variation with peak inverse voltage rating of the design, and
(b) Variation of conducting forward voltage drop with thickness of *n* base region. (After Herlet.[40] *Proc. of the European Semiconductor Device Research Conference*, copyright 1971, Inst. of Physics, London.)

that the forward voltage drop in conduction increases. The surge current and operating current capabilities therefore decline with increase of the voltage rating. The forward voltage drop might be lessened by increasing the conductivity modulation of the base regions but this would require higher minority carrier lifetimes and thus increase the thyristor turn-off time.

The effect of wafer diameter on the power capability is illustrated in Fig. 5.16(a). The curves for the 44-mm and 60-mm diameters do not peak because the chamfer area effect is proportionately less significant in the larger

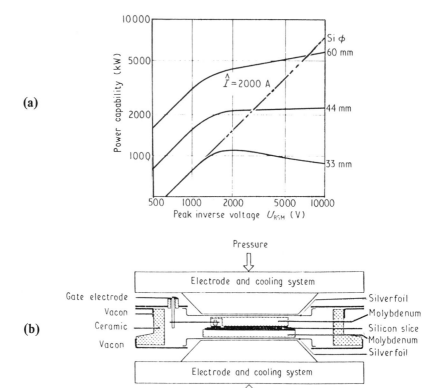

Fig. 5.16.
(a)  Power capability vs peak inverse thyristor voltage for three wafer diameters, and
(b)  Typical high power thyristor construction. (After Herlet.[40] *Proc. of the European Semiconductor Device Research Conference*, copyright 1971, Inst. of Physics, London.)

wafers. The chain dot line represents a current amplitude of 2000 A which is about the maximum that can be allowed in an air cooled 33-mm-diameter design. For higher current ratings, disk type cells are used as in Fig. 5.16(b) to facilitate heat removal from both sides of the wafer with liquid or forced air heat sinking.

Cooling fin design for natural and forced air heat dissipation is discussed by Gentry et al.[31]

## 5.3 TRIACS AND OTHER MULTILAYER STRUCTURES

The thyristor structures described so far conduct load current only in one direction, i.e., from the anode to the cathode. However, in some circuits it may be desirable and economical to have a device that will conduct in both directions so that both halves of the ac supply can provide load current in response to gate control. The first approach is to consider two normal thyristors connected in parallel with opposite conduction directions as shown in Fig. 5.17(a). If these are combined on a single wafer the result is as shown in Fig. 5.17(b). The presence of two gates $G$ and $G'$ is inconvenient for circuit reasons. Fortunately by the use of lateral bias effects, involving the division of $n_3$ into two sections suitably placed, it is possible to dispense with the second gate $G'$ and trigger the thyristor into conduction in either direction with a positive pulse applied to gate $G$. The structure with the required division and overlapping of the $n_3$ region is shown in Fig. 5.18(a).

The action is as follows: Assume $T_2$ is negative with respect to $T_1$, which is at ground potential, and that a positive pulse of voltage is applied to the gate $G$ that causes electron injection from $n_1$ into $p_1$. These electrons when collected by region $n_2$ lower the potential of this region and cause $p_1$ to inject holes into region $n_2$. These holes diffuse towards junction $J_3$ where they are collected by the ohmic contact on $p_2$. But in doing so they cause a lateral bias in overlap $O$ that starts $n_3'$ injecting electrons which further adds to the device turn-on process. As the turn-on region spreads to the right the lateral bias effect of current flow in the overlap region $O$ helps turn on the right-hand side of the

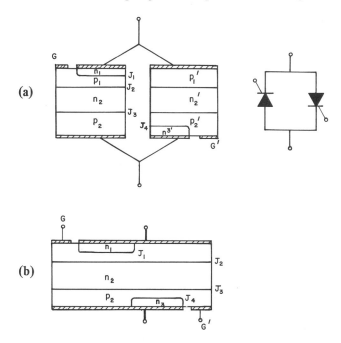

Fig. 5.17. Hypothetical double thyristor construction for full-wave controlled conduction:
(a)   Two thyristors connected in parallel, and
(b)   Single wafer construction.

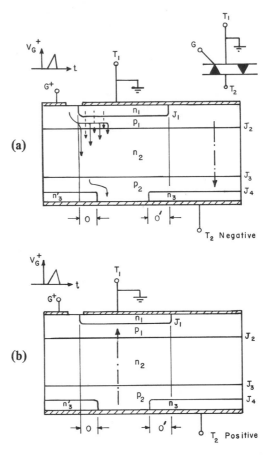

Fig. 5.18. Double thyristor structure for full-wave controlled conduction with positively pulsed single gate.
(a)  With $T_2$ negative, and
(b)  With $T_2$ positive.

device by causing $n_3$ to inject and the current shown by the chain-dot arrow develops.

For the other half cycle of the ac supply voltage the terminal $T_2$ is positive with respect to $T_1$ at ground potential. The application of a positive pulse to the gate $G$ causes the device to turn on much as a conventional $p_2 n_2 p_1 n_1$ thyristor with main current flow as shown by the chain-dot arrow in Fig. 5.18(b). The gate terminal is seen to be at a relatively low potential with respect to ground potential which is convenient in most circuits.

In the structure that has been described a positive gate pulse turns on either direction of current flow. However, in some bilateral thyristor switches (usually called triacs) it is convenient to have turn-on by either positive or negative gate pulses. This is accomplished by providing a partial $n$ region under the gate electrode as shown in Fig. 5.19(a). If the gate is pulsed negatively the region $n_5$ injects electrons and begins the turn-on process. A positive pulse on $G$ would cause the gate contact overlap region $O_G$ to conduct and turn on the device by raising the $\alpha_{1s}$ of the $n_1 p_1 n_2$ section of the structure. But the $O_G$ region is rather remote and the overlap area is small so this trigger action is not as sensitive as the negative-pulse turn-on action.

If the alternating voltage half cycle has terminal $T_2$ positive, the application of a positive gate pulse causes the region $n_1$ to inject electrons to the contact region $O_G$ and the thyristor section formed by $n_1 p_1 n_2 p_2$ turns on. Turn-on is also possible for a negative gate pulse but the sensitivity is usually lower. Triacs are now widely used for low and moderate power handling circuit applications of thyristors.

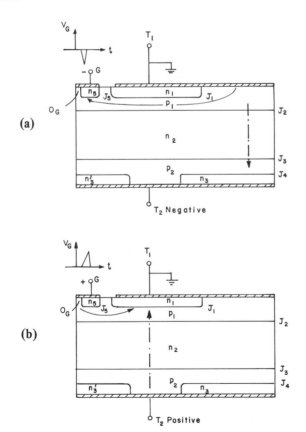

Fig. 5.19. Triac construction:
(a)   Negatively pulsed gate is used when $T_2$ is negative, and
(b)   Positively pulsed gate is used when $T_2$ is positive.

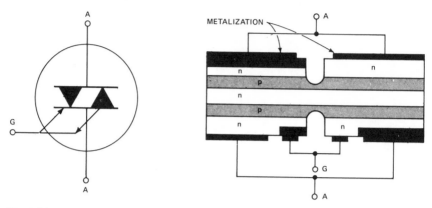

Fig. 5.20. A seven-layer triac type structure. (After Mattera. Reprinted from *Electronics*, July 10, 1975; copyright McGraw-Hill, Inc., NY, 1975.)

Many variations of the triac structure are possible. For instance the seven-layer structure shown in Fig. 5.20 has good rating characteristics.[60]

However, let us now return to a consideration of thyristor gate structures rather than further develop the triac concept. Some important gate structures are shown in Fig. 5.21. In the "field-initiated" gate a strip of the $n$ emitter region adjacent to the gate is free of contact metal. This causes a lateral voltage drop in the $n$ emitter region when turn-on initially begins near the inner edge of the $n$ region. The lateral voltage drop then causes rapid turn-on of the whole of the $p$ region up to the metallization inner edge of the $n$ emitter. This fast spread reduces hot-spot formation and assists in the complete turn-on of the device. This concept is taken one step further in the regenerative gate structure of Fig. 5.21(b). Here an inner contact ring $R_1$ picks up some of the lateral voltage drop in the $n$ region after initial turn-on and transfers it to an outer ring $R_2$ which acts as a transfer gate and initiates turn-on at the outer perimeter of the cathode. To ensure that the initial turn-on always occurs near the inner perimeter of the $n$ emitter region the diffusion there may be made deeper as shown in the right-hand half of Fig. 5.21(b). The deeper diffusion, known as a dip-emitter, produces a higher alpha in this region thus determining the location of the initial turn-on.

The structure in Fig. 5.21(c) is one in which an amplifying gate action is provided by a small pilot thyristor built around the signal gate. The output of the auxiliary thyristor then is the triggering current drive for the main thyristor. An experimental amplifying-gate thyristor in which the auxiliary thyristor is outside the main cathode is shown in Fig. 5.22. The curves for the turn-on of a 300-V, 300-A load show the switching losses to be fairly low. However, the auxiliary thyristor takes up a significant amount of wafer area and some current carrying capacity is lost. Therefore, if a high-power gate device source is available it may be better to use a direct-fired thyristor.

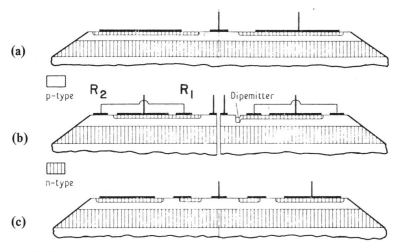

Fig. 5.21. Thyristors with various forms of gate structures to improve turn-on:
(a) Field aided gate,
(b) Regenerative gate dip-emitter shown on the right, and
(c) Amplifying gate with auxiliary thyristor section. (After Herlet.[40] *Proc. of the European Semiconductor Device Research Conf.*, copyright 1971, Inst. of Physics, London.)

When a pilot thyristor provides gate amplification it is important that the auxiliary unit turns on before partial turn-on of the main thyristor so that the main unit is hit with a large base drive. One way to achieve this is by a dip-emitter but another is by heavier doping of the $n$ base region under the auxiliary thyristor as shown in Fig. 5.23. The field strength is higher in the heavier doped region since the depletion region is thinner. The multiplication of carriers in the higher field region is therefore greater and so ensures that the effective alpha is greater under the auxiliary thyristor. The $n$ base region resistivity may be 50 ohm-cm in this region instead of about 70 ohm-cm elsewhere on the wafer. This difference is sufficient to take care of any random resistivity variations across the wafer that might otherwise cause premature turn-on of the main cathode. The breakdown voltage of the depletion region has been shown to be approximately dependent on $(\rho_n)^{0.45}$.

Although the gate in a normal thyristor makes contact to the $p_2$ base region, in some designs an additional $n$ region is provided under the gate as shown in Fig. 5.24. The equivalent circuit is as shown in Fig. 5.24(b) and the firing is by a negative pulse on the inverse gate that causes $n_3$ to inject electrons. When switching takes place at the $p_2 n_1$ junction the voltage drop across $R_1$ causes the $n_2 p_2$ junction to be forward biased and the whole structure turns on with a gate amplification or regeneration effect. The gate structure may be distributed in location to provide more rapid turn-on of the main structure if high $dI/dt$ ratings (300-1000 A/$\mu$sec) are required. A 1200-V inverse gate thyristor will

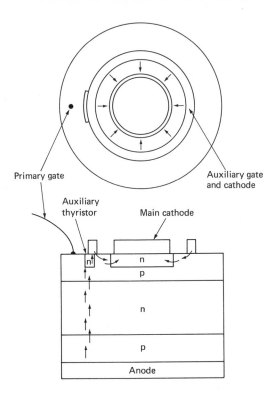

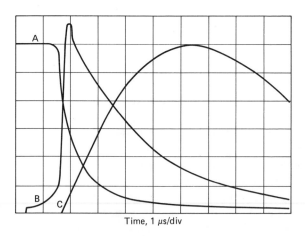

Time, 1 μs/div

Fig. 5.22. Amplifying gate thyristor structure and turn-on waveforms. In the waveform, *A* is the anode-cathode voltage, 50 V/div., *B* is the auxiliary gate-cathode voltage, 5 V/div., and *C* is the anode current 50 A/div. (After Cordingley.[19] Reprinted with permission from *Journal of Science and Technology.*)

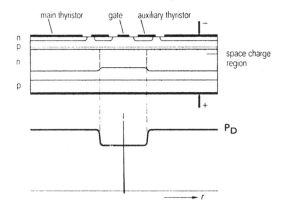

Fig. 5.23. Thyristor with amplifying gate and increased doping under the auxiliary thyristor to ensure that triggering occurs there first. This protects against possible partial triggering of the main thyristor. (After Voss.[102] Reprinted with permission from *Solid-State Electronics*, Pergamon Press, Ltd.)

handle a half-sine current pulse of 3000 A peak of 100 μsec half-width at a repetition rate of 400 pulses/sec. The upper operating frequency limit is about 10 kHz for such a unit.

Reduction of the turn-off time of thyristors is desirable. Less power is dissipated during rapid turn-off and the commutation circuit costs are reduced. The faster units may be operated at a higher power level, higher ambient temperature or at a higher frequency. In the pulsed operation of a thyristor many factors affect the turn-off time. They include: 1) pulse current amplitude, width and voltage when conducting, 2) pulse repetition rate, 3) $dV/dt$ of the reapplied forward blocking voltage and its amplitude, 4) the device case temperature and 5) the magnitude of the gate forward-on drive.[27]

One way of reducing the turn-off time is by Au doping of the base but this has limitations (as previously discussed in Sec. 5.2). Speeding up the desaturation time is possible by sweeping out the excess carriers stored during forward conduction by use of a gate to the *n* base region (in addition to the cathode gate needed for turn-on triggering). Figure 5.25(a) shows a structure with a gate for charge extraction from the *n* base region during turn-off. The transverse flow path to the saturated regions was undesirably long and therefore in another experimental design a base region gate was provided through the center of the anode region as shown in Fig. 5.25(b). With an "anode-gate" improvements in turn-off times of up to 50% were obtained [see Fig. 5.25(c)] for moderate current 50-A thyristors. However, this was at the expense of some increased device and circuit complexity.

Schottky diode clamps are used in transistors to prevent saturation of base regions that would slow up the turn-off time. There is the possibility of applying

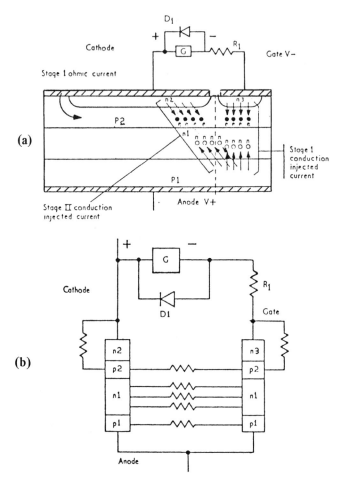

Fig. 5.24. Thyristor with inverse gate having $n$ type doping under most of its metallization:
(a)   Turn-on with gate negatively pulsed, and
(b)   Equivalent circuit representation. (After Oakman and Smallbonet.[72])

a somewhat similar technique to low power *pnpn* thyristors although some degree of base saturation is required to give low forward drop voltages in a thick-base thyristor. Figure 5.26 shows the application of a Schottky diode to a planar *pnpn* switch. The equivalent circuit shows that the Schottky barrier diverts some of the collector current of the *pnp* section of the switch that would otherwise flow into the base of the *npn* transistor. Useful improvements in turn-off time are obtained in this way.

Optical triggering of thyristors is possible.[94, 112, 113] This is useful where the thyristor is floating at a high voltage that would present circuit isolation

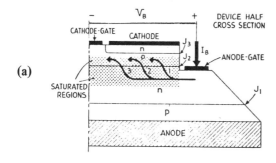

(a)

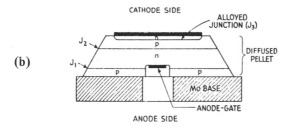

(b)

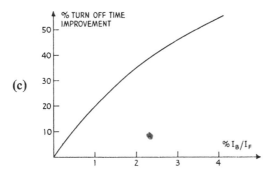

(c)

Fig. 5.25 Thyristor with a gate added to the *n* base region to remove stored carriers during turn-off:

(a)  Extra base gate located at the side is not very effective because of lateral effects,

(b)  Anode gate centrally located is more effective,

(c)  Improvement of turn-off time with ratio of biasing current $I_B$ to forward current $I_F$ for the structure (b). The turn-off time was 10-15 $\mu$sec without improvement. (After Assalit and Studtmann, Borg-Warner Research Center. Reprinted with permission from *IEEE Trans. Electron Devices*, ED-21, 1974, pp. 418, 419.)

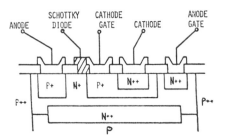

p-n-p-n switch with a Schottky diode clamp.

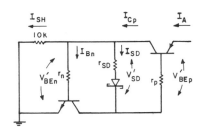

Fig. 5.26. Planar *pnpn* thyristor with Schottky diode clamp to control saturation and subsequent turn-off time. (After Schade and Lin. Reprinted with permission from *IEEE Trans. Electron Devices*, **ED-20**, 1973, pp. 1159, 1160.)

difficulties in supplying gate-drive current. The light may be introduced by optical-fiber coupling of radiation from a diode injection laser or a light emitting diode. The optical power that is available is normally only a few tens of mW and, therefore, high trigger sensitivity is desirable. A 1000-V, 200-A thyristor fired by 20-50 mW light power at the gate region, from a GaAs (Si doped) light emitting diode, is shown in Fig. 5.27. The $n^+$-emitter in the window region is about 5-μm thick to avoid too much absorption of the light which has a penetration depth of a few tens of μm. The sheet resistivities of the $n^+$ region and $p$ base region were chosen to give good transverse field emitter action. The $dI/dt$ rating was 150 A/μsec and the turn-on delay less than 1.5 μsec. The $dV/dt$ rating was greater than 500 V/μsec. This was achieved by providing a ring electrode around the cathode perimeter to reduce lateral biasing effects that might cause $dV/dt$ triggering. If the compensation ring was not provided and the gate $n^+$-zone was directly connected to the main cathode the $dV/dt$ rating became less than 50 V/μsec.

Among variations of thyristor structures there is a non-trigger device termed the field controlled thyristor (FCT). It is a $pn^-n^+$ rectifier with a $p^+$ grid forming a collar around the $n^-$ region as in Fig. 5.28(a). In the on-state the diode is forward biased and conductivity modulation provides a low forward voltage drop. For instance at 10 A and 1300 A cm$^{-2}$ the forward voltage may be 1.45 V,

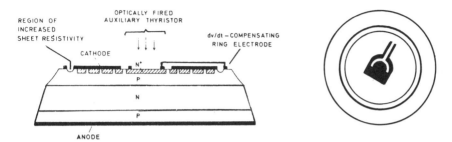

Fig. 5.27. Thyristor (1000 V, 200 A) with amplifying gate designed for optical trigger action with fiber optics LED or injection laser illumination. (After Silber and Fullmann. Reprinted with permission from *International IEEE Electron Devices Meeting*, 1975, p. 374.)

but with a grid bias of −150 V with respect to the cathode ground a pinch-off occurs as shown in Fig. 5.28(b) and 650 V may be sustained on the anode with less than 10 μA leakage current. The FCT has none of the plasma spreading problems of a thyristor since turn-on is accomplished by removing the gate bias and so the $dI/dt$ rating is good. Also, there is no problem of $dV/dt$ or voltage-spike triggering as in a thyristor. In the turn-off action of the FCT there is no regenerative cycle to be interrupted and the grid depletion regions help to remove stored charge (although there is a region in front of the anode where recombination must be relied on).

The field controlled thyristor therefore has promise in the medium power (500-800 V, 20-100 A) range for solid state power conditioning systems up to 15-30 kHz.

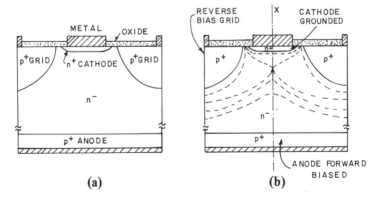

Fig. 5.28. Field controlled thyristor:
(a) The structure is a $p^+n^-n^+$ diode rectifier with $n^-$ region pinch-off provided by the depletion region created by a $p^+$ grid, and
(b) In the off-mode, 150 V applied to the $p$ grid depletes the $n^-$ region for the forward blocking state. (After Houston, et al. Reprinted with permission from *International IEEE Electron Devices Meeting*, 1975, pp. 381, 382.)

Compared with a switching transistor, a thyristor has the ability to handle large currents with lower forward voltage drops and requires no significant base current to maintain in the on-state. If a forward voltage and current surge occurs on a transistor the device may come out of saturation and the surge power dissipation increases considerably. Combating this effect by providing a large excess transistor base current may create a slower turn-off transient because of increase of saturation.

The general rule is that transistors are more convenient than thyristors in low power circuits and that thyristors provide the best performance at least cost in high power applications. The boundary between these two areas is suggested in Fig. 5.29.

## 5.4 COMPUTER AIDED DESIGN MODEL FOR A THYRISTOR CIRCUIT

Computer aided design of thyristor circuits must begin with a model that simulates parameters such as turn-on spreading time, turn-on rise and delay time, turn-off time, anode holding and latching current, gate turn-on voltage and current, shorted-emitter resistance, nonlinear static dynamic on characteristics and gate isolation when the thyristor is conducting.

Formulation of a model that takes account of these effects and of geometrical effects and is not too involved for practical use is not easy.

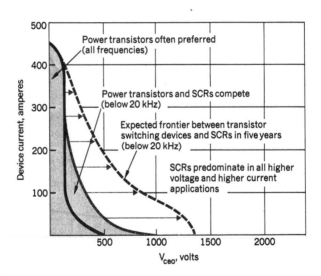

Fig. 5.29. The "moving frontier" (dotted line), behind which designers must choose carefully between thyristors and transistors for best cost/performance trade-offs, will advance deeply into "SCR-only" territory from 1977-1982. (After Harnden. Reprinted with permission from *IEEE Spectrum*, August, 1977, **14**, p. 44.)

Some success has been obtained with the circuit model of Fig. 5.30. The three junctions in a *pnpn* structure are represented by the three diodes where the $\theta$ terms are the thermally dependent coefficients $q/nkT$ with

$$J_A = I_{A(SAT)}(e^{\theta_A V_{CA}} - 1)$$
$$J_C = I_{C(SAT)}(e^{\theta_C V_{CC}} - 1) \tag{5.13}$$
$$J_K = I_{K(SAT)}(e^{\theta_K V_{CK}} - 1)$$

$1 < n < 2$. Turn-on must occur, for the anode positive to the cathode, when a positive gate to cathode voltage is applied of suitable magnitude. This forward biases diode $K$ and a fraction $\alpha_1$ of the total injection current $J_K$ diffuses into the center collection junction. This is represented by $J_M$. Similarly the current component received at the center collection junction from the injection at the anode junction is $J_N = \alpha_2 J_A$. Other current components needed are shown in Fig. 5.30. The junction capacitances are modeled as diffusion components that are proportional to the injection currents and as depletion layer components which are of lesser importance and may be treated as constants. Hence

$$C_A = K_{dA} J_A + C_{tA}$$
$$C_C = K_{dC} J_C + C_{tC} \tag{5.14}$$
$$C_K = K_{dK} J_K + C_{tK}$$

The current $J_H$ is a fraction of the cathode current that varies with the gate current and $J_P$ is a current required for simulation of the thyristor's nonlinear on-resistance.

The description by Bowers and Nienhaus follows:

The effect of spreading (the reduction in on-voltage as the conducting area of the current increases from an initially confined cross section near the gate) is simulated by the charging time of capacitor $C_B$. During turn-on, the current through resistor $R_B$ is equal to the current required to charge capacitor $C_A$ minus the current in $C_B$, which produces a larger voltage drop across $R_B$ than in the static case. Therefore, the collector diode $C$ will forward-bias (representing the turn-on) at lower values of anode current in the dynamic situation. As $C_B$ discharges, the voltage drop across $R_B$ slowly decreases to a steady-state value. The time required for this to occur is the spreading time and is primarily dependent on the values of $R_B$, gain factor $\alpha_q$, and $C_B$. The charge on capacitor $C_B$ is a nonlinear increasing function of injection current $J_A$.

The anode and cathode diffusion-capacitance constants, $K_{dA}$ and $K_{dK}$, help simulate the turn-on delay and rise time. In addition, the ratio of $K_{dA}/K_{dK}$ simulates the breakpoint between the rise time and the spreading

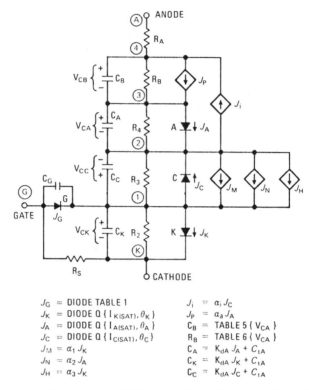

$$J_G = \text{DIODE TABLE 1}$$
$$J_K = \text{DIODE Q}\{I_{K(SAT)}, \theta_K\}$$
$$J_A = \text{DIODE Q}\{I_{A(SAT)}, \theta_A\}$$
$$J_C = \text{DIODE Q}\{I_{C(SAT)}, \theta_C\}$$
$$J_M = \alpha_1 J_K$$
$$J_N = \alpha_2 J_A$$
$$J_H = \alpha_3 J_K$$

$$J_i = \alpha_i J_C$$
$$J_P = \alpha_a J_A$$
$$C_B = \text{TABLE 5}\{V_{CA}\}$$
$$R_B = \text{TABLE 6}\{V_{CA}\}$$
$$C_A = K_{dA} J_A + C_{tA}$$
$$C_K = K_{dA} J_K + C_{tA}$$
$$C_C = K_{dA} J_C + C_{tA}$$

Fig. 5.30. Computer-aided-design model for a thyristor. Diode currents are derived from diode models in subroutines, and values for $J_G$, $C_B$, and $R_B$ are obtained from piecewise-linear tables. Resistors $R_2$, $R_3$, $R_4$, and $R_A$, as well as capacitor $C_G$, are included for programming convenience and do not directly affect model behavior. (After Bowers and Nienhaus. Reprinted from *Electronics*, Apr. 14, 1977, copyright McGraw-Hill, Inc., NY, 1977.)

time. The initial portion of the turn-on delay is simulated by the constant component of the cathode junction capacitance, $C_{tC}$. The collector diffusion-capacitance constant, $K_{dC}$, determines the turn-off time.

To account for the SCR's anode latching current being greater than its holding current, a third dependent current source, $\alpha_3 J_K$, has been included in the model as $J_H$. The current gain $\alpha_3$ is zero when the gate injection current $J_G$ is positive, while it has a small finite value for all other values of $J_G$. The holding current occurs at current levels such that $\alpha_1 + \alpha_2 = 1$. At low current levels, $\alpha_1$ and $\alpha_2$ are increasing functions of current, so latching occurs at a higher current level than that of holding.

The turn-on current and voltage characteristics of the SCR's gate are simulated by the resistance $R_s$, the low-current values of $\alpha_1$ and $\alpha_2$, and the nonlinear characteristics of the cathode and gate diodes. Besides their role in modeling the static characteristics of the SCR, the low-current values of

$\alpha_1$ and $\alpha_2$ help simulate the turn-on delay, while their high current values help simulate the rise time.

The resistance $R_s$ represents the relatively small resistance between the gate and cathode electrodes. This resistance is attributable to the shorted-emitter construction used in the manufacture of most high-power thyristors.

The thyristor's nonlinear static on-resistance between the anode and cathode terminals is represented by the model's nonlinear resistance $R_B$ in shunt with the linear-dependent current source, $\alpha_a J_B$, shown as $J_P$ in Fig. 5.30. This resistance decreases as the anode current increases, because a larger portion of the total cross-sectional area of the SCR is conducting.

In the dynamic case, the spreading effect causes the on-voltage to be less than in the static case. This nonlinear behavior is effectively simulated by the inter-dependence of the current source and the voltage across the anode capacitor. The upshot is a higher on-resistance for the dynamic case than for the static case.

The diode labeled $G$ is not an actual junction. It is a convenient way to represent the nonlinear resistance of a thyristor's gate region between the gate electrode and the active portions of the cathode junction. (In a device with an amplifying gate, the pseudo-diode also will take into account the voltage drop across the pilot cathode junction.) The relatively high back resistance of this diode simulates the thyristor's invulnerability to turn-off if the gate is reversed-biased.

It can be shown that $\alpha_1 + \alpha_2$ is less than unity for currents below the anode latching level, and $\alpha_1 + \alpha_2$ is greater than unity in the current range from latching to maximum surge-current rating. Once a gate pulse drives the thyristor into the region where $\alpha_1 + \alpha_2 > 1$, the device's regenerative behavior will cause it to continue to turn on and stay on, even if the gate pulse is reduced to zero.

The computer aided design program that has been adapted for use with this thyristor circuit model is termed Super-Sceptre (from System for Circuit Evaluation and Prediction of Transient Radiation Effects). It is in FORTRAN language and suitable for running on an IBM 360 computer, or equivalent unit.

An example of the program is shown in Fig. 5.31 for a G.E. thyristor type C358E in a commutation circuit. The result was a representation of the voltage across $R_1$ vs time that agreed closely with the observed waveform. The run time required was 113 sec of IBM 360 central processor time.

## 5.5 THYRISTOR APPLICATIONS (BRIEF COMMENTS)

Applications of thyristors are covered in many books and in the technical journal literature. The purpose of this section is little more than to lead into this information. Books that are available include: a) Gentry, et al., *Semiconductor Controlled Rectifiers*, Prentice-Hall, 1964; b) Atkinson, *Thyristors and Their*

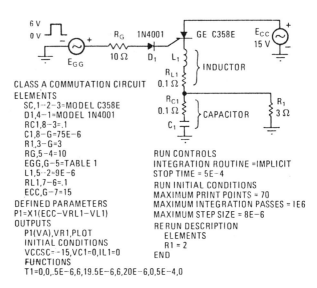

```
CLASS A COMMUTATION CIRCUIT
ELEMENTS
  SC,1-2-3=MODEL C358E
  D1,4-1=MODEL 1N4001
  RC1,8-3=.1
  C1,8-G=75E-6
  R1,3-G=3
  RG,5-4=10
  EGG,G-5=TABLE 1
  L1,5-2=9E-6
  RL1,7-6=.1
  ECC,G-7=15
DEFINED PARAMETERS
  P1=X1(ECC-VRL1-VL1)
OUTPUTS
  P1(VA),VR1,PLOT
  INITIAL CONDITIONS
  VCCSC=-15,VC1=0,IL1=0
FUNCTIONS
  T1=0,0,.5E-6,6,19.5E-6,6,20E-6,0,.5E-4,0
```

```
RUN CONTROLS
INTEGRATION ROUTINE =IMPLICIT
STOP TIME = 5E-4
RUN INITIAL CONDITIONS
MAXIMUM PRINT POINTS = 70
MAXIMUM INTEGRATION PASSES = 1E6
MAXIMUM STEP SIZE = 8E-6
RERUN DESCRIPTION
  ELEMENTS
  R1 = 2
END
```

Fig. 5.31. Thyristor commutation circuit and simulation program. The thyristor is turned on by a 6-V, 20-μsec gating pulse and is turned off by LC ringing. Program listing includes initial-conditions specifications and a rerun command that disclosed failure to turn off when $R_1$ was reduced from 3 to 2 ohms. (After Bowers and Nienhaus. Reprinted from *Electronics*, April 14, 1977, copyright McGraw-Hill, Inc., NY, 1977.)

*Applications*, Crane, Russak Co., 1971; c) Davis, *Power Diode and Thyristor Circuits*, Cambridge University Press, 1971; d) Pelly, *Thyristor Phase-Controlled Converters and Cycloconverters*, Wiley, 1971; e) Mazda, *Thyristor Control*, Halsted Press, 1973; f) Blicher, *Thyristor Physics*, Springer, 1977.

Other excellent sources of information on circuits and thyristor ratings are the handbooks by General Electric Co. and Westinghouse Electric Co. of the USA and the application notes issued by companies such as RCA, TI, Motorola and others. Useful information may be found in the collection of papers published by IEEE (New York) *Power Semiconductor Applications*, Vols. I and II, edited by Harnden and Golden, 1972; also in the *IEEE Transactions on Power* and the *IEEE Power Electronic Specialists Conference Proceedings*.

The principal applications of power thyristors are in the areas of electrical power conversion, frequency changing, control of dc motors and other power regulation and switching functions. Thyristor rectifier systems provide controlled transformation of ac to dc power. Inverters or choppers transform dc to ac. Converters transform dc to dc with a change of voltage, usually to a higher level. Cycloconverters change a higher frequency ac to a lower frequency without a dc link.

Single-phase controlled SCR rectifying circuits are shown in Fig. 5.32 and some three-phase circuits are given in Fig. 5.33. The rectified output may be

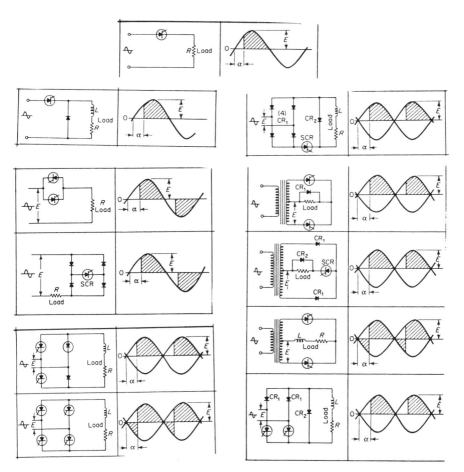

Fig. 5.32. Single-phase thyristor rectifying circuits with typical output wave forms. When the load is partly inductive the use of a free-wheeling rectifier to prevent voltage buildup during turn-off is usually desirable. (After Gentry, et al., from *Semiconductor Controlled Recifiers; Principles and Applications of p-n-p-n Devices*, copyright 1964, Prentice-Hall, Inc., Englewood Cliffs, NJ, pp. 296-301.)

used for electrochemical and battery charging applications or for the control of dc motors (which is perhaps 80% of the market).

DC motors may be regulated in speed by control of the field or the armature currents. The speed is sensed with a tachometer and a closed loop feedback system is generally used. Position control is also a common application in servo systems for machine tool operation, elevators, steering control for ships, control of radar antennas and various military applications. The solid state control of electric drives has been reviewed by Schieman et al.[87]

The thyristor has opened the way to the development of commutatorless dc

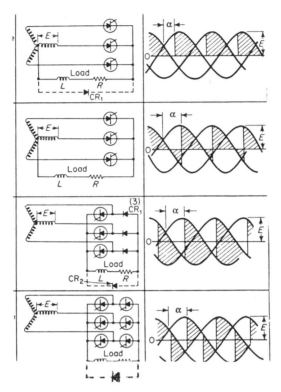

Fig. 5.33. Three-phase controlled rectifying thyristor circuits. (After Gentry et al., from *Semiconductor Controlled Rectifiers: Principles and Applications of p-n-p-n Devices*, copyright 1964, Prentice-Hall, Englewood Cliffs, NJ, p. 298.)

motors that are free of the usual commutator and brush maintenance problems. A basic arrangement is shown in Fig. 5.34. The currents through the three-phase stator winding are switched to match the position of the rotor. The firing sequence determined by the gate controller GC is shown in the lower part of Fig. 5.34. A more practical arrangement is given in Fig. 5.35. The three phase ac power supply is first converted to dc by a three phase thyristor bridge and smoothing reactor $L_s$. The output dc voltage is adjusted to provide the desired motor speed by a phase-shift gate controller in a servo loop that operates on the difference between the output voltage of a tachometer generator and a speed-set control voltage. The thyristor commutator is sometimes provided with a forced commutation system to provide more satisfactory performance, particularly on overload. Natural commutation uses the generated negative-going emfs to turn-off the thyristors that have completed their conduction period. Natural commutation has been used for quite high-power motors, with, for instance, dc ratings of 250 kW, 3000 V and 88 A. In this application each power thyristor leg consisted of four series-connected thyristors of 2500-V, 400-A rating.

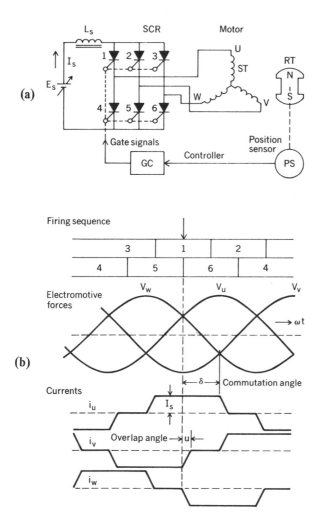

Fig. 5.34. Thyristor commutatorless dc motor system:
(a)  Basic configuration, and
(b)  Thyristor firing sequence and induced emf and line current. (After Inagaki, Kuniyoshi, Tadakuma. Reprinted with permission from *IEEE Spectrum*, June 1973, **10**, p. 57.)

Paralleling of thyristors is possible. Figure 5.36 shows a master thyristor (of inverse gate construction for good *dI/dt* performance) turning on two or three standard slave thyristors. $I_1$ is the initiating trigger current that turns on the master thyristor and $I_2$ then provides a hard gate drive to turn on the slave units. This arrangement requires considerably less initial gate drive than does a parallel assembly of three or four standard units.

Many possible circuits are known for trigger pulsing of thyristors. One com-

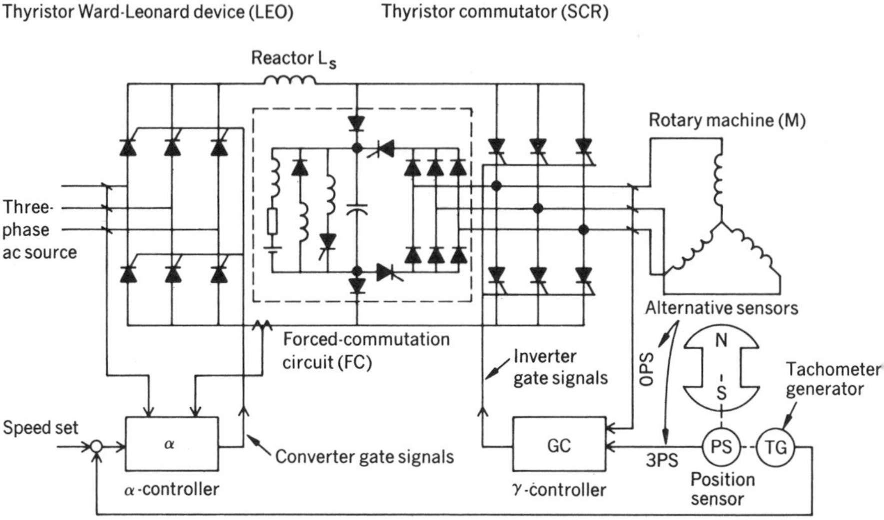

Thyristor Ward-Leonard device (LEO)          Thyristor commutator (SCR)

Reactor $L_s$

Rotary machine (M)

Three-
phase
ac source

Alternative sensors

Forced-commutation
circuit (FC)

Inverter
gate signals

Tachometer
generator

Speed set

$\alpha$

Converter gate signals

GC

3PS

Position
sensor

$\alpha$-controller

$\gamma$-controller

Fig. 5.35. Commutatorless dc motor control system with speed-set controller and forced-commuation circuit. (Cyclo-converter systems are also possible.) (After Inagaki, Kuniyoshi, Tadakuma. Reprinted with permission from *IEEE Spectrum*, June, 1973, **10**, p. 53.)

ponent that has been found useful in this respect is the unijunction transistor. In a simple circuit using a unijunction transistor the junction is connected to a capacitor that charges through a resistor until the voltage reaches a level that causes the *pn* junction on the transistor to inject. Then a negative resistance effect occurs that rapidly discharges the capacitor through the device and

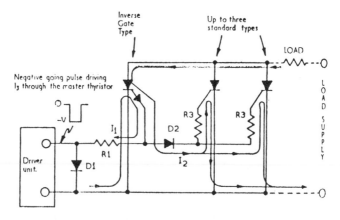

Fig. 5.36. Parallel operation of thyristors is possible. This circuit shows an inverse-gate thyristor of high $dI/dt$ rating turning on first and then triggering two standard thyristors in parallel. (After Oakman and Smallbonet.[72])

another resistor and produces a voltage spike capable of triggering a thyristor. Another component that has been found valuable for generating the turn-on trigger pulse is a *pnpn* diode.

One problem with thyristors is that of achieving immunity from voltage spikes induced during switching of load current in inductive loads. Many thyristor systems have switching actions occurring that may produce voltage transients that are capable of $dV/dt$ triggering of other thyristors. Spiked waveforms that may be produced during various circuit switching actions are shown in Fig. 5.37. Some measure of protection may be obtained by providing a path in parallel with the thyristor to limit the voltage spike. The path may contain a Zener diode pair or equivalent varistor [see Fig. 5.38(a)] or capacitive circuit [see Fig. 5.38(b)] with resistance and some line inductance to limit $dI/dt$ and $dV/dt$. If the thyristor is turning off a load current, particularly with a partially inductive load, the snubber circuit of Fig. 5.38(c) is useful in absorbing some of the load current while the device is recovering its blocking voltage. Another method of protection against surges caused in the turn-off of current through an inductive load $(R_L, L)$ is shown in Fig. 5.38(d). If the capacitor is made equal to $L/R^2$ and $R$ is equal to $R_L$, the ac steady-state equivalent impedance is then purely resistive. Protection of thyristors by the use of high speed fuses is possible but not easy.[64, 92]

In a few circuit instances, such as radar modulator or laser pulse applications, $dV/dt$ triggering may be the chosen method of turning on the thyristor. The device is made without a gate and is sometimes then called a reverse switching rectifier (RSR). Recovery is obtained by allowing the load current to fall below the holding current $I_H$ as in Fig. 5.39(a). The characteristics of a typical unit are:

| | |
|---|---|
| Breakover voltage | 1200 V |
| Pulse current | 5000 A in 20 $\mu$sec pulses at 250 pps |
| Forward voltage drop | < 10 V at 5000 A |
| Turn-on time | < 100 nsec |

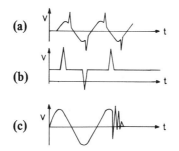

Fig. 5.37. Some typical spiked waveforms of voltage that may occur in switching circuits. (After Schlicke and Struger. Reprinted with permission from *IEEE Spectrum*, June, 1973, **10**, p. 33.)

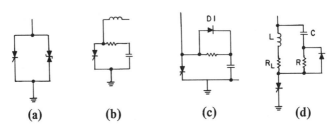

**Fig. 5.38.** Some protective circuits for thyristors to control voltage and current buildup.

Recovery time                     $<200\ \mu$sec
Rate of current rise ($dI/dt$)    2500 A/$\mu$sec to 5000 A
Required trigger $dV/dt$          5000 V/$\mu$sec

The $dI/dt$ rating is excellent. The current capability of the trigger pulse circuit must be 20 A. Figure 5.39(b) shows a series arrangement of three switching rectifiers with parallel capacitors of three different values. The trigger voltage first triggers the RSR with the smallest parallel capacitance. When this fires the other units receive greater rates of voltage rise and fire in rapid succession. A stack of six units of the rating above will switch 5000 V at 5000 A with a duty cycle of 0.005.

## 5.6 QUESTIONS

1. In a Si thyristor assume the $p$ anode region to be doped $10^{18}$ cm$^{-3}$, the 50 $\mu$m $n$ base $3 \times 10^{14}$ cm$^{-3}$, the 50 $\mu$m $p$ base $10^{15}$ cm$^{-3}$ and the $n$ cathode region $10^{18}$ cm$^{-3}$. What are the expected forward and reverse breakover voltages?

2. Assume the current flow across the base regions of a Si thyristor is by diffusion only and the $p$ and $n$ dopings are $10^{15}$ cm$^{-3}$ and $3 \times 10^{14}$ cm$^{-3}$. Also assume that the recombination mechanisms are identical for the two regions (that implies the same density of recombination centers $N_R$ and the same capture cross-sections whether for electron or holes). If the dc $\alpha_1$ is to be 0.25 and $\alpha_3$ is to be 0.65 what must be the relative thicknesses of the $p$ and $n$ base regions?

3. Consider a Si thyristor with $n$ base doping $3 \times 10^{14}$ cm$^{-3}$ and a $p$ base doping of $10^{15}$ cm$^{-3}$. Suppose the thyristor has $\alpha_1$ and $\alpha_3$ values of 0.25 and 0.65 (independent of current flow and applied voltage) but that those values undergo avalanche multiplication in the center junction depletion region. What will be the turn-on voltage inferred from information on avalanche in Si junctions (25°C) and from Eq. (5.3)?

4. Consider $dV/dt$ triggering by displacement current flow. Assume a thyristor to have doping profiles somewhat similar to those of Fig. 5.7 (you may simplify if you wish) and the $n^+pn$ transistor to have $\alpha_{3S}$ small signal alpha characteristics as for $W/L = 0.5$ in Fig. 5.4 and to have $\alpha_{1S}$ as a constant value 0.3. If the device is of area 0.1 cm$^2$ and is forward biased at 30 V what value of $dV/dt$ is likely to trigger the device with a resistive load?

5. If the cathode shunt resistance $R_S$ in Fig. 5.5 cannot be greater than $10^3$ ohms, how does this modify the effective alpha?

6. Design a Si thyristor rated at 600 V, 20 A for 300°K stud temperature. What do you estimate the $dI/dt$ and $dV/dt$ ratings to be?

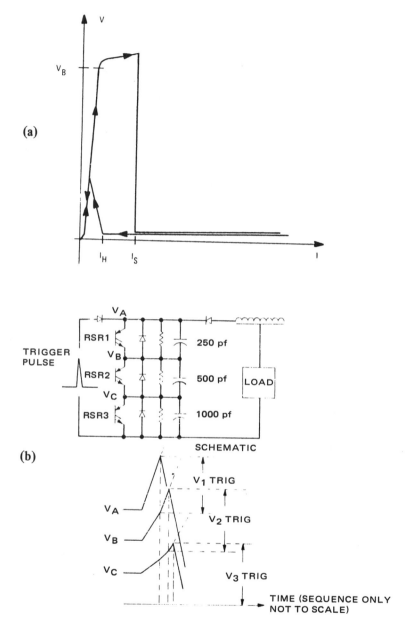

Fig. 5.39. A reverse switching rectifier (RSR) is a thyristor which depends on $dV/dt$ triggering to achieve a high $dI/dt$ rating.

(a)  Device characteristic, and

(b)  Three RSR's in a series arrangement can be triggered in rapid sequence. (After Gardenghi, Westinghouse Electric Corp. Reprinted with permission from *International IEEE Electron Devices Meeting*, 1975, p. 369, 370.)

7. What power could be handled by the design of problem 6 if six thyristors of this kind are used in a 60-Hz three-phase circuit?

8. Design an LED light-triggered Si thyristor of rating 500 V, 10 A. What light power input is needed for turn-on? Did you consider the use of an auxiliary turn-on thyristor structure?

9. Review the concepts involved in thyristors with turn-off gain. Outline a design for a Si unit rated at 200 V, 20 A with a turn-off gain of about 10.

10. A thyristor data sheet states that a single unit is listed at 500 V, 50 A with a $dI/dt$ rating of 50 A/$\mu$sec and a $dV/dt$ rating of 50 V/$\mu$sec. What may be inferred about the design? Such a thyristor is used in conjunction with four rectifiers (see mid-left diagram of Fig. 5.32) to control the power to a resistive load $R$ of 100 ohms with 500 V peak alternating voltage applied. What is the power dissipation it must handle as a function of the control angle $\alpha$ if the supply frequency is 60 Hz? What is the highest frequency that the circuit might be able to handle with this thyristor?

11. Discuss the factors that control the dc alpha and small signal alpha of the junctions in a thyristor with doping profiles similar to those shown in Fig. 5.7.

12. Review the assumptions that went into the calculations of alphas shown in Fig. 5.4. To what extent are they appropriate to thyristor design?

13. If the $p$ and $n$ base regions of a thyristor are doped with Au,[108] how would you estimate the Au concentration in each region?

14. Review various methods of observing the position and propagation of partial turn-on regions, or hot spots, in thyristors.

15. Review the factors controlling the probable ultimate voltage and current ratings of thyristors and estimate what the limits might be. Attempt in a simple kind of analysis to estimate the relationship between size and power rating of thyristors and turn-on and turn-off times.

16. Discuss thyristor regenerative gate design and compare its merits with those of other gates intended to improve $dI/dt$ ratings.

17. Review various classes of gate turn-on and turn-off circuits for thyristors in controlled single-phase rectifier circuits for the general power range 300 V, 100 A.

18. Discuss practical methods of using thyristors in series and in parallel.

19. Apply charge control concepts to the switching of a thyristor. Discuss delay time, rise time, storage time and fall time.

20. Discuss the concept of integral Schottky diode control of the saturation of a thyristor to improve the turn-off time.

21. Discuss four relatively different uses of transverse voltage drops in thyristor or triac operation.

22. Describe five useful low power (<10 A, 200 V) applications of thyristors and make it clear why thyristors are used instead of transistors.

23. Discuss the performance and design of amplifying gate thyristors.

24. Discuss the behavior of a thyristor with an inductive-resistive load.

25. Use the program shown in Fig. 5.31 to compute the response of this circuit with a model for a thyristor of your own choice. If necessary estimate the piecewise linear tables for the model of Fig. 5.30.

26. Discuss the temperature dependences of thyristor characteristics including effects on voltage ratings, current ratings and gate trigger needs.

27. Discuss in greater detail than in the text, the gate switching behavior of triacs.

28. Discuss the seven-layer structure shown in Fig. 5.20 and compare it with a triac.

29. Discuss in greater detail than in the text, the transient turn-on and turn-off characteristics of a practical triac.

30. What are the problems in constructing a thyristor of GaAs instead of Si? From what

you can find out about GaAs in the literature, what do you believe to be a possible thyristor rating that might be aimed at?

**31.** Explain ways of increasing the sensitivity of LED light-triggered thyristors of 200-A, 600-V current rating.

**32.** A nontriggering structure known as a field controlled thyristor has been described. Discuss the merits and limitations of this device relative to triggered thyristors and to transistors.

**33.** Develop the kind of gate control circuits that might be expected to be in the box marked *GC* in the motor control circuit of Fig. 5.31.

**34.** Describe forced commutation as it applies to the control of a thyristor dc motor control system as in Fig. 5.32.

**35.** Discuss the problem of circuit and fuse protection of thyristors from voltage and current transients.

## 5.7 REFERENCES AND FURTHER READING SUGGESTIONS

1. Anheier, W., et al., "Rigorous numerical analysis of a planar thyristor," *IEEE International Electron Devices Meeting Technical Digest,* Washington, D.C., 1975, p. 363.

2. Adler, M.S., and V.A.K. Temple, "Accurate calculations of the forward drop of power rectifiers and thyristors", *IEEE International Electron Devices Meeting Technical Digest,* Washington, D.C., 1976, p. 499.

3. Adler, M.S., and V.A.K. Temple, "The dynamics of the thyristor turn-on process, *IEEE International Electron Devices Meeting Technical Digest,* Washington, D.C., 1977, p. 300.

4. Adler, M.S., "Factors determining the forward drop and the forward blocking gain of the field terminated diode (FTD), *IEEE International Electron Devices Meeting Technical Digest,* Washington, D.C., 1977, p. 42.

5. Aldrich, R.W., and N. Holonyak, Jr., "Two-terminal asymmetrical and symmetrical silicon negative resistance switches," *J. Appl. Phys.* 30, 1954, p. 1819.

6. Anheier, W., W.L. Engl, and R. Sittig, "Numerical analysis of gate triggered SCR turn-on transients," *IEEE International Electron Devices Meeting Technical Digest,* Washington, D.C., 1977, p. 303A.

7. Arai, M., "Magnetosensitive thyristor," *Japan J. Appl. Phys.,* **12**, 1973, p. 1278.

8. Assalit, H.B., and G.H. Studtmann, "Description of a technique for the reduction of thyristor turn-off time," *IEEE Trans. Electron Devices,* **ED-21**, 1974, p. 416.

9. Atkinson, P., *Thyristors and Their Applications,* Crane, Russak Co., NY, and Mills and Boon, London, 1972.

10. Bakowski, M., and K.I. Lundstrom, "Depletion layer characteristics at the surface of beveled high-voltage *p-n* junctions," *IEEE Trans. Electron Devices,* **ED-20**, 1973, p. 550.

11. Baliga, B.J., and E. Sun, "Lifetime control in power rectifiers and thyristors using gold, platinum and electron irradiation," *IEEE International Electron Devices Meeting Technical Digest,* Washington, D.C., 1976, p. 495.

12. Bassett, R.J., C.A. Hogarth, and J.P. Newman, "An investigation into the mode of switching of inverse gate thyristors," *Int. J. Electronics,* **31**, 1971, p. 453.

13. Becke, H., et al., "Gate turn-off silicon controlled rectifiers,—A user's guide," RCA Solid State Division, Application Note AN-6357, 1974.

14. Bedford, B.D. and R.G. Hoft, *Principles of Inverter Circuits,* Wiley-Interscience NY 1964.

15. Bergman, D., "Gate isolation and commutation in bi-directional thyristors," *Int. J. Electronics,* **21**, 1966, p. 17.

16. Bergman, G.D., "The gate-triggered turn-on process in thyristors," *Solid-State Electronics*, **8**, 1965, p. 757.
17. Blicher, A., *Thyristor Physics*, Springer-Verlag, New York, Heidelberg, 1977.
18. Bowers, J.C., and H.A. Nienhaus, "Model for high power SCRs extends range of computer-aided design," *Electronics*, April 14, 1977, p. 100.
19. Cordingley, B.V., "Improving the turn-on performance of high-power thyristors," *J. Science and Technology*, **38**, 1971, p. 2.
20. Cornu, J., "Field distribution near the surface of beveled *p-n* junctions in high voltage devices," *IEEE Trans. Electron Devices*, ED-20, 1973, p. 347.
21. Cornu, J., S. Schweitzer, and O. Kuhn, "Double positive beveling: a better edge contour for high voltage devices," *IEEE Trans. Electron Devices*, ED-21, 1974, p. 189.
22. Cornu, J., R. Sittig, and W. Zimmermann, "Analysis and measurement of carrier lifetimes in the various operation modes of power devices," *Solid-State Electronics*, 17, 1974, p. 1099.
23. Cornu, J., and A.A. Jaecklin, "Processes at turn-on of thyristors," *Solid-State Electronics*, 18, 1975, p. 683.
24. Davis, R.M., *Power Diode and Thyristor Circuits*, Cambridge University Press, London, England. Peter Peregrinus, Stevehage, Herts. 1971.
25. Dewan S.B., and A. Straughen, *Power Semiconductor Circuits*, Wiley-Interscience, NY, 1975.
26. Dodson, W.H., and R.L. Longini, "Skip turn-on of thyristors," *IEEE Trans. Electron Devices*, ED-13, 1966, p. 598.
27. Dyer, R.F., "Concurrent characterization of SCR switching parameters for inverter applications," *Semiconductor Products and Solid State Technology*, April, 1965, p. 15.
28. EIA-NEMA Standard RS 297. Recommended standards for thyristors. Electronic Industry Association, Washington, D.C. 1972.
29. Fukui, H., M. Naito, and Y. Terasawa, "One dimensional analysis of dynamic behavior of a thyristor," *IEEE International Electron Devices Meeting Technical Digest*, Washington, D.C., 1977, p. 304.
30. Gardenghi, R.A., "A super power RSR," *IEEE International Electron Devices Meeting Technical Digest*, Washington, D.C., 1975, p. 367.
31. Gentry, F.E., *Semiconductor Controlled Rectifiers*, Prentice Hall, Englewood Cliffs, NJ, 1964.
32. Ghandhi, S.K., *Semiconductor Power Devices*, Wiley-Interscience, NY, 1977.
33. Gray, D.I., "This SCR is not for burning," *Electronics*, **30**, 1968, p. 96. (Also US Patent 3,486,088.)
34. Grekhov, I.V., L.S. Kostina, and A.E. Otblesk, "Turn-off process in a *pnpn* structure at high injection levels in the base layers," *Soviet Physics—Semiconductors*, 4, 1971, p. 1998.
35. Grekhov, I.V., et al., "Turn-on process in a thyristor," *Soviet Physics—Semiconductors*, 5, 1971, p. 157.
36. Harnden, J.D., and F.B. Golden, "Power semiconductor applications," *Vol. I. General Considerations, Vol. II. Equipment and Systems*, IEEE Press, NY, 1972.
37. Harnden J.D. Jr., "Power semiconductors: looking ahead," *IEEE Spectrum*, August, 1977, p. 40.
38. Hartman, A.R., P.W. Shackle, and R.L. Pritchett, "A junction isolation technology for integrating silicon controlled rectifiers in cross point switching circuits," *IEEE International Electron Devices Meeting Technical Digest*, Washington, D.C., 1976, p. 55.
39. Herlet, A., "The maximum blocking capability of silicon thyristors," *Solid-State Electronics*, **8**, 1965, p. 655.
40. Herlet, A., "High power devices," *Proc. European Semiconductor Device Research Conf. on Solid-State Devices*, 1971, Inst. of Physics, London.

41. Houston, D.E., et al., "Field controlled thyristor (FCT)—A new electronic component," *IEEE International Electron Devices Meeting Technical Digest,* Washington, D.C., 1975, p. 379.
42 Hu, S.P. and B.M Rabinovici, "Two-dimensional current density distribution within three-terminal semiconductor devices," *J. Appl. Phys.,* **45,** 1974, p. 2624.
43. IEEE Power Electronic Specialists Conference Proceedings 1972 thru 1977, IEEE Service Center, Hoes Lane, Piscataway, N.J.
44. IEEE Standard 444—1973 Thyristor Motor Drive, IEEE Service Center, Hoes Lane, Piscataway, N.J. 08854.
45. Ikeda, S., S. Tsada, and Y. Waki, "The current pulse ratings of thyristors," *IEEE Trans. Electron Devices,* ED-17, 1970, p. 690.
46. Ikeda, S., and T. Araki, "The *di/dt* capability of thyristors," *Proc. IEEE,* **55,** 1967, p. 1301.
47. Inagaki, J., M. Kuniyoshi, and S. Tadakuma, "Commutators get the brushoff," *IEEE Spectrum,* June, 1973, p. 52.
48. Kajiwara, Y. et al., "High speed high voltage static induction thyristor," *IEEE International Electron Device Meeting Technical Digest* Washington, D.C., 1977, p. 38.
49. Kao, Y.C., and D.J. Page, "Theoretical limits for high voltage rectifiers and thyristors," *IEEE International Electron Devices Meeting Technical Digest,* Washington, D.C., 1977, p. 313.
50. Köhl, S.G., "A mesa-like edge contour for Si high voltage thyristors," *Solid-State Electronics,* 11, 1968, p. 501.
51. Kokosa, R.A., "The potential and carrier distribution of a *pnpn* device in the on-state," *Proc. IEEE,* **55,** 1967, p. 1309.
52. Kokosa, R.A., and B.R. Tuft, "A high voltage, high temperature reverse conducting thyristor," *IEEE Trans. Electron Devices,* ED-17, 1970 p. 667.
53. Kokosa, R.A., and E.D. Wolley, "Design criteria for amplifying gates on triode thyristors," *IEEE Electron Device Meeting Technical Digest* Washington, D.C., 1974, p. 431.
54. Kurata, M., "One dimensional calculation of thyristor forward voltages and holding currents," *Solid-State Electronics,* **19,** 1976, p. 527.
55. Lietz, M. "Numerical model of the thyristor turn-off," *IEEE International Electron Devices Meeting Technical Digest,* Washington, D.C., 1977, p. 307B.
56. Liniychuk, I.A., and A.I. Polamaazchuk, "Gate turn off of a *pnpn* device at high injection level," *Radio Engineering and Electronic Physics,* 4, 1973, p. 605.
57. Mapham, N., "The rating of silicon-controlled rectifiers when switching into high currents," *IEEE Trans. Communications and Electronics,* 1964, p. 5151.
58. Matsushita, T., H. Hayashi, and H. Yagi, "New semiconductor devices of ultra-high breakdown voltage", *IEEE International Electron Devices Meeting Technical Digest,* Washington, D.C., 1973, p. 109.
59. Matsuzawa, T., "Spreading velocity of the on-state in high speed thyristor," *Trans. IEEE Japan,* **98-C,** 1973, p. 16.
60. Mattera, L., "New type of thyristor challenges triac," *Electronics,* July 10, 1975, p. 122.
61. Mazda, F.F., *Thyristor Control,* Halsted Press, NY 1973.
62. McGhee, J., "A transient model for a three terminal *pnpn* switch and its use in predicting the gate turn-on process," *Int. J. Electronics,* **35,** 1973, p. 73.
63. Morita, K., et al., "Large area high-voltage thyristors for HVDC converters," *IEEE International Electron Devices Meeting Technical Digest,* Washington, D.C., 1977, p. 26.
64. Motto, J.W., "Can fuses protect SCRs and silicon diodes," *Electromechanical Design,* November, 1970, p. 16.
65. Motto, J.W., Jr., (ed), *Introduction to Solid State Power Electronics,* 1977, Westinghouse Electric Corp., Youngwood, PA.

66. Munoz -Yague, A., and P. Leturcq, "Optimum design of thyristor gate-emitter geometry, *IEEE Trans. Electron Devices*, ED-23, 1976, p. 917.
67. Murphy, J.M.D., *Thyristor Control of AC Motors*, Pergamon Press, Oxford, UK 1973.
68. Neilson, J.M., "Light dimmers using Triacs," RCA Application Note AN-3778.
69. New, T.C., and D.E. Cooper, "Turn-on characteristics of beam fired thyristors," *IEEE Conference Record, Industry and Applications*, October, 1973, p. 259.
70. Newell, W.E., "Transient thermal analysis of solid-state power devices–making a dreaded process easy," *IEEE Trans. on Industry Applications* 1A-12, 1976, p. 405.
71. Nield, M.W., and L.W. Davies, "Magnetically switched integrated SCR," *IEEE J. Solid-State Circuits*, SC-8, 1973, p. 175.
72. Oakman, K., and J. Smallbonet, "Regenerative gate thyristors," *Electronic Components*, 12, March 19, 1971.
73. Oka, H., and H. Game, "Electrical characteristics of high voltage, high-power fast-switching reverse conducting thyristor and its application for chopper use," Proc. IEEE Conference on Power Converters, 1973, p. 1.5.1.
74. Otsuka, M., "The forward characteristics of a thyristor," *Proc. IEEE*, 55, 1967, p. 1400.
75. Pelly, B.R., *"Thyristor phase-controlled converters and cycloconverters,"* Wiley-Interscience NY, 1971.
76. Raderecht, P.S., "A review of the "shorted-emitter" principle as applied to pnpn silicon controlled rectifiers," *Int. J. Electronics*, 31, 1971, p. 541.
77. Raderecht, P.S., "The development of a gate arrested turn-off thyristor for use in high frequency," *Int. J. Electronics*, 36, 1974, p. 399.
78. Rahman, S., and W. Shepherd, "Thyristor and diode controlled variable voltage drives for 3-phase induction motors," *Proc. IEE (London)*, 124, 9, September, 1977, p. 784.
79. RCA Solid State Division, *1975 Thyristors/Rectifiers Data Book*, SSD-206C, RCA Somerville, NJ.
80. Read, J.S., and R.F. Dyer, "Power thyristor rating practices," *Proc. IEEE*, 55, 1967, p. 1288.
81. Rice, L.R., "Choosing the best suppression network for your SCR converter, *Electronics*, November 6, 1972, p. 120.
82. Rocher, E.Y. and R.E. Reynier, "Response time of thyristors: theoretical study and application to electronic switching networks," *IBM J. of Research and Development*, 13, July, 1969, p. 447.
83. Rosier, L.L., C. Turrel and W.K. Liebmann, Semiconductor cross-points," *IBM J. of Research and Development*, 13, July, 1969, p. 439.
84. Ruhl, H.J., Jr., "Spreading velocity in the active area boundary in a thyristor," *IEEE Trans. Electron Devices*, ED-17, 1970, p. 672.
85. Sakurai, J., "A new spectroscopic method of measuring transient temperature distribution in high-power thyristors," *Japan J. Appl. Phys.*, Supplement Vol. 42, 1973, p. 97.
86. Schade, P.A., and H.C. Lin, "Characteristics of a *pnpn* switch with an aluminum-silicon Schottky diode clamp," *IEEE Trans. Electron Devices*, ED-20, 1973, p. 1159.
87. Schieman, R.G., E.A. Wilkes, and H.E. Jordan, "Solid-state control of electric drives," *Proc. IEEE*, 62, 1974, p. 1643.
88. Schlegel, E., "Gate assisted turn-off thyristors," *IEEE Trans. Electron Devices*, ED-23, 1976, p. 888.
89. Schlegel, E.S., and D.J. Page, "Gate assisted turn-off thyristor with cathode shunts and dynamic gate," *IEEE International Electron Devices Meeting Technical Digest*, Washington, D.C., 1976, p. 487.
90. Schlegel, E.S., and D.J. Page, "A high-power light-activated thyristor," *IEEE International Electron Devices Meeting Technical Digest*, Washington, D.C., 1976, p. 483.
91. Schlicke, H.M., and O.J. Struger, "Getting noise immunity in industrial controls," *IEEE Spectrum*, June, 1973, p. 38.

92. Schonholzer, E.T., "Fuse protection for power thyristors," *IEEE Conf. Record Industry and General Applications Group,* Chicago, OCtober, 1970.
93. Shimizu, J., et al. "High-voltage high-power gate-assisted turn-off thyristor for high frequency use," *IEEE Trans. Electron Devices,* **ED-32**, 1976, p. 883.
94. Silber, D., and M. Füllmann, "Improved gate concept for light activated power thyristors," *IEEE International Electron Devices Meeting Technical Digest,* Washington, D.C., 1975, p. 371.
95. Silber, D., et al., "Progress in light activated power thyristors," *IEEE Trans. Electron Devices,* **ED-23**, 1976, p. 899.
96. Somos, I., and D.E. Piccone, "Behaviour of thyristors under transient conditions," *Proc. IEEE,* **55**, 1967, p. 1306.
97. Storm, H.F., and J.G. St. Clair, "An involute gate-emitter configuration for thyristors," *IEEE Trans. Electron Devices,* **ED-21**, 1974, p. 520.
98. Sundresh, T.S., "Reverse transient in *pnpn triodes,*" *IEEE Trans. Electron Devices,* **ED-14**, 1967, p. 400.
99. Temple, V.A.K., "Directly light fired inverter thyristors with high di/dt capability," *IEEE International Electron Devices Meeting Technical Digest,* Washington, D.C., 1977, p. 22.
100. Terasawa, Y., "Observation of turn-on action in a gate-triggered thyristor using a new microwave technique," *IEEE Trans. Electron Devices,* **ED-20**, 1973, p. 714.
101. Tserng, H.Q., and H.R. Plumlee, "The turn-on delay time of silicon *pnpn* switches," *Proc. IEEE,* **58**, 1970, p. 792.
102. Voss, P., "A thyristor protected against *dI/dt*—failure at breakover turn-on," *Solid-State Electronics,* **17**, 1974, p. 655.
103. Voss, P., "Observation of the initial phases of thyristor turn-on," *Solid-State Electronics,* **17**, 1974, p. 879.
104. Wessels, B.W., and B.J. Baliga, "A high gain vertical-channel controlled thyristor," *IEEE International Electron Devices Meeting Technical Digest,* Washington, D.C., 1977, p. 30.
105. Williams, B.W., "State-space thyristor computer model," *Proc. IEE* (London), **124**, 9, September, 1977, p. 743.
106. Wolley, E.D., "Gate turn-off in *pnpn* devices," *IEEE Trans. Electron Devices,* **13**, 1966, p. 590.
107. Yang, E.S., and N.C. Voulgaris, "On the variation of small-signal alphas of a *pnpn* device with current," *Solid-State Electronics,* **10** 1967, p. 641.
108. Zimmermann, W., "Spatial dependence of carrier lifetime in the base of an Au-doped silicon controlled rectifier," *J. Electrochem. Soc.,* **120**, 1973, p. 97C.
109. Zucker, O.S.F., et al., "Experimental demonstration of high power fast rise-time switching in silicon junction semiconductors," *Appl. Phys. Lett.,* **29**, 1976, p. 261.
110. Adler, M.S., "Achieving accuracy in transistor and thyristor modeling," *IEEE International Electron Devices Meeting Technical Digest,* Washington, D.C., December 1978, p. 550.
111. Amantea, R., "A measurement technique and algorithm for determining the npn and pnp alphas for a thyristor," *IEEE International Election Devices Meeting, Technical Digest,* Washington, D.C., December 1978, p. 564.
112. Silber, D., M. Fuellmann and W. Winter, "Light activated auxiliary thyristors," *IEEE International Electron Devices Meeting, Technical Digest,* Washington, D.C., December 1978, p. 575.
113. Temple V.A.K., and B. Jackson, "Theoretical and experimental comparison of light triggered and electrically triggered thyristor turn-on," *IEEE International Electron Devices Meeting, Technical Digest,* Washington, D.C., December 1978, p. 579.

# 6
# JFETs and MESFETs – Field Effect Transistors

For their function junction field effect transistors (JFETs) depend on reverse-biased junction depletion regions that control majority carrier conduction in thin channels between ohmic contacts to source and drain regions. The structures require fewer masking steps and fewer diffusions than bipolar transistors and may be made either $n$ channel or $p$ channel. The contact through which the majority carriers enter the channel is termed the source $S$, and the exit is the drain $D$. The circuit symbol and some characteristics for an $n$ channel FET are shown in Fig. 6.1. A MESFET is similar except that the junction is a metal-semiconductor barrier. MOSFETs (metal-oxide-semiconductor FETs) are considered in Chap. 7.

JFETs have uses as small-signal amplifiers, current limiters, voltage-controlled resistors, switches, and in integrated circuits. The small-signal equivalent circuit is shown in Fig. 6.1(b), where $g_m$ the transconductance is typically 0.1-10 mA/V, $C_{gs}$ and $C_{gd}$ are 1-10 pF, $r_d$ is 0.1-1 megohm and $C_{ds}$ is 0.1-1 pF. The resistance between gate and source is determined by the reverse-biased junction leakage current and may be of the order of $10^7$ ohms. At low frequencies the

332

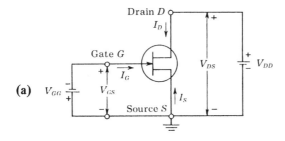

**(a)**

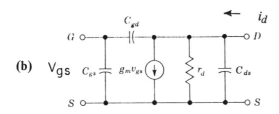

**(b)**

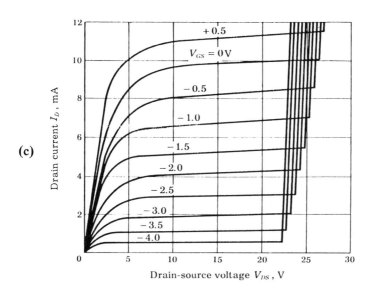

**(c)**

Fig. 6.1. Junction field-effect transistor circuits and characteristics:
(a)  Bias circuit,
(b)  Small signal equivalent circuit, and
(c)  $I_D$ vs $V_{DS}$ characteristics with $V_{GS}$ as parameters. (After Millman and Halkias, *Integrated Electronics*, copyright 1972, McGraw-Hill. Used with permission of McGraw-Hill Book Co., NY.)

capacitances in the equivalent circuit may be small enough to be neglected.

Conventional JFET designs are of low power: The minimum channel resistances are thousands of ohms and the source-drain voltages 10 V or so, hence the power levels are some tens of milliwatts. The shorter the channel length and the greater the width, the lower the channel resistance. FET designs, with width-to-length ratios that are very large, have now been made having tens of watts of output power. Thus, JFETs are no longer regarded as necessarily low in power, although bipolar transistors remain the more usual devices for power handling.

The greater the saturated drift velocity in the channel of a JFET, the higher the frequency response for a given channel length. Since GaAs has a higher peak saturated-drift velocity than Si it is a natural material for microwave frequency FETs. GaAs MESFETs now outperform Si FETs as amplifiers at frequencies of 5 GHz and higher.[141, 142, 143] With gate lengths of 0.5 $\mu$m, GaAs MESFETs give 10-dB gain at 12 GHz with excellent noise figures, for instance 3 dB, as discussed later.

## 6.1 FET MODELING INCLUDING SATURATION VELOCITY EFFECTS

In the part of an FET where the channel is almost pinched-off the field strengths are high and the carriers are moving with saturation drift velocity $v_s$. The characteristics predicted by assuming field-independent mobility do not adequately represent the device behavior, particularly for short-channel structures. The treatment of the FET that follows includes some velocity saturation effects by the use of the expression

$$v = \frac{\mu \mathcal{E}_x}{1 + \mu \mathcal{E}_x / v_s} \tag{6.1}$$

that reduces to $\mu \mathcal{E}_x$ at low field strengths and to $v_s$ at high field strengths.

The FET geometry considered is shown in Fig. 6.2. The thickness of the gate junction depletion layer is given by

$$d = [2\epsilon\epsilon_0(V - V_G - V_b)/Nq]^{1/2} \tag{6.2}$$

where $V$ is the voltage in the channel, $V_G$ is the gate voltage with respect to the source and $V_b$ is the barrier built-in voltage (a negative number). The channel current is

$$I_d = Nq \frac{\mu \mathcal{E}_x}{1 + \mu \mathcal{E}_x / v_s} W(a - d) \tag{6.3}$$

where $\mathcal{E}_x = \partial V / \partial x$ is the field along the channel. If $V_0$ is the gate-to-source voltage for pinch-off, a quantity $(1 - u)$ can be defined that is a measure of the

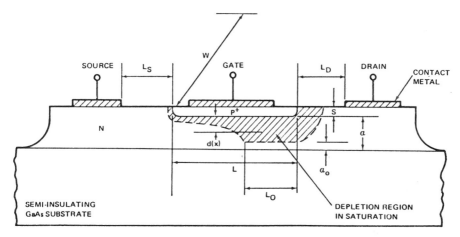

Fig. 6.2. Cross section of a single-gate, epitaxial GaAs junction field-effect transistor and geometrical design parameters. The channel thickness is $a_0$ in the pinch-off region. (After Lehovec and Zuleeg.[73] Reprinted with permission from *Solid-State Electronics*, Pergamon Press, Ltd.)

channel thickness in the saturation region,

$$\frac{(V_D - V_G - V_b)}{V_0} = u^2 < 1 \qquad (6.4)$$

In the region before saturation $V < V_D$ and

$$d^2/a^2 = (V - V_G - V_b)/V_0 \qquad (6.5)$$

Neglecting the distance $L_s$ in Fig. 6.2 the source is at $x = 0$ and integration of Eq. (6.3) along the channel for a distance $x$ gives

$$x/L = \frac{zI_0}{3I_d} \left[ 3 \left( 1 - \frac{I_d}{I_0} \right) \left( \frac{d^2}{a^2} - t^2 \right) - 2 \left( \frac{d^3}{a^3} - t^3 \right) \right] \qquad (6.6)$$

where

$$z = \mu V_0 / v_s L \qquad (6.7)$$

and is a normalized velocity. In Eq. (6.6)

$$t = \frac{d(x = 0)}{a} = \left( \frac{V_G + V_b}{V_0} \right)^{1/2} \qquad (6.8)$$

and is the gate-to-source voltage, normalized and adjusted for the built-in voltage, and

$$I_m = I_0 \left(1 - u_m\right) \tag{6.9}$$

where $I_m$ is the drain saturation current. Then if $x = L, \dfrac{d}{a} = u$ and Eq. (6.6) yields

$$\frac{I_d}{I_p} = \frac{3(u^2 - t^2) - 2(u^3 - t^3)}{1 + z(u^2 - t^2)} \tag{6.10}$$

where

$$I_p = \frac{V_0\, qN\mu a W}{3L} \tag{6.11}$$

is the saturation current for the constant mobility model at $V_G + V_b = 0$; and

$$I_0 = qNv_s aW = 3I_p/z \tag{6.12}$$

is the velocity limited drain current.

Alternatively, Eq. (6.10) may be written as

$$\frac{I_d}{I_p} = \frac{3V_D}{V_0} - 2\; \frac{\left[\left(\dfrac{V_D - V_G - V_b}{V_0}\right)^{3/2} - \left(\dfrac{-V_G - V_b}{V_0}\right)^{3/2}\right]}{1 + \dfrac{\mu V_0}{v_s L}\left(\dfrac{V_D}{V_0}\right)} \tag{6.13}$$

In long channel designs, with low values of $V_D$, there is little or no velocity saturation, $z = \mu V_0/(v_m L)$ tends to zero and the expression becomes the familiar form

$$\frac{I_d}{I_p} = \frac{3V_D}{V_0} - 2\left[\left(\frac{V_D - V_G - V_b}{V_0}\right)^{3/2} - \left(\frac{-V_G - V_b}{V_0}\right)^{3/2}\right] \tag{6.14}$$

Velocity saturation is significant in short-channel designs. Then $z$ may have a value such as 3 and from Eq. (6.13) the effect is a reduced value of $I_d$. Equation (6.13) gives the curves in Fig. 6.3 from which it is seen that velocity saturation produces

- reduction of drain saturation current,
- a drain saturation voltage smaller than the pinch-off voltage,
- reduction of transconductance,
- an almost constant transconductance for small gate voltages, if $z = \mu V_0/v_s L \gg 1$, and
- reduction of drain conductance in the saturation region at a given drain current.

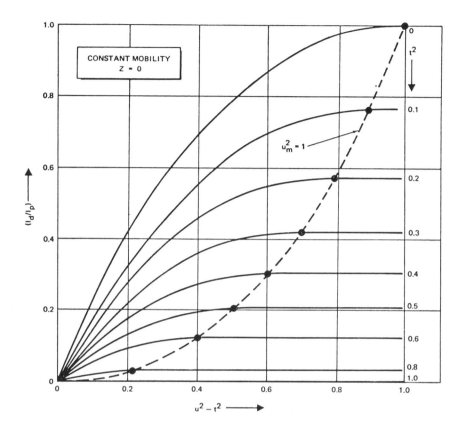

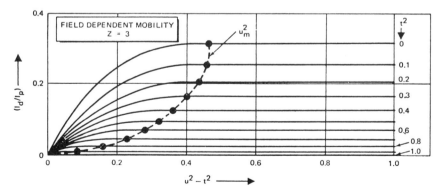

Fig. 6.3. Theoretical JFET characteristics with constant mobility and field dependent mobility (drift velocity saturation). (After Lehovec and Zuleeg.[73] Reprinted with permission from *Solid-State Electronics*, Pergamon Press, Ltd.)

The transconductance in the saturated regime is given by

$$g_m = \frac{\partial I_m}{\partial V_G} = \frac{I_0}{V_0}\frac{\partial u}{\partial t^2} \qquad (6.15)$$

from Eqs. (6.8) and (6.9). Hence

$$g_m = g_{m0}\frac{u_m - t}{z(u_m{}^2 - t^2) + 1} \qquad (6.16)$$

where

$$g_{m0} = I_0 z/V_0 = qNa\mu W/L.$$

and $t^2 = -(V_G + V_b)/V_0$

For the constant mobility model $z = 0$ and $u_m = 1$

$$g_m = g_{m0}(1 - t) \qquad (6.17)$$

The variation of $g_m$ with $(V_G + V_b)/V_0$ for various values of the saturation velocity term $z$ is given in Fig. 6.4. The velocity saturation effect is seen to

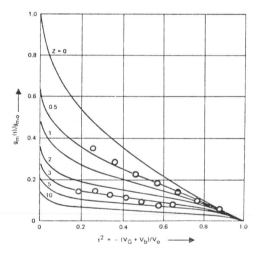

Fig. 6.4.  Transconductance $g_m(t)/g_{m0}$ variation with the gate voltage term $t^2$ for a range of values of the field-dependent mobility saturation term $z$. The experimental points are for $n$ channel GaAs devices. (After Lehovec and Zuleeg.[73] Reprinted with permission from *Solid-State Electronics*, Pergamon Press, Ltd.)

reduce the dependence of $g_m$ on $V_G$ and the transistor becomes more linear in its transfer characteristics.

Consider now the FET gate capacitance, the transit time and the cutoff frequency. The charge on the gate is given by

$$Q_D = W \int_0^L qNd\partial x \tag{6.18}$$

where $d$ is the depletion depth, as shown in Fig. 6.2. The small-signal gate capacitance is then

$$C_G = \frac{\partial Q_D}{\partial V_G} = \frac{1}{V_0} \frac{\partial Q_D}{\partial t^2} \tag{6.19}$$

Evaluation with the aid of Eq. (6.6) gives

$$\frac{C_G}{C_{G0}} = \left(\frac{z}{3}\right) \frac{u-t}{1-u} \left[ \frac{4(u^3 - t^3) - 3(u^4 - t^4)}{(1-u)[u^2 - t^2 + (1/z)]} - 6t \right] \tag{6.20}$$

where

$$C_{G0} = \epsilon\epsilon_0 WL/a = qNWLa/2V_0 \tag{6.21}$$

is the gate capacitance for the fully depleted channel.

The cutoff frequency of the lumped component network is

$$f_{co} = \frac{g_m}{2\pi C_G} = \left(\frac{3}{\pi}\right) \left(\frac{v_m}{Lz}\right) (1-u)^2 \left[ 4(u^3 - t^3) \right.$$
$$\left. -3(u^4 - t^4) - 6t(1-u)[u^2 - t^2 + (1/z)] \right]^{-1} \tag{6.22}$$

The transit time may be obtained as

$$\tau = \int_0^L \frac{\partial x}{v}$$

or as the ratio of the channel charge $(qNWLa - Q_D)$ and the channel current $I_m$:

$$\tau = \frac{qNWLa - Q_D}{I_m}$$
$$= \frac{L}{v_m} \frac{1 - u_m - [(z/6)][u_m^4 - 4u_m t^3 + 3t^4]}{(1-u_m)^2} \tag{6.23}$$

$$= \frac{L}{v_m} \left[ 1 + \frac{z}{6} \frac{(u_m - t)^3}{(1 - u_m)^2} (u_m + 3t) \right] \qquad (6.24)$$

In the limit of a short device so that $z \gg 1$ and at $t = 0$ ($V_G + V_b \to 0$)

$$\frac{g_m}{g_{m0}} \to (3z^2)^{-1/3} \qquad (6.25)$$

and

$$f_{co} \to \frac{1}{4\pi} \frac{v_s}{L} \qquad (6.26)$$

The treatment so far has assumed that pinch-off occurs at $x = L$ and that the field in the channel at pinch-off is infinite. An improvement is to assume a pinched-off channel of length $L_0$ as shown in Fig. 6.2 and a channel field of $\mathcal{E}_0$ at pinch-off. Development of the analysis in this way results in a drain current that does not completely saturate with increasing drain voltage, there is a finite drain conductance $g_d = \partial I_d / \partial V_D$.

Development of the concept gives

$$\frac{V_D}{V_0} = y_m^2 - t^2 + \frac{2a\mathcal{E}_0}{\pi V_0} \sinh\left(\frac{\pi L_0}{2a}\right) \qquad (6.27)$$

where

$$y_m = 1 - \frac{I_d}{I_0} \left( 1 + \frac{v_s}{\mu \mathcal{E}_0} \right) \qquad (6.28)$$

Some manipulation and simplification then gives[73]

$$g_d \simeq \frac{I_d/(\mathcal{E}_0 L)}{1 + z(u_m^2 - t^2)} \qquad (6.29)$$

For a typical Si FET $g_d$ is 50 $\mu$A/V and $\mathcal{E}_0$ determined by matching to the observed $g_d$ value is $5 \times 10^4$ V/cm.

The maximum frequency of oscillation corresponding to this model is

$$f_{max} = \frac{f_{co}}{2} \left( \frac{g_m}{g_d} \right)^{1/2} \qquad (6.30)$$

$$= \frac{f_{co}}{2} \left[ \frac{\mu \mathcal{E}_0}{v_s} \cdot \frac{u_m - t}{1 - u_m} \right]^{1/2} \qquad (6.31)$$

For $z \gg 1$ so that $f_{c0}$ is $v_s/4\pi L$

$$f_{\max} = \frac{v_s}{\gamma L} \left(\frac{3}{z}\right)^{1/6} \tag{6.32}$$

where $\gamma$ is 0.18 if $\mu\mathscr{E}_0/v_s$ is 20.

Equation (6.32) shows the importance of saturation velocity and channel length in determining maximum frequency performance.

The model of the pinch-off process used is still far from complete.[47] Two-dimensional numerical analyses show that departure from space-charge neutrality occurs in the conducting part of the nearly pinched-off channel, as illustrated in Fig. 6.5. At $x_1$ the field strength has reached the critical value $\mathscr{E}_c$ for the drift velocity to be saturated; but the channel voltage increases and the channel thickness continues to narrow as the drain is approached. Therefore the change in channel thickness must be compensated for by an increase in electron concentration to maintain the channel current constant. At $x_2$ the channel cross-section is again $d_1$ and the negative space-charge changes to a positive space-charge, partial electron depletion, to preserve constant current. The electron velocity

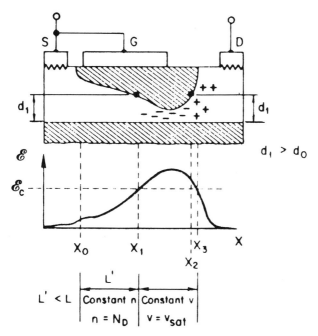

Fig. 6.5. JFET structure showing charge build up and electric field variation in the channel pinch-off region. (After Liechti. Reprinted with permission from *IEEE Trans. Microwave Theory and Techniques*, **MTT-24**, 1976, p. 286.)

remains saturated from $x_2$ to $x_3$ because of the field added by the space-charge. So excess drain voltage forms a dipole space-charge layer in the channel. Shown in Fig. 6.6(a) are curves of constant electron depletion from the equilibrium electron density for the semiconductor in a long-gate structure ($V_{SD}$ = 5.0 V). Figure 6.6(b) shows the space-charge conditions calculated for a short channel JFET ($V_{SD}$ = 5.0 V). The dipole effect is seen to be large in the narrow device and to extend beyond the end of the gate.

## 6.2 GaAs MESFET MODELING

Field-effect transistors in Si or GaAs can be made with diffused or ion-implanted gates. However, there are advantages in the use of Schottky barrier metal gates where the channel is $n$ type and short channel lengths are needed. Metal semiconductor junctions have good barrier heights on $n$ type material. They can be created with very little heat treatment and diffusion widening through oxide windows is not a difficulty. With GaAs the problem of fabricating a suitable

(a)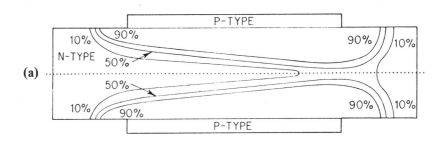

(b)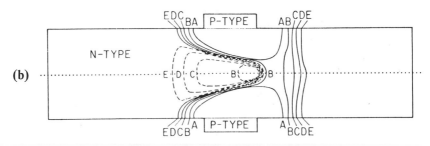

Fig. 6.6. Charge density contours in the channels of long-and short-gate JFETs:
(a) For the long-gate structure the lines represent the percent electron depletion, and
(b) For the short-gate structure the full lines represent electron depletion and the dashed lines represent electron accumulation where $A = 1 \times 10^{17}$, $B = 3 \times 10^{16}$, $C = 1 \times 10^{16}$, $D = 3 \times 10^{15}$, $E = 1 \times 10^{15}$ cm$^{-3}$. (After Kennedy and O'Brien in *Semiconductor Device Modeling for Computer-Aided Design*, Herskowitz and Schilling (eds.), copyright 1972 McGraw Hill. Used with permission of McGraw Hill Book Co., NY.)

oxide with very low interface-state-density at the oxide-semiconductor is avoided. In Si structures the metal gate, if a high temperature material is used, may act as a mask to ensure self-alignment of the source and drain diffusions with respect to the gate.

Since GaAs is a material that is considerably more difficult to diffuse and work with than Si, but makes good Schottky barriers, the MESFET approach is attractive. GaAs is of interest for high frequency FET performance because the peak saturated electron drift velocity is almost double that for Si, and the frequency response is therefore potentially higher [see Eq. (6.32)]. However, at high field strengths in GaAs there is a transfer of high-velocity electrons to a satellite conduction-band valley of lower saturation velocity, with the peaking effect shown in Fig. 6.7. If the velocity-field characteristic is assumed to have equilibrium form, the drift velocity peaks at $x$, in the center of the channel and then falls to the $1 \times 10^7$ cm/sec saturated value and electron accumulation must occur to maintain the current. Between $x_2$ and $x_3$ the field decreases and a second drift-velocity peak occurs in a region of partial electron depletion.

In very-short-channel GaAs FETs the equilibrium velocity-field decay from $2 \times 10^7$ to $1 \times 10^7$ cm/sec does not have the time needed to take place (~1 psec). So, in 0.5-1 $\mu$m gate devices advantage may be taken of the velocity hump in shortening the channel transit time. Analysis of the problem requires Monte Carlo computer simulation studies.

InP MESFETs are also under study[13] since this material has a 50% higher peak drift velocity than GaAs.[13, 144] Also under consideration are ternary materials such as $InP_{1-x}As_x$ and thin films of $In_{.04}Ga_{.96}As$ on GaAs substrates.[28]

## 6.3 DUAL GATE MESFETs

A dual-gate GaAs FET structure is shown in Fig. 6.8 together with its circuit models. The advantages of a dual-gate structure are higher gain and lower feedback capacitance that improves gain stability. Also with variation of the dc bias on the second gate the gain can be controlled over 40 dB for automatic gain control in amplifiers. The gates may each be about 1-$\mu$m long. The performance advantage in making the channel deeper under the second gate, for a larger pinch-off voltage, is that this allows FET2 to provide FET1 with enough current to operate near its maximum drain current and therefore with maximum $g_{m1}$.

Calculated gains for comparable dual- and single-gate structures, neglecting wiring and packaging parasitics, are shown in Fig. 6.9. The maximum stable gain of the dual-gate device is more than 10-dB higher than that of the single-gate unit. This may be expressed as

$$\text{MSG}^{\text{dual}} = \text{MSG}^{(1)} \times \text{MSG}^{(2)} \cong \frac{g_{m1}}{\omega C_{gd1}} \cdot \frac{g_{m2}}{g_{d2}} \qquad (6.33)$$

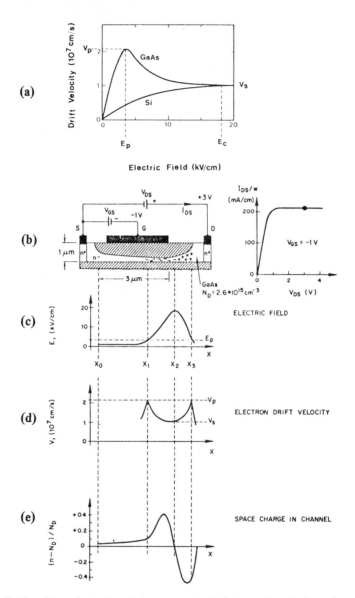

Fig. 6.7. The effect of transferred electron action in GaAs on the velocity and space-charge in the channel of a MESFET:

(a) Drift velocity curve for GaAs,
(b) MESFET structure,
(c) Electric field variation with distance,
(d) Electron drift velocity in the channel, and
(e) Space charge variation. Proceeding from $x_1$ to $x_2$, the channel cross section becomes narrower and, in addition, the electrons "slow down." To preserve current continuity, a heavy electron accumulation has to form. The opposite occurs between $x_2$ and $x_3$. (After Liechti. Reprinted with permission from *IEEE Trans. Microwave Theory and Techniques*, **MTT-24**, 1976, pp. 286, 287.)

344

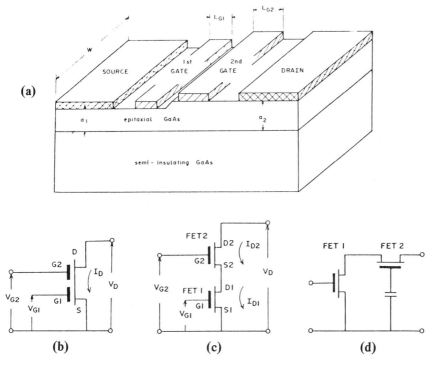

Fig. 6.8. Structure and modeling of a dual-gate GaAs MESFET:
(a)  Structure considered,
(b)  Dual-gate device symbol,
(c)  Decomposition of a dual-gate FET into two series-connected single-gate FETs, and
(d)  Cascade connected common-source and common-gate FETs modeling the ordinary
     application of a dual-gate FET. (After Asai, Murai, Kodera. Reprinted with permission
     from *IEEE Trans. Electron Devices*, **ED-22**, 1975, p. 898.)

where the factor $g_{m2}/g_{d2}$ can be in excess of 10. The stability factor of a dual-
gate FET is approximately given by

$$k^{\text{dual}} \cong k^{(1)} + 2\,\frac{\omega C_{gs2}}{g_{m2}}$$

$$= k^{\text{single}} + 2(f/f_{T2}) \qquad (6.34)$$

where $f_{T2} = g_{m2}/2\pi C_{gs2}$.

If $k > 1$, a good figure of merit is the maximum available gain (MAG) which is
given by

$$\text{MAG} = \text{MSG}\,[k - (k^2 - 1)^{1/2}] \qquad (6.34)$$

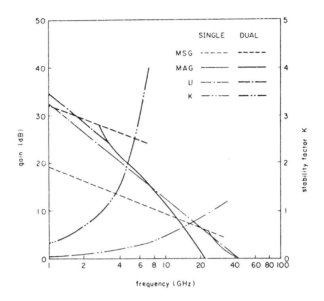

Fig. 6.9. Calculated frequency dependences of power gain and stability factors of single- and dual-gate FETs show the dual-gate unit to have a higher maximum stability gain (MSG). (After Asai, Murai, Kodera. Reprinted with permission from *IEEE Trans. Electron Devices*, **ED-22**, 1975, p. 890.)

For a dual-gate device a high MAG is obtainable over a wide frequency range (3 GHz in Fig. 6.9). The dual-gate device is attractive for use at frequencies below 10 GHz, and the single-gate device is superior in and above X-band. The cutoff frequency is lower for the tetrode FET because the frequency response is limited by that of the second gate.

## 6.4 MICROWAVE FIELD EFFECT TRANSISTORS

Field-effect transistor structures presently under examination for microwave and high-frequency power applications are illustrated in Fig. 6.10.

Insulated gate structures (IGFETs) operated in the enhancement mode have a transconductance and input capacitance that are almost independent of gate-source voltages and the output capacitances are independent of drain voltages. Thus, IGFETs tend to have advantages over JFETs and MESFETs in class A linear power applications with low amplitude- and phase-distortion. Also, the ability to operate from depletion into enhancement allows a large gate-voltage range. The IGFET structure tends at present to be limited to Si devices because of high interface-state densities associated with GaAs-oxide interfaces.

At 700 MHz an *n* channel planar depletion-type Si MOSFET with a 20-mm-

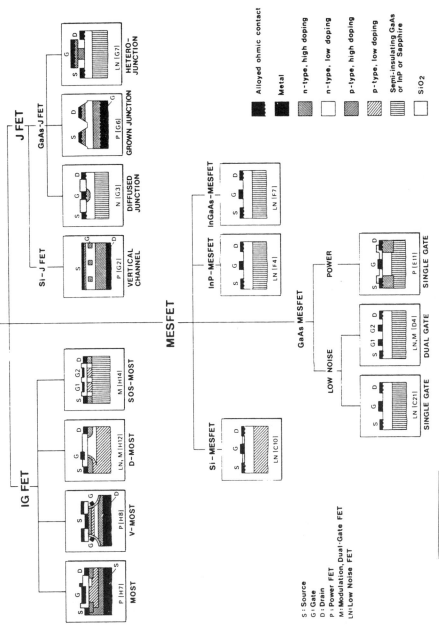

Fig. 6.10  This "family tree" of microwave FETs shows the cross-sections of various FET structures. (After Liechti. Reprinted with permission from *IEEE Trans. Microwave Theory and Techniques*, MTT-24, 1976, p. 280.)

wide channel, 5-$\mu$m long, has given 16-W output power (measured at 4-dB gain compression) with 6-dB power gain and 26% power-added efficiency. (Power-added efficiency is the RF output power minus the RF input power divided by the dissipated dc power.) For an 18 mm $\times$ 1 $\mu$m gate MOSFET a CW output power of 4 W has been measured with 5-dB gain at 2 GHz. Figure 6.11 shows a vertical MOS geometry that is considered to have prospects for useful performance up to 4 GHz. MOSFET behavior is discussed in detail in Chap. 7.

In JFET structures there are fabrication difficulties in obtaining very narrow junction gates by diffusion. Ion implantation provides better control and vertical structures with 1.8 mm $\times$ 1 $\mu$m channels have given 0.2-W output power at 2.7 GHz with 6-dB power gain.[69]

For a GaAs diffused junction structure of 2 $\mu$m gate length a 10-dB gain has been obtained at 4 GHz with the noise figure a low 2.5 dB. This structure had high tolerance to a 1 MeV neutron flux of $5 \times 10^{16}$ cm$^{-2}$.[15]

The heterojunction JFET shown in Fig. 6.10 has a $p$ gate of $Ga_{0.5}Al_{0.5}As$ grown on an $n$-GaAs layer and is selectively etched to give a 1-$\mu$m gate strip and self-alignment of the source and gate.[128]

GaAs MESFETs have produced the highest frequency and highest frequency times power product for microwave transistors to date. Furthermore, they are of remarkably low noise levels. For instance a 0.5-$\mu$m single-gate GaAs MESFET may have 10-dB gain at 12 GHz with a noise figure of only 3 dB.[97] The gain and noise figure of this GaAs MESFET plotted vs frequency is given in Fig. 6.12 together with curves that represent the state of the art for bipolar transistors. At 90°K a GaAs FET may have a noise figure of 0.8 dB ($T_e = 60°K$) and an associated gain of 8 dB. A three-stage amplifier for the 11.7-12.2 GHz commu-

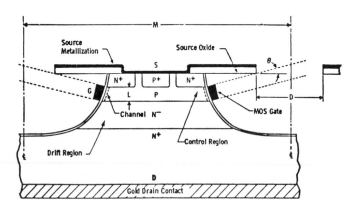

Fig. 6.11. Topology of vertical power MOS transistor geometry. (After Oakes et al. Reprinted with permission from *IEEE Trans. Microwave Theory and Techniques*, **MTT-24**, 1976, p. 305.)

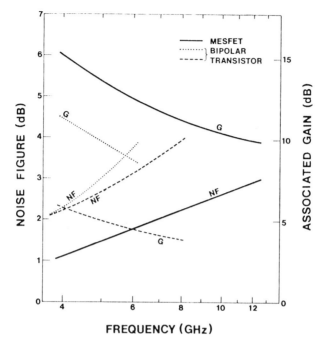

Fig. 6.12. Lowest reported noise figures and associated gains of microwave transistors are plotted vs frequency. The GaAs-MESFET reported by Ogawa et al. has 0.5-$\mu$m gate length. The dashed line represents the Si bipolar transistor with the lowest published noise figure. The bipolar transistor with the dotted line has a considerably higher gain. (After Liechti. Reprinted with permission from *IEEE Trans. Microwave Theory and Techniques*, **MTT-24**, 1976, p. 281.)

nication band cooled to 60°K has shown a 1.6-dB noise figure ($T_e = 130°K$) for 31-dB gain.[76]

Short-gate GaAs MESFETs have very low noise figures. This may be for various reasons, not all of which are fully understood. One possibility is that for a suitable drain current ($I_{DS}/I_{DSS} \approx 0.15$), the amplified input-noise current may destructively interfere with the correlated noise current if the MESFET's gain and transmission phase is properly adjusted with an optimized input termination.[120]

In comparison with Si bipolar transistors, GaAs MESFETs have better performance above 4 GHz. The gains are higher, the noise figures are lower, the amplifier efficiencies are higher, the input impedance is higher, they are self-ballasting (since with rising temperature the channel resistance increases and prevents thermal runaway) and free from second breakdown. They also have better reverse isolation and lower third-order intermodulation distortion.

The development of high power from a MESFET depends on the achievement of a wide gate so that a large drain current may be controlled. However, there is a maximum gate width beyond which the voltage along the gate becomes nonuniform. The maximum width can be estimated from the criterion that the transmission line formed by the source and gate electrodes should be a very small fraction of a wavelength. For example, the maximum gate width for a 1-$\mu$m-long gate is calculated to be about 50 to 250 $\mu$m at 10 GHz. Since uniform gate voltage is desirable for efficient operation, a number of 50 to 250-$\mu$m-wide gates are connected in parallel to obtain a total gate width necessary for controlling a large drain current.

Parallel connection of gates for a total gate width of 2600 $\mu$m/cell is illustrated in Fig. 6.13. The power output characteristic of this GaAs MESFET is shown in Fig. 6.14 as a function of frequency and represents about the present state of the art. One figure of merit for an FET design expresses power and frequency in terms of W(GHz)$^2$/mm of gate width. Recent GaAs designs have values of 20-70 W(GHz)$^2$/mm.

Electron beam lithography offers the possibility of sub-micron gate lengths and therefore higher frequency performance, but the channel thickness $D$ must also be decreased to maintain $L/D > 1$. The thinner channel implies a heavier doping level to maintain a proper channel impedance. But the doping cannot exceed about $4 \times 10^{17}$ cm$^{-3}$ in Si without voltage breakdown effects. So there is no advantage in fabricating Si MESFETs with gate lengths below 0.1 $\mu$m and the limit of current-gain Si bandwidth is likely to be about 70 GHz.[110]

Collections of papers on microwave FETs appear in the *IEEE Transactions on Microwave Theory and Techniques*, **MTT-24**, June 1976, in the *Transactions on Electron Devices*, ED-25, June 1978 and in the *International Electron Devices Meeting Technical Digest*, Washington, D.C., Dec. 1978.

## 6.5 SOME APPLICATIONS OF JFETs AND MESFETs

### 6.5.1 Voltage-Controlled Resistors

At low values of drain current and voltage a JFET has a channel resistance that is increased by increase of the gate-source reverse-bias voltage since this controls the effective channel depth. The characteristics for a typical device, showing resistance variation from about 400 ohms at $V_{GS} = 0$ to about 30 kilohms at $V_{GS} = -3$ V, are given in Fig. 6.15(a). RC circuits based on an FET and a capacitor may be used to give voltage-controlled phase-advance and phase-retard circuits as shown in Fig. 6.15(b), (c).

A cascaded VCR attenuator is shown in Fig. 6.16. Also shown is an automatic gain-control circuit and a voltage-controlled variable gain amplifier. Two filter circuits incorporating JFETs as VCRs are given in Fig. 6.17.

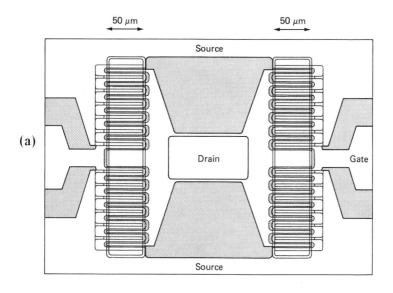

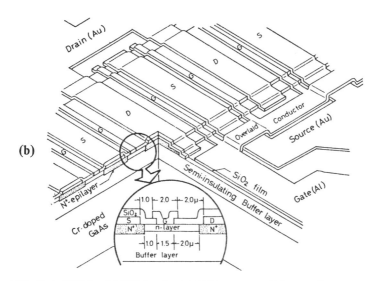

Fig. 6.13. GaAs FET for 4-8 GHz band, with 2600-μm gate width per cell.
(a)  View of the two cells, and
(b)  Cut-away view of the MESFET. (After Fukuta, et al. Reprinted with permission from
     (a) *International IEEE Devices Meeting*, 1974, p. 1 and (b) *IEEE Trans Microwave
     Theory and Techniques*, **MTT-24**, 1976, p. 5.)

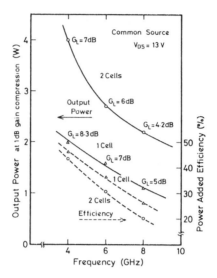

Fig. 6.14. Output power and power added efficiency at operating frequencies from 4 to 8 GHz for the GaAs MESFET of Fig. 6.13. (After Fukuta. Reprinted with permission from *IEEE Trans. Microwave Theory and Techniques*, **MTT-24**, 1976, p. 12.)

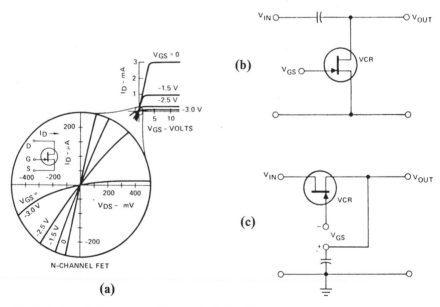

Fig. 6.15. Use of a JFET as a voltage-controlled resistor:
(a)   Section of the characteristics that show approximate behavior as a resistor controlled by the gate voltage,
(b)   A VCR phase-advance circuit, and
(c)   A VCR phase-retard circuit. (Courtesy Siliconix Inc., Applications Note AN 73-7.[117])

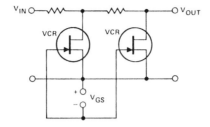

Cascaded VCR Attenuator

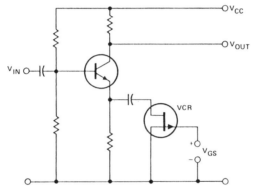

Wide Dynamic Range AGC Circuit. No Gain Through FET
With Distortion Proportional to Input Signal Level.

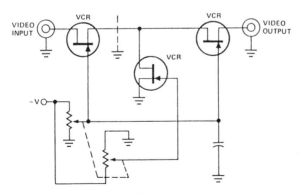

Voltage Controlled Variable Gain Amplifier. The Tee Attenuator
Provides For Optimum Dynamic Linear Range Attenuation.

Fig. 6.16. Some voltage-controlled resistor applications of JFETs. (Courtesy Siliconix Inc.,
Applications Note AN 73-7.[117])

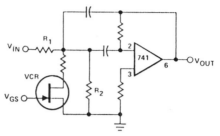

Voltage–tuned Filter Octave Range With Lowest Frequency
at JFET $V_{GS(OFF)}$ and Tuned by $R_2$. Upper Frequency is
Controlled by $R_1$.

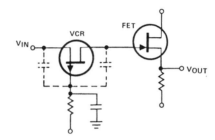

Tunable Low–Pass RC Pi–Filter. Voltage–Tunable Over 100:1 Range.
Frequency Range Changed By Shunting Gate And Source With
Fixed Capacitors.

Fig. 6.17. Applications of JFETs in filter circuits for voltage tuning. (Courtesy Siliconix Inc., Applications Note AN 73-7.[117])

### 6.5.2 FET Switching Circuits

JFET switches have some advantages over bipolar transistor switches. The ON to OFF ratio is high and there is no offset current so JFETs may be used to switch microvolt signals without error. The drive power is lower than for BJTs because of the high gate-input impedance.

Some common switching functions are shown in the circuits of Fig. 6.18. In Fig. 6.18(a) $Q_1$ through $Q_3$ perform as analog switches, $Q_4$ is a sample-and-hold switch, $Q_5$ functions as a signal clamp with zero reference, and $Q_6$-$Q_7$ are in a gain-switching circuit. In Fig. 6.18(b) the two FETs are connected as the modulator and demodulator of a chopper-type dc amplifier, driven in anti-phase so as to simulate the action of a double-pole electromechanical chopper.

### 6.5.3 Amplifiers

The common-gate amplifier of Fig. 6.19(a) may be used as a receiver front-end at 450 MHz with a 3-dB bandwidth of 6 MHz. It has excellent reverse-isolation characteristics that reduce local oscillator radiation from the antenna. Applica-

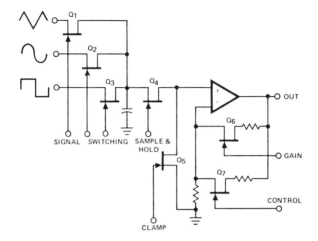

**FET Switching Circuits**

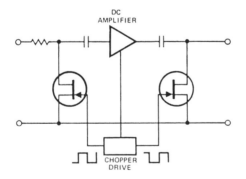

**DC Amplifier Using FET Choppers**

Fig. 6.18. Switching uses of JFETs. (Courtesy of Siliconix, Inc., Applications Note AN 73-7.[117])

tions include fixed and mobile communication systems, paging systems, UHF-TV, and police and fire services.

The cascode-coupled circuit of Fig. 6.19(b) operates at 30 MHz in a 50-ohm system with a 5.5-MHz (3 dB) bandwidth, a gain of 20 dB and a noise-figure of 1.5 dB. Some FET cascode techniques provide excellent differential amplifier performance with voltage protection.[137]

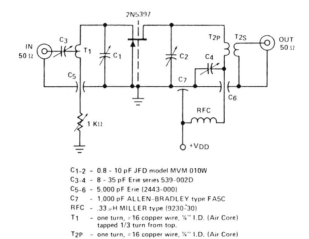

C1-2 - 0.8 - 10 pF JFD model MVM 010W
C3-4 - 8 - 35 pF Erie series 539-002D
C5-6 - 5,000 pF Erie (2443-000)
C7   - 1,000 pF ALLEN-BRADLEY type FA5C
RFC  - .33 μH MILLER type (9230-30)
T1   - one turn, ≈16 copper wire, ¼″ I.D. (Air Core)
       tapped 1/3 turn from top.
T2P  - one turn, ≈16 copper wire, ¼″ I.D. (Air Core)
T2S  - one turn, ≈16 copper wire, ¼″ I.D. (Air Core)

**Common–Gate 450 MHz Amplifier**

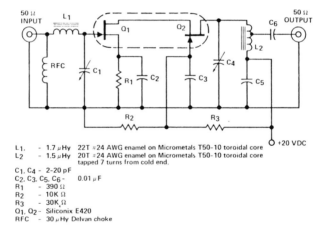

L1.  - 1.7 μHy  22T #24 AWG enamel on Micrometals T50-10 toroidal core
L2   - 1.5 μHy  20T #24 AWG enamel on Micrometals T50-10 toroidal core
                tapped 7 turns from cold end.
C1, C4 - 2-20 pF
C2, C3, C5, C6 -  0.01 μF
R1   - 390 Ω
R2   - 10K Ω
R3   - 30K Ω
Q1, Q2 - Siliconix E420
RFC  - 30 μHy Delvan choke

**Cascode 30 MHz Intermediate Frequency Amplifier**

Fig. 6.19. Amplifier applications of JFETs. (Courtesy Siliconix Inc., Applications Note AN 73-7.[117])

In the circuit of Fig. 6.20(a) the JFETs are used as buffers between a low-impedance 50-ohm input and near-unity $Q$ ratio resonator. The circuit is a conventional VHF amplifier, capacitively-coupled to a two-pole helical coil resonator. The compact resonators offer unusually-high unloaded $Q$, often exceeding 1,000 in the VHF spectrum. Impedance match between resonator input and output is achieved by tapping up on the helix. The amplifier is

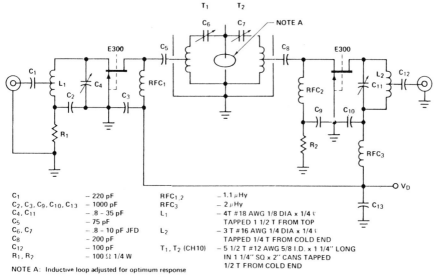

| | | | |
|---|---|---|---|
| $C_1$ | – 220 pF | $RFC_{1,2}$ | – 1.1 μHy |
| $C_2, C_3, C_9, C_{10}, C_{13}$ | – 1000 pF | $RFC_3$ | – 2 μHy |
| $C_4, C_{11}$ | – .8 – 35 pF | $L_1$ | – 4T #18 AWG 1/8 DIA x 1/4 ᶜ |
| $C_5$ | – 75 pF | | TAPPED 1 1/2 T FROM TOP |
| $C_6, C_7$ | – .8 – 10 pF JFD | $L_2$ | – 3 T #16 AWG 1/4 DIA x 1/4 ᶜ |
| $C_8$ | – 200 pF | | TAPPED 1/4 T FROM COLD END |
| $C_{12}$ | – 100 pF | $T_1, T_2$ (CH10) | – 5 1/2 T #12 AWG 5/8 I.D. x 1 1/4" LONG |
| $R_1, R_2$ | – 100 Ω 1/4 W | | IN 1 1/4" SQ x 2" CANS TAPPED |

NOTE A: Inductive loop adjusted for optimum response
1/2 T FROM COLD END

### Selective VHF Amplifier Circuit

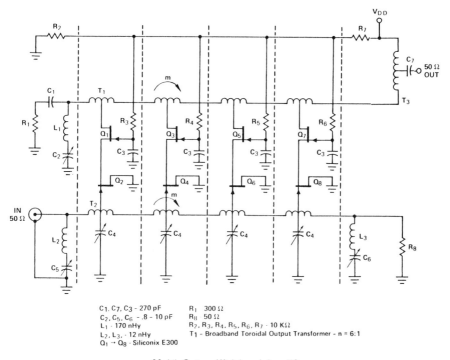

| | | |
|---|---|---|
| $C_1, C_7, C_3$ – 270 pF | $R_1$ | 300 Ω |
| $C_2, C_5, C_6$ – .8 – 10 pF | $R_8$ | 50 Ω |
| $L_1$ – 170 nHy | $R_2, R_3, R_4, R_5, R_6, R_7$ – 10 KΩ |
| $L_2, L_3$, – 12 nHy | $T_1$ – Broadband Toroidal Output Transformer – n = 6:1 |
| $Q_1 \rightarrow Q_8$ – Siliconix E300 | |

### Multi–Octave Wideband Amplifier

Fig. 6.20. VHF and wideband amplifiers using JFETs. (Courtesy of Siliconix Inc., Applications Note AN 73-7.[117])

stagger-tuned to provide a 4.5-MHz color TV bandwidth at about 200 MHz and has 15-dB gain with a noise figure less than 4 dB.

The design of distributed amplifiers basically involves the configuration chosen for the input and output transmission (delay) lines. A number of designs are possible, but the simplest is the image-parameter $m$-derived low-pass filter shown in Fig. 6.20(b), which offers 5-dB gain and a bandpass of 50-300 MHz with excellent group delay characteristics across a major area of the passband.

The use of JFETs in operational amplifiers gives excellent performance[31, 41] as discussed briefly in Chap. 9.

### 6.5.4 MESFET Microwave Amplifiers

Cascading of microwave FET amplifier stages is possible if care is taken. Figure 6.21 shows the block schematic and output characteristic of a 6-GHz, four-stage GaAs MESFET power amplifier which has a power output of 1 W with a gain of 26 dB and a noise figure of 8 dB. The efficiency is 22% and the third-order intermodulation characteristic is 31 dB below the carrier at an output of 1 W.

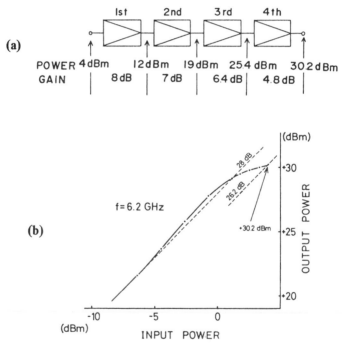

Fig. 6.21. Four stage 6 GHz JFET amplifier:
(a) Stage power levels and gains, and
(b) Output versus input power. (After Arai et al. Reprinted with permission from *IEEE Trans. Microwave Theory and Techniques*, **MTT-24**, 1976, p. 4,5.)

The 3-dB bandwidth is 200 MHz. This amplifier was developed for a microwave FM radio relay. The package size was 19 cm X 4 cm X 2.6 cm.

In microwave applications there are advantages in using balanced amplifiers in conjunction with 3 dB quadrature hybrid couplers as shown in Fig. 6.22(a). Such balancing gives improvements in:

- input- and output-impedance matching in an amplifier optimized for output power or low noise figure,
- short- and open-circuit stability,
- linearity and low intermodulation characteristics,
- temperature insensitivity

and offers the possibility of increasing the output power by combining transistors as shown in Fig. 6.22(b).

If balancing is not used the number of MESFETs needed is halved but performance is somewhat more difficult to achieve and isolators are usually needed as in Fig. 6.22(c) to meet the VSWR requirements.

Broadband amplification is possible by the provision of interstage coupling networks, as shown in Fig. 6.23, that level the gain over the required band by attenuating at the low end and passing at the high frequency end where the MESFET gain is diminished.

Broadband amplification is also possible in GaAs transistors that use the growth of traveling space-charge waves in a thin film of $n$-GaAs biased above the transferred electron threshold.[25, 26] In the 8-18 GHz range gains of 12 dB were obtainable with 32 dB reverse attenuation. The noise figure was 18 dB.

### 6.5.5  FETs as Microwave Oscillators

GaAs MESFET oscillators give excellent performance at frequencies as high as 10 GHz.[78] The voltage needed is low (10 V) and powers of 80 mW or more have been obtained with efficiencies of about 20%.

A typical circuit configuration is shown in Fig. 6.24 where the impedance $jX_3$ is the series feedback element. The oscillation frequency is determined by the gate input capacitance and the external circuit.

The noise figures are comparable to or better than for Gunn oscillators, BARITT diodes (see Chap. 11) and IMPATT diodes (see Table 6.1).

Further development of microwave FET oscillators with increased power output and low noise performance is expected in the next few years.

### 6.5.6  GaAs MESFETs as Mixers

GaAs FETs have the potential for low-noise operation as microwave mixers with gain and a large dynamic range.

The equivalent circuit and characteristics of a MESFET used in one mixer study are shown in Fig. 6.25. Analysis shows that the maximum available

**BALANCED AMPLIFIER**

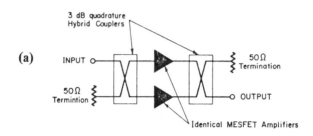

(a)

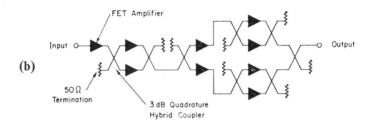

(b)

**UNBALANCED AMPLIFIER**

(c)

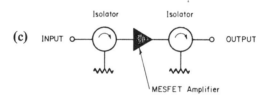

Fig. 6.22. Balanced and unbalanced MESFET amplifiers:
(a) Balanced amplifier with 3-dB hybrid couplers at input and output,
(b) Cascade stages of balanced amplifiers, and
(c) Single amplifier stages with isolators at the input and output ports may be cascaded. (After Liechti. Reprinted with permission from *IEEE Trans. Microwave Theory and Techniques*, **MTT-24**, 1976, pp. 293, 294.)

conversion gain is given by an expression of the form

$$G_c = \frac{g_1^2}{4\omega^2 \overline{C}^2} \frac{\overline{R}_d}{R_{in}} \qquad (6.35)$$

where $g_1$, $\overline{C}$ and $\overline{R}_d$ are time-averaged quantities. If the MESFET is used in a

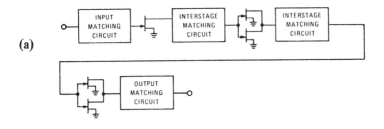

(a)

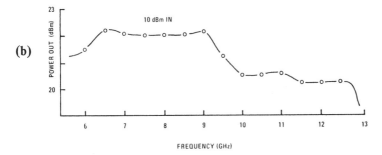

(b)

Fig. 6.23. Three-stage 100-mW 5.9-12.4-GHz amplifier:
(a)  Configuration, and
(b)  Output power vs frequency at $P_{in}$ = 10 dBm. (After Hornbuckle and Kuhlman. Reprinted with permission from *IEEE Trans. Microwave Theory and Techniques*, **MTT-24**, 1976, p. 341.)

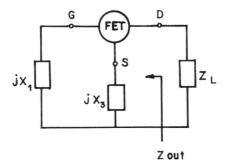

Fig. 6.24. Typical FET oscillator circuit with $jX_3$ the feedback element. (After Okazaki et al.[98])

**Table 6.1. Comparison of Oscillation Performances.**

|  | $f$(GHz) | $P_{out}$ | $\eta$ (%) | NF (dB) |
|---|---|---|---|---|
| GaAs FET | 10.5 | 80 mW | 21 | 4 |
| BARITT | 6.5 | 100 mW | 2 | 15 |
| GaAs IMPATT (conventional) | 10 | 3 W | 19 | 18 |
| GaAs IMPATT (high-low) | 9.5 | 9.25 W | 36 | 38 |
| Gunn | 10 | 2 W | 9 | 15 |

After Okazaki et al., 1974[98].

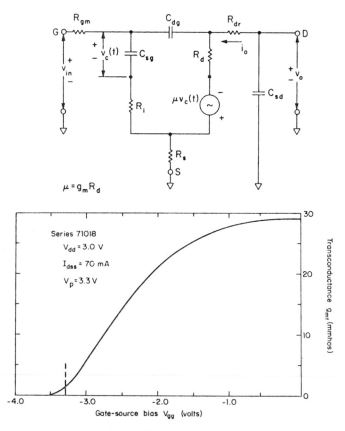

Fig. 6.25. Small-signal equivalent circuit and transconductance of a GaAs MESFET used for the mixer circuit of Fig. 6.26. (After Pucel, Massé, Bera. Reprinted with permission from *IEEE Trans. Microwave Theory and Techniques*, **MTT-24**, 1976, p. 351.)

straight amplifier circuit the maximum available gain is similar in form, namely,

$$G_a = \frac{g_m^2}{4\omega^2 C^2} \frac{R_d}{R_{in}} \qquad (6.36)$$

where $g_m$, $C$ and $R_d$ are values corresponding to the bias condition instead of time-averaged values. The conversion gain $G_c$ (for example, 6.4 dB) can be greater than $G_a$ (for example, 4.7 dB) since the ratios $C/\overline{C}$ and $\overline{R}_d/R$ are greater than unity (although $g_1 < g_m$).

The circuit configuration used is shown in Fig. 6.26(a) for a single FET mixer and in Fig. 6.26(b) for a balanced arrangement. In diode mixers cancellation of the local oscillator-introduced noise can be achieved by reversing the terminals of one of the diodes. This cannot be done with FETs, but a 180° phase shift between the two IF branches can be accomplished by use of leading and lagging phase shifters as shown. These phase shifters are constructed of lumped elements and can be incorporated into the IF matching network.

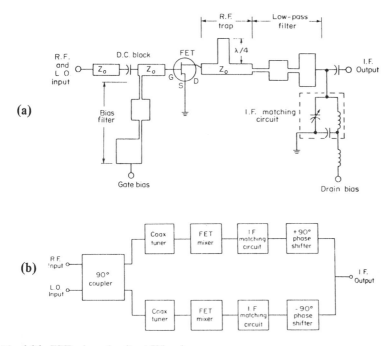

Fig. 6.26.  FET mixer circuits at X band:
(a)   Single mixer circuit, and
(b)   Block diagram of balanced FET mixer showing phase shifters in output circuit for cancellation of LO noise. (After Pucel, Massé, Bera. Reprinted with permission from *IEEE Trans. Microwave Theory and Techniques*, **MTT-24**, 1976, p. 356.)

The conversion gain for a single mixer varies with local oscillator power as shown in Fig. 6.27(a). The balanced mixer gain and noise figure variations are shown in Fig. 6.27(b).

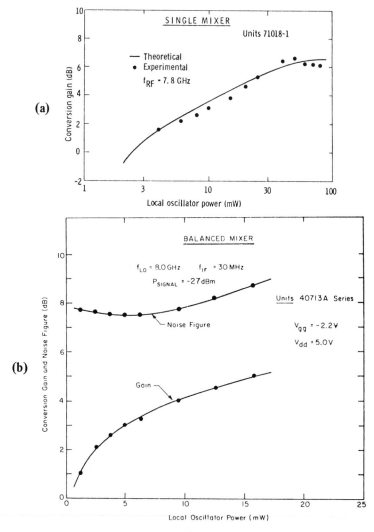

Fig. 6.27. Conversion gains for single-ended and balanced MESFET mixers at X band vs local oscillator power:
(a)  Single-mixer experimental and theoretical conversion gains, and
(b)  Measured conversion gain and double channel noise figure for a balanced mixer. (After Pucel, Massé, Bera. Reprinted with permission from *IEEE Trans. Microwave Theory and Techniques*, **MTT-24**, 1976, pp. 357, 358.)

The conversion loss of a simple Schottky barrier mixer is at least 5 dB (see Fig. 3.4). Therefore, the MESFET mixer is 10 dB better even in its present early stage of development. Other performance figures are given in Table 6.2 for comparison.

### 6.5.7 Other FET Applications

A GaAs MESFET with dual gates may be used for very high speed switching. For instance, the modulation of an 8-GHz carrier with a rise-time of 100 psec, an RF burst-length of 7 cycles and a fall-time of 65 psec is shown in Fig. 6.28, for a 10-GHz FET having two gates, each 1-$\mu$m long. Such devices are of interest for subnanosecond pulsed amplitude modulation (PAM) and for phase- or frequency-shift keyed carrier modulation (PSK) and (FSK).

Regeneration and amplification of fast pulses in the 50-psec range have been established using GaAs MESFETs: The sharpening effect is caused mainly by the variation of gate capacitance with input pulse amplitude. Sharpening factors, $t_{rise\ in}/t_{rise\ out}$, of 3 and voltage amplification factors of 2 have been achieved.[17]

In digital-transmission systems at microwave frequencies the distances between expensive linear repeater relay stations can be increased by the use of low cost regenerator units. An FET circuit for direct regeneration of phase-shift keying signals is shown in Fig. 6.29. This is simpler than the use of tunnel diode parametric amplifier regenerators.

The use of a GaAs MESFET for direct modulation of a double heterojunction GaAlAs laser has been reported with the circuit of Fig. 6.30. The laser required a current of 750 mA at 3 V. The FET was capable of an 8 dB gain at 1 GHz. The rise and fall times of the light with the FET control are about 200 psec and the bit rate of the available word generator limited the performance. GaAs MESFETs will be useful for gigabit modulation of injection lasers in fiber-optics communications.

Table 6.2. Noise Performance and Signal-Handling Capabilities of a GaAs MESFET Mixer and a Low-Level Diode Mixer.

| Mixer | Maximum Gain | Minimum Noise Figure | Output Third-order IM Intercept | Output (1 dB Gain Compression Level) |
|---|---|---|---|---|
| GaAs FET | +6 dB | +7.4 dB | +20 dBm | +5.5 dBm |
| Diode (low-level) | −5 dB | 5-7 dB | +5 dBm | −6 to −1 dBm |

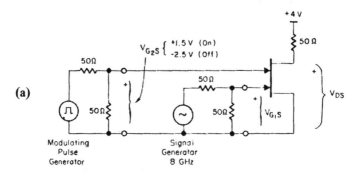

(a)

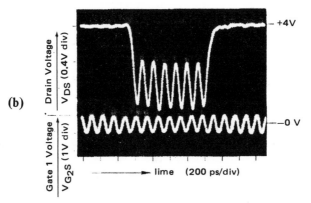

(b)

Fig. 6.28.  Use of a dual-gate GaAs MESFET for high speed switching of an 8-GHz signal:
(a)   Switching circuit, and
(b)   Pulsed-amplitude modulation of an 8-GHz carrier. The RF input voltage at the first
gate is shown in the lower trace and the output-voltage waveform at the drain is shown
in the upper trace. (After Liechti. Reprinted with permission from *IEEE Trans. Microwave Theory and Techniques*, **MTT-23**, 1975, p. 468.)

High speed integrated logic should also be possible at clockrates of 2-3 GHz
with GaAs MESFETs that have propagation delays of less than 100 psec.[129]
Bistable switching action has been observed in GaAs Schottky gate field-effect
transistors with transition times of less than a microsecond associated with deep
trap action.[112] Other JFET charge storage or memory structures are known.[1]

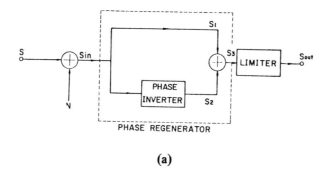

**(a)**

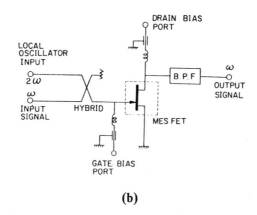

**(b)**

Fig. 6.29. Phase regenerator circuits:
(a) Typical block diagram of a regenerator, and
(b) Circuit diagram of a MESFET phase regenerator. (After Komaki, Kurita, Memita. Reprinted with permission from *IEEE Trans. Microwave Theory and Techniques*, **MTT-24**, 1976, pp. 367, 369.)

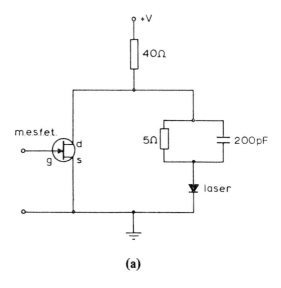

**(a)**

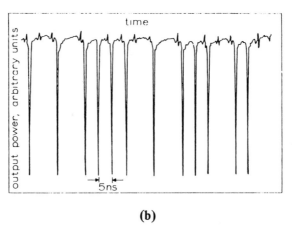

**(b)**

Fig. 6.30. GaAs MESFETs may be used to modulate double-heterojunction GaAlAs laser diodes:

(a) Circuit used for a laser diode requiring about 750 mA at 3 V, and

(b) Detected light output against time for 200 Mbit/sec pseudorandom, return-to-zero bit stream. (Estimated peak power = 40 mW.) (After Ostoich, Jeppesen, Slaymaker.[100] Reprinted with permission from *Electronics Lett.*, **11**, 1975, p. 515, 516, Institution of Electrical Engineers, London.)

## 6.6 QUESTIONS

1. Consider the design of an $n$ channel Si MESFET fabricated on a 5-$\mu$m-thick $n$ epi-layer dopd $10^{15}$ cm$^{-3}$, grown on a sapphire substrate. The Al gate is 100-$\mu$m wide and the channel gate is 10-$\mu$m long. The source and drain contacts are 10-$\mu$m long and spaced 2 $\mu$m from the gate edges. The velocity saturation in the channel may be assumed to be given by $\mu\mathscr{E}/(1 + \mu\mathscr{E}/v_s)$. Calculate the characteristics of the device and estimate the small signal parameters for a bias point in the saturated region.

2. How will the characteristics change if the temperature is (a) 77°K and (b) 450°K?

3. Select a bias condition for the transistor of problem 1 that allows a large ac signal swing. Calculate the voltage along the channel for this bias condition and the transit time of the carriers.

4. If the device of problem 1 is used as a voltage controlled resistor, what will be its characteristics?

5. Consider again the basic structure of problem 1, but with two gates 4-$\mu$m long, spaced 2-$\mu$m apart, to form a dual gate Si MESFET. As suggested by Fig. 6.8 the epi-thickness may be reduced for the first gate if you see any advantage in doing so. Estimate the characteristics that may be achieved from the dual-gate structure under several conditions of gate bias.

6. An $n$ channel JFET silicon on sapphire is to be designed to operate from a 50-V power supply and feed a 150-ohm load in a common-source circuit application as a power device to match the performance of a bipolar transistor. What do you propose for the gate width and length and the layout geometry? What should be the channel doping and thickness? What will be the device capacitance? What will be the voltage and power gain and the linearity, the maximum output power to the load and the maximum device power dissipation? What do you see as the deficiencies of your design?

7. The GaAs FET with the characteristic shown in Fig. 6.25 is to be used to mix a 10-mV, 980-MHz signal with a 750-MHz local oscillator in a single mixer circuit. What must be the design considerations in the circuit and what may be expected for the conversion gain? How would the performance compare with that of a Schottky barrier mixer?

8. Consider the GaAs MESFET characteristic of Fig. 6.25. If the geometry of the device was unchanged but the material was changed from $n$ GaAs to $n$Si of similar doping, what change would be expected in the characteristic?

9. Review the factors controlling the linearity (or nonlinearity) of FET input/output transfer curves.

10. Normally the channel of a JFET is assumed to be in a region of uniform doping. Would there be any advantage in grading the channel doping?

11. Discuss temperature dependencies of JFETs and MESFET parameters.

12. Review the factors controlling the voltage rating of JFETs and MESFETs and the power dissipation.

13. Discuss the low-temperature performances of JFETs and MESFETs.

14. Discuss vertical channel FETs.

15. Review the merits of ion-implantation in the fabrication of JFETs and MESFETs.

16. Discuss typical $S$ parameter variation with frequency in MESFETs.

17. Discuss the effects deep-impurity traps may have in Si and GaAs FETs.

18. Review the microwave-frequency equivalent circuits of MESFETs including package parasitics.

19. Consider linearity of amplification and phase distortion in JFETs and IGFETs for high-frequency amplification.

20. What is the experimental evidence for the velocity overshoot effect in the channel of a GaAs MESFET aiding the frequency response?

21. Review the design of any three-stage FET amplifier.
22. What does *third-order intermodulation* mean in an FET circuit?
23. Consider the use of 3-dB hybrid couplers in conjunction with FET amplifiers for improving VSWR performance and for power combining.
24. Discuss the techniques for obtaining 1 $\mu$m or submicron gate lengths in MESFETs using optical photolithography.
25. Discuss how the various parameters of a Si MESFET will change as the dimensions are reduced.
26. What progress has been made recently in fabricating oxides or other insulating films on GaAs with low densities of interface states suitable for use in JFETs or IGFETs?
27. Consider the merits of ternary III-V compound MESFETs or heterojunction FETs.
28. A figure of merit for FET designs may be W GHz$^2$/mm (Ref. Liechti, 1976). Discuss this relationship of power and frequency and gate width for GaAs. How does the performance of Si FETs compare? Can bipolar Si and GaAs transistors be expressed in a similar way and with what results?
29. It has been stated that a MESFET with a complex-conjugate matched input port becomes unstable with decreasing frequency because a larger fraction of the output voltage is fed back to the input.[75] Discuss the problem of conditional stability and gain-frequency dependence in MESFETs.
30. Review the noise theory for Si JFETs. How must it be modified for GaAs MESFETs?
31. Discuss the tolerance of FETs to neutron and other radiation. How does it compare with the tolerance of comparable BJTs?
32. A JFET has characteristics somewhat resembling those of a grid-controlled vacuum tube triode and pentode. Discuss the attempts that have been made to develop transistors that are analogs of vacuum tubes, or replacements for vacuum tubes in existing circuits.
33. Review recent developments in the field of FET oscillators, and compare with competing high-frequency diode oscillators.
34. Review the uses of FETs as voltage-controlled resistors for applications such as attenuators, electronic gain control, phase advance and retard circuits and voltage-tuned filters.
35. By integrating two field-effect transistors something termed a LAMBDA transistor has been produced (*Electronics*, April 18, 1974, p. 52). Discuss its possible uses.
36. A device has been made that is termed a gate-modulated bipolar transistor (GAMBIT).[10] It may be modeled as a combination of field effect transistor and a bipolar transistor. Discuss its performance and possible usefulness. Compare with the LAMBDA transistor.
37. Compare FET amplifier performance at high frequencies with that of competing devices such as traveling wave tubes, tunnel diodes, parametric amplifiers, oscillator diodes and bipolar transistors.
38. Discuss the performance of GaAs MESFET mixers at 5 GHz or higher.

## 6.7 REFERENCES AND FURTHER READING SUGGESTIONS

1. Arai, M., "Charge-storage junction field-effect transistor," *IEEE Trans. Electron Devices,* ED-22, 1975, p. 181.
2. Arai, Y., et al., "A 6-GHz four-stage GaAs MESFET power amplifier," *IEEE Trans. on Microwave Theory and Techniques,* MIT-24, 1976, p. 381.
3. Asai, S., S. Ishioka, H. Kurano, S. Takahashi, and H. Kodera, "Effects of deep centers

on microwave frequency characteristics of GaAs Schottky barrier gate FET," *Proceedings of the Fourth Conference on Solid State Devices, Tokyo, 1972. Supplement to the Journal of the Japan Society of Applied Physics,* 42, 1973, p. 71.

4. Asai, S., H. Kurono, S. Takahashi, M. Hirao, and H. Kodera, "Single-and dual-gate GaAs Schottky-barrier FET's for microwave frequencies," *Proceedings of the Fifth Conference (1973 International) on Solid State Devices, Tokyo, 1973. Supplement to the Journal of the Japan Society of Applied Physics,* 43, 1974, p. 442.

5. Asai, S., F. Murai, and H. Kodera, "GaAs Dual-gate Schottky-barrier FET's for microwave frequencies," *IEEE Trans. Electron Devices,* ED-22, 1975, p. 897.

6. Asai, S., S. Okazaki, and H. Kodera, "Optimized design of GaAs FET's for low noise microwave amplifiers," *Solid-State Electronics,* 19, 1976, p. 463.

7. Baechtold, W., "Noise behavior of GaAs field-effect transistors with short gate lengths," *IEEE Trans. Electron Devices,* ED-19, 1972, p. 674.

8. Baechtold, W., "Noise behavior of Schottky barrier gate field-effect transistors at microwave frequencies," *IEEE Trans. Electron Devices,* ED-18, 1971, p. 97.

9. Baechtold, W., K. Daetwyler, T. Forster, T.O. Mohr, W. Walter, and P. Wolf, "Si and GaAs 0.5 $\mu$m gate Schottky-barrier field-effect transistors," *Electronics Lett.,* 9, 1973, p. 232.

10. Baliga, B.J., D.E. Houston, and S. Krishna, "GAMBIT: Gate Modulated Bipolar Transistor," *Solid-State Electronics,* 18, 1975, p. 937.

11. Bandy, S.G., and J.G. Linvill, "The design, fabrication, and evaluation of a silicon junction field-effect photodetector," *IEEE Trans. Electron Devices,* ED-20, 1973, p. 793.

12. Barrera, J.S., "Microwave transistor review, Part 1, GaAs field effect transistors," *Microwave Journal,* February, 1976, p. 28.

13. Barrera, J., and R. Archer, "InP Schottky-gate field-effect transistors," *IEEE Trans. Election Devices,* ED-22, 1975, p. 1023.

14. Benjamin, J.A., and E.T. Casterline, "Trends in microwave power transistors," *Solid State Tech.,* April 1975, p. 51-56.

15. Behle, A., and R. Zuleeg, "Fast neutron tolerance of GaAs JFET's operating in the hot electron range," *IEEE Trans. Electron Devices,* ED-19, 1972, p. 993.

16. Beneking, H., J. Jahncke, J. Naumann, "Procedures for the measurement of s-parameters and the construction of stripline connected GaAs MSFET's in the 10 GHz range," *Third Biennial Cornell Electrical Engineering Conference,* Ithaca, NY, 1971, p. 427.

17. Beneking, H., and W. Filensky, "The GaAs MESFET as a pulse regenerator on the Gigabit per second range," *IEEE Trans. on Microwave Theory and Technique,* MTT-24, 1976, p. 385.

18. Blocker, T.G., H.M. Macksey, and R.L. Adams, "X-band RF power performance of GaAs FET's," *IEEE International Electron Devices Meeting Technical Digest,* Washington, D.C., 1974, p. 288.

19. Brewer, R.J., "The 'Barrier Mode' behavior of a junction FET at low drain currents," *Solid-State Electronics,* 18, 1975, p. 1013.

20. Burman, B., "Vacuum tubes yield sockets to hybrid JFET devices," *Electronics,* April 10, 1972, p. 85.

21. Camisa, R.L., J. Goel, and I. Drukier, "GaAs MESFET linear power-amplifier stage giving 1 W," *Electronics Lett.,* 11, 1975, p. 572.

22. Cho, A.Y., and D.R. Chien, "GaAs MESFET prepared by molecular beam epitaxy (MBE)," *Appl. Phys. Lett.,* 28, 1976, p. 30.

23. Dacey, G.C., and I.M. Ross, "Unipolar 'field-effect' transistor," *Proc. IRE,* 41, 1973, p. 970.

24. Das, M.B., and P. Schmidt, "High-frequency limitations of abrupt-junction FET's," *IEEE Trans. Electron Devices,* ED-20, 1973, p. 779.

25. Dean, R.H., "Reflection amplification in thin layers of $n$-GaAs," *IEEE Trans. Electron Devices*, **ED-19**, 1972, p. 1148.
26. Dean, R.H., A.B. Dreeben, J.J. Hughes, R.J. Matarese, and L.S. Napoli, "Broad-band microwave measurements on GaAs traveling wave transistors," *IEEE Trans. on Microwave Theory and Techniques*, **MTT-21**, 1973, p. 805.
27. Dean, R.H., and R.J. Matarese, "Submicrometer self-aligned dual-gate GaAs FET," *IEEE Trans. Electron Devices*, **ED-22**, 1975, p. 358.
28. Decker, D., R. Fairman, and C. Nishimoto, "Microwave InGaAs Schottky-barrier-gate field-effect transistors–prelimary results," in: *Proc. 1975 Cornell Conf. on Active Semiconductor Devices for Microwaves and Integrated Optics*, p. 305.
29. DeKold, D.F., "Diodes stabilize FET gain to 1% over 100°C range," *Electronics*, June 7, 1971, p. 82.
30. DeMassa, T.A., and S.R. Iyer, "Closed form solution for the linear graded JFET," *Solid-State Electronics*, **18**, 1975, p. 931.
31. Diamond, L., and A.V. Siefert, "Designing differential FET inputs with overall performance in mind," *Electronics*, June 21, 1971, p. 76.
32. Drangeid, K.E. and R. Sommerhalder, "Dynamic performance of Schottky-barrier field-effect transistors," *IBM Journal of Research and Development*, **14**, 1970, p. 82.
33. Driver, M.C., H.B. Kim, and D.L. Barrett, "Gallium arsenide self-aligned gate field-effect transistors," *Proc. IEEE*, **50**, 1971, p. 1244.
34. Engelmann, R.W.H., and C.A. Liechti, "Gunn domain formation in the saturated current region of GaAs MESFETs," *IEEE International Electron Devices Meeting Technical Digest*, Washington, D.C., 1976, p. 351.
35. Evans, A.D., "Characteristics of unipolar field-effect transistors," *Electronic Industries*, **22**, 1963, p. 99.
36. Fair, R.B., "Graphical design and iterative analysis of the DC parameter of GaAs FET's," *IEEE Trans. Electron Devices*, **ED-21**, 1974, p. 357.
37. Fairman, R.D., and R. Solomon, "Submicron epitaxial films for GaAs field effect transistors," *J. Electrochem. Soc.* **120**, 1973, p. 541.
38. Filensky, W., H-J. Klein, and H. Beneking, "The GaAs MESFET as a pulse regenerator, amplifier, and laser modulator in the Gbit/s range," *IEEE J. Solid-State Circuits*, **SC-12**, 1977, p. 276.
39. Fukuta, M., H. Ishikawa, K. Suyama, and M. Maeda, "GaAs 8 GHz-band high power FET," *IEEE International Electron Devices Meeting Technical Digest*, Washington, D.C., 1974, p. 285.
40. Fukuta, M., et al., "Power GaAs MESFET with a high drain-source breakdown voltage," *IEEE Trans. on Microwave Theory and Techniques*, **MTT-24**, 1976, p. 312.
41. Fullagar, D., "Better understanding of FET operation yields viable monolithic JFET op amp," *Electronics*, November 6, 1972, p. 89.
42. Gelnovatch, V.G., "ECOM M/W contracts," *Microwave Journal*, **18**, January, 1975, p. 30.
43. Gelnovatch, V.G., and C. Liechti, "FET Fortran IV," *Microwave Journal*, **18**, January 1975, p. 22.
44. Gosling, W., "The pre-history of the transistor," *Radio and Electronic Engineer*, **43**, 1973, p. 10.
45. Graffevil, J., and J. Caiminade, "Low-frequency noise in GaAs Schottky gate FET's," *Electronics Lett.*, **10**, 1974, p. 266.
46. Graham, E.D., Jr., and C.W. Gwyn, *Microwave Transistors*, Artech House, Inc., Dedham, Massachusetts, 1975.
47. Grebene, A.B., and S.K. Ghandhi, "Pinched-mode operation of field-effect transistors," *Proc. IEEE*, **57**, 1969, p. 230.
48. Grove, A.S., *Physics and Technology of Semiconductor Devices*, John Wiley and Sons, NY, 1967.

49. Hardeman, L.J., "FET low-noise R&D heats up," *Electronics,* January 17, 1972, p. 90.
50. Haslett, J.W., E.J.M. Kendall, and F.J. Scholz, "Design considerations for improving low-temperature noise performance of silicon JFET's," *Solid-State Electronics,* **18,** 1975, p. 199.
51. Hault, D.I., and R.E. Richards, "UHFFET preamplifier with a 0.3 dB noise figure," *Electronics Letters,* **11,** 1975, p. 596.
52. Heald, R.A., and D.A. Hodges, "Multilevel random access memory using one JFET per cell," *IEEE International Electron Devices Meeting Technical Digest,* Washington, D.C., 1975, p. 324.
53. Herskowitz, G.J., and R.B. Schilling, eds., *Semiconductor Device Modeling for Computer-Aided Design,* McGraw-Hill, NY, 1972.
54. Hewitt, B.S., et al., "Low noise GaAs MESFETs," *Electronic Lett.,* **12,** June 10, 1976, p. 309.
55. Higgins, J.A., B.M. Welsh, F.H. Eisen, and C.D. Robinson, "Performance of ion-implanted GaAs MESFETs," *IEEE International Electron Devices Meeting Technical Digest,* Washington, D.C., 1975, p. 5.
56. Higgins, J.A., R.L. Kuvas, and D.R. Chen, "Modeling, fabrication and performance of ion implanted low-noise GaAs FETs," *IEEE International Electron Devices Meeting Technical Digest,* Washington, D.C., 1977, p. 506.
57. Himsworth, B., "A two-dimensional analysis of gallium-arsenide junction field-effect transistors with long and short channels," *Solid-State Electronics,* **15,** 1972, p. 1353.
58. Hogeboom, J.H., and R.S.C. Cobbold, "Etched Schottky-barrier MOSFET's using a single mask," *Electronics Lett.,* **7,** 1971, p. 133.
59. Holt, A.J., Jr., and R.L. Berger, "Hybrid integrated networks," *Western Electric Engineer,* **19,** 1975, p. 3.
60. Hornbuckle, D.P., and L.J. Kuhlman, Jr., "Broad-band medium-power amplification in the 2-12.4 GHz range with GaAs MESFETs," *IEEE Trans. on Microwave Theory and Techniques,* **MTT-24,** 1976, p. 338.
61. Hunsperger, R.G., and N. Hirsch, "Ion-implanted microwave field-effect transistors in GaAs," *Solid-State Electronics,* **18,** 1975, p. 349.
62. Johnson, E.O., "Physical limitations on frequency and power parameters of transistors," *RCA Review,* **26,** 1965, p. 163.
63. Katz, G., "FET voltage regulator eliminates ripple feedthrough and permits self-starting," *Electronic Engineering,* **44,** December, 1972, p. 57.
64. Kellermann, K.J., "Gross-signalregelung mit feldeffecht transistoren," *Internationale Elektronische Rundschau,* **NR-7,** 1973, p. 159.
65. Kohn, E., "GaAs-MESFET for digital application," *Solid-State Electronics* **20,** 1977, p. 29.
66. Komaki, S., O. Kurita and T. Memita, "GaAs MESFET regenerator for phase-shift keying signals at the carrier frequency," *IEEE Trans. on Microwave Theory and Techniques,* **MTT-24,** 1976, p. 367.
67. Kurita, O., and K. Morita, "Microwave MESFET Mixer," *IEEE Trans on Microwave Theory and Techniques,* **MTT-24,** 1976, p. 361.
68. Lazarus, M.J., "The short-channel IGFET," *IEEE Trans. Electron Devices,* **ED-22,** 1975, p. 351.
69. Lecrosnier, D., and G. Pelous, "Ion implanted FET for power applications," *IEEE Trans. Electron Devices,* **ED-21,** 1974, p. 113.
70. Lehovec, K., and R.S. Miller, "Field distribution in junction field-effect transistors at large drain voltages," *IEEE Trans. Electron Devices,* **ED-22,** 1975, p. 273.
71. Lehovec, K., and W.G. Seeley, "On the validity of the gradual channel approximation for junction field effect transistors with drift velocity saturation," *Solid-State Electronics,* **16,** 1973, p. 1047.
72. Lehovec, K., and W.G. Seeley, "Photo-effects in junction field effect transistors," *Solid-State Electronics,* **14,** 1971, p. 1077.

73. Lehovec, K., and R. Zuleeg, "Voltage-current characteristics of GaAs J-FET's in the hot electron range," *Solid-State Electronics*, **13**, 1970, p. 1415.

74. Liechti, C.A., "Performance of dual-gate GaAs MESFET's as gain-controlled low-noise amplifiers and high-speed modulators," *IEEE Trans. on Microwave Theory and Techniques*, **MTT-23**, 1975, p. 461.

75. Liechti, C.A., "Microwave field-effect transistors—1976," *IEEE Trans. on Microwave Theory and Techniques*, **MTT-24**, 1976, p. 279.

76. Liechti, C.A., and R.B. Larrick, "Performance of GaAs MESFETs at low temperatures," *IEEE Trans. on Microwave Theory and Techniques*, **MTT-24**, 1976, p. 376.

77. Macken, W.J., "FETs as variable resistances in op amps and gyrators," *Electronic Engineering*, **44**, December 1972, p. 60.

78. Maeda, M., S. Takahashi, and H. Kodera, "CW oscillation characteristics of GaAs Schottky-barrier gate field-effect transistors," *Proc. IEEE*, **63**, 1975, p. 320.

79. Maloney, T.J., and J. Frey, "Frequency limits of GaAs and InP field-effect transistors," *IEEE Trans. Electron Devices*, **ED-22**, 1975, p. 357.

80. Martin, T.B., "Circuit applications of the field effect transistor, Part 1," *Semiconductor Products*, **5**, February, 1962, p. 33.

81. Martin, T. B., "Circuit applications of the field effect transistors, Part 2," *Semiconductor Products*, **5**, March, 1962, p. 30.

82. Mayer, D.C., N.A. Masnari, and R.J. Lomax, "A submicron-channel vertical junction field-effect transistor," *IEEE International Electron Devices Meeting Technical Digest*, Washington, D.C., 1977, p. 532.

83. McIlvenna, J.F., and C.J. Sletten, "Contractual and in-house research at AFCRL," *Microwave Journal*, **18**, May, 1975, p. 24.

84. Medley, M.W., "Stretch FET amp design beyond octave bandwidth," *Microwaves*, **16**, May, 1977, p. 55.

85. Middleback, S., "Metalization processes in fabrication of Schottky- barrier FET's," *IBM Journal of Research and Development*, **14**, 1970, p. 148.

86. Millman, J., and C.C. Halkias, *Integrated Electronics Analog and Digital Circuits and Systems*, McGraw-Hill Book Company, NY, 1972, p. 310.

87. Mohr, Th.O., "Silicon and silicon-dioxide processing for high-frequency MESFET preparation," *IBM Journal of Research and Development*, **14**, 1970, p. 142.

88. Mok, T.D., and C.A.T. Salama, "The characteristics and applications of a V-shaped notched-channel field-effect transistor (VFET)," *Solid-State Electronics*, **19**, 1976, p. 159.

89. Moline, R.A., W.C. Gibson, and L.D. Heck, "An ion-implanted Schottky-barrier gate field-effect transistor," *IEEE Trans. Electron Devices*, **ED-20**, 1973, p. 317.

90. Morkoc, H., "An AlGaAs gate heterojunction microwave GaAs FET," *IEEE International Electron Devices Meeting Technical Digest*, Washington, D.C., 1977, p. 334.

91. Napoli, L.S., J.J. Hughes, W.F. Reichert, and S. Jolly, "GaAs FET for high power amplifiers at microwave frequencies," *RCA Review*, **34**, 1973, p. 608.

92. Neumark, G.F., "Theory of the influence of hot electron effects on insulated gate field effect transistors," *Solid-State Electronics*, **10**, 1967, p. 169.

93. Nishizawa, J.I., T. Terasaki, and J. Shibata, "Field-effect transistor versus analog transistor (static induction transistor)," *IEEE Trans. Electron Devices*, **ED-22**, 1975, p. 185.

94. Notthoff, J.K., and R. Zuleeg, "High speed, low power GaAs JFET integrated circuits," *IEEE International Electron Devices Meeting Technical Digest*, Washington, D.C., 1975, p. 624.

95. Nuzillat, G., C. Arnodo, and J.P. Puron, "A subnanosecond integrated switching circuit with MESFET's for LSI," *IEEE J. Solid-State Circuits*, **SC-11**, 1976, p. 385.

96. Oakes, J.G., et al., "A power silicon microwave MOS transistor," *IEEE Trans. on Microwave Theory and Techniques*, **MTT-24**, 1976, p. 305.

97. Ogawa, M., et al., "Submicron single-gate and dual-gate GaAs MESFETs with improved

low noise and high gain performance," *IEEE Trans. on Microwave Theory and Techniques*, MTT-24, 1976, p. 300.

98. Okazaki, S., S. Takahashi, M. Maeda, and H. Kodera, "Microwave oscillation with GaAs FET," *Proc. of the Sixth Conf. on Solid State Devices, Tokyo, 1974, Suppl. to the J. of the Japan Soc. of Appl. Phys.*, 44, 1975, p. 157.

99. Olsen, D.R., "Equivalent circuit for a field-effect transistor," *Proc. IEEE*, 51, 1963, p. 254.

100. Ostoich, V., P. Jeppesen, and N. Slaymaker, "Direct modulation of D.H. GaAlAs lasers with GaAs MESFET's," *Electronic Lett.*, 11, October 16, 1975, p. 515.

101. Ozawa, O., Y. Sasaki, H. Iwasaki, and H. Ikoma, "A 5000-channel power FET with a new diffused gate structure," *IEEE International Electron Devices Meeting Technical Digest*, Washington, D.C., 1975, p. 163.

102. Ozawa O., H. Iwasaki, and K. Muramoto, "A vertical channel JFET fabricated using silicon planar technology," *IEEE J. Solid-State Circuits*, SC-11, 1976, p. 511.

103. Pancholy, R.K., and W.W. Grannemann, "Gallium arsenide phosphide Schottky barrier field effect transistor," *J. Electrochem. Soc.*, 124, 1977, p. 430.

104. Pengelley, R.S., and J.A. Turner, "Monolithic broadband GaAs FET amplifiers," *Electronic Lett.*, 12, May 13, 1976, p. 251.

105. Pucel, R.A., H. Haus, and H. Statz, "Signal and noise properties of gallium arsenide microwave field-effect transistors," in: *Advances in Electron Physics*, Vol. 38, Academic Press, NY 1975, p. 195.

106. Pucel, R.A., D.J. Massé and C.F. Krumm, "Noise performance of gallium arsenide field-effect transistors," *IEEE J. Solid-State Circuits*, SC-11, 1976, p. 243.

107. Pucel, R.A., D. Massé, and R. Bera, "Performance of GaAs MESFET mixers at X-band," *IEEE Trans. on Microwave Theory and Techniques*, MTT-24, 1976, p. 351.

108. Pullen, K.A., Jr., "Comments on the bibiliography on field-effect transistors," *IEEE Trans. on Electron Devices*, ED-17, 1970, p. 1014.

109. Regier, F.A., "Channel edge location and potential distribution in a junction field effect transistor," *Solid-State Electronics*, 19, 1976, p. 969.

110. Reiser, M., and P. Wolf, "Computer study of submicrometre FET's," *Electronics Lett.*, 8, May 1972, p. 254.

111. Rode, D.L., B. Schwartz, and J.V. DiLorenzo, "Electrolytic etching and electron mobility of GaAs for FET's," *Solid-State Electronics*, 17, 1974, p. 1119.

112. Rossel, P., J.J. Cabot, and J. Graffeuil, "Bistable switching on gallium arsenide Schottky gate field-effect transistors," *Appl. Phys. Lett.*, 25, 1974, p. 510.

113. Sah, C.T., "Theory of low-frequency generation noise in junction-gate field-effect transistors, *Proc. IEEE*, 52, 1964, p. 795.

114. Salama, C.A.T., "V-groove power field effect transistors," *IEEE International Electron Devices Meeting Technical Digest*, Washington, D.C. 1977, p. 412.

115. Sechi, F.N., and R.W. Paglione, "Design of a high-gain FET amplifier operating at 4.4-5.0 GHz," *IEEE J. Solid-State Circuits*, SC-12, 1977, p. 285.

116. Sevin, L.J., Jr., *Field-effect Transistors*, McGraw-Hill, NY, 1965.

117. Siliconix, Inc., "An introduction to FETs," Application Note AN73-7, December, 1973, Siliconix, Inc., 2201, Laurelwood Rd., Santa Clara, CA 95054.

118. Sitch, J.E., and P.N. Robson, "The performance of GaAs field-effect transistors as microwave mixers," *Proc. IEEE*, 61, 1973, p. 399.

119. Slaymaker, N.A., R.A. Soares, and J.A. Turner, "GaAs MESFET small signal X-band amplifiers," *IEEE Trans. on Microwave Theory and Techniques*, MTT-24, 1976, p. 329.

120. Statz, H., H.A. Huas, and R.A. Pucel, "Noise characteristics of gallium arsenide field-effect transistors," *IEEE Trans. Electron Devices*, ED-21, 1974, p. 549.

121. Stoneham, E., T.S. Tan, and J. Gladstone, "Fully ion implanted GaAs power FETs,"

*IEEE International Electron Devices Meeting Technical Digest*, Washington, D.C., 1977, p. 330.

122. Sze, S.M., *Physics of Semiconductor Devices*, Wiley-Interscience, NY, 1969.

123. Takahashi, S., et al., "Reproducible submicron gate fabrication of GaAs FET by plasma etching," *IEEE International Electron Devices Meeting Technical Digest*, Washington 1976, p. 214.

124. Tarui, Y., Y. Komiya, and T. Yamaguchi, "Self-aligned GaAs Schottky barrier gate FET using preferential etching," *Proc. Fourth Conference on Solid State Devices, Tokyo, 1972, Supplement to the J. Japan. Soc. of Appl. Phys.*, 42, 1973, p. 78.

125. Teszner, S., "Gridistor development for the microwave power region," *IEEE Trans. Electron Devices*, ED-19, 1972, p. 355.

126. Tolar, N.J., and D.L. Ash, "Silicon MESFETs for improved VHF and UHF mixer performance," *IEEE International Electron Devices Meeting Technical Digest*, Washington, D.C., 1977, p. 382.

127. Turner, J.A., et al., "Dual-gate gallium-arsenide microwave field-effect transistor," *Electronics Lett.*, 7, 1971, p. 661.

128. Umebachi, S., et al., "A new heterojunction gate GaAs FET," *IEEE Trans. Electron Devices*, ED-22, 1975, p. 613.

129. Van Tuyl, L. and C.A. Liechti, "High-speed integrated logic with GaAs MESFET's," *J. of Solid-State Circuits*, SC-9, 1971, p. 269.

130. Vendelin, G.D., W. Alexander, and D. Mock, "Computer analyses rf circuits with generalized Smith charts," *Electronics*, March 21, 1974, p. 102.

131. Vodicha, V.W., and R. Zueleeg, "Ion implanted GaAs enhancement mode JFET's," *IEEE International Electron Devices Meeting Technical Digest*, Washington, D.C., 1975, p. 625.

132. Wallmark, J.T., "The field-effect transistor–a review," *RCA Review*, 24, 1963, p. 641.

133. Watson, B., "Audio-frequency noise characteristics of junction FETs," Application Note AN74-4, Siliconix, Inc., Santa Clara, CA 95054.

134. Watson, H.A., et., *Microwave Semiconductor Devices and Their Circuit Applications*, McGraw-Hill, NY, 1969.

135. Wisseman, W.R., "Power GaAs FETs," *IEEE International Electron Devices Meeting Technical Digest*, Washington, D.C. 1977, p. 326.

136. Wolf, P., "Microwave properties of Schottky-barrier field-effect transistors," *IBM Journal of Research and Development*, 14, 1970, p. 125.

137. Wyland, D.C., "FET cascode technique optimizes differential amplifier performance," *Electronics*, January 18, 1971, p. 81.

138. Yau, L.D., "A simple theory to predict the threshold voltage of short-channel IGFET's," *Solid-State Electronics*, 17, 1974, p. 1059.

139. Zuleeg, R., "Development in GaAs FET's and IC's," *IEEE International Electron Devices Meeting Technical Digest*, Washington, D.C., 1976, p. 347.

140. Zuleeg, R., et al., "Femto-joule, high-speed planar GaAs E-JFET logic," *IEEE International Electron Devices Meeting Technical Digest*, Washington, D.C., 1977, p. 198.

# 7

# Insulated Gate Field-Effect-Transistors: MOSFETs, IGFETs and Related Devices

## CONTENTS

## 7.1 INTRODUCTION

Modulation of the conductance of a thin layer by surface charges, a concept that goes back to 1930, was revived for semiconductors in 1948.[233] Practical devices did not ensue until about 1960, when diffusion and oxide technologies for Si were developed. MOSFET structures manufactured during the next few years are illustrated in Fig. 7.1. The current flows by majority carrier movement from the source to drain and is controlled by the voltage on the insulated metal gate which determines the channel charge. In the $n$-channel enhancement type structure [see Fig. 7.1(a)] the channel is not initially present and must be induced by a positive gate voltage $V_{GS}$. A threshold voltage $V_T$ is needed before any usable characteristic is obtained. In the $n$-channel depletion type structure [see Fig. 7.1(b)] the channel is already present (as a result of a shallow $n$-type diffusion or implant or as a result of interface conditions). This channel can be partially depleted by a negative gate voltage or enhanced by a positive gate voltage.

377

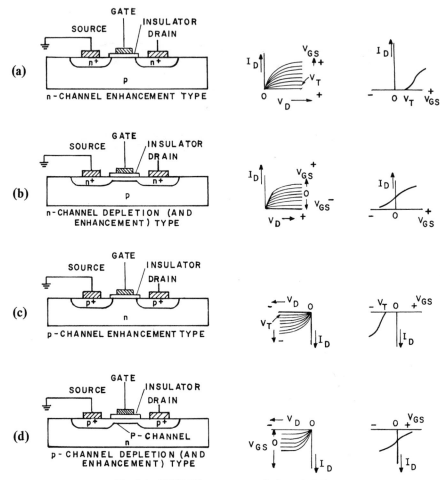

Fig. 7.1.  MOSFET structures and characteristics.

In Fig. 7.1(c), (d) $p$-type channel structures are shown. The MOSFETs first commercially available were $p$ channel since stable characteristics were easier to obtain for $p$ rather than for $n$ channel FETs. In those times $SiO_2$ grown contained $Na^+$ ions in uncontrollable amounts (they are mobile under the influence of an electric field in the insulator). In an $n$-channel enhancement type structure the gate is normally positive with respect to the substrate and, hence, the positively charged ions collect along the interface between the $SiO_2$ and the Si substrate. The positive charge from this layer of ions attracts free electrons in the channel which tends to make the transistor turn-on prematurely. In $p$-channel devices the gate voltage is normally negative and the $Na^+$ ions move to the gate-$SiO_2$ interface where they do not affect the channel.

In the last decade undesirable contaminants in gate insulating oxides have been eliminated and $n$-channel MOSFET technology has become reliable. The electron mobility in Si is more than twice the hole mobility, so an $n$ channel MOSFET may be narrower in channel width for the same ON resistance and similar operating conditions. Hence, a greater packing density can be achieved with $n$-channel than with $p$-channel designs. The operating speed, assuming near-saturation drift velocity conditions in the channel, is determined primarily by the internal resistance-capacitance time constants and $n$-channel has an advantage here since the capacitances are smaller for the smaller width devices.

The substrate may be connected to the source, or provided with an independent lead so that it can be separately biased. Control of the substrate voltage allows the effective threshold voltage $V_T$ to be adjusted to suit the circuit needs.

The $SiO_2$ layer of the gate may be 1000 Å or thinner. Hence, if the breakdown strength of the oxide is on the order of $10^7$ V/cm, a static charge induced on the gate due to careless handling can easily generate the voltage needed to puncture the insulating layer, particularly at local defects (weak spots).[151] MOSFET structures, therefore, may be fabricated with a Zener diode between the gate and source to protect against voltage damage. The presence of the Zener diode lowers the input impedance from MOSFET to JFET levels.

Some source, gate and drain geometries for MOS transistors are shown in Fig. 7.2 to illustrate a few of the arrangements used. Many other patterns are used depending on the skill and inclination of the design group involved.

MOSFETs have certain advantages over JFETs. The dc input resistance may be higher because of the insulated gate, and the input capacitance for an enhancement device does not vary as much with voltage as for a depletion device. The possibility of variation from depletion to enhancement modes, or of mixing such types, is a useful feature in some circuits.

A small-signal equivalent circuit for a MOSFET in the common-source arrangement is shown in Fig. 7.3. The dominant parameters at low frequencies are the transconductance $g_m$ for which typical values may be 0.1-20 mA/V, and the source-drain resistance $r_d$ for which normal values are 1-50 kilohms. $C_{ds}$ may be 0.1-1 pF and $C_{gs}$ and $C_{gd}$ may be 1-10 pF.

The voltage-gain expression for the circuit of Fig. 7.3 is

$$A_V = \frac{-g_m + Y_{gd}}{Y_L + Y_{ds} + g_d + Y_{gd}} \tag{7.1}$$

which at low frequencies reduces to

$$A_V = \frac{-g_m}{Y_L + g_d} = \frac{-g_m r_d Z_L}{r_d + Z_L} = -g_m Z_L' \tag{7.2}$$

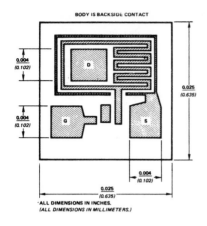

(a)                                                           (c)

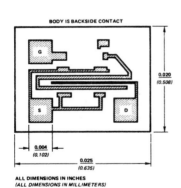

(b)                                                           (d)

Fig. 7.2. Some Si MOSFET geometries:

(a)  $n$-channel enhancement unit with integrated Zener-clamp protecting the gate;

(b)  $n$-channel depletion type, ultrahigh input impedance (no gate protective diode);

(c)  $p$-channel enhancement type, with Zener clamp, $10^{10}$ ohms input resistance, square law characteristics; and

(d)  $p$-channel enhancement type, with Zener clamp. $C_{gs}$ less than 0.5 pF. Suitable for switching applications requiring low OFF currents. (Courtesy Siliconix, Inc.)

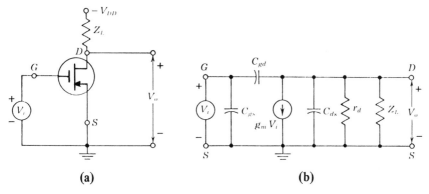

Fig. 7.3. MOSFET common source amplifier circuit:
(a)  $p$ channel transistor (holes flow from source to drain); and
(b)  Small-signal equivalent circuit (the bias network is not shown).

where $Z'_L \equiv Z_L \| r_d$. If $g_m$ is 10 mA/V and $Z'_L$ is 5 kilohms, the voltage gain is 50. The input admittance is

$$Y_i = Y_{gs} + (1 - A_V)Y_{gd} \qquad (7.3a)$$

and the output admittance, with $Z_L$ excluded, is

$$Y_0 = \frac{1}{r_d} + Y_{ds} + Y_{gd} \qquad (7.3b)$$

## 7.2 FIRST ORDER THEORY OF A MOSFET

In a simple model the enhanced channel is assumed to be thin so that variation of carrier density in the depth of the semiconductor channel need not be considered. The carrier mobility in the inversion layer is taken as a constant, and leakage current effects are neglected. In the absence of bias voltage across the insulator, the band edges are assumed to be flat as shown in Fig. 7.4(a). This neglects work function differences between the metal and the semiconductor and neglects inherent surface charge effects at the insulator-semiconductor interface. Application of a positive voltage to the gate causes depletion of the surface layer of the $p$-semiconductor. At a voltage $V_T$ the bending down of the conduction band is $\phi_B$ and the surface is just intrinsic [see Fig. 7.4(b)]. This voltage $V_T$ for the emergence of a channel in an enhancement FET is usually termed the threshold voltage. Further increase in gate-channel voltage creates an $n$-channel of substantial conductivity as in Fig. 7.4(c). Consider the variation of the energy $\phi$ with distance in Fig. 7.4(c). Poisson's equation in one-dimensional form is

$$\frac{d^2\phi/q}{dx^2} = \frac{-\rho(x)}{\epsilon_{Si}\epsilon_0} = -q\frac{(p_p - n_p + N_D^+ - N_A^-)}{\epsilon_{Si}\epsilon_0}$$

$$= \frac{-q}{\epsilon_{Si}\epsilon_0}\left[p_{po}\left(\exp\frac{(-\phi)}{kT} - 1\right) - n_{po}\left(\exp\frac{(-\phi)}{kT} - 1\right)\right] \qquad (7.4)$$

Integration of this equation gives the electric field at the surface $\mathcal{E}_S$, and Gauss' law gives for the charge required to produce this field

$$Q_S = \epsilon_{Si}\epsilon_0\mathcal{E}_S = \frac{2\epsilon_{Si}\epsilon_0 kT}{qL_D}\left[\left(\exp\left(\frac{-\phi}{kT}\right) + \frac{\phi}{kT} - 1\right) + \right.$$

$$\left. \frac{n_{po}}{p_{po}}\left(\exp\frac{(-\phi)}{kT} - \frac{\phi}{kT} - 1\right)\right]^{1/2} \qquad (7.5)$$

where $L_D$ is the extrinsic Debye length for holes $(2kT\epsilon_{Si}\epsilon_0/(p_{po}q^2))^{1/2}$

The variation of $Q_S$ with the surface potential $\psi_s$ is shown in Fig. 7.4(d) for $4 \times 10^{15}$ cm$^{-3}$ $p$ type Si. With the semiconductor surface increasingly positive with respect to the bulk there is first a depletion of holes from the surface region, then a weak inversion and finally the onset of strong inversion at a surface potential of about $2\psi_B$.

The voltage between the metal and the semiconductor bulk for a channel charge $Q_S$ is the sum of the voltage across the semiconductor channel region $\psi_S$ and the voltage across the insulator (corresponding to the electric-field lines that cross the insulator thickness $t$). Here

$$V_{ms} = \psi_S + \frac{t}{\epsilon_i}Q_S = \psi_S + \frac{Q_S}{C_G} \qquad (7.6)$$

$C_G$ is the capacitance per unit area for an insulator of thickness $t$. For the condition of $\psi_S = \phi_B/q$, shown in Fig. 7.4(b), the charge in the channel is caused by negatively-charged acceptor atoms and the density per unit area is $qN_A x$ where $x$ is the depleted depth of the channel which from Poisson's equation is $(2\epsilon_{Si}\epsilon_0\phi_B/N_A)^{1/2}$. Hence,

$$V_T = \frac{\phi_B}{q} + \frac{t}{\epsilon_i}\sqrt{2\epsilon_{Si}\epsilon_0 N_A \, \phi_B} \qquad (7.7)$$

This is the threshold voltage for the very start of formation of the $n$ channel. However, as seen from Fig. 7.4(d) a semiconductor surface voltage of about

$2\phi_B/q$ is usually needed to achieve significant electron charge in the channel. Also, there may be charge $Q_{ss}$ at the SiO$_2$-Si interface and work function differences between the metal and Si that were not considered in Fig. 7.4(a).

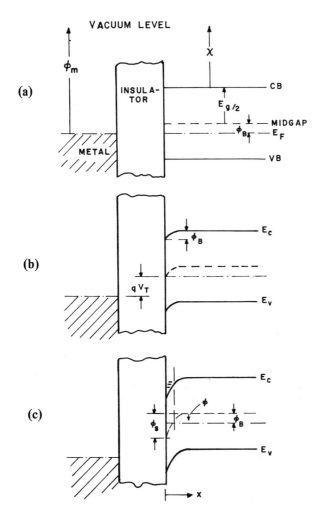

Fig. 7.4. Energy band conditions at a $p$-semiconductor insulator interface:
(a) Flat band condition assumed in the absence of charge at the interface. $\phi_m = \chi + \dfrac{E_g}{2} + \phi_B$;

(b) At a voltage $V_T$ the semiconductor at the interface has just become intrinsic;
(c) $n$ channel forms in the presence of a voltage $V_G > V_T$; and
(d) Channel charge as a function of the semiconductor surface potential $\Psi_s$ ($= \phi_s/q$). (After Garret and Brattain.[91] Reprinted with permission from *Physics of Semiconductors*, by S.M. Sze, copyright 1969, John Wiley & Sons, Inc., NY p. 433.)

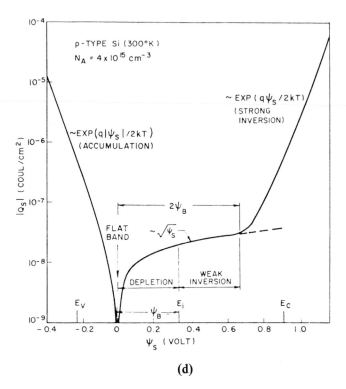

**(d)**

Fig. 7.4. (continued)

A more complete expression for threshold voltage is given later [see Eq. (7.13)].

Consider now the device channel conditions. Let the capacitance of the insulated gate electrode be $C_G$ for the geometry of Fig. 7.5 and the voltage at some distance along the channel be $V(z)$. Then, the free electron charge per unit area is given by

$$q \, \Delta n(z) = \frac{C_G}{WL} (V_{GS} - V(z) - V_T)$$

where $C_G = \epsilon_{ox}\epsilon_o WL/t_{ox}$ and $W$ is the gate width. A length $\delta z$ of the channel is of conductivity

$$\sigma(z) = q \, \Delta n(z) \, \mu_n \, W/\delta z \qquad (7.9)$$

Therefore, the drain current may be written as

$$I_D = \sigma(z) \, \delta \, V$$

$$= \frac{\mu_n C_G}{L} (V_{GS} - V(z) - V_T) \frac{dV}{dz} \qquad (7.10)$$

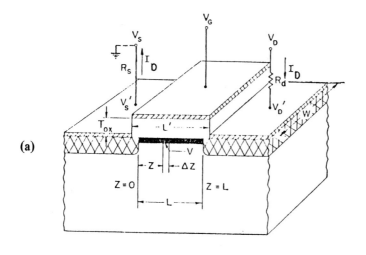

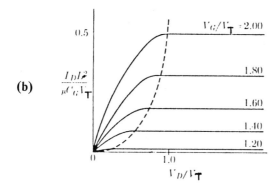

Fig. 7.5. MOSFET model and characteristics. (a) (After Hofstein and Heiman. Reprinted with permission from *Proc. IEEE*, **51**, 1963, p. 1192.)

The total voltage drop $V_{DS}$ along the channel may then be found by integration with the result

$$I_D = \frac{\mu C_G}{L^2}\left(V_{GS} - V_T - \frac{V_{DS}}{2}\right)V_{DS} \tag{7.11}$$

This may also be expressed as

$$\frac{I_D L^2}{\mu C_G V_T^2} = \left(\frac{V_{GS} - V_T}{V_T} - \frac{V_{DS}}{2V_T}\right)\frac{V_{DS}}{V_T} \tag{7.11a}$$

Equation (7.11a) gives the curves to the left of the dashed line in Fig. 7.5(b). At high values of $V_{DS}/V_T$ the current saturates and becomes constant with $V_D$. The channel at the drain end absorbs any increase in $V_{DS}$ since the gate-to-channel voltage there, $(V_{GS} - V_T) - V_{DS}$, is insufficient to sustain the channel conductivity. This is the pinch-off condition analogous to pinch-off in a JFET. The nonlinear spacing of the FET characteristics [see Fig. 7.5(b)] with variation of $V_{GS} - V_T$ shows that the transistor sensitivity is not constant with gate voltage variation.

By differentiation of Eq. (7.11) the device transconductance is

$$g_m = \frac{dI_D}{dV_{GS}} = \frac{\mu_n C_G}{L^2}(V_{GS} - V_T) \tag{7.12}$$

This equation may be tested by substitution of some typical values; the $g_m$ value obtained is 2 mA/V if $L$ is 10 $\mu$m; $\mu_n = 500$ cm$^2$V$^{-1}$sec$^{-1}$; $C_G = 2$ pF; $V_T = 2$ V and $V_{GS} = 4$ V.

## 7.3 FURTHER CONSIDERATION OF MOSFET CHARACTERISTICS

Typical grounded-source characteristics for an $n$-channel Si MOSFET with an oxide of thickness 500 Å and an effective gate length of 4.5 $\mu$m are shown in Fig. 7.6 and the pinch-off effect is illustrated. The characteristics of $I_D$ in the saturation region increase with increase of $V_{DS}$ because the pinch-off point moves towards the source and slightly lowers the effective channel resistance. The effective mobility that matches the characteristics is 510 cm$^2$V$^{-1}$sec$^{-1}$. This is considerably less than 1300 cm$^2$V$^{-1}$sec$^{-1}$, the electron mobility in bulk Si, because of the influence of the thin channel and the surface interface on mobility. MOSFETs with ion implanted buried channels may offer larger mobilities.[223]

The transfer characteristics of $I_D$ vs $V_{GS}$ are shown in Fig. 7.6(b) and are seen to be reasonably linear for low $V_D$ and the substrate bias conditions used. The theoretical variation of threshold voltage $V_T$ with substrate bias voltage $V_S$ is shown in Fig. 7.7 from Eq. (7.13)

$$V_T = \frac{t_{ox}}{\epsilon_0 \epsilon_{ox}} \left[ -Q_{ss} - Q_{ox} + \sqrt{2\epsilon_0 \epsilon_{Si} q N_{a,eff}(V_{s\text{-}sub} + \psi_s)} \right] \tag{7.13}$$

$$+ \Delta W_F + \psi_s + \Delta V_{DT}$$

This is of the same basic form as Eq. (7.7) but with extra terms to include interface state charge and other effects as defined below:

$V_T$ = minimum gate-to-source voltage to turn on the device;

$\epsilon_{ox}$ = relative dielectric constant of oxide under gate under electrode ($\epsilon_{ox}$ = 4 for $SiO_2$);

$\epsilon_{Si}$ = relative dielectric constant of Si ($\epsilon_{Si}$ = 12);

$\epsilon_0$ = dielectric constant of free space (8.85 $\times$ $10^{-14}$ F/cm);

$t_{ox}$ = thickness of oxide under gate;

$q$ = electronic charge (1.6 $\times$ $10^{-19}$ coulombs);

$Q_{ox}$ = oxide charge (considered to be located at $Si$-$SiO_2$ interface);

$Q_{ss}$ = surface state charges at $Si$-$SiO_2$ interface ($Q_{ox}$ + $Q_{ss}$ was taken as 0.5 $\times$ $10^{11}$ positive electronic charges $cm^{-2}$);

$N_{a,eff}$ = effective impurity concentration level of substrate including allowance for B depletion (5 $\times$ $10^{15}$ atoms $cm^{-3}$);

$V_{s\text{-}sub}$ = voltage applied between source and substrate;

$\Delta W_F$ = work function difference, expressed in volts, between Al and Si ($-0.8$ V for $p$-type substrate);

$\psi_s$ = voltage across depletion layer at the onset of conduction in the absence of a substrate potential (0.75 V for 2.3 ohm-cm $p$-type Si);

$\Delta V_{DT}$ = shift in threshold due to B depletion effect (taken as $-0.07$ V).

The threshold voltage for the 6000 Å oxide is of interest because this thickness of oxide is used to allow cross-over Al interconnections and underpass diffusions. The metal cross-overs may create unwanted thick-oxide devices in integrated circuits but these are not of serious consequence in logic circuits provided the threshold voltages are large enough.

The characteristics of two MOSFETs having oxide thicknesses of 500 Å and 1000 Å are shown in Fig. 7.8. At the mask level the source-drain spacings were 8.9 and 12.5 μm, respectively, but the effective spacings after diffusion were 4.8 and 7.1 μm because of lateral spreading action.

The effects of temperature on the characteristics for a 500 Å device are shown in Fig. 7.9. The term $\gamma_m$ is the normalized transconductance $\mu_{eff}$ ($\epsilon_0 \epsilon_{ox}/t_{ox}$) and this decreases with increase of temperature because of decrease of mobility. The threshold voltage also decreases slightly.

The effect of lowering the temperature to 77°K for a short-channel ($L$ = 0.8 μm) MOSFET design is shown in Fig. 7.10. MOSFETs are capable of acceptable performance down to 4.2°K.[163] Some deterioration of characteristics occurs at temperatures at which deionization of the source and drain diffusions begins.

The drain voltage in an MOS transistor is limited by two processes, namely avalanche multiplication in the drain depletion region, and punch-through of the drain depletion region to the source. The punch-through voltage $V_{PT}$ is the

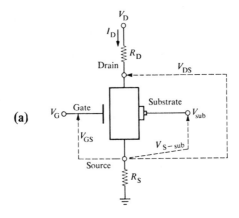

Grounded-source characteristics

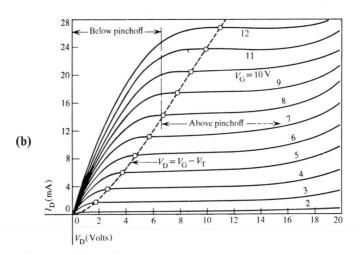

Fig. 7.6. Grounded-source and transfer characteristics for a Si $n$ channel enhancement MOSFET. Oxide thickness 500 A; effective channel length $L_{eff}$ = 4.6 $\mu$m (0.18 mil); channel width $W$ = 100 $\mu$m (4.0 mil); $V_{sub}$ = -7 V. (After Critchlow, Dennard, Schuster; *IBM J. of Research and Development*, **17**, 1973, p. 432 copyright 1973 by International Business Machine Corporation; reprinted with permission.)

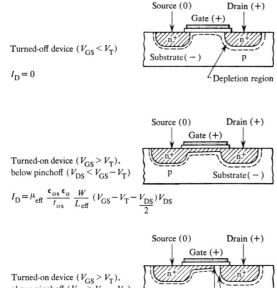

Turned-off device ($V_{GS} < V_T$)

$I_D = 0$

**(c)**    Turned-on device ($V_{GS} > V_T$), below pinchoff ($V_{DS} < V_{GS} - V_T$)

$$I_D = \mu_{eff} \frac{\epsilon_{ox} \epsilon_o}{t_{ox}} \frac{W}{L_{eff}} \left( V_{GS} - V_T - \frac{V_{DS}}{2} \right) V_{DS}$$

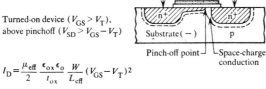

Turned-on device ($V_{GS} > V_T$), above pinchoff ($V_{SD} > V_{GS} - V_T$)

$$I_D = \frac{\mu_{eff}}{2} \frac{\epsilon_{ox} \epsilon_o}{t_{ox}} \frac{W}{L_{eff}} (V_{GS} - V_T)^2$$

**(d)**

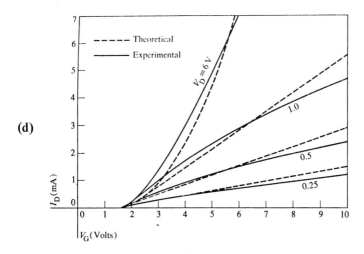

Fig. 7.6. (continued)

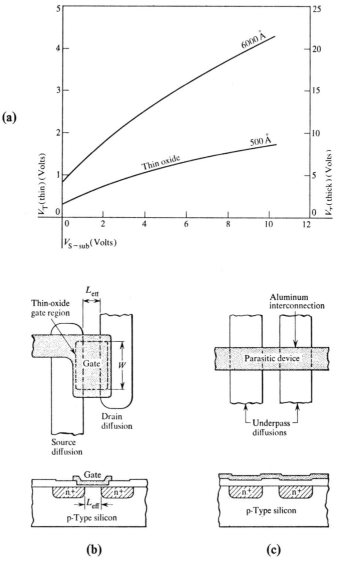

Fig. 7.7. Threshold voltage for an $Al\text{-}SiO_2\text{-}pSi$ structure as a function of source-substrate voltage:

(a) $V_T$ for two oxide thicknesses (plotted for $Q_{ox} + Q_{ss} = 0.5 \times 10^{11}\,cm^{-2}$, $N_{a,eff} = 5.0 \times 10^{15}\,cm^{-3}$, $\Delta V_{DT} = 0.07$ V);

(b) Thin oxide (500 Å) MOSFET; and

(c) Cross-over interconnection creates an unwanted parasitic thick oxide (6000 Å) device. (After Critchlow, Dennard, Schuster; *IBM J. of Research and Development*, 17, 1973, pp. 432, 431, copyright 1973 by International Business Machine Corporation; reprinted with permission.)

**(a)**

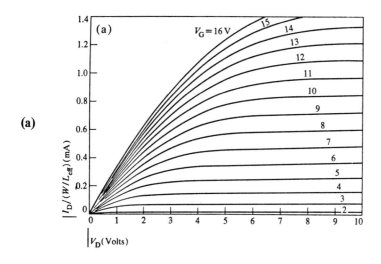

**(b)**

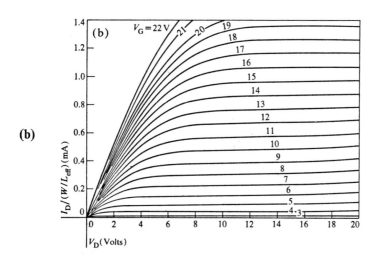

Fig. 7.8. Normalized grounded source characteristics for 500 Å and 1000 Å Si $n$ channel MOSFETs:
(a)  500-Å device, $L = 4.8$ μm ($R_D W/L_{eff} = 110$ ohms, $R_S W/L_{eff} = 75$ ohms; and
(b)  1000-Å device, $L = 7.1$  μm ($R_D W/L_{eff} = 400$ ohms, $R_S W/L_{eff} = 60$ ohms). (After Critchlow, Dennard, Schuster; *IBM Journal of Research and Development*, **13**, 1973, p. 434, copyright 1973 by International Business Machines Corporation, reprinted with permission.)

**(a)**

**(b)**

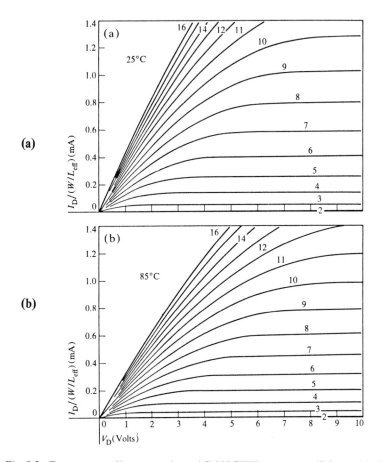

Fig. 7.9. Temperature effects on $n$ channel Si MOSFET parameters ($SiO_2$ = 500 Å):
(a)  Grounded-source characteristics at 25°;
(b)  Grounded-source characteristics at 85°C; and
(c)  Threshold voltage and transconductance variations vs temperature. (After Critchlow, Dennard, Schuster; *IBM Journal of Research and Development*, **17**, 1973, p. 441; copyright 1973 by International Business Machines Corporation; reprinted with permission.)

drain-to-source voltage at which the longitudinal field at any point along the edge of the source region inverts in sign to permit the drift of minority carriers from source to drain. The voltage $V_{PT}$ decreases as the channel length and substrate doping concentration decrease, and as the oxide thickness and diffusion depth increase. These effects are illustrated in Fig. 7.11.

The breakdown voltage associated with drain avalanche multiplication falls as the gate bias rises above threshold because of charge injected from the channel into the drain depletion region. The result is that MOS devices exhibit a "soft

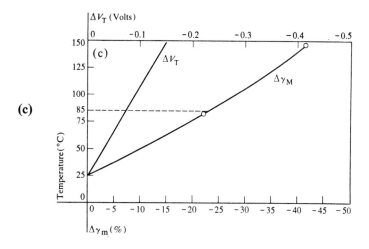

Fig. 7.9. (continued)

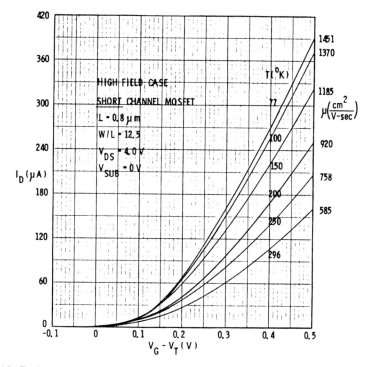

Fig. 7.10. Drain current, $I_D$, vs ($V_G$-$V_T$) in the square law region (i.e., $V_{DS} > V_G$) for a Si $n$ channel device ($L = 0.8$ μm) as a function of operating temperature $T$. The right-hand scale indicates high field mobilities at a gate voltage of 0.5 V above threshold. (After Gaensslen, Rideout, Walker. Reprinted with permission from *IEEE International Electron Devices Meeting Technical Digest*, 1975, p. 46.)

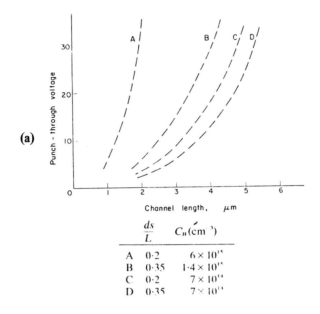

**(a)**

| | $\dfrac{ds}{L}$ | $C_B\,(\text{cm}^{-3})$ |
|---|---|---|
| A | 0·2 | $6 \times 10^{15}$ |
| B | 0·35 | $1·4 \times 10^{15}$ |
| C | 0·2 | $7 \times 10^{14}$ |
| D | 0·35 | $7 \times 10^{14}$ |

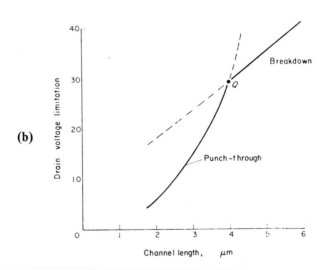

**(b)**

Fig. 7.11. Drain voltage limitations in a Si MOS transistor:
(a) Variation of punch-through voltage $V_{PT}$, with channel length $L$ for different substrate doping concentrations $C_R$ and normalized diffusion depths $ds/L$; $T_{ox} = 0.12\ \mu\text{m}$; and
(b) Limitation on drain voltage as a function of channel length $L$; Q defines point of transition from punch-through limited to breakdown limited; $T_{ox} = 0.12\ \mu\text{m}$, $ds/L = 0.35$, $C_B = 1.4 \times 10^{15}\,\text{cm}^{-3}$. (After Bateman, Armstrong, Magowan.[17] Reprinted with permission from *Solid-State Electronics*, Pergamon Press Ltd.)

breakdown" when operated in the deep saturation region even though the actual drain $p$-$n$ junction breakdown is sharp.[3]

Several structures have been proposed to achieve short channel lengths with higher drain breakdown voltages. One such structure has an enhancement and depletion mode region under the same gate, and is fabricated by ion implantation as shown in Fig. 7.12. The device is roughly equivalent to three MOS transistors connected in series as shown in Fig. 7.13(a). A typical $I_D - V_D$ characteristic may be regarded as having four bend points as in Fig. 7.13(b). The curve at point A shows the usual saturation of Tr 2. But drain current $I_D$ rises, as $V_D$ is increased beyond point A, due to the punch-through current of Tr 2. Point B indicates the saturation point of Tr 3. The region between points C and D suggests the "channel breakdown" due to impact ionization of carriers in the channel. The sharp increase beyond D corresponds to the "hard breakdown" of a drain junction between drain and substrate.

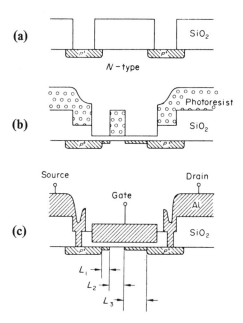

Fig. 7.12. Enhancement and depletion regions may be provided in the same channel and give good drain breakdown characteristics:
(a) Diffusion step for the $p^+$ source and drain regions of a $p$ channel device;
(b) $B^+$ implantation through the gate oxide; and
(c) Fabricated $E/D$ gate MOSFET after Al-metallization. (After Sasaki.[221] Reprinted with permission from *Solid-State Electronics*, Pergamon Press Ltd.)

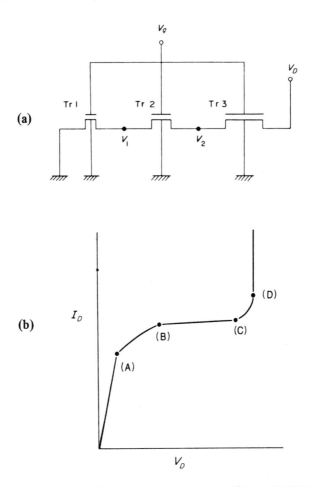

Fig. 7.13. Representation and $I_D$-$V_D$ characteristics of an $E/D$ gate MOSFET as in Fig. 7.12:
(a)  Representation as two depletion mode transistors Tr 1 and Tr 3, and one enhancement mode transistor Tr 2; and
(b)  Bend points in the $I_D$-$V_D$ characteristic (see text). (After Sasaki.[117] Reprinted with permission from *Solid-State Electronics*, Pergamon Press Ltd.)

The stacked gate structure shown in Fig. 7.14(a) also has a good drain-breakdown voltage, and a low drain-gate feedback capacitance. The control gate $G_1$ is of molybdenum and the insulator is formed as an oxide-nitride sandwich. The offset gate is located on the thick insulator layer between the control gate and the drain and is usually operated at a constant voltage $V_{G2}$. This allows a higher drain breakdown voltage because of the rearrangement of field lines and current flow suggested in Fig. 7.14(b). There is some tendency in this structure for high-energy channel electrons to enter the oxide and

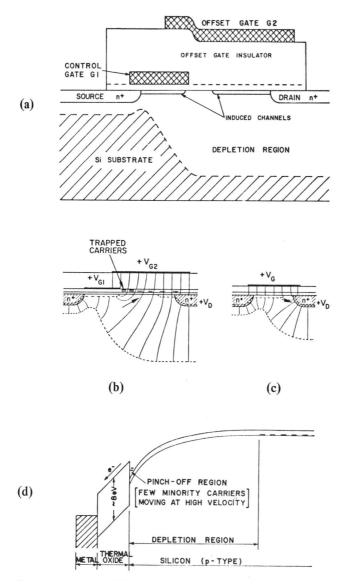

Fig. 7.14. A stacked gate MOS transistor structure may have a high drain voltage and a small drain-gate feedback capacitance:

(a)  Structure showing the two gates;

(b)  Direction of carriers in the pinch-off region for a stacked gate tetrode tends to be towards the oxide;

(c)  The flow towards the oxide is less for a full gate device; and

(d)  Energy band diagram of a section through the pinch-off region illustrating electron injection into the oxide. (After Erb, Dill, Toombs. Reprinted with permission from *IEEE Trans. Electron Devices*, **ED-18**, 1971, p. 105.)

become trapped there and produce threshold-shifts. However, by careful design, including control of oxide traps, structures can be made that will operate at drain voltages of 100-200 V.

A problem with $n$ channel devices is that positive charge, which may be present in the $SiO_2$ insulating layer, can induce thin inversion layers that provide surface leakage to the substrate lead. The leakage paths shown in Fig. 7.15(a) are:

- Leakage directly to the substrate by generation current in the $np$ junctions formed by the parasitic inversion layers.
- The inversion layer may connect an active source or drain region to inactive source or drain regions that increase the leakage current possibilities.
- The inversion channels may reach crystal defect areas at the edge of the chip and so provide leakage to the substrate.

One solution to the problem of weak $n$-inversion layers is to surround the $n$-channel FET with a $p^+$ guard-ring diffusion to isolate it from neighboring devices. Another solution, less costly in chip area, is to use the substrate bias to advantage in conjunction with an $n^+$ guard ring. This is illustrated in Fig. 7.15(b). With substrate bias applied the region between the drain and the guard ring may be depleted and the leakage path "turned-off." Outside the guard ring, the surface is still inverted (if the charge is sufficiently large). In this area, the guard ring behaves as a drain while any other diffusion that is not deliberately tied to another potential will seek the substrate potential and act as a source.

In a densely packed array each junction has only a very limited vacant area around it, except around the outside edge of the array. Furthermore, each of the surrounding junctions is also biased above substrate potential (at ground or higher) thereby effectively serving as a guard ring. The net result is that special guard rings are needed only around the edge of the chip.

## 7.4 MOSFET SATURATION MODELS

The finite output conductance of an MOSFET is caused in part by the spreading of the depletion region near the drain, which results in a reduction of the channel length. Allowance must also be made for drift velocity saturation, as in the study of JFET channel flow, and for the effect of the gate electrode on the electric field in the drain region. The output conductance is therefore a function of the insulator thickness under the gate as well as of the impurity concentration in the substrate.[87]

In general the problem must be studied by numerical solution of semiconductor equations for two-dimensional models. Sometimes the channel is treated as a source region and a drain region, with suitable matching of equations provided between the two sections, as shown in Fig. 7.16(a). The mathematical

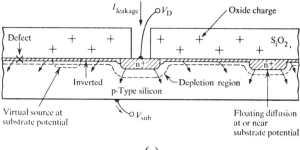

**(a)**

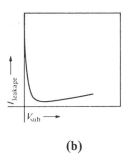

**(b)**

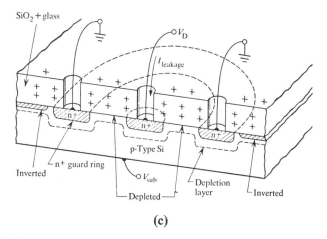

**(c)**

Fig. 7.15.  Leakage current problems in MOS transistors:
(a)  Various leakage current paths for an *n* channel device;
(b)  Leakage current vs. substrate voltage for a device with a large surface charge; and
(c)  Structure of guard ring using biased *n*⁺ diffusion. (After Critchlow, Dennard, Schuster. *IBM J. of Research and Development,* **17,** 1973, p. 437, copyright 1973 by International Business Machine Corporation; reprinted with permission.)

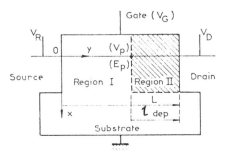

Fig. 7.16. Modeling of a MOSFET as a two region structure to allow calculation of the output conductance in saturation. (After Rossel, Martinot, Vassilieff.[218] Reprinted with permission from *Solid-State Electronics*, Pergamon Press, Ltd.)

development is too long to present here in detail, but results in the expression

$$I_D \left( \frac{\partial V_D}{\partial I_D} \right) V_G V_R \simeq [L^2 E_p^2 + 2A(V_D - V_P)(1 + BI_D)]^{1/2}$$

$$\cdot \left\{ 1 - \frac{\ell_{dep}}{L} + \frac{BI_D}{1 + BI_D} \left[ \frac{\ell_{dep}}{L} \right. \right. \tag{7.14}$$

$$\left. \left. \cdot \frac{V_D - V_P}{[L^2 E_p^2 + 2A(V_D - V_P)(1 + BI_D)]^{1/2}} \right] \right\}$$

where $E_p$ is the longitudinal field at the transition between the two regions, $A$ is $qN_A L^2/\epsilon$, $B = \lambda/(qN_A)$ and the charge density and $\lambda$ are defined by $p(y) = -(qN_A + \lambda I_D)$.

Eq. (7.14) may be further simplified to the form

$$(R_D I_D)^2 \simeq [L^2 E_p^2 + 2A(V_D - V_P)(1 + BI_D)] \tag{7.15}$$

This equation shows that for the saturation region the product $(R_D I_D)^2$ is a linear function of the drain voltage and the slope of the relation is a linear function of the average drain current. Experiments that confirm these relationships are shown in Fig. 7.17. The quantities $A$ and $B$ of the model may be determined by such measurements.

Other parameters such as $\mu_o$, $(Z/L)$, $C_{ox}$, $(Q_{ss}/C_{ox})$, $\phi_{ms}$, $\phi_B$, $\phi_F$, $\psi$, and $LE_o$ may be determined in conventional ways. These then allow testing of the model against experimental characteristics. As shown in Fig. 7.18 the agreement is good.

Acceptable agreement between theory and experiment has also been obtained for other models, some similar and some not so similar.[73,85,195] In the USA

**(a)**

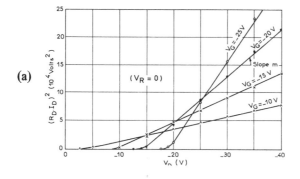

**(b)**

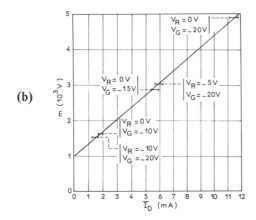

Fig. 7.17. For the saturation region of a MOSFET the product $(R_D I_D)^2$ is predicted to be a linear function of the drain voltage and the slope of the relation to be a linear function of drain current. The experimental curves shown above for a $p$ channel device confirm these relationships. $V_R$ is the substrate-to-source voltage. (After Rossel, Martinot, Vassilieff.[218] Reprinted with permission from *Solid-State Electronics*, Pergamon Press Ltd.)

extensive use is made of the IGFET model discussed in Sec. 7.9, since this has been made the basis of a computer-aided design program known as SPICE.

## 7.5 THE TRANSITION FROM IGFET TO BIPOLAR TRANSISTOR PERFORMANCE

A bipolar transistor is characterized by $I_c = \alpha I_e$, where

$$I_e \simeq I_{e0} \exp\left(\frac{q V_{eb}}{kT}\right) \tag{7.16}$$

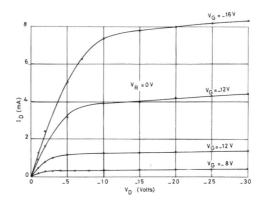

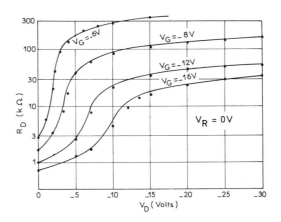

Fig. 7.18. The two-region MOSFET model (of Rossel et al.) is capable of theoretical results that match experimental points:
(a) $I_D$ vs $V_D$ theoretical curves and observed points; and
(b) Output resistance variation with $V_D$ and $V_G$. (After Rossel, Martinot, Vassilieff.[218] Reprinted with permission from *Solid-State Electronics*, Pergamon Press Ltd.)

and the small signal input resistance of the emitter-base diode is given by

$$r_d = \frac{kT/q}{I_e} \qquad (7.17)$$

At high frequencies the input impedance is determined by the diffusion capacitance which for a uniform-doped base model is given by

$$C_D = \frac{\tau_{base}}{r_d} = \frac{\tau I_e}{kT/q} \qquad (7.18)$$

The small-signal low-frequency input voltage is therefore $r_{die}$ and the transconductance $i_c/v_{eb}$ is

$$g_{m_{BJT}} = \frac{I_e q}{kT} \qquad (7.19)$$

The dominant role played by $q/kT$ in the control action is apparent.

If $I_e$ is 10 mA the transconductance from Eq. (7.19) at $300°K$ is $390\ mA/V$. This is large compared with typical values of 0.1 to 20 mA/V for IGFETs. It is usual to dismiss the difference by stating that IGFETs and BJTs are intrinsically different in the details of their operation. However, physical arguments can be developed so that when the IGFET specific gate capacitance is increased, such as by decreasing insulator thickness or by increasing the dielectric constant, device performance will change toward the low impedance, high transconductance characteristics typical of bipolar transistors.

The standard formula for $g_m$ in an MOSFET is

$$g_m = \frac{\bar{\mu}_n C_G}{L^2} (V_G - V_\tau) \qquad (7.20)$$

whence
$$g_m = \frac{\bar{\mu}_n \epsilon_{ox} \epsilon_0 WL}{L^2 t_{ox}} (V_G - V_T) \qquad (7.21)$$

Hence, it might be inferred that for a very short channel length the transconductance can become arbitrarily large. However, study of assumptions built into Eq. (7.20) reveals that the $g_m$ expression for a short channel length approaches the same limit, $qI/(kT)$, as for a BJT.

The analysis depends upon recognizing that the source-drain potential has the form shown in Fig. 7.19 where the barrier $V_s$ across the high-low source junction can be significant. This is particularly true if the channel is short and the gate-insulator capacitance is very large so that the charge induced in the channel is large and the voltage-drop in the main part of the channel is small. Then by a Boltzmann relationship

$$\frac{n_c}{n_s} = \exp\left(-\frac{qV_s}{kT}\right) \qquad (7.22)$$

where for an $n$-channel device $n_c$ is the electron density on the channel side of the interface and $n_s$ is the electron density on the source side.

The drain current $I$ is related to the channel charge $Q_c$ by

$$I = \frac{Q_c}{\tau} \qquad (7.23)$$

where $\tau$ is the carrier transit time in the channel. Also, the capacitance is related to the channel charge density by

$$C_1 = \frac{dQ_c}{dV_s} = \frac{d}{dV_s}\left(\text{const} \cdot \exp\frac{qV_s}{kT}\right) \qquad (7.24)$$

assuming $V_s$ is the dominant voltage in the circuit. At this extreme limit the transconductance is given by

$$g_m = \frac{dI}{dV_g} = \frac{dI}{dV_s} = \frac{1}{\tau}\frac{dQ_c}{dV_s} = \frac{Q_c}{\tau}\frac{q}{kT} \qquad (7.25)$$

which, from Eq. (7.23) becomes

$$g_m = \frac{I_e q}{kT} \qquad (7.26)$$

Hence, in this hypothetical limit the MOSFET has the same transconductance as a bipolar transistor.

This explains the experimental observation that field-effect transistors tend to exhibit a transconductance/current ratio of $q/(kT)$ for very low drain currents.

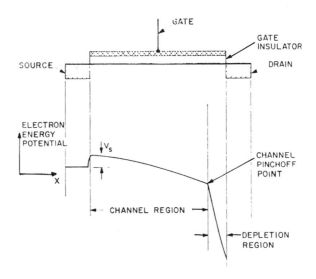

Fig. 7.19. General form of the source-drain potential distribution in an $n$-channel IGFET operating in drain saturation. (After Johnson.[128] Reprinted with permission from *RCA Review*; **34**, 1973, p. 82.)

A somewhat more complete analysis[146] uses for the channel transit time

$$\tau_1 = \frac{4}{3}\frac{L^2}{\bar{\mu}_n(V_G - V_T)}.$$ (7.27)

which leads to the transconductance formula

$$g_m = \frac{qI}{kT}\frac{C}{C + (2qI/kT)\cdot\frac{4}{3}[L/\bar{\mu}_n(V_G - V_T)]}$$ (7.28)

Then if $L$ is very small, $g_m$ tends to $qI/kT$, and if $L$ is large, $g_m$ tends to a form similar to Eq. (7.21).

The input impedance level is $(\omega C)^{-1}$, but $C$ depends on $Q_C$ and $g_m$ from Eq. (7.25) also depends on $Q_C$. Therefore there is a trade-off between input impedance and transconductance in an FET.

Practical weight is hard to assign to this broad-concept study, but it is certainly interesting to consider an FET as a bipolar transistor in disguise.

## 7.6 SEMICONDUCTOR-INSULATOR AND INSULATOR-METAL INTER-FACES

In a typical MOSFET the semiconductor is Si, the insulator is $SiO_2$ and the gate metal is Al. Alternatively, polycrystalline heavily-doped Si is used as the gate to allow a self-aligning mask process and to give a lower threshold voltage. The threshold difference occurs because of a change of the work-function between the gate and the oxide. The results to be expected for well-prepared 1000 Å oxides are shown in Fig. 7.20.

The threshold voltage depends on the Si gate-doping, the surface state density, the substrate doping $N_S$ and the substrate bias voltage.

$$V_T = \phi_{GS} - q\frac{N_{SS}}{C_{ox}} + 2\phi_F + \frac{\sqrt{2\epsilon_{Si}\epsilon_0 qN_S(2\phi_F + V_R)}}{C_{ox}}$$ (7.29)

where

$\phi_{GS}$    = gate-background work function difference = $\frac{kT}{q}\ln\frac{4\times 10^{20}\,N(x)}{n_i^2}$,
(for an $n$-doped Si gate with an assumed effective doping concentration of $4\times 10^{20}$ atoms/cm$^3$), expressed in volts;

$n_i$    = intrinsic concentration = $1.5\times 10^{10}$ carriers/cm$^3$ at $27°C$;

$N_S$    = substrate doping, atoms/cm$^3$;

$N_{SS}$    = surface-state density; $10^{11} - 10^{12}$ cm$^{-2}$; positively charged interface states, as in Fig. 7.25;

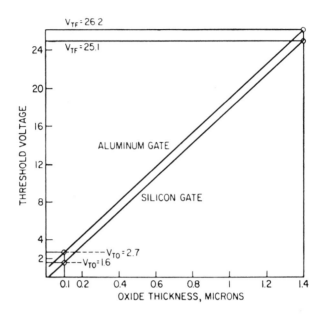

Fig. 7.20. A Si gate gives a lower threshold voltage than for Al for any oxide thickness: Moreover, it gives a higher ratio of parasitic threshold to device threshold ($V_{TF}/V_{TO}$). The values are for (111) Si substrate orientation. For (100) Si the values are lower–perhaps as low as 0.4 V for a poly-Si gate on 0.1 $\mu$m oxide thickness. The lines are for a $p$Si substrate capable of breakdown voltages of 35-45 V. (After Faggin & Klein in *Large and Medium Scale Integration*, S. Weber (ed.); copyright 1974, McGraw-Hill, NY. Reprinted with permission from McGraw-Hill Book Co.)

$\phi_F$     = Fermi level associated with a given doping concentration =
$\dfrac{kT}{q} \ln \dfrac{N(x)}{n_i}$ ;

$C_{ox}$    = gate capacitance per unit area $\epsilon_{ox}\epsilon_0/t_{ox}$;

$\epsilon_{ox}$     = dielectric constant of oxide = $(4 \times 8.85 \times 10^{-14}$ F/cm);

$t_{ox}$     = gate oxide thickness;

$\epsilon_{Si}\epsilon_0$    = dielectric constant of Si = $(11.7 \times 8.85 \times 10^{-14}$ F/cm);

$V_R$     = reverse bias of the substrate relative to the source.

Calculated values of $V_T$ vs the $p$-substrate concentration are shown in Fig. 7.21 to illustrate the range to be expected.

Oxides for Si FETs or MOS capacitors may be grown thermally in dry oxygen, or (less usually) in wet oxygen, with temperatures in the 950-1250°C range. This is followed by an annealing process. A fabrication flow diagram for a typical MOSFET process is shown in Fig. 7.22. The oxide is grown in two stages under dry conditions.

In general the presence of water in the oxygen increases the growth rate and the rate is also greater for (111) Si substrates than for (100) surfaces

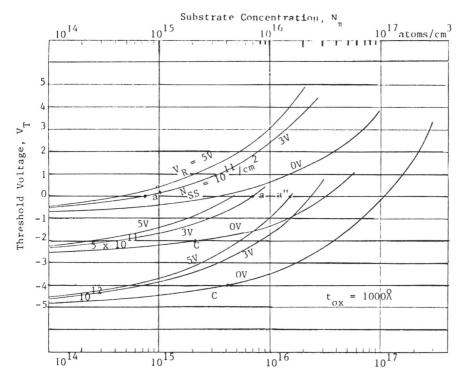

Fig. 7.21. Threshold voltage calculated for poly-Si gate as a function of $p$ substrate doping, with $N_{SS}$ and substrate-source bias $V_R$ as parameters. [After Lin, Halsor and Benz, (NASA sponsorship); reprinted with permission from *IEEE International Electron Devices Meeting Technical Digest*, 1975, p. 55.]

(see Fig. 7.23). The latter orientation is sometimes preferred because it produces lower threshold voltages. The $SiO_2$ thickness vs time data fits a linear-parabolic equation if the region of initial rapid $SiO_2$ growth is eliminated. Oxygen must diffuse through the oxide layer already present to reach the Si surface. The equation that results is of the form

$$t = A(d-d_0) + B(d-d_0)^2 \qquad (7.30)$$

Ellipsometer measurements may be used to determine the thickness and refractive index ($\bar{n}$ = 1.48) of the $SiO_2$. For thick layers the depth can be roughly estimated from the interference colors seen by direct reflected illumination. Table 7.1 lists these colors for $SiO_2$ and also for $Si_3N_4$, another insulator used for IGFETs, observed vertically with tungsten filament light.

In addition to good insulating properties, the most important requirement for the oxide is a minimum density of charges and trapping centers both in the

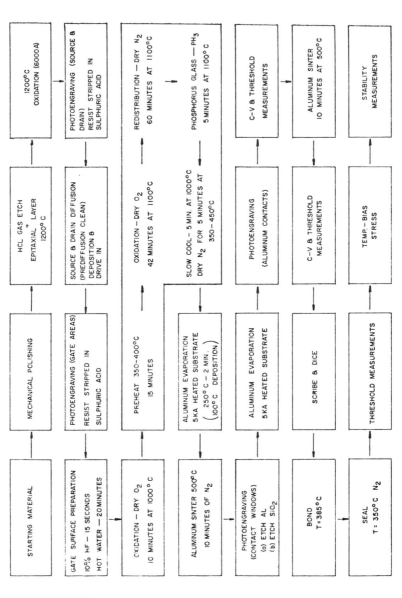

Fig. 7.22.  Fabrication flow diagram for a Si p-channel Al-gate MOS field-effect transistor. (After White and Cricchi.[276] Reprinted with permission from *Solid-State Electronics*, Pergamon Press Ltd.)

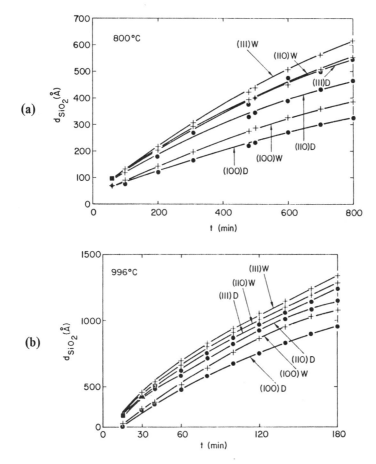

**(a)**

**(b)**

Fig. 7.23. The growth of $SiO_2$ on Si (100),(110) and (111) orientations for dry $O_2$ (D) and 25 ppm $H_2O$(W) in $O_2$:

(a) $SiO_2$ thickness vs time at 800°C; and

(b) At 996°C. For each curve the closed circles are for dry oxygen (< 1 ppm $H_2O$); the crosses are for 25 ppm $H_2O$ in oxygen. The symbols ✦, △, and ■ represent 2, 3, and 4 coincident points, respectively. (After Irene; reprinted from *Journal of the Electrochemical Society*, **121**, 1974, p.1614 with permission of the publisher, The Electrochemical Society.)

bulk and at the $SiO_2$-Si interface: care must be taken to eliminate impurities, particularly $Na^+$ ions, from the $SiO_2$.[143] Cleaning and gettering procedures involving HCl are often used. The $Na^+$ ions move readily in $SiO_2$ at temperatures of a few hundred degrees C with the field strengths applied across the gate insulator. This causes band-bending as shown in Fig. 7.24 and MOSFET threshold voltages that are unstable.

Table 7.1.  Color Comparison of $SiO_2$ and $Si_3N_4$ Films

| Order | Color | $SiO_2$ Thickness Range* ($\mu$m) | $Si_3N_4$ Thickness Range ($\mu$m) |
|---|---|---|---|
| | Silicon | 0-0.027 | 0-0.020 |
| | Brown | 0.027-0.053 | 0.020-0.040 |
| | Golden Brown | 0.053-0.073 | 0.040-0.055 |
| | Red | 0.073-0.097 | 0.055-0.073 |
| | Deep Blue | 0.097-0.010 | 0.073-0.077 |
| 1st | Blue | 0.10-0.12 | 0.077-0.093 |
| | **Deep Blue** | 0.12-0.13 | 0.097-0.10 |
| | Very Pale Blue | 0.13-0.15 | 0.10-0.11 |
| | Silicon | 0.15-0.16 | 0.11-0.12 |
| | Light Yellow | 0.16-0.17 | 0.12-0.13 |
| | Yellow | 0.17-0.20 | 0.13-0.15 |
| | Orange Red | 0.20-0.24 | 0.15-0.18 |
| 1st | Red | 0.24-0.25 | 0.18-0.19 |
| | Dark Red | 0.25-0.28 | 0.19-0.21 |
| 2nd | Blue | 0.28-0.31 | 0.21-0.23 |
| | Blue-Green | 0.31-0.33 | 0.23-0.25 |
| | Light Green | 0.33-0.37 | 0.25-0.28 |
| | Orange Yellow | 0.37-0.40 | 0.28-0.30 |
| 2nd | Red | 0.40-0.44 | 0.30-0.33 |

*The ratio of refractive index $\dfrac{\bar{n}(Si_3N_4)}{\bar{n}(SiO_2)} = \dfrac{1.97}{1.48} = 1.33 = \dfrac{SiO_2 \text{ thickness}}{Si_3N_4 \text{ thickness}}$  for similar color

(From Reizman and van Gelder.[204] Reprinted with permission from *Solid State Electronics*, Pergamon Press, Ltd.)

A more dense insulator is $Si_3N_4$ and movement of $Na^+$ ions is much less in this material. However, its interface-state density is larger than for the $SiO_2$-Si interface and it is not acceptable as the insulator in an FET if used alone, but it may be used in oxide-nitride sandwich structures.

A thin layer of $SiO_2$ on the surface of Si tends to form an electron accumulation region as shown in Fig. 7.25 because of donor type interface states. At a cleaved Si face a dangling Si bond density on the order of $10^{15}$ cm$^{-2}$ [about $(5 \times 10^{22})^{2/3}$] may be expected from simple geometrical considerations. The reduction of this state density by replacement of unsaturated Si bonds at the surface with Si-O bonds is considerable. For instance oxidation of a Si surface at 1200°C in dry $O_2$ and annealing in very pure He gives interface state densities as low as $10^9$-$10^{10}$ cm$^{-2}$ provided the steps are performed in place with radio-frequency heating so that the reactor tube is relatively cold and impermeable to $H_2O$. The roles of $H^+$, OH and $H_2O$ in $SiO_2$ are quite complicated.[123, 177, 205]

Hydrogen in various forms can be incorporated in thermally or anodically grown $SiO_2$ films during the growth process and various postoxidation treatments. Depending on the incorporation process, the two main forms of H

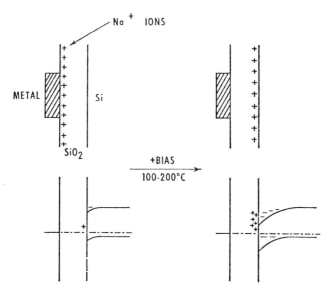

Fig. 7.24. Ion drift instability. Originally, the $Na^+$ ions are piled up at the oxide-metal inter-face. They induce a slight negative charge in Si as their net electrostatic effect on the Si-SiO$_2$ interface is relatively small (see lower left part). After applying a positive bias to the metal electrode at elevated temperature, the $Na^+$ ions are concentrated at the Si-SiO$_2$ interface. Their effect on the Si space-charge region is now much larger than before. After the removal of the bias, the large negative space-charge in Si consists mainly of electrons; this corresponds to the inversion regime in $p$-type silicon. (After Revesz. Reprinted from *Journal of Non-crystalline Solids*, **11**, 1973, p. 311, North Holland Publishing Co.)

incorporated are Si-OH and Si-H but interstitial H and $H_2O$ are also possible. The Si-H bonds are less stable than the Si-OH bonds, and they may act as electron donors after dissociation, whereas Si-OH bonds act as electron acceptors. The Si-H bonds at the Si-SiO$_2$ interface can reduce the apparent density of interface states but are also responsible for various negative bias instability and radiation effects. Since Si-OH bonds are electron acceptors, they decrease the radiation-induced positive charge in SiO$_2$.

The variation of an MOS capacitance with bias voltage and with frequency, provides useful information on the band-bending at the SiO$_2$-Si interface. For an oxide on a $p$-Si substrate the application of a positive gate voltage repels holes and decreases the capacitance. Further voltage increase on the gate induces an electron layer at the Si surface and the capacitance rises again to the nominal oxide value $C_{ox}$. However, if the frequency for the capacitance measurement is very high the substrate does not communicate fast enough with the surface induced layer and the capacitance remains low. Calculated curves of $C/C_{ox}$ that illustrate this for $p$-type Si with a 1000 Å of oxide are shown in Fig. 7.26. The surface state density $N_{SS}$ was assumed to be zero and the work-function

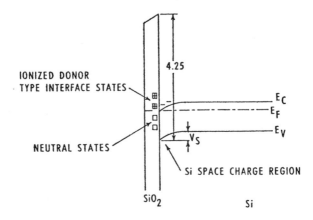

Fig. 7.25. Thin SiO$_2$ film on Si. Energy is plotted vertically, distance (not on scale) horizontally. The SiO$_2$ film is so thin (20-40 Å) that the whole film is considered as part of the Si surface. $E_C$, $E_V$, and $E_F$ designate the conduction band, valence band, and Fermi energy, in this order; $V_S$ is the Si surface potential. Donor type surface states are shown. The Si is $n$-type, and since there are more electrons in the space-charge region than in the bulk, the saturation depicted here corresponds to the accumulation regime. The energy barrier has been determined from electron injection by photoemission measurements. (After Revesz. Reprinted from *Journal of Non-crystalline Solids*, 11, 1973, p. 311, North Holland Publishing Company.)

difference of the metal and the semiconductor $\phi_{ms}$ was also taken as zero. If these are not negligible the $C$-$V$ curve is shifted by an amount,

$$\Delta V = \phi_{ms} - q \frac{N_{SS} t_{ox}}{\epsilon_{ox} \epsilon_0} \qquad (7.31)$$

For MOS structures prepared by the processing steps of Fig. 7.22, the displacement is about 4.4 V as shown in Fig. 7.27 and corresponds to a surface-state density of $7 \times 10^{11}$ cm$^{-2}$. The values of $\phi_{ms}$ for aluminum on $p$Si have been measured as about $-0.9$ to 1.0 V.[56] The values of $\phi_{ms}$ for polycrystalline Si gates on SiO$_2$ have also been determined.[301]

Numerous electrical techniques have been devised for studying the interface state densities in MOS structures and the surface recombination velocity and the bulk lifetime.[92] On the application of a step of gate voltage that causes a depletion region to form in the semiconductor there is a fast capacitance decrease. However, as hole-electron pairs are generated in the depletion region the minority carriers, electrons in Fig. 7.28, flow to the surface to form an inversion layer and there is a transient increase of capacitance. Analysis shows that the plot of $\dfrac{-d(C_o/C)^2}{dt}$ vs $(C_F/C - 1)$ has a straight line portion the slope of which gives the bulk lifetime and the intercept of which is proportional

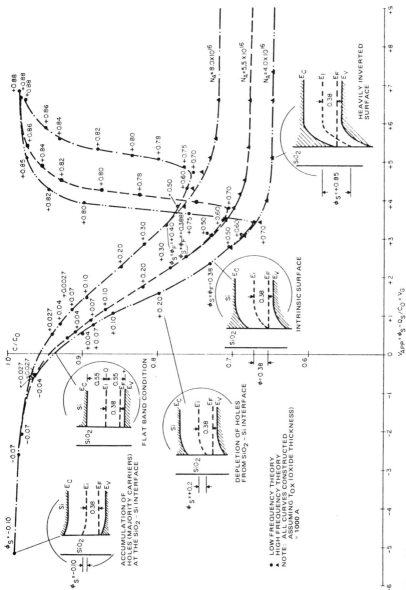

Fig. 7.26. Theoretical *C-V* curves for an Al-SiO₂ (1000 Å)-*p*Si system. (After White & Cricchi.[276] Reprinted with permission from *Solid-State Electronics*, Pergamon Press Ltd.)

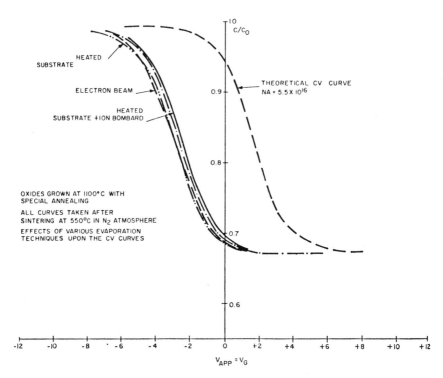

Fig. 7.27. The $C$ vs $V_G$ curve for an $SiO_2$-$p$Si gate capacitance is usually displaced considerably from the theoretical location that neglects the presence of interface charge. The displacement seen here corresponds to a surface-state density of $7 \times 10^{11}$ cm$^{-2}$. (After White & Cricchi.[276] Reprinted with permission from *Solid-State Electronics*, Pergamon Press Ltd.)

to the interface recombination velocity $S$. This is known as a Zerbst plot.[46]

Other important techniques exist for the study of the surface-state density and its energy distribution within the Si bandgap. One involves study of the conductance in parallel with the capacitance.[187] Another technique involves measurement of the variation of the capacitance flat-band voltage with temperature.[97] Generally the interface state density is found to have peaks at about 0.1 or 0.2 eV from the conduction and valence band edges and to be lower by an order of magnitude in the center of the bandgap.

The interface state density may also be determined for an MOS transistor by study of the slopes of the $I_D$ vs $V_D$ curves for small drain voltages[259] or from the charge pumping current that flows to the substrate when gate pulses are applied.[72]

Consider now the problems of insulator formation on GaAs. Deposited insulators such as $SiO_2$ and $Si_3N_4$ usually give imperfect results.[263] Unless grown with great care, $SiO_2$ on a GaAs surface tends to have a much higher interface-

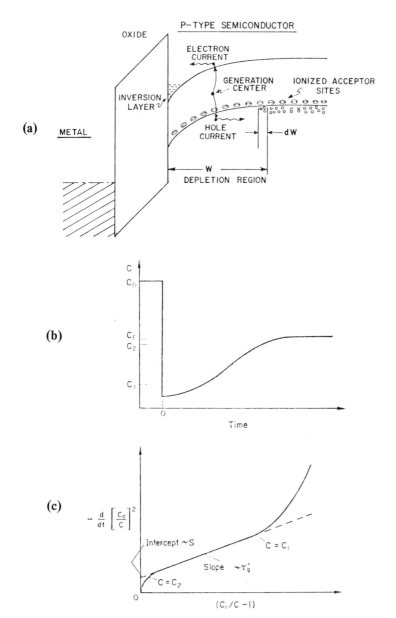

Fig. 7.28. After the depletion edge is moved in an MOS capacitor by a step of voltage, generation in the new depletion region causes a relaxation of capacitance with a transient that can be related to the bulk lifetime and the interface recombination:

(a) Generation process; (After Heiman. Reprinted with permission from *IEEE Trans. Electron Devices*, **ED-14**, 1967, p. 781)

(b) Capacitance transient after a voltage pulse; and

(c) A Zerbst plot of the pulsed response showing the linear portion and the intercept s-value. (After Schroder & Nathanson.[225] Reprinted with permission from *Solid-State Electronics*, Pergamon Press Ltd.)

state density than for a thermal oxide on Si.[140] There are also problems of thermal coefficient of expansion mismatches that cause strain and cracking of $SiO_2$ and $Si_3N_4$ layers during subsequent diffusion processes or lateral diffusion effects along the interface. The addition of $P_2O_5$ to the $SiO_2$ has been found to help to reduce cracking.[15]

Films of $Al_2O_3$ may be deposited on GaAs from the decomposition of aluminum-isopropoxide in Ar at 300°C. This may be followed by a coating of $SiO_2$ from tetraethoxy-silane at 680°C. The $SiO_2$ layer improves the stability and with a dual-layer structure reasonable GaAs IGFET characteristics are obtained if source and drain regions are formed by alloying processes.[124]

Amorphous native oxides may be grown on GaAs by anodization in aqueous solutions of 30% $H_2O_2$ pH 2.0. Baking at 250°C is necessary to remove included water. Such layers successfully mask against zinc diffusions at 612°C for 2 hr, which is sufficient for good surface concentrations ($\sim 10^{20}$ $cm^{-3}$) and adequate junction depths ($>0.5$ $\mu$m) in the unmasked regions.[241, 242] Other anodizing solutions that produce denser native oxides have been proposed.[102, 179, 239, 246] Thermal oxidation in air at 500°C has recently given well-adhering layers at a moderate growth rate of about 500 Å/hr.

The temperature of native oxide layers on GaAs must be limited to 650°C or less, to avoid the formation of brittle crystalline forms of $\beta$-$Ga_2O_3$ or $GaAsO_4$. Thus, high temperature diffusions in GaAs are not practical with such oxide masks. However, low temperature diffusions and ion-implantation processes are possible.

IGFETs have been successful on many semiconductors including Ge,[268] InAs,[134] InSb,[230] CdS,[274] and GaAs.[299]

## 7.7 FABRICATION PROCESSES FOR IGFETs

The fabrication of insulated-gate FETs involves many processes that may be reviewed from the reading suggestions given. The processes include wafer preparation, epitaxial growth, insulator growth, photolithography, etching, diffusion, ion implantation, metallization and contacts, and packaging.

Some of the usual geometries and processes are:

- $n$ and $p$ channel planar FETs with Al gate metallization added after the source and drain diffusions;
- planar structures with poly-Si gates (see Fig. 7.29) to provide self-alignment of the gate during the source and drain diffusion and to give low threshold voltages;
- planar structures with refractory metal gates (Mo or W) to withstand the diffusion temperatures and provide self-masking;[270]
- ion-implantation of the channel to control the threshold voltage, followed by implantation for source and drain extension (see Fig. 7.30). Enhance-

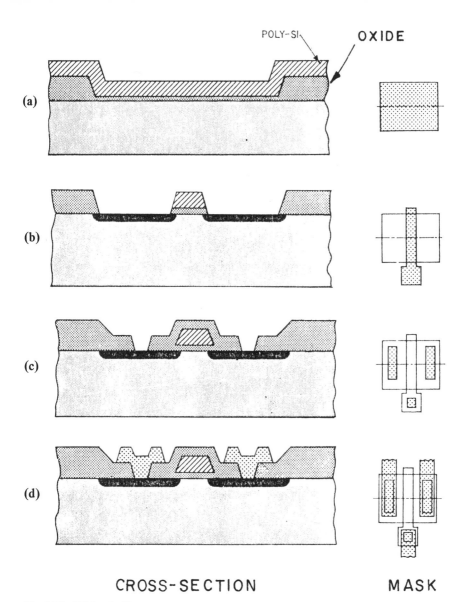

POLY-SI

OXIDE

(a)

(b)

(c)

(d)

CROSS-SECTION                    MASK

Fig. 7.29. MOS poly-Si gate process:
(a) An oxide is grown on an *n* wafer and a window made for each MOS transistor and then poly-Si is grown;
(b) The poly-Si is masked to define the gate and the poly-Si interconnection strip and the boron diffusion made for the source and drain regions. This also dopes the poly-Si gate *p* type to provide the desired conductivity and work function;
(c) A layer of $SiO_2$ is vapor deposited and patterned; and
(d) Slice is covered with Al and then masked and etched to provide the source and drain region contacts and the second interconnection plane. (After Faggin & Klein in: *Large and Medium Scale Integration*, edited by S. Weber, copyright 1974 McGraw-Hill, NY. Reprinted with permission from McGraw-Hill Book Co. p. 28.)

417

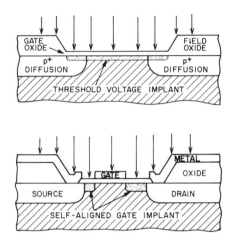

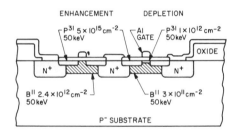

Fig. 7.30. Ion implantation used in the fabrication of MOSFET structures:
(a) For threshold-voltage adjustment;
(b) For source and drain extension; and
(c) In combination with electron-beam lithography for the fabrication of enhancement and depletion mode devices with 1-$\mu$m channel widths. (After Lee and Mayer. Reprinted with permission from *Proc. IEEE*, **62**, 1974, p. 1242, 1250.)

ment and depletion mode transistors may be made in this way on the same wafer as shown in Fig. 7.31. A channel-stop (chanstop) region is provided by a heavy boron implant so that a spurious $n$ channel that links the two transistors cannot form. In the layout of integrated circuits, care must always be taken that spurious channels do not occur. In Fig. 7.32 parasitic FETs may be formed by the cross-over of the metallization lines on the thick oxide regions, but may not be harmful if their threshold voltages are high. Occasionally parasitic bipolar geometries may occur that provide harmful signal paths;[65, 254]

• in general optical lithography on large wafers is limited to line separations of 2-5 $\mu$m and electron-beam lithography is used for finer resolution. However, by subtle mask-etching and oblique-metallization procedures,

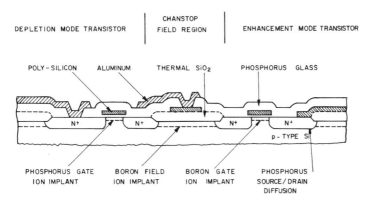

Fig. 7.31. *N*-channel enhancement and depletion mode Si-gate MOS transistors made by ion implantation.

1-μm gate widths can be obtained with optical lithography—if high yield is not essential;[186, 212]

- short-channel devices may be made by a double-diffused process (DMOS)[40, 59, 170] in which the channel length is determined by the difference in lateral diffusion of two impurity distributions [see Fig. 7.33(a)]. An *n* channel is induced (enhancement mode) in the 1-μm *p*-channel region. This is linked to the $n^+$ drain region by an $n^-$ conducting drift region about 5-μm long. Figure 7.33(b) shows a D-MOS logic inverter that studies show to have excellent low-power and high-speed capabilities;

- SOS (silicon on sapphire) is a technology in which the fabrication of the FETs is on small islands of Si that have been defined by etching from a thin Si layer grown on a sapphire substrate. The island structure provides insulation that greatly reduces parasitic capacitances as shown in Fig. 7.34. Certain manufacturers are using SOS successfully but others have encountered yield problems;

- VMOS uses V-shaped grooves obtained by a directional etch on (100) wafers with $n^+$ substrates and *p* (1-μm) and π (1-μm) epi-layers as shown in Fig. 7.35. The *p* layer has the higher threshold and so determines the effective transistor threshold and channel length for most operating conditions. The π region reduces the drain source capacitance and raises the punch-through voltage determined by a depletion region extending from drain to source. The $n^+$ source being below the drain requires no surface area in this form of VMOS and also serves as the ground plane needed in memory applications. Compared with *n*-MOS, VMOS technology saves about 40% in random-logic area because of the savings in source and ground areas, and also because, for a given ratio of load device to pull-down device, *n*-MOS requires a larger surface area. Compared with $I^2L$, VMOS is more dense because $I^2L$ has a base stripe width that must

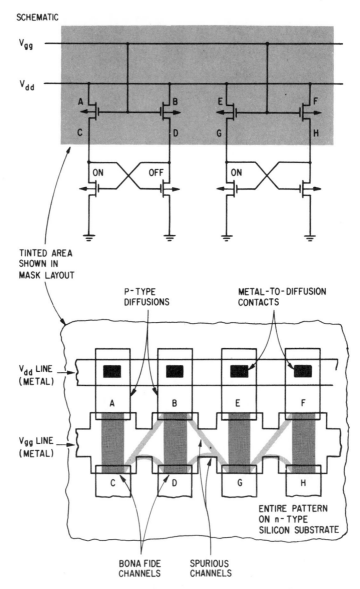

Fig. 7.32. Parasitic transistors formed by spurious channels. In the four transistors that serve as load impedance for these two static flip-flops, there are parasitic transistors between points as shown by the spurious channels (medium shading). When the flip-flop transistor connected to point $C$ is conducting, point $C$ is at or near ground, whereas points $B$ and $D$ are both at the supply voltage. Metal strip forming $V_{gg}$ line is continuous across the entire area, forming structure essentially identical to bona fide transistor between points $A$ and $C$ (dark shading). (After Bridwell. Reprinted from *Electronics*, April 13, 1970; copyright McGraw-Hill, Inc., NY, 1970.)

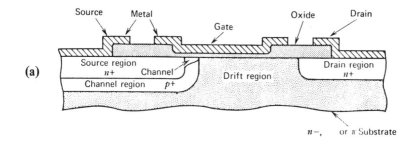

(a)

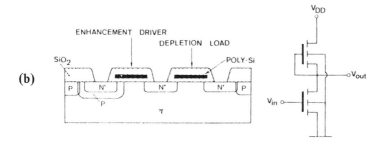

(b)

Fig. 7.33. DMOS technology:
(a) DMOS transistor with channel formed by double diffusion in the source region; and
(b) DMOS inverter with enhancement driver and depletion load (After Declereq and Laurent. Reprinted with permission from *IEEE J. of Solid-State Circuits*, **SC-12**, 1977, p. 265).

accommodate a concentric collector diffusion as well as a contact hole, whereas the VMOS diffusion line must accommodate only a contact hole; the channel width $W$ is large for given design rules because it extends completely around the V groove. A drive current comparison can be made with TTL current-sinking output drivers of a VMOS and an $n$-MOS product, both using rules of about 6 $\mu$m. The VMOS driver is less than half the size of the $n$-MOS driver (45 mil$^2$ vs 94 mil$^2$) but at the same time can sink 28 mA or 0.6 mA/mil$^2$, while the $n$-MOS driver can sink 16 mA or 0.2 mA/mil$^2$.

Gain of the VMOS transistor is good. When short-channel devices are operating in electron velocity saturation, their gain $g_m$, has an upper limit, given by:

$$g_m = WCV_s \qquad (7.32)$$

where $W$ is the channel width and $C$ is the channel capacitance per unit area. For a given area of active gate region the channel width is greater for VMOS than for other $n$-MOS technologies.

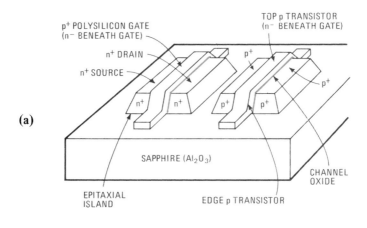

(a)

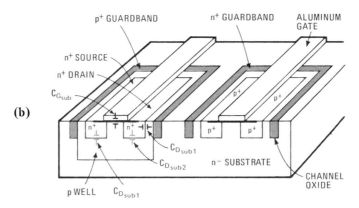

(b)

Fig. 7.34. Comparison of silicon-on-sapphire (SOS) MOSFETS.
(a) Conventional structures
(b) (The parasitic capacitances $C_{D_1}$ and $C_{D_2}$ and the size of $C_{GD}$ in the conventional structure result in lower speed than for the SOS structure. (After Eaton. Reprinted from *Electronics*, June 12, 1975; copyright McGraw-Hill, Inc., NY, 1975.)

However, the VMOS technology is somewhat more complicated than that of planar *n*-MOS technology. The steps required are shown in Fig. 7.36. Figure 7.37 illustrates how several different types of devices may be formed during VMOS processing. In some versions of VMOS where a common ground plane is not required the source and drain are both on the top surface.

The structure and characteristics of a power VMOS (50 V, 2 A) transistor are shown in Fig. 7.38. Designs are now available for ratings of 450 V and 6 A and on-resistances of a few ohms. Such units may be connected in parallel to increase current handling capability since the positive temperature coefficient of the source-drain resistance ensures equal load sharing, or in series to increase

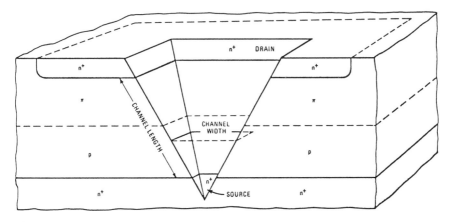

Fig. 7.35. Cross-section of a VMOS transistor. A V-shaped groove is etched into the surface of the Si down through all the $n^+$ and $p$ layers, and the channels formed on the slopes of the groove. To save surface area, the source is in the body of the Si, beneath the drain. (After Jenné; courtesy of American Microsystems, Inc. Reprinted from *Electronics*, August 18, 1977; copyright McGraw-Hill, Inc., NY, 1977.)

the voltage rating (see Fig. 7.39). An 80-W audio-frequency amplifier circuit with 3 paralleled FETs in each output push-pull limb is shown in Fig. 7.40. Other high power VMOS transistors are capable of operation to 200 MHz with greater than 10-W output.

There is considerable activity in the development of high-power high-frequency MOSFET amplifiers and power levels of 50 W at 1 GHz are within reach.[178, 285] Power MOSFETs with large gate widths are capable of controlling many amperes of load current with switching times of about ten nanoseconds (since they are majority carrier devices) compared with a few hundred nanoseconds for BJTs of comparable power handling ability. Thus the role of FET technology is becoming increasingly dominant.[294, 295]

## 7.8 CMOS STRUCTURES AND LOGIC

Various classes of FET logic circuits have developed over the years, one of the most interesting is CMOS that uses the complementary symmetry of $n$ channel and $p$ channel units (see Fig. 7.41) to allow a voltage swing from one transistor to the other with very low power dissipation per gate at standby. The propagation delay is 20-50 nsec. The current rating of an MOSFET at a given voltage is linearly proportional to the ratio of gate width $W$ to channel length $L$. The operating frequency is limited by the ability of the transistors to supply current to charge the node capacitances, and these are primarily the device gate

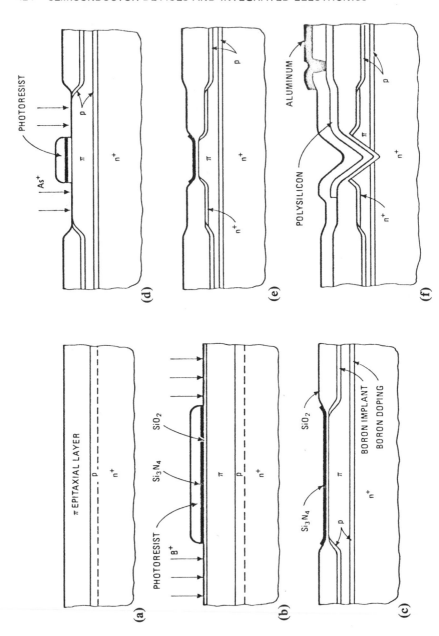

Fig. 7.36. The VMOS process. An $n^+$ substrate, which forms the source, is overlaid with a nonuniformly doped $p$ layer, to which is added an $n$ layer for the drain. Then a V-shaped groove is etched through the epitaxial layer, the exposed surface of which forms the channel. (After Jenné; courtesy of American Microsystems, Inc. Reprinted from *Electronics* August 18, 1977; copyright McGraw-Hill, Inc., NY, 1977.)

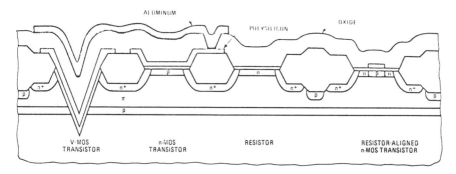

Fig. 7.37.  Several types of devices can be formed during VMOS process; a VMOS transistor, an *n*-MOS transistor, an *n*-channel resistor, and a resistor-aligned *n*-MOS transistor. The last is useful as the pass gate in a random-access memory cell. (After Jenné; courtesy of American Microsystems, Inc. Reprinted from *Electronics*, August 18, 1977; copyright McGraw-Hill, Inc., NY, 1977.)

capacitances.* Therefore, $W \times L$ should be small and $W/L$ should be large for high speed of operation. Hence, the maximum frequency of operation is determined by $L^{-2}$ so $L$ should be small.

The dynamic power dissipation of CMOS is proportional to the load (or node) capacitance, the square of the operating voltage and the operating frequency. So if the first two parameters are set, the power dissipation is directly proportional to the clock frequency.

CMOS has low quiescent power and good noise immunity. A large fan-out implies a larger load capacitance and thus a longer propagation delay. The performance of CMOS ensures a place in future integrated circuit technology. However, the logic circuit packing density is not as good as can be achieved by *n*-MOS and certain other logic families (see Chaps. 8 and 9).

The process steps by which *n* and *p* channel MOSFETs may be produced on the same wafer are shown in Fig. 7.42. Three different CMOS technologies are shown in Fig. 7.43. The planar structure of CMOS II allows a tightening of the dimensions, and the self-alignment of the $p^-$ well allows closer spacing of the *p* and *n*-channel devices. The design rules for the two designs may differ therefore as shown below:

|  | CMOS I | CMOS II |
|---|---|---|
| Si-gate channel length | 7½ μm | 6½ μm |
| Si-gate separation | 6½ μm | 5 μm |
| Al width/separation | 7½ μm | 5 μm |
| Diffusion width/separation | 6½ μm/7½ μm | 5 μm/5 μm |
| Contact windows | 6½ μm × 6½ μm | 5 μm × 5 μm |

*In LSI circuits this may not be true because the interconnecting lead capacitances may play an important role.

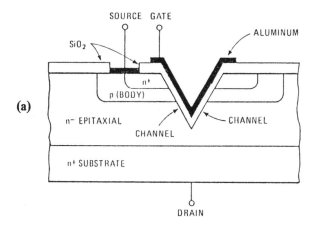

(a)

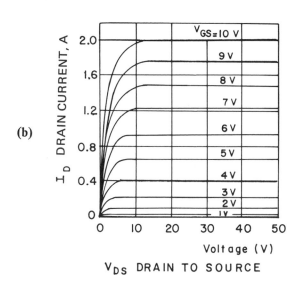

(b)

Fig. 7.38.  High power VMOS transistor:
(a)  Structure; and
(b)  Characteristics. The short channel length of VMOS produces a carrier saturation effect, which, in turn, results in a very linear drain-current vs gate voltage relationship. (After Van Der Kooi and Ragle. Reprinted from *Electronics*, June 24, 1976; copyright McGraw-Hill, Inc., NY, 1976.)

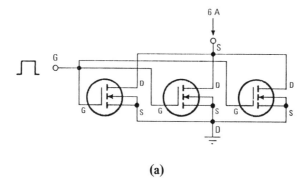

(a)

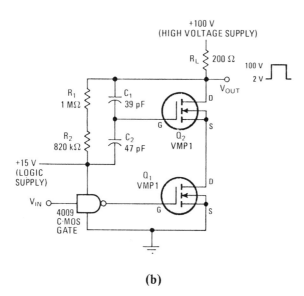

(b)

Fig. 7.39. MOSFETS may be connected in parallel (a) or in series (b) to increase the power controlled. (After Van Der Kooi and Ragle. Reprinted from *Electronics*, June 24, 1976; copyright McGraw-Hill, Inc, NY, 1976.)

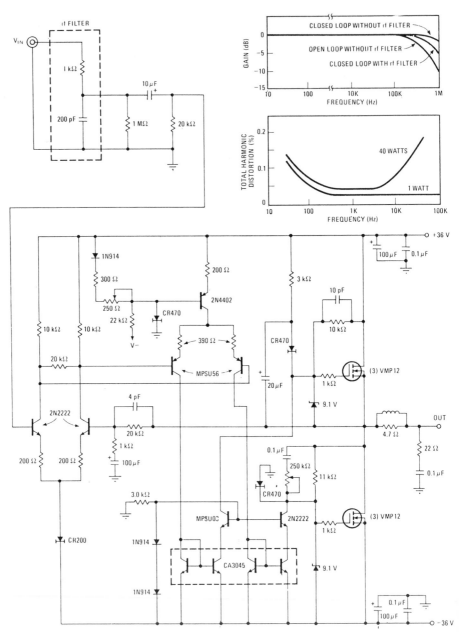

Fig. 7.40. High quality audio FET amplifier. One channel of an 80-watt stereo amplifier uses six power FETs in a push-pull arrangement. A total harmonic distortion of less than 0.04% is achieved with only 22 dB of negative feedback, compared to about 40 dB usually used with bipolar transistor stages. (After Van Der Kooi and Ragle. Reprinted from *Electronics*, June 24, 1976; copyright McGraw-Hill, Inc., NY, 1976.)

$V_{DD}$ = + I. 5 V

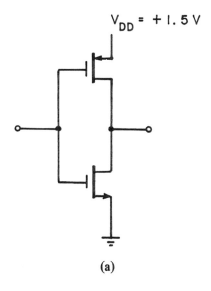

**(a)**

+ $V_{DD}$

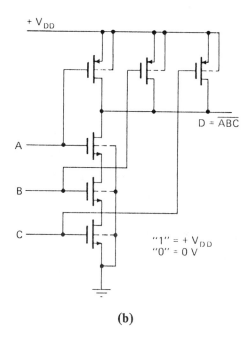

D = $\overline{ABC}$

"1" = + $V_{DD}$
"0" = 0 V

**(b)**

Fig. 7.41. CMOS LOGIC
(a)   Building block
(b)   Typical NAND Gate (After Weber, *Large and Medium Scale Integration*, copyright 1974, McGraw-Hill. Reprinted with permission of McGraw-Hill Book Company, NY.)

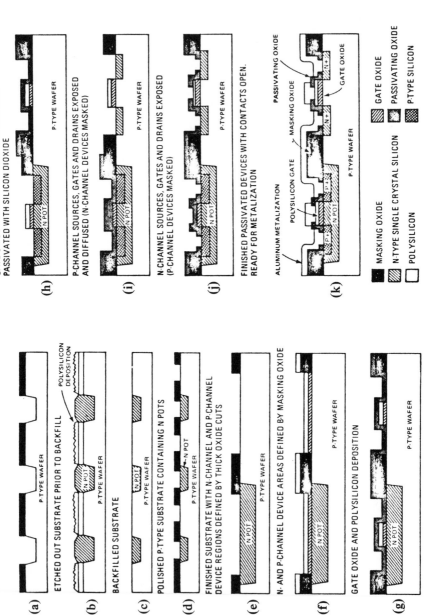

Fig. 7.42. How $n$ and $p$ channel Si gate MOSFETs may be achieved on the same wafer. (After Burgess and Daniels, in: *Large and Medium Scale Integration*, S. Weber (ed.), copyright 1974, McGraw-Hill. Reprinted with permission of McGraw-Hill Book Company, NY.)

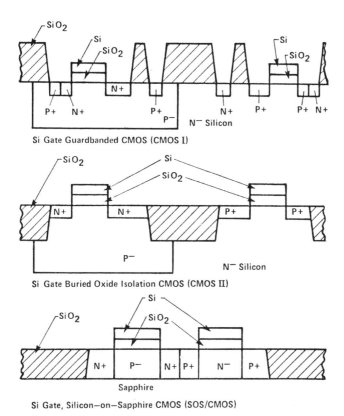

Si Gate Guardbanded CMOS (CMOS I)

Si Gate Buried Oxide Isolation CMOS (CMOS II)

Si Gate, Silicon–on–Sapphire CMOS (SOS/CMOS)

Fig. 7.43. Comparison of three complementary-symmetry CMOS technologies. (After Aitken, et al. Reprinted with permission from *International IEEE Electron Devices Meeting Technical Digest*, 1976, p. 329.)

In CMOS II there is greater freedom of layout and the result is that in a typical logic circuit it requires less than half the area of CMOS I. More recently there has been further miniaturization of CMOS towards 3½ $\mu$m design rules. The parasitic capacitances of CMOS II are lower than for CMOS I but not as low as for SOS/CMOS. However, SOS/CMOS does not generally outperform CMOS II with regard to speed except to a marginal extent because the electron mobility in the 0.6 $\mu$m thin SOS structure tends to be lower (typically 450 cm$^2$/volt-sec compared with 550 cm$^2$/volt-sec) and there are floating substrate effects that influence the threshold voltage. Bipolar transistors may be produced on the same wafer as CMOS by the 8-mask process illustrated in Fig. 7.44. CMOS integrated circuits may also be made with short double-diffused (DMOS) channels.[170]

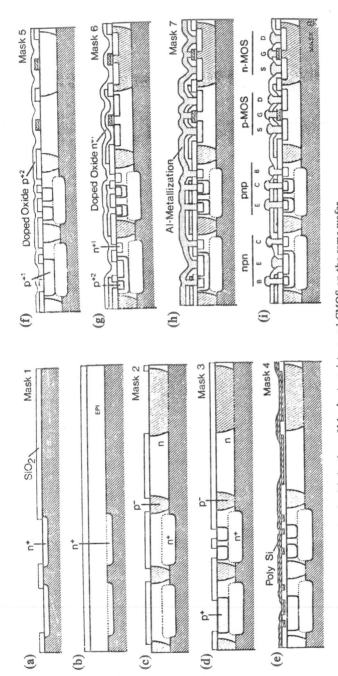

Fig. 7.44.  An 8 mask process for the fabrication of bipolar transistors and CMOS on the same wafer.

(a)  Window etching and buried layer diffusion;
(b)  $n$ epitaxial layer deposition and oxide growth;
(c)  Window etching and $p^-$ diffusion;
(d)  $p^+$ diffusion.
(e)  Opening of $n^{+1}$ and $p^{+2}$ regions, gate oxidation and polycrys-talline Si deposition;
(f)  Poly-Si delineation, window opening $n^{-1}$ and $p^{+1}$; Doped oxide deposition $p^{+2}$;

(g)  Window etching for $n^{+1}$ regions, doped oxide deposition $n^{+1}$, simultaneous diffusion of $p^{+2}$ and $n^{+1}$ regions;
(h)  Etching of contact windows and evaporation of Al;
(i)  Delineation of contacts and interconnections. (After Darwish and Tauberest. Reprinted by permission of the publisher, The Electrochemical Society, from *J. of the Electrochemical Society*, **121**, 1974, p. 1120.)

A technique known as C$^2$MOS is shown in Fig. 7.45. The circuit operates as a NAND gate when clock pulses are high and retains previous levels when the clock pulses are low. This technique differs from standard CMOS in being static, in the sense that cell contents are retained statically, yet dynamic, in the sense that peripheral circuits are cycled during each transient time. (To do this, row decoders, sense circuits, and input/output controllers are designed with clocked gates and half-bit inverters.)

This technique, developed at Toshiba, yields a fully static 4-k RAM (random access memory) chip that, at only 4.7 mm$^2$, is no larger than equivalent $n$-MOS 4-k RAMs yet dissipates one fifth the power. The standby current of 0.1 $\mu$A is a fraction of that needed by $n$-MOS RAMs, and the delays are in the low nanoseconds.

Another development is the back-gated MOSFET that can operate at input voltage levels as low as 0.1 V with application of the signal to the substrate and biasing the normal gate to a suitable potential. An Hitachi B-MOS inverter is constructed out of an $n$-channel B-MOS driver and a conventional $p$-MOS current source (see Fig. 7.46). In configuration, it resembles a C-MOS inverter, in which an $n$-channel element acts in conjunction with a $p$-channel element. But it is more compact than conventional low-power C-MOS, where $n$ and $p$ transistors lie side-by-side. The $p$-type well of the back-gate input terminal slips into the $n$-type substrate needed for the output transistor. Just as in integrated injection logic, the B-MOS transistor pair merges into the space of a single transistor, keeping the level of power consumption low and the level of circuit integration high.

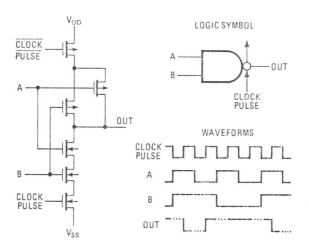

Fig. 7.45. Clocked CMOS logic NAND gate: the concept is known as C$^2$MOS. (Courtesy *Electronics*; Reprinted from *Electronics*, June 9, 1977; copyright McGraw-Hill, Inc., NY, 1977.)

Any MOS logic family is likely to contain the following components (and more):

- NAND gate
- NOR gate
- AND-OR gate
- Full adder with parallel carry
- Buffer/converter (inverting or noninverting)
- Static shift register
- Presettable divide-by-N counter
- JK master-slave flip-flop
- D flip-flop with set/reset
- BCD-to-decimal decoder
- Latch circuits
- Multiplexers
- Liquid crystal display driver
- Light-emitting diode display driver
- Random access memories (RAM)
- Read only memories (ROM)
- Programmable memories (PROM)
- Analog-to-digital encoders.

Although available space does not allow discussion of a complete logic family, the circuit schematics and diagrams of conventional CMOS, NOR and AND-OR gates are given in Fig. 7.47 to illustrate the practical realization of such functions. The JK flip-flop illustrated in Fig. 7.48 is seen to involve about 40 transistors.

## 7.9 COMPUTER AIDED DESIGN OF MOSFET CIRCUITS FOR LARGE SCALE INTEGRATION

In the design of MOSFET integrated circuits, computer studies must be made of the steady-state and transient characteristics of circuits containing many tens of transistors. It is also important to be able to predict changes of these characteristics with scaling of the device in size and with temperature. The device model for a single transistor therefore must be sophisticated enough to include second-order effects such as short-channel charge effects, overlap capacitance, side-wall and substrate capacitances and diodes, mobility dependence on field strength, sub-threshold leakage weak-inversion effects and surface-state effects.

In the USA a computer-aided-design program known as SPICE is widely used and the IGFET model in this is shown in Fig. 7.49. This model (sometimes termed CSIM for compact short-channel IGFET model) is the result of contributions from many workers including Thomasco, Chawla, Poon, and Scharfetter. The gate is represented by three capacitances $C_{GD}$, $C_{GS}$ and $C_{GB}$ modeled by

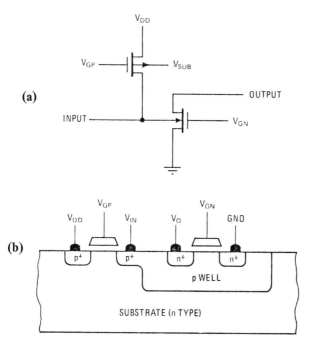

Fig. 7.46. A BMOS inverter uses a back-gated MOS transistor driver with a p-MOS current source. This allows a merged structure with low threshold voltages and microwatt power dissipation, but at the expense of extra voltage supplies. (Courtesy *Electronics*; reprinted from *Electronics*, June 9, 1977; copyright McGraw-Hill, Inc., NY, 1977.)

parameters such as the overlap capacitance or distance, $C_{ov}$ or $D_{ov}$; the oxide thickness, entered as $C_{ox}$, per unit area; gate length $L$ and gate width $W$. The substrate capacitors $C_{BD}$ and $C_{BS}$ are depletion-layer capacitances derived from the zero-bias substrate capacitance, $C_{BG}$ per unit area, and other parameters that take account of its variation with voltage. The two substrate diodes are modeled by equations of the usual form,

$$I_{BD} = I_{SD} \left[\exp\left(V_{BD}/V_t\right) - 1\right] \qquad (7.33a)$$

$$I_{BS} = I_{SS} \left[\exp\left(V_{BS}/V_t\right) - 1\right] \qquad (7.33b)$$

where $I_{SD}$ and $I_{SS}$ are selected to fit the junction leakage current rather than the forward-bias diode voltage drop and $V_t$ is the thermal voltage $kT/q$.

The non-linear current source $I_D$ in the triode region of the characteristics

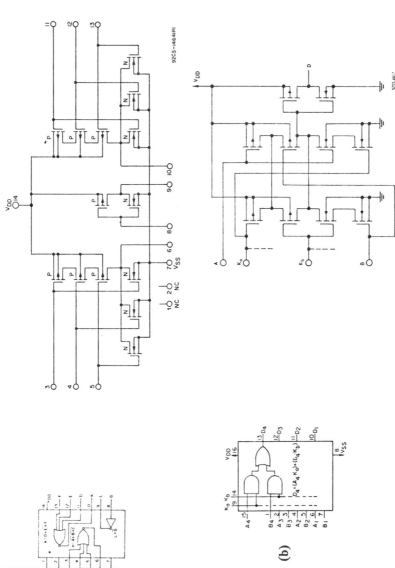

Fig. 7.47. CMOS NOR and AND-OR gates:
(a) NOR gate;
(b) AND-OR SELECT gate. (Selection is by the control bit $K_a$, $K_b$.) (Courtesy RCA Solid State Division.)

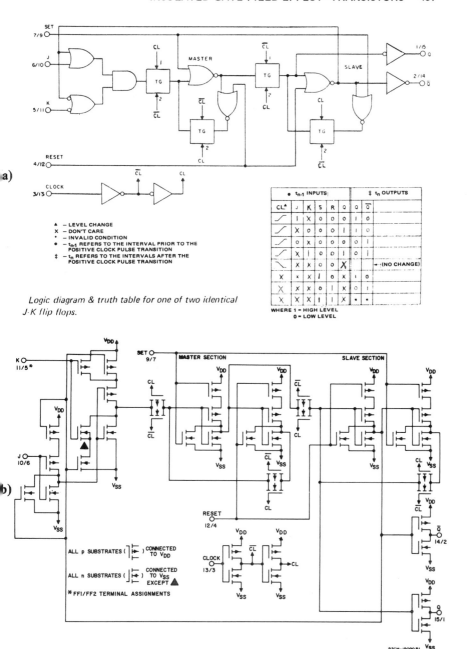

Logic diagram & truth table for one of two identical J-K flip flops.

| • $t_{n-1}$ INPUTS | | | | | | ‡ $t_n$ OUTPUTS | |
|---|---|---|---|---|---|---|---|
| CL▲ | J | K | S | R | Q | Q | $\overline{Q}$ |
| ⟋ | I | X | 0 | 0 | 0 | I | 0 |
| ⟋ | X | 0 | 0 | 0 | I | I | 0 |
| ⟋ | 0 | X | 0 | 0 | 0 | 0 | I |
| ⟋ | X | I | 0 | 0 | I | 0 | I |
| ⟍ | X | X | 0 | 0 | X | →·(NO CHANGE) | |
| X | x | X | I | 0 | x | I | 0 |
| X | X | X | 0 | I | X | 0 | I |
| X | X | X | I | I | X | • | • |

WHERE 1 = HIGH LEVEL
0 = LOW LEVEL

▲ – LEVEL CHANGE
X – DON'T CARE
* – INVALID CONDITION
• – $t_{n-1}$ REFERS TO THE INTERVAL PRIOR TO THE POSITIVE CLOCK PULSE TRANSITION
‡ – $t_n$ REFERS TO THE INTERVALS AFTER THE POSITIVE CLOCK PULSE TRANSITION

Schematic diagram for one of two identical J-K flip flops.

Fig. 7.48.  Logic diagram and circuit diagram for a CMOS *JK* flip flop. (Courtesy RCA Solid State Division.)

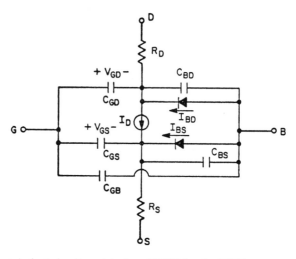

Fig. 7.49. The equivalent circuit model of an IGFET for the SPICE computer-aided-design program. (The model for a JFET is similar except that B takes on the function of $G$ and the left-hand part of the circuit is omitted.) (After Poon. Published with permission of the American Telephone and Telegraph Company.)

before saturation is modeled by an equation of the form

$$
I_D = \beta \left\{ \left[ V_{GS} - V_{FB} - 2\phi_f - \frac{V_{DS}}{2} \right] V_{DS} \right.
$$

$$
\left. - \frac{2}{3} K_1 \left[ \left( 2\phi_f - V_{DS} - V_{BS} \right)^{3/2} - \left( 2\phi_f - V_{BS} \right)^{3/2} \right] \right\} \qquad (7.34)
$$

where

$$
\beta = \mu C_{ox} \frac{W}{L} \qquad (7.35)
$$

and

$$
K_1 = \frac{(2\epsilon\epsilon_0 q N_A)^{1/2}}{C_{ox}} \qquad (7.36)
$$

In the saturated region of the characteristic the equation for $I_D$ becomes

$$
I_D = I_{sat} \left\{ 1 + \sqrt{\frac{2\epsilon\epsilon_0}{qN_A}} \left[ \left( V_{DS} - V_{BS} \right)^{1/2} - \left( V_{sat} - V_{BS} \right)^{1/2} \right] \right\} \qquad (7.37)
$$

The variation of channel mobility with voltage is represented as dependent on a mobility degradation parameter $V_0$ where

$$\mu = \frac{\mu_0}{1 + \dfrac{V_{GS} - V_{TH}}{2V_0}} \tag{7.38}$$

Short channel effects may be included by taking

$$V_{TH} = V_{FB} + 2\phi_f + \frac{(2\epsilon\epsilon_0 q N_A)^{1/2}\,(2\phi_f - V_{BS})^{1/2}}{HC_{0x}} \tag{7.39}$$

with

$$H = 1 + \frac{0.8\,D_j}{L}\left\{\left[\left(\frac{2\epsilon\epsilon_0}{qN_A}\right)\left(\frac{L}{D_j}\right)^2\left(\phi_B - V_{BS}\right)\right]^{2/5} - 0.43\right\} \tag{7.40}$$

where $D_j$ is a parameter chosen to correct for the dependence of $V_{th}$ on $L$ and on $(V_{BG} + 2\phi_f)$ in a short channel device. Because of the shape of the electric field distribution in a short device, crowding of the field lines may produce the effect of a lowering of the threshold voltage when $L$ is decreased.

The expression for $I_D$ in the weak inversion subthreshold leakage condition [(see Fig. 7.4(d)] follows the treatment of Swanson and Meindl.[247] Parameters $M$ and $N$ are introduced where

$$I_D = \frac{Z}{L}\,\mu_n\,C_0\,\frac{1}{m}\left(n\frac{kT}{q}\right)\exp\left\{q\left[(1-\eta)(V_G - V_T)\right.\right.$$
$$\left.\left. + \eta V_D - \frac{nkT}{q}\right]\Big/nkT\right\}\cdot\left\{1 - \exp(-mqV_D/nkT)\right\} \tag{7.41}$$

where $\qquad \eta = \dfrac{\epsilon_s\,t_{ox}}{\pi\epsilon_{ox}L}\qquad n = \dfrac{C_0 + C_d, + C_{fs}}{C_0}\qquad m = \dfrac{C_0 + C_d}{C_0}$

for $\qquad\qquad V_{GS} \leq V_T + n\left(\dfrac{kT}{q}\right) \tag{7.42}$

A set of SPICE parameters for an IGFET model, with some typical default values, follows in Table 7.2.

The agreement that may be achieved in representing the static characteristics of a transistor by such parameters may be seen in Fig. 7.50. The SPICE model is also successful in predicting the transient response of FET logic circuits and

Table 7.2.  SPICE Parameters for an IGFET

| Parameter | Definition | Default | Unit |
|---|---|---|---|
| L | Channel length | $1 \times 10^{-3}$ | cm |
| W | Channel width | $1 \times 10^{-2}$ | cm |
| BETA | Transconductance parameter | $1 \times 10^{-5}$ | $a/v^2$ |
| VTO[a] | Zero-bias threshold voltage | 0 | volt |
| DOP | Substrate doping | $1 \times 10^{15}$ | $cm^{-3}$ |
| COX | Oxide capacitance, per unit area | $1.7 \times 10^{-7}$ | $f/cm^2$ |
| VO | Mobility degradation parameter | 20 | volt |
| CBG | Zero-bias substrate junction capacitance, per unit area | 0 | $f/cm^2$ |
| CJD | Zero-bias drain-substrate capacitance | * | farad |
| CJS | Zero-bias source-substrate capacitance | * | farad |
| PB | Substrate junction potential | 0.75 | volt |
| MB | Substrate junction grading coefficient | 0.5 | − |
| BB | Maximum substrate junction depletion capacitance parameter | $1 \times 10^{-3}$ | − |
| COV | Source or drain overlap capacitance | * | farad |
| ISD | Drain junction saturation current | $1 \times 10^{-15}$ | amp |
| ISS | Source junction saturation current | $1 \times 10^{-15}$ | amp |
| RD | Extrinsic drain resistance | 0 | ohm |
| RS | Extrinsic source resistance | 0 | ohm |
| M | Subthreshold leakage-current coefficient | 1000 | − |
| N | Subthreshold leakage-current exponent | 3 | − |
| VTDS | Subthreshold leakage-current drain-source threshold voltage | $1 \times 10^6$ | volt |
| DJ | Drain and source junction depth | $5 \times 10^{-5}$ | cm |
| EA | Activation energy | 1.21 | eV |
| TO | Temperature at which the model parameters are measured | 25 | °C |

For a 2N4869 depletion mode n-channel Si JFET a limited set of parameters might be: BETA = 0.77 × $10^{-3}$, VTO = −1.8, LAMBDA (channel-length modulation parameter) = 4 × $10^{-3}$, CJD = 25pF, CJS = 25pF, PB = 0.6, $I_s$ = 0.25 × $10^{-9}$ A (bulk junction reverse−saturation current), $R_D$ = $R_s$ = 0 − 100 ohms.

[a] A positive value of VTO designates an enhancement mode device, and a negative value of VTO designates a depletion mode device.

the amplitude and phase response of analog FET circuits. Noise analysis can also be built into the program. The SPICE program can be run well on a computer such as the Honeywell HIS 6070 and analyze circuits with up to 1200 nodes and 600 devices. On a smaller computer such as the CDC CYBER-72 the implementation may be limited to 200 nodes and 100 devices.

Another computer-aided-design program in operation in the USA is the MSINC program of Young and Dutton.[287] The program is structured modularly so that the built-in model can be readily changed. A version of the program Mini-MSINC has been devised that runs on the HP 2100-series minicomputer

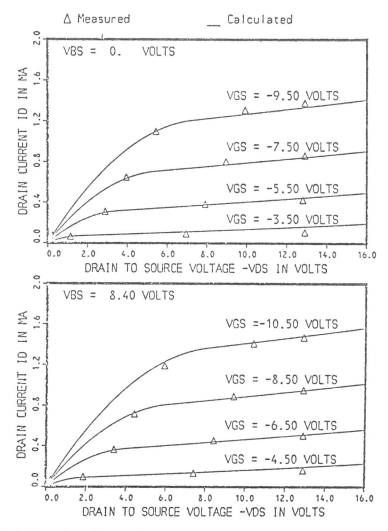

Fig. 7.50. Comparison of measured and calculated Si IGFET characteristics using the SPICE computer model. (After Poon. Published with permission of the American Telephone and Telegraph Company.)

with 32K words of memory and this is capable of handling circuits with many tens of MOS transistors.[288] The second-order effects implemented include 1) bulk-charge effects, 2) channel length modulation, 3) channel carrier mobility reduction resulting from large normal electric field, 4) carrier velocity saturation, and 5) voltage-dependent gate capacitances.

Simulations of MOS circuits require both process and geometry dependent parameters. In MOS circuit design, the process dependent parameters are considered separately from the geometry dependent parameters. For a given process, the designer changes device scaling to alter circuit performance objectives separately from specific circuit topologies. With this in mind, geometry-dependent parameters and process-dependent parameters are entered in Mini-MSINC via separate inputs.

An input program for Mini-MSINC is given in Fig. 7.51(a). The transistor cards $Q_1$, $Q_2$ and $Q_3$ enter the transistor parameters that are geometry dependent and indicate the nodes between which the drain, gate and source are connected.

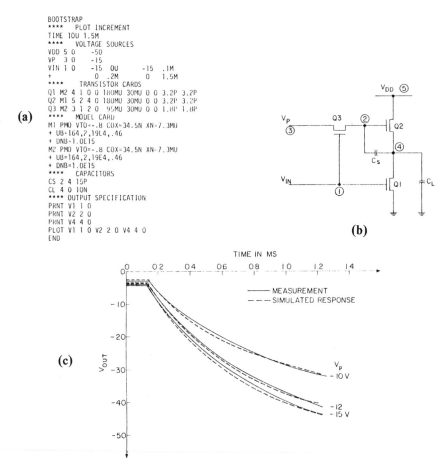

(a)
```
BOOTSTRAP
****    PLOT INCREMENT
TIME  10U  1.5M
****    VOLTAGE SOURCES
VDD 5 0    -50
VP   3 0    -15
VIN 1 0    -15   0U      -15  .1M
+            0  .2M       0  1.5M
****    TRANSISTOR CARDS
Q1 M2 4 1 0 0 180MU 30MU 0 0 3.2P 3.2P
Q2 M1 5 2 4 0 180MU 30MU 0 0 3.2P 3.2P
Q3 M2 3 1 2 0 95MU 30MU 0 0 1.8P 1.8P
****    MODEL CARD
M1 PMO VTO=-.8 COX=34.5N XN-7.3MU
+ UB=164,2,19L4,.46
+ DNB-1.0E15
M2 PMO VTO=-.8 COX=34.5N XN-7.3MU
+ UB=164,2,19E4,.46
+ DNB=1.0E15
****    CAPACITORS
CS 2 4 15P
CL 4 0 10N
****    OUTPUT SPECIFICATION
PRNT V1 1 0
PRNT V2 2 0
PRNT V4 4 0
PLOT V1 1 0 V2 2 0 V4 4 0
END
```

(b)

(c)

Fig. 7.51. Circuit simulation using the Mini-MSINC program:
(a)  An input deck of program cards;
(b)  The bootstrap circuit being analyzed;
(c)  Measured and simulated responses for different precharge voltages. (After Young and Dutton; reprinted with permission from *IEEE Journal of Solid-State Circuits*, **SC-11**, 1976, p. 731.)

The model cards $M_1$ and $M_2$ enter the process dependent parameters. Figure 7.51(b) shows the circuit being analyzed, a bootstrap circuit driving a large capacitance load ($C_L$ = 10,000 pF), and Fig. 7.51(c) shows the measured and simulated transient responses for three precharge voltages. The simulation requires less than 1 min. of minicomputer time. A test circuit of 26 transistors with 150 time-points can be simulated in 10 min.

### 7.10 MOSFET SWITCHES

Turn-on behavior of MOSFETs depends on the gate drive, the transistor channel length, the substrate doping, the parasitic capacitances and the drain load. Figure 7.52(a) illustrates the buildup of hole density in a $p$-channel MOSFET during switching-on in the saturated region of the $I_C - V_{DS}$ characteristics. There is a delay time in the growth of $I_D$ [see Fig. 7.52(b)]. The spike shown in the $V_{DS}$ curve occurs because part of the gate pulse is transferred to the drain via the overlap capacitance.

Several types of MOS switches or driver gates are shown in Fig. 7.53. The TTL logic signals (0.8 V to 2.0 V) are amplified in a level-shifting stage to provide the −20 V to −10 V gate signals required to turn the MOSFET ON or OFF. These circuits use $p$-channel devices and the ON resistance depends on the drive voltage. Two enhancement mode MOSFETs in the CMOS configuration of Fig. 7.54, however, have an ON resistance that is fairly constant. The $p$-MOS and $n$-MOS FETs are in parallel and may be turned ON by driving the $p$ channel gate negative and the $n$ channel gate positive simultaneously [see Fig. 7.54(a)]. In Fig. 7.54(b) only the $p$-MOS is ON and in 7.54(c) only the $n$-MOS is ON. All three ON conditions have about the same resistance, since the drive $V_{GS}$ is 30 V when a single transistor is ON and only 15 V when the two transistors are ON in parallel.

The turn-on transient of a 1-A V-MOS FET switch is shown in Fig. 7.55. The response is seen to be quite fast for the power levels involved and the drive from CMOS logic circuits.

MOS switches are also used in 2-GHz microwave systems in place of $p$-$i$-$n$ diodes since no bias current is needed to drive them.

### 7.11 NOISE IN MOSFETs

MOS transistors have an extensive region of $f^{-1}$ noise that is related to fluctuations caused by carriers in the channel tunneling into and out of traps distributed in the oxide-semiconductor interface region. The noise power expression is

$$v_n^2 = \left( \frac{q \, V_{DS} \bar{\mu}_p W}{g_m L} \right)^2 \left( \frac{kT \bar{N}_{SS}}{WL \alpha D_t f} \right) \tag{7.43}$$

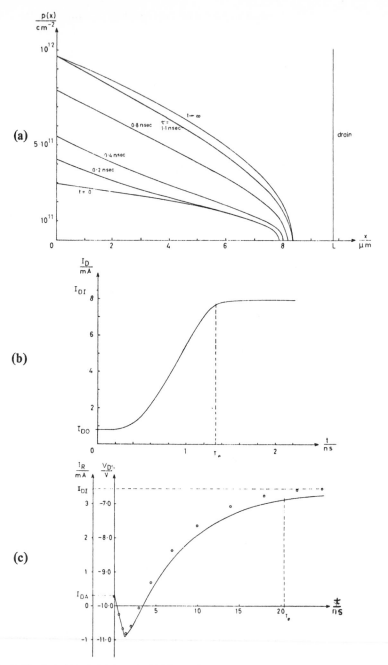

Fig. 7.52. Switching behavior of a MOS transistor:
(a) The calculated carrier concentrations in the inversion layer, if a MOST is turned on in the saturation region ($L$ = 0.8 μm, $W$ = 0.8 mm, $d_{ox}$ = 1290 Å, $N_D$ = 6 × $10^{14}$ cm$^{-3}$, $\mu_p$ = 230 cm$^2$ V$^{-1}$ sec$^{-1}$, $V_T$ = −3.2 V, $R$ = 100Ω, $V_B$ = −10 V, $V_{G0}$ = −5 V, $V_{G1}$ = −9V);

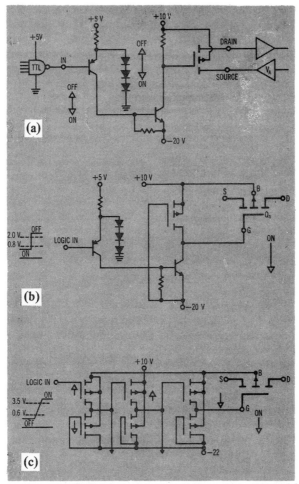

Fig. 7.53. Three types of *p*-MOS driver gates or switches: a basic FET and driver (a), FETs combined with bipolars (b), and an all-*p*-MOS unit (c). (After Boag. Reprinted with permission from *Electronics Industry* and *Electronics Products Magazine*, Jan. 15, 1973, p. 91.)

(b)   The drain current during the switching process; and

(c)   Solid line: calculation of the transient of the MOST with consideration of the parasitic capacitances. The circles correspond to measurements. (After Zahn.[290] Reprinted with permission from *Solid-State Electronics*, Pergamon Press Ltd.)

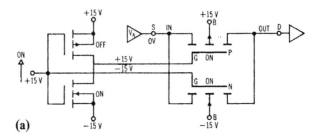

**(a)**

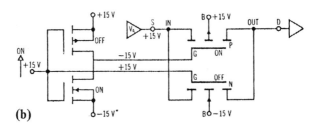

**(b)**

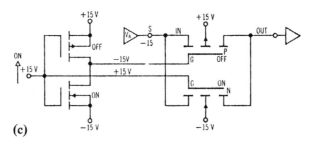

**(c)**

Fig. 7.54. This *p*-MOS and *n*-MOS parallel driver gate switch has the advantage of remaining constant (about 70 ohms) in ON resistance for the three different drive conditions shown. (After Boag. Reprinted with permission from *Electronics Industry* and *Electronics Products Magazine*, Jan. 15, 1973, p. 91.)

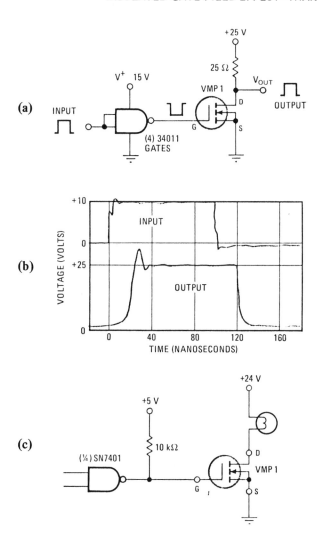

Fig. 7.55. Logic compatible. Four paralleled CMOS logic gates provide sufficient drive to turn on the 1-ampere VMOS FET switch (a) in 20 nsec; (b) Using only a single logic gate increases switching time to about 50 nsec. Open-collector TTL logic can easily turn on the VMOS FET driving the lamp circuit in (c). In such applications the TTL pull-up resistor can range from 1 kilohm to 100 kilohms. (After Vander Kooi and Ragle. Reprinted from *Electronics*, June 24, 1976; copyright McGraw-Hill, Inc., NY, 1976.)

where $\alpha$ is a tunnel probability parameter, $D_t$ is the maximum depth of traps in the oxide and $W$ is the channel width.

There is an explicit dependence on bias through the $V_{DS}\bar{\mu}_p/g_m$ term, and an implicit dependence on bias through $\bar{N}_{SS}$. The quantity $V_{DS}\bar{\mu}_p/g_m$ is constant when the following drain current approximation is valid,

$$I_D \simeq \frac{W}{L}\,\bar{\mu}_p\,C_0\left[(V_{GS} - V_T)\,V_{DS} - \frac{V_{DS}^2}{2}\right] \qquad (7.44)$$

Evaluation of $g_m$ from Eq. (7.44) then gives[13]

$$v_n^2 = \left(\frac{q}{C_0}\right)^2 \left(\frac{kT\bar{N}_{SS}}{WL\alpha D_t f}\right)$$

Generation-recombination noise may show at midband frequencies.[279] Ion implantation can contribute to noise by generation-recombination at residual damage sites. This is apparent at low drain currents (see Fig. 7.56).

Radiation effects in MOSFETs (associated with fast neutrons, high energy electrons and $\gamma$ rays) are partially understood[98] but problems remain.

## 7.12  SPECIAL PURPOSE MOSFETs

Semiconductor memory circuits are the subject of much specialized technology. This is covered briefly in Chap. 9. The discussion that follows is confined to two unusual versions of MOSFETs.

MOSFETs may be used as infrared sensing devices if the $p$ substrate contains a deionized deep impurity, such as Ga, In, or Au, in addition to the background (B) doping. The structure is shown in Fig. 7.57 and is operated as an integrating photon detector at low temperatures (24-77°K). The energy level of Ga in Si is 0.072 eV above the valence band.

The device is reset, or first preset, by applying a large negative gate voltage to accumulate at the surface and fill all impurity centers with holes. For In- and Ga-doped devices the impurity centers are in the neutral charge state following the reset operation. The application of infrared radiation causes photoionization of an electron from the valence band to the impurity (this may be stated in an alternative way as the photoemission of a hole). The change of the impurity to the negative charge-state means that for a given gate voltage the number of free electrons in the channel must decrease. The conductivity of the device therefore decreases in response to the illumination, as seen in Fig. 7.57(b).

For wavelengths in the 8-14-$\mu$m range a Ga-doped Si self-scanning infrared imaging array is capable of a $3 \times 10^{10}$ photon cm$^{-2}$sec$^{-1}$ sensitivity level for a 10 sec integration time.[83] This is better than the sensitivity of photoconductor

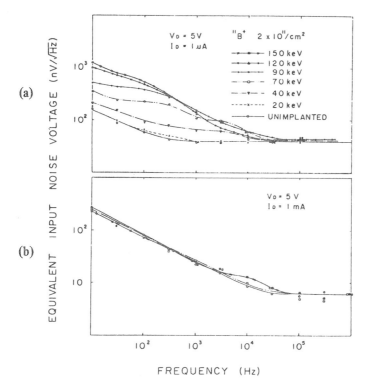

Fig. 7.56. Frequency characteristics of equivalent input noise voltages of $^{11}$B$^+$-implanted p-channel MOS transistors annealed at 1000°C for 10 min. measured at drain currents of (a) 1 $\mu$A and (b) 1 mA with acceleration energies as parameters. (After Nakamura, Kudoh, Kamashida. Reprinted with permission from *J. Appl. Phys.*, 46, 1975, p. 3189.)

systems for this wavelength range. Indium doped MOSFETs have given sensitivities of 1 mA/microwatt for photon wavelengths of about 4 $\mu$m (the energy level of In in Si being at $E_V + 0.16$ eV). The Au donor level in Si is at $E_V + 0.35$ eV and the wavelength range of interest for Au-doped MOSFETs is 1.4-1.8 $\mu$m.

Long integration times are needed for low photon flux conditions and thermally generated $f^{-1}$ noise may be a limitation under these conditions. Under shot noise or background limited operation the maximum attainable signal-to-noise ratio is determined only by the device area and surface charge storage capability and is independent of flux, integration period, and quantum efficiency.

MOSFETs may be used as strain sensing transducers, as Hall effect devices, and as gas sensitive devices. These applications are discussed in Chap. 15.

In general, semiconductor devices are not used in conjunction with moving

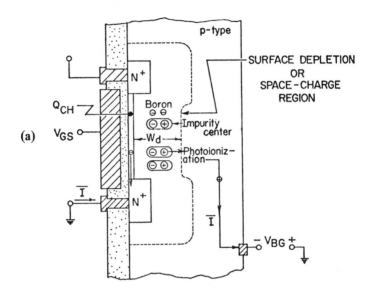

(a)

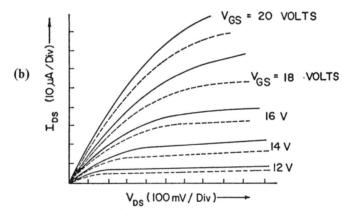

(b)

Fig. 7.57. An infrared sensing MOSFET (IRFET):
(a) The MOSFET structure contains a deep impurity in the depletion region the charge state of which is changed by infrared radiation so that the transistor current changes; and
(b) IRFET with Ga as the deep impurity undergoes characteristic changes such as shown in the presence of 8-14 μm radiation (24.5°K). (After Wittmer et al. Reprinted with permission from *International IEEE Electron Devices Meeting Technical Digest*, 1975, p. 512.)

parts (except, perhaps, for Hall sensitive semiconductor devices moving relative to a magnetic field). Therefore, the MOSFET geometry of Fig. 7.58 with a vibrating gate electrode is an interesting concept. Cantilever gate electrodes of Au about 0.1-cm long, can be made resonant at 3 kHz with $Q$s of about 150 (20-Hz bandwidth). Such structures may be used in tone detection circuits, for filter circuits, as micropower electrostatic relays and as detectors of movement in accelerometers.[82, 183, 296]

## 7.13 QUESTIONS

1. Consider an MOS capacitor (Al − SiO$_2$ − $n$ Si $10^{14}$ cm$^{-3}$) of area 1 cm$^2$. The Al is 1000-Å thick, the SiO$_2$ is 2000-Å thick, undoped and free of charge, and the Si is an epi-layer doped $10^{14}$ cm$^{-3}$ and 20-$\mu$m thick on an $n^+$ $10^{18}$ cm$^{-3}$ substrate. The work function of Al may be taken as 4.2 eV and the $\phi_{ms} = V_{ms}$ barrier may be found from *J. Phys. Chem. Solids*, **27**, 1966, p. 1873. Neglect any interface charge between the SiO$_2$ and the Si. Sketch the capacitance variation as a function of voltage applied to the Al (both polarities) with $C$ in $\mu$F and $V$ in volts. If the oxide contains $10^{12}$ Na$^+$ ions resketch $C$ vs $V$ for (a) all the ions at the SiO$_2$-Si interface, (b) the ions uniformly distributed through the oxide and (c) all the ions at the Al-SiO$_2$ interface.

2. An Al gate, enhancement-mode, $n$ channel Si MOSFET has an insulator 1000-Å thick of clean SiO$_2$ (charge free). The gate is 10-$\mu$m long and 100-$\mu$m wide. Make any assumptions that seem appropriate for the SiO$_2$/Si interface charge. The $p$ region in which the $n$ channel is formed is doped $4 \times 10^{14}$ cm$^{-3}$. What is the threshold voltage of the transistor? What is a suitable drain to source voltage? Sketch the transfer characteristics and transconductance under typical conditions. What is the frequency limitation for the transistor?

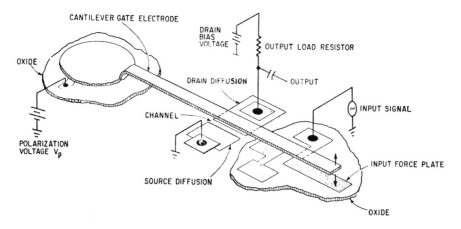

Fig. 7.58. A MOSFET with a mechanically resonant Au-beam gate which is set vibrating by the input signal to the force plate. (After Nathanson, Newell, Wickstrom. Reprinted from *Electronics*, Sept. 20, 1965; copyright McGraw-Hill, Inc., NY, 1965.)

3. The gate material of the transistor in problem 2 is changed to heavily doped poly-Si— what is the effect on the characteristics? If the $p$ region is also changed to a doping of $4 \times 10^{15}$ cm$^{-3}$ what additional changes occur in the characteristics?

4. An Al gate enhancement mode $p$ channel Si MOSFET is desired with characteristics (low frequency to 10 kHz) as similar as possible to the $n$ channel unit of problem 2. How will this be designed? In which significant respects will the $p$-channel and $n$-channel transistors differ in behavior?

5. A dry oxide is grown at 1000°C on (110) Si for 3 hr. What color is seen if the oxide is illuminated at 90° with daylight? If the daylight is collimated and applied at 45° to the surface? If tungsten light is used?

6. A capacitance vs voltage curve for $5.5 \times 10^{16}$ cm$^{-3}$ $p$ type Si is shown in Fig. 7.27. Verify the position of the theoretical curve and also show that the observed curve(s) correspond to a surface state density of $7 \times 10^{11}$ cm$^{-2}$.

7. Calculate the Zerbst plot that might be expected for the oxide of Fig. 7.27 if the bulk electron lifetime in the $p$ type Si is $10^{-6}$ sec. and if the SiO$_2$-Si interface recombination velocity is $10^3$ cm/sec. Does the recombination velocity result in a conductance in parallel with the capacitance?

8. It is stated that a V-MOS driver may be less than half the size of an $n$-MOS driver (45 mil$^2$ vs 94 mil$^2$) but V-MOS can sink 0.6 mA/mil$^2$ compared with 0.2 mA/mil$^2$ for $n$-MOS. Explain.

9. The diagram of an SR flip-flop consisting of NAND gates and provided with means for presetting or clearing is shown below.

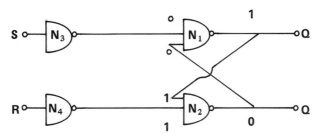

On squared paper draw a layout of this flip-flop using typical 7-$\mu$m design rules for (a) for Al gate $n$ channel logic, and (b) for Al gate CMOS. Compare the size of the two designs and the number of major process needed.

10. If a clocked SR flip-flop is required so that reset is in synchronism with the clock pulses, $N_3$ and $N_4$ are modified to be two-input NAND gates with the clock train applied to the paralleled extra input leads. Explain the mode of operation of the flip-flop, and draw the truth table. Design a layout for this in CMOS. How relevant is this to the C$^2$MOS of Fig. 7.45?

11. Draw a three or four-stage IGFET shift register for which the stages form a closed loop. Using the SPICE parameters shown in Table 7.2 and making a reasonable allowance for load capacitance estimate the frequency of oscillation of the closed loop.

12. Verify the simulated response for the circuit shown in Fig. 7.51 using either the Mini-MSINC program or any other convenient computer-aided-design program.

13. Discuss the problem of MOSFET design for good linearity of the $I_D$ vs $V_{GS}$ relationship.

14. Review the small-signal equivalent circuits of MOSFETs.

15. Discuss channel mobility in Si MOSFETs.

16. Discuss the factors controlling threshold voltages in enhancement Si MOSFETs.

17. Sketch some wafer masking procedures that result in $n$ channel MOSFETs with Zener diode gate protection.

18. Review the relationships between output power, power gain and frequency at 1 GHz for IGFETs.

19. Design an $n$-channel Si MOSFET capable of $V_{DS}$ = 100 V, and 10 W of output at 500 MHz.

20. Review stacked gate concepts in IGFETs.

21. Can MOSFETs be made with buried channels or buried gates?

22. Discuss substrate bias voltage as a means of adjusting MOSFET characteristics in circuit uses.

23. Review the studies of source-drain resistance in the saturated region of MOSFET behavior.

24. Review the drain voltage limitations of MOS transistors[17] in planar and VMOS forms.

25. Discuss the multilevel interconnect and parasitic transistor problem (MOS and BJT) in MOS large scale integrated circuits.

26. Achievement of short-channel MOSFETs by double diffusion procedures (DMOS) is an important technology. Discuss the range of channel lengths that is possible in Si $n$ channel units by specifying the diffusants and diffusion times.

27. Discuss poly-Si gates in MOSFET technology. Gates made of poly-Si have higher resistance than metal gates. What effect is this likely to have on the frequency response?[157]

28. Discuss the role of insulator thickness in determining the characteristics of MOSFETs.

29. Discuss the common-gate circuit configuration of MOSFETs and its uses.

30. Discuss the use of transistors rather than resistors as loads in intermediate stages of amplifiers.[117]

31. Discuss the temperature dependence of the large and small signal parameters of MOSFETs.

32. Review MOSFET behavior at cryogenic temperatures.[163]

33. Review the effects of fast neutrons, high energy electrons and gamma rays on the behavior of MOSFETs.

34. Discuss the use of ion-implantation in the fabrication of MOSFETs. What are the effects on threshold voltage and noise? Can double implant procedures produce short channel structures as in DMOS?

35. Discuss subthreshold drain-source leakage current in MOSFETs.[95]

36. Compare the performance of Si on sapphire MOSFETs with that of Si MOSFETs on Si substrates.

37. Review the problem of testing MOS/LSI arrays.[209]

38. VMOS integrated circuit technology has some attractive features but so does planar technology. Compare the two.

39. Review the turn-on behavior of an MOSFET. What is the relationship of the turn-on time to the cut-off frequency response? What is the role of circuit parasitics? Can a charge storage model be used to characterize it?

40. A great deal of information may be obtained from capacitance vs voltage curves for MOS structures at various frequencies. Explain.

41. An MOS capacitor has three distinct capacitance states. Describe an integrated circuit for tuning RF circuits that allows digital switching between the high frequency inversion capacitance and the deep depletion capacitance.[31]

42. Discuss the Zerbst plot and related techniques for studying channel lifetime and interface recombination velocity.[225,226]

43. What information can be obtained from capacitance transient curves of Si MOS transistors? What temperatures are needed for these studies and what can be learned by the application of light?

44. Review the interface state density and energy distribution at typical Si-SiO$_2$ interfaces.
45. Discuss the leakage current across the insulator in a MOSFET at $100°C$.[184,185]
46. Discuss the growth of SiO$_2$ that is free of traps and has a low interface state density at the Si surface. Compare the characteristics of dry oxygen grown SiO$_2$ and steam-grown oxides.[135]
47. Discuss the effects of sodium and other contaminants in the insulators (SiO$_2$, Si$_3$N$_4$, Al$_2$O$_3$) of IGFETs.
48. Discuss electron trapping at the Si-SiO$_2$ interface in MOSFETs.
49. Discuss hole and electron injection and transport in SiO$_2$ and Si$_3$N$_4$.
50. Discuss the measurement of substrate lifetime ($10^{-3}$-$10^{-10}$ sec) in a metal-oxide-semiconductor capacitor by transient biasing from strong inversion to deep depletion.[77]
51. Discuss the use of charge pumping currents to measure surface state densities in MOS transistors.[72]
52. Describe the weak-inversion technique for determining the number of surface states in MOSFETs.[259]
53. Review the parallel conductance method of obtaining information about interface states in MOS capacitors.
54. Discuss noise in MOSFETs.
55. Discuss power consumption of CMOS circuits and noise-voltage margins.[106]
56. Describe the geometry and performance of MOSFET shift registers and consider some applications.
57. What are the fundamental performance limits of Si MOSFETs (a) in logic circuits, (b) in high frequency performance, (c) in power?
58. Review the present state of knowledge of insulator films on GaAs for possible IGFET devices.
59. Discuss the characteristics and possible uses of InSb and InAs insulated-gate field effect transistors.
60. What is the merit in the use of vacuum-deposited thin-film CdS insulated gate FETs? Why is CdS used and not chemical vapor deposited Si polycrystalline films?
61. Discuss the electrically variable channel length effects that can be seen in NMOS transistors.[159]
62. Review what has been learned about the possibility of charge storage at SiO$_2$-Si$_3$N$_4$ interfaces for memory purposes.
63. Compare tunnel-injection and floating-gate concepts for IGFET memory cells. Discuss memory IGFETs that involve more than the usual junctions.[262]
64. A particular NMOS structure is stated to have a charge retention time of 10 years at $300°K$ and 3 years at $350°K$. How would the decay processes have been measured and what energy can be inferred from the measurement?[161,165,262]
65. Describe random-access write-read electron beam memories that use metal-oxide-semiconductor capacitors as targets.[45]
66. Review the characteristics of an MOS transistor that has a continuously variable threshold voltage (CVTD).[283]
67. In one modification of an $n$-channel enhancement IGFET an $n$ type region underlies the shallow $p$ substrate layer and is connected to the gate through a resistor.[190] Describe the effects that result in negative resistance in the device. Does the device have any circuit merit?
68. Describe the characteristics of metal-ferroelectric-semiconductor, IGFETs.
69. Compare the performance of infrared sensing MOSFETs with that of cooled photoconductive sensors of deep-impurity doped Si and lower bandgap semiconductors. Discuss gain and noise level and speed of response.
70. Discuss the fabrication of resonant-gate MOSFET structures and their frequency limitations and potential uses.

## 7.14  REFERENCES AND FURTHER READING SUGGESTIONS

1. Abbas, S.A., "Substrate current, a device and process monitor," *IEEE International Electron Devices Meeting Technical Digest,* Washington, D.C., December, 1974, p. 404.
2. Abbas, S.A., and R.C. Doeherty, "Hot-carrier instability in IGFET's," *Appl. Phys. Lett.,* **27**, 1975, p. 147.
3. Abbas, S.A., and R.C. Doeherty, "*N*-channel IGFET design limitations due to hot electron trapping," *IEEE International Electron Devices Meeting Technical Digest,* Washington, D.C., December, 1975, p. 35.
4. Agajanian, A.H., "A bibliography on MOSFET devices," *IEEE Trans. Electron Devices,* **ED-20**, 1973, p. 757.
5. Ahuja, B.K., and A.R. Boothroyd, "Modelling of V-channel MOS transistor," *IEEE International Electron Devices Meeting Technical Digest,* Washington, D.C. 1976, p. 573.
6. Aitken, J.M., and D.R. Young, "Electron trapping by radiation-induced charge in MOS devices," *J. Appl. Phys.,* **47**, 1976, p. 1196.
7. Aitken, A., et al., "The relative performance and merits of CMOS techniques," *IEEE International Electron Devices Meeting Technical Digest,* Washington, D.C., 1976, p. 327.
8. Alessandrini, E.I., D.R. Campbell, and K.N. Tu, "Surface reactions on MOS structures," *J. Appl. Phys.,* **45**, 1974, p. 4888.
9. Armstrong, L., "Ion implantation coming of age," *Electronics,* June 19, 1972, p. 69.
10. Armstrong, W.E., and J.R. Lidke, "An analog system for measurement of effective surface mobility in MOS devices," *IEEE J. Solid-State Circuits,* SC-8, 1973, p. 180.
11. Arnold, E., and M. Poleshuk, "Carrier generation at the Si-SiO$_2$ interface under pulsed conditions," *J. Appl. Phys.,* **46**, 1975, p. 3016.
12. Baccarini, G., et al., "Majority- and minority-carrier lifetime in MOS structures," *Solid-State Electronics,* **18**, 1975, p. 1115.
13. Backensto, W.V., and C.R. Viswanathan, "An improved 1/f noise model of an MOS transistor," *IEEE International Electron Devices Meeting Technical Digest,* Washington, D.C., December, 1975, p. 469.
14. Backensto, W.V., and C.R. Viswanathan, "The utilization of charge pumping techniques to evaluate the energy and spatial distribution of interface states of an MOS transistor," *IEEE International Electron Devices Meeting Technical Digest,* Washington, D.C., 1976, p. 287.
15. Baliga, B.J. and S.K. Ghandhi, "Doped mask solves some GaAs woes," *Electronics,* September 25, 1972, p. 36.
16. Barker, R.W.J., "Small-signal subthreshold model for IGFET's," *Electronics Lett.,* May 13, 1976, **12**, p. 260.
17. Bateman, I.M., G.A. Armstrong, and J.A. Magowan, "Drain voltage limitations of MOS transistors," *Solid-State Electronics,* **17**, 1974, p. 539.
18. Beeke, H., R. Hall, and J. White, "Gallium arsenide MOS transistors," *Solid-State Electronics,* **8**, 1965, p. 813.
19. Berger, J., "Deep-channel MOS transistor," *IEEE Trans. Electron Devices,* **ED-22**, 1975, p. 314.
20. Berglund, C.N., and R.J. Powell, "Injection into SiO$_2$: Electron scattering in the image force potential well," *J. Appl. Phys.,* **42**, 1971, p. 537.
21. Bergueld, P., "Development, operation, and application of the ion-sensitive field-effect transistor as a tool for electrophysiology," *IEEE Trans. Biomedical Engineering,* **BME-19**, 1972, p. 342.
22. Biancomano, V., "Logic-simulator programs set pace of computer-aided design," *Electronics,* October 13, 1977, p. 98.

23. Bhatti, I.S., T.J. Rodgers and J.R. Edwards, "Minimization of parasitic capacitances in VMOS transistors," *IEEE International Electron Devices Meeting Technical Digest*, Washington, D.C., 1976, p. 565.
24. Boag, T.R., "Designing with monolithic FET switches," *Electronic Products*, January 15, 1973, p. 85.
25. Boleky, E.J., "The performance of complementary MOS transistors on insulating substrates," *RCA Rev.*, 31, 1970, p. 372.
26. Boothroyd, A.R., and Y.A. El-Mansy, "A new accurate model of the IGFET for CAD applications," *IEEE International Electron Devices Meeting Technical Digest*, Washington, D.C., December, 1974, p. 31.
27. Breed, D.J., "A new model for the negative voltage instability in MOS devices," *Appl. Phys. Lett.*, 26, 1975, p. 116.
28. Bridwell, W., "Three ways to build low-threshold MOS," *Electronics*, April 13, 1970, p. 118.
29. Britton, J., J.R. Cricchi, and L.G. Ottobre, "Metal-nitride-oxide IC memory retains data for meter reader," *Electronics*, Oct. 23, 1972, p. 119.
30. Brotherton, S.D., "Dependence of MOS transistor threshold voltage on voltage substrate resistivity," *Solid-State Electronics*, 10, 1967, p. 611.
31. Brown, D.M., et al., "*P*-channel refractory metal self-registered MOSFET," *IEEE Trans. Electron Devices*, ED-18, 1971, p. 931.
32. Brown, D.M., et al., "High frequency MOS digital capacitor," *IEEE Trans. Electron Devices*, ED-22, 1975, p. 938.
33. Brown, W.L., "*n*-Type surface conductivity on *p* type germanium,"*Phys. Rev.*, 91, 1953, p. 518.
34. Bulucea, C., "Avalanche injection into the oxide in silicon gate-controlled devices— I theory," *Solid-State Electronics*, 18, 1975, p. 363.
35. Burghard, R.A. and C.W. Gwyn, "Radiation failure modes in CMOS integrated circuits," *IEEE Trans. Nuclear Sci.*, NS-20, 1973, p. 300.
36. Burns, J.R., "Switching response of complementary-symmetry MOS transistor logic circuits," *RCA Rev.*, 25, 1964, p. 627.
37. Burns, J.R., "High-frequency characteristics of the insulated-gate field-effect transistor," *RCA Rev.*, 28, 1967, p. 385.
38. Butler, W.J., and C.W. Eichelberger, "Temperature-stable MOSFET reference voltage source," *IEEE International Electron Devices Meeting Technical Digest*, Washington, D.C., 1976, p. 587.
39. Card, H.C., and A.G. Worrall, "Reversible floating-gate memory," *J.Appl. Phys.*, 44, 1973, p. 2326.
40. Cauge, T.P., et al., "Double-diffused MOS transistor achieves microwave gain," *Electronics*, February 15, 1971, p. 99.
41. Chang, L.L., and H.N. Yu, "The germanium insulated-gate field-effect transistor (FET),"*Proc. IEEE*, 53, 1965, p. 316.
42. Churchill, J.N., T.W. Collins, and F.E. Holmstrom, "Electron irradiation effects in MOS systems," *IEEE Trans. Electron Devices*, ED-21, 1974, p. 768.
43. Clemens, J.T., R.H. Doklan, and J.J. Nolen, "An *N*-channel Si-gate integrated circuit technology," *IEEE International Electron Devices Meeting Technical Digest*, Washington, D.C., December, 1975, p. 299.
44. Clemens, J.T., et al., "A *P*-channel Si-gate integrated circuit technology," *IEEE International Electron Devices Meeting Technical Digest*, Washington, D.C., December 1975, p. 291.
45. Cohen, M.S., and J.S. Moore, "Physics of the MOS electron-beam memory," *J. Appl. Phys.*, 45, 1974, p. 5335.
46. Collins, T.W., and J.N. Churchill, "Exact modeling of the transient response of an MOS capacitor," *IEEE Trans. Electron Devices*, ED-22, 1975, p. 90.

47. Combs, S.R., D.C. D'Avanzo, and R.W. Dutton, "Characterization and modeling of simultaneously fabricated DMOS and VMOS transistors," *IEEE International Electron Devices Meeting Technical Digest*, Washington, D.C. 1976, p. 569.
48. Cricchi, J.R., et al., "Space charge effects in MNOS memory devices and endurance measurements," *IEEE International Electron Devices Meeting Technical Digest*, Washington, D.C., December, 1975, p. 459.
49. Critchlow, D.L., "The $N$-channel IGFET for logic memory," *Nerem Record*, 1968, p. 142.
50. Critchlow, D.L., R.H. Dennard, and S.E. Schuster, "Design and characteristics of $n$-channel insulated-gate field-effect transistors," *IBM Journal of Research and Development*, 17, 1973, p. 430.
51. Curran, L., "Readers reply on MOS/LSI testing," *Electronics*, March 1, 1971, p. 65.
52. Curtis, O.L., Jr., J.R. Srour, and K.Y. Chiu, "Hole and electron transport in $SiO_2$ films," *J. Appl. Phys.*, 45, 1974, p. 4506.
53. Dang, L.M., "A one-dimensional theory on the effects of diffusion current and carrier velocity saturation on E-type IGFET current-voltage characteristics," *Solid-State Electronics*, 20, 1977, p. 781.
54. Darwish, M., and R. Tauberest, "C-MOS and complementary isolated bipolar transistor monolithic integration process," *J. Electrochem. Soc.*, 121, 1974, p. 1119.
55. D'Avanzo, D.C., S.R. Combs, and R.W. Dutton, "Effects of the diffused impurity profile on the DC characteristics of VMOS and DMOS devices," *IEEE J. Solid-State Circuits*, SC-12, 1977, p. 356.
56. Deal, B.E., E.H. Snow, and C.A. Mead, "Barrier energies in metal-silicon dioxide-silicon structures," *J. Phys. Chem. Solids*, 27, 1966, p. 1873.
57. Deal, B.E., "The current understanding of charges in the thermally oxidized silicon structure," *J. Electrochem. Soc.*, 121, 1974, p. 198C.
58. Declercq, M.J., "Applications of the anisotropic etching of silicon to the development of complementary structures," *IEEE International Electron Devices Meeting Technical Digest*, Washington, D.C., 1974, p. 519.
59. Declercq, M.J., and T.Laurent, "A theoretical and experimental study of DMOS enhancement/depletion logic," *IEEE J. Solid-State Circuits*, SC-12, 1977, p. 264.
60. Declerek, G.J., et al., "Some effects of 'trichloroethylene oxidation' on the characteristics of MOS devices," *J. Electrochem. Soc.*, 122, 1975, p. 436.
61. Dennard, R.H., "Design of ion-implanted MOSFETs with very small physical dimensions," *IEEE J. Solid-State Circuits*, SC-9, 1974, p. 256.
62. DiMaria, D.J., "Effects on interface barrier energies of metal-aluminum oxide-semiconductor (MAS) structures as a function of metal electrode material, charge trapping, and annealing," *J. Appl. Phys.*, 45, 1974, p. 5454.
63. DiMaria, D.J., and D.R. Kerr, "Interface effects and high conductivity in oxides grown from polycrystalline silicon," *Appl. Phys. Lett.*, 27, 1975, p. 505.
64. Dingwall, A.G.F., and R.E. Stricker, "$C^2$ L: a new high-speed high density bulk CMOS technology," *IEEE J. Solid-State Circuits*, SC-12, 1977, p. 344.
65. Dishman, J.M., "Limitation on the maximum operating voltage of CMOS integrated circuits," *IEEE International Electron Devices Meeting Technical Digest*, Washington, D.C., December, 1975, p. 551.
66. Doeherty, R.C., S.A. Abbas, C.A. Barile, "Low-leakage $n$- and $p$-channel silicon-gate FETs with an $SiO_2$-$Si_3N_4$-gate insulator," *IEEE Trans. Electron Devices*, ED-22, 1975, p. 33.
67. Dorey, A.P., "A high sensitivity semiconductor strain sensitive circuit," *Solid-State Electronics*, 18, 1975, p. 295.
68. DuBow, J., "Process yields low-cost FET op amp," *Electronics*, Dec. 20, 1973, p. 128.
69. Eaton, S.S., "Sapphire brings out the best in C-MOS," *Electronics*, June 12, 1975, p. 115.

70. Edwards, J.R., and G. Marr, "Depletion-mode IGFET made by deep ion implantation," *IEEE Trans. Electron. Devices,* ED-20, 1973, p. 283.
71. EerNisse, E.P., and C.B. Norris, "Introduction rates and annealing of defects in ion-implanted $SiO_2$ layers in Si," *J.Appl. Phys.,* 45, 1974, p. 5196.
72. Elliot, A.B.M., "The use of charge pumping currents to measure surface state densities in MOS transistors," *Solid-State Electronics,* 19, 1976, p. 241.
73. El-Mansy, Y.A., and A.R. Boothroyd, "A simple two-dimensional saturation model of short-channel IGFETs for CAD applications," *IEEE International Electron Devices Meeting Technical Digest,* Washington, D.C., December, 1974, p. 35.
74. El-Mansy, Y.A., and D.M. Caughey, "Modelling weak avalanche multiplication currents in IGFETs and SOS transistors for CAD," *IEEE International Electron Devices Meeting Technical Digest,* Washington, D.C., December 1975, p. 31.
75. Erb, D.M., H.G. Dill, and T.N. Toombs, "Electron gate currents and threshold stability in the *n*-channel stacked gate MOS tetrode," *IEEE Trans. Electron Devices,* ED-18, 1971, p. 105.
76. Evans, S.A., I.H. Morgan, and T.N. Jernigan, "A model for the accurate prediction of the current voltage behavior of ion-implanted MOS transistors at low drain voltages," *IEEE International Electron Devices Meeting Technical Digest,* Washington, D.C., December, 1974, p. 47.
77. Fahrner, W.R., and C.P. Schneider, "A new fast technique for large-scale measurements of generation lifetime in semiconductors," *J. Electrochem. Soc.,* 123, 1976, p. 100.
78. Farmer, J.W., and R.S. Lee, "Photocurrents and photoconductive yield in MOS structures during k irradiation," *J.Appl. Phys.,* 46, 1975, p. 2710.
79. Farzan, B., and C.A.T. Salama, "Depletion V-groove MOS (VMOS) power transistors," *Solid-State Electronics,* 19, 1976, p. 297.
80. Ferris-Prabhu, A.V., "Theory of MNOS memory device behavior," *IBM Journal of Research and Development,* 17, 1973, p. 125.
81. Fischer, W., "Equivalent circuit and gain of MOS field effect transistors," *Solid-State Electronics,* 9, 1966, p. 71.
82. Flinn, I., G. Bew, and F. Berz, "Low frequency noise in MOS field effect transistors," *Solid-State Electronics,* 10, 1967, p. 833.
83. Forbes, L. and J.R. Yeargan, "Design for silicon infrared sensing MOSFET," *IEEE Trans. Electron Devices,* ED-21, 1974, p. 459.
84. Fripp, A.L., "Dependence of resistivity on the doping level of polycrystalline silicon," *J. Appl. Phys.,* 46, 1975, p. 1240.
85. Frohman-Bentchkowsky, D., "On the effect of mobility variation on MOS device characteristics," *Proc. IEEE,* 56, 1968, p. 217.
86. Frohman-Bentchkowsky, D., "FAMOS-a new semiconductor charge storage device," *Solid-State Electronics,* 17, 1974, p. 517.
87. Frohman-Bentchkowsky, D., and A.S. Grove, "Conductance of MOS transistors in saturation," *IEEE Trans. Electron Devices,* ED-16, 1969, p. 108.
88. Gabler, L. et al., "Extraction of implantation profiles from the differential body effect of ion-implanted MOS transistors," *Electronics Lett.,* 12, 1976, p. 257.
89. Gaensslen, F.H., V.L. Rideout, and E.J. Walker, "Design and characterization of very small MOSFETs for low temperature operation," *IEEE International Electron Devices Meeting Technical Digest,* Washington, D.C., December, 1975, p. 43.
90. Gannon, J.J., and C.G. Nuese, "A chemical etchant for the selective removal of the GaAs through $SiO_2$ masks," *J. Electrochem. Soc.,* 121, 1974, p. 1215.
91. Garrett, C.G.B., and W.H. Brattain, "Physical theory of semiconductor surfaces," *Phys. Rev.,* 99, 1955, p. 376.

92. Goetzberger, A., and S.M. Sze, "Metal-insulator-semiconductor (MIS) physics," in: *Applied Solid State Science*, Vol. 1, R.Wolfe and C.J. Kreissman (eds.), Academic Press, NY, 1969, p. 15.
93. Goodman, A.M., "A useful modification of the technique for measuring capacitance as a function of voltage," *IEEE Trans. Electron Devices*, ED-21, 1974, p. 753.
94. Gosch, J., "Nitride makes its mark," *Electronics*, Jan. 24, 1975, p. 72.
95. Gosney, W.M., "Subthreshold drain leakage currents in MOS field-effect transistors," *IEEE Trans. Electron Devices*, ED-19, 1972, p. 213.
96. Gosney, W.M., and L.H. Hall, "The extension of self-registered gate and doped-oxide diffusion technology to the fabrication of complementary MOS transistors," *IEEE Trans. Electron Devices*, ED-20, 1973, p. 469.
97. Gray, P.V., and D.M. Brown, "Density of $SiO_2$-Si interface states," *Appl. Phys. Lett.*. 8, 1966, p. 31.
98. Gregory, B.L., and C.W. Gwyn, "Radiation effects on semiconductor devices," *Proc. IEEE, 62*, 1974, p. 1264.
99. Hamilton, D.J., and W.G. Howard, *Basic Integrated Circuit Engineering*, McGraw-Hill, NY, 1975.
100. Hariuchi, S., "Electrical characteristics of boron diffused polycrystalline silicon layers," *Solid-State Electronics, 18*, 1975, p. 659.
101. Hariuchi, S., and R. Blanchard, "Boron diffused in polycrystalline silicon layers," *Solid-State Electronics, 18*, 1975, p. 529.
102. Hasegawa, H., K.E. Forward, and H.L. Hartnagel, "New anodic native oxide of GaAs with improved dielectric and interface properties," *Appl. Phys. Lett., 26*, 1975, p. 567.
103. Heiman, F.P., "On the determination of minority carrier lifetime from the transient response of an MOS capacitor," *IEEE Trans. Electron Devices, ED-14*, 1967, p. 781.
104. Heiman, F.P. and H.S. Miller, "Temperature dependence on n-type MOS transistors," *IEEE Trans. Electron Devices, ED-12*, 1965, p. 142.
105. Hemmert, R.S., "Invariance of the Hall effect MOSFET to gate geometry," *Solid-State Electronics, 17*, 1974, p. 1039.
106. Henderson, R., "Let the designer beware," *IEE (London) Electronics and Power*, May 29, 1975, p. 629.
107. Hess, D.W., and B.E. Deal, "Effect of nitrogen and oxygen/nitrogen mixtures on oxide charges in MOS structures," *J. Electrochem. Soc., 122*, 1975, p. 1123.
108. Hnatek, E.R., *Applications of Linear Integrated Circuits*, Wiley-Interscience, NY, 1975.
109. Hodges, D.A. (ed.), "*Semiconductor Memories*," IEEE Press, NY, 1972.
110. Hoffait, A.H., and R.D. Thornton, "A simple derivation of field-effect transistor characteristics," *Proc. IEEE, 51*, 1963, p. 1146.
111. Hofstein, S.R., "Minority carrier lifetime determination from inversion layer transient response," *IEEE Trans. Electron Devices, ED-14*, 1967, p. 785.
112. Hofstein, S.R., and F.P. Heiman, "The silicon insulated-gate field-effect transistor," *Proc. IEEE, 51*, 1963, p. 1190.
113. Hofstein, S.R. and G. Warfield, "Carrier mobility and current saturation in the MOS transistor," *IEEE Trans. Electron Devices, ED-12*, 1965, p. 129.
114. Hogeboom, J.G., "Etched Schottky-barrier MOSFETs," *Electronics Lett., 7*, 1971, p. 133.
115. Holmes, F.E., and C.A.T. Salama, "VMOS- a new MOS integrated circuit technology," *Solid-State Electronics, 17*, 1974, p. 791.
116. Holmes, F.E., "V-groove MOS (VMOS) enhancement load logic," *Solid-State Electronics, 20*, 1977, p. 775.
117. Hsu, S.T., "A COS/MOS linear amplifier stage," *RCA Rev., 27*, 1976, p. 136.

460    SEMICONDUCTOR DEVICES AND INTEGRATED ELECTRONICS

118. Hsu, S.T., "Drain characteristics of thin-film MOSFET's," *RCA Rev.*, 38, 1977, p. 139.
119. Hughes, R.C., "Hole mobility and transport in thin $SiO_2$ films," *Appl. Phys. Lett.*, 26, 1975, p. 436.
120. Ihantola, H.K.J., "Design theory of a surface field-effect transistor," *Solid-State Electronics*, 7, 1964, p. 423.
121. Ipri, A.C. and J.C. Sarace, "CMOS/SOS semi-static shift registers," *IEEE J. Solid-State Circuits*, SC-11, 1976, p. 337.
122. Ipri, A.C. and J.C. Sarace, "Low-threshold low-power CMOS/SOS for high frequency counter applications," *IEEE J. Solid-State Circuits*, SC-11, 1976, p. 329.
123. Irene, E.A., "The effects of trace amounts of water on the thermal oxidation of silicon in oxygen," *J. Electrochem. Soc.*, 121, 1974, p. 1613.
124. Ito, T., and Y. Sakai, "The GaAs inversion-type MIS transistors," *Solid-State Electronics*, 17, 1974, p. 751.
125. Jenkins, F.S. et al., "MOS device modelling for computer implementation," *IEEE Trans. Circuit Theory*, CT-20, 1973, p. 649.
126. Jenné, F.B., "Grooves add new dimension to VMOS structure and performance," *Electronics*, August 18, 1977, p. 100.
127. Jeppson, K.O., and J.L. Gates, "The effects of impurity distribution on the subthreshold leakage current in CMOS *n*-channel transistors," *Solid-State Electronics*, 19, 1976, p. 83.
128. Johnson, E.O., "The insulated-gate field-effect transistor—a bipolar transistor in disguise," *RCA Rev.*, 34, 1973, p. 80.
129. Jones, D., and R.W. Webb, "Chopper-stabilized op.amp combines MOS and bipolar elements on one chip," *Electronics*, Sept. 27, 1973, p. 110.
130. Jutzi, W., A MESFET distributed amplifier with 2 GHz bandwidth," *Proc. IEEE*, 57, 1969, p. 1195.
131. Kabayashi, K., and Y. Haneta, "Memory performance of MOS transistors with $SiO_2$ films prepared by $SiH_4$-$H_2O$ system," *Japan. J. Appl. Phys.*, 12, 1973, p. 715.
132. Kahng, D., and E.H. Nicollian, "Physics of Multilayer-Gate IGFET Memories," in: *Applied Solid State Science*, Vol. 3, R. Wolfe (ed.), Academic Press, NY, 1972.
133. Kamins, T.I., and R.S. Muller, "Statistical considerations in MOSFET calculations," *Solid-State Electronics*, 10, 1967, p. 423.
134. Kano, K., and H. Kunig, "Effect of thermal instability on ultra-high frequency performance of insulated-gate field effect transistors," *Solid-State Electronics*, 12, 1969, p. 719.
135. Kar, S., "Interface charge characteristics of MOS structures with different metals on steam grown oxides," *Solid-State Electronics*, 18, 1975, p. 723.
136. Katsube, T., Y. Adachi, and T. Ikoma, "Trap centers in MNOS memory devices measured by thermally stimulated currents," *Japan. J. Appl. Phys.*, 12, 1973, p. 1633.
137. Kazan, B. (ed.), *Advances in Image Pickup and Display*, Vol. 3, Academic Press, NY, 1977.
138. Kendall, E.J.M. and J.W. Haslett, "On the removal of the memory properties of MNOS devices," *IEEE Trans. Electron Devices*, ED-19, 1972, p. 287.
139. Kennedy, E.J., "Gate leakage currents in MOS field effect transistors," *Proc. IEEE*, 54, 1966, p. 1098.
140. Kern, W., and J.P. White, "Interface properties of chemically vapor deposited silica films on gallium arsenide," *RCA Rev.*, 31, 1970, p. 771.
141. Kohn, E. et al., "High-speed 1 μm GaAs MESFET," *Electronics Lett.*, 11, April 17, 1975, p. 171.
142. Kotecha, H.N., and K.E. Beilstein, "Currents and capacitances in narrow width MOSFET structures," *IEEE International Electron Devices Meeting Technical Digest*, Washington, D.C., December 1975, p. 47.
143. Kriegler, R.J., "High quality $SiO_2$ for integrated circuits," *1975 IEEE International Solid-State Circuits Conf. Abstracts*, IEEE, NY, p. 56.

144. Kudok, O., K. Nokamura, and M. Kamoshida, "Implant dose profile dependence of electrical characteristics of ion-implanted MOS transistors," *J.Appl. Phys.*, **45**, 1974, p. 4514.
145. Lagel, L.W., "SPICE 2: A computer program to simulate electronic circuits," University of California, Berkeley Memo, ERL-M520, May 1975.
146. Lazarus, M.J., "The short-channel IGFET," *IEEE Trans. Electron Devices*, **ED-22**, 1975, p. 351.
147. Learn, A.J., "Effects of MOS metallization geometry and processing on mobile impurities," *J. Electrochem. Soc.*, **122**, 1975, p. 1127.
148. LeBoss, B., "C-MOS gets a rise out of LSI," *Electronics*, October 13, 1977, p. 65.
149. Lee, D.H., and J.W. Mayer, "Ion-implanted semiconductor devices," *Proc. IEEE*, **62**, 1974, p. 1241.
150. Lehovec, K., and R. Zuleeg, "Negative resistance of a modified insulated gate field-effect transistor," *Proc. IEEE*, **62**, 1974, p. 1163.
151. Lenzlinger, M., "Gate protection of MIS devices," *IEEE Trans. Electron Devices*, **ED-18**, 1971, p. 249.
152. Lepselter, M.P., and S.M. Sze, "SB-IGFET: An insulated-gate field-effect transistor using Schottky barrier contacts as source and drain," *Proc. IEEE*, **56**, 1968, p. 1400.
153. Lepselter, M.P., A.U. MacRae, and R.W. MacDonald, "SB-IGFET, II: An ion implanted IGFET using Schottky barriers," *Proc. IEEE*, **57**, 1969, p. 812.
154. Leuenberger, F., "Dependence of threshold voltage of silicon p-channel MOSFET's on crystal orientation," *IEEE Proc.* **54**, 1966, p. 1985.
155. Liechti, C.A., "Recent advances in high-frequency field-effect transistors," *IEEE International Electron Devices Meeting Technical Digest*, Washington, D.C., December, 1975, p. 6.
156. Lile, D.L., A.R. Clawson, and D.A. Collins, "Depletion-mode GaAs MOSFET, *Appl. Phys. Lett.*, **29**, 1976, p. 207.
157. Lin, H.C., et al., "Effect of silicon-gate resistance on the frequency response of MOS transistors," *IEEE Trans. Electron Devices*, **ED-22**, 1975, p. 255.
158. Lin, H.C., J.L. Halsor, H.F. Benz, "Optimized load device for DMOS integrated circuits," *IEEE International Electron Devices Meeting Technical Digest*, Washington, D.C., December, 1975, p. 547.
159. Lonky, M.L., and A.P. Turley, "Electrically variable channel-length changes in MNOS transistors," *Appl. Phys. Lett.*, **28**, 1976, p. 162.
160. Lukes, Z., "Characteristics of the metal-oxide-semiconductor transistor in the common-gate electrode arrangement," *Solid-State Electronics*, **9**, 1966, p. 21.
161. Lundkvisk, L., C. Svensson, and B. Hansson, "Discharge of MNOS structures at elevated temperatures," *Solid-State Electronics*, **19**, 1976, p. 221.
162. Lundstrom, I., et al., "A hydrogen-sensitive MOS field-effect transistor," *Appl. Phys. Lett.*, **26**, 1975, p. 55.
163. Maddox, R.L., "pMOSFET parameters at cryogenic temperatures," *IEEE Trans. Electron Devices*, **ED-23**, 1976, p. 16.
164. Maes, H.E., and R.J. Van Overstraeten, "Low-field transient behavior of MNOS devices," *J. Appl. Phys.*, **47**, 1976, p. 664.
165. Maes, H.E., and R.J. Van Overstraeten, "Memory loss in MNOS capacitors," *J. Appl. Phys.*, **47**, 1976, p. 667.
166. Mar, H.A., and J.G. Simmons, "Determination of the energy distribution of interface traps in MIS systems using non-steady-state techniques," *Solid-State Electronics*, **17**, 1974, p. 131.
167. Mar, H.A., and J.G. Simmons, "Determination of bulk-trap parameters in MIS structures," *Appl. Phys. Lett.*, **25**, 1974, p. 503.
168. Mar, H.A., and J.G. Simmons, "Determination of the energy distribution of interface traps in MIS systems using non-steady-state techniques," *Solid-State Electronics*, **17**, 1974, p. 131.

169. Masuda, H., et al., "Device design of E/D gate MOSFET," *Proc. Fourth Conf. on Solid State Devices*, Tokyo, *1972 Suppl. to J. Japan. Soc. Appl. Phys.*, 42, 1973, p. 167.

170. Masuhara, T., and R.S. Muller, "Complementary DMOS process for LSI, *IEEE International Electron. Devices Meeting Technical Digest*, Washington, D.C., December, 1975, p. 543.

171. Mavar, J., *M.O.S.T. Integrated Circuit Engineering*, Peter Peregrinus, Ltd. Herts., England, 1973.

172. Mayer, J.W., L. Erikson, and J.A. Davies, *Ion Implantation in Semiconductors*, Academic Press, NY, 1970.

173. McCaughan, D.V., and B.C. Wonsiewicz, "Effects of dislocations on the properties of metal SiO$_2$ silicon capacitors," *J. Appl. Phys.*, 45, 1974, p. 4982.

174. Meyer, J.E., "MOS models and circuit simulation," *RCA Rev.*, 32, 1971, p. 42.

175. Mohsen, A.M., and F.J. Morris, "Measurements on depletion-mode field effect transistors and buried channel MOS capacitors for the characterization of bulk transfer charge-coupled devices," *Solid-State Electronics*, 18, 1975, p. 407.

176. Moline, R.A., and G.W. Reutlinger, "Self-aligned maskless chan stops for IGFET integrated circuits," *IEEE Trans. Electron Devices*, ED-20, 1973, p. 1129.

177. Montillo, F., and P. Balk, "High-temperature annealing of oxidized silicon surfaces," *J. Electrochem. Soc.*, 118, 1971, p. 1463.

178. Morita, Y., et al., "Si UHF MOS high-power FET," *IEEE Trans. Electron Devices*, ED-21, 1974, p. 733.

179. Muller, H., F.H. Eisen, and J.W. Mayer, "Anodic oxidation of GaAs as a technique to evaluate electrical carrier concentration profiles," *J. Electrochem. Soc.*, 122, 1975, p. 651.

180. Murarka, S.P., "Thermal oxidation of GaAs," *Appl. Phys. Lett.*, 26, 1975, p. 180.

181. Nakamura, K., O. Kudoh, and M. Kamoshida, "Noise characteristics of ion-implanted MOS transistors," *J. Appl. Phys.*, 46, 1975, p. 3189.

182. Nathanson, H.C., W.E. Newell, and R.A. Wickstrom, "Tuning forks' sound a hopeful note," *Electronics*, Sept. 20, 1965, p. 84.

183. Nathanson, H.C., and J. Guldberg, "Topologically structured thin films in semiconductor device operation," in: *Physics of Thin Films*, Vol. 8, Academic Press, NY. 1975.

184. Negro, V.C. and L.H. Goldstein, "An analytical expression for MOSFET gate leakage current," *Proc. IEEE*, 61, 1973, p. 1509.

185. Negro, V.C., and L. Parrone, "Self-heating and gate leakage current in a guarded MOSFET," *Proc. IEEE*, 60, 1971, p. 342.

186. Nicholas, K.H., H.E. Brockman, and I.J. Stemp, "Fabrication of submicron polysilicon lines by conventional techniques," *Appl. Phys. Lett.*, 26, 1975, p. 398.

187. Nicollian, E.H., and A. Goetzberger, "The Si-SiO$_2$ interface—electrical properties as determined by the metal-insulator-silicon conductance techniques," *Bell System Tech. J.*, 46, 1967, p. 1055.

188. Ning, T.H., and H.N. Yu, "Optically induced injection of hot electrons into SiO$_2$," *J. Appl. Phys.*, 45, 1974, p. 5373.

189. Notthoff, J.K., "Radiation responses of matched silicon junction field-effect transistors," *IEEE Trans. Nuclear Science*, 18, 1971, p. 397.

190. Oakley, R.E. "MNOS: a new non-volatile store," *Component Technology* (GB), 4, 1970, p. 17.

191. Oguey, H. And E. Vittoz, "Resistance-CMOS circuits," *IEEE J. Solid-State Circuits*, SC-12, 1977, p. 283.

192. Okabe, T., et al., "A complementary-pair high-power MOSFET," *IEEE International Electron Devices Meeting Technical Digest*, Washington, D.C., 1977, p. 416.

193. Osburn, C.M., and D.W. Ormond, "Sodium-induced barrier-height lowering and dielectric breakdown on SiO$_2$ films on silicon," *J. Electrochem. Soc.*, 121, 1974, p. 1195.

194. Park, J.K., and W.W. Grannemann, "Fabrication of a ferroelectric gate memory transistor," *IEEE International Electron Devices Meeting Technical Digest*, Washington, D.C., December, 1975, p. 463.
195. Pao, H.C., and C.T. Sah, "Effects of diffusion current on characteristics of metal-oxide (insulator)-semiconductor transistors," *Solid-State Electronics*, 9, 1966, p. 927.
196. Penney, W.M. (ed.), *MOS Integrated Circuits*, Van Nostrand Reinhold, NY, 1972.
197. Perkins, C.R., "Complex MOS circuit arrays," *Proc. of 20th Electronics Components Conf.*, Washington, D.C., May 1970, p. 475.
198. Pierret, R.F., "A linear-sweep MOS-C technique for determining minority carrier lifetimes, *IEEE Trans. Electron Devices*, ED-19, 1972, p. 869.
199. Plummer, J.D., and J.D. Meindl, "A monolithic 200-V CMOS analog switch," *IEEE J. Solid-State Circuits,"* SC-4, 1976, p. 809.
200. Rapp, A.K., and E.C. Ross, "Silicon-on-sapphire substrates overcome MOS limitations," *Electronics*, Sept. 25, 1972, p. 113.
201. Rapp, K., "Silicon-on-sapphire: inception, implementation, application," *Electronic Products*, 15, 1973, p. 83.
202. Raymond, R.K., and M.B. Das, "Fabrication and characteristics of MOS-FET's incorporating anodic aluminum oxide in the gate structure," *Solid-State Electronics*, 19, 1976, p. 181.
203. Reddi, V.G.K., and C.T. Sah, "Source to drain resistance beyond pinch-off in metal-oxide-semiconductor transistors (MOST)," *IEEE Trans. Electron Devices*, ED-12, 1965, p. 139.
204. Reizman, F. and W. VanGelder, "Optical thickness measurement of $SiO_2$ and $Si_3N_4$ films on silicon," *Solid-State Electronics*, 10, 1967, p. 625.
205. Revesz, A.G., "Noncrystalline silicon diode films on silicon: a review," *J. Non-Crystalline Sol.*, 11, 1973, p. 309.
206. Reynolds, F.H., and W.D. Morton, "Metal-oxide-semiconductor (MOS) integrated circuits. Part 2—Simple logic circuits," *Post Office Electrical Eng. J.*, 63, 1970, p. 105.
207. Richman, P., *MOS Field Effect Transistors and Integrated Circuits*, Wiley-Interscience, NY 1973.
208. Rideout, V.L., F.H. Gaensslen, and A. LeBlanc, "Device design considerations for ion implanted *n*-channel MOSFETs," *IBM Journal of Research and Development*, 19, 1975, p. 50.
209. Robinton, M.A. , "A critique of MOS/LSI testing," *Electronics*, February 1, 1971, p. 62.
210. Rodgers, T.J., et al., "An experimental and theoretical analysis of double diffused MOS transistors," *IEEE J. Solid-State Circuits*, SC-10, 1975, p. 312.
211. Rodgers, T.J., et al., "DMOS Experimental and theoretical study," *1975 IEEE International Solid-State Circuits Conf. Digest*, p. 122, IEEE, NY.
212. Rodriquez, A., et al., "Fabrication of short-channel MOSFET's with refractory metal gates using RF sputter etching," *Solid-State Electronics*, 19, 1976, p. 17.
213. Ronen, R.S., "Low-frequency 1/f noise in MOSFET's," *RCA Rev.*, 34, 1973, p. 280.
214. Ronen, R.S., and L. Strauss, "The silicon-on-sapphire MOS tetrode-some small-signal features LF to UHF," *IEEE Trans. Electron Devices*, ED-21, 1974, p. 100.
215. Ronen, R.S., et al., "High voltage SOS/MOS device structures," *IEEE International Electron Devices Meeting Technical Digest*, Washington, D.C., December, 1975, p. 4.
216. Ronen, R.S., M.R. Splinter, R.E. Tremain, Jr., "High-voltage SOS/MOS devices and circuit elements: Design, fabrication, and performance," *IEEE J. Solid-State Circuits*, SC-11, 1976, p. 431.
217. Root, C.D., and L. Vadasz, "Design calculations for MOS field effect transistors," *IEEE Trans. Electron Devices*, ED-11, 1964, p. 294.
218. Rossel, P., H. Martinot, and G. Vassilieff, "Accurate two sections model for MOS transistor in saturation," *Solid-State Electronics*, 19, 1976, p. 51.

219. Sah, C.T., "Characteristics of the metal-oxide-semiconductor transistors," *IEEE Trans. Electron Devices*, **ED-11**, 1964, p. 324.

220. Salama, C.A.T., "A new short channel MOSFET structure (UMOST), " *Solid-State Electronics*, **20**, 1977, p. 1003.

221. Sasaki, N., "Characteristics of enhancement/depletion (E/D) gate MOSFET fabricated using ion implantation," *Solid-State Electronics*, **18**, 1975, p. 777.

222. Scheff, P.S., "Destruction of transistors by static electricity," *Solid State Tech.*, May 1973, p. 22.

223. Schemmert, W., L. Gobler, and B. Hoefflinger, "Sub-threshold and active-region characterization of ion-implanted buried-channel MOS transistors, *IEEE International Electron Devices Meeting Technical Digest*, Washington, D.C., December, 1974, p. 546.

224. Schickman, H., and D.A. Hodges, "Modeling and simulation of insulated-gate field-effect transistor switching circuits," *IEEE J. Solid-State Circuits*, **SC-3**, 1968, p. 285.

225. Schroder, D.K. and H.C. Nathanson, "On the separation of bulk and surface components of lifetime using the pulsed MOS capacitor," *Solid-State Electronics*, **13**, 1970, p. 577.

226. Schroder, D.K., and J. Guldberg, "Interpretation of surface and bulk effects using the pulsed MIS capacitor," *Solid-State Electronics*, **14**, 1971, p. 1285.

227. Schroen, W., R.D. Woodruff, and D. Farrington, "Influence of non-equilibrium carriers on the surface breakdown of diodes and MOS-structures," *IEEE Trans. Electron Devices*, **ED-13**, 1966, p. 570.

228. Sesnic, S.S., and G.R. Craig, "Thermal effects in junction field effect transistors and metal oxide semiconductors field effect transistors devices at cryogenic temperatures," Texas Univ. Austin Electronics Res. Center, AFOSR-TR-72-2376.

229. Sewell, Jr., F.A., "The light-sensitive MNOS memory transistor, *IEEE Trans. Electron Devices*, **ED-20**, 1973, p. 563.

230. Shappir, J., S. Margalit, and I. Kidron, "p-channel MOS transistor in indium antimonide," *IEEE Trans. Electron Devices*, **ED-22**, 1975, p. 960.

231. Shibata, H., et al., "A new fabrication method of short channel MOSFET—multiple walls self-aligned MOSFET," *IEEE International Electron Devices Meeting Technical Digest*, Washington, D.C., 1977, p. 395.

232. Shine, M.C., "A simplified technique for measuring fast surface states," *Solid-State Electronics*, **18**, 1975, p. 1135.

233. Shockley, W., and G.L. Pearson, "Modification of conductance of thin films of semiconductors by surface charges," *Phys. Rev.*, **74**, 1948, p. 232.

234. Shoji, M., "Analysis of high-frequency thermal noise of enhancement mode MOS field-effect transistors," *IEEE Trans. Electron Devices*, **ED-13**, 1966, p. 520.

235. Six, P., W. Sansen, "Comparison of MOS-gate protection networks," *IEEE International Electron Devices Meeting Technical Digest*, Washington, D.C., 1977, p. 159.

236. Sigmon, T.W., and R. Swanson, "MOS threshold shifting by ion implantation," *Solid-State Electronics*, **16**, 1973, p. 1217.

237. Simmons, J.G., and G.W. Taylor, "Theory of non-steady-state interfacial thermal currents in MOS devices, and the direct determination of interfacial trap parameters," *Solid-State Electronics*, **17**, 1974, p. 125.

238. Simmons, J.G., and L.S. Wei, "Theory of transient emission current in MOS devices and the direct determination of interface trap parameters," *Solid-State Electronics*, **17**, 1974, p. 117.

239. Singh, R., and H. Hartnagel, "New method of passivating GaAs with $Al_2O_3$," *IEEE International Electron Devices Meeting Technical Digest*, Washington, D.C. December, 1974, p. 576.

240. Small, D.W., and R.F. Pierret, "Defect-controlled generation in deeply depleted MOS-C structures," *Appl. Phys. Lett.*, **27**, 1975, p. 148.

241. Spitzer, S.M., B. Schwartz and G.D. Weigle, "Native oxide mask for zinc diffusion in gallium arsenide," *J. Electrochem. Soc.*, **121**, 1974, p. 820.
242. Spitzer, S.M., B. Schwartz, and G.D. Weigle, "Preparation and stabilization of anodic oxides on GaAs," *J. Electrochem. Soc.*, **122**, 1975, p. 397.
243. Sproul, M.E., and A.G. Nassihian, "Field-effect mobilities of gold-doped MOS structures," *J. Appl. Phys.*, **46**, 1975, p. 1255.
244. Sproul, M.E., and A.G. Nassihian, "Temperature dependence of turn-on voltage of gold-doped MOSFETs," *IEEE Trans. Electron Devices*, **ED-22**, 1975, p. 8.
245. Stern, R.J., and Y.C.M. Yeh, "A 15% efficient antireflection-coated metal-oxide-semiconductor solar cell," *Appl. Phys. Lett.*, **27**, 1975, p. 95.
246. Stoneham, E.B., "A nonaqueous electrolyte for anodizing GaAs and GaAsP," *J. Electrochem. Soc.*, **121**, 1974, p. 1382.
247. Swanson, R.M., and J.D. Meindl, "Fundamental performance limits of MOS integrated circuits," *1975 IEEE International Solid-State Circuits Conf. Digest*, IEEE, NY, p. 110.
248. Sze, S.M., *Physics of Semiconductor Devices*, Wiley-Interscience, NY, 1969.
249. Takagi, H., and G. Kano, "Dual depletion CMOS ($D^2$CMOS) static memory cell," *IEEE J. Solid State Circuits*, **SC-12**, 1977, p. 424.
250. Tandon, J.D., and D.J. Roulston, "Low noise photodetector circuits using FET's," *Solid-State Electronics*, **17**, 1974, p. 607.
251. Tarui, Y., et al., "Diffusion self-aligned MOST-a new approach for high speed devices," in: *Proc. First Conf. Solid-State Dev.*, suppl. to *J. Japan Soc. Appl. Phys.*, **39**, pp.105-110.
252. Tarui, Y., Y. Hayoshi, and T. Sekigawa, "Diffusion self-aligned enhance-depletion MOS-IC(DSA-ED-MOS-IC), *Proc. Second Conf. on Solid State Dev.*, Tokyo 1970, Suppl. to *J. Japan Soc. Appl. Phys.*, **40**, 1971, p. 193.
253. Temple, V.A.K., and J. Shewchun, "The exact low-frequency capacitance-voltage characteristics of metal-oxide-semiconductor (MOS) and semiconductor-insulator-semiconductor (SIS) structures," *IEEE Trans. Electron Dev.*, **ED-18**, 1971, p. 235.
254. Tihanyi, J. and H. Schlotterer, "Properties of ESFI MOS transistors due to the floating substrate and the finite volume," *IEEE Trans. Electron Dev.*, **ED-22**, 1975, p. 1017.
255. Tihanyi, J. et al., "DIMOS; a novel IC technology with submicron channel MOSFETs," *IEEE International Electron Devices Meeting Technical Digest*, Washington, D.C., 1977.
256. Uchida, Y. et al., "A 1024 bit MNOS RAM using avalanche-tunnel characteristics below saturation," *IEEE International Solid-State Circuits Conference Digest*, 1975, p. 108.
257. Vadasz, L., and A.S. Grove, "Temperature dependence of MOS transistor characteristics below saturation," *IEEE Trans. Electron Devices*, **ED-13**, 1966, p. 863.
258. Valsamakis, E.A., "A short channel MOSFET model," *IEEE International Electron Devices Meeting Technical Digest*, Washington, D.C., 1977, p. 516.
259. Van Overstraeten, R.J., G.J. Declerck, and P.A. Meels, "Theory of the MOS transistor in weak inversion–new method to determine the number of surface states," *IEEE Trans. Electron Dev.*, **ED-22**, 1975, p. 282.
260. Vander Kooi, M., and L. Ragle, "MOS moves into higher power applications," *Electronics*, June 24, 1976, p. 98.
261. Vendelin, G.D., et al., "MOS transistors for microwave receivers," *Third Biennial Cornell Electrical Engr. Conf.*, Ithaca, NY, 1971, p. 417.
262. Verwey, J.F., and R.P. Kramer, "ATMOS–an electrically reprogrammable read-only memory device," *IEEE Trans. Electron Devices*, **ED-21**, 1974, p. 631.
263. von Munch, W., "Gallium arsenide planar technology," *IBM J. of Research and Development*, **10**, 1966, p. 438.

264. von Munch, W., H. Statz, and A.E. Blakeslee, "Isolated GaAs transistors on high-resistivity GaAs substrate," *Solid-State Electronics*, 9, 1966, p. 826.

265. Wada, A., "Estimation of the impurity redistribution in silicon substrates from MOSFET characteristics," *IEEE Trans. Electron Devices*, ED-20, 1973, p. 602.

266. Walker, L.G., and G.W. Pratt, Jr., "Long-term charge storage in interface states of ZnS MOS capacitors," *J. Appl. Phys.*, 46, 1975, p 2992.

267. Wallmark, J.T., and L.G. Carlstedt, *Field-Effect Transistors in Integrated Circuits*, John Wiley & Sons, NY, 1974.

268. Wang, K.L. and P.V. Gray, "Fabrication and characterization of germanium ion-implanted IGFET's," *IEEE Trans. Electron Devices*, ED-22, 1975, p. 353.

269. Wang, R., and T.A. DeMassa, "Silicon-aluminum gate complementary IGFET's," *IEEE Trans. Electron Devices*, ED-21, 1974, p. 130.

270. Weber, S. (ed), *Large and Medium Scale Integration*, McGraw-Hill, NY, 1974, p. 24.

271. Wedlock, B.D., "Direct determination of the pinch-off voltage of a depletion-mode field-effect transistor," *Proc. IEEE*, 57, 1969, p. 75.

272. Wegener, H.A.R., et al., "Radiation resistant MNOS memories," *IEEE Trans. Nuclear Sci.*, NS-19, 1972, p. 291.

273. Wei, L.S., and J.G. Simmons, "Trapping, emission and generation in MNOS memory devices," *Solid-State Electronics*, 17, 1974, p. 591.

274. Weimer, P.K., "The insulated-gate thin film transistor," in: *Physics of Thin Films*, Vol. 2, edited by G. Hass and R.E. Thun, Academic Press, NY, 1964.

275. Weinberg, Z.A., and R.A. Pollok, "Hole conduction and valence-band structures of $Si_3N_4$ films on Si," *Appl. Phys. Lett.*, 27, 1975, p. 254.

276. White, M.H., and J.R. Cricchi, "Complementary MOS transistors," *Solid-State Electronics*, 9, 1966, p. 991.

277. Wittmer, L.L. et al., "The infrared sensing MOSFET," *IEEE International Electron Devices Meeting Technical Digest*, Washington, D.C., December, 1975, p. 510.

278. Woods, M.H., and R. Williams, "Hole traps in silicon dioxide," *J. Appl. Phys.*, 47, 1976, p. 1082.

279. Wu, S.-Y., "Theory of the generation-recombination noise in MOS transistors," *Solid-State Electronics*, 11, 1968, p. 25.

280. Wu, S.Y., "A new ferroelectric memory device, metal-ferroelectric semiconductor transistor," *IEEE Trans. Electron Devices.*, ED-21, 1974, p. 499.

281. Wu, S.H. and R.L. Anderson, "MOSFET's in the $0°K$ approximation," *Solid-State Electronics*, 17, 1974, p. 1125.

282. Yamada, T., Y. Hirata, and S. Sato, "New field effect devices with linear V-I characteristics," *IEEE International Electron Devices Meeting Technical Digest*, Washington, D.C., December, 1974, p. 533.

283. Yamaguchi, T., and S. Sato, "Continuously variable threshold voltage devices," *IEEE International Electron Devices Meeting Technical Digest*, Washington, D.C., 1974, p. 537.

284. Yang, P. Ou, "Double ion implanted V-MOS technology," *IEEE J. Solid-State Circuits*, SC-12, 3, 1977.

285. Yoshida, I., M. Kubo, and S. Ochi, "A high power MOSFET with a vertical drain electrode and meshed gate structure," *IEEE International Electron Devices Meeting Technical Digest*, Washington, D.C., December, 1975, p. 159.

286. Yoshida, I., M. Kubo, and S. Ochi, "A high power MOSFET with a vertical drain electrode and a meshed gate structure," *IEEE J. Solid-State Circuits*, SC-11, 1976, p. 472.

287. Young, T.K., and R.W. Dutton, "MSINC—an MOS simulator for integrated nonlinear circuits with modular built-in model," Stanford Electronics Lab Tech Rept., SU SEL 74-038, TR 5010-1, July 1974.

288. Young, T.K., and R.W. Dutton, "Mini-NSINC–a minicomputer simulator for MOS circuits with modular built-in model," *IEEE J. Solid-State Circuits,* SC-11, 1976, p. 730.

289. Yu, K.K., T.P. Brody, and P.C.Y. Chen, "Experimental realization of floating-gate-memory thin-film transistor," *Proc. IEEE,* 63, 1975, p. 826.

290. Zahn, M.E., "Calculation of the turn-on behavior of MOST," *Solid-State Electronics,* 17, 1974, p. 843.

291. Zechnall, W., and W.M. Werner, "Determination of generation lifetime from the small-signal transient behavior of MOS capacitors," *Solid-State Electronics,* 18, 1975, p. 971.

292. Zemel, J.N., "Ion-sensitive field effect transistors and related devices," *Analytical Chemistry,* 47, 1975, p. 255A.

293. Zohta, Y., "Frequency dependence of $\Delta V/\Delta(C^{-2})$ of MOS capacitors," *Solid-State Electronics,* 17, 1974, p. 1299.

294. Evans, A.D. et al., "Higher power ratings extend VMOS FETs' domain," *Electronics,* 51, June 22, 1978, p. 105.

295. Lisiak, K.P. and J. Berger, "Optimization of nonplanar power MOS transistors," *IEEE Trans. on Electron Devices,* ED-25, 1978, p. 1229.

296. Petersen, K.E., "Micromechanical voltage controlled switches and circuits," *IEEE International Electron Devices Meeting Technical Digest*, Washington, D.C., Dec. 1978, p. 100.

297. Estreich, D.B., A. Ochoa, Jr. and R.W. Dutton, "An analysis of Latch-up prevention in CMOS IC's using an epitaxial-buried layer process," *IEEE International Electron Devices Meeting Technical Digest*, Washington, D.C., Dec. 1978, p. 230.

298. Sun, E. et al., "Breakdown mechanism in short channel MOS transistors," *IEEE International Electron Devices Meeting, Technical Digest*, Washington, D.C., Dec. 1978, p. 478.

299. Sugano, T. et al., "30-40 GHz GaAs insulated gate field effect transistors," *IEEE International Electron Devices Meeting Technical Digest*, Washington, D.C., Dec. 1978, p. 148.

300. Hoefflinger, B. et al., "Model and performance of hot-electron MOS transistors for high-speed, low power LSI," *IEEE International Electron Devices Meeting Technical Digest*, Washington, D.C., Dec. 1978, p. 463.

301. W. M. Werner, "The work function difference of the MOS-system with aluminium field plates and polycrystalline silicon field plates," *Solid-State Electronics,* 17, 1974, p. 769.

# 8
# Integrated
# Circuit Fundamentals

## CONTENTS

### 8.1 LARGE SCALE INTEGRATION

Integrated circuit technology has made possible instruments and applications that were inconceivable on the grounds of cost, size and reliability a decade ago.

The batch processing nature of IC fabrication is the feature that results in low cost. A large area integrated circuit is typically $0.5 \times 0.5$ cm$^2$ and may contain 5,000-10,000 transistors and be equivalent to 1,000 circuit functions. A user is of course interested in the number of circuit functions obtained for a given cost. A slice of Si of diameter about 7.5-cm fully processed, may cost about $50 after processing with seven or eight masking steps and 50-100 individual steps. If the slice contains 120 chips and if the yield of perfectly functioning integrated circuits is 10%, the cost per circuit is $4. If there are 800 circuit functions per chip, the cost per circuit function is $.005. A macro-circuit of many transistors and other discrete components to perform this function might require a price two orders of magnitude higher.

The problem of yield is a critical one. If the yield of perfect chips is 20% instead of 10% the cost per circuit is correspondingly reduced. Other factors of considerable importance concern interconnections and packaging costs. As the number of circuit functions per chip is increased the number of external interconnections per system is decreased, and the ratio of circuits to pins is increased. Both of these are factors that improve reliability. Since 1950, the

468

number of circuit failures per 1000 hr in computers has been decreased by a factor of 10 every 5 years largely because of the trend from discrete to small scale to medium scale to large scale integration.

Large scale integration brings other advantages such as decreased system size, lower power per circuit, reduced assembly costs, smaller transmission delays because of compactness and reduced circuit interconnection capacitance (or reactance). Large scale integration is of course limited to applications having a high volume of sales to recover the considerable costs of design and to aid in achievement of satisfactory yields from the process.

## 8.2 YIELD: THE DESIGNER'S DILEMMA

In planning an integrated circuit family, systems analysts partition the system and relate these parts to integrated circuit chips. The past experience of the company and of the industry as a whole is borne in mind in choosing particular logic forms. The whole circuit logic may then be constructed and tested in hardware form using readily available discrete gates or smaller scale integrated chips. The system performance will also be simulated on a computer so that the role of parasitics may be determined.

The designer must choose between general technology types such as MOS, bipolar and surface-charge transfer (CCD). If MOS is involved decisions must be made on matters such as $n$ channel, $p$ channel, CMOS, VMOS and the use of bipolar in conjunction with MOS. If bipolar technology is used the logic circuits must be selected from such efficient forms as $T^2L$, $T^3L$, ECL, EFL and $I^2L$. The logic levels of voltage must be decided on and clock frequencies chosen and compatibility ensured between all parts of the system. Then, estimates can be made of power dissipation and logic speed, and a chip layout can be roughed out. A typical flow pattern is shown in Fig. 8.1. The polycell library of step 5(b) relates to the set of standard logic subcircuits such as inverters, logic gates, dividers, memory cells and driver transistors maintained by the manufacturer in process files.

The polycells are usually designed to be rectangular and of standard heights and widths, say, 105-$\mu$m high by widths that are multiples of 15 $\mu$m. IC chips are then composed by positioning the required cells in rows and interconnecting them in routing channels between the rows. Mapping is simplified if a consistent style is adopted for bringing out the connections, such as from the width dimensions. The interconnections may be a continuation of diffusion lines or poly-Si gate lines or metallization lines to windows on the polycells. Busbars for the supply voltages may be arranged to be along the rows of polycells.

Computer aided layout programs usually are necessary to decrease the design time and to reduce errors. Such programs involve four main steps, namely, Input, Mapping, Routing and Output. Most layout programs can be run in a step-by-step mode with the interim results displayed on an oscilloscope screen so

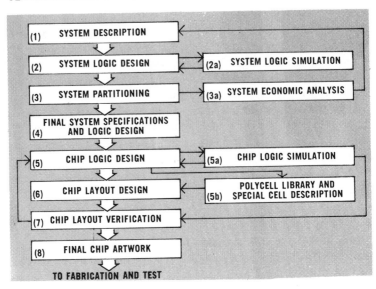

Fig. 8.1. Flow diagram for a typical integrated circuit system development.

that the user can interrogate the system and make on-line manual modifications at various stages. The effectiveness of the layout may be judged by the chip size that results and by the total length of each kind of interconnect. Interchanging of polycell positions in the rows or grosser changes may have to be tried in a man-machine interaction to arrive at a satisfactory arrangement.

Many layout programs have been constructed and details may be found in the IEEE Design Automation Conference Proceedings. A few of them with the originating company and the class of computer on which they have been run, are listed below:

| | | |
|---|---|---|
| LTX | Bell Laboratories[94] | H.P. 2100 Computer |
| PR2D | RCA Corporation[35] | Unspecified |
| FLOSS | RCA Corporation[17] | |
| FAMOS | GTE Sylvania[53] | IBM 370 |
| CALMOS | Catholic University of Leuvens[8] | Any with FORTRAN Compiler |
| DA4 | International Computers Ltd.[21] | Unspecified |
| AVESTA | Siemens AG[72] | Siemens 40004/151 |

The work days needed for an LSI design cycle are shown in Table 8.1. However, designs may take longer if they are not variations or extensions of previous successes.

The design cost, including prototype production, tends to vary between $20 and $100 per gate depending on the complexity of the chip. Thus a 5,000 gate IC design may cost between $100,000 and $500,000.

With the technology chosen and the chip area determined the question of

Table 8.1.  Automated LSI Design Cycle Time Intervals

| Function | Step Time (Work Days) | Cumulative Time (Work Days) |
|---|---|---|
| Network Description | 5-6 | 5-6 |
| Unit Delay Simulation and Vertification | 2-5 | 7-11 |
| Network Array Design | 3-5 | 10-16 |
| Real Delay Simulation and Verificiation | 2-4 | 12-20 |
| Graphic Editing | 3-5 | 15-25 |
| Interconnect Check | 3 | 18-28 |
| Pattern Generation | 11 | 29-39 |
| Artwork Review | 2 | 31-41 |
| Mask Generation | 13 | 44-54 |
| Wafer Processing | 18 | 62-72 |
| Design Verification | 3-12 | 65-84 |
| Packaging | 3 | 68-87 |
| Final Testing | 1 | 69-88 |

Total Elapsed Time:  14-18 weeks.

(After Herrick and Sims. Reprinted with permission from Design Automation Conference. 13th, 1976, p. 74, #76 CH1098-3C.)

yield arises. The designer asking for finer dimensional or electrical tolerances than photolithography or multiple mask alignments can readily provide is living dangerously. Circuits should not demand transistor betas, mutual conductance values or resistance or capacitance values held within unusually close tolerances. The circuit should not be dependent on precision interrelation of clocking pulses or neglect parasitic effects or temperature effects. Judgment of when to break new ground and when to back off and play safer comes from intuition and experience due to learning from one's own successes and mistakes and those of others.

Whether the circuit is made on bulk wafer material or epitaxial layers, the starting material always contains defects and, as the processing continues, other defects will be introduced by heat or chemical interaction. Some of these defects, associated with crystal imperfections, inclusions or precipitates (for instance, Cu on a stacking fault), or doping variations, may be serious enough to ruin the circuit performance. A single defect may give trouble, but sometimes the deterioration is more gradual and depends on the density or magnitude of the typical defect.

Obviously the greater the chip area the lower the yield. If the defects are uniformly distributed over the wafer the yield $Y$ for area $A$ is given by $Y = (Y_0)^{A/A_0}$ where $Y_0$ is the yield for area $A_0$. For instance the probability of no defects in area $2A_0$ is the product of the probabilities of no defects in two areas $A_0$, namely $Y = Y_0^2$. If $A_0$ is chosen so that $Y_0 = 1/e$ then we have

$Y = e^{-A/A_0}$ and the logarithm of yield is inversely proportional to chip area. The observed yield decrease is however usually at less than an exponential rate as suggested by Fig. 8.2(a). This comes about because the defects tend to be clustered on the wafer instead of uniformly distributed. Temperature non-uniformities during processing may cause one segment of a wafer to incur more strain and therefore more precipitation than the rest of the wafer. There are also other factors that can lead to clustering of defects.

Although the yield falls at less than an exponential rate, the loss is very substantial with increasing area. Various attempts have been made to model the situation by introducing the concept of an average deleterious defect density $D$, per unit area. One approach assumes that the defect density probability value is higher for zero defect density and decreases exponentially with increase in the density.

Tightening of the design rules from, say, 5 $\mu$m to 3.5 $\mu$m allows a smaller chip size for a given IC and, therefore, more chips per wafer. However, closer spacing of the diffusion or metal lines causes more defects to become deleterious, hence $D$ is a function of the design rules. Hence, reduction of IC size by closer packing of features if carried too far results in a lower yield. Too large a chip area, however, results in fewer circuit starts per wafer. Consequently, there may be an optimum size for lowest cost per working chip.

Under constant design rule conditions, larger area chips allow more circuit functions per IC, but the chance of including a deleterious defect increases.[133] One company, for instance, finds that yield decreases exponentially with the square root of increasing area. Such a yield curve is shown in Fig. 8.2(b), but the second line also shows that as time passes and experience with the processing is gained, the yield increases. The improvement may be expressed as a rate of decrease of $D$, perhaps a 40% decrease per year.

One conclusion to be drawn from such modeling and experience is that over a period of years a low-yield large-area chip increases proportionately more in yield than a smaller chip. A significant question, though, is whether the larger chip will generate lower cost per function in a few years time.

To illustrate this matter, consider a processed wafer costing $100 that may contain either 400 chips of an MSI design of area $A$ cm$^2$ and $n$ circuit functions per chip (such as a 4K memory) or 100 chips of an equivalent LSI design of area $4A$ cm$^2$ and $4n$ circuit functions per chip (such as a 16K memory). In 1975 let the yields be 5% and 10%, respectively, and in 1978 suppose the respective yields to be 20% and 30%. Thus, for the larger chip a fourfold increase in yield has been assumed and for the smaller chip a threefold increase in yield during the time span considered.

Consider the 1978 figures. For MSI there are 120 useful chips per wafer and $120n$ circuit functions per wafer. For the LSI design there are 20 useful chips per wafer and this corresponds to $80n$ circuit functions. Thus the evidence seems to favor MSI since this yields more useful circuit-functions per wafer.

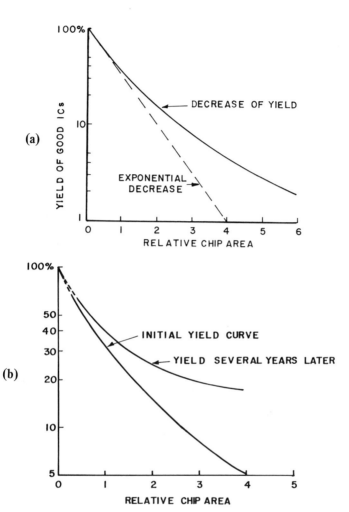

Fig. 8.2.  Typical yield curves for ICs:
(a)  Yield of ICs decreases rapidly with increase of chip area but usually at less than an
     exponential rate; and
(b)  Illustration of how IC yield may increase with time.


However, now consider packaging and test costs. Suppose such costs for an
LSI chip to be $2 and for an MSI to be $1. Then Table 8.2(a) shows the overall
result (1978) to be in favor of LSI in terms of cost per circuit functions. Since
one IC package can be used in place of 4 MSI packages, the socket assembly
and interconnection costs weigh matters much more strongly in favor of the LSI
version than might appear from first inspection of Table 8.2.
A designer therefore must decide on the level of integration that represents

Table 8.2.   Comparison of Cost per Circuit Function When
Yields and Packaging Costs Are Considered[a]

(a)   *For 1978 yields of 20% and 30%*

|  | Circuit Functions per Wafer | Useful Chips | Cost of Packaging | Total Cost Wafer Plus Packaging | Cost per Circuit Function (when packaged) |
|---|---|---|---|---|---|
| LSI | $80n$ | 20 | $ 40 | $140 | $1.75/n$ |
| MSI | $120n$ | 120 | $120 | $220 | $1.83/n$ |

(b)   *For 1975 yields of 5% and 10%*

| LSI | $20n$ | 5 | $10 | $110 | $5.50/n$ |
| MSI | $40n$ | 40 | $40 | $140 | $3.50/n$ |

[a]The numbers used in this example are hypothetical and not necessarily representative of practice.

the best balance between the needs of the immediate vs the more remote future. Usually this must be done on the basis of inadequate information about yields, costs and market demand. If too conservative in the scale of integration, the product may fail to be priced competitively with that of a bigger thinking rival. On the other hand, if the company goes for the long return and encounters unexpected yield problems, the short term losses may be considerable.

Observation of price declines over a decade of semiconductor experience and development has produced some evidence that each time the cumulative production doubles the unit price decreases to a certain percentage of the previous price. For IC gates, doubling the production gives a price that is 70% of the previous price. The log-log plot of price per gate vs accumulated production of units is a straight line and in this instance is referred to as a 70% learning curve. This 70% value is a useful working rule, but must not be taken too rigidly. A major technological breakthrough may drop the price level significantly below one learning curve and start production costs tracking down a lower parallel curve. The transition from minicomputers to microprocessor computers is one example.

IC reliability is an essential goal. Testing, which is expensive and difficult to do fully for a complex circuit, can eliminate the results of obvious processing and handling mistakes. Component burn-in may also be necessary to weed out units with incipient faults.[76]

Extended high-temperature-stress life tests are needed to determine if more subtle problems such as electromigration exist. It is usual to conduct operating life and storage tests at temperatures such as 125°C or 150°C, and to predict from these the expected failure rate at a lower ambient temperature such as 55°C. A failure rate of 0.05%/1000 hr (quoted at a 60% upper confidence level) from $125°C$ and $150°C$ data extrapolated to $55°C$, might be representative of CMOS. For large MOS memory systems this is hardly adequate since failure rates of less than one in $10^8$ hr are needed, and various error correction and detection circuits must be provided. For telephone system use in electronic switching stations, the reliability level of components aimed for is one failure per $10^9$ operating hr.

## 8.3 BIPOLAR IC TECHNOLOGY

Selection of the logic family for various parts of a system requires good information, experience and judgment. Only the student aspiring to be an integrated circuit designer need understand all of the circuit trade-offs and refinements. However, a general understanding of the kinds of problems involved is readily obtained, and the references at the end of this chapter lead to the more extensive literature when this is needed for advanced study.

In general bipolar technology is selected over MOS technology when substantial output power is needed. However recent developments in VMOS power devices are beginning to erode this advantage. A TTL gate is shown in Fig. 8.3. This logic family has been used for more than a decade and is well discussed and modeled.[44, 46] TTL is high in power dissipation and not very high in speed since in its original form transistor saturation may occur. Schottky TTL in which a Schottky metal-semiconductor junction is provided across the base-collector junction is shown in Fig. 8.4. This prevents the transistor from being driven into saturation forward bias since excessive base drive is diverted by conduction of the Schottky diode. Figure 8.4 also shows a buried $n^+$ collector layer that reduces collector resistance and allows shallower diffusions that produce faster transistor action. A variation of TTL known as $T^3L$ that improves noise immunity and circuit convenience also exists.[36]

Emitter coupled logic is an important high-speed bipolar technology. A basic ECL OR gate is shown in Fig. 8.5. If all the inputs are near ground potential, no current flows in $R_1$ and $T_5$ turns on and gives a logical 1 output across its emitter resistor. The voltage $V_{bb}$ sets the threshold level for a logical 1 at the input. Since the transistors do not become saturated the switching speed is high and switching does not generate noise glitches but the impedances are such that the power consumption is fairly large.[46, 111]

Gate propagation delays, power dissipations and speed-power products for various technologies are shown in Fig. 8.6 and Table 8.4. Recent versions of TTL may be compared with ECL and EFL (an emitter follower logic variation) in performance. $I^2L$ is a merged-transistor bipolar logic that will be discussed later.

Isolation of bipolar transistors and associated resistors is possible by the use of deep diffusions to box off sections of the wafer into islands, as shown in Fig. 8.7(a). A triple diffusion technique is shown in Fig. 8.7(b). Both are somewhat wasteful of wafer area ("real-estate") and parasitics can be a problem. Dielectric isolation is possible but is not easy or low cost to achieve. Figure 8.8 shows how regions of semiconductor grade Si with $SiO_2$ isolation may be created in a poly-Si matrix. However, it is more practical to achieve some degree of dielectric isolation with an original wafer that has deep regions of $SiO_2$ grown into it as in the Fairchild Isoplanar II process illustrated in Fig. 8.9.

Another dielectric isolation technology is the etching of V shaped grooves in the wafer so that air is essentially the insulator.[85] However, interconnections pass up and down the sides of grooves and this, although practical, requires careful control of the metallization process to avoid strain and thickness fluctuations that may ultimately affect reliability. Some technologies therefore coat the sides of the grooves with $SiO_2$ or $Si_3N_4$ and fill with poly-Si to provide a flat surface.[102]

Following this brief review of bipolar fabrication techniques, consider the bipolar technology known as integrated injection logic, $I^2L$. This system is economical of wafer area since there is some merging of the transistor functions. (Merged transistor logic might be a better name for it.) It is reasonably fast in speed and quite low in power, so that the speed-power product is very low. The development of an $I^2L$ gate is illustrated by the transitions shown in Fig. 8.10 and a typical construction is illustrated in Fig. 8.11. The packing density is better than $T^2L$ by a factor of 10 and the speed-power product about 100 times better for a propagation delay per gate of the order of 10 nsec. The use of dielectric isolation or ion implantation in recent $I^2L$ designs has further improved performance to delays of the order of 5 nsec at a speed-power product of 0.7 pJ/gate. Using a second substrate as an injector has also led to a higher

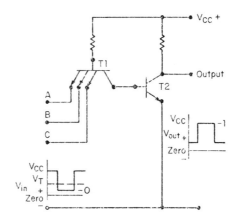

Fig. 8.3. A typical TTL gate. (After Morris and Miller, from *Designing with TTL Integrated Circuits*, copyright 1971, McGraw-Hill, NY. Reprinted with permission of McGraw-Hill Book Company and Texas Instruments, Inc.)

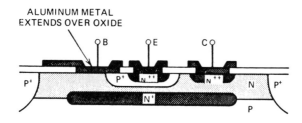

SCHOTTKY EXTENDED-METAL TRANSISTOR

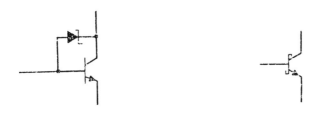

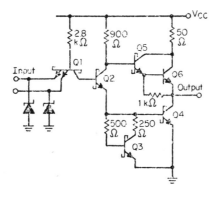

Fig. 8.4. Structure and symbols for a Schottky clamped transistor and a Schottky clamped TTL gate circuit. (After Morris and Miller, from *Designing with TTL Integrated Circuits*, copyright 1971, McGraw-Hill, NY. Reprinted with permission of McGraw-Hill Book Company and Texas Instruments, Inc.)

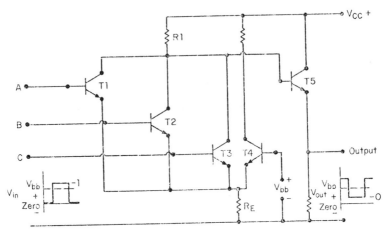

Fig. 8.5. An emitter-coupled logic (ECL) gate. (After Morris and Miller, from *Designing with TTL Integrated Circuits*, copyright 1971, McGraw-Hill, NY. Reprinted with permission of McGraw-Hill Book Company and Texas Instruments, Inc.)

packing density and a lower speed-power product. $I^2L$ is seen to plot very well in comparison with other logic families in Fig. 8.6.

Modifications of the $I^2L$ structure have included the use of Schottky diodes to reduce the logic swing and hence increase the speed of operation.[134] A power delay improvement factor of 5 and a speed limit improvement factor of 2 have been reported at a reduced logic swing of 150 mV.

The minimum propagation delay of an $I^2L$ gate is an increasing function of the upward transit time and the upward and downward current gains. The power-delay product, on the other hand, decreases with decreasing voltage swing and decreasing gate capacitance. By substituting the collector-base junction of the switching transistor by a Schottky contact, it is possible to reduce both the voltage swing and the downward current gain and thus to increase the speed of the gate over the entire power range. For a suitable collector Schottky barrier height the base region of the transistor must be $n$ type hence the structure is a *pnm* transistor. Figure 8.12 then shows the evolution of a Schottky-collector $I^2L$ gate.

The large saturation current of the Schottky causes the forward voltage drop across the base-collector junction to be small and the resulting absence of substantial minority carrier injection results in a very low value of downward current gain.

Hence, the Ebers-Moll model for a transistor simplifies into the form shown on the right-hand side of Fig. 8.13(a) by eliminating the down current term.

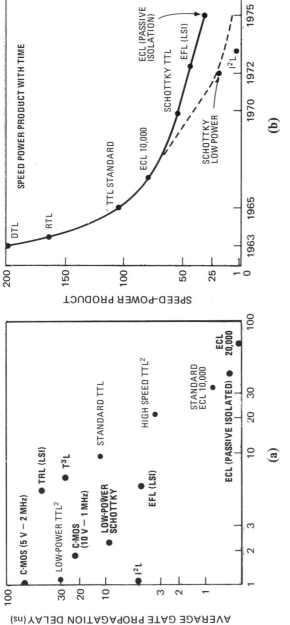

Fig. 8.6. The propagation delay, gate dissipation and speed-power product of various logic families. (After Altman. Reprinted from *Electronics*, February 21, 1974; copyright McGraw-Hill, Inc, NY, 1974.)

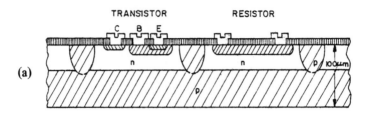

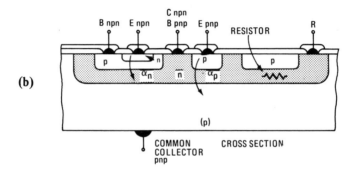

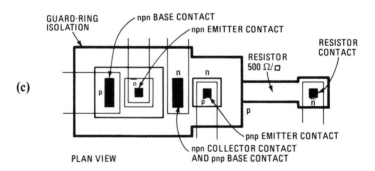

Fig. 8.7. Isolation techniques for bipolar devices on a wafer:
(a)   The *n* well is created from the original *n* substrate by the deep *p* diffusions;
(b)   The *n* well is created by diffusion and the *npn* transistor is created by triple diffusion; and
(c)   Plan view of (b). (After Buie. Reprinted from *Electronics*, August 9, 1975; copyright McGraw-Hill, Inc., NY, 1975.)

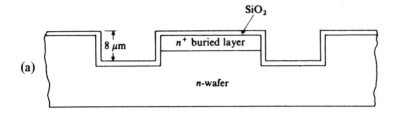

(a)

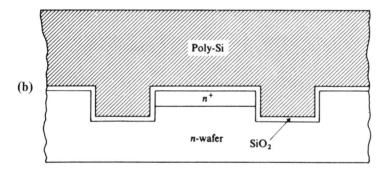

(b)

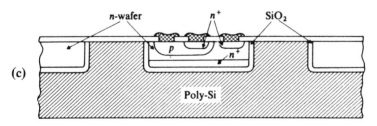

(c)

Fig. 8.8. Device quality $n$ wells are shown imbedded, and $SiO_2$ isolated, in a polycrystalline substrate created by an inversion process:
(a)  Original $n$ wafer with pockets etched is $SiO_2$ coated;
(b)  After addition of poly-Si to form new substrate; and
(c)  After inversion, removal of original substrate and fabrication of devices in the $n$ wells. (After Hamilton and Howard, from *Basic Integrated Circuit Engineering*, copyright 1975 by McGraw-Hill, Inc., NY. Used with permission of McGraw-Hill Book Company.)

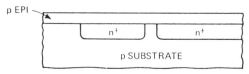

**(a)** EPITAXIAL LAYER

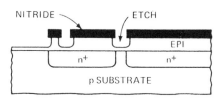

**(b)** ISOLATION ETCH

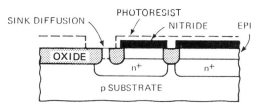

**(c)** ISOLATION OXIDATION AND SINK

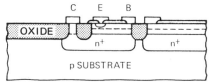

**(d)** FINAL METAL

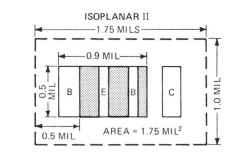

**(e)**

Fig. 8.9.  The Isoplanar II Process of Fairchild:
(a)   A $p$ epi-layer is grown after the buried $n^+$ collector regions have been formed;
(b)   Silicon nitride masking used to open wells for oxide growth;
(c)   Isolation oxidation and sink diffusion to the $n^+$ region;
(d)   Possible final structure after devices created by other diffusions and metallization; and
(e)   Typical plan geometry of a transistor. (After Baker et al., Reprinted from *Electronics*, March 29, 1973; copyright McGraw-Hill, Inc., NY, 1973.)

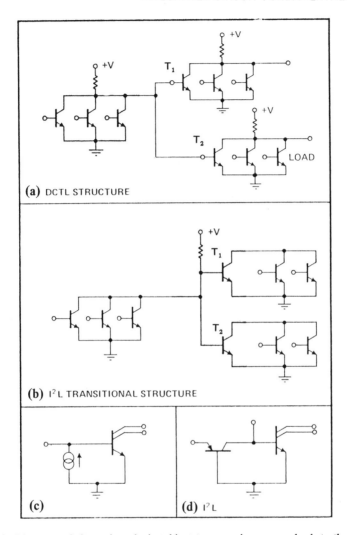

**(a)** DCTL STRUCTURE

**(b)** I²L TRANSITIONAL STRUCTURE

**(c)**

**(d)** I²L

Fig. 8.10. Direct coupled transistor logic subject to a merging process leads to the concept of integrated injection logic (I²L) gates. The transistors $T_1$ and $T_2$ with common bases in (a) are placed in a common region of the wafer in (b). It is then apparent that $T_1$ and $T_2$ can be merged and replaced by a multicollector transistor, with a current-source base device as in (c). This current source can then be provided by a *pnp* transistor as in (d). Note the collector of the *pnp* unit is common with the base of the *npn* unit and the base of the *pnp* is common with the emitter of the *npn* transistor. Hence, considerable merging action and efficient use of wafer space is possible. (After Hart, Slob and Wulms. Reprinted from *Electronics*, October 3, 1974; copyright McGraw-Hill, Inc., NY, 1974.)

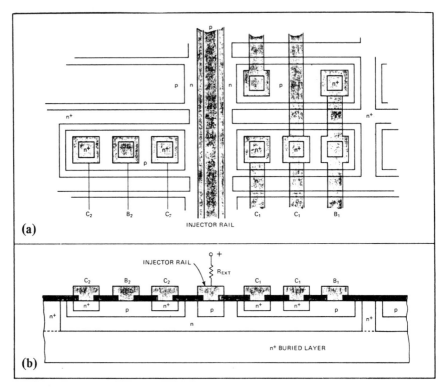

Fig. 8.11. Gates of an $I^2$ L chip (a) are situated on both sides of an injector rail, which forms the emitter of the lateral *pnp* current source transistor. A heavily doped $n^+$ isolation region increases the current amplification factor of the *npn* transistor and kills the parasitic effects of the *pnp* transistors between two adjacent gates. In (b) are shown the space-saving features of $I^2$ L. (After Hart, Slob and Wulms. Reprinted from *Electronics*, October 3, 1974; copyright McGraw-Hill, Inc., NY, 1974.)

Addition of the current-source transistor then gives Fig. 8.13(b) for the equivalent circuit of the gate. From Kirchhoff's current law at node $A$:

$$I_{FN} - I_{SP}\phi(-V_{in}) - I_{SS}\phi(V_{in} - V_{out}) - \frac{I_{SP}}{\beta_U}\phi(-V_{out})$$

$$+ I_{DS}\phi(V_{out} - V_{in}) = 0 \tag{8.1}$$

The symbol $\phi_V$ stands for $\exp[(q/kT)V] - 1$; $I_{SP}$ is the saturation current of the *pnp*; $I_{SS}$ is the saturation current of the Schottky barrier; $\beta_U$ is the beta of the vertical *pnp* with injector grounded; and $I_{FN}$ is the injector current = $I_{SN}\phi(-V_{CC})$. Numerical solution of Eq. (8.1) gives the transfer characteristic of the gate, as shown by the example in Fig. 8.13(c).

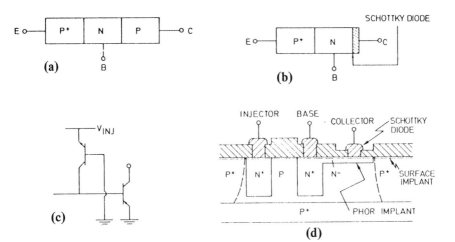

Fig. 8.12. Schottky collector $I^2L$ inverter gate:
(a, b) $p^+np$ transistor evolving into Schottky collector $p^+nm$ structure;
(c)    $I^2L$ gate; and
(d)    Structure of the gate. (After Blackstone and Mertens. Reprinted with permission from *IEEE Journal of Solid-State Circuits*, **SC-12**, 1977, pp. 271-272.)

The low downward beta of Schottky collector $I^2L$ involves some limitation of both the fan-out and the fan-in. However, delays of less than 10 nsec have been measured using a 10-$\mu$m technology and a divide-by-two circuit with a maximum toggle frequency of 12.5 MHz has been built.

Other means of controlling the saturation of the transistors in $I^2L$ structures have been proposed.[135] The inherent saturation control of $I^2L$ structures can be used. This is done by controlling the back injection of the common base region. Another method of controlling the saturation of the *npn* transistors is based on adding an extra "dummy collector" to the multicollector *npn* transistor and folding this back to the base.

Figure 8.14(a) shows three $I^2L$ inverters where $Q_1$, $Q_2$, and $Q_3$ are multi-collector *npn* transistors. When $Q_1$ is turned off, $Q_2$ is on and $Q_3$ is off. The transistor $Q_2$ is heavily saturated in its conduction state because of the excessive base current supplied to its base. In this case, $I_B=I_C=I_O$. The excess stored charge contributes to the minimum delay of the inverter. To improve this delay, the saturation of $Q_2$ has to be controlled. This can be achieved by adding an extra dummy collector to the *npn* transistor and folding this back to the input base, as shown in Fig. 8.14(b). Similar circuit techniques have been used success-fully in controlling the saturation of the output transistor of $T^2L$ structures.

The minimum delay time of an $I^2L$ structure is proportional to $(\beta_U)^{1/2}f_T^{-1}$

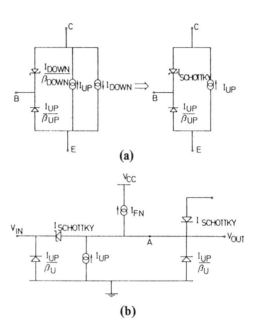

**(a)**

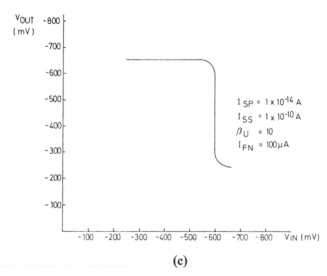

**(b)**

$I_{SP} = 1 \times 10^{-14}$ A
$I_{SS} = 1 \times 10^{-10}$ A
$\beta_U = 10$
$I_{FN} = 100 \mu$A

**(c)**

Fig. 8.13. Analysis of a Schottky collector $I^2L$ gate:
(a)  Schottky transistor model from Ebers-Moll form;
(b)  Equivalent circuit of the inverter gate; and
(c)  Calculated transfer characteristics of the gate. (After Blackstone and Mertens. Reprinted with permission from *IEEE Journal of Solid-State Circuits*, **SC-12**, 1977, p. 271.)

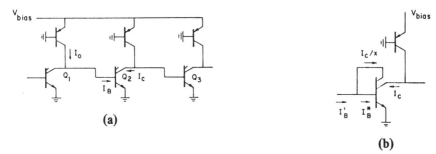

(a)

(b)

Fig. 8.14. Folded-collector integrated injection logic:
(a) Three $I^2L$ inverters, illustrating saturation of $Q_2$.
(b) Folded collector $I^2L$ structure. The ratio of the areas of the output collector and the "dummy" collector controls the saturation of the $npn$ transistor and improves the minimum delay of the basic $I^2L$ gate. (After Elmasry. Reprinted with permission from *IEEE Journal of Solid-State Circuits*, SC-11, 1976, p. 644.)

where $\beta_U$ is $I_C/I_B$ and $f_T$ is proportional to the ratio of the collector to the base area. From Fig. 8.14(b)

$$\beta_U = \frac{I_C}{I'_B} = \frac{I_C}{I_B^* + I_C/x} \simeq \frac{I_C}{I_B + I_C/x} = \frac{x\beta_U}{x + \beta_U}. \qquad (8.2)$$

where $x$ is the ratio of the areas of the output and dummy collectors. The minimum delay time may then be shown to be reduced by the dummy collector structure in a suitable current operating range.

One variation of $I^2L$ termed vertical injection logic (VIL) uses a vertical *pnp* transistor in place of the lateral *pnp* transistor of conventional $I^2L$. The current gain of the vertical transistor is higher than for a lateral transistor which leads to an improved power-delay product. The intrinsic delay time is also improved by the action of the bottom injector as a hole sink. The fabrication process is illustrated in Fig. 8.15 and compared with that for $I^2L$. In one study the alpha of the vertical transistor was 0.6-0.9 for currents of 0.1 to 300 $\mu A$ compared with 0.3-0.5 for $I^2L$. The results in a 13-stage closed-loop inverter chain showed a minimum delay time of 8.8 nsec and a power-delay product of 0.07 pJ at low power level below 1 $\mu W$ for VIL compared to 25 nsec and 0.18 pJ for standard $I^2L$.

Computer simulation of $I^2L$ logic circuits is possible by models that use a modified Ebers-Moll approach that includes lateral and current redistribution effects.[33] Lateral current transfer between adjacent gates (because of parasitic transistor effects) and injector current redistribution effects reduce gate propagation delay times. A macromodel for an $I^2L$ gate with a fan-out of three is shown in Fig. 8.16.

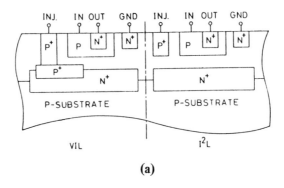

(a)

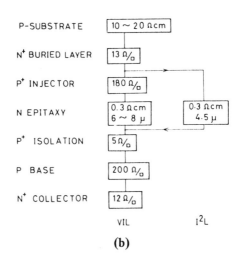

(b)

Fig. 8.15. Vertical injection logic compared with $I^2$ L:
(a)  Schematic cross sections. The $p^+$ horizontal layer in VIL is the emitter of the vertical
*pnp* transistor. The *n* epitaxial layer acts both as the base of the *pnp* transistor and
as the emitter of an *npn* transistor; and
(b)  Fabrication steps and process parameters of VIL and $I^2$ L. The $p^+$ injector step is added
in VIL to conventional $I^2$ L process. (After Tomisawa et al. Reprinted with permission
from *IEEE Journal of Solid-State Circuits*, **SC-11**, 1976, p. 638.)


A merged transistor dynamic bipolar memory cell is shown in Fig. 8.17(a).
The logic ones and zeros are stored in the shared collector-base junction of the
merged transistors. Although the storage capacitor is quite small, on the order
of 0.1 pF, the relatively high doping concentrations provide both high coupling
capacitance and low leakage per unit area. In addition, the *npn* cell transistor
provides a β-gain of 70 during readout, so that the effective coupling capacitance
is about 7 pF and the signal available at the output is large enough to drive

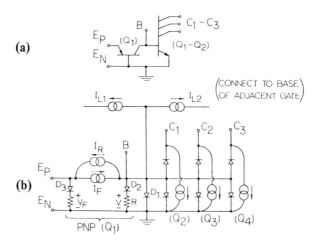

Fig. 8.16.  $I^2$ L gate with a fan-out of three:
(a)  Circuit schematic; and
(b)  Circuit model for use with SPICE program. (After Estreich, Dutton, and Wong. Reprinted with permission from *IEEE Journal of Solid-State Circuits,* **SC-11**, 1976, p. 649.)

high-capacitance bit lines. The double diffused *pnp* transistor used is seen from Fig. 8.17(c). The isoplanar topology has led to the term $I^3L$. The cell size is only 0.7 mil$^2$ for a 16K bit memory.

The performance of an $I^3L$ 16K bit dynamic memory is shown in Table 8.3 and is seen to be competitive with that of recent *n*-MOS 16K bit memories. Hence, bipolar technology is still very much a contender in some areas of the integrated circuit field.

## 8.4 MOS INVERTERS

MOS logic circuits in the last decade have tended to replace bipolar logic in many classes of applications. The reasons for this are:

● MOS fabrication processes usually have less than half the operator handling steps of bipolar processes and there are fewer masking and high temperature steps.

● Although in the smaller slices good mask alignment control is needed, the yield is usually better than for comparable bipolar circuit functions and the packing density is high. The cost, therefore, is lower.

● Both *n* and *p* channel technology are now available, so complementary symmetry circuits are possible for low power dissipation.

● Furthermore, *n*-channel MOS has the advantage of greater channel mobility and therefore improved performance and greater packing density for the same impedance level.

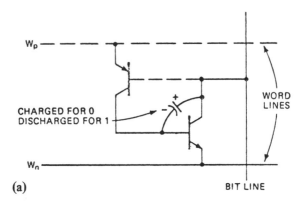

(a)

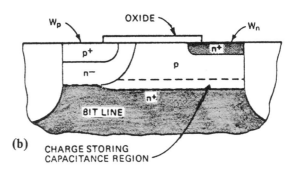

(b)

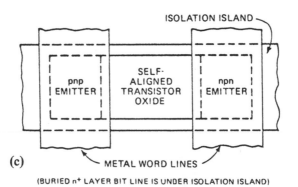

(c)

(BURIED n⁺ LAYER BIT LINE IS UNDER ISOLATION ISLAND)

Fig. 8.17. Memory cell using isoplanar $I^2L$. (After Sander, Early and Longo. Reprinted from *Electronics*, August 18, 1977; copyright McGraw-Hill, Inc., NY, 1977.)

Table 8.3.  Comparison of 16K nMOS and $I^3L$ Memory Chips

| Memory Characteristics | nMOS 4027-2 | $I^3L$ 93481A |
|---|---|---|
| Access time (ns) | 150 | 100 |
| Cycle time (ns) | 320 | 240 |
| Page access time (ns) | 90 | 65 |
| Page cycle time (ns) | 160 | 65 |
| Maximum power (at max cycle time) (mW) | 462 | 450 |
| Maximum power (standby) (mW) | 27 | 70 |
| Power supply (V) | +12; +5; −5 (10%) | +5 (5%) |
| Chip selects | 1 | 2 |
| Timing inputs | 2 | 1 |
| Data latch | always | user-controlled |
| Input capacitance (pF) | 4-8 typical | 2 typical |
| Output drive low/high (mA) | 3.2/5 | 16/5 |
| Chip size (mil²) | 14,500 | 11,700 |
| Refresh | 64 lines; 2 ms | 32 lines; 2 ms |

(After Sander, Early and Longo, 1977.)

- The technology of self-aligning poly-crystalline Si gates has reduced mask alignment problems and has also lowered the threshold voltage for better logic voltage swings.
- The impedance levels tend to be higher than for bipolar circuits and this contributes to a lower power dissipation than for bipolar logic. The lower impedance of bipolar circuits makes them more suitable for driving large capacitance loads and high power peripheral output functions. Interfacing between MOS logic and bipolar driver circuits is usually not too inconvenient.
- MOS transistors do not have the minority carrier charge storage effects that often limit the switching speeds of bipolar logic circuits although they may have RC time constant limitations.
- An MOS device has the possibility of short time charge-storage in its inherent gate capacitance or long term memory by charge storage in a complex insulator region. Both effects are made use of in semiconductor memory array structures.
- MOS transistor type structures can serve as load resistors and therefore the process is highly self-compatible, and under certain conditions the performance is fairly insensitive to temperature effects.

Some important features of MOS logic may be understood by considering the most elementary logic gate which is the inverter. There are several versions of such circuits that must be considered, depending on the relationship of the driver to the load. An inverter circuit consists of a transistor in series with a load and the output is taken from the common node. If the driver transistor is conducting perhaps 0.1 mA of current flows through the load and so for a 10-V swing the effective resistance of the load must be 100 kilohms. In an integrated circuit the ohms per square of a diffused resistor region may be 200 ohms/square and so for a width of 10 $\mu$m and a resistance of 100 kilohms the length would have to be 500 $\times$ 10 $\mu$m, or 5,000 $\mu$m, and the area 50,000 $\mu$m$^2$. However, an MOS transistor of suitable impedance may be obtained from an area that is substantially less than 1000 $\mu$m$^2$. Hence, transistor loads are used in inverter circuits.

An MOS inverter circuit with an enhancement-mode driver and a depletion mode load is illustrated in Fig. 8.18 and shows the load line (curve for $V_{GS(load)}$ equal to zero) crossing the characteristics of the driver transistor. The transfer characteristic, $V_{out}$ vs $V_{in}$, that results depends on whether the load characteristic is adjusted to be in the saturated region of the curve or in the "triode" region as it crosses the driver characteristics. The positioning of the load line depends on bias and on the ratio of the transconductance of the driver to that of the load. This is termed the electrical beta ratio $\beta_R$ and is related to a

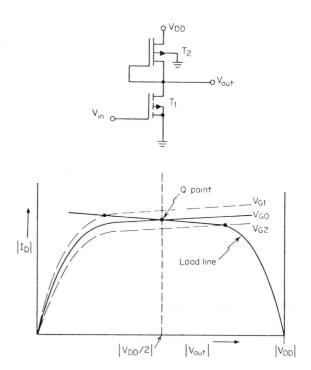

Fig. 8.18. MOS Inverter with enhancement-mode driver and depletion-mode load in the saturated region. (After Carr and Mize, from *MOS/LSI Design and Application*. Reprinted with permission of Texas Instruments, Inc., and McGraw-Hill, Inc., copyright 1972.)

geometrical beta ratio which is derived from the mask geometry. If the width to length ratio of the driver is 5 and for the load is 0.25 then the geometrical beta ratio is 20 and the electrical beta will be comparable in value. The effect of various beta ratios on transfer characteristics is shown in Fig. 8.19 for a $p$ enhancement driver and a depletion load in the triode region (which may be termed a PDLT combination). The transfer curves for high beta ratios are seen to result in sharper transitions and greater logic swings than for $\beta_R = 1$. The results for an inverter $p$-channel with an enhancement-type load in the triode region are shown in Fig. 8.20 for comparison. Again, high beta ratios are preferable. If the enhancement load is operated in the saturated region (PELS

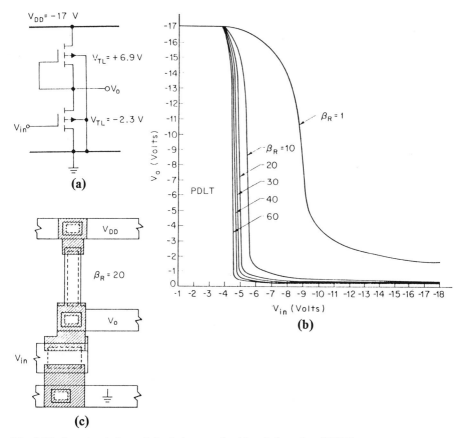

Fig. 8.19. Inverter, $p$-channel, depletion-type load in triode region (PDLT):
(a) Circuit schematic;
(b) Transfer characteristic curves; and
(c) Mask geometry for a masking $\beta_R = 20$. (After Carr and Mize, from *MOS/LSI Design and Application*. Reprinted with permission from Texas Instruments, Inc., and Mc-Graw-Hill, Inc., copyright 1972.)

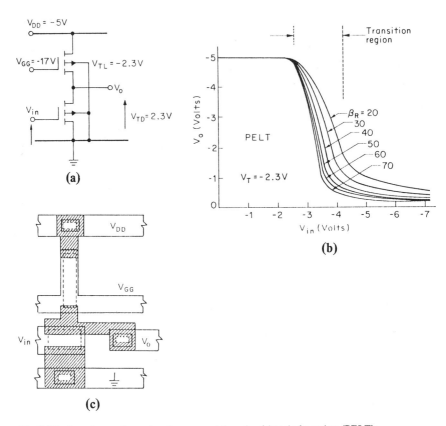

**Fig. 8.20.** Inverter, *p*-channel, enhancement-type load in triode region (PELT):
(a)  Circuit schematic;
(b)  Transfer characteristic curves; and
(c)  Mask geometry for a masking $\beta_R = 20$. (After Carr and Mize, from *MOS/LSI Design and Application.* Reprinted with permission from Texas Instruments, Inc., and Mc-Graw-Hill, Inc., copyright 1972.)

mode) it has a higher effective load resistance than in the PELT mode and so charging of the load capacitance is slower.

The transfer characteristics for a CMOS inverter are shown in Fig. 8.21. For this inverter a masking ratio as low as $\beta_R = 1$ is seen to produce a transition of high gain. Steepness of the transfer characteristics with a square knee form is desirable for an inverter since it increases the stability of quiescent points in the presence of noise spikes in the input voltage. In Fig. 8.22(a) the noise margins $NM_A$ and $NM_B$ are shown as the voltage spikes that can be tolerated from the points A and B to the unity-gain points on the transfer curve. The two noise margins should preferably be equal in size (since the smaller determines the effective margin) and on the order of several volts. Some normalized transfer

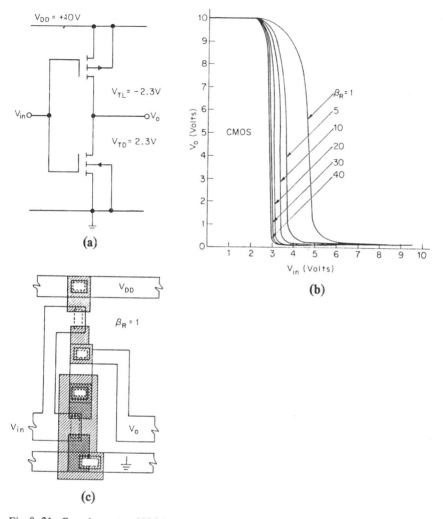

Fig. 8. 21. Complementary MOS inverter (CMOS):
(a) Circuit schematic;
(b) Transfer characteristic curves;
(c) Geometry for a masking $\beta_R = 1$. (After Carr and Mize, from *MOS/LSI Design and Application.* Reprinted with permission from Texas Instruments, Inc., and McGraw-Hill, Inc., copyright 1972.)

characteristics are shown in Fig. 8.22(b) to allow comparison of various inverter combinations. Many factors besides noise margin must be considered in the choice of the inverter logic, including ease of fabrication, number of masking steps and packing density.

Although a beta ratio of between 5 and 20 is desirable for steep transfer

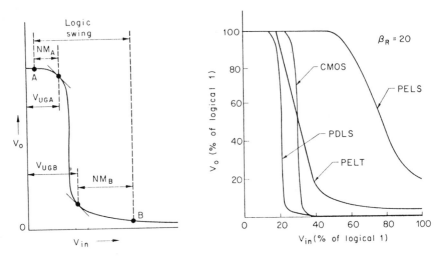

Fig. 8.22. Inverter transfer characteristics:
(a) Transfer curve with noise margins, logic swing, and unity gain points defined; and
(b) Comparison of normalized transfer curves for various types of inverters. (After Carr and Mize, from *MOS/LSI Design and Application*. Reprinted with permission from Texas Instruments, Inc. and McGraw-Hill, Inc., copyright 1972.)

curves with DLS and ELT type loads, there is the disadvantage that high load resistances result in slow charging of the output node capacitance. Hence, there is a marked difference between the voltage rise and fall times, as seen in Fig. 8.23. The equivalent circuits used in the calculation of static and transient characteristics are shown in Fig. 8.24.

The power dissipation of most logic inverters is determined mainly by the quiescent power. In CMOS, however, the stand-by power is not large enough to be significant and the loss is determined by that generated during logic switching. The dynamic power dissipation of CMOS is proportional to the load (or node) capacitance, the square of the operating voltage and the operating frequency. Thus, if the first two parameters are set, the power dissipation is directly proportional to the data input frequency as shown in Fig. 8.25.

Typical performances of some logic families are summarized in Table 8.4 which shows the low quiescent power and the good noise immunity of CMOS. A large fan-out implies a large load capacitance and a longer propagation delay. The performance of CMOS ensures a place in future integrated circuit technology. However, the logic circuit packing density is not as good as can be achieved by skillful *n* channel logic design.

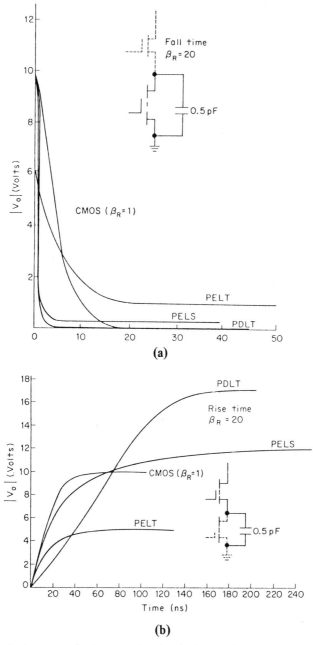

**(a)**

**(b)**

Fig. 8.23. Transient response for inverters showing the longer rise time seen for high beta ratio logic:
(a) Capacitance load discharges through the driver; and
(b) Charges through the load device. (After Carr and Mize, from *MOS/LSI Design and Application*. Reprinted with permission from Texas Instruments, Inc., and McGraw-Hill, Inc., copyright 1972.)

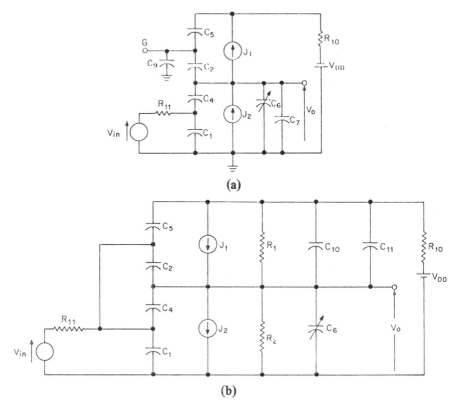

(a)

(b)

Fig. 8.24. Equivalent circuit for inverters:
(a) PELT, PELS and PDLT configurations; and
(b) CMOS configurations. (After Carr and Mize, from *MOS/LSI Design and Application.*
Reprinted with permission from Texas Instruments, Inc., and McGraw-Hill, Inc.,
copyright 1972.)

### 8.5 MOS LOGIC CIRCUITS AND SCALING

Many MOS logic circuits are available and extensive review books exist for
detailed study. Only a few matters need be remarked on before considering
scaling effects in the reduction of MOS circuits in size for future large scale
integration.

The function of a logic circuit depends on whether positive or negative logic
is chosen. For positive logic a logic 1 is the most positive voltage. PELT and
CMOS implementations of positive NAND gates are shown in Fig. 8.26. With
negative logic the circuit of Fig. 8.26(b) would become a negative NOR and that
of Fig. 8.27(b) a negative NAND.

In the CMOS form six transistors are needed for a three-input gate, whereas
in $p$ or $n$ channel logic only four transistors would be needed because the load

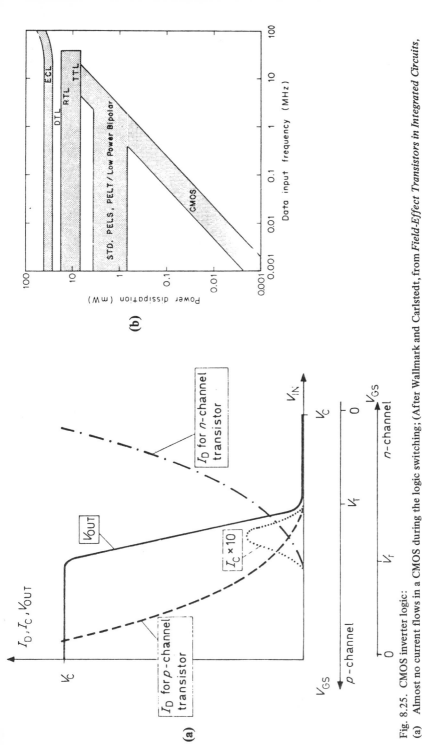

Fig. 8.25. CMOS inverter logic::
(a) Almost no current flows in a CMOS during the logic switching; (After Wallmark and Carlstedt, from *Field-Effect Transistors in Integrated Circuits*, copyright 1974, MacMillan Press Ltd., London.)
(b) The power dissipation in CMOS is therefore proportional to the data input rate. (After Carr and Mize, from *MOS/LSI Design and Application*. Reprinted with permission from Texas Instruments, Inc., and McGraw-Hill, Inc., copyright 1972.)

499

**Table 8.4. Comparison Chart of the Major IC Digital Logic Families (1970).**

| Parameters | RTL | Low-Power RTL | DTL | HTL | 12-ns TTL | 6-ns TTL | 4-ns ECL | 2-ns ECL | 1-ns ECL | P-MOS | CMOS |
|---|---|---|---|---|---|---|---|---|---|---|---|
| 1. Circuit form | resistor-transistor | | diode-transistor | diode-Zener transistor | transistor-transistor | | emitter-coupled current mode | | | p-channel MOS | complementary MOS |
| 2. Positive logic function of basic gate | NOR | | NAND | | AND-OR-INVERT | | OR/NOR | | | NAND | NOR or NAND |
| 3. Wired positive logic function | implied AND (some functions) | | implied AND | implied AND (A-O-I) | | | implied OR (all functions) | | | none | |
| 4. Typical high-level $Z_0$, ohms | 640 | 3.6 k | 6 k or 2 k | 15 k or 1.5 k | 70 | 10 | 15 | 6 | 6 | 2 k | 1.5 k |
| 5. Typical low-level $Z_0$ | $R_{sat}$ | $R_{sat}$ | $R_{sat}$ | $R_{sat}$ | $R_{sat}$ | $R_{sat}$ | 15 ohms or 2.7 mA | 6 ohms or 6.7 mA | 6 ohms or 21 mA | 25 k | 1.5 k |
| 6. Fanout | 5 | 4 | 8 | 10 | 10 | 10 | 25 | 25 inputs or 50 ohms | 10 low-Z inputs or 50 ohms | 20 | 50 or higher |
| 7. Specified temperature range, °C | -55 to 125 0 to 75 15 to 55 | -55 to 125 0 to 75 15 to 55 | -55 to 125 0 to 75 | -30 to 75 | -55 to 125 0 to 75 | -55 to 125 0 to 75 | -55 to 125 0 to 75 | -55 to 125 0 to 75 | 0 to 75 | -55 to 125 0 to 75 | -55 to 125 |
| 8. Supply voltage | 3.0 V ± 10% 3.6 V ± 10% | 3.0 V ± 10% 3.6 V ± 10% | 5.0 V ± 10% | 15 ± 1 V | 5.0 V ± 10% 5.0 V ± 5% | 5.0 V ± 10% 5.0 V ± 5% | -5.2 V + 20% -10% | -5.2V + 20% -10% | -5.2 V ± | -27 ± 2 V -13 ± 1 V | 4.5 to 16 V |
| 9. Typical power dissipation per gate | 12 mW | 2.5 mW | 8 mW or 12 mW | 55 mW | 12 mW | 22 mW | 40 mW | 55 mW plus load | 55 mW plus load | 0.2 to 10 mW | 0.01 mW static ≈ 1 mW at 1 MHz |
| 10. Immunity to external noise | nominal | fair | good | excellent | very good | very good | good | good | good | nominal | very good |
| 11. Noise generation | medium | low-medium | medium | medium | medium-high | high | low | low-medium | medium | medium | low-medium |
| 12. Propagation delay per gate, ns | 12 | 27 | 30 | 90 | 12 | 6 | 4 | 2 | 1 | 300 | 70 |
| 13. Typical clock rate for flip-flops, MHz | 8 | 2.5 | 12 to 30 | 4 | 15 to 30 | 30 to 60 | 60 to 120 | 200 | 400 | 2 | 5 |
| 14. Number of functions; family growth rate | high; growing | high; growing | fairly high; new functions in TTL | nominal; growing | very high; growing | very high; growing | high; growing | high; growing | high; growing | low; growing | low; growing |
| 15. Cost per function | low | low | low | medium | low | medium | low | medium | high | medium to high | medium to high |

After Garrett IEEE Spectrum December 1970.
(Some upgrading of performance has been achieved in the last few years. For instance 10 MHz, 25 ns CMOS is available (Ref. Walker 1975). See also Table 8.6.

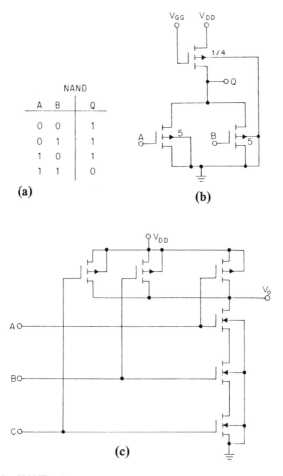

| NAND | | |
|---|---|---|
| A | B | Q |
| 0 | 0 | 1 |
| 0 | 1 | 1 |
| 1 | 0 | 1 |
| 1 | 1 | 0 |

**(a)**

**(b)**

**(c)**

Fig. 8.26. Positive NAND gates:
(a) Truth table for NAND with two inputs;
(b) PELT implementation positive NAND gate (two inputs); and
(c) CMOS positive NAND gate (three inputs). (After Carr and Mize, from *MOS/LSI Design and Application*. Reprinted with permission from Texas Instruments, Inc., and Mc-Graw-Hill, Inc., copyright 1972.)

transistor is shared. A comparable situation exists with NOR gates as seen from Fig. 8.27. This partly explains the packing problem observed with CMOS and shows why $n$ (or $p$) channel logic is preferred in many systems even though it involves beta-ratio-logic and therefore differences in rise and fall times.

The use of dynamic logic circuits involving clocking of the gates is an option open to the designer. A two-phase clocked ratio shift register is shown in Fig. 8.28. The evaluation of the relative merits of static and dynamic logic

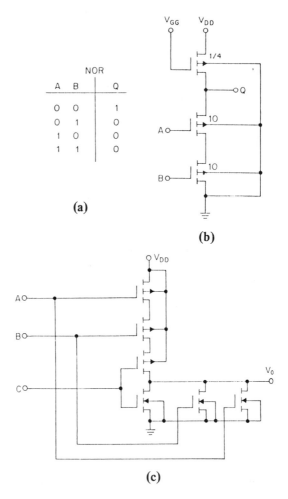

NOR

| A | B | Q |
|---|---|---|
| 0 | 0 | 1 |
| 0 | 1 | 0 |
| 1 | 0 | 0 |
| 1 | 1 | 0 |

(a)

(b)

(c)

Fig. 8.27. Positive NOR gates:
(a)  Truth table for NOR;
(b)  PELT implementation (two inputs); and
(c)  CMOS implementation (three inputs). (After Carr and Mize, from *MOS/LSI Design and Application*. Reprinted with permission from Texas Instruments, Inc., and Mc-Graw-Hill, Inc., copyright 1972.)

circuits is complicated and dealt with in technical studies included in the reading list for this chapter. Rather than pursue these matters it is important that we now consider the fundamental effects of scaling of MOS transistors in size.

In MOS transistors the separation between the source and the drain must be larger than the sum of the depletion thicknesses at the junctions. The field

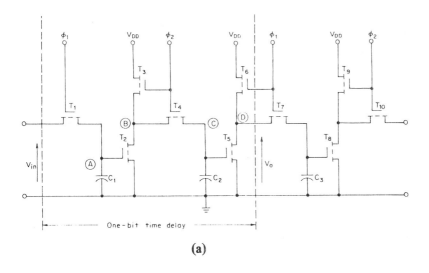

**(a)**

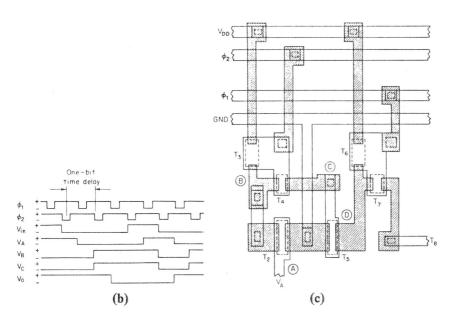

**(b)**                                                    **(c)**

Fig. 8.28.  Two-phase ratio shift register:
(a)  Circuit schematic;
(b)  Timing sequence; and
(c)  Layout geometry for 1 bit. (After Carr and Mize, from *MOS/LSI Design and Application.* Reprinted with permission from Texas Instruments, Inc., and McGraw-Hill, Inc., copyright 1972.)

at the surface end of a depleted layer of thickness $w$ is

$$\mathcal{E}_0 = N\,qw/\epsilon \tag{8.3}$$

and the electric field in the oxide must match $\mathcal{E}_0$

$$\mathcal{E}_{0x} = \mathcal{E}_0\,\epsilon/\epsilon_{0x} \tag{8.4}$$

So the breakdown field in the insulator limits $\mathcal{E}_0$ and this sets a limit to $Nw$. But the field is related to the applied voltage $V$ (plus a small inversion potential Fermi-level separation term, $2\phi$) by expressions of the form

$$\mathcal{E}_0 = 2\left(\frac{V + 2\phi}{w}\right) \tag{8.5}$$

where

$$w = \sqrt{(2\epsilon(V+2\phi)/qN)^{1/2}} \tag{8.6}$$

So for a chosen circuit voltage the doping $N$ must not exceed a certain value determined by the oxide breakdown field. From this value of $N$, the depletion thickness $w$ can be determined and the minimum source-drain separation inferred. In a typical circuit with a voltage $V_{DD}$ of 4 V and a doping of $5 \times 10^{16}$ cm$^{-3}$ the minimum source drain distance is just under 1 $\mu$m.

If all the dimensions of an FET including the oxide thickness are reduced by a factor $S$ and the voltages also reduced by this factor, then Eq. (8.6) shows that the doping $N$ must be increased by a factor $S$. Table 8.5 shows the influence of the scaling factor on other device and circuit parameters. A large advantage from scaling down in size may be expected in the power-delay product, $S^{-3}$, and in power dissipation, $S^{-2}$.

**Table 8.5. MOS Device Scaling**

| Device/Circuit Parameter | Scaling Factor |
|---|---|
| Device dimension, $T_{OX}$, L, $L_D$, W, $X_J$ | 1/S |
| Substrate doping, $C_B$ | S |
| Supply voltage, V | 1/S |
| Supply current, I | 1/S |
| Parasitic capacitance, WL/$T_{OX}$ | 1/S |
| Gate delay, VC/I ($\tau$) | 1/S |
| Power dissipation, VI | 1/S$^2$ |
| Power-delay product | 1/S$^3$ |

(After Pashley et al. Reprinted from *Electronics*, August 18, 1977; copyright McGraw-Hill, Inc., NY, 1977.

Obviously straightforward size reduction according to the scaling of Table 8.5 has some limits. The supply voltage would become inconveniently low and the gate oxide too thin for reliability. Furthermore, interconnection capacitances and contact effects begin to set limits to size reduction.

The effects of reducing channel length from 6 $\mu$m to 3.5 $\mu$m and oxide thickness from 1200 Å to 700 Å are shown in Fig. 8.29 together with predictions for $L = 2$ $\mu$m and $T_{ox} = 400$ Å. Table 8.6 shows the progressive improvements achieved as MOS technology has advanced from enhancement-mode $n$-MOS to the high density H-MOS of 1977. The results of experimental studies on 2-$\mu$m channel length logic are also given.

## 8.6 FUTURE LIMITS IN DIGITAL ELECTRONICS

Large scale integration (LSI) began to develop in the period 1972-1975 and is now in a dominant position in sophisticated electronic systems because of the cost and performance advantages. Moreover progress is being made towards very large scale integration (VLSI).[138] Table 8.7 shows some industry projections for 1980 compared with the situation in 1973/74. The movement from LSI to VLSI

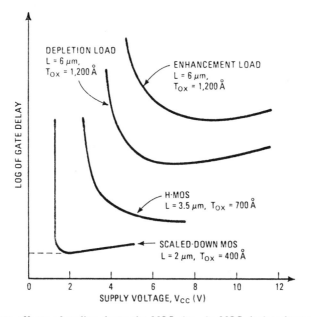

Fig. 8.29. Some effects of scaling down the MOS size. As MOS devices become smaller, substantial improvements in gate delay and other factors are possible but the supply voltage must be smaller. For 2-$\mu$m channel lengths, a 3-V supply yields the best gate-delay performance and process reliability. (After Pashley et al. Reprinted from *Electronics*, August 18, 1977; Copyright McGraw-Hill, Inc., NY, 1977, p. 94.)

Table 8.6. Evolution of MOS Device Scaling

| Device-Circuit Parameter | Enhancement-mode n-MOS 1972 | Depletion-mode n-MOS 1976 | H-MOS 1977 | MOS 1980 |
|---|---|---|---|---|
| Channel length, L ($\mu$m) | 6 | 6 | 3.5 | 2 |
| Lateral diffusion, $L_D$ ($\mu$m) | 1.4 | 1.4 | 0.6 | 0.4 |
| Junction depth, $X_j$ ($\mu$m) | 2.0 | 2.0 | 0.8 | 0.8 |
| Gate-oxide thickness, $T_{OX}$ (A) | 1,200 | 1,200 | 700 | 400 |
| Power supply voltage, $V_{CC}$ (V) | 4-15 | 4-8 | 3-7 | 2-4 |
| Shortest gate delay, $\tau$ (ns) | 12-15 | 4 | 1 | 0.5 |
| Gate power, $P_D$ (mW) | 1.5 | 1 | 1 | 0.4 |
| Speed-power product (pJ) | 18 | 4 | 1 | 0.2 |

(After Pashley et al. Reprinted from *Electronics*, August 18, 1977; copyright McGraw-Hill, Inc., NY, 1977.

Table 8.7. Integrated Circuit Performance, LSI and VLSI

| Integrated Circuit | Industry Capability 1973/1974 (LSI) | 1980 (VLSI) |
|---|---|---|
| Random-access memory | 2,048 bits | 64,000 bits |
| Serial memory | 30,000 bits (CCD) | $5 \times 10^6$ bits (CCD) |
| (Random) logic | 500 gates | 10,000 gates |
| Digital correlator | 64 bits | 2,048 bits |
| Rf analog circuits (bipolar) | — | S-band rf circuits |
| Secure code generator | 32 bits | 1,024 bits |
| Sensor mosaics (CCD) | 100-by-100 elements | 1,000-by-1,000 elements |
| Minicomputer CPU (bipolar) | 10 chips at 5 MHz | 1 chip at 50 MHz |
| Performance | | |
| Clock rate (maximum) (MHz) | 300 | 2,000 |
| Transistor bandwidth (GHz) | 1 | 6 |
| Speed-power product (pJ) | 3-10 | 0.1-1 |
| Complexity | | |
| Chip size (maximum) (mil) | 250 | 500 |
| Device area (mil²) | 2-5 | 0.1-0.3 |
| Transistors per chip (maximum) | 5,000 | 200,000 |

(After Altman. Reprinted from *Electronics*, February 21, 1974; copyright McGraw-Hill Inc., 1974.)

is possible because of decreases in device circuit function area, improvements in the speed-power product and better technology control that allows reasonable yields with larger chip sizes. $N$ channel MOS will be the dominant technology in many fields and VMOS, $I^2L$ and CCD in others. There are some fundamental limits to this progress and it is important to consider what they are likely to be.

### 8.6.1  Lithography Limitations

Most of the miniaturizations that can be hoped for, with 3.5 $\mu$m design rules, by skillful layouts and improved interconnection technology have been achieved. Therefore, problems of scaling down devices to increase circuit function packing density must be considered. Photolithography is limited by the wavelength of the light used.[130] In contact printing with UV irradiation ($\lambda$ = 2000 to 2500 Å) with hard surface masks, resolution of features as small as 2 $\mu$m is possible with satisfactory edge control. Flexible masks and wafers allow conformal printing and 1 $\mu$m resolution. Proximity printing and projection printing to avoid mask wear have resolutions that are about 3 $\mu$m because of lens problems and the fact that large area Si wafers become non-flat (bowed) during processing. An f2 lens system can, if perfect, project 2.0-$\mu$m lines and spaces with 70% contrast but with a depth of focus less than the typical distortion of wafer flatness. If scanning projection printing is used the wafer area exposed at a time is reduced and resolution is improved, but at the expense of throughput.

Electron lithography offers the possibility of much shorter wavelengths but introduces problems of scanning and throughput.[130] It is possible to make very high quality Cr master masks by electron beam pattern generation. Scanning systems have also been used to make special fine-line devices, such as 0.3-0.6 $\mu$m gate-length FETs for microwave transistors. However, widespread use of electron beam lithography for direct device writing of integrated circuits either by scanning or electron image projection is not very practical at this time. A scanning exposure for a typical single integrated circuit may require one second or more. This is an order of magnitude too long and too expensive for general production use at this time and the initial equipment cost is typically in excess of $1 million. Much work is in progress to improve the technology and lower the cost. One approach that is being studied is step-and-reduction electron beam projection but this at present has distortion and other problems (the masks for instance must be transparent to electrons).

Lithography by X-rays of wavelength 4-8 Å has also been studied in a proximity printing form with the mask a heavy metal such as gold or platinum on a thin substrate such as Si or Mylar.[130,131] X-ray systems provide excellent image resolution in typical resist thicknesses (several $\mu$m) and allow high aspect ratios of line width (0.3 $\mu$m) to resist thicknesses. X-ray exposures do not suffer from scattering effects as do electron beam exposures or from diffraction pattern degradation as do optical methods. However, there remain at present, problems of large-area mask stability and beam divergence; or of photo-resist sensitivity if a high-speed step-and-repeat system is desired.

Also under study are ion-beam exposure systems but these are limited by the beam technology available at present.

### 8.6.2 More on Device Miniaturization

Scaling down of device size for MOS transistors has been discussed with reference to Table 8.5. More comments are needed, however, on general concepts. The Si bandgap and the value of $kT/q$ at normal ambient operating temperatures oblige us to work with on-voltages in the 0.5-1 V range and reverse- or off-voltages of at least a few volts. For given voltages, then, as the device dimensions are reduced the depletion regions can only be allowed to shrink to the limits set by the avalanche breakdown field of about $5 \times 10^5$ Vcm$^{-1}$. Thus as the scaling down continues, the depletion regions cannot be allowed to continue to decrease, so they become proportionally larger relative to the rest of the device.

In a bipolar transistor in which the base region is lightly doped, the depletion region that spreads in from the collector side (and to some extent from the emitter side also) sets a limit to the minimum base thickness that is practical. Thus for 10 V on an abrupt junction of doping $10^{17}$ cm$^{-3}$ the depletion thickness is about 0.4 $\mu$m which corresponds to a peak field strength of $5 \times 10^5$ Vcm$^{-1}$, the breakdown field strength.

In a planar bipolar transistor the collector may be more lightly doped than the base and the depletion region extends into the collector but also curves and spreads along the surface. The minimum area of a planar transistor needs to be some hundreds of times greater than $w^2$ where $w$ is the depletion region thickness. If $10^{18}$ cm$^3$ is the maximum convenient collector doping level in a planar transistor and corresponds to a collector voltage rating of a few volts and a $w$ of about 0.1 $\mu$m, then $500w^2$ gives 5 $\mu$m$^2$ for the minimum transistor area. This is an order of magnitude less than the smallest device area considered in Table 8.7.

When device active volumes are very small the number of doping atoms involved is small. For instance, a volume of 1 $\mu$m$^3$ contains only 1,000 atoms for a doping of $10^{15}$ cm$^{-3}$, and therefore, the random statistical spread of impurities in a Poisson distribution might be expected to have some effect on the breakdown voltage. Actually, because of the dependence of field on the square root of $N$ the effect is not large. Typically the effect degrades the reverse voltage rating of a diode by a volt or so from the value predicted by the average doping.[68]

In another consideration of statistical effects, namely the range of turn-on voltages for FETs because of doping fluctuations, Hoeneisen and Mead[57] find the voltage range for a typical doping to be only 0.07 V. This, of course, takes no account of nonrandom doping variations that can be significant across a large wafer.

In IC layouts by the best computer-aided-layout programs it is not unusual to find that the active transistor cells occupy only one-third to one-half of the chip area and the remaining area is needed for interconnection channels. The interconnection parasitic capacitance may then become as much as the device node capacitance. This reduces the speed of the circuit to the point where one may well wish to rearrange the logic flow even at the expense of extra logic gates in order to reduce the interconnection parasitics.

If device linear dimensions are reduced by a factor of 5, from Table 8.5 the current is a fifth but the current density is 5 times greater. If there are then 25 times as many circuits on the chip the total current in the chip busbar may increase by a factor of 5 and the ratio of IR voltage drop to signal voltage may be increased. Such factors must be considered as the packing density is increased.

The fastest speed of response in a semiconductor device is related to fundamental material properties such as the saturation drift velocity and the avalanche field strength. For an electron moving in Si at the saturation velocity of $10^7$ cm/sec and a field strength of $3 \times 10^5$ Vcm$^{-1}$, the energy input to the electron is 5 ergs/sec$^{-1}$. At this rate of energy input, the time for an electron to acquire a voltage of 100 $kT/q$ (regarded as necessary for logic in semiconductor junction devices) is $10^{-12}$ sec. In an actual semiconductor device the response time is more usually in the nanosecond range than in the picosecond range because only in a very small portion of a junction device can the electrons be moving under these limiting field conditions.

Signal propagation time need not be a serious speed limitation if interconnection layout technology is efficient.[68] If there are 20,000 gates on a 1-cm$^2$ chip the lattice spacing is about 60 $\mu$m between gates. If a gate interacts, on the average, over five gate-lattice constants and the propagation velocity of the signal is $5 \times 10^9$ cm/sec$^{-1}$, the propagation delay on the interconnection is about 1 psec. This is considerably less than probable future system speeds of 1-0.1 nsec and clock rates of $10^8$-$10^9$ Hz.

There is also the hope that in years to come the many technological disadvantages of the compound semiconductor GaAs may be brought under control, and that large scale integrated circuits of this material (that take advantage of its high electron mobility and special saturation velocity characteristics) may become possible. Some predictions for its performance are shown in Fig. 8.30. Enhancement MESFETs are low voltage devices which have low power dissipations.

### 8.6.3 The Power Dissipation Limitations

Since increasing the power drive has been shown to increase speed and since future technology holds out the possibility of 50,000 gates per chip of area

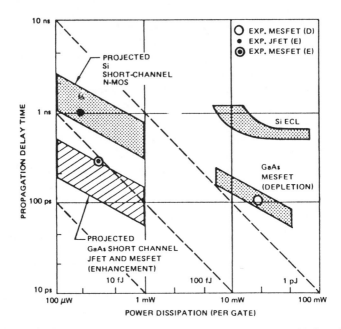

Fig. 8.30. Propagation delay vs gate power for Si technology of various kinds and for a projected GaAs technology. (After Zuleeg et al. Reprinted with permission from *IEEE International Electron Devices Meeting Technical Digest*, 1977, p. 200.)

1 cm$^2$ with each logic gate dissipating an average of 0.1 mW (more when switching but less while on standby) it is important to consider the heat transfer problem.

For heat transfer from a surface to free air, a useful rule is that 1 mW can be transferred per cm$^2$ of surface area for each °C temperature difference between the surface and the ambient air. If the air is blown over the surface with a fan the dissipation can be increased 10 to 30 times to 0.03 W cm$^{-2}$ °C$^{-1}$. Replacing the air by a fluid such as Freon and circulating the liquid can give power transfer rates of 10 W cm$^{-2}$ for a 20°C temperature differential between the surface and the bulk of the coolant. A further temperature differential exists between the integrated circuit chip and the cooled surface of the module because the heat must be carried from the chip to the surface by conduction. The thermal conductivity is likely to be in the range 10-25°C W$^{-1}$. So unless special design efforts are made, the power dissipation for a 50°C rise in temperature is not likely to exceed 5 W cm$^{-2}$. A power dissipation of 100 mW would not be unusual for a chip 0.25 cm × 0.25 cm: this corresponds to 1.6 W cm$^{-2}$. As the packing of circuit modules increases it becomes increasingly difficult to maintain a good dissipation-to-area ratio. For a circuit card of 100 cm$^2$ the permissible dissipation may be only 20 W.[11, 70, 105]

The permissible power dissipation $Q_m$ W cm$^{-2}$ imposes a limitation on the packing density of circuit functions per cm$^2$. For instance, 50,000 gates each dissipating 0.01 mW represents an average power of 0.5 W and if this must be dissipated from a chip of area 1 cm$^2$, the dissipation density is 0.5 W cm$^{-2}$ and this is pushing the state of the art for air cooling of large arrays. The search for logic with low average power dissipation per system must be continued with emphasis on low stand-by power.

The simple concepts presented above do not fully characterize the effects of device miniaturization and for a more complete understanding further reading must be done. Other matters must also be considered including the advantages to be gained by operating at temperatures as low as 77°K.

Progress in integrated circuit technology continues and in the USA important sources of information include:

- The *IEEE International Solid-State Circuits Conferences* held in February each year in Philadelphia or San Francisco. A digest of technical papers is available for purchase from the IEEE;
- the *IEEE International Electron Devices Meeting* held in December each year in Washington, D.C. Again the technical digest of papers is available from the IEEE;
- *IEEE Journal of Solid State Circuits*;
- *IEEE Transactions on Electron Devices*;
- *Electronics*, McGraw-Hill, N.J. This journal presents frequent state-of-the-art reviews from both technical and industrial viewpoints.

## 8.7 QUESTIONS

1. Consider the problem of yield vs chip area for integrated circuit wafers in which the distributions of deleterious defects have the forms shown in Fig. P 8.1. Calculate yield vs chip size and illustrate with numerical examples. Name some deleterious defects and suggest how they might be reduced in density.

2. Consider in a design problem what level of integration to choose. Postulate that yield decreases as the square root of the number of circuit functions per chip ($n$) and that the chip area is proportional to the number of circuit functions. The yield is initially 5% for $n$ functions per chip (the lowest level of integration) and the chip size is 0.25 cm × 0.25 cm. The yield (irrespective of the level of integration) increases each year by 30% of its previous value. The maximum level of integration that can be contemplated is $5n$ functions per chip. The initial cost for development is $100,000 and thereafter the cost of processing an 8-cm-diameter wafer is $50, and packaging and testing is $1 per chip (irrespective of chip size). Assume that the total market for the device is $10^7$ a year for each of 10 years and then becomes zero. You have a competitor who chooses the lowest level of integration $n$ and stays with this for 10 years. His price $P_c$ which includes a 50% profit sets the price level. It is initially $P_1$ per chip (of $n$ functions) and decreases by 30% per year to reflect his improved yield. Your fraction of the market in any year depends on your price level $P$ by the expression $2(P_c/P + P_c)^2$ which is limited to 100% when your price is 41% of the competitor's price. Interest rates for borrowing (for start-up costs) and savings (as the profits come in) are 10%. What should be your level of integration and your

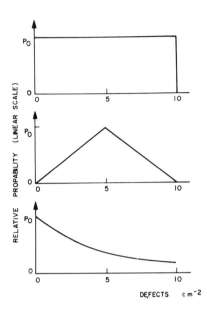

relative price level $P/P_C$ (both assumed constant) to maximize your return? How could your return be further improved if each year you are allowed to readjust your price relative to your competitor?

3. Assume that an IC product in its first year of introduction costs $10 per unit and $10^5$ are sold for a yearly market of $10^6$. Thereafter the product conforms to the 70% learning rule—that is doubling the total accumulated production gives a price of 70% of the previous level. If the market per year remains at $10^6$ how many units will have been sold at the end of the 7th year after introduction? What will be the unit price?

4. A thousand ICs are tested at $100°C$ for 1,000 hr and a further thousand at $150°C$ for 1,000 hr. From the evidence, the quality control group estimated the failure rate to be $0.05\%/1,000$ hr at $50°C$ (quoted at a 60% upper confidence level). What was the evidence?

5. Apply the SPICE transistor parameters given in Chap. 4 (see Table 4.1) to calculate the $V_{out}/V_{in}$ transfer characteristic of the TTL circuit of Fig. 8.4. Select a suitable $V_{cc}$ and adjust the circuit components only if necessary to obtain proper circuit behavior.

6. Apply the SPICE transistor parameters given in Chap. 4 (see Table 4.1) to calculate the $V_{out}/V_{in}$ transfer characterization of the ECL gate shown in Fig. 8.5. Select $V_{cc}$ and the resistors to suit the transistor parameters. How much does the total current drawn vary with the logic state? Explain how you would calculate the response time and power involved in switching.

7. Calculate the transfer characteristic of the $I^2L$ inverter gate shown in Fig. 8.13(c) for the data given (assume any other values that may be needed).

8. For the inverter circuits of Fig. 8.19, 8.20 or 8.21 with $\beta_R = 1$ and the MOSFET defined by the SPICE values in Table 7.2, or equivalent, calculate the $V_{out}/V_{in}$ inverter characteristic.

9. The characteristics of a Si $n$-channel MOSFET (oxide thickness 500 A, $L = 4.6$ $\mu$m) are shown in Figs. 7.6 and 7.8. If all the dimensions except the oxide thickness are

reduced by a factor of two, predict the effect on the characteristics. What changes might you wish to make in the doping to improve the curves?

10. The effects of scaling down MOSFETs ($L$ = 6 $\mu$m, $T_{ox}$ = 1200 Å) to H-MOS ($L$ = 3.5 $\mu$m, $T_{ox}$ = 700 Å) and to $L$ = 2 $\mu$m and $T_{ox}$ = 400 Å are shown in Fig. 8.29. But the vertical axis has no divisions marked on it. Estimate a scale and justify it from MOSFET theory and scaling principles.

11. Design a package for a chip 0.5 cm $\times$ 0.5 cm $\times$ 0.01 cm thick, that will allow it to dissipate 1 W to a free air ambient without blower cooling for an active region rise in temperature above ambient air temperature of only 50°C. What would be the approximate derating needed if 1000 such packages were closely stacked to form a cube of dimension 10 packages per edge?

12. Discuss bipolar and MOS logic systems from the viewpoint of noise immunity and threshold levels.

13. Explain techniques of interfacing between bipolar and MOS logic and vice versa.

14. Discuss the relative merits of the $T^2L$, $T^3L$, ECL, EFL and $I^2L$ bipolar logic families.

15. Discuss the relative merits of MOS logic $p$-channel, $n$-channel, CMOS, D-MOS and VMOS for particular applications.

16. Discuss various methods of circuit isolation on a chip in terms of space required and effectiveness.

17. Review the logic families in terms of fan-out capability and the effect of the fan-out ratio on speed.

18. Discuss the problems of scale down of size of bipolar transistors and the probable ultimate limits.

19. Consider the scale down of size of MOS transistors and the factors that become important in the limit.

20. The breakdown field in Si and the saturated limit of the drift velocity determine the rate of energy transfer. Discuss the relevance of this concept to bipolar and MOS transistors in integrated circuits.

21. Consider analytically the effects of a possible 20% variation of doping uniformity from one side to the other across the diameter of a wafer of bipolar or MOS circuits.

22. Discuss the limits of resolution of optical lithography and the economics of replacing it with electron beam lithography.

## 8.8 REFERENCES AND FURTHER READING SUGGESTIONS

1. Aitken, A., et al., "The relative performance and merits of CMOS technology," *IEEE International Electron Devices Meeting Technical Digest*, Washington, D.C., 1976, p. 327.

2. Allstot, D.J., et al., "A high voltage analog-compatible $I^2L$ process," *IEEE International Electron Devices Meeting Technical Digest*, Washington, D.C., 1977, p. 175.

3. Altman, L., "Logic's leap ahead creates new design tools for old and new applications," *Electronics*, February 21, 1974, p. 81.

4. Altman, L., "MOS makers worry about $I^2L$ progress," *Electronics*, July 24, 1975, p. 70.

5. Altman, L., "Five technologies squeezing more performance from LSI chips," *Electronics*, August 18, 1977, p. 91.

6. Baker, W.D., et al., "Oxide isolation brings high density to production bipolar memories," *Electronics*, March 29, 1973, p. 67.

7. Becker, P.W., and F. Jensen, *Design of Systems and Circuits for Maximum Reliability or Maximum Production Yield*, McGraw-Hill, NY, 1977.

8. Beke, H., W.M.C. Sansen, and R. Van Overstraeten, "CALMOS: A computer-aided layout program for MOS/LSI," *IEEE J. Solid-State Circuits*, SC-12, 1977, p. 281.
9. Bishop, R.A., "Complementary MOS offers many advantages to the digital systems designer," *Electronic Engineering*, November, 1972, p. 67.
10. Blackstone, S.C., and R.P. Mertens, "Schottky collector I²L," *IEEE J. Solid-State Circuits*, SC-12, 1977, p. 270.
11. Bolvin, R., "Thermal characteristics of IC's gain in importance," *Electronics*, October 31, 1974, p. 87.
12. Brothers, J.S., "Integrated circuit development," *Radio and Electronic Engineer*, 43, 1973, p. 39.
13. Buie, J., "Improved triple diffusion means densest IC's yet," *Electronics*, August 7, 1975, p. 103.
14. Camenzind, H.R., *Electronic Integrated Systems Design*, Van Nostrand Reinhold, NY, 1972.
15. Carr, W.N., and J.P. Mize, *MOS/LSI Design and Application*, McGraw-Hill, NY, 1972.
16. Chang, T.H.P., et al., "Electron-beam lithography draws a finer line," *Electronics*, May 12, 1977, p. 89.
17. Cho, Y.E., A.J. Korenjak, and D.E. Stockton, "FLOSS: an approach to automated layout for high-volume designs," *1977 Design Automation Conference*, p. 138, IEEE Cat. 77CH1216-1C.
18. Cook, P.W., D.L. Critchlow, and L.M. Terman, "Comparison of MOSFET Logic Circuits," *IEEE J. Solid-State Circuits*, SC-8, 1973, p. 348.
19. Crawford, B., "Implanted depletion loads boost MOS array performance," *Electronics*, April 24, 1972, p. 85.
20. Crippen, R.E., et al., "High-performance integrated injection logic: a microprogram sequencer built with I³L," *IEEE J. Solid-State Circuits*, SC-11, 1976, p. 662.
21. Crocker, N.R., R.W. McGuffin, and A. Micklethwaite, "Automatic ECL LSI Design," *1977 Design Automation Conference*, p. 158, IEEE Cat. 77CH1216-1C.
22. C. Crook, "Comparing the power of C-MOS with TTL," *Electronics*, May 16, 1974, p. 132.
23. Davies, R.D., and J.D. Meindl, "Poly I²L—a high speed linear-compatible structure," *IEEE J. Solid-State Circuits*, SC-12, 1977, p. 367.
24. Davies, R.D., and J.D. Meindl, "Device characteristics for poly I²L," *IEEE International Electron Devices Meeting Technical Digest*, Washington, D.C., 1977, p. 170.
25. Davies, D.E., et al., "Automatic registration in an electron-beam lithographic system," *IBM Journal of Research and Development*, 21, 1977, p. 498.
26. Dean, K.J., "Trends in semiconductor digital circuits," *Radio and Electronic Engineer*, 43, 1973, p. 67.
27. Declereq, M.J., and T. Laurent, "A theoretical and experimental study of DMOS enhancement/depletion logic," *IEEE J. Solid-State Circuits*, SC-12, 1977, p. 264.
28. Dennard, R.H., et al., "Design of ion-implanted MOSFETS with very small physical dimensions," *IEEE J. Solid-State Circuits*, SC-9, 1974, p. 256.
29. Dill F.H., J.A.Tuttle, and A.R. Neureuther, "Limits of optical lithography," *IEEE Electron Devices Meeting Technical Digest*, Washington, D.C., 1974, p. 13.
30. Eaton, S.S., "Sapphire brings out the best in C-MOS," *Electronics*, June 12, 1975, p. 115.
31. Elliott, M.T., and M.R. Splinter, "Size effects in e-beam fabricated MOS devices," *IEEE International Electron Devices Meeting Technical Digest*, Washington, D.C., 1977, p. 11A.
32. Elmasry, M.I., "Folded-collector integrated logic," *IEEE J. Solid-State Circuits*, SC-11, 1976, p. 644.
33. Estreich, D.B., R.W. Dutton, and B.W. Wong, "An integrated injection logic (I²L)

macro-model including lateral and current redistribution effects," *IEEE J. Solid-State Circuits,* **SC-11,** 1976, p. 648.

34. Evans, S.A., et al., "High Speed I²L fabricated with e-beam lithography and ion implantation," *IEEE International Electron Devices Meeting Technical Digest,* Washington, D.C., 1977, p. 266.

35. Feller, A., "Automatic layout of low-cost quick-turnaround random-logic custom LSI devices," *1976 Design Automation Conference,* p. 79, IEEE Cat. 76CH1098-3C.

36. Fleishhammer, W., G. Schneider, and G. Koppe, "T³L achieves C-MOS noise immunity while retaining TTL speed," *Electronics,* March 7, 1974, p. 95.

37. Fogiel, M., *Microelectronics: Principles, Design Techniques, Fabrication Processes,* Research and Education Association, NY, 1968.

38. Fukahori, K., and P.R. Gray, "Computer simulation of integrated circuits in the presence of electrothermal interaction," *IEEE J. Solid-State Circuits,* **SC-11,** 1976, p. 834.

39. Fuller, C.E., P.A. Gould, and D.J. Vinton, "Device processing using electron image projection lithography," *IEEE International Electron Devices Meeting Technical Digest,* Washington, D.C. 1977, p. 7L.

40. Gaensslen, F.H., "Geometry effects of small MOSFET devices," *IEEE International Electron Devices Meeting Technical Digest,* Washington, D.C., 1977, p. 512.

41. Gary, P., "I²L: it's getting faster and smaller," *IEEE Spectrum,* June 1977, p. 30.

42. Ghandi, S.K., *The Theory and Practice of Microelectronics,* J.Wiley & Sons, NY, 1968.

43. Gordon, E.I., and D.R. Herriott, "Pathways in device lithography," *IEEE Trans. Electron Devices,* **ED-22,** 1975, p. 371.

44. Greenbaum, J.R., "Digital IC models for computer-aided design," *Electronics,* December 6 and 20, 1973, pp. 121, 107, and January 24, 1974, p. 98.

45. Grundy, D.L., J. Bruchez, and B. Down, "Collector diffusion isolation packs many functions on a chip," *Electronics,* July 3, 1972, p. 96.

46. Garrett, L.S., "*Integrated circuit digital logic families, Parts I, II, and III, IEEE Spectrum,* 7, October, November, December, 1970.

47. Hamilton, D.J., and W.G. Howard, *Basic Integrated Circuit Engineering,* McGraw-Hill, NY, 1975.

48. Hart, C.M., A. Slob, and H.E.J. Wulms, "Bipolar LSI takes a new direction with integrated injection logic," *Electronics,* October 3, 1974, p. 115.

49. Hebenstreit, E., and K. Horninger, "High-speed programmable logic arrays in ESFI SOS technology," *IEEE J. Solid-State Circuits,* **SC-11,** 1976, p. 370.

50. Hennig, F., et al., "Isoplanar integrated injection logic: a high-performance bipolar technology," *IEEE J. Solid-State Circuits,* **SC-12,** 1977, p. 101.

51. Herman, J.M., "High performance output Schottky I²L MTL," *IEEE International Electron Devices Meeting Technical Digest,* Washington, D.C., 1977, p. 262.

52. Herman, J.M., III, S.A. Evans, and B.J. Sloan, Jr., "Second generation I²L/MTL: a 20 ns process/structure," *IEEE J. Solid-State Circuits,* **SC-12,** 1977, p. 93.

53. Herrick, W.V., and J.R. Sims, "A successful automated IC design system," *1976 Design Automation Conference,* p. 74, IEEE Cat. 76CH1098-3C.

54. Hershberger, W.D., *Topics in Solid State and Quantum Electronics,* Wiley, NY, 1971, "Advances in LSI technology," A.S. Grove. Ch. 10.

55. Hewlett, F.W., Jr., and W.D. Ryden, "The Schottky I²L technology and its application in a 24 × 9 sequential access memory," *IEEE J. Solid-State Circuits,* **SC-12,** 1977, p. 119.

56. Hnatek, E., *Applications of Linear Integrated Circuits,* Wiley-Interscience, NY, 1975. (See also, *A User's Handbook of Integrated Circuits,* Wiley-Interscience, NY, 1973.)

57. Hoeneisen, B., and C.A. Mead, "Fundamental limitations in microelectronics," *Solid-State Electronics,* **15,** 1972, pp. 819 and 981.

58. Hodges, D.A., *Semiconductor Memories,* IEEE Press, NY, 1972.

59. Hoefflinger, B., J. Schneider, and G. Zimmer, "Advanced compatible LSI process for N-MOS and bipolar transistors," *IEEE International Electron Devices Meeting Technical Digest*, Washington, D.C., 1977, p. 261 A.
60. Hsu, S.T., "A COS/MOS linear amplifier stage," *RCA Review*, 37, 1976, p. 136.
61. Huijsing, J.H., and F. Tol, "Monolithic operational amplifier design with improved HF behavior," *IEEE J. Solid-State Circuits*, SC-11, 1976, p. 323.
62. "Japan presses innovations to reach VLSI goal," *Electronics*, June 9, 1977, p. 110.
63. Jenné, F.B., "Grooves add new dimension to V-MOS structure and performance," *Electronics*, August 18, 1977, p. 100.
64. Johnson, E.O., "Physical limitations in frequency and power parameters of transistors," *RCA Review*, 26, 1965, p. 163.
65. Kaneko, K., T. Ojabe, and M. Nagata, "Stacked $I^2 L$ circuit," *IEEE J. Solid-State Circuits*, SC-12, 1977, p. 210.
66. Kawagoe, H., and N. Tsuji, "Minimum size ROM structure compatible with silicon-gate E/D MOS LSI," *IEEE J. Solid-State Circuits*, SC-11, 1976, p. 360.
67. Keonjian, E., *Microelectronics*, McGraw-Hill, NY, 1963.
68. Keyes, R.W., "Physical limits in digital electronics," *Proc. IEEE*, 63, 1975, p. 740.
69. Klaassen, F.M., "Device physics of integrated injection logic," *IEEE Trans. Electron Devices*, ED-22, 1975, p. 145.
70. Kokkas, A.G., "Thermal analysis of multiple-layer structures," *IEEE Trans. Electron Devices*, ED-21, 1974, p. 674.
71. Koike, S., and G. Kambara, "Electrically erasable nonvolatile optical MNOS memory device," *IEEE J. Solid-State Circuits*, SC-11, 1976, p. 303.
72. Koller, K.W., and U. Lauther, "The Siemens-Avesta-system for computer-aided design of MOS standard cell circuits," *1977 Design Automation Conference*, p. 153, IEEE Cat. 77CH1216-1C.
73. Laermer, L., "Air through hollow cards cools high power LSI," *Electronics*, June 13, 1974, p. 113.
74. Lin, H.C., *Integrated Electronics*, Holden-Day, San Francisco, 1967.
75. Lindgren, N., "Semiconductors face the '80s," *IEEE Spectrum*, October 1977, p. 42.
76. Loranger, J.S., "The case for component burn-in: the gain is well worth the price," *Electronics*, January 23, 1975, p. 73.
77. Lynch, W.T., "The reduction of LSI chip costs by optimizing the alignment yields," *IEEE International Electron Devices Meeting Technical Digest*, Washington, D.C., 1977, p. 7G.
78. Maitland, D., "*N* or *p* channel MOS: take your pick," *Electronics*, August 3, 1970, p. 79.
79. Masuhara, T., N. Nagata, and N. Hashimoto, "A high performance *n*-channel MOS LSI-using depletion-type load elements," *IEEE J. Solid-State Circuits*, SC-7, 1972, p. 224.
80. Mashuhara, T., and R.S. Muller, "Complimentary DMOS process for LSI," *IEEE J. Solid-State Circuits*, SC-11, 1976, p. 453.
81. Mavor, J., *MOST Integrated Circuit Engineering*, Peregrinus, England, 1973.
82. Meindl, J.D., *Micropower Circuits*, J. Wiley & Sons, NY, 1969.
83. Meusburger, G., "A new circuit configuration for a single-transistor cell using Al-gate technology with reduced dimensions," *IEEE J. Solid-State Circuits*, SC-12, 1977, p. 253.
84. Morris, R.L., and J.R. Miller, *Designing with TTL Integrated Circuits*, McGraw-Hill, NY, 1971.
85. Mudge, J., and K. Taft, "V-ATE memory scores a new high in combining speed and bit density," *Electronics*, July 17, 1972, p. 65.
86. Mulder, C., and H.E.J. Wulms, "High speed integrated injection logic ($I^2 L$)," *IEEE J. Solid-State Circuits*, SC-11, 1976, p. 379.
87. Müller, R., "Current hogging injection logic: new functionally integrated circuits,"

*IEEE International Solid-State Circuits Conference*, 1975, p. 174.

88. Müller, R., H.-J. Pfleiderer, and K.-U. Stein, "Energy per logic operation in integrated circuits: definition and determination," *IEEE J. Solid-State Circuits*, SC-11, 1976, p. 657.

89. Murakami, K., et al., "Folded-collector integrated injection logic," *IEEE J. Solid-State Circuits*, SC-11, 1976, p. 644.

90. Panousis, P.T., and R.L. Pritchett, "GIMIC-O—A low cost non-epitaxial bipolar LSI technology suitable for application to TTL circuits," *IEEE International Electronic Devices Meeting Technical Digest*, Washington, D.C., 1974, p. 515.

91. Pashley, R., et al., "H-MOS scales traditional devices to high performance levels," *Electronics*, August 18, 1977, p. 95.

92. Pease, R.L., K.F. Galloway, and R.A. Stehlin, "Gamma radiation effects on integrated injection logic cells," *IEEE Trans. Electron Devices*, ED-22, 1975, p. 348.

93. Penney, W.M., and L. Lau, (American Micro-Systems, Inc.), *MOS Integrated Circuits*, Van Nostrand Reinhold, NY, 1972.

94. Persky, G., D.N. Deutsch, and D.G. Schweiket, "LTX—a system for the directed automatic design of LSI circuits," *1976 Design Automation Conference*, p. 399, IEEE Cat. 76CH1098-3C.

95. Ragonese, L.J. and N.T. Yang, "Enhanced integrated injection logic performance using novel symmetrical cell topography," *IEEE International Electron Devices Meeting Technical Digest*, Washing, D.C., 1977, p. 166.

96. Rein, H.M., and R. Ranfft, "Improved feedback ECL gate with low delay-power product for the subnanosecond region," *IEEE J. Solid-State Circuits*, SC-12, 1977, p. 80.

97. Remshardt, R., and U. Baitinger, "A high performance low power 2048-bit memory chip in MOSFET technology and its application," *IEEE J. Solid-State Circuits*, SC-11, 1976, p. 352.

98. Roesner, B.B., and D.J. McGreivy, "A new high speed $I^2L$ structure," *IEEE J. Solid-State Circuits*, SC-12, 1977, p. 114.

99. Sakurai, J., "A new buried-oxide isolation for high-speed, high-density MOS integrated circuits," *IEEE International Electron Devices Meeting Technical Digest*, Washington, D.C., 1977, p. 388.

100. Saltich, J.L., W.L. George, and J.G. Soderberg, "Processing technology and AC/DC characteristics of linear compatible $I^2L$," *IEEE J. Solid-State Circuits*, SC-11, 1976, p. 478.

101. Sander, W., J. Early, and T. Longo, "Injection logic boosts bipolar performance while dropping cost," *Electronics*, August 18, 1977, p. 107.

102. Sanders, T.J., and W.R. Morcom, "Polysilicon-filled notch produces flat, well-isolated bipolar memory," *Electronics*, April 12, 1973, p. 117.

103. Schwartz, S., *Integrated Circuit Technology, Instrumentation and Techniques for Measurement, Process and Failure Analysis*, McGraw-Hill, NY, 1967.

104. Schmidt, B., "Using C-MOS in systems designs—those all-important details," *Electron. Des. News*, 18, January 5, 1973, p. 5.

105. Scott, A.W., *Cooling of Electronic Equipment*, Wiley-Interscience, NY, 1974.

106. Scrupski, S.E., "Plastic cavity packages for MOS/LSI fill a very present need—but for how long?" *Electronics*, May 22, 1972, p. 102.

107. Spaanerburg, L., "Circuit implications of the metal-gate polysilicon source-and-drain MOST process," *IEEE J. Solid-State Circuits*, SC-12, 1977, p. 258.

108. Stern, L., *Fundamentals of Integrated Circuits*, Hayden, NY, 1968.

109. Suran, J.J., "A perspective on integrated electronics," *IEEE Spectrum*, January, 1970, p. 67.

110. Takagi, H., and G. Kano, "Dual depletion CMOS ($D^2$CMOS) static memory cell," *IEEE J. Solid-State Circuits*, SC-12, 1977, p. 424.

111. Taub, H., and D. Schilling, *Digital Integrated Electronics*, McGraw-Hill, NY, 1977.
112. Tokumaru, Y., et al., "I² L with a self-aligned double diffused injector," *IEEE J. Solid-State Circuits*, SC-12, 1977, p. 109.
113. Tomisawa, O., Y. Horiba, and S. Kato, "Vertical injection logic," *IEEE J. Solid-State Circuits*, SC-11, 1976, p. 637.
114. Torrero, E.A., "Focus on C-MOS," *Electron. Des.*, 6, March 15, 1974, p. 86.
115. Vacca, A.A., "The case for emitter-coupled logic," *Electronics*, April 26, 1971, p. 48.
116. VanTuyl, R., and C. Liechti, "Gallium arsenide spawns speed," *IEEE Spectrum*, 1977, p. 41.
117. Vittoz, E., and J. Fellrath, "CMOS analog integrated circuits based on weak inversion operation," *IEEE J. Solid-State Circuits*, SC-12, 1977, p. 224.
118. Walker, R., "C-MOS specifications: Don't take them for granted," *Electronics*, January 9, 1975, p. 103.
119. Walsh, P.S., and G.W. Sumerling, "Schottky I² L (substrate fed logic)–an optimum form of I² L," *IEEE J. Solid-State Circuits*, SC-12, 1977, p. 123.
120. Warner, R.M., and J.N. Fordemwalt, *Integrated Circuits–Design Principles and Fabrication*, McGraw-Hill, NY, 1965.
121. Warner, R.M., "Applying a composite model to IC yield problem," *IEEE J. Solid-State Circuits*, SC-9, 1974, p. 86.
122. Weber, S., *Large and Medium Scale Integration*, McGraw-Hill, NY, 1974.
123. Weber, E.V., and H.S. Yourke, "Scanning electron-beam system turns out IC wafers fast," *Electronics*, November 10, 1977, p. 96.
124. Wickes, W.E., *Logic Design with Integrated Circuits*, Wiley-Interscience, NY, 1968.
125. Wittenzellner, E., "Computer-aided design of large-scale integrated I² L logic circuits," *IEEE J. Solid-State Circuits*, SC-12, 1977, p. 199.
126. H. Wolff, "How much will SOS help?," *Electronics*, January 23, 1975, p. 66.
127. Yang-Ou, P., "Double ion implanted V-MOS technology," *IEEE J. Solid-State Circuits*, SC-12, 1977, p. 3.
128. Zuleeg, R., "Developments in GaAs FET's and IC's," *IEEE International Electron Devices Meeting Technical Digest*, 1976, p. 347.
129. Zuleeg, R., et al. "FEMTO-JOULE, high-speed, planar GaAs E-JFET logic," *IEEE International Electron Devices Meeting Technical Digest*, Washington, D.C., 1977, p. 198.
130. Broers, A.N., "Fine line lithography systems for VLSI," *IEEE International Electron Devices Meeting, Technical Digest*, Washington D.C., Dec. 1978, p. 1.
131. Hughes, G.P. and R.C. Fink, "X-ray lithography breaks the VLSI cost barrier," *Electronics*, 51, Nov. 9, 1978, p. 99.
132. Eden, R.C., "GaAs integrated circuits: MSI status and VLSI prospects," *IEEE International Electron Devices Meeting, Technical Digest*, Washington D.C., Dec. 1978, p. 6.
133. Glaser, A.B. and G.E. Subak-Shappe, *Integrated Circuit Engineering Design, Fabrication and Applications*, Addison Wesley, NY, 1977.
134. Bahraman, A. and S. Chang, "Design trade-offs in Schottky-base I² L–an advanced bipolar technology," *IEEE International Electron Devices Meeting Technical Digest*, Washington, D.C., Dec. 1978, p. 205.
135. Agraz-Guerena, J., R.L. Pritchett and P.T. Panousis, "High performance "upward" bipolar technology for VLSI," *IEEE International Electron Devices Meeting Technical Digest*, Washington, D.C., Dec. 1978, p. 209.
136. Kato, S. et al., "A new I² L memory cell–double diffused base structure," *IEEE International Electron Devices Meeting Technical Digest*, Washington, D.C., Dec. 1978, p. 213.
137. Handy, E.Z. and M.I. Elmasry, "Bipolar structures for BIMOS VLSI," *IEEE International Electron Devices Meeting, Technical Digest*, Washington D.C., Dec. 1978, p. 217.
138. Mead, C.A. and L.A. Conway, "Introduction to VLSI Systems," Addison-Wesley, Reading, Mass. 1979.

# 9
# Integrated
# Circuit Applications

## CONTENTS

The worldwide sales of integrated circuits (ICs) exceeded sales of discrete semiconductor devices in 1977 by a considerable margin (the figures being $3.5 billion and $3 billion, respectively). About $1.8 billion was spent for MOS ICs and the balance for bipolar circuits. About $750 million of the total was for circuits classified as linear rather than digital. This chapter begins with a discussion of linear circuit applications.

## 9.1 LINEAR INTEGRATED CIRCUITS

### 9.1.1 Operational Amplifiers

Operational amplifiers (op amps) are one of the key linear components of ICs and hundreds of millions of units are used each year. Prices for op amp monolithic circuits consisting of 20 to 50 transistors range from considerably less than a dollar to several dollars.

The dc parameters that define op amp performance include voltage gain, input current, input-offset voltage and current, and input-offset drift with temperature. The ac performance is measured primarily in terms of slew rate

(V/μsec rise time in attempting to follow a step input) and of bandwidth. To keep the dc input-offset small it is desirable to have low input currents. Then even with high beta transistors the operating current may not charge the circuit capacitances fast enough for good ac response. The problem is compounded if the design contains lateral transistors since these tend to have wide base thickness and poor frequency response. However, recent monolithic designs use concepts such as feed-forward and achieve excellent dc and ac parameters.

One of the earliest commercially successful op amps IC design is the μA 741, which has a voltage gain of over 100,000, an input offset voltage of 5 mV, a maximum slew rate of about 0.5 V/μsec, and a unity-gain frequency of about 1 MHz. This slew rate roughly corresponds to the ability of an op amp to provide an output voltage of 10 V peak at 10 kHz without large distortion: not a very impressive frequency performance. Another popular design of this general class is the LM 101 op amp, the basic circuit of which is shown in Fig. 9.1. The two voltage-gain stages drive a class AB emitter-follower pair to provide a low impedance output to the load $R_L$. In practice, a Darlington output transistor may be used, and a transistor added to the base level of $Q_3$ and $Q_4$ to provide a current "mirror" that reduces dc offset caused by the base currents.

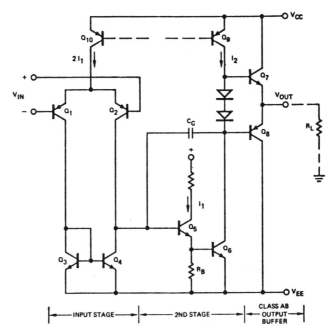

Fig. 9.1. A basic operational amplifier circuit. (After Solomon. Reprinted with permission from *IEEE Journal of Solid-State Circuits*, SC-9, 1974, p. 314.)

If these circuit refinements are neglected and a simple $\pi$ transistor model is used, the dc voltage gain is

$$A_v(0) = \frac{V_{out}}{V_{in}} \cong \frac{g_{m1} \, \beta_5 \beta_6 \beta_7 R_L}{1 + r_{i2}/r_{o1}} \qquad (9.1)$$

where

$$r_{i2} = \beta_5(r_{e5} + \beta_6 r_{e6}) \text{ and } r_{o1} = r_{o4}/r_{o2}.$$

The $r_e$ values are $kT/gI_e$ and the $r_0$ values are the output impedances in the $\pi$ model. The subscript numbers relate to the transistor numbers. If $\beta_5 = \beta_6 = 150$ and $\beta_7 = \beta_8 = 50$, the voltage gain for a load $R_L$ of 2 kilohms is 625,000. Discrete components give this gain, but the monolithic amplifier gives only a fraction of this gain because of thermal feedback effects. The problem is that transistors $Q_1$ and $Q_2$ cannot be located so that their temperatures are identical when the output stage is delivering power; thus, the base-emitter voltages of $Q_1$ and $Q_2$ are not the same and a spurious thermal feedback voltage exists at the input. The effect is significant even when care is taken in the design layout.

The small signal high-frequency gain of the op amp of Fig. 9.1 can be modeled as shown in Fig. 9.2 with the result

$$A_v(\omega) = \left| \frac{v_{0(s)}}{v_i} \right| = \left| \frac{g_{m1}}{sC_c} \right| = \frac{g_{m1}}{\omega C_c} \qquad (9.2)$$

Hence, the open-loop unity-gain frequency $\omega_u$, which is equivalent to the closed-loop gain-bandwidth product, is

$$\omega_u = \frac{g_{m1}}{C_c} \qquad (9.3)$$

where $C_c$ is the compensation capacitor. Here, $C_c$ may be 30 pF but cannot be much smaller because various excess phase shifts cannot be allowed to become too large. If the first stage bias current $I_1$ is 10 $\mu$A then $g_{m1}$ is 0.192 mA/V from $g_{m1} = qI_1/2kT$. Hence, from Eq. (9.3) the unity gain frequency $fu = (\omega_u/2\pi)$ is 1 MHz.

The slew rate of the amplifier is also related to $C_c$ as shown in Fig. 9.3. If the input voltage is large enough to switch the input differential circuit fully, the tail current $2I_1$ is, in effect, switched into the integrator formed by $C_c$ and the amplifier, A. The slew rate is then

$$\left. \frac{dV_0}{dt} \right|_{\substack{\text{slew} \\ \text{max}}} = \frac{2I_1}{C_c} \qquad (9.4)$$

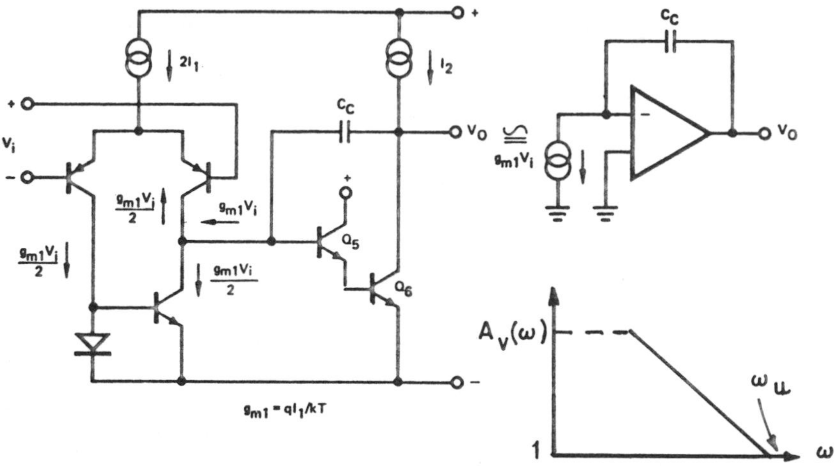

Fig. 9.2. First-order model of op amp used to calculate small signal high frequency gain. At frequencies of interest the input impedance of the second stage becomes low compared to first stage output impedance due to $C_c$ feedback. Because of this, first stage output impedance can be assumed infinite, with no loss in accuracy. (After Solomon. Reprinted with permission from *IEEE Journal of Solid-State Circuits*, SC-9, 1974, p. 9.)

So, if $I_1$ is 10 $\mu$A and $C_c$ is 30 pF the maximum slew rate is 0.67 V/$\mu$sec as found for the 741 op amp design. Alternatively, the slew expression can be written as

$$\left.\frac{dV_0}{dt}\right|_{\substack{\text{slew} \\ \text{max}}} = \frac{2\omega_u I_1}{g_{m1}} \tag{9.5}$$

For a simple bipolar input stage the ratio $I_1/g_{m1}$ is $kT/q$ and the slew rate is determined by $\omega_u$. The highest sinusoidal frequency that can be amplified without distortion is given by

$$\frac{dV_0}{dt} = \omega V_p \cos \omega t$$

hence

$$\omega_{\max} = \left.\frac{1}{V_p} \frac{dV_0}{dt}\right|_{\substack{\text{slew} \\ \text{max}}} \tag{9.6}$$

Above this frequency the onset of distortion is severe. Hence a slew rate of 7 V/μsec would be required to amplify a 100-kHz signal of amplitude 10 V.

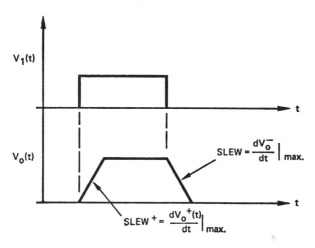

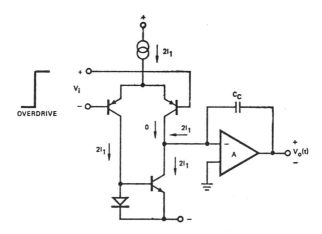

Fig. 9.3. Model for calculation of the slew rate. (After Solomon. Reprinted with permission from *IEEE Journal of Solid-State Circuits*, SC-9, 1974, pp. 11, 12.)

One method of increasing $I_1/g_{m1}$ for a bipolar transistor stage is to add feedback resistors $R_E$ in series with the emitters of $Q_1$ and $Q_2$ of Fig. 9.1. The equation then becomes

$$\frac{I_1}{g_{m_1}} = \frac{kT}{q}[1 + R_E I_1 / kT/q)] \tag{9.7}$$

Thus, if $R_E I_1$ is 500 mV the improvement is a factor of about 20. For the LM 118 op amp the $R_E$ value is 2 kilohms. Therefore, 1 mV of extra input-offset voltage occurs for a mismatch of 4 ohms (0.2%), so careful control of the $R_E$ values is essential.

An alternative method of increasing the slew rate is the use of an FET input stage. For a JFET

$$I_D = I_{DSS} \left( \frac{V_{GS}}{V_T} - 1 \right)^2 \tag{9.8}$$

where

$$\frac{g_m}{I_D} \cong \frac{2}{|V_T|} \left( \frac{I_{DSS}}{I_D} \right)^2 \tag{9.9}$$

and this is a minimum when $I_D = I_{DSS}$. Hence, the ratio of the JFET slew rate to that of a basic bipolar stage is given by

$$\frac{SLEW_{\text{JFET}}}{SLEW_{\text{BJT}}} = \left( \frac{W_{u\text{JFET}}}{W_{u\text{BJT}}} \right) \frac{|V_T|}{2kT/q} \tag{9.10}$$

So, for a JFET threshold of 2 V, the slew rate is better by about a factor of 40 than for a BJT input stage. A design using $p$ channel JFETs is shown in Fig. 9.4. This has a slew rate of 33 V/$\mu$sec, a unity-gain frequency of 10 MHz and an input current of 10 pA. The offset voltage is 3 mV and its drift is 3 $\mu$V/°C.

Other matters of importance in op amp design are covered in the tutorial paper by Solomon and this and Millman's book are recommended for further study.[136,185]

A number of IC op amps are compared in Table 9.1 with regard to performance. The chip area of a type 741 op amp design was originally about 0.15 × 0.15 cm (56 × 56 mils). However, areas of comparable units are now a quarter of this and 8000 units can be fabricated on a single wafer of diameter 7.5 cm. The type 088 op amp design is a two chip design but interesting because of the very low input-offset voltage and the high slew rate. The signal is divided into two parts as shown in Fig. 9.5. This allows the dc component to be chopped for amplification and feedback against the input for cancellation of the dc voltage offset and drift.

Another basic technique is that of feed forward of the ac component of the signal so that it bypasses transistors that are slow in response, such as lateral $pnp$ transistors that would introduce excess phase. Unity-gain bandwidths greater than 70 MHz and slew rates of 100-300 V/$\mu$sec have been achieved by such techniques.

Although a high slewing rate is desirable it results in overshoot of the final output level and a settling time is required before the desired accuracy is

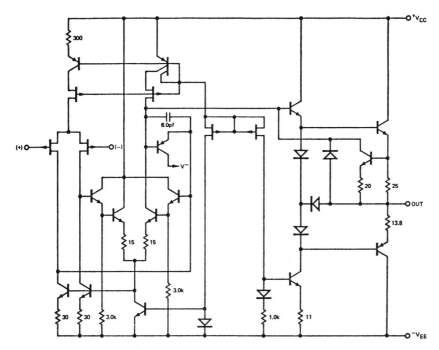

Fig. 9.4.  Monolithic op amp employing compatible *p*-channel JFETs on the same chip with normal bipolar components. (After Solomon. Reprinted with permission from *IEEE Journal of Solid-State Circuits*, SC-9, 1974, p. 15.)

## Table 9.1.  Comparison of Popular IC Operational Amplifiers

| Device | Slew Rate (V/$\mu$s) | Input Current (nA) | Input-Offset Voltage (mV) | Figure[a] of Merit (nA-ns)$^{-1}$ |
|--------|-----------|---------|----------|------------|
| 741  | 0.5   | 500 | 5     | 0.0002   |
| 101A | 0.5   | 75  | 2     | 0.00333  |
| 108A | 0.1   | 2   | 0.5   | 0.1      |
| 725A | 0.005 | 75  | 0.5   | 0.000133 |
| 770  | 2.5   | 15  | 4     | 0.04166  |
| 531  | 30    | 500 | 5     | 0.012    |
| 740  | 6     | 0.2 | 20    | 1.5      |
| 118  | 70    | 250 | 4     | 0.07     |
| 088  | 25    | 5   | 0.075 | 66.67    |

(After Callahan) Reprinted from Electronics, August 16, 1973; copyright © McGraw-Hill, Inc., 1973)

[a]The figure of merit is:

$$FM = \frac{\text{Unity-gain slew rate}}{\text{input current} \times \text{input-offset voltage}}$$

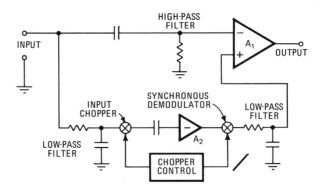

Fig. 9.5. *Two paths.* In one path (upper), high-frequency components are amplified directly, and in a second path (lower), the d.c. signal is chopped and fed back against the input. (After Callahan. Reprinted from *Electronics*, August 16, 1973; copyright McGraw-Hill, Inc., NY, 1973.)

acquired. The settling time is a function of the pole-zero configuration of the complete amplifier transfer function, and careful circuit design is needed to minimize the acquisition time. For an op amp with 12.5 MHz bandwidth and a unity-gain slew rate of 65 V/μsec the total time (slewing plus settling) may be 650 nsec to 0.01% accuracy (or 450 nsec to 1% accuracy) for a 10 V input signal. For this particular amplifier the input offset voltage was 2 mV, the open-loop gain was 100 dB and the common-mode rejection ratio was 95 dB.

In another design with complementary bipolar transistors, the op amp slews at 500 V/μsec and settles to within 0.1% in 200 nsec when used as a pulse inverter.

The uses of op amps are extensive since they are high-gain differential input direct-coupled amplifiers of fairly high input impedance and well suited to the addition of external feedback circuits. Op amps easily perform many mathematical functions including inversion (sign changing), addition, subtraction, differentiation, integration and phase changing. Therefore, they are the building blocks of electronic analog computers, active RC circuit filters, analog-to-digital (A/D) converters, voltage regulators and other circuits.[57,107]

### 9.1.2  Analog/Digital Converters

Transducers for electrical sensing of properties such as position, velocity, acceleration, pressure and temperature are usually analog in output rather than digital. The processing of analog signals for telemetering and measurement is difficult since voltage drops in lines and nonlinear effects result in loss of accuracy. Conversion, therefore, of analog signals to digital form for transmission and computer processing has become common practice. Numerical display of the information is also common practice since integrated circuit technology and light emitting diode or liquid crystal displays have made digital read-out instruments much less expensive than a few years ago.

The reverse process, namely digital-to-analog conversion, is needed when the processed digital information must operate an analog output controller such as a proportional electromagnetic-controlled flow valve. Digital control systems are now used in most industrial manufacturing processes, including those for chemicals, metal processing, machine tools, paper and textiles, and also in power generating stations and in military applications.

An important feature of most digital telemetering or communication systems is time-sharing multiplexing so that many signals can be sent along one communication channel. Disturbance of a digital signal by noise during transmission can affect accuracy but is usually controllable by proper setting of threshold on-off levels and pulse reshaping when needed. So, digital transmission accuracy is likely to be better than for analog transmission. However, the overall system performance depends on the A/D and D/A conversion accuracies.

(A) *Circuit Concepts of A/D and D/A*     Techniques of A/D and D/A conversion are numerous, and only a few are outlined here. Since many A/D converter circuits use a D/A converter in a feedback path it is appropriate to first consider digital-to-analog conversion.

Two broad classes of D/A conversion are parallel digital-to-analog converters and serial digital-to-analog converters. The first class is faster since the digital information bits are handled in parallel, but the component count is high.

The binary signal bits—$B$ in number—are fed in on $B$ lines, each with an analog switch that connects a reference voltage to a register. The current that flows is of a magnitude determined by the resistance in each bit line involved and the currents for all bit lines are summed in an op amp to give the analog output voltage. (Good bipolar current switches can be made without too much difficulty but it is difficult to make good bipolar voltage switches; therefore it is appropriate to make current the conserved quantity.) In a weighted-resistor D/A converter as shown in Fig. 9.6 currents $I/2, I/4, I/8 \ldots I/2n$ are generated by using resistors of magnitude $R, 2R, 4R \ldots 2^n R$. A 12-bit converter therefore requires 12 analog switches, 14 resistors, one reference voltage and one op amp. If $R$ is 10 kilohms and $V_{ref}$ is 5 V the current for the most significant bit (MSB) is 0.5 mA and for the least significant bit (LSB) is 0.25 $\mu$A. For low voltage offset the voltage across each switch must be less than a few millivolts and the on resistance (for the MSB) must be only a few ohms if 0.1% accuracy is to be achieved. The reference supply must be stable to about 0.01% and the ratio of the resistors must be better than 0.05%.

Another resistive circuit arrangement that is even more important than the weighted resistor design is known as $R$-$2R$ ladder. This, as illustrated in Fig. 9.7, requires only two values of resistance instead of the wide range of values required by the previous method. The property of the resistance ladder that is

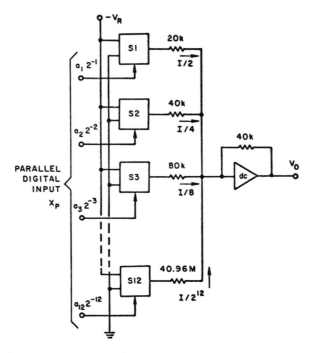

Fig. 9.6. A 12-bit weighted-resistor D/A converter. (After Schmid, from *Electronic Analog/ Digital Conversions*; copyright 1970 by Litton Educational Publishing, Inc. Reprinted by permission of Van Nostrand Reinhold Company.)

important is that every time a current enters a node such as $B_1$ it is divided in half. This is shown in Fig. 9.7(a) where bit 1 is on and the current $I$ is $V_{ref}/3R$ and the output voltage is $IR$. If bit 2 is turned on, the currents flowing as a result of this bit are as shown in Fig. 9.7(b). Therefore, if all bits are on together corresponding to a digital signal 111 . . . . . . 111 the output voltage is

$$IR + \frac{IR}{2} + \frac{IR}{2^2} + \cdots \cdot \frac{IR}{2^{11}}$$

The resistance ratios of an $R$-$2R$ ladder may be close enough, switch voltage drops small enough and the op amp output linear enough to provide an overall system accuracy of better than 0.05%.

Other parallel D/A converters use resistance potential dividers to provide a weighted voltage mode of operation. There are also many variations of resistance ladder circuits.

Serial D/A converters operate at the clock rate of the input signal. Some analog storage is needed and usually this is in capacitance form. Such converters

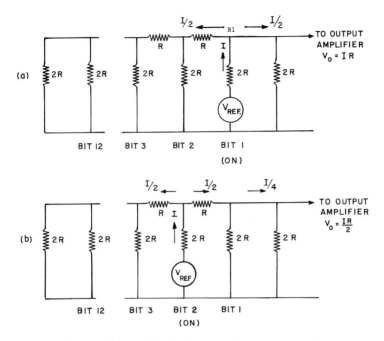

Fig. 9.7. An $R$-$2R$ resistive ladder network for D/A conversion.
(a)   Output is IR with bit 1 on.
(b)   Output is IR/2 with bit 2 on.

are very appropriate for time multiplexed signals since the digital information is already in serial form. The books of Schmid[130] and Hoeschele[70] provide information on serial techniques.

Consider now analog-to-digital converters that use a parallel D/A converter in the feedback path. As shown in Fig. 9.8 the error voltage $V_E$ is the sum of the input voltage $V_X$ and the negative feedback voltage $(-V_F)$. If $V_E$ is larger than $+V_{TH}$ the threshold circuit instructs the logic and storage circuits to increase the digital output. Similarly, if $V_E$ is negative and below the $-V_{TH}$ threshold the digital output is decreased. Some amplification of $V_E$ is desirable before it is compared with the threshold voltages.

In a successive-approximation A/D converter, $n$ successive comparisons are made between the input signal $V_X$ and the feedback signal $V_F$ where $n$ is the number of bits in the output. In the first comparison step the feedback voltage is $\frac{1}{2}V_R$ and on the next comparison the feedback voltage is reset to $(\frac{1}{2} \pm \frac{1}{4})V_R$, and so on for all $n$ steps. Although numerous switching operations must be performed in the clock period and the op amp must settle on a new level and new storage conditions be established, the accuracy and speed achievable can be quite good. For instance, a 12-bit successive-approximation A/D converter can give an accuracy of $\pm 0.05\%$ of full scale at $10^4$ conversions/sec.

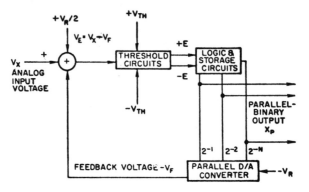

Fig. 9.8. Generalized parallel-feedback d.c. to digital converter. (After Schmid, from *Electronic Analog/Digital Conversions*, copyright 1970 by Litton Educational Publishing, Inc. Reprinted by permission of Van Nostrand Reinhold Company.)

The output from a successive-approximation A/D converter appears with the MSB first. In the feedback path this does not suit a serial D/A converter since the input to this must first be the least-significant bit. However, the difficulty has led to the development of newer concepts of serial feedback that are known as circulation or charge equalizing converters.

Consider now a ramp-comparison A/D converter as shown in simple form in Fig. 9.9. A sawtooth voltage is generated and compared with the input analog voltage $V_X$. The time $t$ for the ramp voltage to exceed the input voltage is determined by counting a clock signal and the count is converted to a parallel binary number. At a clock frequency of 10 MHz a 12-bit converter with an amplifier of comparison sensitivity ±1 mV can provide 2500 conversions/sec with an accuracy of ±0.05% of full scale over a limited temperature range. The resistor $R$, about 20 kilohms, and the capacitor $C$, about $10^4$ pF, must maintain precise RC values in this simple integrator configuration.

However, another mode of operation exists known as dual slope integration in which this dependence on RC is avoided. The technique is illustrated in Fig. 9.10. The input voltage $V_X$ is applied to the capacitor for a half cycle $t_1$ to obtain a voltage

$$V_0(t_1) = V_i + \frac{1}{RC} \int_0^{t_1} V(dt) = V_i + V_X \left(\frac{t_1}{RC}\right) \tag{9.11}$$

where the capacitor may have a small initial voltage $V_i$ on it. In the second half cycle the reference voltage $V_R$ is applied to discharge the capacitor in time $t_2$ to the starting level $V_i$. Hence,

$$V_X \frac{t_1}{RC} = V_{ref} \left(\frac{t_2}{RC}\right) \quad \text{or} \quad t_2 = t_1 \frac{V_X}{V_{ref}} \tag{9.12}$$

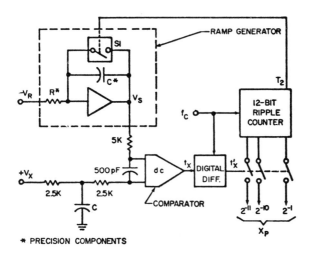

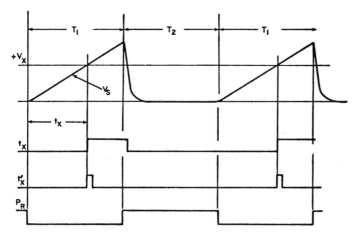

Fig. 9.9. Basic ramp-comparison A/D converter circuit and waveforms. (After Schmid, from *Electronic Analog/Digital Conversions*, copyright 1970 by Litton Educational Publishing, Inc. Reprinted by permission of Van Nostrand Reinhold Company.)

From this it is seen that $t_2$, a time that can be easily counted, is linearly proportional to the input voltage and is independent of $V_i$ and of the value of RC or its temperature drift. Accuracy depends on the precision determination of $t_2/t_1$ and $V_R$. Offset corrections are easy to make in the circuit.

(B) *Some Integrated Circuit A/D Converters*    The use of dual slope integration for a low cost 3½ digit multimeter application is shown in Fig. 9.11. One chip carries both *p* channel MOS and bipolar transistors in a voltage controlled

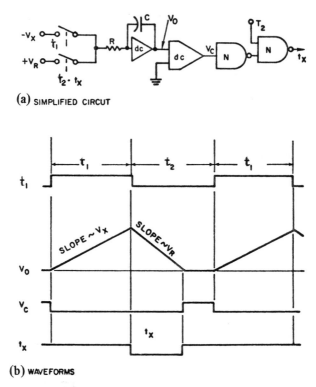

(a) SIMPLIFIED CIRCUT

(b) WAVEFORMS

Fig. 9.10. Dual slope ramp A/D converter. (After Schmid, from *Electronic Analog/Digital Conversions*, copyright 1970 by Litton Educational Publishing, Inc. Reprinted by permission of Van Nostrand Reinhold Company.)

oscillator (VCO) running at 40 kHz. In a 100 msec counting time this results in a count of 4000 in the register of the second chip. The voltage-to-frequency converter has a sensitivity of −10 kHz/V so an input signal of say 1.500 V applied in a second 100 msec counting period results in a countdown at a rate of 25 kHz and after the count is complete 1,500 remain in the register. This is read out by a decoder/driver chip and appears on an LED display as the measured voltage, 1.500 V.

In a digital panel meter it may be important to preserve the free floating capability of a simple analog meter so that voltages and currents can be measured at any reasonable voltage, for example ±300 V, with respect to ground potential. This can be done by transformer coupling of the pulse output from a free floating input chip.[55]

Although Zener diode circuits are normally used for voltage reference circuits,[21] in low cost A/D conversion the reference may be provided by a lightly-loaded mercury battery such as those used in wristwatches.[54]

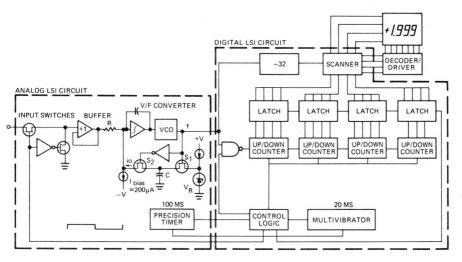

Fig. 9.11. Dual slope A/D conversion in a low cost 3½ digit multimeter. (After Strong. Reprinted from *Electronics*, September 11, 1972; copyright McGraw-Hill, Inc., 1972.)

Good switching of information is an essential feature of all A/D and D/A converter designs and attention must be paid to temperature gradients in the chips.[85, 96] In some designs thin film NiCr resistor networks that can be trimmed and are very stable are used. This hybrid approach with 12-bit resolution is capable of 0.02% accuracy over the −55°C to +125°C temperature range but is relatively costly.[147] Generally one or more bipolar chips are used for the analog and display driver functions and an MOS chip for the digital functions.

Several approaches, however, are possible to an all-MOS A/D system. In one serial D/A system, based on charge-redistribution concepts, an op amp is not needed and operation depends upon only two capacitors. An advantage of the MOS charge conservation approach over bipolar current conservation approaches is that capacitors in IC chips can be made with greater accuracy than can resistors. This is mainly because resistors of suitable values tend to be long and narrow in shape and so suffer from line-width errors, whereas capacitors can be square in shape.

A basic charge redistribution circuit is shown in Fig. 9.12(a). If the LSB $b_0$ is one, $S_2$ is closed and $C_2$ charges to $V_R$. Switch $S_1$ is then closed momentarily giving an output voltage $V_{out} = b_0 V_R/2$. The charge is left on $C_1$ but discharged from $C_2$ and then bit $b_1$ is applied to $S_2$ so that $C_2$ is recharged if the bit is one and through switch $S_1$ a charge redistribution is again produced. The output voltage is then

$$V_{out} = \frac{b_0 V_R}{4} + \frac{b_1 V_R}{2} \tag{9.13}$$

After $n$ bits the result is

$$V_{out} = \frac{b_n\, V_R}{2} + \frac{b_{n-1}\, V_R}{2^2} + \cdots\cdots \frac{b_0\, V_R}{2^n} \qquad (9.14)$$

which is the desired D/A form.

This serial D/A converter can be used as the basis of a successive approximation A/D converter. In contrast to the D/A conversion, the MSB must first be determined for A/D. The control logic takes on a particularly simple form since the D/A input string at any given point in the conversion is just the previously encoded word LSB first. For example, consider a point during the A/D conversion in which the $k$ MSBs have been decided. To decide the $(k+1)$th bit, a $(k+1)$ bit D/A conversion is carried out assuming the bit under consideration is a 1. If the bit is a zero, the D/A output voltage will exceed the analog voltage being encoded, and this will be indicated by the comparator. The correct value of the bit is stored, and the next serial D/A is started. Since the D/A conversion for the $k$th bit requires $k$ redistributions, the total number of redistributions required for an $M$-bit A/D conversion is $M(M+1)$.

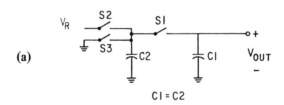

(a)

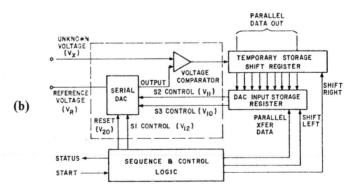

(b)

Fig. 9.12. MOS A/D charge-redistribution converter.
(a) Basic circuit principle.
(b) Complete analog to digital converter. (After Suarez, Gray and Hodges. Reprinted with permission from *International IEEE Solid-State Circuits Conference*, 1974, p. 194.)

The principal factors limiting the linearity of this technique are the error charges coupled onto $C_1$ and $C_2$ by the switch parasitics, the capacitor matching and the capacitor nonlinearity. In another approach to MOS A/D conversion, an array of binary weighted capacitors is used. In a prototype 10-bit design (see Fig. 9.13) the capacitors ranged in size from 120 pF to 0.24 pF. The transistors were made by $n$ channel metal-gate technology. Parasitic capacitance effects such as the gate-drain capacitance of the switches have little effect on accuracy. The sample mode acquisition time was 25 $\mu$sec and for a 0-10 V input voltage range the input offset voltage was −4 mV and the linearity ±½LSB.

MOS circuitry can also be applied to a constant-slope ramp type A/D converter with a resolution of 11 bits and a sample rate of 50/sec in a chip that has the potential of being a very low-cost design.[135] In other MOS studies, storage charge has been varied at the interface between aluminum oxide and silicon dioxide to produce continuously variable voltage threshold structures that form part of clockless A/D converters.[173]

Since there are many principles of A/D and D/A conversion, IC design work is likely to continue in this field for a number of years. The relative status of monolithic and hybrid types (1977) is shown in Table 9.2, from which it is seen that the IC approach tends to have a price advantage.

### 9.1.3  Consumer Use of Linear ICs

About a third of the linear IC market is in consumer oriented equipment such as radio receivers, TV sets and automobiles, instead of in the industrial and professional instrumentation field. A few circuits are presented to illustrate some typical performance levels achieved.

A monolithic 5-W integrated audio amplifier circuit is shown in Fig. 9.14. This was designed for use in stereo phonographs, tape systems and FM receivers. The output impedance is 0.6 ohms and the input signal level for 5-W power output is 185 mV. The open-loop voltage amplification is 62 dB but in a typical application the resistor $R6$ provides about 28 dB of negative feedback. The −3 dB power points occur at 22 Hz and 100 kHz. The general principles of audio frequency amplifiers are discussed in numerous books.

A high power op amp chip may be used as an audio amplifier. In one power op amp design the open-loop voltage gain is 50,000 and the unity-gain bandwidth is 750 kHz. Outputs of ±22 V and 1.2 A can be obtained with an output resistance of 2 ohms. In a class A-B audio amplifier 6 W rms of power can be obtained in an 8-ohm load with a distortion of 0.3% at 1 kHz. The chip size is 2500 $\mu$m $\times$ 1750 $\mu$m (100 $\times$ 70 mils). Built into the chip are two protective features shown in the block diagram of Fig. 9.15. The maximum output current is adjustable by the value of the external resistor $R_{Lim}$. A thermal shutdown circuit, based on the voltage drop across a diode, limits the output current if the chip temperature tends to exceed 150°C. Thus, protection is provided against load faults, inadequate heat sinking and excessive ambient temperature.

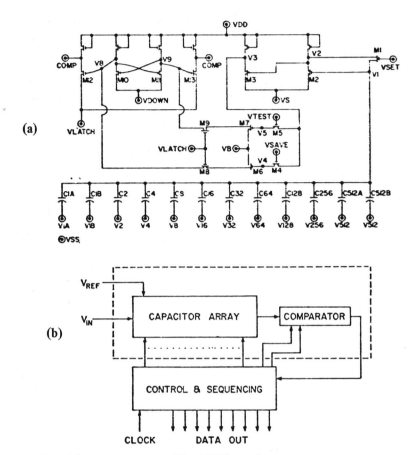

Fig. 9.13. A/D bit array binary-weighted MOS capacitor converter.
(a) Capacitor array, offset compensated comparator and output latch circuit.
(b) Complete A/D converter. (After McCreary and Gray. Reprinted with permission from *IEEE International Solid-State Circuits Conference*, Technical Abstracts, 1975, p. 38.)

In the TV industry solid-state chassis have become generally available but these are usually mixtures of discrete and integrated circuits. Progress is being made towards reducing the number of discrete circuits and increasing the functions per chip of the integrated circuits. The rate of progress in this direction depends primarily on the economics of such conversion. Functions of the receiver that are particularly amenable to integration include the IF stage, audio, chroma, sync. and automatic gain control.

The schematic of a monochromatic TV receiver[16] in which nearly all the components are on a single chip is shown as Fig. 9.16. The chip size is 220 $\mu$m $\times$ 175 $\mu$m (86 $\times$ 70 mils) and it is in a 24 pin package. Bipolar transistors are used throughout. The performance of the chip is summarized below and it gives good visual display and sound.

**Table 9.2.  D/A and A/D Converters of IC and Hybrid Form Compared**

| Type | | Specification | Monolithic IC | Hybrid |
|---|---|---|---|---|
| Digital-to-analog | | Resolution<br>Settling time<br>Price<br>Notes | 6 – 12 bits<br>100 ns – 2 μs<br>$5 – $20<br>Most are current output; microprocessor-compatible 8-bit devices emerging. | 6 – 16 bits<br>0.5 – 100 μs<br>$20 – $150<br>Most are fully self-contained; microprocessor-compatible versions emerging. |
| Analog-to-digital | Integrating | Resolution<br>Conversion time<br>Price<br>Notes | 10 – 13 bits<br>30 – 300 ms<br>$7 – $25<br>Two-chip sets are giving resolutions up to 16 bits. | –<br>–<br>–<br>Very little available. |
| | Successive-approxima-tion | Resolution<br>Conversion time<br>Price<br>Notes | 8, 10 bits<br>20 – 40 μs<br>$10 – $30<br>Microprocessor-compatible versions emerging; complete 8-bit micro-processor-compatible data-acquisition systems coming. | 8 – 12 bits<br>1 – 50 μs<br>$40 – $300<br>Microprocessor-compatible versions emerging; complete 8- and 12-bit analog I/O systems available. |

(After Mattera) Reprinted from *Electronics*, October 12, 1977; copyright © McGraw-Hill, Inc., NY, 1977.)

*TV Chip Performance*:

- video IF amplifier and detector conversion gain (with recommended external components), 15 μV rms input per volt output;
- AGC range, 65 dB;
- tuner AGC (*n-p-n*), 1.5-10.5 V;
- tuner AGC (*p-n-p*), 0-6 mA;
- AFC carrier output, 1 mA peak-to-peak;
- video output (zero carrier), 7.0 ± 0.4 V; (sync. tip), 2.1 ± 0.25 V;
- intermodulation products (50 μV to 10 mV rms input), less than −60 dB; (50 mV rms input), less than −50 dB;
- FM detector sensitivity, 100 μV rms;
- AM rejection (10 mV rms), 55 dB;
- volume control range, 70 dB;
- audio output, 2 V peak-to-peak.

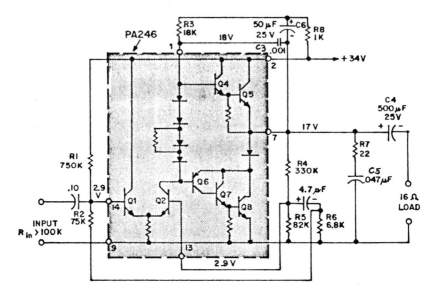

Fig. 9.14. A 5-watt integrated circuit audio amplifier. (After Jones. Reprinted with permission from *Proceedings of the National Electronics Conference*, 1968, *V.24*, p. 784.)

The present trend in television is towards more use of UHF channels. This is facilitated by preset tuning that can be selected by a push-button keyboard and LED display system. Figure 9.17 shows what is typically involved in such a system.

A consumer product is the video tape recorder/player to extend TV programming. Integrated circuits are being used to lower the cost of such systems to bring them into more widespread use.

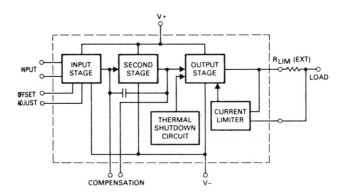

Fig. 9.15. Block diagram of a power operational amplifier chip with current limiter and thermal shutdown features. (After Gray. Reprinted with permission from *IEEE Journal of Solid-State Circuits*, SC-7, 1972, p. 475.)

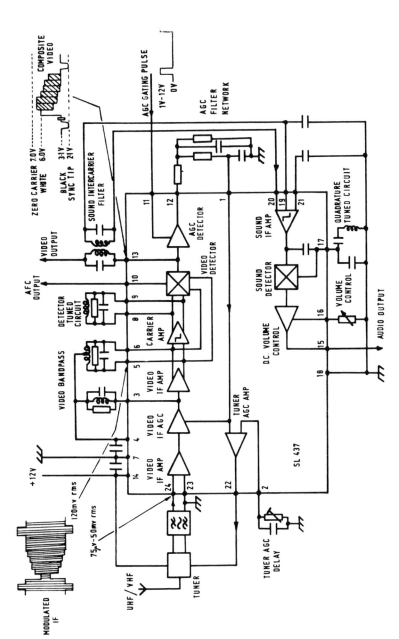

Fig. 9.16. Block diagram of chip showing the low number of external components required for a complete television IF system. (After Baskerville. Reprinted with permission from *IEEE Journal of Solid-State Circuits, SC-7*, 1972, p. 456.)

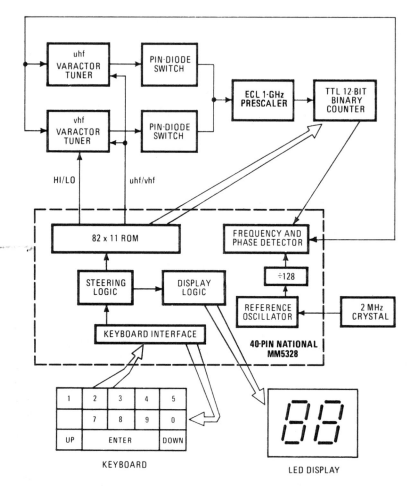

Fig. 9.17. Push button television UHF keyboard and LED display system. (Reprinted from *Electronics*, December 26, 1974, copyright McGraw-Hill, Inc., 1974.)

In automobiles there are numerous applications for linear and digital ICs but cost is the barrier to their increased use. Since digital processing is the primary feature here the topic will be dealt with in a later section of this chapter.

Voltage regulators are systems that involve large scale use of linear ICs. Voltage regulator principles are discussed in general circuit textbooks and in more detail in specialized IC engineering books. The basic concept is that a series transistor reduces an unregulated voltage to a lower regulated voltage (see Fig. 9.18) under the command of a differential amplifier which acts on the error signal difference between a scaled-down sample of the output voltage and a preset reference voltage. The differential amplifier may be a standard

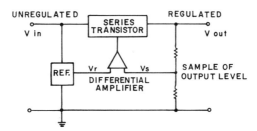

Fig. 9.18. Block diagram of a voltage regulator.

power op amp. A typical performance is an input voltage range of 8-40 V, an output voltage range of 5-37 V (depending on the input voltage and the permissible chip dissipation of 500 mW), a maximum output current of 150 mA, an output impedance of 50 milliohms, line regulation 0.02%/V and temperature stability of 30 ppm/°C.

## 9.2 COMMUNICATIONS APPLICATIONS

Communications include voice telephone systems, digital telephone line transmission, microwave relay systems, satellite relay systems, facsimile transmission, data telemetry, and mobile communications. The range of circuit needs is large and only a few examples are discussed here.

Electronic switching is now used in new telephone installations, but some of the designs predate the era of integrated circuits and are therefore based on discrete or hybrid circuits. A telephone company with large amounts of equipment installed is concerned with problems of maintenance and stocking of spare modules for servicing. Therefore, design changes tend to be introduced in big steps rather than as a continuous improvement process.

Reliability is of considerable importance. The Bell Telephone No. 1 Electronic Switching System, for example, is expected to have a total down-time of only 11 hr in 40 years of life. Only half of this is expected to be the result of hardware failure since human error is still a factor. A typical ESS installation contains 122,700 transistors, 379,100 diodes, 35 million bits of memory (magnetic) and 514,900 passive components such as resistors, capacitors and transformers. The component reliability is defined in units of FITs which are failures in $10^9$ device hr. The observed rates are for transistors 3.0 FITs, for diodes and resistors 0.2 FITs, for capacitors 4.0 FITs and for transformers 1.0 FITs. The ultimate aim is to achieve a system uptime of 99.99943% (the 1972 value was 99.997%).[28]

As further examples of telecommunications systems involving semiconductor devices and integrated circuits, consider briefly four other applications in the Bell System. The T-Carrier system handles about a million voice channels in the USA (1972) for short interoffice trunk lines and is being extended for long-haul service. A channel bank is used to pulse-code-modulate 24 voice-frequency signals and associated signaling information into a 1.5 Mbits/sec digital stream for transmission over a repeater line. Voice signals are coded in nonlinear digital codes to conserve digital transmission bandwidth with a good signal to noise or distortion ratio. Integrated circuits based on arrays of binary weighted MOS capacitors have recently been shown to have potential as PCM voice coders.[61,156]

The TH-3 Microwave Radio System is based on the 6-GHz band and is capable of a 4,000-mile haul in 150 hops. There is an 1800 message load capability and the base-band signals in the multiplexers cover the frequency range 0.564-8.524 MHz. Figure 9.19 shows a block diagram of the system. The integrated circuits in the system are of hybrid strip-line form. Another radio system is the DR 18 which is for short-range major communication arteries in metropolitan areas. This digital radio system at 18 GHz will carry up to 28,224 simultaneous telephone conversations with a data rate of 274 Mbits/sec.[18]

Coaxial cables are however still important. The L5 coaxial-carrier transmission system is a solid state broad-band system designed to transmit 10,800 long-haul message channels on a standard 9.5 mm (3/8 inch) coaxial cable. It also has high-speed digital signal capability. Repeaters are at 1 mile intervals and main power feed repeater stations and transmitters are every 75 miles.[86] Low-distortion *npn* transistors with a 3-GHz gain-bandwidth product are used to handle the highest message frequency of 60.5 MHz. The structure of these transistors is shown in Fig. 9.20. The emitter fingers are 2.5-$\mu$m wide by 105-$\mu$m long, $h_{fe}$ is 90 and third harmonic distortion is $-90$ dB below 1 mW and the noise figure is 2.7 dB. The thermal impedance is $30°C/W$ and the transistors may dissipate 1.7 W in a peak ambient of $85°C$.

Video communications require a large bandwidth, and noise or crosstalk can be a problem that degrades picture quality. Digital transmission provides better noise immunity. This requires electrical crosspoint switches for exchanging digitized video signals in a space division switching network. The switches should be of logic circuit form so that pulse waveform regeneration is achievable. For a low-power design full power must be supplied only to the crosspoint switches that must be operated and almost zero power to the idle switches. An integrated circuit for a 4 × 4 matrix has been designed on a chip (2.0 × 2.6 mm) recently for study in a 100 Mb/sec system. The general configuration is shown in Fig. 9.21. The power dissipation is 50 mW per crosspoint and the peripheral circuits consume 500 mW, but further power reduction seems possible.

At least 100 dB of crosstalk attenuation is desirable at 1.5 kHz in a crosspoint switch. The use of silicon-on-sapphire CMOS technology has been suggested to provide exceptional isolation between devices.[40]

In another area of telephone applications there is the task of driving telephone relays at 30-60 V. If bipolar circuits are used the current drain is 10-20

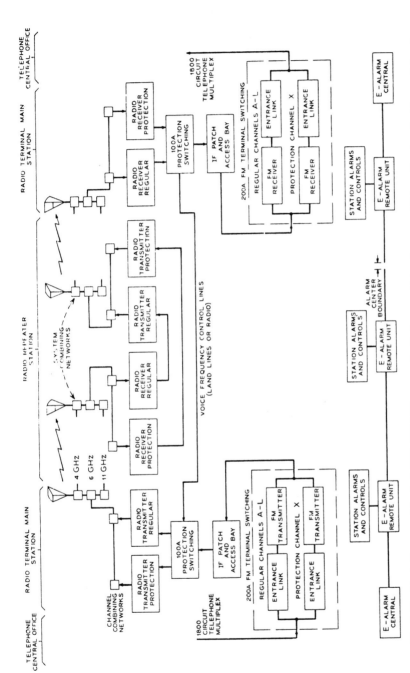

Fig. 9.19.  Block diagram of the Bell 6GHz TH-3 Microwave Radio System. (After Janson and Prime. Reprinted with permission from the *Bell System Technical Journal*, **V.50**, p. 2088; copyright 1971, The American Telephone and Telegraph Company.)

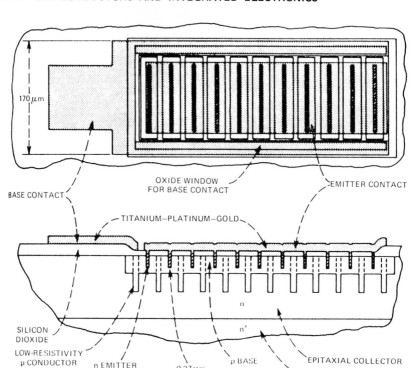

Fig. 9.20. Structure of the low-distortion npn L5 transistors used in the Bell System to handle a 60.5 MHz message frequency in a co-axial cable system. (After D'Altroy et al. Reprinted with permission from *Bell System Technical Journal*, **V.53**, p. 2195; copyright 1974, American Telephone and Telegraph Company.)

mA at standby. This can be greatly reduced by use of two CMOS NAND gates driving an emitter follower Darlington pair on a $2110 \times 1830$ μm ($83 \times 72$ mil) monolithic chip. In other approaches thyristor (SCR) switches with dielectric isolation are used.[132]

In telecommunications extensive use is made of active filters to obtain high $Q$ networks at low frequencies. Since an active filter basically involves one or more op amps, with RC networks connected around, the hybrid approach was the first to be used since trimming of resistor values is then easy. Figure 9.22 shows the components of such a filter which removes frequencies over 4 kHz from telephone lines by attenuation of 40 dB. The system arrangement is shown in Fig. 9.23. After filtering, the voice-frequency signal is chopped at an 8-kHz rate for digital encoding and multiplexing before transmission. In the receive channel the active filter smoothes the incoming signal to voice-frequency form.

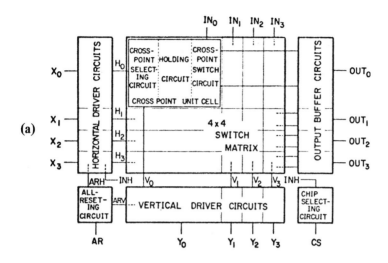

(a)

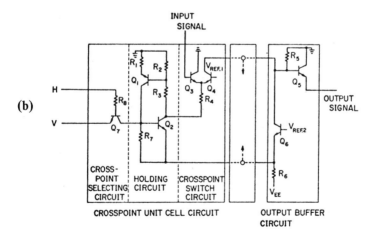

(b)

Fig. 9.21. 4 × 4 Matrix crosspoint switch for a 100 Mb/s digital transmission system.
(a)  Block diagram of the 4 × 4 matrix.
(b)  Unit cell and buffer circuit. (After Sunazawa et al. Reprinted with permission from
*IEEE Journal of Solid-State Circuits,* SC-10, 1975, p. 119.)

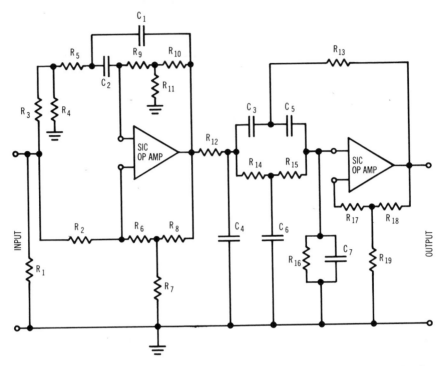

Fig. 9.22. Hybrid thin film active filter, with two operational amplifiers, for filtering frequencies over 4 KHz from telephone lines. (After Dupcak and DeGroot. Reprinted with permission from *Western Electric Engineer*, **V.18**, 1974, p. 24.)

The high frequency performance of an active filter is limited by the frequency behavior of the op amp used.[20, 98, 109, 119]

Future telecommunications systems may use millimeter waveguides. A 6.2-cm (2.5-inch) diameter circular waveguide has a potential capacity of 500,000 conversations in the 40-110 GHz frequency band. The transmission distance is tens of miles. This is a substantial improvement over the 108,000 channel coaxial cable system discussed earlier, that requires repeaters at 1 mile intervals. Present day satellites operate in the 4 and 6-GHz bands which are also used for terrestrial microwave links. Designs, however, are complete for the use of the 12 and 14-GHz bands and in the future satellites may even operate in the 30-GHz range.

Looking further into the future the use of fiber optics for communications can be predicted for wide bandwidth low cost transmission. The attenuation of optical waveguides is now acceptably low, for instance 1-2 dB/km, and extensive laboratory and field studies are in progress on the modulating, demodulating and multiplexing components needed for complete systems.

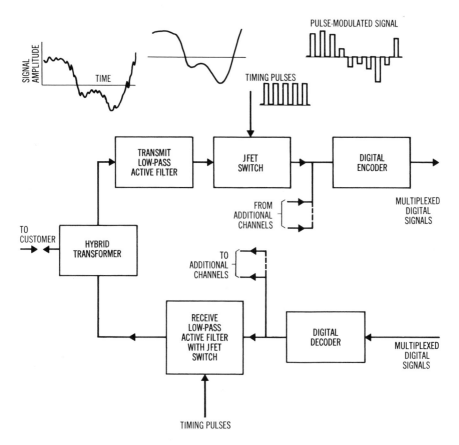

Fig. 9.23. One low-pass filter is required for transmission and another for reception in one channel of a pulse modulation system. (After Dupcak and DeGroot. Reprinted with permission from *Western Electric Engineer*, **V.**18, 1974, p. 24.)

## 9.3 APPLICATIONS IN WATCHES, CAMERAS AND AUTOMOBILES

### 9.3.1 Electronic Watches

About 200 million watches are sold each year (worldwide) for a total market of about $2 billion. A third of the sales consists of high grade watches costing $50 or more. For some hundreds of years the mechanical spring balance movement of accuracy 5 min./month in typical use has sufficed. The introduction of battery driven tuning fork designs in the mid 1950s raised the accuracy to 1 min./month. The next improvement was in 1969 with the development of a quartz crystal oscillator and a CMOS frequency divider circuit movement to provide timing for a stepping motor driving a conventional hands display.

This increased the accuracy to a few seconds a month. In recent models the mechanical display function has been superseded by liquid crystal or light-emitting-diode displays.

The components of a watch with a mechanical display are: a) the battery; b) the quartz oscillator; c) the frequency divider; d) the driver decoder; and e) the stepping motor. For an all-electronic watch the driver decoder is con-siderably changed and (e) becomes the display.

The battery cells are of mercuric oxide (1.3 V) or silver oxide (1.5 V) and have ratings of 100 to 200 mAh. For one year of continuous use the average current drain is limited to 10-20 $\mu$A.[44]

The quartz crystal is usually a 32.768 kHz flexure-mode bar trimmed to the desired frequency by capacitance adjustment. However, frequencies as low as 8.192 kHz and as high as 5 MHz have been used.[50,145] If the frequency is low the quartz crystal is inconveniently large and if the frequency is high the divider circuit may have too many stages and consume too much power. A conventional binary stage uses 16 or more transistors to divide by 2, and to scale down in binary stages from 1.5 MHz, for instance, would require 80 $\mu$A of current. In one high frequency design at 1.548288 MHz the initial dividing is by 7 (using 28 transistors), followed by three stages of division by 3 and 14 stages of division by 2 to give the desired 0.5 Hz output for a current drain of only 3 $\mu$A. The output transistors produce square wave voltages at 0.5 Hz to give current impulses of 300 $\mu$A peak to the stepping motor. However, the mean current is only about 7.5 $\mu$A and the overall current drain for the complete watch is 11.5 $\mu$A.

The block diagram of a watch operating from a 32 KHz crystal oscillator is shown in Fig. 9.24. A typical CMOS watch may contain between 300 and 1,000 transistors. A circuit for scale-of-two division is shown in Fig. 9.25 and involves 19 transistors. This large number of transistors is necessary because of the Boolean equation that must be satisfied. The dynamic power consumption per stage is 1.6 nW/kHz and the divider is capable of working up to 2 MHz at 1.35 V. The power dissipation is of course greatest in the first few stages since these have the highest switching rates. The search for minimum power circuitry for electronic watches has led to the development of many sophisticated circuit systems as can be discovered by exploring the patent literature.[139] Although CMOS dominates the circuits used presently, integrated injection logic circuits are also possible.

Two types of digital display are presently available. An LED display in a watch is a large consumer of power since it requires from 5-100 mA, depending on the size and brightness required. Since the mean battery drain must be 10 $\mu$A or so for a one-year life, continuous activation of LED displays is not possible. Liquid crystal displays however require much less power, although they require 4-15 V from a voltage converter for good operation. A CMOS chip with drive for an LCD as well as the oscillator and divider functions may have a power

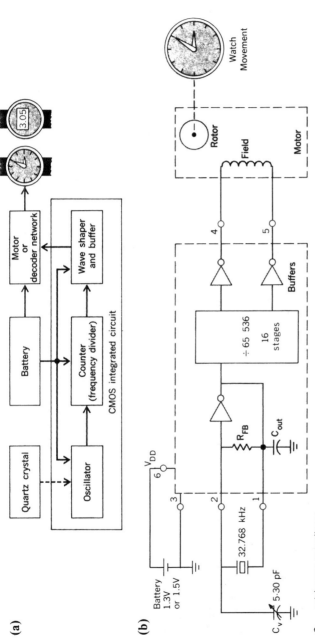

$C_v$ = trimmer capacitance
$C_{out}$ = integrated oscillator output capacitance
    $\approx$ 20 pF
$R_{FB}$ = integrated oscillator feedback resistance
    $\approx$ 40 M$\Omega$

Fig. 9.24. An electronic watch with either an analog (mechanical hands) output or a digital display.
(a) Block diagram.
(b) Analog display with 32 KHz crystal and 16 CMOS divider stages. (After Eleccion. Reprinted with permission from *IEEE Spectrum*, April 1973, p. 25.)

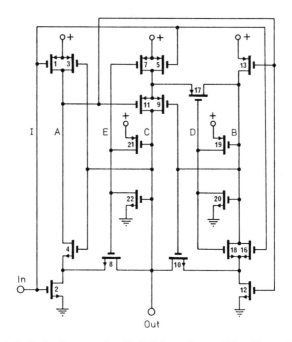

Fig. 9.25. A divide by 2 stage using 19 CMOS transistors. (After Vittoz, Gerber and Leuen-berger. Reprinted with permission from *IEEE Journal of Solid-State Circuits,* SC-7, 1972, p. 101.)

demand of less than 10 $\mu$A and cost not much more than a $1 in quantity. LCDs are not readable in the dark and the contrast is not good under certain lighting conditions. However, because of the convenience of a constant display, they have tended to replace LED displays in most watch designs.

### 9.3.2 Electronics in Cameras

About 15 million cameras are sold each year and the market for electronic components in cameras is from some tens of millions of dollars to $100 million. The systems in use vary from CdS photoconductor light-sensors operating needle displays, to sophisticated exposure-control and photo-flash systems.

An automatic exposure system of the simplest kind is one in which the user selects the shutter speed and the aperture is set by a CdS photocell, driving a galvanometer movement that mechanically adjusts the iris. In another approach the user chooses the aperture (or as in the Minox-C the aperture is fixed) and the shutter speed is electronically determined by an RC time constant in which the photocell is the variable resistance. In the Nikkormat EL the photocell charges a capacitor to a reference voltage corresponding to the light level. When the shutter is opened a timing capacitor begins to charge and when its voltage reaches the reference level the shutter is closed. In another approach known as a programmed shutter, the aperture and speed are interrelated and the combination is selected by the photocell response (Minolta Hi-matic E).

In the Kodak Instamatic cameras the circuit takes the form shown in Fig. 9.26. The aperture is controlled by a positioning servo represented by solenoids $L_1$, $L_2$ and $L_3$, which are controlled by the amplifier $A_1$ whose input is the difference between the photocell voltage and a reference voltage across $R_2$. If the light level is too low the indicator $I_1$ comes on and the user can release the shutter button without an exposure. If the indicator remains off the user completes his depression of the shutter button which closes the shutter switch and capacitance $C_3$ discharges through the lower photocell element to provide the timing action needed. If more light is needed a flash cube may be used and then $R_1$ limits the exposure time to 1/30 sec.

Electronic flash units use a Xe tube discharge from a capacitor (200-500 $\mu$F) charged to 300-500 V. This voltage is obtained from a battery of a few volts by a charging step-up circuit. In sophisticated flash systems the light reflected from the subject is sensed during the exposure, by a light sensitive thyristor for instance, and when the integrated exposure reaches a preset level a hidden Xe quench tube is fired to extinguish the main flash.

The block diagram for a single lens reflex camera with an aperture preset system and an LED array indicating shutter speed is given in Fig. 9.27. A Si photodiode is used for light sensing although it must be adjusted for blue response. Alternatively a GaAsP diode may be used so that there is a better match to the film sensitivity without the need for filtering.

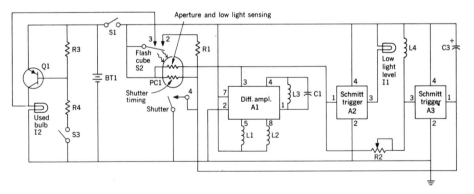

Fig. 9.26. In the Kodak Instamatic 60 camera three custom ICs provide both aperture and shutter speed control. A1 is an IC differential amplifier with high input impedance. A2 and A3 are IC Schmitt triggers, operated by a voltage ratio between input and supply voltage, rather than absolute voltage levels. To fit the components in the limited space available they are distributed around the camera and interconnected by flexible printed circuits that are packed around the mechanical and optical components. (After Lapidus. Reprinted with permission from *IEEE Spectrum*, June, 1973, p. 23.)

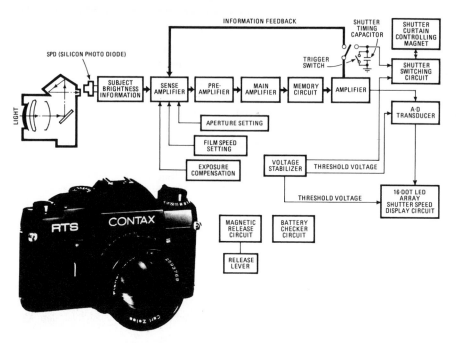

Fig. 9.27. Block diagram of the Contax RTS camera electronics. (After Schneiderman. Reprinted from *Electronics*, August 21, 1975; copyright McGraw-Hill, Inc., 1975.)

The Polaroid SX-70 folding instant picture camera uses 400 transistors in 7 integrated circuits and 4 nonintegrated discrete passive components. Figure 9.28(a) shows the block diagram of the exposure control module that uses 4 of the ICs. The taking lens aperture is programmed to provide a good compromise between f stop and exposure time. The photometer integrates photocurrents between 10 and 200 pA and gives exposure times between $10^{-3}$ and 22 sec with accuracies of $\pm 1\%$ for a $V_{cc}$ variation from 3.0 to 5.8 V and a temperature range 0-50°C. The overall exposure is accurate to $\pm 5\%$ and higher precision is not needed because the film and processing variables introduce greater variations.

The battery is contained in the film pack and controls the IC motor control module which ejects the film. The motor requires 1-2 A of current and dynamic braking is provided. About 22 transistors are present in the block shown in Fig. 9.28(b).

Extensive use of integrated circuits in 8-mm movie cameras is developing since automated circuit functions and low weight are important. The latest Agfa-Gevaert cameras contain chips with 1800 MOS transistors.

Automatic focusing cameras working on the principle that an in-focus image has more contrast than a non-focused image are under study. Systems are also available in which a split-image is reduced to coincidence automatically instead of manually as in a split-image rangefinder. Reflection of sound pulses may also be used to control the focusing action.

### 9.3.3 Automobile Electronics

Cost and reliability are primary considerations in the acceptance of new electronic systems in cars. Estimates of the volume of US business involved for 1978 vary from $200-400 million for the electronic semiconductor components, to over $1 billion for the complete systems including entertainment equipment and test gear. About 15 million cars are sold each year in the USA and each car carries between $15 and $30 worth of components. The predictions are for a further 40% increase by 1980.

Automobile electronic systems are located in inhospitable environments. The temperature range in the vicinity of the engine is from $-40°C$ to $+125°C$ and vibration, humidity, and corrosion conditions are severe. Electric conditions may also be bad because of surges and voltage glitches and transient effects. Nevertheless, more than fifty electronic control areas or systems for use in cars have been identified and studied. As seen in Table 9.3 they range from control of major engine functions to simple convenience items for vehicle users. The future will probably involve grouping of functions and multiplexing of digital information and microprocessor control as shown in Fig. 9.29. The lack of reliable low cost sensors and actuators is the limitation in many instances.[181] The schematic of a closed-loop electronic fuel-injection system given in Fig. 9.30 shows the sensing and control points required for this function. The oxygen concentration in the exhaust gases is used to determine the air/fuel ratio to $±0.1\%$ to give the optimum combustion conditions. This gives fuel economy and assists in pollution control.

Two electronic controls that have already entered the automobile market are electronic regulation of alternator voltage and electronic ignition systems. With the electronic regulator, much better battery charging conditions and dependability are obtained. The electronic ignition system provides more accurate timing of the spark and a rapid rise of voltage that allows firing on fouled spark plugs at low engine turnover speeds.

Another control that is gaining acceptance is that for antispin braking since this provides better action than manual control. Guidance and safety concepts involving radar or sonics are at present only experimental.[88]

Studies are in progress of the microprocessor concept in automobiles in major companies. General Motors has been examining a four-bit system for display of information that relates to preprocessed data on matters such as: speedometer, odometer, trip odometer, time of day, elapsed time, engine speed, four-wheel-lock control, cruise control, traction control, ignition timing, ignition dwell, automatic door locks, speed warning, speed limiting, and antitheft. Other GM processors are concerned with automobile function diagnostics both in motion and during servicing.

**Table 9.3. Existing and Proposed Electronic Systems for Automobiles**

| | | |
|---|---|---|
| Alternator | Voltage regulator | Electronic fuel injection |
| Electronic ignition | Intermittent windshield wipers | |
| Wheel lock control | Traction control | Headlamp dimming |
| Climate control | Air cushion restraint system | Digital clock |
| Automatic door locks | Alcohol detection systems | Sleep detectors |
| Flasher control systems | Programmed driving controls | High speed warning |
| High speed limiting | Lamp monitor systems | Electronic horn |
| Crash recorder | Traffic controls | Tire pressure monitor |
| Tire pressure control | Automatic seat positioner | Automatic mirror control |
| Automatic icing control | Road surface indicator | Four-wheel antilock |
| Vehicle guidance | (radar, infrared, laser, sonic) Station keeping Automatic brakes, Predictive crash sensors | Electronic timing |
| Multiplex harness systems | Electronic transmission control | Electronic cooling system control |
| Closed loop emission control | Accessory power control | Cruise control |
| Theft deterrent systems | On board diagnostic systems | Off-board diagnostic systems |
| Leveling controls | Radio frequency display | Digital speedometers |
| Digital tachometers | Elapsed time clock | Electronic odometer |
| Trip odometer | Destination mileage | Miles per gallon |
| Estimated arrival time | Trip fuel consumption | Average speed |
| Average miles per gallon | Digital fuel gauge | Service interval |
| Digital temperature gauges | Digital pressure gauges | Digital voltmeter |
| Acceleration gauge | | |

## 9.4 SEMICONDUCTOR MEMORIES

The need for memory in a computer system falls into several categories depending on the capacity needed and the access time. Very large bulk archival stores may extend to $10^{10}$ bits and are based on magnetic tape or photooptical recording. In the $10^{6}$-$10^{10}$ range bits magnetic tape, disk and drum files predominate. Magnetic core arrays for smaller faster memory action have access times on the order of 250 nsec to 10 $\mu$sec. Magnetic-plated wire systems have been designed with speeds down to 75 nsec. In the last few years, however, semiconductor memories have begun to take over very large sections of the memory field—up to $10^{7}$ bits—in new system designs. The cost and access times of various technologies for 1977, and prediction for 1980, are shown in Fig. 9.31.

Moving-surface memory systems are low in cost but slow in access time and typically have a mean time before failure (MTBF) of $10^{3}$-$10^{4}$ hr even with good preventive maintenance service. Although a semiconductor memory cell may

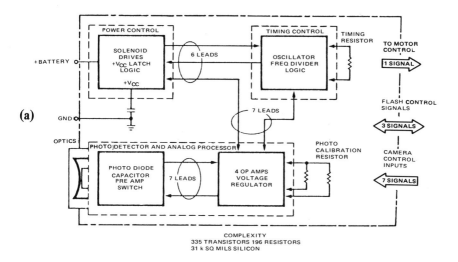

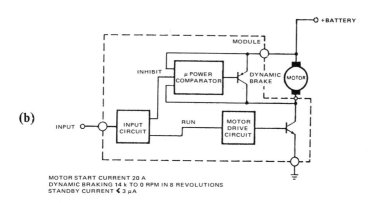

Fig. 9.28.  Block diagrams for the Polaroid SC camera.
(a)  Exposure Control module.
(b)  Motor Control module. (After Murphy and Steffe. Reprinted with permission from *IEEE Journal of Solid-State Circuits,* SC-8, 1973, pp. 428, 432.)

have an estimated mean failure time of $10^8$ hr, the MTBF of a complete system of these cells is much less. Error correcting techniques that ensure prompt maintenance by replacement of defective sections of memory every few thousand hr, are needed to achieve an MTBF of $10^5$ hr for a complete system. The extra circuits and components needed for error correction increase the cost of the system by about 15%. Also, extra care must be taken in testing and by burn-in of some hundreds of hr to eliminate early failures.

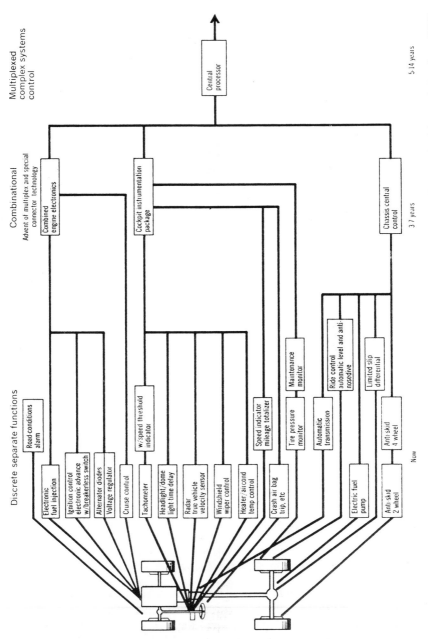

Fig. 9.29. Possible grouping of automobile control functions for processing. (After Jurgen. Reprinted with permission from *IEEE Spectrum*, June, 1975, p. 73.)

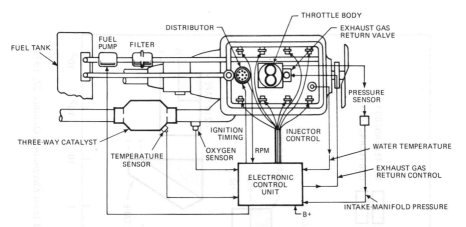

Fig. 9.30. The Bendix closed-loop electronic fuel-injection system mounted on an engine (top) provides precise fuel management based on engine operating requirements signaled to a control unit (bottom) that computes maximum fuel-burning efficiency. (After Walker. Reprinted from *Electronics*, June 21, 1973; copyright McGraw-Hill, Inc., 1973.)

Early semiconductor memories were unnecessarily complicated in basic cell design and wasteful of Si wafer space. MOS memory technology has now moved in the direction of $n$ channel designs with Si or metal gates and six or three transistors per memory cell. The design rules imposed by the photolithographic processing tend to be minimum feature sizes for windows and lines of 3.5-5 $\mu$m and alignment tolerance of 2 $\mu$m. The number of interconnect lines and contacts through to the Si are measures of circuit complexity. Unsuitable circuit concepts can be detected by such counts. For instance, CMOS static read-write memory cells require 1½-4 array interconnect lines and 3½-10 contacts per storage cell. These counts are much higher than for $n$ channel MOS dynamic memory cells. Dynamic cells are clock pulse controlled and the information is stored on a capacitance and must be refreshed at intervals. The contact and line counts for Si gate 3 and 1 transistor memory cells are illustrated in Fig. 9.32. (The count is fractional because of sharing with adjacent cells.) The line count for CCD or bucket-brigade memories (discussed in Chap. 10) is 2 lines and zero contacts per cell so this also is a high-packing density technology. Information from a CCD or similar memory must be read out in serial form from rows that can be randomly accessed. Whereas in a normal random access memory (RAM) the individual bit can be X,Y accessed. However, there are many instances where a serial read-out from a row, or part of a row, is the required mode of operation. The serial read-out nature of the CCD approach to memory is then not a disadvantage, or not much of a disadvantage. Of course a row of bits can be read out quite readily from a RAM by suitable addressing. Close-packing densities for CCD memories and $n$ channel MOS RAMs show that CCDs have greater

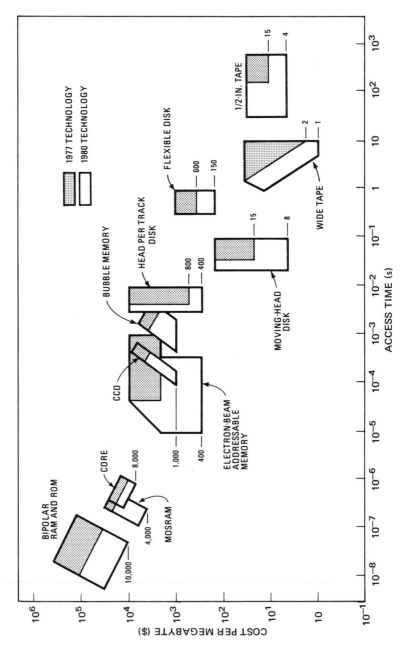

Fig. 9.31. Cost and access time for various memory technologies. (After Capece. Reprinted from *Electronics*, October 27, 1977; copyright McGraw-Hill, Inc., 1977.) The numbers on the blocks are $ per megabyte.

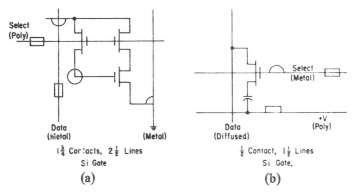

Fig. 9.32. Contact and line counts for MOS silicon gate 3 and 1 transistor memory cells. (After Hodges. Reprinted with permission from *Proceedings of the IEEE*, V.63, 1975, p. 1138.)

densities, but the factor (2 or so) is not so great that CCDs are likely to quickly displace $n$ channel MOS designs.[179]

There is no space in this volume to review systematically the many cell layouts that have been developed over the last decade. A few illustrations may, however, provide the general ideas and lead to the detailed treatments in some important technical papers and books.[26, 43, 73, 94, 106, 124] Figure 9.33 shows the layout and circuit schematic for a read-only memory (ROM). A six transistor static RAM of a flip-flop cross-coupled form is shown in Fig. 9.34. A VMOS version of this is shown in Fig. 9.35 and Fig. 9.36 shows a Mostek version that uses ion-implanted poly-Si resistors.

Single transistor memory cells are possible by the use of dynamic clock-driven memories. The single-transistor storage cell layout of a recent Intel 16K bit memory is shown in Fig. 9.37. A diffused bit line is used with a second-level poly-Si access gate. As may be seen from the cross section, the cell is charge coupled as in CCD technology. The signal consists of electrons stored in the inversion layer under first-level poly-Si. They are transmitted directly to and from the bit line, with no intermediate $n^+$ diffusion. Features of the cell include small size (455 $\mu m^2 \sim 0.7$ mil$^2$); high efficiency ($\sim 30\%$ storage area); small word line pitch (13 $\mu$m); and limited contact (1/2 contact per cell, none to diffusion).

The storage cell shown is 13-$\mu$m high by 35-$\mu$m wide, and yields a storage to bit line capacitance ratio $C_s/C_b$ between 3 and 6. As in all single-transistor cell memories, a read operation refreshes the dynamic storage cell. The memory is TTL compatible and the performance observed is $T_{acc}$ 150-nsec typical; $T_{cycle}$ 350 nsec to 100 ms; $I_{DDavg}$ 40-mA typical at 500-nsec $T_{cycle}$; and $I_{DDstandby}$ 1-mA typical.

Dynamic RAMs with access times of 50-80 nsec are in development and even faster times are seen in laboratory prototypes.[179]

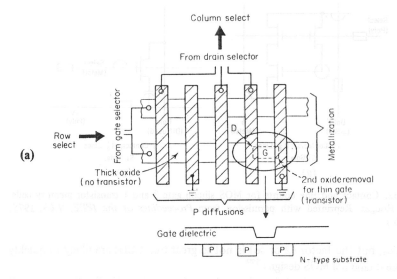

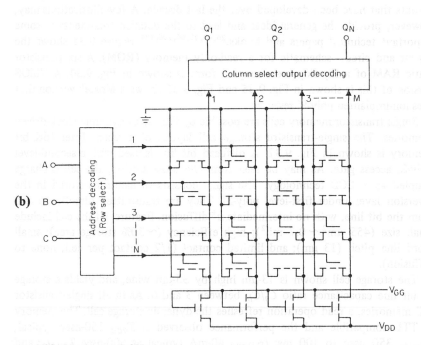

Fig. 9.33. Matrix-type ROM:
(a) geometry layout;
(b) circuit schematic and input-output decoding. (After Carr and Mize, from *MOS/LSI Design and Application.* Reprinted with permission from Texas Instruments, Inc. and McGraw-Hill Book Company, 1972, p. 197.)

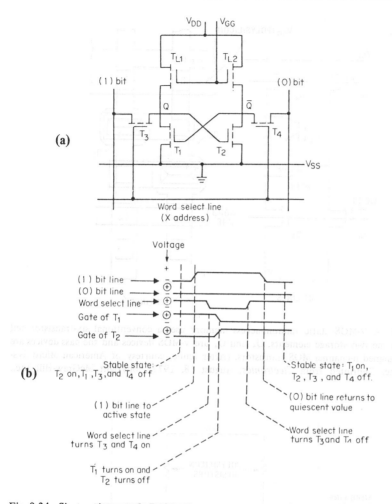

Fig. 9.34. Six-transistor static RAM cell:
(a) circuit schematic;
(b) timing for write operation. (After Carr and Mize, from *MOS/LSI Design and Application*. Reprinted with permission from Texas Instruments, Inc. and McGraw-Hill Book Company, 1972, p. 198.)

The organization of a RAM of $W$ words and $B$ bits is illustrated in Fig. 9.38. In general a square arrangement with $W=B$ is preferred since in an $N$ bit memory ($=W \times B$) the number of address decoders ($W+B$) and other peripheral circuits is minimized and the off-chip connections, the number of pins required, is low. The address lines needed are $\log_2 W + \log_2 B$. The timing and control functions include the clock system, the refresh counter and control, the read-write control, the chip enable circuit and other features. Typically 30% of the chip area is needed for these functions.

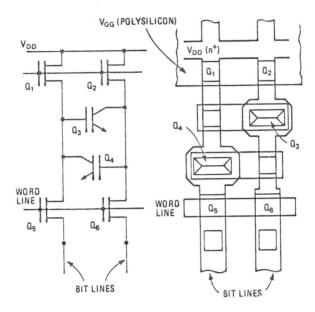

Fig. 9.35. A V-MOS static random-access memory uses a conventional six-transistor cell in which the two storage elements, $Q_3$ and $Q_4$ are V-MOS devices and the pass devices are resistor-aligned n-channel MOS transistors. (After Jenné, courtesy of American Micro Systems, Inc. Reprinted from *Electronics*, August 18, 1977; copyright McGraw-Hill Inc., 1977.)

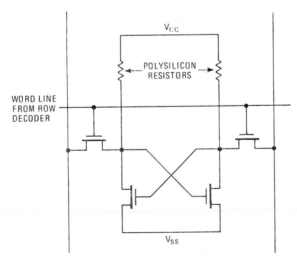

Fig. 9.36. A static RAM cell design (Mostek 4104) that uses resistors as loads saves space and reduces power consumption. Each 5000 megohm resistor is an ion-implanted polysilicon device that draws less than 1 nanoampere of current. (After Young. Reprinted from *Electronics*, May 12, 1977; copyright McGraw-Hill, Inc., 1977.)

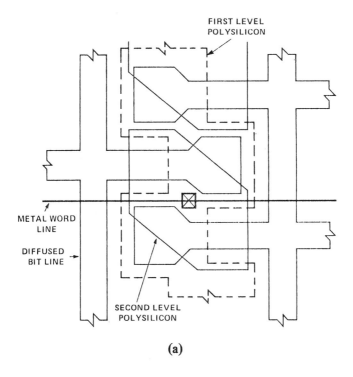

**(a)**

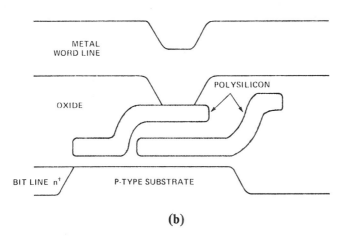

**(b)**

Fig. 9.37. Storage cell in an Intel memory (a) layout and (b) cross section. The storage capacitor is formed by first-level polysilicon biased to +12 V to form an inversion layer. Overlapping second-level polysilicon electrode forms the access gate. (After Ahlquist et al. Reprinted with permission from *IEEE Journal of Solid-State Circuits,* **SC-11**, 1976, p. 571.)

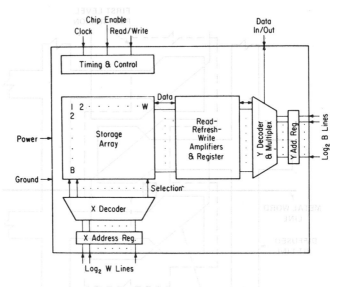

Fig. 9.38. Block diagram of RAM component organization. (After Hodges. Reprinted with permission from *Proceedings of the IEEE*, **V.63**, 1975, p. 1139.)

Consider now read-only memory technology in computers. Some fraction of the memory function may be in the form of permanent information, such as look-up tables of mathematical functions or other set routines, and then a read-only action is all that is required. In other applications the file information is subject to change, such as lists of addresses and routes, but only infrequently. This need led to the development of ROMs and read-mostly memories. Some read-only memories are preprogrammed by the manufacturer but others are programmed by the user by burn-out of selected fuse links prior to installation. The possibility of electronic writing for read-only memories has been explored in technologies such as MNOS (metal, nitride, oxide, silicon) and FAMOS (floating-gate avalanche-injection MOS). In the MNOS structures, charge is stored at the interface between the silicon nitride and the silicon oxide to provide the long-term memory action. Such designs have not become generally available because the manufacturing control problems are severe. Floating-gate avalanche-injection ROMs, however, are in general use. In Fig. 9.39(a) the Si gate is isolated by 1000 Å of $SiO_2$ from the $p$-type channel and by 1 $\mu$m of oxide from the surface. The application of $-30$ V between the 8 ohm-cm $n$ substrate and the channel gives an avalanche action at the $pn$ junction and high energy electrons cross the 1000 Å oxide and become trapped on the floating gate. This charging effect proceeds at the rate of about $10^{-7}$ A/cm$^2$ and gives a stored charge on the order of a few times $10^{12}$ electrons/cm$^2$. This gives a turned-on gate condition when the circuit is put to use. The charge induced field

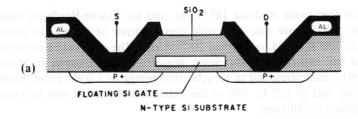

(a)

FLOATING SI GATE

N-TYPE SI SUBSTRATE

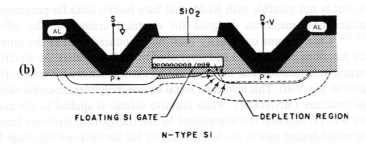

(b)

FLOATING SI GATE          DEPLETION REGION

N-TYPE SI

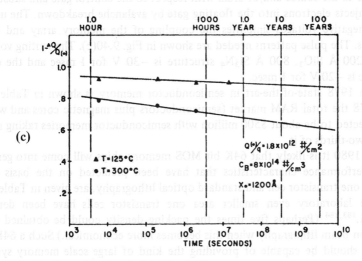

(c)

Fig. 9.39.  Floating-gate avalanche-injection ROM.
(a)  Cross-section of the structure, FAMOS.
(b)  Structure under bias with charge on the gate.
(c)  Charge decay as a function of time at 125° and 300°C. (After Frohman-Bentchkowsky. Reprinted with permission from *IEEE Journal of Solid-State Circuits,* SC-6, 1971, pp. 302, 303.)

across the thin oxide is about $10^6$ V/cm, and for Fowler-Nordheim emission from the poly-Si gate into the oxide over a barrier of 3.2 eV the discharge current at $300°K$ would be less than $10^{-40}$ A/cm$^2$. The observed barrier is not so large but still entirely adequate. Some measured discharge curves are shown in Fig. 9.39(b) and correspond to an activation energy of 1.0 eV. From this it can be inferred that at $125°C$, 70% of the initial induced charge may be expected to be retained for 10 years.

Removal of the charge for reprogramming can easily be accomplished by the current produced by X-ray irradiation ($> 5 \times 10^4$ rads) of sealed packages, or by ultraviolet illumination if the package is provided with a quartz window. Thorough testing of floating-gate ROMs is practical because of the ease of erasure, but is not possible with ROMs that have fusable links for programming.

Consider now memories designed for electrical erasure of the information.[138,182] Usually some complexity must be introduced such as the provision of extra junctions in the device and reversal of the voltage pulses. An electrically re-programmable memory cell that requires no extra junctions for erase action is illustrated in Fig. 9.40. This involves a metal-nitride-molybdenum-oxide-silicon $p$ channel structure (MNMoOSi). When negative voltage is applied to the control gate the WRITE function is accomplished by the transport of electrons from the floating molybdenum gate to the Si substrate by the barrier-lowering image force effect at the $SiO_2$-Mo interface. For the ERASE function a negative voltage is applied to the source and/or drain with respect to the control gate and substrate. This injects electrons into the floating gate by avalanche breakdown. The use of only negative voltages simplifies the coupling of the memory array and logic circuits. The pulse patterns needed are shown in Fig. 9.40(c). The writing voltage for a 200 Å $SiO_2$, 800 Å $Si_3N_4$ structure is $-30$ V for 1 msec and the erase voltage is $-20$ V for 1 msec.

The 1978 state-of-the-art in semiconductor memory is shown in Table 9.4. In 1978 the total RAM market (semiconductors plus magnetic cores and wires) is expected to be about $500 million with semiconductor memories taking more than two-thirds of it.

By 1980 it is likely that 64K bit MOS memory chips will come into general use. Performance characteristics that have been predicted on the basis of a 1 mil$^2$ one transistor cell and standard optical lithography, are given in Table 9.5. In the laboratory even smaller area one transistor cells have been demonstrated.[183,184] (Perhaps five times the packing density could be obtained with electron beam lithography when this becomes more economical.) Such a 64K bit design should be capable of providing the kind of large scale memory system suggested in Table 9.6.

Electron beam addressable memories are also under development with promising results. A 32 Mbit MOS system has been constructed with an access time of 30 $\mu$sec and 10 Mbit/sec transfer rate. The size is 43 cm (17 in.) by 10 cm (4 in.) in diameter and the cost at the system level is 0.02 cents and likely to be

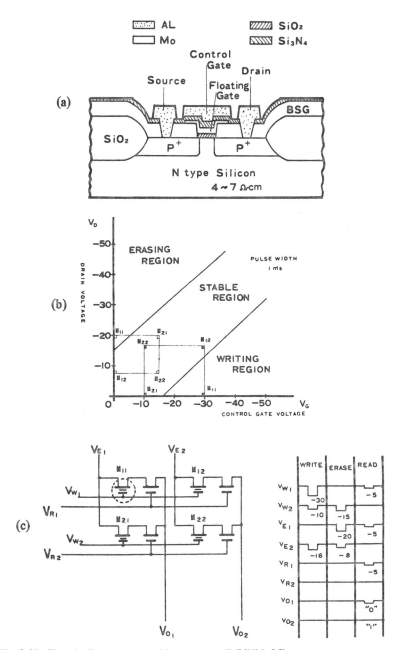

Fig. 9.40. Electrically reprogrammable memory cell (MNMoOS)
(a)  Negative voltage (WRITE) on the control gate transports electrons to the floating Mo
     gate. Negative voltage to the source or drain provides the ERASE action.
(b)  Bit selection method.
(c)  Circuit configuration of a MNMoOS memory array and its associated pulse patterns.
     (After Rai, Sasami and Hasegawa.  Reprinted with permission from *International IEEE
     Electron Devices Meeting Technical Digest*, 1975, pp. 314, 316.)

Table 9.4.  1978 State-of-the-Art Memory Technology

| Device Type | Access Time (ns) | Active Power Dissipation per Chip (mW) | Application |
|---|---|---|---|
| 4-k to 16-k MOS dynamic RAMs | 150 – 300 | 400 – 600 | large mainframe and microcomputer-based systems |
| 4-k fully static 2114-type MOS statics | 150 – 300 | 500 | peripheral and microcomputer systems |
| 4-k clocked 4104-type MOS statics | 150 – 200 | 120 (30 standby) | peripheral and microcomputer systems |
| 1-k to 4-k fast MOS statics | 50 – 100 | 500 | fast mainframe, cache, or microcomputer-based systems |
| 4-k to 16-k $I^3$ L dynamic RAMs | 90 – 125 | 500 | fast mainframe and microcomputer systems |
| 4-k to 16-k $I^2$ L static RAMs | 70 – 125 | 500 (20 standby) | fast mainframe, peripheral, and microcomputer systems |
| 1-k to 4-k TTL/ECL static RAMs | 30 – 70 | 500 – 1 watt | cache |
| 32-k to 64-k MOS ROMs | 200 – 500 | 500 | fixed memory and program storage |
| 1-k to 8-k TTL ROMs | 50 – 100 | 700 | fixed storage |
| 8-k to 16-k EPROMs (UV-erasable) | 500 | 500 | prototype program storage |
| 8-k EAROMs | | | nonvolatile RAM; reprogrammable ROM |
| nitride types | 1 $\mu$s | 500 | |
| Famos types | 500 | 500 | |
| 64-k CCD memory | 1 $\mu$s | 500 | disk replacement; auxiliary serial memory |
| 100-k to 1,000-k bubble memory | 10 $\mu$s | 500 | nonvolatile disk replacement; microcomputer storage |

(After Altman and Capece) reprinted from Electronics, Oct. 27, 1977; copyright © McGraw-Hill, Inc. 1977)

much less in $10^8$-$10^9$ bit systems presently under study. Reliability of electron beam systems, however, must be fully examined.

## 9.5 MICROPROCESSORS

With the advent of LSI and the progress in memory chips and integrated logic in the late 1960s, the concept of a powerful computer consisting of five or six ICs became a feasible goal to work towards. By the early 1970s the first central processor unit (CPU) chips were produced and RAMs and ROMs and peripheral chips were made available to match them. The cost of a computer on a single

**Table 9.5. Projected Characteristics of 64K Memory Component in 1980**

| | |
|---|---|
| Externally-viewed organization | 64K × 1 bit |
| Electrical parameters | |
| Access and cycle times | 1 $\mu$sec |
| Active power | 64 mW |
| Standby and refresh power | 7 mW |
| Refresh interval | 2 msec |
| Die Size | 9.2 × 7 mm |
| Package-dual inline (1.0 × 2.5 cm) | 18 pin |
| $\left(\frac{16}{2}\right.$ Address, 3 power, 2 data, 2 clock, 2 enable, 1 read/write $\left.\right)$ | |
| Economic parameters | |
| Die sites per 7.6 cm wafer | 60 |
| Good die/wafer (25% yield) | 15 |
| Die cost | $ 3.00 |
| Cost of packaged, tested component | 5.00 |
| Selling price of component | 10.00 |
| Component price per bit | .016 cent |
| Reliability (MTBF per component) | $10^7 - 10^8$ hr. |

(After Hodges, reprinted with permission from Proc. IEEE, 1975, V. 63, p. 1144)

**Table 9.6. Projected Characteristics of 32-Million Byte Storage Module in 1980**

| | |
|---|---|
| Externally-viewed organization | 4 million words, 64 bits each |
| Electrical parameters | |
| Access and cycle times | 2 $\mu$sec |
| Active power | 40 W. |
| Standby and refresh power | 30 W. |
| System physical dimensions (without power supply) | 0.5 × 0.4 × 0.5 m |
| System selling price per bit | .04¢ |
| System selling price total | $100,000. |
| Reliability (at $10^7$ hours MTBF per component) | |
| MTBF (components) | 2200 hrs. |
| MTBF (system with ECC) | $6 \times 10^5$ hours |

(After Hodges, reprinted with permission from Proc. IEEE, 1975, V. 63, p. 1145)

circuit board (excluding teletypes and printers) became less than $1,000 and whole new classes of use in process control were practical.

In the last 5 years, progress has been made towards single chip micro-computers and prices have dropped to a few tens of dollars.[180] At this level incorporation of microprocessors in almost all analytical instruments becomes possible. By 1980 the market for microprocessors is expected to be $300 million per year. The U.S. analytical instrument market is presently at $500 million and appears headed for $1000 million in 1980.

The first generation of microprocessors (circa 1972) were 4-bit units and required many associated ICs and expensive input/output equipment to provide complete computing action. By 1974, the CPUs available were 8-bit (or more) units with much more of the associated circuitry incorporated on the chip or related chips of the family. The microprocessors available in 1974/75 included the Intel 4040, Rockwell PPS4, Intel 8080, Intel 3000, RCA COSMAC, Motorola MC 6800, National IMP 8 and 16, Fairchild FP8 and Toshiba TLCS-12.

By 1977, microprocessors available included the Intel 8085 (which is the basis of a 5 MHz 3 chip computer) and the 8048; the Texas Instrument 9940 (8 bit, 1 chip); the Fairchild 9440; the Mostek 3871; the Zilog Z-80; the AMI S2000 (equivalent to 13,000 transistors), Advanced Micro Devices 2900, General Instruments 1650, Signetics' 2650, Rockwell's 6500, RCA's 1802 Cosmac to mention only a few of the models. New chips and microcomputer families were being announced every few months. By 1978, these included many 16-bit-wide designs such as the Intel 8086, the Zilog Z-8000 and the Motorola 6809.

There is a trend towards 2 and 4 kilobytes of on-chip ROM and 128 bytes of on-chip RAM as scratch-pad memory, and towards direct connection to external chips without interfacing chips. Table 9.7 lists a few of the peripheral chips that are used in microcomputer systems. Such peripheral chips are quite complex themselves (for instance a programmable cathode ray-tube controller may represent 20,000 transistors) but their roles are to make the overall circuit functions more convenient and lower in cost.

A microprocessor chip contains typically the following component functions: (a) instruction decoding and control; (b) an arithmetic and logic unit (ALU); (c) input/output buffer circuits; (d) register and accumulator memories needed by the ALU and (e) address buffers to supply the location of the next instruction and the addresses to supply or deposit the data needed or generated. The flow paths are indicated in Fig. 9.41. One or two phase clock is used and the rate may be 1 to 4 MHz.

The MC6800, an early microprocessor family, includes among many peripherals a RAM, a ROM, a peripheral interface chip, a communications interface chip and a 600 BPS low speed modem chip to communicate with a voice grade telephone channel. The organization around an 8-bit bidirectional data bus and a 16-bit address bus for direct addressing of up to 65536 memory locations is shown in Fig. 9.42. The 128 × 8 static RAM needs no clocks or

## Table 9.7.  Microcomputer Peripherals

### General-Purpose Peripherals

decoders
latches
priority interrupt
bus drivers
serial communication interface
parallel communication interface
interval timer
direct-memory-access controller
interrupt controller
analog-to-digital converter
digital-to-analog converter

### Memory-Oriented Peripherals

random access-memory with input/output
read only memory with I/O
erasable programmable ROM with I/O

### Dedicated Peripherals

floppy-disk controller
synchronous data-link control protocol controller
programmable cathode-ray-tube controller
programmable keyboard/display interface
arithmetic units
media encryption converters

(After Altman, Reprinted from *Electronics* Dec. 8, 1977; copyright © McGraw-Hill, Inc., 1977)

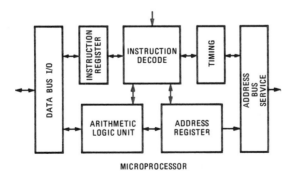

MICROPROCESSOR

Fig. 9.41. Flow paths in a microprocessor. (After Young, Bennett and Lavell, Reprinted with permission of Motorola Semiconductor Products Group and *Electronics*, April 18, 1974; copyright McGraw-Hill, Inc., 1974.)

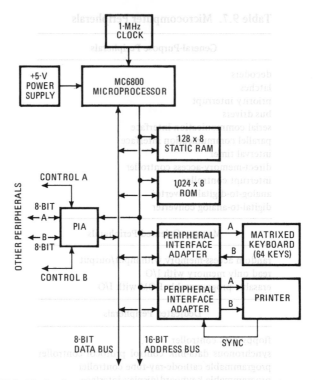

Fig. 9.42. Eight-bit family. Motorola's M6800 family of components is organized around the concept of the parallel data bus. Consequently, all memory and peripheral interface adapter (PIA) chips are simply designed to hang on its CPUs eight bi-directional data lines. (After Young, Bennett and Lavell. Reprinted with permission of Motorola Semiconductor Products Group and *Electronics*, April 18, 1974; copyright McGraw-Hill, Inc., 1974.)

refreshing and the maximum access time is 575 nsec. The ROM is mask-programmable and has a similar access time.

Another early central processor unit (CPU) is the Intel 8080. This has a chip of dimensions 4200 $\mu$m $\times$ 4850 $\mu$m (165 $\times$ 191 mil) and contains about 5000 $n$-channel MOS transistors. Assembly language fetch and execute times of 2-9 $\mu$sec are achieved (comparable to many minicomputers) and it has a large instruction set that provides for convenience and speed. In a typical operation six or more TTL packages may be needed, as suggested in Fig. 9.43. The 8080 takes four control inputs from external devices (ready, hold, interrupt, reset), generates six control outputs (sync, data-bus-input, wait, write, hold-acknowledge and interrupt-enable) and provides eight status bits on the data bus at sync time.

Designers of microprocessors have been under constant pressure to reduce the need for auxiliary packages by adding more logic and address functions

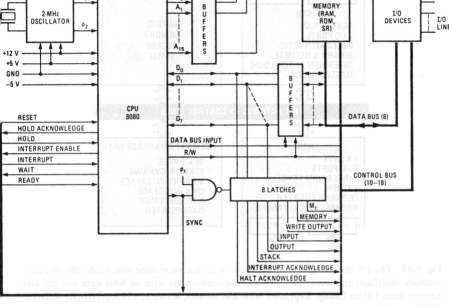

Fig. 9.43. 8080 at work. When connected in a microprocessor system, the 8080 requires only six external TTL packages, as against the 20 needed by the 8008. The address bus can access up to 64 kilobytes of memory and up to 256 input and 256 output ports. (After Shima and Faggin. Reprinted from *Electronics*, April 18, 1974; copyright McGraw-Hill, Inc., 1974)

to the basic chips. The Fairchild F8 system has the four basic chips shown in Fig. 9.44. An address bus between chips is not necessary thus saving 16 pins for input/output uses. Use of three chips in a traffic-light controller system is shown in Fig. 9.45.

The need for processors with word widths of any desired number of bits has led to the adoption in some microprocessor families of "bit-slice" organization. An example of this is the Schottky bipolar technology of the Intel 3000 family. Each central processing element (CPE) represents a 2 bit slice. An array of 8 such units with a microprogram control unit, a look-ahead carry generator and 8 memory units provides a complete 16 bit processor on a 15 × 15 cm (6 × 6 inch) circuit board. The register-to-register add time for 16 bits is less than 125 nsec: This is 15 times faster than the same process in the 8 bit Intel 8080.

The architecture and software support for a microprocessor family are both extremely important in determining its market acceptance. The TMS 9940 (Texas Instruments) is a 16-bit one-chip microcomputer organized in a memory-to-memory configuration (see Fig. 9.46), with program and data registers maintained in external memory as in minicomputer architecture.

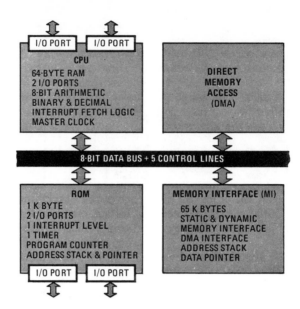

Fig. 9.44. The F8 system. The four F8 chips—a central processing unit, read-only memory, memory interface, and direct memory access-communicate over an 8-bit data bus and five control lines. (After Chung. Reprinted from *Electronics*, March 6, 1975; copyright McGraw-Hill, Inc., 1975.)

Consider now some of the uses of microprocessors. In addition to the automotive developments already discussed, they are utilized in oscilloscopes and other electrical instruments, chemical analyzers, process controllers, floppy disk controllers, medical instruments, machine tool controllers, cash registers, communication line controllers, printer controllers, traffic light controllers, copiers, paging and telephone board systems, film processing, automated gas

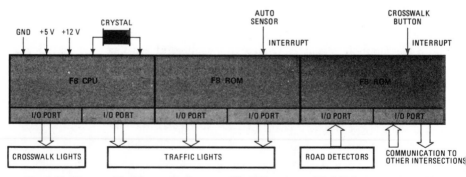

Fig. 9.45. As a traffic-light controller, one F8 CPU and two F8 ROMs can count passing autos, provide a pedestrian crosswalk interrupt input, and also time the traffic-light changes. An external crystal precisely sets the clock frequency. (After Chung. Reprinted from *Electronics*, March 6, 1975; copyright McGraw-Hill, Inc., 1975.)

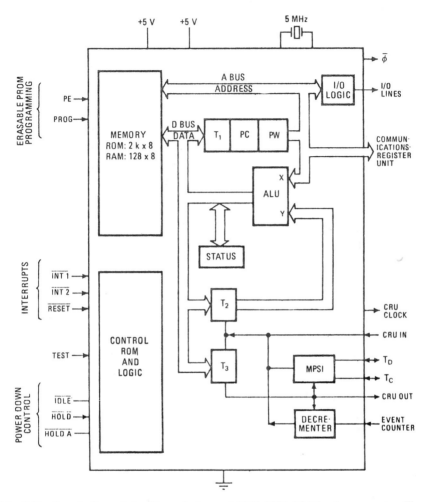

Fig. 9.46. Organization of the Texas instrument TMS 9940 16 bit microcomputer. The 9940 can address 32 bits of I/0 and this may be expanded by another 256 bits. (After Bryant and Longley. Reprinted from *Electronics*, June 23, 1977; copyright McGraw-Hill, Inc., 1977.)

pumps, tape handlers, vending machines, electronic scales, specialty calculators, electronic games and home appliances.

In some applications the input signal is an analog voltage and the output must also be analog in nature, such as power to control a proportional valve. Use must then be made of A/D and D/A peripheral converter chips.

Figure 9.47 shows the organization of a microprocessor for control of the cycles of a washing machine as an example of memory requirements in a

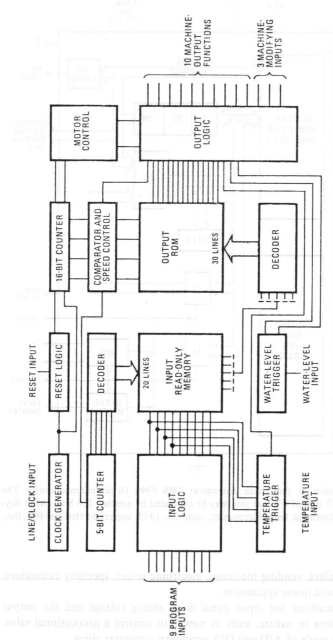

Fig. 9.47. Microprocessor control of a washing machine with an ITT 7150 chip requires two ROMs. The input ROM is split into nine sections, each of which is controlled by a separate program. The output ROM routes data from first ROM into 18 lines that control operations. (After Walker. Reprinted from *Electronics*, April 4, 1977; copyright McGraw-Hill, Inc., 1977.)

dedicated application. The use of a 4 bit microcomputer in a programmable microwave oven is shown as an example in Fig. 9.48.

In electronic instruments, compensation and calibration problems can be readily handled by microprocessor systems with improvement in accuracy. Automated zero compensation is important in all weighing operations. If a product is on a moving belt with individual weighing pans or shackles these may vary in weight and a microprocessor can keep track of each and provide the correct product weight. (Correction for temperature can be made by computing from the power polynomial that represents the temperature sensor curve.) Instruments with self-calibrating functions can be made that tune themselves to conditions of high sensitivity. Instruments that integrate, differentiate, invert, compute products and ratios and sums now are on the market in the price range of a few thousand dollars. This sophistication will be available at even lower cost as the microprocessor costs decrease and the market expands.

## 9.6 SMALL CALCULATORS

The hand-held calculator, discussed briefly below, is an example of a product that has rapidly developed a wide market range.

The block diagram for an early Hewlett Packard calculator (HP-35) is shown in Fig. 9.49. The three main components are the arithmetic and register chip, the ROM chip and the timing and control chip. An interesting account of the design and procurement of these has been given by Walker.[164]

In the Texas Instruments calculator line a single chip (TMS0102) contains all the calculator functions—arithmetic and logic, register RAM (182 bits), program ROM (3520 bits), input encoding, control and timing and output peripheral action. The components needed to complete a calculator are merely the clock generator, the display drivers and the LED display.

A simple account of the functioning of small electronic calculators has been given by McWhorter.[103] Much more detailed accounts of the architectures used may be found by inspection of the patent literature. As starting points one may take US Patents: 3,755,806; 3,781,852; 3,819,921; 3,855,577; 3,904,863; 3,921,142; 3,932,846; 3,934,233; 3,955,181; 3,987,416; 3,988,604; 3,989,939; 3,991,305; 3,991,306 and 4,014,013.

The trend more recently in calculators has been towards programmable systems and print-out features.

The achievements of microelectronics in recent years match the greatest accomplishments of man in engineering, science and medicine.

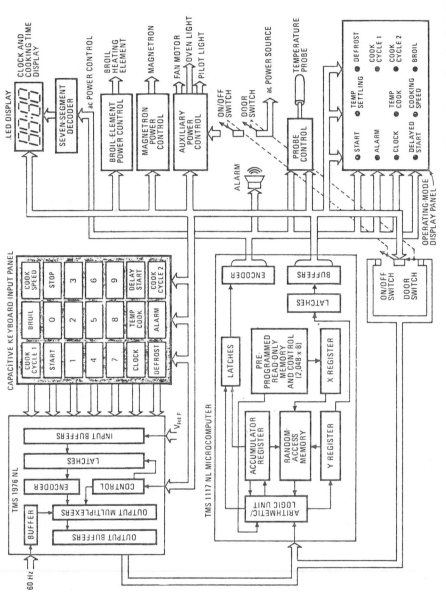

Fig. 9.48. Microprocessor (TMS 1117) control of a microwave oven. (After Walker. Reprinted from *Electronics*, April 4, 1977; copyright McGraw-Hill, Inc. 1977.)

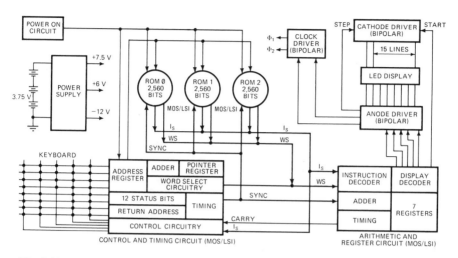

Fig. 9.49. The HP-35 block diagram above shows the MOS LSI circuits three ROMs, a control and timing circuit, and an arithmetic and register circuit. Three bipolar ICs drive the LED display. (After Walker. Reprinted from *Electronics*, February 1, 1973; copyright McGraw-Hill, Inc., 1973.)

## 9.7 QUESTIONS

1. Complete the design of the differential amplifier shown in Fig. P9.1 by selecting the values of $R_1$, $R_2$, $R_3$, $R_4$, $R_5$ to give suitable bias conditions. The on-voltage of an emitter-base junction or of a diode may be taken as 0.7 V. Assume each transistor has a basic gain $\beta$ of 50. What is the voltage gain $V_O/(V_1 - V_2)$ of your circuit? What is the dc offset voltage at the output? What is the maximum output voltage?

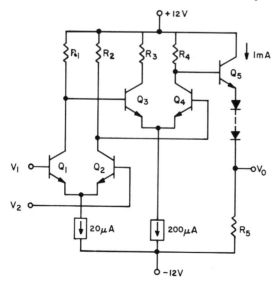

2. An op amp with a unity gain slew rate of 50 V/$\mu$sec and a 10-MHz bandwidth is reported to have a settling time of 500 nsec to 1% accuracy for a 5-V input signal. Show by analysis that this settling time is reasonable (assume any further parameters needed for the analysis).

3. Show that an op amp with a diode characterized by $I = I_0 \exp(qV/kT)$ connected in the feedback path provides an output voltage that is the logarithm of an input voltage fed through a resistor $R$. What is the corresponding circuit for the antilog function? Design a circuit using op amps that provides the function $V_O = C \log\left(\dfrac{V_{1A}}{V_{1B}}\right) + V_{1C}$. Estimate the permitted range of the input voltages $V_{1A}$, $V_{1B}$ and $V_{1C}$ in your circuit.

4. For the circuit shown in Fig. P9.4 where $R_2$ represents the input impedance of the op amp of voltage gain $A_V$, obtain an expression for $V_O/V_1$. Suppose the input offset voltage of an op amp of open loop gain $10^4$ is 1 mV at 300°K and that this changes by 5 $\mu$V/°K, what will the output offset voltage be at 400°K in the circuit shown if $R_F/R_1 = 5$ or 100?

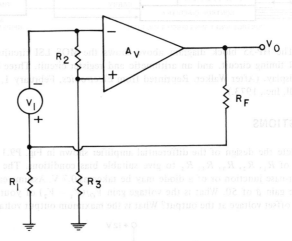

5. Discuss serial D/A converter technology and compare the component counts for some typical designs.

6. D/A and A/D technology usually depends for accuracy on the precision of a reference voltage. Discuss the design of such precision reference voltages.[21]

7. A 12-bit weighted-resistor D/A converter is shown in Fig. 9.6. How does the output resistance of the network to the left of the op amp vary with the digital word? Discuss how well this layout could be achieved in an integrated circuit chip. Compare this with the $R$-$2R$ network of Fig. 9.7 and with the capacitor charge redistribution system of Fig. 9.12.

8. Outline the concepts that might be used for the design of analog multipliers of better than ±1% full scale accuracy, and consider integrated circuit implementation[75] (also J. Solid-State Circuits, December, 1974).

9. Analog-to-digital converters require precise switching. Make a study of the performance properties required of such switches for a 12-bit A/D converter of accuracy 0.05%.

10. Consider the integrated circuit concepts that might be used to produce the square root of a voltage in the 1 to 100-V range with an accuracy of 1%. Would such a circuit have application to an RMS voltmeter?

11. A dual slope ramp A/D converter is shown in Fig. 9.10. Accuracy to $10^{-4}$ is required

for an input voltage of 10 V, and $10^{-2}$ for an input of 0.1 V, in a reading cycle response time of not more than 10 msec. The time ratio $t_2/t_1$ is to be determined by counting the pulses of a crystal oscillator. What suggestions do you have for the circuit design and the oscillator frequency?

12. Design a transformerless push-pull class B, complementary symmetry *pnp/npn* driver transistor. The power supply is to be +24 V and one side of the load is grounded. The maximum power into an 8-ohm speaker coil load is to be 1.5 W or more. The signal frequency may be from 100 Hz to 10 kHz. The $h_{fe}$ of the transistors may be taken to be 100 for each unit. What will be the power dissipation of the amplifier when delivering full power to the load? Having completed the design, attempt an integrated circuit layout (with the exception of any large capacitors). Discuss the problems that occur in the design.

13. Some integrated voltage regulator chips incorporate a thermal cut-out or control of the output current, so that the junction temperature does not exceed 150°C. Describe ways this could be done.[169]

14. Discuss pulse-code-modulation of voice signals from a systems viewpoint and at the circuit block level.

15. Discuss all classes of transistor signal choppers and consider in particular the integrated circuit choppers that might be used for pulse-code modulation or input chopping for an op amp to reduce drift.

16. Compile a set of 5 to 10 circuits for active filters suitable for integrated circuit implementation. Include characteristics if possible. Which of the circuits would be voltage tunable, how would this be done, and over what range? (A starting reference might be *J. Solid-State Circuits*, February 1974.)

17. Discuss the frequency, voltage and current limitations of active filters that use a particular op amp (characteristics known) such as the LM 101.

18. Explain why a scale of two divider circuit in a wristwatch may require between 18 and 24 transistors per stage.[44]

19. In a wristwatch circuit it has been stated[145] that a binary divide-by-two system consumes about 80 $\mu$A to scale from 1,548,288 to 0.5 Hz. However, the current required is only 3 $\mu$A if the division is done in one stage of 7, three stages of 3 and 14 binary stages.
    (a) Describe a dynamic divider stage that will divide accurately by 7 or by 3.
    (b) Show that the reduction from 80 $\mu$A to 3 $\mu$A, or thereabouts, is what might be expected.

20. Tabulate information on the range of op amps presently available. Types which are essentially duplications of others need not all be listed. Then assume that your company, a large telecommunications system, requires you to choose only five standard designs that will have to meet all their circuit needs for the next 5 years. What units would you select and why? (A starting reference might be the *Bell Laboratory Record*, 1974, p. 122.)

21. Consider an integrated circuit chip for a small motor drive, as in the Polaroid SX70 camera, and discuss the problem of providing dynamic braking.

22. Explain how an MOS one-transistor per cell memory works and consider its properties relative to a three-transistor per cell design.

23. Prepare a study of address decoder circuits for 4K bit MOS *n*-channel RAMs.

24. Consider a specific C-MOS memory cell array and also an *n*-channel MOS RAM 3 transistor cell array, and identify and count the array interconnect lines and the contacts through to the Si per memory cell. Compare your counts with the typical ranges given by Hodges.[71]

25. Consider how a dynamic RAM chip could be operated as a kind of static RAM.[110]

26. Discuss the memory refresh procedures in two different 4K bit *n*-channel MOS RAMs.

27. Discuss error correcting techniques and mean time before failure as applied to a 4K bit $n$-channel MOS RAM. (Take the discussion beyond the introduction found in Hodges, *Proc. IEEE*, August 1975.)

28. In general 4K bit dynamic $n$ MOS RAMs are not fast enough for main frame memory systems operating at 100 nsec. Recently one company has claimed that a charge-pumped 4K bit pseudo-static RAM which is ECL compatible and has an access time of 80 nsec can be built. (See *Electronics*, February 20, 1975.) Explain what is involved.

29. What test routines would you recommend to test effectively, but economically a 16K bit $n$-channel RAM?[126]

30. Discuss the relative merits of dual-in-line (DIP) packaging, solder reflow flip-chip packaging and beam lead packaging of integrated semiconductor memories.[73]

31. Discuss the package pin requirements for a 4K bit MOS $n$-channel RAM (taking account of the varying amounts of peripheral circuitry that might be on the chip) for a square organization and a rectangular 4 to 1 block organization.

32. Discuss avalanche-injection programming of ROMs and compare with alternative methods of electrical writing for nonvolatile memory.

33. Select three families of microprocessors, having the same bit length, and compare their relative merits under the headings of performance, convenience and cost. To be specific choose a particular application such as a traffic light system with pedestrian signals.

34. The control signals associated with the READ cycle of the Motorola MC 6800 micro-processor are given in Fig. P9.34. During a READ operation the READ/WRITE (R/W) line must be at a logic 1 state ($\simeq 2.4$ V). In terms of the times shown in the figure, state the maximum allowable access time of a memory chip which can be used with the MC 6800. If $T_R = 25$ nsec, $T_{DSU} = 100$ nsec and $T_{ASR} = T_{ASC} = T_{VS} = 300$ nsec, calculate the maximum access time for a RAM to be used with a 1 MHz MC 6800. Is this time compatible with most MOS RAMs? Is it compatible with FAMOS EPROMs?

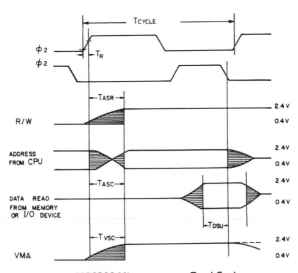

MC6800 Microprocessor Read Cycle

$T_{ASR}$ =  R/W line setting time       $T_{VSC}$ =  Valid Memory Address line settling time
$T_{ASC}$ =  Address settling time       $T_{DSU}$ =  Data set up time during which data must be
$T_R$   =  $\phi 1$ clock rise time                        stable for MPU read
Note: $T_{ASR} = T_{ASC} = T_{VSC}$

35. If a "slow memory" is to be used with an MC 6800 microprocessor which does not meet the maximum access time limitation, the $\phi_2$ clock can be lengthened to provide the additional time required to READ. The maximum time by which $\phi_2$ can be increased is 5 $\mu$sec in the MC 6800. Calculate the percent decrease in program execution time if the $\phi_2$ clock stretching technique is used to accommodate a 900 nsec access time memory.
36. Discuss the merits of bit-slice microprocessor architecture.
37. Discuss the design of a memory for a pocket size scientific calculator[125] (such as the TI SR50) that sells for between $50 and $100.

## 9.8 REFERENCES AND FURTHER READING SUGGESTIONS

1. Ahlquist, C.N., et. al., "A 16384-Bit Dynamic RAM," IEEE, *J. Solid-State Circuits*, SC-11, 1976, p. 570.
2. Albarran, J.F., and D.A. Hodges, "A charge-transfer multiplying digital-to-analog converter," IEEE, *J. Solid-State Circuits*, SC-11, 1976, p. 772.
3. Aldridge, D., "Building your own digital voltmeter," *Electronics*, December 6, 1973, p. 132.
4. Allan, R., "Semiconductor memories," *IEEE Spectrum*, August, 1975, p. 40.
5. Altman, L., "Single-chip microprocessors open up a new world of applications," *Electronics*, April 18, 1974, p. 81.
6. Altman, L., "Solid State," *Electronics*, October 17, 1974, p. 139.
7. Altman, L., "Microprocessors," *Electronics Book Series*, McGraw-Hill, NY, 1975.
8. Altman, L., "New MOS processes set speed, density records," *Electronics*, October 27, 1977, p. 92.
9. Altman, L., "Microcomputer families expand, part 1: the new chips," *Electronics*, Dec. 8, 1977, p. 89.
10. Altman, L., and R.P. Capece, "One-chip controllers and 4-K static RAMS star," *Electronics*, Oct. 27, 1977, p. 96.
11. Apfel, R.J., and P.R. Grey, "A fast-settling monolithic operational amplifier using doublet compression techniques," *IEEE J. Solid-State Circuits*, SC-9, 1974, p. 332.
12. Barna, A., and D. Porat, "*Integrated Circuits in Digital Electronics*," Wiley-Interscience, NY, 1973.
13. Barna, A., and D. Porat, *Introduction to Microcomputers and Microprocessors*, Wiley Interscience, NY, 1976.
14. Barnes, J.J., and F.B. Jenne, "The buried-source VMOS dynamic ram device," *IEEE International Electron Devices Meeting Technical Digest*, Washington, D.C., 1977, p. 272.
15. Barnes, J., J. Linden, and J. Edwards, "Operations and characterization of $N$ channel EPROM cells," *IEEE International Electron Devices Meeting Technical Digest*, 1976, p. 173.
16. Baskerville, G., "A single-chip television IF system," *IEEE J. Solid-State Circuits*, SC-7, 1972, p. 455.
17. Beason, J., "Better bipolar MOS process yields linear ICs with good a.c. and d.c. specs," *Electronics*, October 31, 1974, p. 65.
18. Bell Laboratories, (Ed.), "New digital radio transmission system," *Bell Laboratories Record*, 52, 1974, p. 362.
19. Blakeslee, T.R., "Digital Design with Standard MSI and LSI," Wiley-Interscience, NY, 1975.
20. Brandt, R., "Active resonators save steps in designing active filters," *Electronics*, Apr. 24, 1972, p. 106.
21. Brokow, A.P., "A simple three-terminal IC bandgap reference," *IEEE J. Solid-State Circuits*, SC-9, 1974, p. 388.

22. Bryant, J.D., and R. Longley, "16-bit microcomputer is seeking a big bite of low-cost controller tasks," (TI-TMS 9940), *Electronics*, June 23, 1977, p. 118.
23. Callahan, M., "Chopper-stabilized IC op amps achieve precision, speed, economy," *Electronics*, Aug. 16, 1973, p. 85.
24. Capece, R.P., "Memory-oriented designs maximize throughput," *Electronics*, Oct. 27, 1977, p. 104. "Microcomputer families expand, part 2: the new boards," *Electronics*, Dec. 22, 1977, p. 65.
25. Card, H.C. and E.L. Heasell, "Modelling of channel enchancement effects on the write characteristics of FAMOS devices," *Solid-State Electronics*, 19, 1976, p. 965.
26. Carr, W.N., and J.P. Mize, *MOS/LSI Design and Application*, McGraw Hill, NY, 1972.
27. Chung, D., "Four-chip microprocessor family reduces system parts counts," *Electronics*, March 6, 1975, p. 87.
28. Clement, G.F., W.C. Jones, and R.J. Walters, "No. 1 ESS processors: How dependable have they been?" *Bell Laboratories Record*, 52, 1974, p. 21.
29. Cohen, L., et al., "Single-transistor cell makes room for more memory on an MOS chip," *Electronics*, Aug. 2, 1971, p. 69.
30. Coker, D., "10-K RAM eases memory design for main frames and microcomputers," *Electronics*, April 28, 1977, p. 115.
31. Cole, B., "4 bit controller system upgraded," *Electronics*, November 14, 1974, p. 167.
32. Cripps, L.G., "Electronics in cars," *IEEE Electronics and Power*, 16, 1970, p. 394.
33. Curran, L., "Seat-belt interlock deadline nears," *Electronics*, March, 1973, p. 68.
34. D'Altroy, F.A., et al., "Ultralinear transistors," *BSTJ*, 53, 1974, p. 2195.
35. Davis, P.C., S.F. Moyer, and V.R. Saari, "High slew rate monolithic operational amplifier using compatible complementary PNP's," *IEEE, J. Solid-State Circuits*, SC-9, 1974, p. 340.
36. DeMan, H.J., R.A. Vanparys, and R. Cuppens, "A low input capacitance voltage follower in a compatible silicon gate MOS-bipolar technology," *IEEE, J. Solid-State Circuits*, SC-12, 1977, p. 217.
37. Dingwall, A.G.F., and R.E. Stricker, "$C^2$ L: a new high speed, high density bulk CMOS technology," *IEEE International Electron Devices Meeting Technical Digest*, Washington, D.C., 1976, p. 188.
38. Draper, D.A. et al., "Fabrication and characterization of a VMOS EPROM," *IEEE International Electron Devices Meeting Technical Digest*, Washington, D.C., 1977, p. 277.
39. Dupcak, J., and R.H. DeGroot, "The manufacture of thin-film active filters," *Western Electric Engineer*, 18, 3, July, 1974, p. 18.
40. Eaton, S.S., "Sapphire brings out the best in C-MOS," *Electronics*, June 12, 1975, p. 115.
41. Editor, "C-MOS/Darlingtons sense phone relays," *Electronics*, January 9, 1975, p. 36.
42. Editor, "C-MOS chip drives display," *Electronics*, April 3, 1975, p. 119.
43. Eimbinder, J., (ed.), *Semiconductor Memories*, Wiley-Interscience, NY, 1971.
44. Eleccion, M., "The electronic watch," *IEEE Spectrum*, April, 1973, p. 24.
45. Embree, M.L., and D.G. March, "Family planning for Bell System op amps.," *Bell Laboratories Record*, 52, 1974, p. 122.
46. Fagan, J.L., M.H. White, and D.R. Lampe, "A 16-kbit nonvolatile charge addressed memory," *IEEE J. Solid-State Circuits*, SC-11, 1976, p. 631.
47. Falk, H., "The microprocessor: jack-of-all trades," *IEEE Spectrum*, November, 1974, p. 46.
48. Falk, H., "Self-contained microcomputers ease system implementation," *IEEE Spectrum*, December, 1974, p. 53.
49. Feth, G.G., "Memories: smaller, faster and cheaper," *IEEE Spectrum*, 13, 1976, p. 37.
50. Forrer, M.P., "Survey of circuitry for wristwatches," *Proc. IEEE*, 60, 1972, 1047.
51. Frohman-Bentchkowsky, D., "A fully decoded 2048 bit electrically programmable

FAMOS read only momory," *IEEE J. Solid-State Circuits,* SC-6, 1971, p. 301.
52. Gersbach, J.E., "Current steering simplifies and shrinks 1K bipolar RAM," *Electronics,* May 2, 1974, p. 110.
53. Gilbert, B., "A versatile monolithic voltage-to-frequency converter," *IEEE J. Solid-State Circuit,* SC-11, 1976, p. 852.
54. Gillette, G.C., and J. Crosby, "C-MOS teams with liquid crystals to make a reliable digital VOM," *Electronics,* December 6, 1973, p. 99.
55. Gordon, B.M., "Digital panel meter matches analog performance," *Electronics,* October 23, 1972, p. 108.
56. Gosling, W., "Intergrated circuits for analogue systems," *The Radio and Electronic Engineer,* 43, 1973, p. 58.
57. Graeme, J.G., C.E. Tobey, and L.P. Huelsman, *Operational Amplifiers, Design and Applications* McGraw Hill, NY, 1971.
58. Gray, J.S., "Recent advances in high bit rate technology," *Microwave Systems News,* January, 1973, p. 21.
59. Gray, P.R., "A 15 W monolithic power operational amplifier," *IEEE J. Solid-State Circuits,* SC-7, 1972, p. 474.
60. Gray, P.R., and R.G. Meyer, "Recent advances in monolithic operational amplifier design," *IEEE Trans. Circuits and Systems,* CAS-21, 1974, p. 317.
61. Gray, P.R., et al., "Companded pulse-code-modulation voice code C using monolithic weighted capacitor arrays," *IEEE J. Solid-State Circuits,* SC-10, 1975, p. 497.
62. Grebene, A.B., *"Analog Integrated Circuit Design,"* Van Nostrand, NY, 1972.
63. Gundlach, R., "Large-scale integration is ready to answer the call of telecommunications," *Electronics,* April 28, 1977, p. 93.
64. Gurtler, R., and C. Maze, "Nematic liquid crystals and their application to the electronic watch," *Microtecnic,* 27, 1973, p. 353.
65. Hamakawa, Y., et al., A nonvolatile memory FET using PLT thin film gate," *IEEE International Electron Devices Meeting Technical Digest,* Washington, D.C., 1977.
66. Hamilton, D.J., and W.G. Howard, *Basic Integrated Circuit Engineering,* McGraw-Hill, NY, 1975.
67. Heald, R.A., and D.A. Hodges, "Multilevel random-access memory using one-transistor per cell," *IEEE J. Solid-State Circuits,* SC-11, 1976, p. 519.
68. Hewlett, F.W., Jr., and W.D. Ryden, "The Schottky $I^2$ L technology and its application in a 24 x 9 sequential access memory," *IEEE International Electron Devices Meeting Technical Digest,* 1976, p. 304.
69. Hillburn, J.L., and D.E. Johnson, *Manual of Active Filter Design,* McGraw-Hill, NY, 1973.
70. Hoeschele, D.F., *Analog-to-Digital/Digital-to-Analog Conversion Techniques,* Wiley-Interscience, 1968.
71. Hodges, D.A., "A review and projection of semiconductor components for digital storage," *Proc. IEEE,* 63, 1975, p. 1136.
72. Hodges, D.A., "Microelectronic Memories," *Scientific American,* September 1977, p. 130.
73. Hodges, D.A., (ed.), *Semiconductor Memories, IEEE Press,* NY, 1972.
74. Hoffmann, K., "The behavior of the continuously charge-coupled random access memory ($C^3$ RAM)," *IEEE J. Solid-State Circuits,* SC-11, 1976, p. 591.
75. Holt, J.G., "A two-quadrant analog multiplier integrated circuit," *IEEE J. Solid-State Circuits,* SC-8, 1973, p. 434.
76. Huelsman, L.P., *Active Filters: Lumped, Distributed, Integrated, Digital and Parametric,* McGraw-Hill, NY, 1970.
77. Iizuka, H., et al., "Stacked-gate avalanche-injection type MOS (SAMOS) memory," Proc. Fourth Conference Solid State Devices, Tokyo, 1472: Supplement *J. Japan. Applied Physics,* 42, 1973, p. 158.

78. Itoh, K., et al, "A high-speed 16 k bit n-MOS random-access memory," *IEEE J. Solid-State Circuits*, SC-11, 1976, p. 585.
79. Jansen, R.M., and R.C. Prime, "TH-3 Microwave radio system: system considerations," *Bell System, Technical J.*, 50, 1971, p. 2085.
80. Jenné, F.B., "Groves add new dimension to V-MOS structure and performance," *Electronics*, Aug. 18, 1977, p. 100.
81. Jones, D.V., "A monolithic 5 watt integrated amplifier," *Proc. National Electronics Conference*, 1968, p. 781. (Obtainable from University Microfilms, Ann Arbor, Michigan 48107.)
82. Jones, D.V., and R.F. Shea, *Transistor Audio Amplifiers*, Wiley, NY, 1968.
83. Juliussen, J.E., D.M. Lee, and G.M. Cox, "Bubbles appearing first as microprocessor mass storage," *Electronics*, Aug. 4, 1977, p. 81.
84. Jurgen, R.K., "The microprocessor: in the driver's seat?" *IEEE Spectrum*, June, 1975, p. 73.
85. Karp, H.R., "Digital-to-analog converters: trading off bits and bucks," *Electronics*, March 13, 1972, p. 84.
86. Kelcourse, F.C., and J.J. Herr, "L5 system: overall description and system design," *Bell System Technical Journal*, 53, 1974, p. 1901.
87. Klaassen, F.M., "Physics of and models for $I^2 L$," *IEEE International Electron Devices Meeting Technical Digest*, Washington, D.C., 1976, p. 299.
88. Kondoh, T., K. Ban, and M. Kiyoto, "Radar in automobile on collision course tightens passengers' seat belts," *Electronics*, February 7, 1974, p. 107.
89. Lapidus, G., "Electronics in cameras," *IEEE Spectrum*, June, 1973, p. 22.
90. Lee, H.S., and W.D. Pricer, "Merged charge memory (MCM) a new random access cell," *IEEE International Electron Devices Meeting Technical Digest*, Washington, D.C., 1976, p. 15.
91. Lewandowski, R., "Preparation: the key to success with microprocessors," *Electronics*, March 20, 1975, p. 101.
92. Lodi, R.J., et al., "MNOS-BORAM memory characteristics," *IEEE J. Solid-State Circuits*, SC-11, 1976, p. 622.
93. Luce, N.A., "C-MOS digital wristwatch features liquid crystal display," *Electronics*, April 10, 1972, p. 93.
94. Luecke, G.J., P. Mize, and W.N. Carr, *Semiconductor Memory Design and Application*, McGraw-Hill, NY, 1973.
95. Maas, M.A., "Food industry takes a bit out of waste by tightening process control," *Electronics*, August 21, 1975, p. 86.
96. Maddox, E., "Current-steering chip upgrades performance of d-a converter," *Electronics*, Apr. 4, 1974, p. 125.
97. Martinex, A., "Transistor circuits in television: some evolutionary aspects," *The Radio and Electronic Engineer*, 43, 1973, p. 103.
98. Mattera, L., "Active filters get more of the action," *Electronics*, June 19, 1972, p. 104.
99. Mattera, L., "Converters adjust to LS1, Bifet op-amps emerge," *Electronics*, October 12, 1977, p. 112.
100. Mattera, L., "Data converters latch onto microprocessors," *Electronics*, Sept. 1, 1977, p. 81.
101. McCreary, J., and P.R. Gray, "A high speed, all MOS successive-approximation weighted capacitor A/D conversion technique," 1975 *IEEE International Solid-State Circuits Conference, Technical Abstracts*, NY, p. 38.
102. McGreivy, D.J., and B.B. Roesner, "Up-diffused $I^2 L$, a high speed bipolar LS1 process," *IEEE International Electron Devices Meeting Technical Digest*, Washington D.C., 1976, p. 308.

103. McWhorter, E.W., "The small electronic calculator," *Scientific American*, March 1976, p. 88.
104. Mennie, D., "Self-contained electronic games," *IEEE Spectrum*, December 1977, p. 21.
105. Meusburger, G., "A new circuit configuration for a single transistor cell using Al-gate technology with reduced dimensions," *IEEE J. Solid-State Circuits*, SC-12, 1977, p. 253.
106. Middelhoek, S., et al., *Physics of Computer Memory Devices*, Academic Press, NY, 1976.
107. Millman, J., and C.C. Halkias, *Electronic Devices and Circuits*, McGraw Hill, NY, 1967.
108. Mitra, S.K., *Analysis and Synthesis of Linear Active Networks*, Wiley, NY, 1969.
109. Mitra, SK., (ed.), *Active Inductorless Filters*, IEEE Press, NY, 1971.
110. Mrazek, D., "Dynamic RAM elements can yield static RAM systems," *Electronic Design News*, June 20, 1973, p. 46.
111. Murphy, H.E. and W.C. Steffe, "Monolithic integration for a camera control system," *IEEE J. Solid-State Circuits*, SC-8, 1973, p. 427.
112. Nemec, J., "One-chip bipolar microcontroller approaches bit-slice performance," *Electronics*, September 1, 1977, p. 91.
113. Ohmori, Y., M. Sunazawa, and Y. Fujimura, "MOS IC crosspoint switch for space division digital switching networks," *IEEE J. Solid-State Circuits*, SC-9, 1974, p. 142.
114. Pashley, R., et al., "H-MOS scales traditional devices to higher performance levels," *Electronics*, August 18, 1977, p. 95.
115. Pashley, R., et al., "Speedy RAM runs cool with power-down circuitry," *Electronics*, Aug. 4, 1977, p. 103.
116. Petritz, R.L., "The pervasive microprocessor: trends and prospects," *IEEE Spectrum*, July 1977, p. 18.
117. Plassche van de, R.J., "A wideband monolithic instrumentation amplifier," 1975 *IEEE International Solid-State Circuits Conference Abstracts*, p. 194.
118. Rai, Y., T. Sasami, and Y. Hasegawa, "Electrically reprogrammable non-volatile semiconductor memory with MNMoOS (metal-nitride-molybdenum-oxide-silicon) structure," *IEEE International Electron Devices Meeting Technical Digest*, Washington, D.C., 1975, p. 313.
119. Rao, K.R., and S. Srinivasan, "A bandpass filter using the operational amplifier pole," *IEEE J. Solid-State Circuits*, SC-8, 1973, p. 245.
120. Rattner, J., J-C Cornet, and M.E. Hoff, "Bipolar LSI computing elements usher in new era of digital design," *Electronics*, September 5, 1974, p. 89.
121. Renwick, W., and A.J. Cole, *Digital Storage Systems*, 2nd ed., Chapman and Hall, London, 1971.
122. Reyling, G.F., "Single-chip microprocessor employs mini-computer word length," *Electronics*, December 20, 1974, p. 87.
123. Rideout, V.L., J.J. Walker, and A. Cramer, "A one-device memory cell using a single layer of polysilicon and self-registering metal-to-polysilicon contact," *IEEE International Electron Devices Meeting Technical Digest*, Washington, D.C., 1977.
124. Riley, W.B., *Electronic Computer Memory Technology*, McGraw-Hill, NY, 1975.
125. Riley, W.B., "Semiconductor memories are taking over data-storage applications," *Electronics*, August 2, 1973, p. 75.
126. Riley, W.B., "4096-bit memories pose test woes," *Electronics*, December 20, 1973, p. 65.
127. Rodgers, T.J., et al., "VMOS ROM," *IEEE J. Solid-State Circuits*, SC-11, 1976, p. 614.
128. Rutkowski, G.E., *Handbook of Integrated Circuit Operational Amplifiers*, Prentice Hall, Englewood Cliffs, 1975.
129. Schlageter, J.M. et al., "Two 4K static 5-V RAM's," *IEEE J. Solid-State Circuits*, SC-11, 1976, p. 602.
130. Schmid, H., *Electronic Analog/Digital Conversions* Van Nostrand Reinhold, NY, 1970.

131. Schneiderman, R., "Smart cameras clicking with electronic functions," *Electronics*, August 21, 1975, p. 74.
132. Scrupski, S.E., "Communications," *Electronics*, October 17, 1974, p. 83.
133. Shiga, K., T. Itoh, and I. Anbe, "A monostable CMOS RAM with self-refresh mode," *IEEE J. Solid-State Circuits*, SC-11, 1976, p. 609.
134. Shima, M. and F. Faggin, "In switch to *n* MOS microprocessor gets a $\mu$s cycle time," *Electronics*, April 18, 1974, p. 95.
135. Smarandoiu, G. et al., "An all-MOS analog-to-digital converter using a constant slope approach," *IEEE J. Solid-State Circuits*, SC-11, 1976, p. 408.
136. Solomon, J.E., "The monolithic Op amp: a tutorial study," *IEEE J. Solid-State Circuits*, SC-9, 1974, p. 314.
137. Soucek, B., *Microprocessors and Microcomputers*, Wiley-Interscience, NY, 1976.
138. Stewart, R.G., "A CMOS/SOS electrically alterable read only memory," *IEEE International Electron Devices Meeting Technical Digest*, Washington, D.C., Dec. 1978, p. 344.
139. Strocka, R.L., and D.F. Broxterman, Electronic Watch, US Patent 3,815,354, June 11, 1974.
140. Strong, N., "LS1 converts an old technique into low-cost a-d conversion," *Electronics*, September 11, 1972, p. 102.
141. Strong, N., "Rough life of digital multimeter puts tough demands on design," *Electronics*, June 23, 1977, p. 106.
142. Suarez, R.E., P.R. Gray, and D.A. Hodges, "An all-MOS charge-redistribution A/D conversion technique," *1974 IEEE International Solid State Circuits Conference Technical Abstracts*, NY, p. 194.
143. Sunazawa, M., et al., "Low power CML IC crosspoint switch matrix for space division digital switching networks," *IEEE J. Solid-State Circuits*, SC-10, 1975, p. 117.
144. Sutherland, I.E., and C.A. Mead, "Microelectronics and computer science," *Scientific American*, September, 1977, p. 210.
145. Swinbank, E., T.R. Heeks, and J.H. Shelly, "Quartz crystal watches," *IEEE(London) Electronics and Power*, October 31, 1974, p. 887.
146. Tarui, T., K. Namimoto, and Y. Takahashi, "Twelve-bit microprocessor nears minicomputer's performance level," *Electronics*, March 21, 1974, p. 111.
147. Tatro, R.D., "Which hybrid converter: single switch or quad?" *Electronics*, July 25, 1974, p. 89.
148. Taylor, G.W., and I. Lefkowitz, "Techniques for decreasing power and increasing legibility of electronic watches," *Optics Communications*, 8, 1973, p. 426.
149. Terman, L.M., "MOS FET memory circuits," *Proc. IEEE*, 59, 1971, p. 1044.
150. *Texas Instruments, Circuit Design for Audio, AM/FM and TV*, McGraw-Hill, NY, 1967.
151. *Texas Instruments, Designing with TTL Integrated Circuits*, McGraw-Hill, NY, 1971.
152. Tillman, J.R., "Expanding Role of Semiconductor Devices in Telecommunicators," *Radio and Electronic Engineering*, 43, p. 82, 1973.
153. Toong, H.M.D., "Microprocessors," *Scientific American*, September 1977, p. 146.
154. Tremaine, H.M., *Audio Cyclopedia*, 2nd ed., H.A. Sams Co., Indianapolis, 1969.
155. Tsividis, Y.P., and P.R. Gray, "An integrated NMOS operational amplifier with internal compensation," *IEEE J. Solid-State Electronics*, SC-11, 1976, p. 748.
156. Tsividis, Y.P., et al., "A segmented $\mu$-255 law PCM voice encoder utilizing NMOS technology," *IEEE J. Solid-State Circuits*, SC-11, 1976, p. 740.
157. Tucci, P.A., and L.K. Russell, "An $I^2L$ watch chip with direct LED drive," *IEEE J. Solid-State Circuits*, SC-11, 1976, p. 847.
158. Vacroux, A.G., "Microcomputers," *Scientific American*, May, 1975, p. 32.
159. Vadasz, L.L., et al., "Silicon-gate technology," *IEEE Spectrum*, 6, October 19, 1969, p. 28.

160. Vittoz, E., B. Gerber, and F. Leuenberger, "Silicon gate CMOS frequency divider for the electronic wristwatch," *IEEE J. Solid-State Circuits,* SC-7, 100, 1972. (See also Vittoz, E. "LSI in watches," ESSCIRC, 1976, Toulouse, France.)
161. Volpe, G.T., and L.M. Freeman, "Design of a hybrid integrated circuit elliptic active RC filter," *IEEE J. Solid-State Circuits,* SC-9, 1974, p. 76.
162. Wada, T., "Electrically reprogrammable ROM using $n$-channel memory transistors with floating gate," *Solid-State Electronics,* 20, 1977, p. 623.
163. Walker, G.M., "Inside electronic watches: a microprocessor movement," *Electronics,* April 12, 1971, p. 97.
164. Walker, G.M., "The HP-35: a tale of teamwork with vendors," *Electronics,* February 1, 1973, p. 102.
165. Walker, G.M., "LSI controls gaining in home appliances," *Electronics,* April 14, 1977, p. 91.
166. Walker, G.M., "Automotive electronics gets the green light," *Electronics,* September 29, 1977, p. 83.
167. Weber, S., *Large and Medium Scale Integration,* McGraw-Hill, NY, 1974.
168. Weissberger, A.J., "Microprocessors expand industry applications of data acquisition," *Electronics,* September 5, 1974, p. 107.
169. Widlar, R.J., "New developments in IC voltage regulators," *IEEE J. Solid-State Circuits,* SC-6, 1971, p. 2.
170. Widlar, R.J., "Design techniques for monolithic operational amplifiers," *IEEE J. Solid-State Circuits,* SC-4, 1969, p. 184.
171. Wilnai, D., and P.W.J. Verhofsbabt, "One-chip CPU packs power for general purpose minicomputers," *Electronics,* (Fairchild 9440), June 23, 1977, p. 113.
172. Wooley, B.A., S.Y.J. Wong, and D.O. Pederson, "A computer-aided evaluation of the 741 amplifier," *IEEE J. Solid-State Circuits,* 1974, p. 357.
173. Yamaguchi, T. and S. Sato, "Monolithic clockless A/D converter integrated circuits," *IEEE Trans. Electron Devices,* ED-22, 1974, p. 295.
174. Young, L., T. Bennett, and J. Lavell, "$N$-channel MOS technology yields new generation of microprocessors," *Electronics,* April 18, 1974, p. 88.
175. Yoshida, K., "A 16 bit LSI minicomputer," *IEEE J. Solid-State Circuits,* SC-11, 1976, p. 696.
176. Young, S., "Uncompromising 4K static RAM runs fast on little power," *Electronics,* May 12, 1977, p. 99.
177. Zicko, C.P., "New applications open up for the versatile isolation amplifier," *Electronics,* March 7, 1974, p. 96.
178. Zuch, E.L., "Where and when to use which data converter," *IEEE Spectrum,* June 1977, p. 39.
179. Capece, R.P., "Memories," *Electronics,* October 26, 1978, 51, p. 126.
180. Capece, R.P., "Microprocessors and microcomputers," *Electronics,* October 26, 1978, 51, p. 138.
181. Wolber, W.G., "Sensor development in the microcomputer age," *IEEE International Electron Devices Meeting Technical Digest,* Washington, D.C., Dec. 1978, p. 12.
182. Guterman, D.C. et al., "Electrically alterable hot-electron injection floating gate MOS memory cell with series enhancement," *IEEE International Electron Devices Meeting Technical Digest,* Washington, D.C., Dec. 1978, p. 340.
183. Koyanagi, M. et al., "Novel high density, stacked capacitor MOS RAM," *IEEE International Electron Devices Meeting Technical Digest,* Washington, D.C., Dec. 1978, p. 348.
184. Sasaki, N. et al., "Charge pumping SOS-MOS transistor memory," *IEEE International Electron Devices Meeting Technical Digest,* Washington, D.C., Dec. 1978, p. 356.
185. Millman, J., *Microelectronics,* McGraw-Hill, New York, 1979.

# 10
# Charge-Transfer Devices

## CONTENTS

## 10.1 GENERAL CONCEPTS

Shift registers of simple form have been known since the days of vacuum tube circuits. Bipolar and MOS transistor versions of static and dynamic registers using discrete components followed in due course. The concept of a shift register involves the passage of charge along a line of capacitors by the sequential switching of transistors in response to clock pulses. This class of circuit has acquired the name bucket brigade since the action is reminiscent of the transfer of water buckets down a line of people. In 1970 integrated versions of these circuits were shown to be practical for delay and other applications. A MOS FET version of an integrated bucket-brigade circuit is shown in Fig. 10.1(a). The storage regions are $p$ islands in the $n$ substrate and the metal gates are offset. When the gates are negatively pulsed, conducting channels are created that link adjacent $p$-storage regions. The storage condition ($V_1 = V_2$) is shown in Fig. 10.1(b) and the transfer condition ($V_2$ more negative than $V_1$) during which time positive charge flows from well 1 to well 2 is shown in Fig. 10.1(c). The equivalent circuit, considered as discrete components, is given in Fig. 10.1(d) and consists of a line of $p$-channel MOS FETs with the storage capacitors between the gates and drains.

590

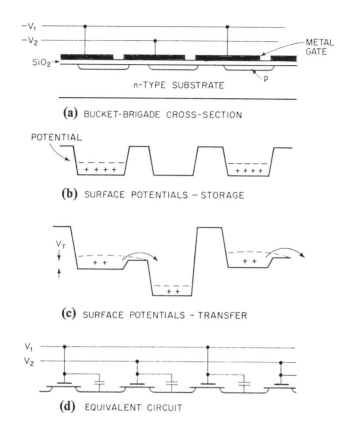

Fig. 10.1. Bucket-brigade shift register concepts. (After Altman. Reprinted from *Electronics*, June 21, 1971; Copyright © McGraw-Hill, Inc., N.Y., 1971.)

The next step, a major one in integration, was the concept by Boyle and Smith,[20] verified by Amelio, Tompsett, and Smith,[6] that the charge could be handed on in the form of minority carriers by manipulation of potential wells and that the $p$ islands of the bucket brigade were not necessary. This leads to a true integrated charge-transfer device that cannot be represented by discrete components.* A three-phase clock version of a charge-coupled shift register is shown in Fig. 10.2.

The voltage on $V_2$ in Fig. 10.2(a) is large enough to create an inversion region with minority carriers (holes) confined in the potential well at the interface between the oxide and the silicon. However, when $V_3$ is made more negative than $V_2$ the holes move to the right. The role of $V_1$ is to prevent movement of the charge to the left, hence the need for a three-phase clock.

---

*The term charge-transfer device (CTD) is a little broader in meaning than charge-coupled device (CCD) since it includes bucket-brigade circuits and other variations. However, CCD is in general use.

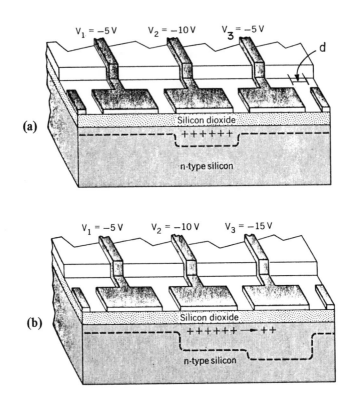

Fig. 10.2. A charge-coupled device in the storage mode (a) and in the transfer mode (b). (After Boyle and Smith. Reprinted with permission from *IEEE Spectrum*, 8, 1971, p. 21.)

A simpler way in which directionality can be achieved is through the use of a stepped oxide as shown in Fig. 10.3. The region under the thicker oxide acts as a potential barrier that ensures transfer in only one direction (from left to right in the diagram) even though only a two-phase clock is used. Typically, the packing density is improved by having the array constructed in a serpentine fashion as shown in Fig. 10.4.

Almost all applications of CCDs call for many hundreds of transfer steps and therefore the reduction of charge losses during transfer has been the aim of much of the recent work in this field. The early versions of CCDs had charge losses of 1 to 0.1% for each transfer. From the beginning of the CCD concept it was clear that the spacing between the plates [distance $d$ in Fig. 10.2(a)] needed to be small if potential barriers that would impede full transfer of the charge were not to occur. The effects of the substrate doping density and interface charge on such barriers were examined and improvements were achieved by the use of multiple layers of metallization or polysilicon to effectively eliminate the gaps without loss of insulation.

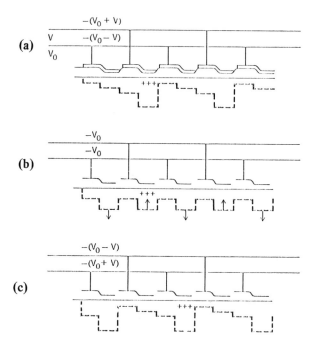

Fig. 10.3. A stepped oxide in a two-phase CCD provides the barriers that ensure directionality of flow. (After Boyle and Smith. Reprinted with permission from *IEEE Spectrum,* 8, 1971, p. 22.)

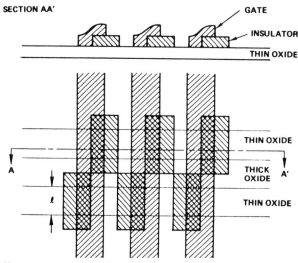

Fig. 10.4. Offset gate structure and serpentine array CCD. (After Bower, Zimmerman, and Mohsen. Reprinted with permission from *IEEE Trans. Electron Devices,* **ED-22,** 1975, p. 72.)

Sealed metallization and sealed oxides are desirable to prevent charge buildup on thin oxide layers altering the CCD characteristics. Fig. 10.5 shows various sealed structures with polysilicon gates. The cell dimensions and tolerances needed for such structures are shown in Fig. 10.6 and Table 10.1. The projections for the future envisage electron-beam exposure techniques and plasma and ion-beam processing.

The moving carriers may be electrons instead of holes. There is some advantage in this because electrons have higher mobility and a larger diffusion coefficient. Such a structure with a $p$ substrate is shown in Fig. 10.7 with an input diode to inject the pulse at the start of the line and a sensing diode to extract it at the end of the CCD. Injection is also possible by pulsing the voltage on the first gate of a CCD to create avalanche-charge buildup and this was used in early devices. Sensing at the end of the CCD can be by injection of the charge into the substrate or by having the final gate floating in potential.

Various charge-injection schemes are shown in Fig. 10.8. In the dynamic current-injection scheme [Fig. 10.8(a)] the electrical input signal is applied to the input diode (ID) referenced to a dc potential on an input gate (IG) which is the first separately accessible transfer electrode. However, with this scheme the magnitude of the injected packet has poor linearity with the signal voltage and is dependent on the threshold voltage of the gate and on the available injection time and thus on the clock frequency.

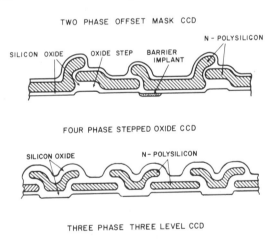

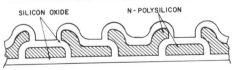

Fig. 10.5. CCD electrode structures with polysilicon gates. (After Mohsen and Retajczyk. Reprinted with permission from *IEEE J. Solid-State Circuits,* **SC-11,** 1976, p. 186.)

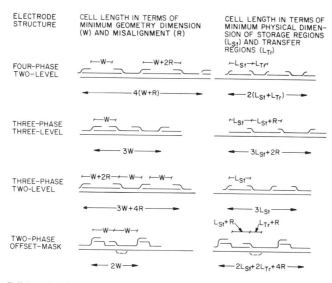

Fig. 10.6. Cell length of polysilicon-gate CCD electrode structures. (After Mohsen and Retajczyk. Reprinted with permission from *IEEE J. Solid-State Circuits,* **SC-11**, 1976, p. 187.)

## Table 10.1. CCD Cell Dimensions and Tolerances

| Electrode Type | Photolithographic Limitations | Physical Limitations | State of the Art (μm) | On the Horizon (μm) | Future (μm) |
|---|---|---|---|---|---|
| Two-phase | $4(W + R)$ | $2(L_{tr} + L_{st})$ | 32 | 20 | 10 |
| Two-phase offset gate | $2W$ | $2(L_{tr} + L_{st} + 2R)$ | 14 | 10 | 6 |
| Three-phase three level | $3W$ | $3(L_{st} + R)$ | 18 | 12 | 6 |
| $L_{st}$ (μm) | | minimum storage region | 2 | 2 | 1 |
| $L_{tr}$ (μm) | | minimum transfer region | 1 | 1 | 1 |
| $R$ (μm) | alignment tolerance | | 2 | 1 | 0.5 |
| $F$ (μm) | minimum feature size | | 6 | 4 | 2 |

(After Mohsen et al, reprinted with permission from IEEE Journal of Solid-State Circuits, 1976, SC-11, p. 53)

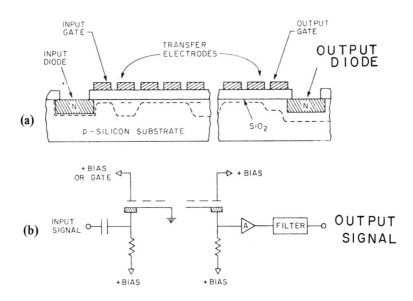

Fig. 10.7. (a) Schematic cross section of a CCD showing the input and output configuration. (b) Electrical connections for input and output of charge. (After Tompsett and Zimany. Reprinted with permission from *IEEE J. Solid-State Circuits*, **SC-8**, 1973, p. 152.)

In a better arrangement Fig. 10.8(b)

the signal is applied to the input diode (ID) and fills the potential well under the first regular transfer electrode (P2) to a corresponding value when the input gate (IG) is pulsed fully open. Subsequently, the gate is turned off and thus isolates this charge packet from the input diode before it is transferred to the following electrode. Input signal voltages as high as the pulse amplitude $V_p$ can be applied to the input diode in this voltage input or diode cutoff method. However, there is still a nonlinear term in the relation between the input signal voltage and the charge packet size due to the variable depletion capacitance of the metering potential well. A somewhat better performance can be expected if the input diode is held at a fixed low potential and the input signal is applied to a second separately addressable input gate at the location of the first P2 electrode. The well underneath is then always filled to the same potential value given by the voltage on ID and some of the nonlinear terms cancel out.

Considerably better linearity and lower noise can be obtained with the potential equilibration method, also called "fill and spill," or charge preset method, of Fig. 10.8(c) as explained by Séquin and Tompsett.[94]

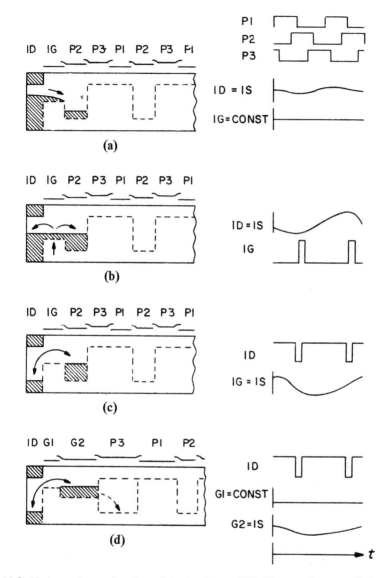

Fig. 10.8. Various schemes for charge injection into a CCD. The wave forms applied to the electrodes are shown on the right:
(a) dynamic current injection,
(b) signal voltage applied to the input diode and input gate pulsed to cut off the input diode,
(c) potential equilibration method with a pulsed-metering electrode, and
(d) improved equilibration method with input signal applied to enlarged metering electrode. (After Séquin and Tompsett, from *Charge Transfer Devices*, Academic Press, N.Y., © 1975, p. 49.)

The signal is applied to the single input gate (IG) and the first regularly pulsed transfer electrode (P2) forms the metering potential well. The input diode (ID) is pulsed low to inject charge across IG when the first transfer electrode (P2) is turned on. The charge packet size is then determined by the active area of the metering electrode and by the potential difference $(V_{P2} - V_{IG})$, independently of the substrate doping and the depletion capacitance.

In another version, Fig. 10.8(d),

the first gate (G1) is held at a low fixed reference level, while the input signal is applied to a second gate (G2) under which the metering well is formed. Potential equilibration then always takes place at the same value of the interface potential and the effects of fringing fields or of a different oxide thickness under the barrier electrode (G1) are constant. If the potential of the signal electrode is never raised above $(V_R + V_P/2)$, all carriers will discharge into the subsequent transfer electrode (P3) when that phase turns on. Thus in this method the metering electrode G2 does not have to be pulsed, which results in a substantial reduction in noise. In a proper design the metering gate and the following two transfer electrodes have to be made at least twice as large as the regular transfer electrodes to match the signal handling capability of the input to that of the rest of the device.

Charge-sensing circuits used at the output ends of CCDs are illustrated in Fig. 10.9. In Fig. 10.9(a), the output charge passes into the depletion region of the diode OD and is detected by the current sensitive preamplifier. The on-chip amplifier shown in Fig. 10.9(b) is preferable because it introduces less parasitic capacitance. The output obtained is a few volts. Nondestructive sensing is possible from potential swings induced on the gate of a MOSFET.

## 10.2 LOSS MECHANISMS IN CCDs

Modeling of CCD action has been attempted in many ways, including one-dimensional closed-form simplifications, traveling-wave models, and two-dimensional numerical treatments. Our treatment is limited to a discussion of some of the physical causes of charge loss.

Transfer inefficiency is defined as

$$\epsilon = \frac{\Delta N}{N_{\text{sig}} - N_{fz}} \tag{10.1}$$

where $\Delta N$ is the net charge lost from the packet in a transfer cycle and $(N_{\text{sig}} - N_{fz})$ is the amount by which the signal exceeds the background charge (usually termed "fat-zero"). The physical mechanisms that are responsible for loss include:

(a)

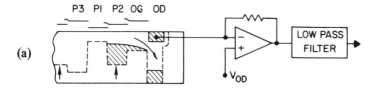

(b)

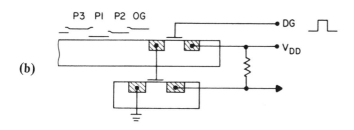

(c)

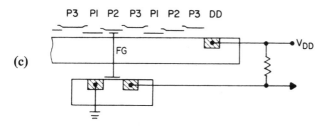

Fig. 10.9. Basic charge-sensing circuits using:
(a) an external preamplifier connected to an output diode,
(b) on-chip sense and reset MOSFET's form a gated-charge integrator, and
(c) a floating-gate sense MOSFET which measures the charge packet nondestructively. (After Séquin and Tompsett, from *Charge Transfer Devices*, Academic Press, N.Y., © 1975, p. 53.)

- Inadequate time for the carriers to move the required distance under the influence of diffusion or bulk field sweep out.
- Potential barrier humps that may exist between wells because of inter-electrode gaps.
- Trapping of carriers at interface states between the silicon and the $SiO_2$ and later release of these carriers during subsequent clock pulses.
- Trapping of carriers at bulk states and later release.

The transfer inefficiency, in part, is independent of the magnitude of the charge transmitted, but also has components that are charge-amplitude dependent. It is also dependent on the immediate past history of the CCD cell since this determines the occupancy of traps.

The driving pulse amplitude determines the signal charge stored in the well. For a typical Si gate structure, with element length 30 $\mu$m, width 200 $\mu$m, and 1500-Å oxide thickness, the signal charge is about 3 pC for a 10-V amplitude pulse. The transfer inefficiency is then about $1 \times 10^{-4}$ and increases with the decrease of signal charge and gate voltage as shown in Fig. 10.10.

The transfer inefficiency also increases as the clock (drive) frequency is increased. This is shown in Fig. 10.11 for undercut-isolated CCDs with $p$ and $n$ channels. The fairly flat region below about 0.5 MHz has a transfer efficiency of $4 \times 10^{-4}$ and presumably is limited by factors such as interface states. In the measurements, a background charge was injected to provide for filling of the interface states. For a background charge below 20% of a full charge packet, increase of the transfer inefficiency occurred. Between 20 and 80% there was no further improvement.

The charge-transfer process may be modeled to some degree as a discharge of a capacitor through a nonlinear resistance that represents the induced

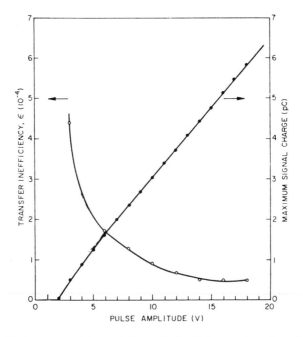

Fig. 10.10. Maximum signal charge and transfer inefficiency $\epsilon$ for a three-phase polysilicon CCD surface-channel device as a function of the driving-pulse amplitude $V_p$. The transfer frequency was 1 MHz. (After Bertram, et al. Reprinted with permission from *IEEE Trans. Electron Devices*, **ED-21**, 1974, p. 761.)

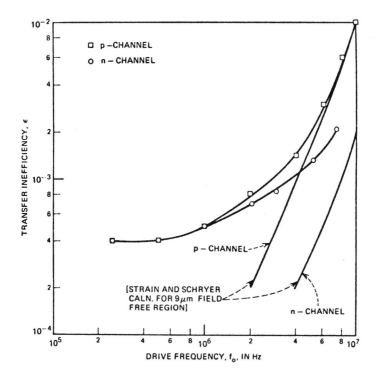

Fig. 10.11. Measurements of transfer inefficiency (per transfer) for both $n$- and $p$-channel undercut-isolated CCDs. The theoretical values for $p$- and $n$-channel devices assuming a 9-$\mu$m field-free region under the transfer electrodes on the thin oxide have also been plotted. (After Tompsett, Kosicki, and Kahng. Reprinted with permission from *Bell Syst. Tech. J.*, **52**, 1973, p. 5; © 1973, American Telephone and Telegraph Company.)

channel.[15,54] However, terms for diffusion and field-aided drift need to be taken into account in more complete modeling.[66]

The inherent frequency response of a CCD with $n$ transfers is given by

$$R(f) = \exp\left\{-n\epsilon\left[1 - \cos(2\pi f/f_0)\right]\right\} \qquad (10.2)$$

where $f$ is the signal frequency and $f_0$ is the drive frequency. The curves that follow are given in Fig. 10.12. If $n$ is 250 and $\epsilon$ is $2 \times 10^{-3}$ at (say) 3 MHz drive frequency, then $n\epsilon$ is 0.5 and the signal is attenuated to about 50% for an input frequency of 1 MHz.

Since CCD performance is influenced by surface or bulk traps, the application of pulses of input to a CCD that has been carrying zeros involves some amplitude attenuation of the first few pulses. When the pulse train is turned off, trailing-edge charge packets are seen as traps empty. This is illustrated in Fig. 10.13 for

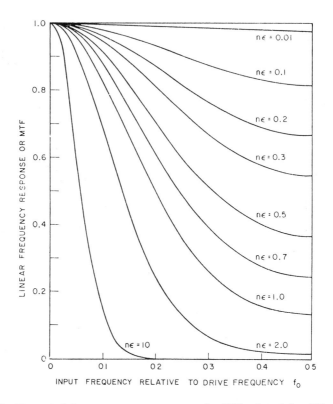

Fig. 10.12. Theoretical frequency-response curves of a CCD, plotted for different values of transfer inefficiency $n\epsilon$. (After Tompsett and Zimany, Reprinted with permission from *IEEE J. Solid-State Circuits*, **SC-8**, 1973, p. 153.)

an input five clock pulses long. If loss is proportional to the input signal amplitude, it may be described by

$$N_{loss} = \epsilon N_{sig} \qquad (10.3)$$

Or the loss may be a fixed amount of charge that is independent of signal amplitude. This may be true for surface-state losses in a surface-channel CCD. The use of a "fat-zero" background charge reduces this loss (Fig. 10.14) by keeping the surface states filled. The net trapping or release of carriers at the $Si\text{-}SiO_2$ interface if the states are not kept filled is illustrated in Fig. 10.15(b). The discharge times of the states are given in Fig. 10.15(c). Accounting for interface state effects is complicated since edge effects must be considered. [76]

The effectiveness of background charge in allowing the CCD to be operated at low-signal level with reasonable transfer efficiency is shown in Fig. 10.16. The rolloff with increasing signal frequency is seen to be almost independent

(a)

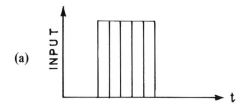

(b)

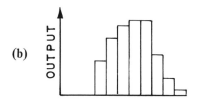

Fig. 10.13. The output of a CCD which has a proportional loss when the input is five clock periods long. (After Brodersen, Buss, and Tasch. Reprinted with permission from *IEEE Trans. Electron Devices*, **ED-22**, 1975, p. 42.)

of signal amplitude over a 60-dB range. The "fat-zero" charge was about 50% of a full charge packet in this study and the clock frequency was 1 MHz. Ability to operate at small signal levels is desirable in the low-light imaging uses of CCDs. The number of transfers for Fig. 10.16 was 300 and the transfer efficiency was inferred to be $2.7 \times 10^{-3}$ by matching to Eq. (10.2).

Another important cause of transfer inefficiency is the existence of potential barriers between wells. These barriers may be associated with the interelectrode

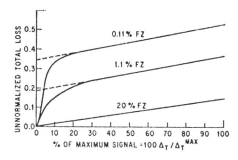

Fig. 10.14. CCD measurements showing that a 20% fat-zero is sufficient to eliminate the part of the transfer loss that does not depend on signal amplitude. (After Brodersen, Buss, and Tasch. Reprinted with permission from *IEEE Trans. Electron Devices*, **ED-22**, 1975, p. 42.)

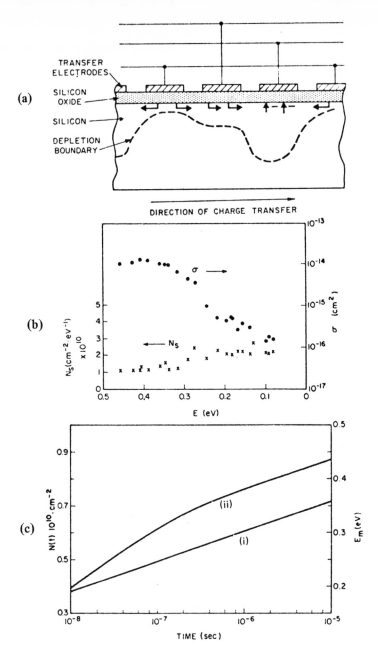

Fig. 10.15. Effects of interface states on CCDs:

(a) Trapping or release of carriers as charge packet moves.

(b) Interface state density and electron-capture cross section as a function of energy below the conduction band for $SiO_2$ on $n$-type substrate.

(c) Number of interface states out of $2 \times 10^{10}\,cm^{-2}\,eV^{-1}$ that empty in time $t$, (i) if capture cross section is $4.5 \times 10^{-16}\,cm^2$ (ii) if capture cross sections range as in Fig. 10.15 (b). $E_m$ is the mean energy to which the interface states have emptied in time $t$ and is shown by the same curves. (After Tompsett. Reprinted with permission from *IEEE Trans. Electron Devices*, **ED-20**, 1973, pp. 47-49.)

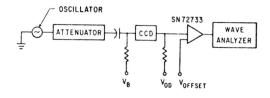

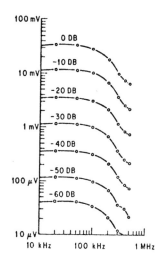

Fig. 10.16. Frequency-response variation for a CCD at various signal amplitudes.
(a)  Test circuit used.
(b)  Frequency-response curves showing the roll-off to be almost independent of signal amplitude. (After Brodersen, Buss, and Tasch. Reprinted with permission from *IEEE Trans. Electron Devices*, **ED-22**, 1975, pp. 42, 43.)

gaps. The trapped charge that can result is illustrated in Fig. 10.17(a). Since the barriers may depend on the charge that has transferred, as shown in Fig. 10.17(b), the loss may be greater for a large signal than for a small signal, so both fixed and proportional losses may occur. The barriers can be reduced by the use of overlapping layers of metallization to avoid effective gaps, and with the proper choice of clock voltages the residual barrier effect is negligible.

An important development in the improvement of transfer efficiency in CCDs, however, has been the elimination of interface state effects by displacing the channel containing the charge packets away from the $SiO_2$-Si interface. In some buried channel designs this has lowered the loss from $10^{-3}$ to $10^{-5}$ per transfer. In a $p$-Si substrate an $n$-type region is created, usually by ion implantation, and this connects to $n^+$ input and output pads at either end of the CCD as shown in Fig. 10.18. Consider some typical numbers. The $n$ layer may be 5 $\mu$m deep and if uniformly doped at $10^{15}$ cm$^{-3}$ in a $p$ substrate of $1 \times 10^{14}$ cm$^{-3}$, the $p$ region can be depleted of its carriers (a total charge of $10^{12}$ cm$^{-2}$) by the application of a gate voltage of about +20 V, and a potential well for electrons formed about 4 $\mu$m from the oxide surface. The charge-

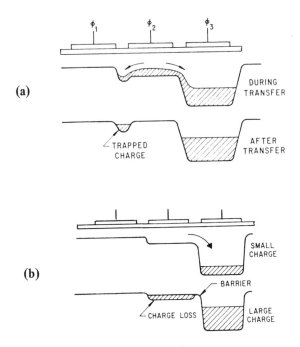

Fig. 10.17. Barrier effects in CCD potential wells.
(a) Barrier associated with interelectrode gap.
(b) Small charge transfers without loss but for a larger charge packet a barrier appears that causes charge loss. (After Brodersen, Buss, and Tasch. Reprinted with permission from *IEEE Trans. Electron Devices,* **ED-22,** 1975, pp. 44, 45.)

carrying capacity of this buried channel is about $4 \times 10^{-8}$ C/cm$^2$ which is adequate since 0.1 pC can be stored in an area 10 $\mu$m $\times$ 250 $\mu$m. However, it is about an order of magnitude less than the $3 \times 10^{-7}$ C/cm$^2$ of a surface-channel CCD.

The effects of bulk traps in buried channel CCDs have been studied. In one such study with no intentionally introduced background charge, a leading-edge deficit of 8.9% was observed in a signal after many "zeros." In the trailing edge of a group of "ones" the charge excess was 2.6%. This corresponds to a transfer inefficiency of $10^{-4}$ from which an effective bulk-trap density of $2 \times 10^{11}$ cm$^{-3}$ may be inferred assuming no other cause for loss.[77] The addition of a background charge lowered the loss to $3 \times 10^{-5}$ per transfer. The channel was an implant of $1.5 \times 10^{12}$ cm$^{-2}$ phosphorous ions, which gave after annealing and oxide growth, a Gaussian distribution with a peak concentration at the surface of $1.6 \times 10^{16}$ cm$^{-3}$ and a channel depth of 2.1 $\mu$m. The $p$-type substrate was doped at $4.5 \times 10^{14}$ cm$^{-3}$.

Bulk-trapping effects in BCCDs differ from interface-trapping effects in SCCDs in two respects.

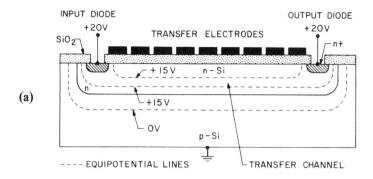

**(a)**

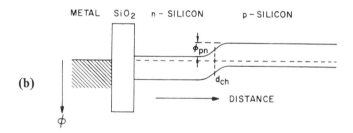

**(b)**

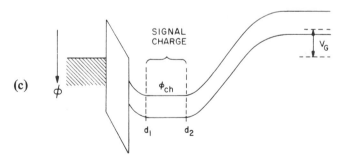

**(c)**

Fig. 10.18. Bulk-(buried) channel CCD.
(a) Cross-sectional view showing the potentials created by the bias on the input and output diodes. (After Walden, et al. Reprinted with permission from *Bell Syst. Tech. J.,* **51**, p. 1636; © 1972, American Telephone and Telegraph Company.)
(b) Energy diagram without bias.
(c) Energy diagram with bias and signal charge present. (b), (c) (After Séquin and Tompsett, from *Charge Transfer Devices,* Academic Press, N.Y., © 1975, p. 14.)

1)    Bulk-trapping effects peak at the clock frequency $f_c = 0.36/\tau_e$, where $\tau_e$ is the emission time of the trap, and drop off exponentially at clock frequencies above or below $0.36/\tau_e$.

2)    The volume occupied by a charge packet decreases as the number of trapping centers with which the charge packet interacts decreases. In typical BCCDs, a bulk-trapping level having a room temperature emission time constant of 900 $\mu$s is observed.[12] Typically, the temporal noise due to bulk trapping in buried-channel devices, operated at $f_c = 0.36/(900 \ \mu s) = 400$ Hz with signal levels of $1 \times 10^4$ electrons/packet, is only 50 electrons per packet after 600 transfers.

The fringing fields in buried-channel CCDs allow lower loss per transfer than in surface-channel CCDs, or operation at higher clock frequencies with some increase in transfer loss. A version of a buried-channel field-swept CCD has been described under the name peristaltic charge-coupled device.[44] Effective operation has been achieved at high frequencies, 135 MHz, and calculations suggest the possibility of operation to the GHz clock frequency range.

The problem of charge-packet degeneration during transfer is of considerable importance in analog uses of CCD, for instance as delay lines, filters, and image sensors. In digital applications it is less serious since signal regeneration stages can be provided at intervals, although the intervals must not be too short since bit density is reduced by the need for such stages. Charge regeneration can be accomplished by extracting the charge after a certain number of transfers have corrupted the logic levels and applying it to a trigger pulse circuit to generate a larger and squarer pulse and then reinserting it in the CCD line either on the same chip or in a following CCD section on another chip. Commonly, regeneration is provided at the end of several CCD rows and the restored charge packet sent back on the next adjacent row in the array. A compact regeneration scheme that uses a floating diffusion to couple and regenerate the charge packet is shown in Fig. 10.19.

As is shown in Fig. 10.19(b), the signal-regeneration cycle is initiated by resetting F-1 to the reference potential $E_2$ by a voltage pulse $V_{c2}$ at the time prior to the beginning of voltage pulse $\phi_3$. The transfer of charge into the last potential well of the shift register during the pulse $\phi_3$ is illustrated in Fig. 10.19(c). The charge signal also begins to accummulate in the F-1 region during the $\phi_3$ pulse. Therefore, the conductive channel previously established between F-1 and D-1 by $V_{c2}$ must now be interrupted. The transfer of charges into the potential well associated with the F-1 region is completed during the fall time of the pulse $\phi_3$. The process of introducing the charge signal to the first potential well of the second circuit by sampling the potential of F-1 by the control voltage pulse $V_{c1}$ is illustrated in Fig. 10.19(d) and (e) for the case of information bits "1" and "0," respectively, detected at the output of the first register. As is shown in Fig. 10.19(e), the signal regeneration inverts the information by filling the first potential well to the voltage level of S-2 if no charge is in F-1. However,

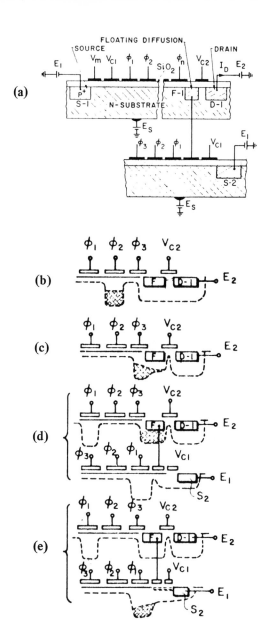

Fig. 10.19.

(a) Operation of a CCD signal-regeneration stage illustrated in terms of the variations of the surface potentials as the charge signal is moved under the influence of the phase-voltage pulses.

(b) F-1 region is reset to the reference potential $E_2$.

(c) Charge is introduced into F region.

(d) Signal regeneration for signal "1" in the first shift register.

(e) Signal regeneration for signal "O" in the first shift register. (After Kosonocky and Carnes. Reprinted with permission from *IEEE Solid-State Circuits,* SC-6, 1971, pp. 314, 315.)

609

if a charge signal is in F-1, as shown in Fig. 10.19(d), it results in positive voltage change of F-1 that cuts off the conduction channel otherwise extending from S-2 to the potential well created by $\phi_1$. Inversion from 1 to 0 and 0 to 1 during regeneration presents no circuit inconvenience. Several variations of this class of regeneration circuit are known.

Digital CCDs may have packing densities of about 600 $\mu m^2$ (1 $mil^2$) per bit of information and for a bit rate of $10^7$ bps a power requirement of 20 $\mu W$ per bit. One potential of digital CCDs is that a coded signal may be loaded at one clock frequency and subsequently transferred out at a higher clock frequency for high-speed data transmission and multiplexing with other signals.

Signal compression or frequency shift is also possible with sampled analog signals although here dispersive loss and noise effects must be kept within bounds.

A circuit for noise studies on CCDs is given in Fig. 10.20(a). The results in Fig. 10.20(b) show the magnitude of the mean-square transfer noise for a three-phase CCD with overlapping polysilicon electrodes (as in Fig. 10.5). This transfer noise corresponds to an effective interface state density in the low $10^9$ $cm^{-2}$ $eV^{-1}$. Since this kind of state density has also been obtained from transfer inefficiency measurements, the transfer noise is of the magnitude expected. However, there are many other sources of noise in CCDs as can be seen from Table 10.2. The extrinsic noise associated with insertion is large and reduction studies are in process. The bulk (buried-channel) CCD is seen to have some noise advantages over the surface-channel structure. The effective signal-to-noise ratio in a 4K CCD memory block is predicted to be about 60 dB for a cell area of 0.8 $mil^2$ [500 $(\mu m)^2$] and $n\epsilon$ = 0.3. This would lead to very small error rates.

## 10.3 CHARGE-COUPLED DELAY LINES AND FILTERS

CCDs are of importance in imaging and memory structures but also in signal-processing functions such as delay lines and filters.

Surface acoustic wave structures on piezoelectric materials are suitable for use from a few gigahertz down to the megahertz range and are capable of delay times from a microsecond to a few hundred microseconds. Bulk acoustic wave devices can give longer delay times and are usual at lower frequencies. Quartz bulk-delay lines may be used in color television receivers based on the PAL and SECAM systems.

CCDs, however, have invaded the radar field[70] and will, with further cost reduction, enter the TV field.[39] In radar, the moving target indicator (MTI) filter discriminates between returns from stationary and moving objects by comparing the return signal from one pulse with the return signal from the next pulse to reveal differences associated with movement. CCD delay lines have applications here. Radar displays may also be improved by the use of video

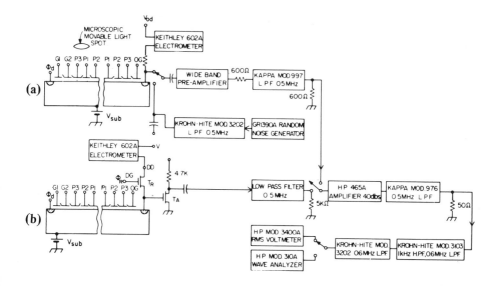

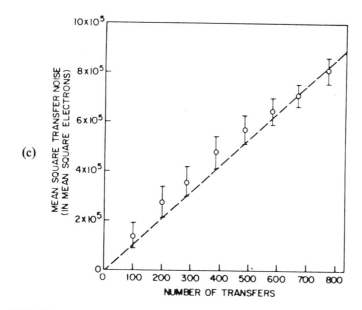

Fig. 10.20.  Mean-square transfer-noise studies for a surface-channel CCD.
(a)  Circuit if CCD has no on-chip preamplifier.
(b)  Circuit if CCD has on-chip preamplifier.
(c)  Noise versus the numbers of transfers from optical injection point to output end of
the device. (After Mohsen, Tompsett, and Séquin. Reprinted with permission from
*IEEE Trans. Electron Devices,* **ED-22**, 1975, pp. 210, 211.)

Table 10.2.  Measured Noise Levels in 256-Element CCD at 1-MHZ Clock
Frequency (Active Element Area 200 × 30 $\mu m^2$)

| Noise Source | Noise Equivalent Signal in SCCD | Noise Equivalent Signal in BCCD |
|---|---|---|
| Electrical Insertion Noise of Background Charge | 750 | Not Required |
| Electrical Insertion of the Signal Charge | 750 | 750 |
| Optical Injection of 1 pC of Signal Charge | 2800 | 2800 |
| Trapping Noise | 700-1000 | <200 |
| | $(N_{SS} = 1 - 2 \times 10^9 \text{ cm}^{-2}\text{eV}^{-1})$ | |
| Dark Current Noise | 160 | 320 |
| On-Chip Amplifier Reset Noise $C_0 = 0.7$ pF | $\simeq 330$ | $\simeq 330$ |
| Pulser Noise with Off-Chip Preamplifier | 440 | 440 |
| Pulser Noise with On-Chip Preamplifier | <30 | |
| Maximum Signal in Electrons for $V_p = 14$ V | $40 \times 10^6$ | $20 \times 10^6$ |

(From Mohsen, Tompsett and Séquin. Reprinted with permission from IEEE Trans. Electron Devices, 1975, ED-22, p. 216)

integrators to enhance the signal-to-noise ratio. In communications systems, signal compression and multiplexing can be obtained readily from CCDs by readout at a higher clock frequency than the input frequency. Functions such as correlation and Fourier and other transformations are also possible.[32, 102]

The signal-to-noise ratio and maximum delay achievable in CCDs has been discussed by Tompsett and Zimany.[106] Signal-to-noise ratios of the order of 60 dB are obtained for 1000 transfers. The ratio is proportional to the size of the charge packet and, to a first approximation, is independent of the drive frequency since the number of packets and the noise bandwidth both rise in direct proportion to the drive frequency.

The delay time in CCDs is readily and precisely controllable by variation of the clock frequency. Another advantage of CCD analog delay lines is that analog-to-digital and digital-to-analog input and output converters are not needed. CCDs have a wide signal-input dynamic range and are simple and flexible enough to become standard items for general use. The maximum delay achievable is limited by the dark current generated in the surface or bulk channel. This restricts the dynamic range and has noise associated with it. A further problem is that the dark current is not the same everywhere in the device because of spatial nonuniformities. Thus if the charge pattern is stationary while temporarily in storage in the device, these nonuniformities show in the analog output signal. At $300°K$ in a well-made CCD the dark current may produce a full packet in about 1 sec but considerably less time is required in high-generation regions associated with wafer nonuniformities. Thus in an analog delay line the maximum reasonable delay time is perhaps $10^{-1}$ sec.

A Bell "Picturephone" video signal has a bandwidth of 0.5 MHz and with a CCD of 500 elements can be delayed up to 242 $\mu$s (2 lines) without significant degradation of spatial resolution or gray-scale discrimination. With a CCD array of 106 × 128 elements a delay of half of a "Picturephone" display was possible for one complete frame, namely 16.66 msec. This represents a considerable delay-bandwidth capability at low cost.[106] The clock frequencies of CCDs are, at present, limited to below 50 MHz in normal designs, but by field-aided charge transfer there is a possibility that this rate may be raised in the next few years.

Let us leave delay lines now and consider sampled data filters. These data filters may be either of the recursive form or the transversal form. In recursive filters, feedback paths are connected between delay stages and the input. In transversal filters the delay stages are tapped, and weighted portions of the delay signal are summed as the output. CCDs are suitable for both filter forms and have significant cost advantages relative to previous approaches.

A transversal filter is shown in Fig. 10.21(a). The delay elements $D$ are tapped and the voltages $V_1 \cdots V_M$ are weighted by the elements $h_1 \cdots h_M$ before summation. In this illustration the unit input pulse is converted to the form shown in Fig. 10.21(b) by the values set for the weighting functions. A CCD transversal filter, designed to respond to a particular signal waveform, is known as a matched filter and will detect this waveform in the presence of substantial noise. If a bandpass filter is required, the impulse response is chosen to be the Fourier transform of the desired bandpass characteristic.

The charge under a CCD electrode at a tapping point must be measured nondestructively and this is done by integrating the current that flows in the clock line during charge transfer.

In order to weight each sampled charge with an arbitrary coefficient $(-1 < h_k < +1)$, each $\phi_3$ electrode is split, as shown in Fig. 10.21(c). The upper portions of each $\phi_3$ electrode are connected together in a common clock line $(\phi_3^{(+)})$, and the lower portions are connected together in a common clock line $(\phi_3^{(-)})$. Identical voltage waveforms are applied to $\phi_3^{(+)}$ and $\phi_3^{(-)}$, but the currents in the two lines are measured separately, and the difference is applied to an external differential amplifier. With the electrode split so that a fraction $0.5(1+h)$ is connected to $\phi_3^{(+)}$ and $0.5(1-h)$ is connected to $\phi_3^{(-)}$, the differential amplifier output is proportional to $h$. If the device is operated with a zero-signal charge, i.e., halfway between the maximum charge packet and the background "fat-zero" charge, the plus and minus range needed for $h$ is readily obtained. A gap midway across the electrode then represents a zero-weighting factor and gaps in the upper and lower halves as in Fig. 10.21(d) correspond to negative and positive weightings, respectively.

The Nyquist sampling theorem requires that a signal having a bandwidth, $W$,

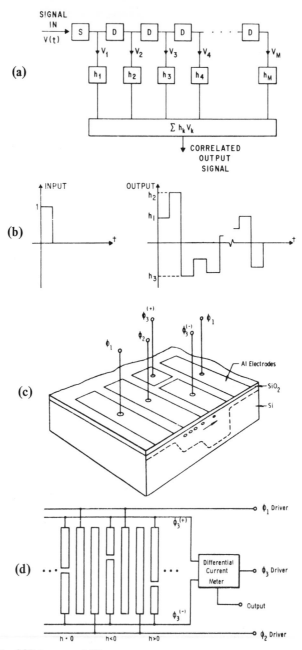

Fig. 10. 21. CCD transversal filter.

(a) Block diagram showning taps of weight $h$ between the delay states $D$.

(b) Response to a unit input has a form determined by the $n$ values.

(c), (d) Electrode-partitioning techniques that give the weighting factors required. (After Buss, et. al. Reprinted with permission from *IEEE J. Solid-State Circuits,* SC-8, 1973, pp. 139-140.)

be sampled at a frequency greater than $2W$. Thus if the CCD clock frequency is 20 MHz the signal bandwidth is limited to 10 MHz. If the charge-transfer inefficiency per stage is $\epsilon$ and the number of delay stages in the filter is $n$, then from calculations (or Fig. 10.12) the limit on $n$ is set by

$$n\epsilon < 2 \qquad (10.4)$$

if the matched filter performance is to be acceptable. So for a bandwidth, $W$, the time duration of the signal that can be processed, $T_d$, is given by

$$T_d W < \frac{1}{\epsilon} \qquad (10.5)$$

Hence if $W$ is 10 MHz and $\epsilon$ is $10^{-3}$, $T_d$ must be less than 100 $\mu$s.

The accuracy with which weighting coefficients can be achieved with the effective geometry and design rules, sets some limits to the precision of CCD filters. In matched filter applications the effect is to add noise to an already high-noise background and slightly increase the probability of nondetection or false recognition.

A CCD performance for a narrow bandpass filter is illustrated in Fig. 10.22. The out-of-band rejection is 29 dB and the $\pm$ 3-dB bandwidth is 110 kHz at a center frequency of 2.5 MHz. There were 101 stages and the center frequency was made a quarter of the clock frequency. The impulse response required for this filter is shown in Fig. 10.22(b). This filter was not particularly sensitive to charge-transfer inefficiency. For instance, a loss per stage of $10^{-3}$ reduced the out-of-band rejection by less than 0.5 dB.

The advantages of such CCD filters are: a) the low-cost performance; b) the tuning that is possible by changing the clock frequency; c) the reasonable signal-to-noise performance; and d) the possibility of exercising phase as well as amplitude control.

Recursive bandpass filters can be realized with high-$Q$ narrow passband response and the out-of-band rejection is not limited by weighting coefficient errors. On the other hand, amplifiers in the feedback paths must have precisely controlled gains and the design of some forms of frequency characteristics presents difficulties. Recursive filter applications have been described by Mattern and Lampe;[70] and time sharing of recursive filters has been discussed by Sealer and Tompsett.[89]

CCD filters can be used to perform many transforms, including frequently used ones, such as Fourier and Hilbert transforms. A transform application has been described that uses the chirp-Z transform (CZT) algorithm[23] for computing the discrete Fourier transform. The standard Fourier expression is

$$X_k = \sum_{n=0}^{N-1} x_n e^{(-i2\pi nk/N)} \qquad (10.6)$$

**(a)**

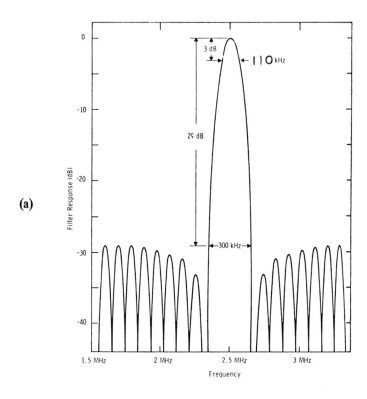

**(b)**

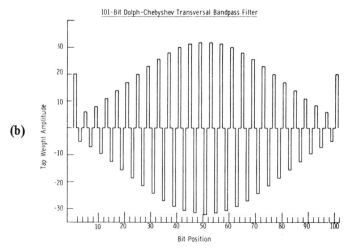

Fig. 10.22. Transversal CCD bandpass filter.
(a) Frequency characteristic (101 stage, CCD filter clocked at 10 MHz).
(b) Impulse response that gives this characteristic. (After Buss, et al. Reprinted with permission from *IEEE J. Solid-State Circuits,* SC-8, 1973, pp. 144.)

and the chirp-Z form, obtained by substituting $k^2 + n^2 - (k-n)^2$ for $2nk$, is

$$X_k = e^{-i\pi k^2/N} \sum_{n=0}^{N-1} (x_n e^{-i\pi n^2/N}) e^{i\pi(k-n)^2/N} \tag{10.7}$$

The algorithm has three steps: 1) premultiplication of the $x_n$ by a complex chirp waveform; 2) convolution in a filter having a complex chirp impulse response; and 3) postmultiplication by a complex chirp waveform.

The CZT algorithm is convenient for CCD implementation because the bulk of the computation is performed in a fixed-weighting-coefficient transversal filter. When only the power-density spectrum is desired, the postmultiplication is replaced by a squaring operation. The resulting block diagram is shown in Fig. 10.23. This was implemented with CCDs of 500 stages with two surface-channel circuits on a $160 \times 100$ mil$^2$ ($0.406 \times 0.254$ cm$^2$) chip. A simple electrode-weighting technique was used for the transversal filters. The charge-transfer efficiency was 99.98% for clock frequencies from 1 kHz to 3 MHz. The chirp generation premultiplication and output squaring were performed off the chip in the first design but complete integration should be possible. In one use of the filter, the Fourier analysis power spectrum of a complex wave was displayed as impulses spaced 9 kHz apart along a scale of frequency from 50 to 100 kHz.

The possibility exists of electrically controlling the weighting functions of transversal CCD filters. This concept leads to charge-transfer device correlators.[102] These may be thought of as programmable transversal filters or as peripheral devices capable of performing hundreds of thousands of multiply-add operations per microsecond. In combination with a microprocessor, a CCD correlator of 128 modules (each of 32 analog samples and 32 binary control signals) at a clock frequency of 3 MHz is capable of performing the equivalent of several thousand 8-bit multiply-add operations per microsecond. Matrix inversion and tomographic reconstruction operations are some of the processing applications that lie ahead for this class of circuitry.

## 10.4 CHARGE-COUPLED MEMORIES

CCD memories are of interest because the cost per bit is low (lower than 0.01 cent per bit is projected) and the access time is good. The cost is low for two reasons: a) the bit size is small—about 200 $(\mu m)^2$ ($1/3$ mil$^2$) with present layout rules, and the possibility exists of reducing this further; and b) the number of contact openings (dig-downs) is small and therefore yields should be high. However, $n$ channel MOS FET random-access-memory technology is now well entrenched and CCD memories, which have line readouts, mainly impact computer memory systems by filling the access time gap illustrated in Fig. 10.24.

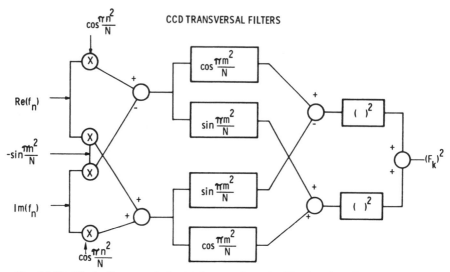

Fig. 10.23. Block diagram of the implementation of a Fourier chirp-Z transform using CCDs. (After Brodersen, et al. Reprinted with permission from *IEEE Int. Solid-State Conference Digest*, 1975.)

Two ways of organizing a CCD memory are presented in Fig. 10.25. The serpentine arrangement in Fig. 10.25(a) is wasteful of power since each bit passes through each storage site in the loop, and must do so at the same high frequency needed for reasonably fast access. In a parallel ac arrangement a reasonable access speed is obtainable at a lower clocking frequency and power is saved. This leads to the series-parallel-series arrangement of Fig. 10.25(b). Here the data stream is introduced into the series high-speed upper register, then transferred to the vertical parallel-channel middle register where the clocking frequency is lower, and extracted from the series lower register.

The organization of an early 9216-bit CCD memory for terminal buffering, display refresh, and microprocessor data storage applications is a 1024-word by 9-bit configuration with the 9-bit bytes handled in byte-serial mode. The serpentine-loop buried-channel structure has nine shift registers each of 1024 bits with refresh turnaround cells every 128 bits. For a clock frequency of 2.5 MHz this gives a worst case access time to a bit of 330 μs and an average access time of 165 μs.

A larger memory aimed at block-oriented cache-buffering, and memory-swapping actions, is structured in a line-addressable format that gives pseudo-random access as shown in Fig. 10.26. The access time, of course, depends on the number of elements in the line. In the 16,284-bit memory under discussion this is 128 bits per line. At a 5-MHz data rate the memory has a worst case access time of 25.6 μs per bit and an average access time of 12.8 μs. Since only one line

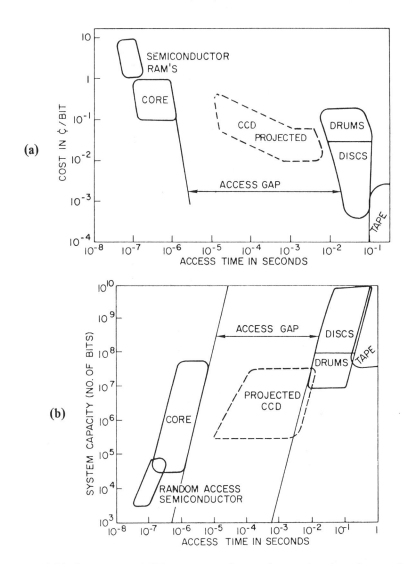

Fig. 10.24. Comparison of different types of memories as a function of access time with respect to: (a) cost per bit, and (b) total storage capacity used in present-day computer systems. (After D.F. Barbe, private communication.)

is operative at any time, the power dissipation is low, in operation 200 mW or in the standby mode only 50 mW.

A 64-kbit (65,536 bits) block-addressed charge-coupled serial-memory layout is shown in Fig. 10.27.

The memory chip is organized as 64 k words by 1 bit in 16 blocks of 4 kbits.

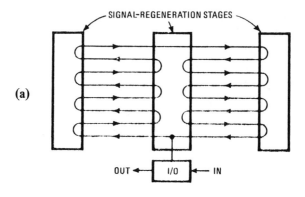

**(a)**

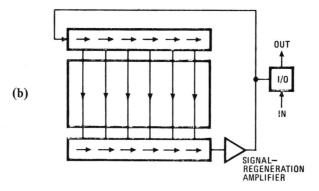

**(b)**

Fig. 10.25. CCD memory organization.
(a)   Serpentine array is wasteful of power.
(b)   Series-parallel-series arrangement has slower clocking rate in the vertical-parallel chan-
      nels and is a good compromise between power access time and bit density. (After
      Altman. Reprinted from *Electronics*, Aug. 8, 1974; © McGraw-Hill, Inc., N.Y., 1974.)

Each 4-kbit block is organized as a serial-parallel-serial (SPS) array. The chip
is fully decoded with write/recirculate control and two-dimensional decoding
to permit memory matrix organization with $X$-$Y$ chip select control. All
inputs and the output are TTL compatible. Operated at a data rate of 1 MHz,
the mean access time is about 2 ms and the average power dissipation is
1 $\mu$W/bit. The maximum output data rate is 10 MHz, giving a mean access
time of about 200 $\mu$s, and an average power dissipation of 10 $\mu$W/bit. The
memory chip is fabricated using an $n$-channel polysilicon gate process. Using
tolerant design rules (8-$\mu$m minimum feature size and ± 2-$\mu$m alignment
tolerance) the CCD cell size is 0.4 mil$^2$ and the total chip size is 218 × 235
mil$^2$.

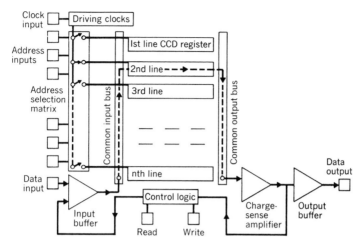

Fig. 10.26. The organization of a CCD line-accessed 16K memory with 128 bits per line and an average bit access time of 12.8 $\mu$s for a data rate of 5 MHz. (Fairchild) (After Allan. Reprinted with permission from *IEEE Spectrum*, Aug. 1975, p. 40.)

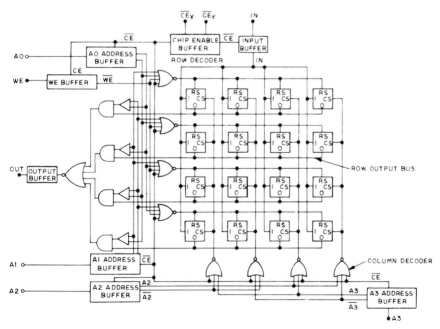

Fig. 10.27. Block diagram of a 64-kbit CCD memory chip. (After Mohsen, et al. Reprinted with permission from *IEEE J. Solid-State Circuits*, **SC-11**, 1976, p. 51.)

Some predictions of future bit densities and speeds for CCD memories are good but some problems remain.[118] The present status (1978) is that CCD memories are available at 2 to 4 times the bit content per chip and at about half to a quarter the price of $n$ MOS RAMs. CCD memories have been assembled into stores of more than a million bits on memory boards of size 30 X 38 cm (12 X 15 in.). The 1980 market for 64 kbit CCD memories is expected to be about 10 million pieces at about $10 a piece. Developments in the direction of 256 kbit CCD memories are also expected by then.

In CCD registers the flow of information is primarily in a one-dimensional fashion. Clearly, it would be desirable to have a two-dimensional array in which the flow could be from any cell to any of its four nearest neighbors. This would allow mass serial-to-parallel conversion, interlace, mixing, and passing maneuvers. Even more limited degrees of freedom would be improvements over the linear flow structures presently used. Some possibilities have been considered by Séquin. Figure 10.28 presents structures that would allow flow in two directions. Only time will tell whether layouts like these will prove to be practical.

The volatility of information stored in CCDs can be prevented by the addition of metal-nitride-oxide-silicon capacitors to the integrated circuit as suggested in Fig. 10.29. There is a factor of two loss in packing density involved, since the MNOS capacitor is alongside the CCD capacitor. Some problems of charge-transfer efficiency and reading and writing are also involved.

## 10.5 IMAGING CCD ARRAYS

CCDs have a bright future in the field of solid-state imaging. Full TV line-scan arrays have already been constructed for special-purpose TV cameras. (The quality is not yet suitable for normal TV studio use.)

The special features that make CCD image arrays attractive include:

- The serial readout provides the desired electrical scanning action in a most convenient way since no electron-beam scanning or $X$-$Y$ addressing is needed.
- The sensitivity is about 500 $\mu$A/lm. The best CCDs considerably exceed the performance of vidicons in quantum efficiency and signal-to-noise ratio.
- The sensitivity can be extended down to wavelengths of 0.9 $\mu$m for Si or 1.6 $\mu$m for Ge CCDs.
- The dynamic range is good and the dark current low so that low-light levels can be handled with buried-channel CCDs.
- The signal-to-noise ratio is good. This is, in part, due to the low-output capacitance, 1 pF for the outout diode, which gives a good signal-to-noise ratio assuming the thermal noise in the input resistor of the preamplifier is limiting.

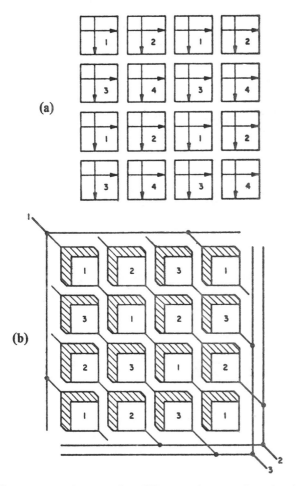

Fig. 10.28. Extension of the two-phase CCD principle to two dimensions leading to: (a) four separate electrodes in each unit cell, and (b) reduction to three independent electrode systems. (After Séquin. Reprinted with permission from *IEEE J. Solid-State Circuits*, SC-9, 1974, p. 135.)

- The transfer efficiencies of recent designs ($\epsilon < 10^{-4}$) are adequate for TV resolution.
- Because of the good packing density and fabrication simplicity (few contact dig-downs) the yield of good CCD arrays is better than for $SiO_2$ array-image surfaces. The cost is therefore lower and the picture quality is less subject to defects.

Early forms of CCD image sensors were linear single-line arrays in which the second dimension was obtained by mechanical means such as a rotating mirror

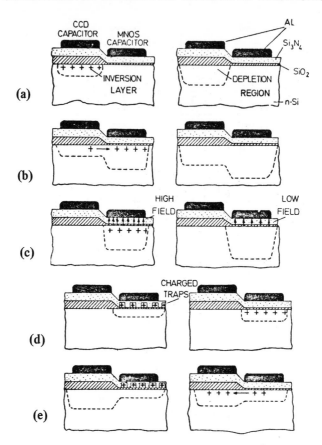

Fig. 10.29. A CCD with MNOS capacitors for memory retention. Two starting conditions are considered: left column, there is an inversion layer below the CCD electrode; right column, there is no inversion layer.
(a)   Starting point.
(b)   The charge is transferred from the CCD to the MNOS capacitors.
(c)   Due to the charge, the electrical field is high in the gate insulator; without charge it is low.
(d)   For readout, an inversion layer is only induced if the traps are neutral.
(e)   The charge is transferred back from the MNOS capacitors to the CCD. (After Goser and Knauer. Reprinted with permission from *IEEE J. Solid-State Circuits,* **SC-9,** 1974, p. 148.)

that scanned the two-dimensional image across the CCD row. But these early forms of CCD image sensors have tended to be replaced by area-image sensors.

The organization of a CCD image array requires that there be an integration period during which light is creating charge packets and that this charge is then moved out rapidly. If charge packets are moved slowly across a continuously illuminated area, integration will be continuing and the image will be smeared.

Two methods of organization are shown in Fig. 10.30(a) and (b). In Fig. 10.30(a) the complete picture from the imaging area is transported rapidly to a comparable size storage area for line-by-line readout during the time the next frame is integrating. The storage area represents a substantial loss of imaging area assuming the chip size is limited, but the address conditions of this organization are fairly simple.

In Fig. 10.30(b) a line-addressed structure is shown. This is a structure that

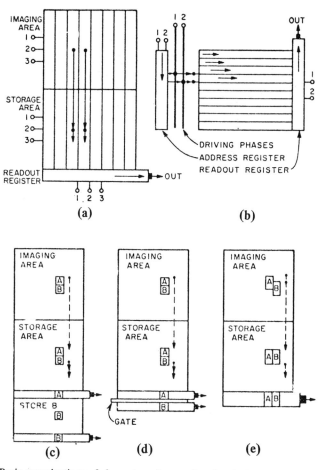

(a)

(b)

(c)        (d)        (e)

Fig. 10.30. Basic organizations of charge-transfer-area imaging devices.
(a)  Frame transfer.
(b)  Line-addressed.
(c)  2:1 interlaced readout with extra store.
(d)  Reduced-signal readout for interlacing.
(e)  Mixed-signal readout for interlacing. (After Séquin. Reprinted with permission from *IEEE Trans. Electron Devices*, **ED-20**, 1973, pp. 535, 536.)

lends itself naturally to line interlacing. In order to reduce flicker in the picture in commercial TV (and in the Bell "Picturephone" system), the information is presented as two interlaced fields, each with half the total number of horizontal scan lines. Unless TV receivers and system concepts are to be modified, it is desirable that CCD cameras conform in their information handling. Since the frame-transfer concept produces a picture that is freer of smearing than the line-addressed concept, it is necessary to consider how frame handling can be adapted to provide interlacing. One method is shown in Fig. 10.30(c) where a half-size store B has been added to collect every other line until they can be read out by a second horizontal register and transmitted as the second field. Since the total device area is now more than 2½ times the imaging section area, this is not a particularly acceptable solution.

A step towards a more attractive arrangement is shown in Fig. 10.30(d). Here the serial readout section consists of two identical CCDs. In two steps a pair of lines is placed into the two readout CCDs: line A is read out and line B is discarded. Thus in one field the odd lines are transmitted and the even lines lost and in the other field the inverse occurs. This is also not a very acceptable solution because half the signal is wasted and the picture quality is degraded except in respect to flicker.

Some loss of signal can be avoided if the outputs of lines A and B are mixed as shown in Fig. 10.30(e). Interlocking is then achieved by using different combinations of line pairs for the two fields. The concepts of such mixing have been described by Séquin who shows that the limiting resolution in the vertical direction is acceptable, but somewhat reduced by the procedure.

A 2-MHz, 220-by-256 cell, three-level polysilicon three-phase CCD imager has been used for Bell "Picturephone" studies.[93] The layout of the chip is shown schematically in Fig. 10.31. In Fig. 10.31 the cell length in the image area (vertical dimension) is 48 $\mu$m and since each cell is a triplet (for the three-phase system), the active electrode length is 16 $\mu$m. In the storage region the electrode length is reduced to 10 $\mu$m. The vertical-transfer channels are spaced 30 $\mu$m apart and separated by 5-$\mu$m-wide channel-stop diffusions. The electrode length in the readout sections is therefore 10 $\mu$m. The diffused bus line (IOD) in conjunction with the input gates (PG1 and PG2) and the overflow drains (OV) inject background charge to improve the efficiency of the vertical-transfer channels. The serial-register output gate (OG) and the output diode (OD) are connected to the gate of the IGFET amplifier (AS,AD). After signal sensing, the charge packet is discharged in the diode (DD) through a dump gate (DG). (The components IG, ID, and CG are for test purposes and the diodes (BD) are to provide busbar isolation.)

The use of polysilicon electrodes (4000 Å thick) allows the imaging region to be illuminated from the front with only 20% loss of light intensity in the yellow region of the spectrum. The dark currents were of the general level 15-150 nA/cm$^2$ but at least an order of magnitude reduction has been observed

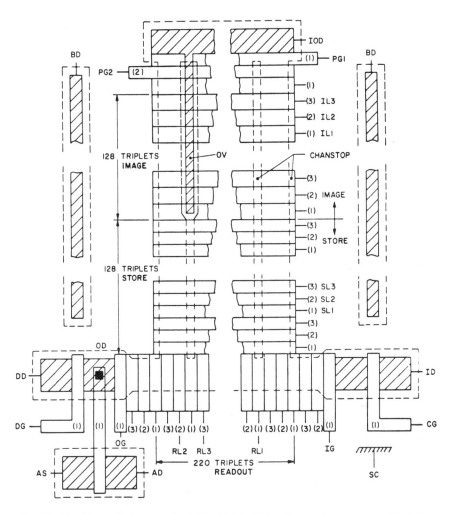

Fig. 10. 31. Schematic layout of a CCD, 220-by-256 cells, area-image sensor. Diffusions are shown hatched. Electrode levels are indicated by number in parentheses (1), (2), and (3). See text for symbolic abbreviations. (After Séquin, et al. Reprinted with permission from *IEEE Trans. Electron Devices,* **ED-21**, 1974, p. 714.)

in later devices. The quantum efficiency under illumination (i.e., electrons per photon) was estimated to be about 25%. The limiting resolution was about 100-line pairs per picture frame and matched the present resolution of the "Picturephone" system. The supporting circuitry included 17 TTL packages for the driving logic and discrete transistors for the pulse generation with a total power demand of about 1.5 W. The camera, complete with batteries, measured about 13 X 8 X 7 cm.

CCD chips with full 525-line (U.S.A.) TV resolution have since been produced by several laboratories. One model has a 256-by-330 element image area, a 256-by-330 storage area, and a 330-stage output register operating at the video data rate of 6 MHz. Interlacing is used to provide the required 512-by-330 picture elements per frame.

In the presence of optical overloads the phenomenon of "blooming" occurs in CCDs (and other semiconductor imaging devices) unless measures are taken to prevent the spreading of overload-generated charge into adjacent regions. Figure 10.32(a) shows the presence of blooming drains in the channel-stop regions of the image area of a chip. Figures 10.32(b) and (c) illustrate a further development in which polysilicon field-shield stop gates are used. The potential well is two-dimensional—having four sides to it. The dashed line shows the surface potential of the transfer gates adjusted to prevent the blooming of charge along the channel. The potential of the blooming barrier set by the polysilicon electrode is slightly lower than the barrier set by the transfer gate. Thus the excess charge will spill from the full well into the $n^+$ blooming bus before it can spill down the channel. Other antiblooming concepts involving ion-inplanted channel regions and stepped-oxide channel-step regions have been devised.[60]

Sophisticated uses of CCD sensors are presently being explored. In the military field infrared image systems are being studied and Ge-based CCDs have been fabricated. Experiments have also been conducted on background subtraction, multiple readout, motion detection, and other video processing achieved by recycling of signals through the sensor.[111]

Charge-injection imaging devices are now considered. In contrast to charge-coupled devices, in which the signal charge is transferred to the edge of an array for sensing, the charge-injection device (CID) approach confines this charge to an image site during sensing. Site addressing may even be by an $X$-$Y$ coincident-voltage technique, not unlike that used in digital memory structures although this is not usual. In early structures, readout was obtained by injecting the charge from individual sites into the substrate and detecting the resultant displacement current, but more recently higher speed has been obtained by a parallel-injection technique that separates the sensing and injection steps.

The organization of a $4 \times 4$ array is illustrated in Fig. 10.33. Each sensing site consists of two MOS capacitors with their surface-inversion regions coupled so that charge can be transferred between them. The level of signal charge at each sensing site is detected by intracell transfer during a line scan and, during the line retrace interval, all charge in the selected line can be injected.

The action from Burke and Michon is as follows.[26]

The voltage applied to the row electrodes is larger than that applied to the column electrodes to prevent the signal charge stored at unaddressed locations from affecting the column lines. At the beginning of a line scan, all rows have voltage applied and the column lines are reset to a reference voltage $V_S$ by means of switches $S_1$ through $S_4$, and then allowed to float.

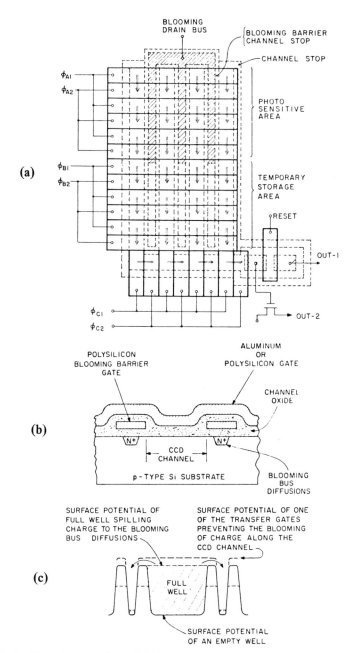

Fig. 10. 32. Blooming control in a CCD image array.
(a) A two-phase frame-transfer charge-coupled imager with blooming-control area.
(b) Cross-sectional view perpendicular to charge flow of static blooming-control structure using polysilicon field-shield gates and resulting surface potential.
(c) As shown by the dotted lines, the blooming-barrier potential determines the size of the well. (After Kosonocky, et al. Reprinted from *RCA Rev.*, **35**, 1974, pp. 5, 7.)

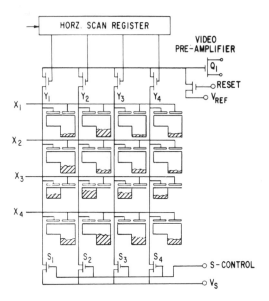

Fig. 10.33. Schematic diagram of a 4 × 4 CID array designed for parallel-injection readout. Silicon-surface potentials and signal-charge locations are included. (After Burke and Michon. Reprinted with permission from *IEEE J. Solid-State Circuits*, SC-11, 1976, p. 122.)

Voltage is removed from the row selected for readout ($X_3$ in Fig. 10.33) causing the signal charge at all sites of that row to transfer to the column electrodes. The voltage on each floating column line then changes by an amount equal to the signal charge divided by the column capacitance. The horizontal scanning register is then operated to scan all column voltages and deliver the video signal to the on-chip preamplifier $Q_1$. The input voltage to $Q_1$ is reset to a reference level prior to each step of the horizontal scan register.

At the end of each line scan all charge in the selected row can be injected simultaneously by driving all column voltages to zero through switches $S_1$ through $S_4$. Alternatively, the injection operation can be omitted and voltage reapplied to the row after readout, causing the signal charge to transfer back under the row electrodes. This action retains the signal charge and constitutes a nondestructive readout operation.

The parallel-injection approach permits high-speed readout for TV-scan formats. A 244-line by 248-element imager, employing this technique and including an on-chip preamplifier, has been designed, fabricated, and evaluated in both the normal and nondestructive readout modes. At 200°K with a light level that corresponded to $6 \times 10^5$ carriers per site the imager was read out more than $3 \times 10^5$ times (3 hr at 30 frames/sec) without significant image degradation. This represents a loss of less than one carrier per site per readout.

The charge-injection approach to image sensing, as for CCDs, is relatively compatible with standard MOS processing; a large fraction of the Si surface can be used for the photon-charge generation; the dark current is low and uniform; and the device is resistant to blooming effects and crosstalk, since each sensing site is electrically isolated from its neighbors. The nondestructive readout operation and the random-image-site addressing possibilities are interesting options available to the CID system designer.

The low-light-level-image resolution of Si CCDs is better than that of Si vidicon tubes as shown by the dashed lines in Fig. 10.34.

The applications of CCDs to infrared image sensing may involve the use of standard CCD structures with the substrate being a narrow bandgap semiconductor sensitive to IR radiation. However, an extrinsic semiconductor with deep impurities sensitive to the infrared may be operated in an accumulation mode. This is possible if the base semiconductor (Si) is semi-insulating or has been made so by freezeout of the shallow dopants at low temperatures. The gate voltages are then used to induce potential wells into which "majority" carriers may be introduced from impurity levels in the semiconductor by IR photon absorption.

Hybrid structures which involve coupling of any one of various types of

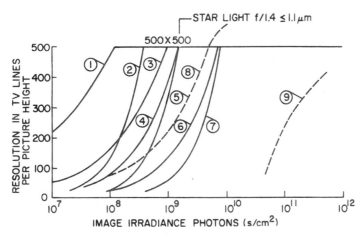

Fig. 10.34. Theoretically observable resolution in TV lines per picture height versus image irradiance for a $500 \times 500$ imaging CCD with element spacing of 25 $\mu$m, frame time of 0.1 sec, quantum efficiency of 1.0, contrast of 0.2, and signal-to-noise ratio $k$. The various curves show the limits imposed by the various noise sources: (1) incomplete transfer of signal, $N_S = 625$, $\epsilon N_g = 0.2$, $k = 1$; (2) MOSFET noise, $\bar{N}_n = 10$, $k = 5$; (3) photoelectron shot noise, $k = 5$; (4) transfer noise of electrically introduced bias charge, $N_b = 10^5$. $\epsilon N_g = 0.2$, $k = 1$; (5) storage noise of electrically introduced bias-charge noise, $\bar{N}_n = 40$, $k = 5$; (6) fast interface trapping noise, $\bar{N}_n = 950$, $k = 1$; (7) thermally generated bias-charge noise, $\bar{N}_n = 200$, $k = 5$; (8) measured resolution for 40 mm I-SIT tube; (9) measured resolution for silicon vidicon. (After Carnes and Kosonocky. Reprinted from *RCA Rev.*, **33**, 1972, p. 607.)

IR photodetectors to a Si CCD shift register are also of considerable interest. These hybrid structures allow the functions of detection and signal processing to be performed in separate regions of an integrated structure.

## 10.6 CCD LOGIC STRUCTURES

Since CCD structures have excellent pipeline-type memory functions it is appropriate to consider the classes of operation in which CCD logic gates may be used on the same chip.

Performing digital-logic functions with the charge-transfer principle requires interaction with the information contained in charge-coupled shift registers. Such interaction can be accomplished in two ways. One way is termed bit-destructive because the original bits lose their individual identities in the process. The other way, designated bit-preserving, detects the presence or absence of charge without disturbing the bit stream. This detection controls charge flow in another register.

Some CCD logic gates are illustrated in Fig. 10.35. In the AND gate of Fig. 10.35(a) the two shift registers $A$ and $B$ are connected to two series-transfer gates. Both gates must be on for the minority charge carrier to pass from the source to the CCD gate marked $C$. In Fig. 10.35(b) the controlled transfer gates are in parallel and the OR function is performed. The controlling bit patterns are not disturbed but merely sensed nondestructively and used to control other registers.

The other gates shown in Fig. 10.35, however, are bit-destructive. In Fig. 10.35(c) shift registers $A$ and $B$ dump their charge packets into well $C$ and any surplus flows into $D$ if the barrier control allows this. If only $A$ or $B$ is full of charge, $C$ will be filled exactly and $D$ will remain empty. So $C$ represents an OR function and $D$ an AND function.

For further development the well $D$ may be provided with a charge-sensing element and drive one of two parallel-transfer gates that control the flow of charge from gate $C$. The charge-sensing element allows the charge under gate $C$ to flow to gate $E$ only if there is no charge under gate $D$. The purpose of transfer gate $F$ is to clear out charge under $C$ if it has not been moved under $E$. Not only does $E$ provide an exclusive-OR function and $D$ an AND function, but the two gates together provide a complete 2-bit half-adder with a sum and a carry. This half-adder circuit can be expanded to a full-adder with additional gates.

Table 10.3 shows a comparison of CCD logic potential relative to that of other 1980 technologies.

## 10.7 BUCKET-BRIGADE CIRCUITS

Charge-coupled device concepts stem from bucket-brigade circuits as discussed in Section 10.1, although CCDs represent a considerable step onwards in

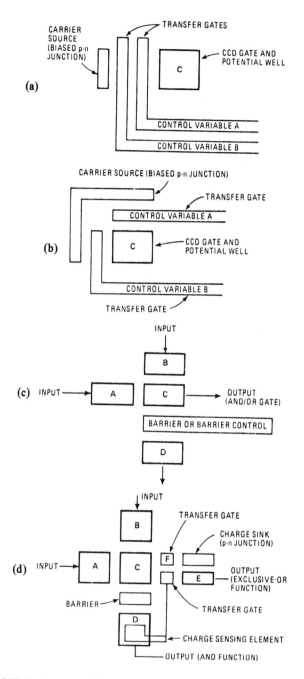

Fig. 10.35. CCD logic gate configurations. Arranging transfer gates, carrier source, and potential wells provides the following logic functions: (a) AND gate, (b) OR gate, (c) AND/OR gate, and (d) exclusive-OR gate where exclusive-OR and AND functions are realized simultaneously. (After Zimmerman and Barbe. Reprinted from *Electronics*, Mar. 31, 1977; © McGraw-Hill, Inc., N.Y., 1977.)

633

## Table 10.3. Comparison of CCD and Other Technologies

| Technology (1980) | Logic Gate | | Shift Register Stage | | | | Chip Size |
| | Speed-Power (pJ) | Area (mil$^2$) | Power/bit @ 1 MHz ($\mu$W) | Area/bit (mil$^2$) | Max. Speed (MHz) | Type of Logic | (20% Yield) (mil$^2$) |
|---|---|---|---|---|---|---|---|
| Charge-coupled-device | 0.2 | 2 | 1 | 0.5 | 50 | dynamic | 250 |
| Triple-diffused VLSI | 3 | 10 | 100 | 30 | 100 | static | 400 |
| Integrated injection logic | 2 | 10 | 50 | 30 | 5 | static | 200 |
| Silicon on sapphire CMOS | 5 | 20 | 100 | 75 | 50 | static | 175 |

Conclusions
1. For mix of repetitive 80% logic + 20% pipeline memory, CCD advantage at 5 MHz is
   power    ×18
   area     ×16
   compared to next best technology (I$^2$ L).

2. Interconnectibility (power, area) degrades more rapidly for CCD than any other technology as circuitry departs from pure arrays to pure random networks.

(After Zimmerman and Barbe, Reprinted from Electronics, March 31, 1977; Copyright © McGraw-Hill, Inc., 1977.)

integration. CCD technology now appears capable of doing everything that BBCs can do, with a greater transfer efficiency, a greater packing density, and higher yields. The BBC articles listed in the further reading section are therefore mainly of interest from a historical development viewpoint. However, at the time valuable experience was obtained with bucket-brigade circuits that hastened the CCD development programs. For instance, 32 × 44 image arrays were constructed and other circuit concepts such as delay lines and filters developed.

### 10.8 QUESTIONS

1. A three-phase Si gate CCD as shown in Fig. 10.2 has an element length of 30 $\mu$m and a width of 200 $\mu$m. The SiO$_2$ oxide thickness is 1500 A and the Si is $p$-type 10 $\Omega$-cm. For a 10-V amplitude pulse, what is the signal charge? The transfer inefficiency is $10^{-4}$ at 10 V, $2 \times 10^{-4}$ at 5 V, and $5 \times 10^{-4}$ at 2 V for a drive frequency $10^6$ Hz. What may be inferred?

2. A Si $p$-channel CCD has an element length of 20 $\mu$m and a width 200 $\mu$m. The oxide thickness is 1500 A and the $n$ substrate is 10 $\Omega$-cm. It is a three-phase undercut-isolated CCD structure. With a 10-V amplitude pulse the transfer inefficiency is $4 \times 10^{-4}$ for a drive frequency of $10^5$ Hz, $5 \times 10^{-4}$ at $10^6$ Hz, $10^{-3}$ at $2 \times 10^6$ Hz, and $2 \times 10^{-3}$ at $4 \times 10^6$ Hz. Can this be satisfactorily explained? At the $10^6$ Hz drive frequency what will be the signal after 400 transfers? How does this depend on signal frequency?

3. A Si CCD has a two-phase drive frequency of 1 MHz and the transfer inefficiency measured conventionally is $10^{-3}$ at this frequency. A signal pulse of 10 V is inserted in the CCD after it has been idle for some time. What is the output waveform observed after 100 transfers? If a block of five pulses as shown in Fig. 10.13(a) is inserted in an idle CCD what is the output waveform observed after 100 transfers?

4. For the Si CCD of Problem 3, the signal inserted is five consecutive pulses ON followed by five pulses OFF and this is repeated constantly. What then is the output waveform after equilibrium has been reached in an array 100 transfers long? If the signal then switches to a cycle of four ON pulses and six OFF pulses what are the waveforms for the next few cycles?

5. A surface-channel Si CCD with electron-charge packets has a loss performance that is dominated by interface states between the $SiO_2$ and the Si. The density of the states is $10^{11}$ cm$^{-2}$ and they have a capture cross section of $10^{-14}$ cm$^{-2}$ for electrons and a depth below the conduction band edge of 0.3 eV. The $p$-type resistivity is 10 $\Omega$-cm, the element dimensions are 20 $\mu$m, and the oxide thickness is 1500 A. If all the interface states are occupied by electrons what is the time for 50% of them to empty at 300°K and appear as spurious charge in otherwise empty pulse packets? If the clock frequency is made equal to this time, what will be the apparent charge-transfer inefficiency for a signal pulse of 5-V amplitude? How will this transfer inefficiency vary with signal amplitudes between 2 and 10 V? If the signal is a continuous sequence of two 5-V pulses ON and two pulses OFF, what will be the output waveform after 200 transfers?

6. Consider a bulk-(buried) channel CCD as shown in Fig. 10.18. The $n$ layer is 5 $\mu$m deep and uniformly doped at $2 \times 10^{15}$ cm$^{-3}$ on a $p$ substrate of $2 \times 10^{14}$ cm$^{-3}$. What gate voltage would be needed to create a reasonable potential well? If the element area is 10 $\mu$m $\times$ 200 $\mu$m what charge can be stored in the well?

7. In a bulk-channel CCD the $n$ surface layer is uniformly doped at $10^{15}$ cm$^{-3}$ and is 4 $\mu$m thick on a $p$ substrate doped at $4 \times 10^{14}$ cm$^{-3}$. The element area is 10 $\mu$m $\times$ 200 $\mu$m. The pulse voltage is 10 V and the clock frequency is 1 MHz. A leading-edge deficit of 18% is observed in a signal after many zeros (300°K) and of about 5% in the trailing edge of a group of "ones." What is the transfer inefficiency? What is the bulk-electron trap density, assuming that the traps are situated at the center of the Si bandgap and have a capture cross section of $10^{-15}$ cm$^{-2}$? If a 20% fat-zero background charge is added, what then is the signal-charge-transfer inefficiency?

8. The 16 kbit CCD memory shown in Fig. 10.26 is stated to have a worst case access time of 25.6 $\mu$s per bit for a 5-MHz data rate. The power dissipation is stated to be 200 mW in operation or 50 mW in standby mode. Show that these numbers are reasonable. How could the memory be reorganized to reduce the access time to about a half? What effect would this have on the power dissipation?

9. Find the best closed-form solution model for a CCD and present and discuss some curves obtained from it and discuss its limitations.[66]

10. Present the most important CCD equations given by Berglund and Thornber[15] and discuss their significance.

11. Consider ion implantation with respect to CCD manufacture. How could ion-implant doping determine the directionality of flow?[63]

12. Review the relative merits of acoustic-wave delay lines (bulk or surface wave) and CCDs for use: (a) below 5 MHz and (b) above 5 MHz.

13. Discuss recent developments in buried-channel CCDs. Is the peristaltic charge-coupled device[44,45] significantly distinct from a normal buried-channel device?

14. How is the charge-carrying capacity of a buried-channel CCD estimated?[110]

15. Design three-phase and two-phase clock circuits for 1-10-MHz CCDs. Specify the actual components you would use.[38]

16. A one-phase clocking system for CCDs has been described by Melen and Meindl.[73] Discuss its merits.

17. Double-junction charge-coupled devices (DJCCD) and Schottky barrier CCDs have been described by Schuermeyer et al.[88] Discuss the merits of these concepts.

18. Consider the effects of interface states in CCDs as causes of transfer inefficiency and noise. Discuss the use of a fat-zero background charge.

19. In the study by Mohsen et al.,[76] mention is made of the advantages of push and drop clocks. Explain.

20. Discuss the role of thermal-charge generation in CCDs.[77,82]

21. Discuss fixed and proportional transfer in efficiency losses in CCD wells and the different effects they have on blocks of signal pulses.[22]

22. Review the sources of noise, intrinsic and extrinsic in a CCD system including drive circuits. How would one design to maximize signal-to-noise ratio?

23. Discuss the dark current density typical of surface and buried-channel CCDs.

24. Discuss bit-regeneration schemes in CCDs.[59,94]

25. A "racetrack" recirculating CCD is oval in shape and closed on itself so that it has no beginning and no end. Discuss input and output structures in such devices and the practical uses of racetrack CCDs.[65]

26. Discuss the sinusoidal frequency response of CCDs and derive the response expression $R(f) = \exp\left\{-n\epsilon\left[1 - \cos\left(2\pi f/f_0\right)\right]\right\}$.

27. Set down the relative merits of recursive and transversal CCD filters and give examples of each.

28. Design a CCD weighting pattern for a transversal filter that will give a bandpass characteristic for a range 1 to 10 kHz.[28]

29. Explore analytically the effect of weighting-coefficient errors caused by masking tolerances in CCD bandpass transversal filters. Assume design rules of 10 $\mu$m minimum feature and 3-$\mu$m edge resolution in a 100-stage filter of 200 $\mu$m width.

30. Claims have been made that amplitude and phase can be controlled somewhat independently in CCD filters. Discuss and show how it may be done with CCDs.

31. Review the application of CCD filters to Fourier transformation.

32. Discuss the purpose of Hilbert signal transformation and show how it may be done with CCDs.

33. Give some instances of matched filter uses that are realized with CCD transversal filter approaches.

34. In CCD matched filters it is claimed that $N\epsilon < 2$ is required for acceptable performance.[28] Consider a simple matched filter under reasonable signal-to-noise ratio conditions and judge the basis for this claim.

35. Discuss electrical methods of controlling tap weight in transversal CCD filters.[5,102]

36. Discuss the relative packing densities and costs of CCD and $n$-channel MOS memories.

37. Review the organization of CCD memories for fast access to bit lines, and to individual bits, with low-power dissipation.

38. For a CCD memory block Mohsen, Tompsett, and Séquin[78] predict a signal-to-noise ratio of 58 dB. See how close you can come to this figure by using some of their numbers.

39. Review nonvolatile CCD memory concepts.[46,113] Deal with such problems as transfer inefficiency, packing density, read write, write-inhibit efficiency, and ease of manufacture.

40. Discuss the minimum light-level sensitivity that may be expected for CCDs at room temperature.[78] Is 500 $\mu$A/lm a high or low number? What are normal classroom illumination levels?

41. In the video processing of CCD image signals, it is suggested[111] that recycling of signals back through the sensor can be used to achieve multiple imaging, background subtraction, and motion detection. Discuss such concepts.

42. In a frame-organized CCD imager, what must be the relationship in size between the photoactive area and the storage area?

43. Discuss frame-interfacing techniques for TV use of CCD image arrays.[91]

44. Blooming-control techniques are important in charge-coupled image arrays. Discuss and rate the known methods.[60]

**45.** Charge-injection imaging devices (CID) have several advantages over conventional CCD imaging structures. Discuss these but also review the disadvantages.

**46.** Accumulation-mode CCDs may be obtained in Si by operation at a temperature such as $4.2°K$ which is below the freezeout temperature of the shallow dopant.[81] Discuss how this mode might be used for IR imaging at wavelengths greater than 3 $\mu$m. What dopants would be used? Assume capture cross sections and make estimates of what sensitivity might be expected.

**47.** Design the layout of a charge-coupled half-adder, with a sum and carry, for a 5-MHz clock pulse. Use typical 5-$\mu$m design rules. How does the size compare with an integrated circuit MOS half-adder logic element made to similar design rules?

**48.** Bucket-brigade circuits can be made either of bipolar transistors or MOSFETS. How much faster is a bipolar array likely to be than a MOSFET using similar design rules? Compare also with CCDs.

## 10.9 REFERENCES AND FURTHER READING SUGGESTIONS

1. Abe, M., et al., "A CCD imager with $SiO_2$ exposed photosensor arrays," *IEEE Int. Electron Devices Meeting, Technical Digest,* Washington, D.C., 1977, p. 542.

2. Allan, R., "Semiconductor memories," *IEEE Spectrum,* **12**, Aug. 1975, p. 40.

3. Altman, L., "The new concept for memory and imaging: charge coupling," *Electronics,* June 21, 1971, p. 50.

4. Altman, L., "Bucket brigade devices pass from principle to prototype," *Electronics,* Feb. 29, 1972, p. 62.

5. Altman, L., "Charge-coupled devices move in on memories and analog signal processing," *Electronics,* Aug. 8, 1974, p. 91.

6. Amelio, G.F., M.F. Tompsett, and G.E. Smith, "Experimental verification of the charge coupled device concept," *Bell Syst. Tech. J.,* **49**, 1970, p. 593.

7. Amelio, G.F., W.J. Bertram, and M.F. Tompsett, "Charge-coupled imaging devices: design considerations," *IEEE Trans. Electron Devices,* **ED-18**, 1971, p. 986.

8. Amelio, G.F., "The impact of large CCD image sensing area arrays," *Proc. CCD 74 Int. Conf., Edinburgh,* 1974, p. 133.

9. Aoki, M., et al., "A 1024 element linear CCD sensor with a new photodiode structure," *IEEE Int. Electron Devices Meeting, Technical Digest,* Washington, D.C., 1977, p. 538.

10. Baertsch, R.D., et al., "The design and operation of practical charge-transfer transversal filters," *IEEE J. Solid-State Circuits,* **SC-11**, 1976, p. 65.

10a. Baker, I.M., and J.D.E. Beynon, "Charge-coupled devices with submicron electrode separations," *Electronics Lett.,* **9**, Feb. 8, 1973, p. 48.

11. Barbe, D.F., "Imaging devices using the charge-coupled concept," *Proc. IEEE,* **63**, 1975, p. 38.

12. Barbe, D.F., "Charge-coupled device and charge-injection device imaging," *IEEE J. Solid-State Circuits,* **SC-11**, 1976, p. 109.

13. Berger, J., J.S. Brugler, and R. Melen, "Measurement of transfer efficiency of charge-coupled devices," *IEEE J. Solid-State Circuits,* **SC-6**, 1971, p. 421.

14. Berglund, C.N., et al., "Two-phase stepped oxide CCD shift register using undercut isolation," *Appl. Phys. Lett.,* **20**, 1972, p. 413.

15. Berglund, C.N., and K.K. Thornber, "A fundamental comparison of incomplete charge transfer in charge transfer devices," *Bell Syst. Tech. J.,* **52**, 1973, p. 147.

16. Bertram, W.J., et al., "A three-level metallization three-phase CCD," *IEEE Trans. Electron Devices,* **ED-21**, 1974, p. 758.

17. Beynon, J.D.E., "Charge-coupled devices," *IEE Electronics and Power,* May 17, 1973, p. 188.

18. Bower, R.W., T.A. Zimmerman, and A.M. Mohsen, "The two-phase offset gate CCD," *IEEE Trans. Electron Devices,* **ED-22,** 1975, p. 70.
19. Bower, R.W., T.A. Zimmerman, and A.M. Mohsen, "Performance characteristics of the offset-gate charge-coupled device," *IEEE Trans. Electron Devices,* **ED-22,** 1975, p. 72.
20. Boyle, W.S., and G.E. Smith, "Charge coupled semiconductor devices," *Bell Syst. Tech. J.,* **49,** 1970, p. 587.
21. Boyle, W.S., and G.E. Smith, "Charge-coupled devices – a new approach to MIS device structures," *IEEE Spectrum,* **8,** 1971, p. 18.
22. Brodersen, R.W., D.D. Buss, and A.F. Tasch, Jr., "Experimental characterization of transfer efficiency of charge coupled devices," *IEEE Trans. Electron Devices,* **ED-22,** 1975, p. 40.
23. Brodersen, R.W., et al., "A 500 point Fourier transform using charge-coupled devices," *1975 IEEE Int. Solid-State Circuits Conference Digest,* p. 144.
24. Brodersen, R.W., C.R. Hewes, and D.D. Buss, "A 500 stage CCD transversal filter for spectral analysis," *IEEE J. Solid-State Circuits,* **SC-11,** 1976, p. 75.
25. Brodersen, R.W., and S.P. Emmons, "Noise in buried channel charge-coupled devices," *IEEE J. Solid-State Circuits,* **SC-11,** 1976, p. 147.
26. Burke, H.K., and G.J. Michon, "Charge-injection imaging: operating techniques and performance characteristics," *IEEE J. Solid-State Circuits,* **SC-11,** 1976, p. 121.
27. Burt, D.J., "Charge-coupled devices and their applications," *IEE Electronics and Power,* Feb. 6, 1975, p. 93.
28. Buss, D.D., et al., "Transversal filtering using charge-transfer devices," *IEEE J. Solid-State Circuits,* **SC-8,** 1973, p. 138.
29. Carnes, J.E., W.F. Kosonocky, and E.G. Ramberg, "Drift-aiding fringing fields in charge-coupled devices," *IEEE J. Solid-State Circuits,* **SC-6,** 1971, p. 322.
30. Carnes, J.E., and W.F. Kosonocky, "Sensitivity and resolution of charge coupled imagers at low light levels," *RCA Rev.,* **33,** 1972, p. 607.
31. Chou, S., "Design of a 16,384-bit serial charge-coupled memory device," *IEEE J. Solid-State Circuits,* **SC-11,** 1976, p. 10.
32. Chowaniec, A., "Charge transfer device analogue delay lines," *IEE Electronics and Power,* Dec. 12, 1974, p. 1122.
33. Chowaniec, A., and G.S. Hobson, "A wide-band quadrature phasing system using charge transfer devices," *Solid-State Electronics,* **19,** 1976, p. 201.
34. Cole, B., and S.E. Scrupski, "CCD memory system stores megabits," *Electronics,* Aug. 21, 1975, p. 109.
35. Collet, M.G., "An experimental method to analyze trapping centers in silicon at very low concentrations," *Solid-State Electronics,* **18,** 1975, p. 1077.
36. Collet, M.G., "The influence of bulk traps on the charge-transfer inefficiency of bulk charge-coupled devices," *IEEE J. Solid-State Circuits,* **SC-11,** 1976, p. 156.
37. Collet, M.G., "A new method to measure very low bulk trap densities in silicon," *IEEE Trans. Electron Devices,* **ED-22,** 1975, p. 1058.
38. Collins, D.R., et al., "Charge-coupled devices fabricated using aluminum-anodized aluminum-aluminum double-level metalization," *J. Electrochem. Soc.,* **120,** 1973, p. 521.
39. Copeland, M.A., D. Roy, and C.C. Chan, "A multiplexed video bandwidth CCD delay line," *1975 IEEE Int. Solid-State Circuits Conference Digest,* p. 146.
40. Elsaid, M.H., and S.G. Chamberlain, "Short-channel effects on the input stage of surface-channel CCD's," *IEEE Trans. Electron Devices,* **ED-24,** 1977, p. 1164.
41. El-Sissi, H., and R.S.C. Cobbold, "One dimensional study of buried-channel charge-coupled devices," *IEEE Trans. Electron Devices,* **ED-21,** 1974, p. 437.

42. El-Sissi, H., and R.S.C. Cobbold, "Potentials and fields in buried-channel CCDs: a two dimensional analysis and design study," *IEEE Trans. Electron Devices*, **ED-22**, 1975, p. 77.
43. Emmons, S.P., A.F. Tasch, and J.M. Caywood, "A low-noise CCD input with reduced sensitivity to threshold voltage," *IEEE Int. Electron Devices Meeting, Technical Digest*, Washington, D.C., 1974, p. 233.
44. Esser, L.J.M., "Peristaltic charge-coupled device: a new type of charge-transfer device," *Electronic Lett.*, 8, Dec. 14, 1972, p. 620.
45. Esser, L.J.M., "The peristaltic charge coupled device for high speed charge transfer," *IEEE Int. Solid-State Circuits Conf.*, 1974 Digest of Tech. Papers, p. 28.
46. Goser, K., and K. Knauer, "Nonvolatile CCD memory with MNOS storage capacitors," *IEEE J. Solid-State Circuits*, **SC-9**, 1974, p. 148.
47. Haken, R.A., et al., "Charge-coupled structures with self-aligned submicron gaps," *IEEE Trans. Electron Devices*, **ED-22**, 1975, p. 289.
48. Hamaoui, H., G. Amelio, and J. Rothstein, "Application of an area-imaging charge-coupled device for television cameras," *IEEE Trans.*, **BTR-20**, 1974, p. 78.
49. Hara, H., "A theoretical analysis on fundamental performances of charge-coupled devices," *Electronics and Communications in Japan*, **54-C**, 1971, p. 133.
50. Hartsell, G.A., and A.R. Kmetz, "Design and performance of a three phase double level metal 160 x 100 element CCD imager," *IEEE Electron Devices Meeting Abstracts*, Washington, Dec., 1974, p. 549.
51. Heller, L.G., and H-S Lee, "Digital signal transfer in charge-transfer devices," *IEEE J. Solid-State Circuits*, **SC-8**, 1973, p. 116.
52. Hinkle, F.E., "C-MOS decade divider clocks bucket brigade delay line," *Electronics*, Aug. 7, 1975, p. 117.
53. Holmes, F.E., and C.A.T. Salama, "A V groove oxide isolated bipolar bucket brigade shift register," *Solid-State Electronics*, **17**, 1974, p. 1193.
54. Jayadevaiah, T.S., and J. Laur, "Simple model for charge-coupled devices," *Electronics Lett.*, 7, Dec. 16, 1971, p. 752.
55. Jespers, P.G., F. Van de Wiele, and M.H. White (Eds.), *Solid State Imaging*, Noordhoff, Leyden, 1976.
56. Kim, C-K, and R.H. Dyck, "Low light level imaging with buried channel charge coupled devices," *Proc. IEEE*, 61, 1973, p. 1146.
57. Kim, C-K, "Two-phase charge coupled linear imaging devices with self-aligned implanted barrier," *IEEE Electron Devices Meeting Abstracts*, Washington, D.C., Dec. 1974, p. 55.
58. Kim. C-K, and M. Lenzlinger, "Charge transfer in charge-coupled devices," *J. Appl. Phys.*, **42**, 1971, p. 3586.
59. Kosonocky, W.F., and J.E. Carnes, "Charge-coupled digital circuits," *IEEE J. Solid-State Circuits*, **SC-6**, 1971, p. 314.
60. Kosonocky, W.F., et al., "Control of blooming in charge-coupled images," *RCA Rev.*, **35**, 1974, p. 3.
61. Kovac, M.G., et al., "Solid state imaging emerges from charge transport," *Electronics*, Feb. 28, 1972, p. 72.
62. Krambeck, R.H., "Zero loss transfer across gaps in a CCD," *Bell Syst. Tech. J.*, **50**, 1971, p. 3169.
63. Krambeck, R.H., R.H. Walden, and K.A. Pickar, "A doped surface two-phase CCD," *Bell Syst. Tech. J.*, **51**, 1972, 1849.
64. Krambeck, R.H., et al., "A 4160 bit C4D serial memory," *IEEE J. Solid-State Circuits*, **SC-9**, 1974, p. 436.

65. Lancaster, A.L., and J.M. Hartman, "A recirculating CCD with novel input and output structures," *IEEE Electron Devices Meeting Abstracts*, Washington, D.C., Dec. 1974, p. 108.

66. Lee, H-S, and L.G. Heller, "Charge-control method of charge-coupled device transfer analysis," *IEEE Trans. Electron Devices*, ED-19, 1972, p. 1270.

67. Leess, A.W., and W.D. Ryan, "A simple model of a buried channel charge coupled device," *Solid-State Electronics*, 17, 1974, p. 1163.

68. Leonberger, F.J., A.L. McWhorter, and T.C. Harman, "PbS MIS devices for charge-coupled infrared imaging applications," *Appl. Phys. Lett.*, 26, 1975, p. 704.

69. Matsumoto, H., et al., "Zig-Zag transfer CCD image sensor," Professional Group on semiconductors and semiconductor device of IECE Japan, SSD 77-3, 1977.

70. Mattern, J., and D. Lampe, "A reprogrammable filter band using CCD discrete analog signal processing," *1975 IEEE Int. Solid-State Circuits Conference Digest*, p. 148.

71. McKenna, J., and N.L. Schryer, "The potential in a charge coupled device with no mobile minority carriers and zero plate separation," *Bell Syst. Tech. J.*, 52, 1973, p. 669.

72. McKenna, J., and N.L. Schryer, "The potential in a charge-coupled device with no mobile minority carriers," *Bell Syst. Tech. J.*, 52, 1973, p. 1765.

73. Melen, R.D., and J.D. Meindl, "One-phase CCD: a new approach to charge-coupled device clocking," *IEEE J. Solid-State Circuits*, SC-7, 1972, p. 92.

74. Melen, R., and D. Buss, *Charge-Coupled Devices: Technology and Applications*, IEEE Press, N.Y., 1977.

75. Mifune, T., et al., "An improvement on structure of charge-coupled devices," Proc. 4th Conf. Solid State Devices, Tokyo 1972 (*J. Japan. Soc. Appl. Phys. Supplement*, 42, 1973, p. 207).

76. Mohsen, A.M., et al., "The influence of interface states on incomplete charge transfer in overlapping gate charge-coupled devices," *IEEE J. Solid-State Circuits*, SC-8, 1973, p. 125.

77. Mohsen, A.M., and M.F. Tompsett, "The effects of bulk traps on the performance of bulk channel charge-coupled devices," *IEEE Trans. Electron Devices*, ED-21, 1974, p. 701.

78. Mohsen, A.M., M.F. Tompsett, and C.H. Séquin, "Noise measurements in charge-coupled devices," *IEEE Trans. Electron Devices*, ED-22, 1975, p. 209.

79. Mohsen, A.M., et al., "A 64 kbit block addressed charge-coupled memory," *IEEE J. Solid-State Circuits*, SC-11, 1976, p. 49.

80. Mohsen, A.M., and T.F. Retajczyk, Jr., "Fabrication and performance of offset-mask charge-coupled devices," *IEEE J. Solid-State Circuits*, SC-11, 1976, p. 180.

81. Nelson, R.D., "Accumulation-mode charge-coupled device," *Appl. Phys. Lett.*, 25, 1974, p. 568.

82. Ong, D.G., and R.F. Pierret, "Thermal carrier generation in charge-coupled devices," *IEEE Trans. Electron Devices*, ED-22, 1975, p. 593.

83. Pike, W.S., et al., "An experimental solid-state TV camera using a 32 x 44 element charge-transfer bucket-brigade sensor," *RCA Rev.*, 33, 1972, p. 483.

84. Powell, R.J., C.N. Berglund, J.T. Clemens, and E.H. Nicollian, "Two-phase stepped oxide CCD shift register using undercut isolation," *Appl. Phys. Lett.*, 20, 1972, p. 413.

85. Rosenblatt, A., "For CCDs, imager is the beginning," *Electronics*, Sept. 27, 1973, p. 71.

86. Sangster, F.L.J., "Integrated MOS and bipolar analog delay using bucket-brigade capacitor storage," *IEEE, ISSCC Digest Tech. Papers*, 1970, p. 74.

87. Schroder, D.K., "A two-phase germanium charge-coupled device," *Appl. Phys. Lett.*, 25, 1974, p. 747.

88. Schuermeyer, F.L., et al., "New structures for charge coupled devices," *Proc. IEEE*, 60, 1972, p. 1444.

89. Sealer, D.A., and M.F. Tompsett, "A dual-differential analog charge-coupled device for time-shared recursive filters," *1975 IEEE Int. Solid-State Circuits Conference Digest*, p. 152.

90. Sekula, J.A., P.R. Prince, and C.S. Wang, "Non-recursive matched filters using charge-coupled devices," *IEEE Electron Devices Meeting Abstracts*, Washington, Dec. 1974, p. 244.

91. Séquin, C.H., "Interlacing in charge-coupled imaging devices," *IEEE Trans. Electron Devices*, ED-20, 1973, p. 535.

92. Séquin, C.H., "Two-dimensional charge transfer arrays," *IEEE J. Solid-State Circuits*, SC-9, 1974, p. 134.

93. Séquin, C.H., et al., "Charge-coupled area image sensor using three levels of poly-silicon," *IEEE Trans. Electron Devices*, ED-21, 1974, p. 712.

94. Séquin, C.H., and M.F. Tompsett, *Charge Transfer Devices*, Academic Press, N.Y., 1975.

95. Séquin, C.H., et al., "All-solid-state camera for the 525-line television format," *IEEE J. Solid-State Circuits*, SC-11, 1976, p. 115.

96. Shortes, S.R., et al., "Characteristics of thinned backside-illuminated charge-coupled device imagers," *Appl. Phys. Lett.*, 24, 1974, p. 565.

97. Steckl, A.J., et al., "Application of charge-coupled devices to infrared detection and imaging," *Proc. IEEE*, 63, 1975, p. 67.

98. Strain, R.J., "Properties of an idealized traveling wave charge coupled device," *IEEE Trans. Electron Devices*, ED-19, 1972, p. 1119.

99. Tanikawa, K., Y. Ito, and H. Sei, "Evaluation of dark-current non-uniformity in a charge-coupled device," *Appl. Phys. Lett.*, 28, 1976, p. 285.

100. Tasch, A.F., et al., "Charge capacity analysis of the charge-coupled RAM-cell," *IEEE J. Solid-State Circuits*, SC-11, 1976, p. 575.

101. Terman, L.M., and L.G. Heller, "Overview of CCD memory," *IEEE J. Solid-State Circuits*, SC-11, 1976, p. 4.

102. Tiemann, J.J., et al., "Charge-transfer devices filter complex communications signals," *Electronics*, Nov. 14, 1974, p. 113.

103. Tompsett, M.F., "A simple charge regenerator for use with charge transfer devices and the design of functional logic arrays," *IEEE J. Solid-State Circuits*, SC-7, 1972, p. 237.

104. Tompsett, M.F., "The quantitative effects of interface states on the performance of charge-coupled devices," *IEEE Trans. Electron Devices*, ED-20, 1973, p. 45.

105. Tompsett, M.F., et al., "Charge-coupled imaging devices: experimental results," *IEEE Trans. Electron Devices*, ED-18, 1972, p. 992.

106. Tompsett, M.F., and E.J. Zimany, Jr., "Use of charge-coupled devices for delaying analog signals," *IEEE J. Solid-State Circuits*, SC-8, 1973, p. 151.

107. Tompsett, M.F., B.B. Kosicki, and D. Kahng, "Measurements of transfer inefficiency of 250-element undercut-isolated charge coupled devices," *Bell Syst. Tech. J.*, 52, 1973, p. 1.

108. Tompsett, M.F., et al., "Charge-coupling improves its image, challenging video camera tubes," *Electronics*, Jan. 18, 1973, p. 162.

109. Ullrich, M., and M. Hegendörfer, "TV receiver puts two pictures on screen at same time," *Electronics*, Sept. 1, 1977, p. 102.

110. Walden, R.H., et al., "The buried channel charge coupled device," *Bell Syst. Tech. J.*, 51, 1972, p. 1635.

111. Weimer, P.K., et al., "Video processing in charge-transfer image sensors by recycling of signals through the sensor," *RCA Rev.*, 35, 1974, p. 341.

112. Wen, D.D., et al., "A distributed floating-gate amplifier in charge-coupled devices," *IEEE Int. Solid-State Circuits Conf.*, 1975, Philadelphia Digest of Tech. Papers. p. 24.

113. White, M.H., et al., "A nonvolatile charge-addressed memory (NOVCAM) cell," *IEEE Electron Devices Meeting Abstracts*, Washington, Dec. 1974, p. 115.

114. Zimmerman, T.A., and D.F. Barbe, A new role for charge-coupled devices: digital signal processing," *Electronics*, Mar. 31, 1977, p. 97.

115. Buss, D.D. et al., "Infrared monolithic HgCdTe IR CCD focal plane technology," *IEEE International Electron Devices Meeting, Technical Digest*, Washington, D.C., Dec. 1978, p. 496.

116. Andrews, A.M., "Hybrid infrared imaging arrays," *IEEE International Electron Devices Meeting, Technical Digest*, Washington, D.C., Dec. 1978, p. 505.

117. Blouke, M.M., J.E. Hall and J.F. Breitzmann, "A 640 kilopixel CCD imager for space applications," *IEEE International Electron Devices Meeting, Technical Digest*, Washington, D.C., Dec. 1978, p. 412.

118. Iversen, W.R., "64-K CCDs face an uncertain future," *Electronics*, Jan. 4, 1979, p. 85.

# 11
# Avalanche–Diode Microwave Oscillators, Amplifiers, and Gunn Devices

## CONTENTS

## 11.1 INTRODUCTION

Microwave systems typically operate in the frequency range from 1 to 100 GHz. Some of the commonly used frequency bands are $L$ band, 1.0-2.6 GHz; $S$ band, 2.60-3.95 GHz; $C$ band, 4.90-7.05 GHz; $X$ band, 8.20-12.40 GHz; $Ku$ band, 15.3-18.0 GHz, and $K$ band, 18.0-26.50 GHz. At $X$ band the wavelength is about 3 cm and above 30 GHz the wavelengths enter the millimeter range.

Reflex klystron oscillator tubes are commonly used for generating power from 20 mW to 1 W in the $X$-band range for commercial and military radars. Klystrons require operating-beam voltages of 200-600 V. They typically have an electronic tuning range of up to 40 MHz with a modulation sensitivity of 1-3 MHz/V, and a good temperature stability $\sim 0.2$ MHz/°C. For operation at higher power levels magnetrons are used, generally in a pulse mode with peak outputs of 0.1 to 1000 kW and pulse durations of 1-2 $\mu$s.

Microwave semiconductor oscillators and amplifiers must compete with klystrons and also with transistor oscillators and amplifiers up to frequencies of 5 GHz. Microwave transistors presently available can provide a few watts of generated power at 4 to 5 GHz with efficiencies of 15-20%. High-power GaAs field-effect transistors under development are predicted to deliver tens of watts of CW power at $X$ band in the next few years.[210]

The status (1976) on GaAs Impatts (avalanche-oscillator diodes) is that about 10 W at 10 GHz has been obtained with 25-30% efficiency under laboratory conditions. A number of such diodes can be put together to give 50 W, or more, of power with a combining efficiency of 80-90%. The efficiencies of Impatts tend to be better than the efficiencies of Gunn oscillator diodes. The Impatt noise levels are low enough for most applications although not as good as for klystrons.

Impatt and Trapatt diodes must be recognized as special limited-purpose devices that will never command the breadth of market of, say, operational-amplifier integrated circuits. In 1973 the total market for microwave diodes of all kinds was about $35 million and oscillator diodes accounted for about $2 million of this.

## 11.2  READ-DIODE OSCILLATOR CONCEPTS

The term Impatt for an oscillator diode is derived from impact-avalanche transit time. The negative resistance action needed to sustain oscillation was first conceived by Read[162] for a diode structure $p^+$-$n$-$i$-$n^+$. Such structures were not easy to make in 1958 and it was not until seven years later that the first Read oscillator was fabricated. Within a year or two it was established that a Read-doping profile was not essential to operation at good oscillator efficiencies. Avalanche-diode oscillators are reviewed in considerable detail in the books by Carroll[21] and Gibbons.[68]

The Read diode, however, remains an excellent starting point for a discussion of transit-time oscillations. The doping profile in a Read diode is that shown in Fig. 11.1(a). Under reverse-voltage bias such that the depletion region reaches across the intrinsic region, the electric field across the diode is as shown in Fig. 11.1(b). If a small ac voltage disturbance occurs so that the peak breakdown field $\mathcal{E}_B$ is reached, a spike of avalanche-induced charge occurs near the junction and develops as the electrons move across the $n$ region towards the interface with the $i$ region. There is a phase delay during this process as suggested by Fig. 11.1(c). The pulse of electrons then enters the intrinsic region of the diode and drifts to the $n^+$ contact (at the saturation drift velocity, if $\mathcal{E}_i$ is high enough). There is a further transit-time delay here. So during the period when the ac voltage is negative, the current flowing through the device is positive and power is being returned to the supply. This is a negative resistance condition and, provided the positive resistance in series with the device is low, oscillations can build up. The amplitude and frequency of the oscillations are limited by phase-shift transit-time conditions and energy-transfer considerations. If the ac voltage swing becomes too large, the field $\mathcal{E}_i$ in the intrinsic region becomes too large and further avalanche ionization may occur which disturbs the desired voltage-current phase relationship. Or if the total voltage is too low during the negative swing of the ac voltage, the drift time across the intrinsic region

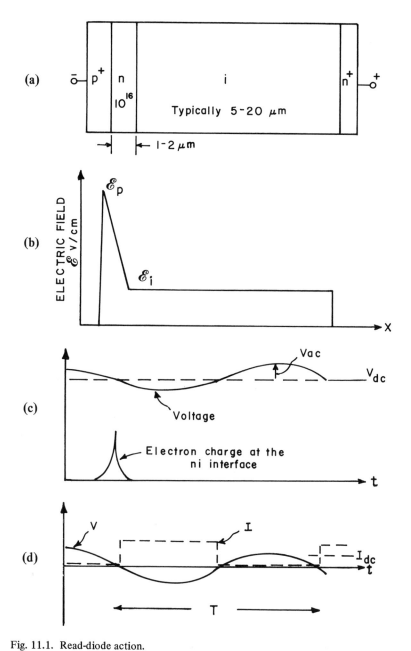

Fig. 11.1. Read-diode action.
(a) Diode $p^+$-$n$-$i$-$n^+$ structure.
(b) Electric field variation across the device under reverse bias.
(c) Voltage and charge variation.
(d) Hypothetical current and voltage relationship corresponding to an efficiency of 30%.

lengthens and again the desired phase relationship is lost. A suitable limit is perhaps that $V_{ac} \leq V_{dc}/2$.

While the electron charge pulse is crossing the drift region a roughly constant current, $I$, of square waveform flows, as shown in Fig. 11.1(d). The dc to ac conversion efficiency of the diode for this condition can be readily estimated. The fundamental component of the fully modulated square wave of current shown is, by Fourier analysis, $(4/\pi)I_{dc} \sin \omega t$.

So for the period $T$ the ac power is

$$P_{ac} = \frac{1}{T} \int \left( \frac{4}{\pi} I_{dc} \sin \omega t \right) (V_{ac} \sin \omega t) dt \qquad (11.1)$$

and the dc power is $I_{dc} V_{dc}$. The power conversion efficiency, $\eta$, therefore becomes

$$\eta = \frac{2}{\pi} \frac{V_{ac}}{V_{dc}} \qquad (11.2)$$

With the concept that $V_{ac}$ will not exceed $V_{dc}/2$, the maximum-power efficiency is therefore just under 30%. This approach has neglected many loss terms in the diode, and efficiencies of half this value are easier to obtain although good diodes now may be 20-25% efficient. The sources of loss include space-charge effects, the detailed nature of the current waveform, series-resistance effects, and possible premature avalanche action arising from stored charge that affects the phase shift.

Negative-resistance concepts are strongly rooted in the approach to oscillators, so it is worth considering the impedance of a Read diode as a function of frequency.

In a small region, $dx$, of the avalanche zone the electron and hole buildup equations are

$$\frac{\partial n}{\partial t} = \alpha v_s(n+p) + \frac{1}{q} \frac{\partial j_n}{\partial x} \qquad (11.3)$$

and

$$\frac{\partial p}{\partial t} = \alpha v_s(n+p) - \frac{1}{q} \frac{\partial j_p}{\partial x} \qquad (11.4)$$

where $\alpha$ is the ionization coefficient at the appropriate field strength and for simplification is assumed to be the same for electrons and holes. The total current density, $J$, is the sum of the electron ($j_n = q v_s n$) and hole ($j_p = q v_s p$) current densities. The avalanche zone is assumed to be of width $x_A$ and the carrier velocity, $v_s$, is taken as constant. Here the transit time across the full zone is $\tau = x_A/v_s$. The boundary conditions are that at $x = 0$, $j_p$ is the thermally generated hole current $j_{ps}$ and at $x = x_A$, $j_n$ is $j_{ns}$. Integration of the avalanche expressions then gives for the current buildup equation

$$\frac{\partial J}{\partial t} = \frac{2J}{\tau}\left(\int_0^{x_A} \alpha \, dx - 1\right) + j_{ns} + j_{ps} \tag{11.5}$$

If the field is assumed uniform in the whole avalanche zone and the leakage current is neglected then

$$\frac{dJ}{dt} = \frac{2J}{\tau}(\alpha x_A - 1) \tag{11.6}$$

The avalanche term may be represented by dc part $\alpha_0$ and a part $(d\alpha/d\mathscr{E})\mathscr{E}_a \, e^{i\omega t}$ where $\mathscr{E}_a$ is the ac field. Substituting this in Eq. (11.6) and selecting out the ac terms gives

$$j_a = \frac{2J_0 \mathscr{E}_a}{i\omega\tau}\left(\left(\frac{d\alpha}{d\mathscr{E}}\right)x_A\right) \tag{11.7}$$

where the impedance per unit area is

$$Z_A = i\omega \left[\frac{\tau}{2J_0}\left(\frac{d\mathscr{E}}{d\alpha}\right)\right] \tag{11.8}$$

This inductive impedance is in parallel with the avalanche-zone depletion-layer capacitance $\epsilon A/x_A$ which carries a displacement current $i\omega\epsilon\mathscr{E}_a$. The resonant avalanche frequency of the parallel $LC$ combination is given by

$$\omega_a^2 = \frac{2J_0 x_A}{\epsilon\tau}\left(\frac{d\alpha}{d\mathscr{E}}\right) \tag{11.9}$$

The total impedance of the $LC$ combination at a frequency $\omega$ is then

$$Z_A = i\omega\tau \left[2J_0 \frac{d\alpha}{d\mathscr{E}}\left(1 - \frac{\omega^2}{\omega_a^2}\right)\right]^{-1} \tag{11.10}$$

To arrive at the total impedance for the Read diode, drift-zone impedance must now be considered and added to the avalanche-zone impedance. The ac current amplitude in the drift zone varies as $j_a \exp(-i\omega x/v_s)$ where $v_s$ is the drift velocity assumed constant. The ac drift field is then given by

$$\mathscr{E}(x) = \frac{1}{i\omega\epsilon}(j - j_a e^{-i\omega x/v_s}) \tag{11.11}$$

where $j$ is the total ac (displacement and conduction) current density. The ac voltage across the drift zone is then

$$\int_0^{x_D} \mathcal{E}(x)dx = \frac{x_D j}{i\omega\epsilon} \left[ 1 - \frac{1}{1-\left(\dfrac{\omega}{\omega_a}\right)^2} \left( \frac{1-\exp(-i\omega x_D/v_s)}{i\omega x_D/v_s} \right) \right] \qquad (11.12)$$

The impedance of the drift zone is therefore

$$Z_D = \frac{1}{i\omega C_D} \left[ 1 - \frac{1}{1-\left(\dfrac{\omega}{\omega_a}\right)^2} \left( \frac{\sin(\omega x_D/v_s)}{\omega x_D/v_s} \right) \right] + \frac{1}{\omega C_D}$$

$$\left[ \frac{1}{1-\left(\dfrac{\omega}{\omega_a}\right)^2} \left( \frac{1-\cos(\omega x_D/v_s)}{\omega x_D/v_s} \right) \right] \qquad (11.13)$$

where $C_D$ is the depletion-layer capacitance and $\omega x_D/v_s$ may be thought of as a transit angle. Addition of $Z_A$ to $Z_D$ gives the total diode impedance. Since $Z_A$ is imaginary the real part of the total impedance is the last term in Eq. (11.13). This term is a negative resistance when the frequency is greater than the avalanche-resonance frequency $\omega_a$ of the diode. The operating frequency is typically chosen to be 20% greater than the resonance frequency.

Having established that the device indeed has a negative resistance and that the magnitude is related to the carrier transit time across the drift zone, let us now consider the space-charge effect of the carriers moving in the drift zone. The space charge, by Poisson's equation, produces a change in the electric field (see Fig. 11.2) from the original profile (dotted line) that exists with negligible charge present. The reduction of the field in the avalanche region tends to extinguish the avalanche prematurely and the device efficiency is lowered. The second effect to note is that the field rises in front of the drifting space-charge pulse. If this increase becomes too large, avalanche occurs at the right-hand side (anode) of the drift region. The hole pulse thus created causes dissipation in the next half cycle. The dc current density $J_{DC}$ is related to the total charge $Q$ flowing in the pulse by

$$J_{DC} = Qf \qquad (11.14)$$

where $f$ is the oscillation frequency. If we assume that the field rise at the anode is limited by $\mathcal{E}_B$, the breakdown field strength [neglecting $\mathcal{E}_i$ in Fig. 11.1(b)] is then

$$Q \leq \epsilon \mathcal{E}_B \qquad (11.15)$$

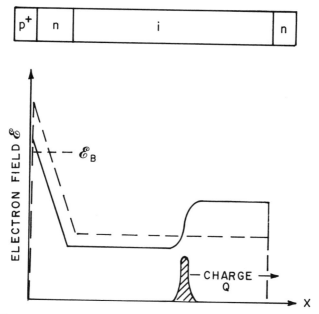

Fig. 11.2. Effect on the electrical-field distribution of a space-charge pulse of carriers in the drift region of a Read diode. The dashed line is the field profile without charge present.

Hence the current density must be limited by

$$J_{DC} \leq \epsilon \mathcal{E}_B f \tag{11.16}$$

If $f$ is 50 GHz and $\mathcal{E}_B$ is $5 \times 10^5$ V-cm$^{-1}$ and $\epsilon$ is $12 \times 8.84 \times 10^{-14}$ for Si, then the maximum current density for avalanche oscillation is about 2500 A-cm$^{-2}$. However, in practical Impatt diodes thermal dissipation considerations tend to limit the current density to some hundreds of amperes per square centimeter.

A familiar effect of the decrease of field through the charge pulse is that if this fall is too large the drift velocity of the carriers in the tail of the pulse may drop below the saturated drift velocity and the bunch will not be well confined. Then the desired current and voltage phase relationship is lost.

In a Read diode there is a power-frequency tradeoff. Such a tradeoff is usual in all transit-time-limited devices and may take the form of a constant $Pf^2$ value for a given device impedance level. The form of the tradeoff relationship depends on the conditions assumed. The $Pf^2$ relationship can be readily demonstrated.

Assume a uniform-field drift zone of width $x_A$ and an avalanche breakdown field of $\mathcal{E}_B$ throughout the zone. Then the voltage applied to the device is

$$V = \mathcal{E}_B x_A \tag{11.17}$$

Further assume that the device dc current is limited by the space-charge effect [Eq. (11.16)] so that for a device of area $A$,

$$I_{max} = \epsilon A \, \mathcal{E}_B f \tag{11.18}$$

The maximum power drain from the supply is then

$$P_{max} = \epsilon A x_A \, \mathcal{E}_B^2 f \tag{11.19}$$

and may be related to the ac output power by the efficiency factor $\eta$. The frequency of oscillation we assume to be related to the transit time across the depletion region by $f/2 = v_s/x_A$. This is not, of course, strictly true because in a Read structure there is a separate drift zone, $x_D$, but since $x_D$ tends to be proportional to $x_A$, the simplification will do for the purpose at hand. Then Eq. (11.19) becomes

$$P_{max} = 4 \frac{\epsilon A}{x_A} \frac{v_s^2}{f} \mathcal{E}_B^2 = 4 C_D \frac{v_s^2}{f} \mathcal{E}_B^2 \tag{11.20}$$

$C_D$, the depletion capacitance of the diode, plays a dominant role in determining the device impedance as may be seen from Eq. (11.13). Hence the circuit reactance, $X_c$, for matching purposes must be proportional to $(f C_D)^{-1}$. Substituting $X_c$ in Eq. (11.20) then gives

$$P_{max} f^2 \propto \frac{v_s^2 \mathcal{E}_B^2}{X_c} \tag{11.21}$$

The tradeoff relationship involving power and frequency squared is therefore obtained.

The relationship shows that a high-saturation drift velocity and a high-breakdown field for avalanche are desirable material properties for a transit-time device. The circuit reactance tends to be limited to values above 10 to 15 $\Omega$ in coaxial or waveguide cavities. As the frequency is raised, the device area must be decreased to maintain the same impedance and this reduction in area reduces the power-handling ability, in concept through Eq. (11.18). In practice, thermal considerations are often dominant and the power-frequency tradeoff tends to be more nearly $Pf$-constant.

The treatment on avalanche and other matters just given is condensed. For fuller discussions it is recommended that the reader consult the references and reading suggestions of Section 11.10. Impatt diodes based on a Read-diode doping structure are sometimes referred to as RIMPATTs.

## 11.3 IMPATT PERFORMANCE

Once the circuits had been set up for Read-diode structures to oscillate, it was found that many other diodes intended for rectifier use would also oscillate at reasonable efficiencies. This led to a period of extensive computer modeling of diode profiles to determine the factors involved in increasing the efficiency. Studies by Misawa,[143, 144] Scharfetter,[168, 169, 170] Gummel,[78, 79] and Clorfeine[26, 27, 28, 29] were impressive in this period and derived performance concepts and efficiency values against which experimental results could be weighed.

The effect of varying the length of the avalanche region relative to the drift-region length is shown in Fig. 11.3 for diodes with 10, 33, and 50% avalanche zones. The 33% curve corresponding to a rather normal $p^+$-$n$ junction is seen to have quite a reasonable negative conductance over a substantial range of transit angle (or frequency).

One useful concept for comparing various structures as oscillators is the $Q$ factor, defined as the energy stored in the device divided by the power dissipation per cycle of the frequency. The $Q$ factor is negative for a diode with negative resistance. The rate of growth of the oscillation and the final amplitude tend to be greatest when the magnitude of $Q$ is small. Narrow-avalanche-region structures, $p^+$-$n$-$i$-$n^+$, have $Q$ factors that tend to degrade more rapidly with increasing current density than for broader avalanche-region structures such as $p^+$-$n$-$n^+$ or $p^+$-$v$-$n^+$. Hence the broad structures are more suitable for substantial power generation.

In the Read diode modeling given earlier the ionization coefficients were assumed equal. But for Si the ionization rate for electrons in an avalanche zone is about 30 times that for holes. One effect of a low-hole ionization coefficient is that in a narrow avalanche region the contribution of holes to the avalanche-electron charge pulse is small since many of the holes pass out of the avalanche region without ionizing collisions. Therefore, the avalanche zone must be made wider if a substantial charge pulse is to flow and there tends to be some loss of device efficiency.

Computer studies of the buildup and transit of pulses of electrons and holes in a Si Read diode $p$-$n$-$v$-$n$ structure are shown in Figs. 11.4 and 11.5. The assumed impurity distribution is given in Fig. 11.4(a) and the carrier concentrations and the electric field strengths are shown in Fig. 11.4(b) for the condition of maximum-voltage swing just before avalanche injection starts (in the inset figure the dot indicates the current and voltage conditions). A quarter of a cycle later [Fig. 11.5(a)] the electron pulse is fully developed and moving to the left and the hole pulse is merging into the cathode. After a half cycle [Fig. 11.5(b)] the electrons fill a large portion of the device but the holes have almost disappeared. After a further quarter cycle the electrons are exiting at the anode.

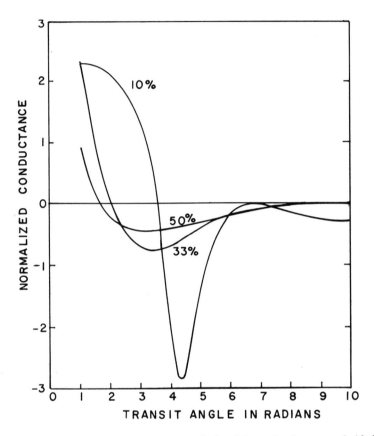

Fig. 11.3.  Read-diode negative conductance calculated for avalanche zones of  10, 33, and 50% of the total device width. (After Misawa. Reprinted with permission from *IEEE Trans. Electron Devices,* **ED-14,** 1967, p. 804.)

The frequency of oscillation inferred from the study was 12.4 GHz and the efficiency 12% for the chosen current density of 200 A-cm$^{-2}$. At 9.4 GHz the same current density gave an efficiency of 18% for this assumed structure.

Several factors may contribute to the fall of efficiency when an Impatt structure is redesigned to work at higher and higher frequencies. An important factor is that depletion-layer modulation becomes more important as the design is made narrower. Another factor that is involved is that the electron ionization rate for Si does not increase linearly as the field strength is increased (see Fig. 11.6). This results in modification of the waveform of the induced avalanche pulse and some falloff in efficiency.

Minority carrier storage is another possible cause of efficiency loss. The mechanism is that in a $p^{+}$-$n$-$i$-$n^{+}$ structure, for instance, electrons may exist as

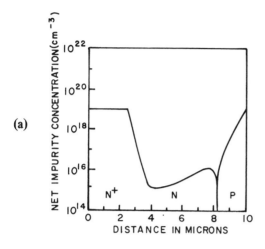

(a)

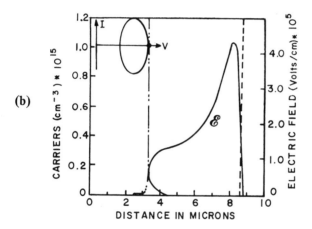

(b)

Fig. 11.4. Read-diode calculations.
(a)  $n^+$-$n$-$p$ impurity concentration assumed.
(b)  Calculated electron concentration ——··—··, hole concentration ---, and electric field
     before avalanche buildup. (After Scharfetter and Gummel. Reprinted with permission
     from *IEEE Trans. Electron Devices*, ED-16, 1969, p. 65.)

stored minority charge in the $p^+$ region near the junction, having been created
there or diffused in during the previous avalanche-pulse formation period, and
if these electrons return to the $n$ region at some later time in the cycle premature
avalanching may occur that disturbs the desired phase relationship.

Series-parasitic resistance can have the effect of decreasing Impatt diode
efficiency, particularly that caused by incomplete depletion of the lightly
doped region. A series resistance of an ohm or two corresponding to perhaps a

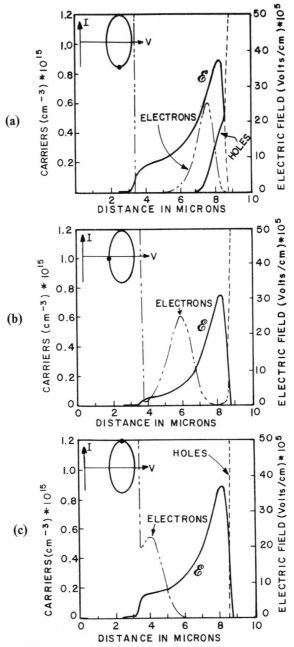

Fig. 11.5. Read-diode carrier and field distributions computed for various stages in the oscillation cycle. (After Scharfetter and Gummel. Reprinted with permission from *IEEE Trans. Electron Devices*, **ED-16**, 1969, p. 70.)

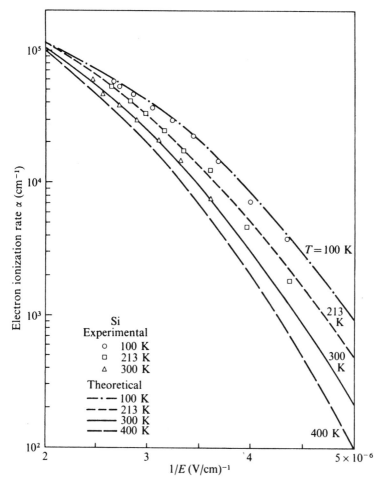

Fig. 11.6. Electron-ionization rate versus reciprocal-electric field for Si. (See also Fig. 1.13.) (After Crowell and Sze. Reprinted with permission from *Appl. Phys. Lett.*, 9, 1966, p. 244.)

micrometer of undepleted material in diodes of typical proportions lowers the efficiency considerably. Diodes, therefore, are designed for complete depletion with punch-through factors (dc voltage applied divided by minimum voltage for depletion) often 1.2 or higher. Excessive ac voltage output can result in a fall of efficiency as seen from the computed curves given in Fig. 11.7. The saturation of efficiency with increasing RF voltage is primarily caused by depletion-layer modulation in diodes operated with a dc voltage corresponding to a punch-through factor of unity. When depletion-layer modulation occurs the electric field near the epitaxial-substrate interface becomes negative and mobile-charge carriers from the substrate are injected into the epitaxial region. The injected

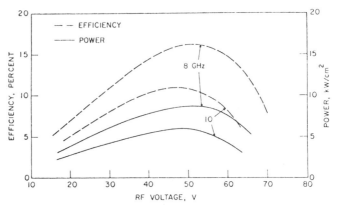

Efficiency and RF power of diode 1-*P* versus RF voltage.
($J_{dc} = 500$ A/cm².)

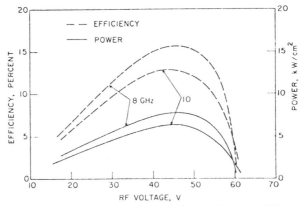

Efficiency and RF power of diode 1-*N* versus RF voltage.
($J_{dc} = 500$ A/cm².)

Fig. 11.7. Theoretical studies of Si Impatt-diode efficiency, power, and RF voltage for *n⁺-p-p* and *p⁺-n-n⁺* structures. (After Lee, Haddad, and Lomax. Reprinted with permission from *IEEE Trans. Electron Devices,* **ED-21,** 1974, p. 138.)

charge induces an external current in phase with the ac voltage and thus degrades the efficiency. At even higher dc biases the efficiency is beyond the saturation hump and is decreasing: this is caused by the space-charge effect of the increased current and ac voltage.

The curves of Fig. 11.7 show that there is not too much difference in maximum efficiency to be expected for $n^+$-$p$-$p^+$ and $p^+$-$n$-$n^+$ structures. Double-drift-diode structures $p^+$-$p$-$n$-$p^+$ have also been studied and have $Q$ values that indicate substantial power advantages over single-drift structures.[175, 211]

The thermal impedance of an Impatt diode on its heat sink is inversely proportional to the diameter, $d$, of the diode (assumed circular in form). The impedance $\theta$ in $°C\ W^{-1}$ is

$$\theta = \frac{2}{\pi dK} \tag{11.22}$$

where the thermal conductivity, $K$, is 3.9 W cm$^{-1}$ $°C^{-1}$ for copper and about three times greater for a type IIa diamond heat sink. Curves relating $\theta$ and $d$ for copper and diamond heat sinks are presented in Fig. 11.8. These curves do not take into account any temperature drop in the device itself or in the solder used to hold the device onto the heat sink so the $\theta$ values shown are low and must be used with caution. In recent designs, ring-shaped structures have been used for high-power devices to improve on the thermal impedance represented by Eq. (11.22).

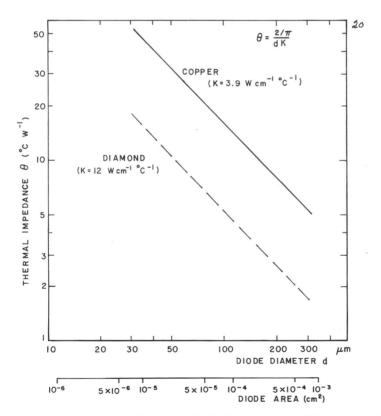

Fig. 11.8. Thermal impedance variation with diode diameter (and area) for copper and diamond heat sinks. Temperature changes in the diode itself and on the solder are not included.

Let us consider some effects of the thermal limitations. For a thermal impedance of $10°C$ $W^{-1}$ on a copper heat sink the diode diameter must be about 163 $\mu$m (6.4 mil). If the maximum junction temperature rise is $300°C$ a dissipation of 30 W is permissible. For a diode of 15% efficiency this corresponds to a power output of 5.3 W. If a larger diameter had been chosen for the diode a larger power output would have followed. However, the diode-depletion capacitance is proportional to the diode area (assuming the doping is unchanged), and the diode impedance is $1/2\pi f C_D$. This impedance cannot be allowed to fall much below 10 to 15 $\Omega$ for convenience of matching, so it follows that there is an inverse relationship between power output and frequency in Impatts. (This relationship differs from the one obtained earlier involving power and frequency squared, since this was derived from the concept of a power limit set by space-charge effect limitations on the dc current density.)

The thermal-based power-frequency relationship has been studied by Scharfetter[169] and some of his scaling results are shown in Fig. 11.9. For a 10-GHz diode of diameter about 220 $\mu$m (8.7 mil) on a copper heat sink the power output at 15% efficiency is limited to about 6 W for continuous wave operation. Under pulsed-operation conditions significantly more power (perhaps three to five times as much) might be expected. A more recent study of scaling conditions relating power-handling ability and frequency is that of O'Hara and Grierson.[150]

Up to frequencies of 6-10 GHz, Impatt-diode circuits are usually of coaxial line form with the diode in a coaxial cavity. Above this frequency level, waveguide cavities are used so that the mode of propagation is a well-defined TEM. The region of reduced height in the waveguide, as shown in Fig. 11.10, lowers the waveguide impedance and improves the matching to the low impedance ($\sim$15 $\Omega$) of the diode.

Although the frequency of an Impatt diode is dependent on the dc current, the resonant circuit must be retuned if a wide frequency range is to be obtained. One way of providing such tuning is by the use of varactor diodes. For example, two voltage-controlled diodes mounted symmetrically on either side of an Impatt diode in a ridged waveguide with two matched outputs, result in a bandwidth of about 30%.[25] At $X$ band this is capable of giving a tuning range from 9.4-10.5 GHz.

If in a $p^+$-$n$-$p^+$ structure the hole-injecting contact is replaced by a Schottky barrier, the $M$-$n$-$p^+$ structure formed is well suited for avalanche injection and transit-time oscillations. Similarly, the other junction may be replaced by a second Schottky contact to give an $M$-$n$-$M$ structure. This barrier-injection transit-time device is known as a Baritt diode.[85] Schottky barrier injectors are extensively used in GaAs Impatts[137] since the technology is simpler than diffusions in GaAs.

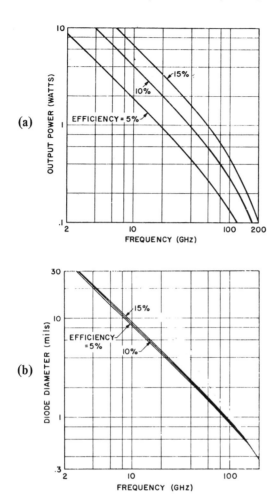

Fig. 11.9. Theoretical studies of Si Impatt-oscillator diodes of 5, 10, and 15% efficiency on copper heat sinks.
(a)  Power output versus frequency.
(b)  Diode diameter versus frequency. (After Scharfetter. Reprinted with permission from *IEEE Trans. Electron Devices*, ED-18, 1971, p. 538.)

## 11.4  TRAPATT OSCILLATIONS

During early experiments on pulsed-mode operation of Impatt diodes a new high-efficiency high-power mode of oscillation was discovered.[161] Typically in these oscillations the waveforms become very nonsinusoidal with a fundamental frequency substantially below the transit-time frequency of the Impatt mode.

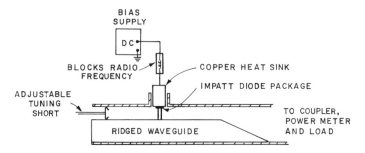

Fig. 11.10.  Schematic Impatt diode mounted in reduced-height waveguide.

The voltage across the diode decreases, the bias current increases, and the RF power efficiency rises to 50-60%.

Computer studies were valuable in explaining this new mode of operation. [107] The oscillation mode discovered by these studies suggested the name Trapatt from "trapped plasma avalanche triggered transient." It occurs when a voltage greater than the breakdown voltage is rapidly applied across the junction and causes an avalanche shock front to traverse the diode at high speed leaving in its wake a trapped plasma. The shock-front velocity is usually several times the saturated drift velocity of carriers. The electric-field variation with distance during the transit is that shown in Fig. 11.11 and the type of current and voltage waveforms are as shown in Fig. 11.11(b) and (c). During the plasma formation the voltage across the device drops rapidly; this is followed by a recovery period as the plasma is extracted. The recovery is slow at first as the plasma in the center of the device is not in a region of high electric field [see Fig. 11.12(a)]. However, as the extraction process continues, the field reaches into the center and the sweep-out action speeds up [see Fig. 11.12(b) and (c)] so the voltage rises rapidly towards the end of the extraction period as shown in Fig. 11.11(b).

The Trapatt oscillation is not self-oscillating but must be triggered for each cycle by a preceding Impatt oscillation. The cavity for the diode must therefore be resonant at one or more frequencies that are harmonics of the desired Trapatt frequency. For instance, a coaxial line cavity with several coaxial tuning slugs suitably spaced along the center conductor can be made resonant at, say, 5 GHz and 1 GHz. The device used is then one capable of Impatt oscillation at 5 GHz in the local cavity surrounding the device. The buildup of the Impatt oscillation provides the voltage pulse that triggers the first plasma oscillation. As the diode voltage drops, a negative voltage pulse travels down the coaxial line to the 1-GHz slug $\lambda/2$ away and the impedance change there reflects it as a positive pulse that arrives one period later to provide the voltage rise needed to trigger the next Trapatt oscillation. Thus a 1-GHz oscillation-repetition rate is obtained.

Several resonant frequencies may be obtained by spacing tuning slugs along the coaxial line (or the ridged waveguide) and, if correctly done, this gives waveforms of improved efficiency.

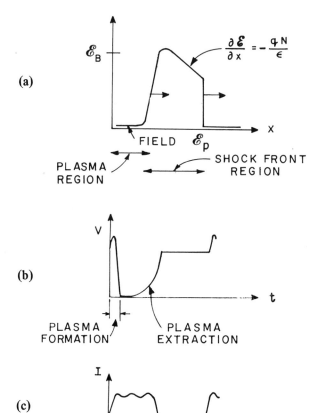

Fig. 11.11.  Trapatt mode of oscillation.
(a)  High-field shock-front transiting the diode.
(b)  Voltage waveform across the device.
(c)  Current waveform through the device.

Consider again the waveforms of Figs. 11.11 and 11.12. As the shock front develops the gradient of the field from Poisson's equation is $\partial \mathcal{E}/\partial x = qN/\epsilon$ where $N$ is the doping level. The displacement current, $J$, that flows is related to $\epsilon \partial \mathcal{E}/\partial t$. The shock-front velocity is given by

$$U = \frac{\partial \mathcal{E}}{\partial t} \bigg/ \frac{\partial \mathcal{E}}{\partial x} \qquad (11.23)$$

which is therefore

$$U = J/qN \qquad (11.24)$$

If $U$ is $3 \times 10^7$ cm-sec$^{-1}$ and $N$ is $2 \times 10^{15}$ cm$^{-3}$, the displacement current density is about $10^4$ A-cm$^{-2}$. For a diode of width 3-5 $\mu$m the shock-front transit time is of the order of $10^{-10}$ sec which accounts for the rapid rise of current and fall of voltage in Fig. 11.11.

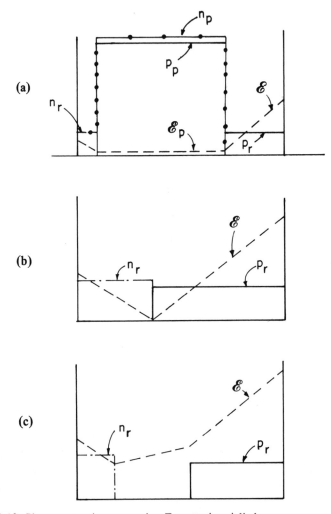

Fig. 11.12. Plasma-extraction process in a Trapatt $p^+$-$n$-$n^+$ diode.

(a)  The recovery is slow during the initial stages of plasma extraction since the field is low in the center region. ($n_D = p_D + N_D$ for an $n$ region.)

(b)  The electric field $\mathscr{E}$ has now fully penetrated the device and extraction proceeds more rapidly.

(c)  Extraction enters its final stage under saturation-drift velocity conditions everywhere. (After Clorfeine, Ikola, and Napoli. Reprinted with permission from *RCA Rev.*, **30**, Sept. 1969, pp. 410, 412.)

The plasma sweep-out transient has been studied by Scharfetter[168] and Gibbons,[68] with the condition that the electric field at the conclusion of the plasma-extraction period must not exceed the breakdown field to avoid premature avalanching. This leads to the conclusion that

$$2 \frac{\epsilon \mathcal{E}_B / q N_A}{W} \geq \frac{U}{v_s} + 1 \tag{11.25}$$

where $N_A$ is the shallow region doping in an $n^+ p p^+$ structure. For Trapatt performance the shock front must propagate faster than the carrier saturation drift velocity $v_s$. Hence if $U/v_s > 1$ we have

$$W < \epsilon \mathcal{E}_B / q N_A \tag{11.26}$$

Therefore, the width (thickness) of the diode must be less than the depletion width that would correspond to a peak-field strength $\mathcal{E}_B$ in an abrupt diode with uniform doping $N_A$. Such a diode is termed a punch-through diode since the entire lightly doped $p$-region is depleted before the avalanche field strength is reached. The depletion distance for breakdown avalanche-field conditions, namely, $\epsilon \mathcal{E}_B / q N_A$, as a ratio of the actual device width $W$ gives the diode punch-through factor $F$. Hence from Eq. (11.25)

$$2F \geq \frac{U}{v_s} + 1 \tag{11.27}$$

In typical Trapatt designs the doping levels and diode widths are chosen so that $F$ is about 2. Hence the shock-wave velocity is about three times the saturation-drift velocity.

The ratio of the nominal Impatt frequency to the Trapatt frequency is known as the subharmonic number. For high efficiency, waveforms rich in harmonics are desirable so the subharmonic number should be large, but this tends to suggest low Trapatt frequencies which are not usually desired. For a subharmonic number of 3, the efficiencies of Si and Ge Trapatts with a punch-through factor of 2.0 are as shown in Fig. 11.13(a) and the power output is as shown in Fig. 11.13(b). The dashed line, CW, indicates the continuous-wave power limit for Si. The corresponding depletion-layer thickness and doping conditions required are shown in Fig. 11.14(a) and (b). From these calculations, a 4-GHz Si design has a possible pulse-power output of 80 W with an efficiency of about 50%. The doping level would be about $6 \times 10^{15}$ cm$^{-3}$ and the device width about 4 $\mu$m. The 4-GHz negative-resistance value was chosen to be 10 $\Omega$ (the minimum value that can be conveniently matched) and for this the device area would have to be about $6 \times 10^{-4}$ cm$^2$.

Guidelines developed for the design of Trapatt-mode oscillators include a

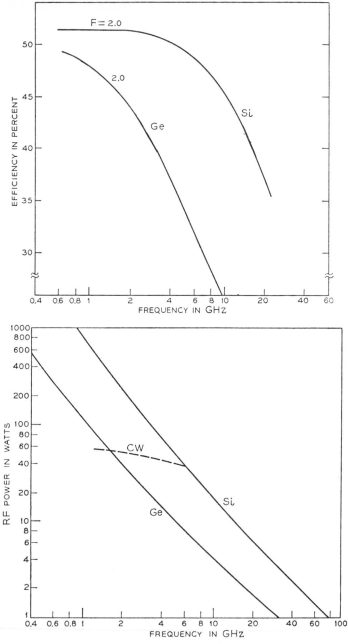

Fig. 11.13. Theoretical studies of Trapatt diodes of Si and Ge with punch-through factors of $F = 2$.

(a)  Efficiency versus frequency.

(b)  Power output versus frequency. (After Scharfetter. Reprinted with permission from *Bell Syst. Tech. J.,* **49**, 1970, pp. 818, 822; © 1970, American Telephone and Telegraph Company.)

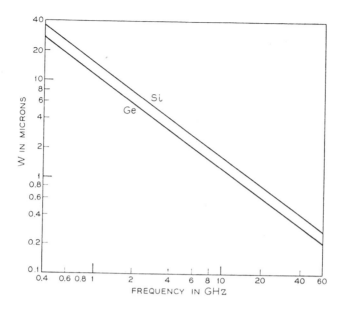

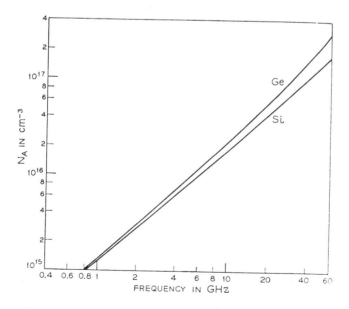

Fig. 11.14. Trapatt width and impurity concentration in the transit zone for the designs of Fig. 11.13. (After Scharfetter. Reprinted with permission from *Bell Syst. Tech. J.*, **49**, 1970, pp. 819, 820; © 1970, American Telephone and Telegraph Company.)

triangle on the plane of $J_{dc}$ versus $N_D W$ [see Fig. 11.15(a)] within which high efficiency should be obtained. Similar considerations have been used by Gibbons to arrive at a design triangle on the $J_{dc}/qNv_s$ versus $F$ plot. Line (1) comes from premature avalanche considerations, line (2) from the condition that the shock-front velocity must exceed the saturated-drift velocity, and line (3) from considerations of avalanche-zone width to total width.

From Fig. 11.13, the use of Ge for Trapatts offers no advantages over the use of Si. Ge has more equal-ionization coefficients but has a lower breakdown field and a lower saturated-drift velocity ($6.5 \times 10^6$ versus $8.9 \times 10^6$ cm/sec for Si). GaAs Impatts and Trapatts, however, may outperform Si structures if the material properties are carefully controlled. GaAs has a high-drift mobility for electrons at low-field strength, and a saturated-drift velocity comparable to that of Si [see Fig. 11.16] with a velocity maximum corresponding to low-mass electrons. The ionization rates for electrons and holes, although not equal in GaAs,[180] are closer in value than for Si, so the avalanche region is more compact. GaAs has the further advantage of a large bandgap which permits high-temperature operation; however, it is a material that is often not as well under control as Si. Diffusions and contacts are more difficult to make and material properties such as lifetime and trapping are sometimes troublesome even in high-grade epitaxially grown material. However, with Schottky-barrier junctions the best GaAs Impatts and Trapatts now surpass Si devices in performance.

Trapatt oscillators are very much influenced by the tuning and triggering of the microwave lines, cavities and waveguides in which they are placed, and the circuits are more complicated to trim than for Impatt oscillators. The circuits must limit the main RF power to the chosen output frequency and provide matching of impedance at this frequency. Furthermore, the reactances must be suitable at the higher harmonics of the output frequency such as the second and third harmonics and this must be accomplished with a small number of relatively independent adjustments. The circuit considerations involved have been reviewed by Clorfeine.[27] At low frequencies, 500 MHz-1 GHz, lumped-element-tuning circuits may be used.[29] At higher frequencies tuned cavities and waveguides are needed. The tuning conditions make it difficult to achieve high-efficiency Trapatt behavior dependably above 7 or 10 GHz.

In general, Trapatt action is observed only after a time representing many Impatt periods from the application of the bias-current pulse. Also there is a randomness about this delay in Trapatts because of the random buildup from noise of the Impatt oscillation. This is known as front-end jitter. Since the avalanche region width in an $n^+$-$p$-$p^+$ Si structure is narrower than for $p^+$-$n$-$n^+$ diodes, the Impatt performance of the former is better. Hence $n^+$-$p$-$p^+$ Trapatts are preferable since the overvoltage needed to trigger the trapped-plasma mode is more readily achieved at the current-density level for CW operation.[83]

Recently, a radial-cavity configuration with a metal disk, as shown in

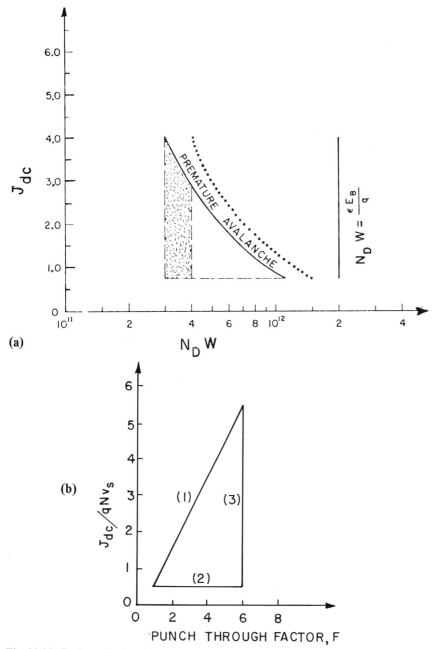

Fig. 11.15. Design triangles for Trapatt diodes: for high efficiency the design should be within the triangle.

(a) $J_{dc}$ versus $N_D W$. (After Clorfeine. Reprinted with permission from *IEEE Trans. Electron Devices*, **ED-18**, 1971, p. 552.)

(b) $J_{dc}/qNv_s$ versus $F$. (After Gibbons, from *Avalanche Diode Microwave Oscillators*, Oxford University Press, 1973.)

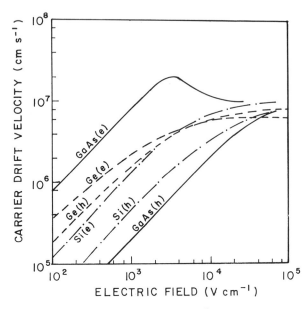

Fig. 11.16. Drift velocities of electrons and holes at 300°K in GaAs, Si, and Ge as a function of electric field.

Fig. 11.17, has been found to result in Trapatt oscillations starting with very little delay. The concept is that the disk, besides being a lumped capacitor with the ground plane at the Trapatt frequency, combines with lead inductance to form a radial cavity with characteristic resonant frequencies higher than the Trapatt frequency. If one of these higher frequencies is excited by the Trapatt current and if it is an integral multiple of the Trapatt frequency (this is done by adjusting the Trapatt frequency by the varying bias current), the voltage components are synchronous. The high-frequency component thus superimposed on the sawtooth waveform produces the necessary overvoltage and Trapatt action can occur. The frequency is controlled by the plasma-extinction time.

## 11.5 AVALANCHE-DIODE AMPLIFIERS

Impatt and Trapatt diodes may be used as amplifiers in suitably chosen applications. To obtain stable unidirectional gain from a negative-resistance device it is necessary to use a circulator-coupled reflectance-type circuit configuration. The basic concept is that if a transmission line of impedance $Z_0$ is terminated in a negative resistance, $R$, the reflected power for an input power $P_{in}$ is

$$P_{out} = \left(\frac{Z_0 + R}{Z_0 - R}\right)^2 P_{in} \qquad (11.28)$$

Hence the output power exceeds the input power and if a circulator is used the output and input may be separated and the circuit acts as a unidirectional amplifier. In most negative-resistance devices, Impatts being no exception, the negative-resistance value is highly dependent on the bias conditions. Therefore, if $(Z_0 - R)$ is made too small, the power gain $P_{out}/P_{in}$, changes considerably

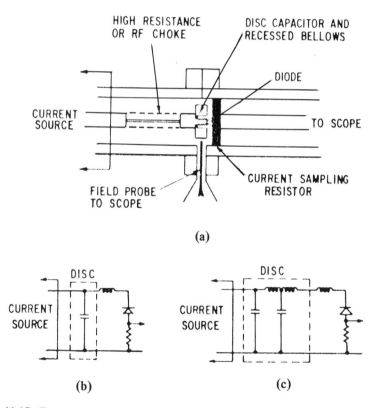

(a)

(b)    (c)

Fig. 11.17.  Trapatt cavity with disk that gives rapid turn-on of oscillation.
(a)  Cavity.
(b)  Equivalent circuit representing the disk as a lumped capacitor.
(c)  Equivalent circuit representing the disk as a capacitor and a resonant cavity. (After Yu and Tantraporn. Reprinted with permission from *IEEE Trans. Electron Devices,* ED-22, 1975, p. 140.)

with variation of $R$. Good design practice suggests that the power gain be limited to a factor of 10 (10 dB) or less per stage.

The circuit for an Impatt-diode-reflectance amplifier is shown in Fig. 11.18(a). Since the negative resistance is dependent on the frequency the output varies with frequency as shown in Fig. 11.18(b). For the frequency of 11.6 GHz the transfer characteristic as shown in Fig. 11.18(c) is quite nonlinear. The falloff of gain with increase of the input signal power is known as gain compression and may or may not be desirable, depending on the application. For an output power of 100 mW the gain is about 14 dB and the efficiency defined in terms of added power, $P_{add}/I_{dc}V_{dc}$, is about 6%.

The performance of a 6-GHz 4-W two-stage Impatt-diode amplifier is shown in Fig. 11.19. The input stage used two GaAs Impatt diodes that were power combined and the output stage used four Si Impatts. The locking bandwidth is of the order of 200 MHz. The power gain is 16 dB and the noise figure is 50 dB.

Several schemes are possible for obtaining high-maximum-power output in Impatt oscillators or amplifiers. One approach is to use individual oscillator circuits with separate current-regulated power supplies and injection lock the oscillators in frequency.[127] Another approach is to connect diodes in series in the same resonator circuit, but the series arrangement involves heat-sink problems.[70] Yet another technique is to connect Impatt (or Trapatt) diodes in parallel[114] in a suitable coaxial or cavity waveguide jig. Generally, the power obtainable from the combination is less than could be obtained from the diodes summed separately. Combining efficiencies are typically in the range of 75-90%.

Consider now Trapatt amplifiers: these handle more power than Impatt amplifiers since they can be operated in a trigger-type class C mode. The diodes are biased below their dc breakdown voltage so that negligible current is drawn in the absence of RF input to the amplifier. When RF is applied to the circulator and filter circuit arrangement of Fig. 11.20 the Trapatts are triggered-on and amplification occurs. Such a configuration centered at 2.5 GHz has given a peak power output of 60 W at 6.5-dB gain with a 14% bandwidth and an efficiency of about 18%. Stagger tuning of Trapatt amplifiers is possible to give broad bandwidths.[110] In typical phased-array radar systems in $S$ band (2.6-3.95 GHz) the bandwidth needed may be 300 MHz and the pulsewidth may have to be 50 $\mu$s (1% duty cycle) with a peak power of 50 W. Trapatt diodes with $n^+$-$n$-$p$-$p^+$ structures have been developed for this class of application.

### 11.6 COMMENTS ON PERFORMANCE

At present, Impatt diodes of about 3-5 W continuous-wave power are readily available in the frequency range of 3-12 GHz. These may be combined to obtain power levels of a few tens of watts (CW) with efficiencies of greater than 10%.

GaAs Impatt diodes with Pt-Schottky-barrier junction and ohmic contacts on diamond heat sinks have been reported with 10-W (CW) power output and

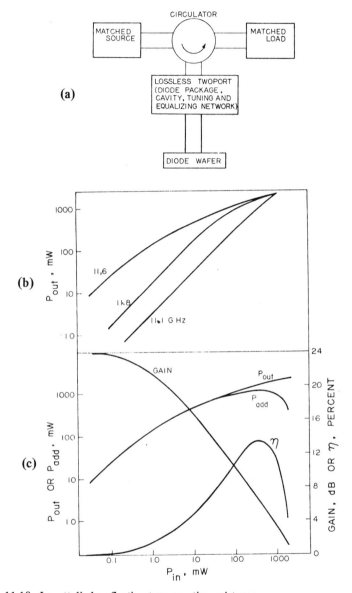

Fig. 11.18.  Impatt-diode reflection-type negative resistance.
(a)  Circulator-coupled circuit to separate the amplifier input and output.
(b)  Variation of output power with input power at different operating frequencies for an Impatt-diode amplifier in a coaxial circuit.
(c)  Variation of the power added, output power, gain, and efficiency with input-power level at a single-operating frequency (11.6 GHz) for the same amplifier. (After Gupta. Reprinted with permission from *IEEE Trans. Microwave Theory and Techniques*, **MTT-21**, 1973, pp. 692, 693.)

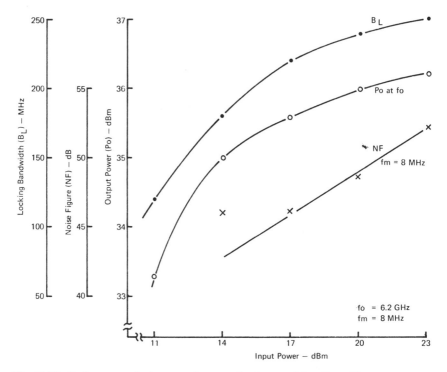

Fig. 11.19. Performance of a two-stage Impatt-reflection amplifier. (After Willing. Reprinted with permission from *IEEE Trans. Microwave Theory and Techniques*, **MTT-21**, 1973, p. 710.)

efficiencies of 20-30% at 6 GHz.[102] These are fabricated by chemical vapor deposition and have hi-lo donor profiles of ratio 20:1 or lo-hi-lo profiles with a charge clump of $3 \times 10^{12}$ cm$^{-2}$ located about 0.2 $\mu$m from the surface. Typical noise values are 52-58 dB. The performance of GaAs Schottky-Read hi-lo Impatt diodes at $X$ band has been discussed by Wisseman, et al.[206] The CW power of a single diode was 2.5 W at 26% efficiency and structures having four such mesas in parallel delivered 7 W at 9 GHz with 18% efficiency.

In the frequency range of 50-150 GHz Impatt diodes are of potential interest for waveguide communication systems and as pump sources for parametric amplifiers. With ion-implanted double-drift Si Impatt structures, a CW power level of 380 mW has been obtained at 92 GHz with an efficiency of 12.5%.[149] At 150 GHz, output powers of over 100 mW have been obtained.

Since Impatt diodes depend upon avalanche ionization and on the transit time of induced carriers, the noise-generating mechanisms may be expected to be significant and to result in AM and FM low-frequency noise. Expressions have been derived by Goedbloed[71] for the ratio of AM to FM noise and for the ratio of single-sideband intrinsic FM noise power to carrier power. The value of

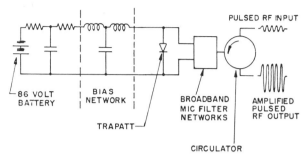

Fig. 11.20. Trapatt class-$C$ amplifier-circuit arrangement. (After Rosen, et al. Reprinted with permission from *RCA Rev.*, **33**, Dec. 1972, p. 733.)

$\Delta f_{\mathrm{rms}}$ depends strongly on the circuit $Q$.[72,199] An FM noise measure may be defined as

$$M_{\mathrm{FM}} = \frac{P_0}{B} \left(\frac{Q_{ex}\Delta f_{\mathrm{rms}}}{f_0}\right)^2 \Bigg/ kT_0 \qquad (11.29)$$

where $\Delta f_{\mathrm{rms}}$ is the rms frequency deviation measured in a window of bandwidth $B$ for a particular output power $P_0$, with an oscillator external quality factor $Q_{ex}$. The noise factor may be interpreted as the ratio of the diode's effective temperature to the thermal noise energy $kT_0$ for an ambient temperature $T_0$ of 290°K.

The FM noise factor for good diodes may be 36 dB for Si and 30 dB for GaAs and the FM and AM noise spectrum is fairly flat to within about 10 kHz of the carrier frequency. A circuit used for the measurement of the AM and FM noise of an Impatt oscillator is shown in Fig. 11.21. The AM and FM spectra observed for a GaAs Impatt-diode oscillator at about 30 GHz are given in Fig. 11.22(a) and (b). The AM noise-to-signal ratio is seen to be nearly as good as that of a klystron oscillator and is independent of the circuit $Q_{ex}$. However, the FM noise, even at a high $Q_{ex}$, is appreciably worse than that of the klystron. The $1/f$ dependence of the AM noise is not expected from avalanche-noise theory and may perhaps be upconverted flicker noise or $1/f$ noise arising from trapping states. The $f^{-1/2}$ dependence of the FM noise is compatible with upconverted $1/f$ noise. For systems using an intermediate frequency much greater than 200 kHz removed from the carrier frequency the excess noise is not likely to be troublesome. However, for Doppler radar systems it is objectionable.

In Impatt-reflection amplifiers, GaAs systems may have noise figures of about 1.0-15 dB less than for Si-based systems. Three Impatt amplifiers developed for Bell Systems microwave communications needs are listed in Table 11.1. The 1-3.5-W units were found to be worthwhile replacements for klystron amplifiers. The 10-W 6-GHz unit was not considered to have performance or economic

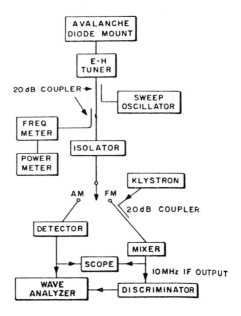

Fig. 11.21. System used for measurement of AM and FM noise of a 30-GHz Impatt oscillator. (After Weller, courtesy of RCA Laboratories. Reprinted with permission from *IEEE Trans. Electron Devices*, **ED-21**, 1973, p. 518.)

advantages over traveling-wave tube (TWT) amplifiers already in existence and delivering lifetimes of 20,000 hours or more.

At frequencies of 4 GHz or less transistor circuits, oscillators, or amplifiers are the most promising approach.

## 11.7 TRANSFERRED-ELECTRON DEVICE (GUNN) OSCILLATORS

The electron-drift velocity in $n$-GaAs peaks at $2.2 \times 10^7$ cm/sec for a field of 3.2 kV/cm and declines to $1.2 \times 10^7$ cm/sec at 10 kV/cm. The reason is the transfer of electrons from the direct bandgap $\Gamma_6^c$ conduction band minimum to the $L_6^c$ <111> minima.[4] This involves a 0.29-eV energy increase (produced by the applied field) and associated with this valley transfer is a higher effective mass. The result is a decrease of mobility from 7350 cm$^2$ V$^{-1}$ sec$^{-1}$ to 920 cm$^2$ V$^{-1}$ sec$^{-1}$. There are also $X_6^c$ minima at 0.35 eV above the $\Gamma_6^c$ minimum, with an effective mobility of 300 cm$^2$ V$^{-1}$ sec$^{-1}$, that may also be involved.

A bar of $n$-GaAs with dc voltage applied along the length develops a non-uniform field distribution associated with bunching of the electrons because of this mobility effect. The result is a depletion-accumulation domain that moves along the bar from cathode to anode and regenerates at the cathode. The current

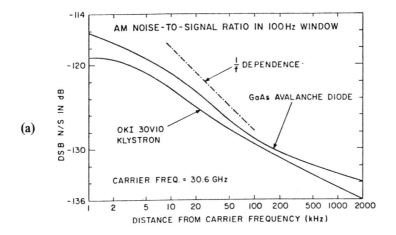

(a)

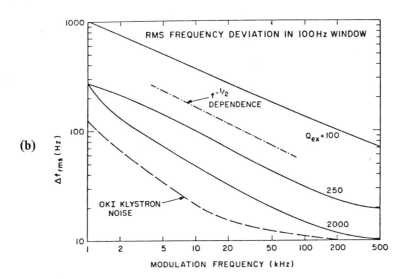

(b)

Fig. 11.22. Noise measurements for a 30-GHz GaAs-Impatt oscillation and comparison with a klystron oscillation at this frequency.

(a)  AM double-sideband noise-to-signal ratio.

(b)  Rms frequency deviation for several values of external cavity $Q$. (After Weller. Reprinted with permission from *IEEE Trans. Electron Devices*, **ED-20**, 1973, p. 520.)

dips with the formation of the domain and the bar develops an ac component that is known as a transferred-electron oscillation observed initially by J. B. Gunn.

Table 11.1.  IMPATT Amplifiers Developed for the
           Bell System

| Characteristics | 6 GHz | 6 GHz | 11 GHz |
|---|---|---|---|
| No. and type of diode | 1 Si | 5 GaAs | 3 GaAs |
| RF output power | 1 W | 10 | 3.5 W |
| dc input power | 24 W | – | 62 W |
| Amplifier noise figure | 50 dB | – | 38 dB |
| Nominal amplifier gain | 20 dB | 30-33 | 23 dB |

The domain electric fields are related to the drift velocity/field curve as shown in Fig. 11.23. The domain develops from a region of locally high field in the negative slope mobility region of Fig. 11.23(c). The electrons ahead of this fluctuation move further ahead of it since their drift velocity is greater than for electrons in the high-field region. Therefore, there is a pile up of electrons in a depletion-accumulation form. The buildup of domain field proceeds until the system acquires a stable domain moving with a drift velocity equal to that of electrons outside the domain.

The expressions to be considered are Poisson's equation:

$$\frac{\partial \mathcal{E}}{\partial x'} = \frac{q}{\epsilon}(n - n_0) \qquad (11.30)$$

where $x' = x - v_{\text{dom}}t$, and the current density equation:

$$J = q n_0 V_n = q n v(\mathcal{E}) - q \frac{\partial D(\mathcal{E})}{\partial x} n + \epsilon \frac{\partial \mathcal{E}}{\partial t} \qquad (11.31)$$

where

$$\frac{\partial \mathcal{E}}{\partial t} = - v_{\text{dom}} \frac{\partial \mathcal{E}}{\partial x'}$$

Division of Eqs. (11.30) and (11.31) to eliminate the variable $x'$, and the assumption that $D(\mathcal{E})$ is independent of $\mathcal{E}$, gives

$$(n - n_0)\frac{q}{\epsilon}\frac{d(Dn)}{d\mathcal{E}} = n\left[v(\mathcal{E}) - v_{\text{dom}}\right] - n_0\left[v_r - v_{\text{dom}}\right] \qquad (11.32)$$

Integration of Eq. (11.32) gives

$$\frac{n}{n_0} - \ln\frac{n}{n_0} - 1 = \frac{\epsilon}{q n_0 D}\int_{\mathcal{E}_r}^{\mathcal{E}}\left(\left[v(\mathcal{E}) - v_{\text{dom}}\right] - \frac{n_0}{n}\left[v_r - v_{\text{dom}}\right]\right)d\mathcal{E} \qquad (11.33)$$

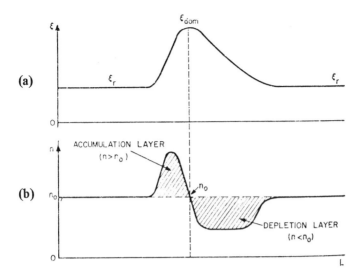

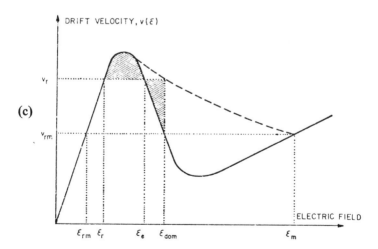

Fig. 11.23. Gunn-domain field variation and depletion-accumulation layer form.
(a)  Electric field versus distance at some domain position.
(b)  Electron-dipole layer constituting a domain that moves from left to right.
(c)  Domain fields related to the drift-velocity field plot by the equal-areas rule. (After Butcher, Fawcett, and Hilsum. Reprinted from *British J. Appl. Phys.*, **17**, 1966, p. 843; © Crown.)

From Fig. 11.23(b), the electron density is $n_0$ if the field is either $\mathscr{E}_r$ or $\mathscr{E}_{dom}$ and the left-hand side of Eq. (11.33) becomes zero. The form of the integration then requires that $v_r = v_{dom}$ and that

$$\int_{\mathscr{E}_r}^{\mathscr{E}_{dom}} [v(\mathscr{E}) - v_r]\, d\mathscr{E} = 0 \qquad (11.34)$$

Equation (11.34) is satisfied if the two areas shown in Fig. 11.23(c) are equal. This equal areas rule results in the dashed line and shows that stable-domain propagation cannot be obtained if the field strength is below $\mathscr{E}_{rm}$.

The frequency of oscillation is given by

$$f = v_{dom}/L_a \qquad (11.35)$$

where $L_a$ is approximately the device length (the qualification is needed because of the finite length of the domain). Hence for an oscillation at 2 GHz, the length must be about 50 $\mu$m.

During the early stages of space-charge growth the buildup of charge is determined by the negative dielectric relaxation time $\tau_{neg} = \epsilon/q\, n_0\, \mu_{neg}$ where $\mu_{neg}$ is the negative electron mobility. The growth that is available on a length $L$ is $\exp(L/v\tau_{neg})$ and if this is to be greater than unity,

$$n_0 L > \frac{\epsilon v}{q\mu_{neg}} \qquad (11.36)$$

$$> 10^{12} \text{ cm}^{-2} \text{ for GaAs}$$

For fully mature domains $n_0 L$ should preferably be $10^{13}$ cm$^{-2}$ or greater. Hence for a bar of length 50 $\mu$m the free-carrier density must be $2 \times 10^{15}$ cm$^{-3}$ or more.

The excess voltage in the domain is given by

$$V_{ex} = V - L\mathscr{E}_r = \int_{-\infty}^{+\infty} (\mathscr{E}(x) - \mathscr{E}_r)\, dx \qquad (11.37)$$

For an excess voltage of 5 V and a doping of $10^{15}$ cm$^{-3}$ the domain width is about 5 $\mu$m.[36]

Consider now the power and efficiency limitations of a transit-time device of frequency $f$ and impedance $R$. The maximum voltage that may be applied is

$$V_{max} = \mathscr{E}_B L = \mathscr{E}_B \frac{v_s}{f} \qquad (11.38)$$

where $\mathscr{E}_B$ is the breakdown (threshold) field and $v_s$ is the saturated drift velocity. The maximum current for an impedance $R$ is

$$I_{max} = \frac{V_{max}}{R} = v_s \frac{\mathscr{E}_B}{Rf} \tag{11.39}$$

The total power assuming square waves and a matched-load resistance is

$$P = \frac{\mathscr{E}_B^2 v_s^2}{4Rf^2} \tag{11.40}$$

Hence the power-impedance product is inversely proportional to the square of the transit-time frequency. (This may be compared with Eq. (11.21) for avalanche diodes.) The impedance at a given frequency is inversely proportional to the device area since the length is set by the frequency.

The waveforms in a practical Gunn oscillator are far from square and, do not represent complete swings of the dc bias current. Usually the load resistance used is many times the low-field resistance of the device since this provides maximum power and efficiency [see Fig. 11.24(a)]. Efficiencies of 4% are typical of early devices, but recently 0.5-W 10-GHz devices have shown efficiencies of 10% or more. To obtain high-power outputs, pulsed modes of operation must be used [see Fig. 11.24(b)].

Recently studies have been made with InP as the material involved instead of GaAs and promising results have been obtained. The impetus to explore transferred-electron effects in InP was based on good velocity/field curves expected from the energy-band structure.

If a transferred-electron device is mounted in a resonant circuit cavity, a high-frequency nontransit mode of operation is possible. This has been termed a limited-space-charge accumulation (LSA) mode since calculations show that apart from an accumulation layer near the cathode, there is little space-charge formation. The mathematical details of operation will not be presented here, but Fig. 11.25 is helpful in visualizing the operating boundaries of this mode. Calculations of dc to RF conversion efficiency suggest 18% or more to be possible in a uniformly doped device, but if doping fluctuations of 20% peak-to-peak exist, the efficiency may be only 10%. Also, the efficiency decreases at frequencies much above 20 GHz because of energy and intervalley relaxation processes. A quenched-domain mode of operation is also possible (see Fig. 11.25) but is lower in efficiency than the LSA mode although less demanding on doping uniformity.

Now consider the tuning of oscillator frequency. Even though the Gunn mode of oscillation is basically domain transit-time-limited the effective resonant $Q$ of a typical device is small (5-10). Therefore, tuning over an octave frequency is possible electrically or by cavity tuning mechanically. For instance, a ferri-magnetic resonator sphere (of diameter 0.030 in.) in a cavity with a Gunn device

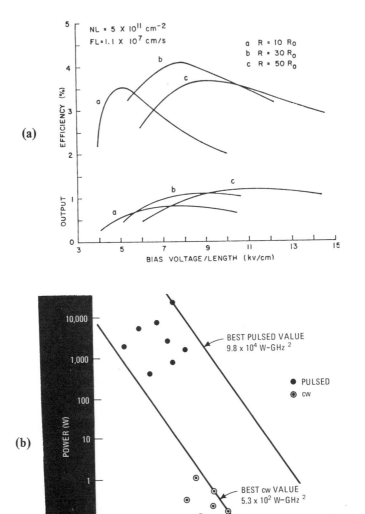

Fig. 11.24. Gunn-diode oscillator sources.
(a)  Calculated efficiency and power curves. Lower curves indicate output power scaled so that one unit on the vertical scale is 200 W/cm². Higher efficiency can be obtained with other doping × length, $nL$, and frequency × length, $fL$, products. (After Copeland. Reprinted with permission from *IEEE Trans. Electron Devices,* **ED-14,** 1967, p. 497.)
(b)  Power obtained for devices of various frequencies under pulse and CW conditions. The product of power and frequency represents a figure of merit. (After Berson. Reprinted from *Electronics,* Dec. 4, 1972; © McGraw-Hill, N.Y., 1972.)

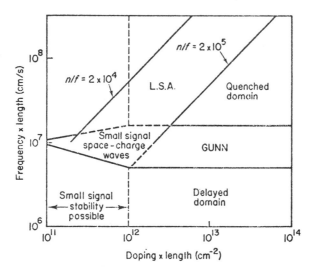

Fig. 11.25. The frequency × length and doping × length products are plotted to show the values appropriate to various modes in GaAs. The positions of the lines vary with material quality, temperature, and circuit conditions, and represent diffuse rather than hard boundaries. (After Copeland. Reprinted with permission from the *J. Appl. Phys.*, **38**, 1967, p. 3096.)

produces a linear tuning range of 6-13 GHz for a magnetic field change of 2 to 4.5 kG.[55] Tuning by a series-connected varactor is also possible.[76]

The frequency of a Gunn oscillator say at 7-10 GHz may change by 0.5 MHz/hr and the temperature dependence may be 0.5-3 MHz/°C change. In a well-designed waveguide cavity, however, the temperature dependence may be only 50 kHz/°C for a range −40° to +70°C. The AM noise level of a Gunn oscillator is quite low and the FM noise level is at least as good as that of a comparable klystron or Impatt oscillator.

Gunn diodes have been designed for use in radar-type circuits as transmitter and local oscillator power sources. Two such applications are illustrated by the schematic circuits of Fig. 11.26.

Gunn-effect devices may also be used as negative-conductance amplifiers as described earlier for Impatt diodes.

## 11.8 TED (GUNN) LOGIC CONCEPTS

Transferred-electron devices (TEDs) operating in the dipole-domain mode have promise for fast combinatorial and sequential-logic circuits. Numerous circuits have been devised involving diode logic TEDs. These suffer, however, from the usual circuit problems of two-terminal devices—namely, the difficulties of isolating input and output and obtaining reflection insensitivity. Therefore, let

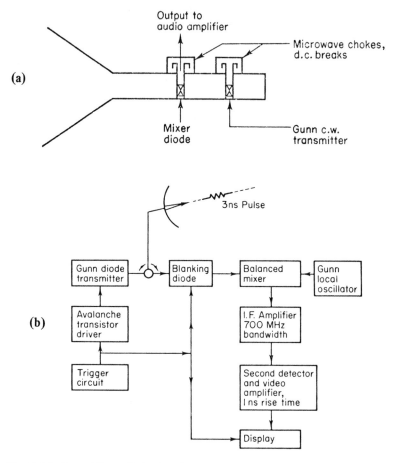

Fig. 11.26.  Uses of Gunn diodes in radar systems.

(a)  Gunn-diode 10-GHz Doppler radar suitable for a burglar alarm or speed meter. A frequency shift of a few tens of hertz may be detected.

(b)  Gunn diode 5-W 14-GHz 3-ns pulse oscillator as part of a high-resolution radar. (After Bulman, Hobson, and Taylor. Reprinted with permission from *Transferred Electron Devices*, Academic Press, Inc., London Ltd., 1972, pp. 305, 307.)

us pass over diode logic concepts in favor of discussing three-terminal structures with Schottky-barrier gates. A typical three-terminal planar TED is shown in Fig. 11.27. The domain is triggered by the application of a negative voltage pulse to the gate near the cathode. This causes the depletion region under the gate to expand and increases the field in the channel to above the critical level for a domain to develop. As the domain forms and moves towards the anode, the current through the TED is decreased and this provides the output signal. The shape of the output pulse is determined by the geometry and doping of

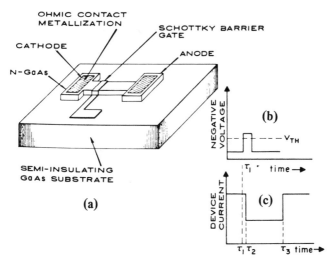

Fig. 11.27. Three-terminal planar TED: (a) device structure; (b) gate voltage; and (c) resulting device current. When the domain reaches the anode, a narrow current spike is usually generated. Such spikes do not, in general, affect the operation of the logic TEDs and these spikes are not shown. (After Sterzer. Reprinted with permission from *RCA Rev.*, 34, March 1973, p. 155.)

the device and is relatively independent of the shape of the input pulse. Hence, pulse regeneration is obtained. TEDs also have gain since the output pulse can be larger than the gate input pulse. This allows fan-outs of up to ten. The device supports only one domain at a time and does not respond to an input signal as long as a domain is moving through it. A further feature of the Schottky-gate structure is that it has a reverse attenuation of as much as 15 dB in blocking response to signals traveling in the reverse direction.

In logic applications it may be desirable to have a delay time between the application of the signal pulse and the appearance of the output pulse. In TEDs, this is easily obtained by including on the structure a notch that reduces the device area near the anode [see Fig. 11.28(a)] and produces a further drop in current as the domain passes through it. Application of such a structure to a three-stage delay line is shown in Fig. 11.28(b). The domain velocity is about $10^7$ cm/sec and so a propagation length before the notch of 100 $\mu$m provides a delay of 1 ns. The regeneration of the pulse at each stage is a feature not obtained with passive-delay circuits such as acoustic lines, and tapped delay line action is also passive. A memory or latch circuit is shown in Fig. 11.28(c). This generates a continuing train of output pulses. It is turned on by the application of a negative set-pulse to the gate of the first TED. This nucleates a domain and when this reaches the notch of the first TED, the voltage in the load resistance drops to a value that triggers a domain in the second TED. When

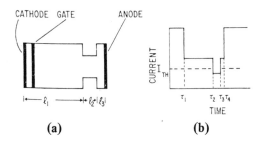

**(a)**                  **(b)**

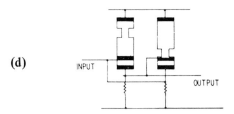

**(c)**

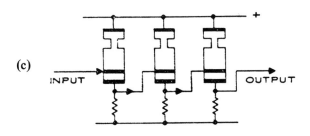

**(d)**

Fig. 11.28. Three-terminal planar TED with notch near anode: (a) device geometry; and (b) device current as a function of time. The domain is nucleated at time $t_1$, reaches the notch at time $t_2$, leaves the notch at time $t_3$, is absorbed at the anode at time $t_4$.
(c) Three-stage delay line using the three-terminal TED.
(d) Schematic diagram of a TED memory element. (After Sterzer. Reprinted with permission from *RCA Rev.*, **34**, March 1973, pp. 157, 158, 160.)

this domain reaches its notch a negative pulse is generated that triggers a second domain in the first device and so on. To turn off the train of pulses, a positive going reset pulse is applied to the gate of the first TED which prevents it being triggered by the second TED.

An inverter or biphase-modulator circuit is shown in Fig. 11.29(a). The

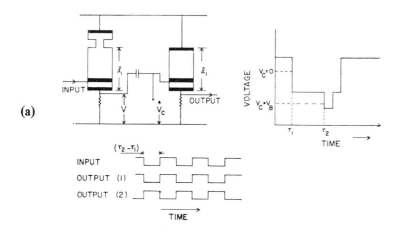

**(a)**

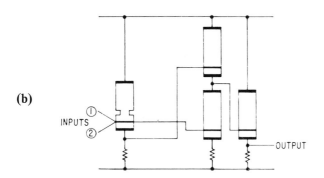

**(b)**

Fig. 11.29. Inverter modulator and exclusive-OR circuits.
(a)  Biphase modulator using two TEDs. Also shown is the voltage across the load resistor
    of the first device as a function of time, and the input and output pulse trains.
(b)  Schematic of an exclusive-OR circuit. (After Sterzer. Reprinted with permission from
    *RCA Rev.*, **34**, March 1973, pp. 160, 161.)

output of the notched TED is applied to the gate of the second shorter TED.
With no applied control voltage, the second TED is triggered at time $t_1$. When a
positive control voltage of appropriate magnitude is applied, the second TED
is triggered at time $t_2$. Biphase modulation is obtained if the time difference
$t_2 - t_1$ is made equal to one-half of the period of the input pulse train.

The second circuit shown in Fig. 11.29 has an exclusive-OR function.
The input pulses are applied to the gate of a notched input TED and to
the gate of the lower of the two series-connected TEDs. The input TED is
triggered if either or both of the negative input pulses are present, while

the lower series-connected TED is triggered only if both input pulses are simultaneously present. The gate of the upper series-connected TED is connected to the cathode of the input TED. The upper TED is triggered whenever a domain passes the notch of the input TED, provided there are no domains traversing the lower TED. If the lower TED carries a domain, its resistance becomes so large that the voltage across the upper TED falls below the threshold value required for domain formation. The anode of the lower TED and the cathode of the upper TED are connected to the gate of an output TED. This output TED is triggered whenever its gate is driven negative.

If there are no input pulses, no domains are triggered in any of the TEDs and no output pulse is generated. If one input pulse is present, the input and the upper TED are triggered, the gate of the output TED is driven negative, and the output pulse is produced. If both input pulses are present, the input TED and the lower TED are triggered, the gate of the output TED is driven positive, and no output pulse is generated.

Hence the circuit performs in accord with the truth table of an exclusive-OR function.

The trigger sensitivity of Schottky-barrier TEDs may be improved by the provision of a subsidiary Schottky anode (see Fig. 11.30) that suppresses the formation of a high-field region near the anode. The $\Delta V_g$ min range may then be 100-200 mV instead of 250-800 mV.

We will now consider further logic functions. Inhibit or NOT action is obtainable from the circuit of Fig. 11.31. When the time difference of the

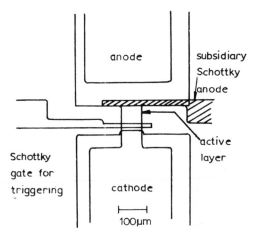

Fig. 11.30. Schottky-gate coplanar Gunn tetrode with subsidiary Schottky anode. (After Kurumada, Mizutani, and Fugimoto. Reprinted from *Electronics Lett.*, **10**, 1974, p. 161, with permission from IEE, London.)

two-input pulse, at $S_1$ and $S_2$, is smaller than the output pulsewidth, the preceding input pulse inhibits the following input pulse. This is because a domain in one device lowers the internal field in the other device below the level at which a gate pulse can trigger it.

AND action can be obtained by requiring trigger pulses on both the anode and the gate to obtain domain action as in Fig. 11.31(b), or by the parallel connection of devices. Circuits with multiple-input gates can perform logic functions such as AND, OR, and INHIBITION.[109] Other functions are possible,

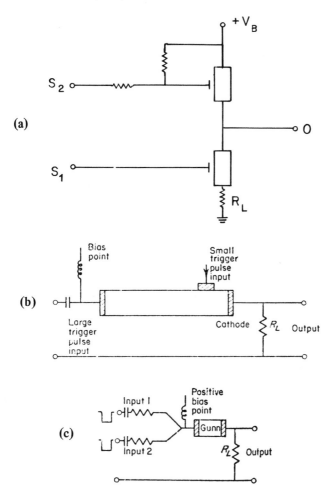

Fig. 11.31.  Other TED logic circuits.
(a)   NOT action depends on the preceding input pulse inhibiting the next input pulse.
(b)   AND action: anode and gate must have trigger pulses for a domain to form.
(c)   NOR or NAND action depends on bias level. (After Hartnagel, from *Gunn-Effect Logic Devices*, Elsevier, N.Y., © 1973.)

for instance, the circuit of Fig. 11.31(c) biased above threshold produces no output for a pulse at either input (NOR function) or no output for pulses at both inputs (NAND function) depending on the bias-adjustment conditions.

A TED memory action has already been described (Fig. 11.28) but a similar action may also be obtained from the TED-capacitor circuit of Fig. 11.32(a). Once a domain is triggered by the gate-input pulse (set pulse), it travels through the device and disappears at the anode. However, while and after the domain

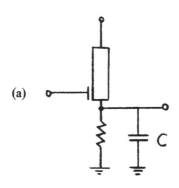

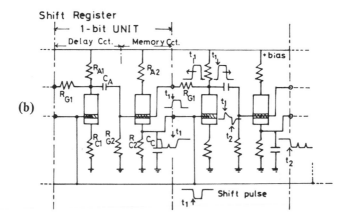

Fig. 11.32. TED and capacitor provides memory-train action.
(a) Basic circuit produces a train of pulses until turnoff by positive gate pulse.
(b) TED shift register. (After Sugeta and Yanai. Reprinted with permission from *Proc. IEEE*, **60**, 1972, p. 238.)

disappears, the cathode voltage returns to the high-bias level with delay due to the charging time of $C$. For this delay, the voltage between the anode and the cathode is higher than the bias level just after a domain disappears. Therefore, the next domain can be triggered, and a train of pulses ensues. Turnoff is by the application of a positive gate pulse.

A shift-register action may be obtained by cascading delay circuits. Such a memory circuit is shown in Fig. 11.32(b). When the first stage has a signal train in its memory unit, a negative shift pulse applied to the shift line and the negative pulse from the anode of the first memory unit, together succeed in nucleating a domain in the diode of the subsequent delay unit. A positive pulse is obtained at the anode of this delay unit and is applied both to the gate of the preceding memory circuit to extinguish the memorized state of the first stage, and to a differentiating capacitance circuit. The differentiated signal contains a positive pulse followed by a negative one, and the latter pulse nucleates a domain in the second memory unit when applied to the gate electrode. In this way, the pulse signal is shifted from the first unit to the subsequent one, and shifting occurs by one step each time a shift signal is applied. TED shift registers capable of operating at clock rates of 1.6 GHz have been reported.

Dynamic-frequency division is possible with TEDs by use of the feature of nonresponse to signals while a domain is in transit. If the input-signal frequency is twice the domain transit-time frequency then every second signal pulse is ignored and division by 2 is obtained. Division by integers up to 6 and for frequencies as high as 16 GHz has been obtained.[195]

Transverse extension of a high-field domain may proceed at a velocity of $10^8$ cm/sec or higher, which is much greater than the normal $10^7$ cm/sec domain transit velocity. An $E$-shaped logic gate that uses this effect consists of two TED branches linked by a center section which has an ohmic gate to control the potential in this region [see Fig. 11.33(a)]. The transverse propagation of the domain in branch 1 into branch 2 is shown by a computer simulation in Fig. 11.33(b) and (c). It occurs only if the ohmic gate is at a suitable potential (3 V in this study). The logic function performed by this device is therefore

$$Z_1 = X + Y.\sim W$$

$$Z_2 = Y + X.\sim W$$

where +, 1, and $\sim$ mean logical sum, logical product, and negation, respectively, and where the terms are applied as shown in Fig. 11.33(d). A delay time of about 50 ps is obtained with a gate width of 100 $\mu$m. Even faster switching times are considered possible.

A 4-bit carry generator using lateral spreading has been fabricated with the comb-shaped structure shown in Fig. 11.34(a). This could be of use in fast adders where the limit to the speed of operation is the carry signal which has to

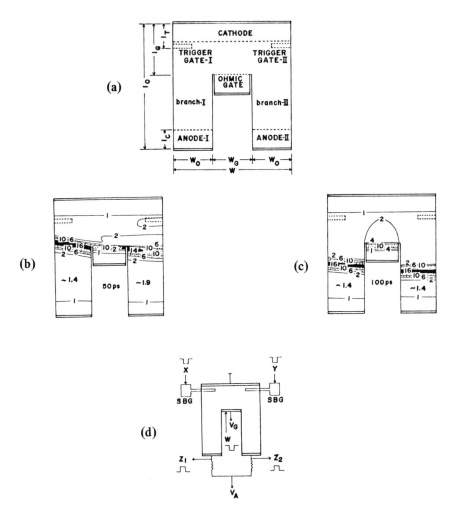

Fig. 11.33. *E*-shaped TED logic gate using transverse extension of the high-feld domain.
(a)  Structure. Typical dimensions may be $L_0 = 24$ μm, $L_g = 10$ μm, $W_0 = W_g = 8$ μm.
(b) and (c) Computer simulation of field strengths in kilovolts per centimeter 50 ps and
      100 ps after the application of voltage to trigger gate No. 1. Transverse extension of
      the domain into branch 2 is seen.
(d)  Logic application. (After Goto, et al. Reprinted with permission from *IEEE Trans.
      Electron Devices,* **ED-23,** 1976, p. 21.)

propagate serially from the least to the most significant bit in the worst case.
The lateral-spreading action provides the traveling domain at each stage which
corresponds to a carry signal in each bit. For a gate-on voltage of 3 V the domain
propagates to the next stage within a delay of 10 ps whereas if the gate voltage
is 2.5 V the domain does not propagate and there is no carry effect. The

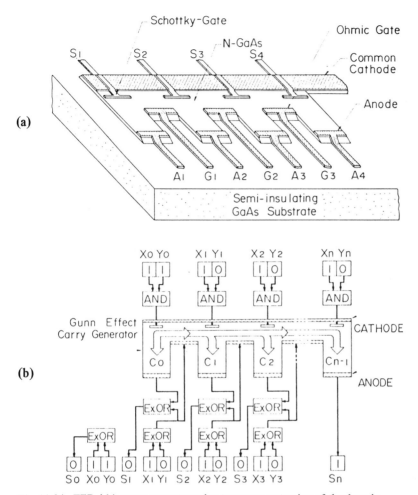

Fig. 11.34. TED 4-bit carry generator using transverse extension of the domain.

(a) Structure.

(b) Principle of $n$-bit parallel adder with the Gunn-effect carry generator. $X1$ and $Y1$ are binary inputs. $S1$ and $C1$ are sum and carry signals, respectively. (After Nakamura, et al. Reprinted with permission from *IEEE Int. Solid-State Circuits Conference, Digest of Technical Papers*, 1975, pp. 166, 167.)

organization of an $n$-bit parallel adder with TED AND gates and exclusive-OR gates is shown in Fig. 11.34(b).

The prospects for further development and ultimate use of TED logic are uncertain at this time for reasons of power dissipation and size. Also GaAs integrated-circuit fabrication is presently not as well developed as for Si and is technically more difficult. The experience that is presently being gained with high-speed (10 GHz) GaAs MESFETs fabricated in thin-epitaxial layers on semi-insulating substrates should be a contributing factor to growth of the logic technology. Work is also in progress with GaAs devices for integrated optics, and optical coupling may be possible as part of TED logic circuits.

In high-speed logic arrays the device areas must be small and the distances between gates short so that signal travel times are low. High-density packing raises the question of the power dissipation requirement. The bias voltage involved in TED logic is $\mathcal{E}_B L$ where $\mathcal{E}_B$ is at least the bias threshold field ($\sim 3$ kV/cm) and $L$ is the device length. The device resistance is $(q\mu n)^{-1} L/A$ where $n$ is the electron density per cm$^3$. Hence the power dissipation when biased is

$$P = \mathcal{E}_B{}^2 \, qA \, \mu \, n \, L \qquad (11.41)$$

For the formation of mature domains the product $nL$ (Fig. 11.25) must be greater or equal to $5 \times 10^{12}$/cm$^2$. However, there is also a limitation on how thin the GaAs layer may be made before surface effects begin to disturb domain formation. This matter has not been fully examined but it appears that the thickness in cm should be greater or equal to $2 \times 10^{11}$/n.[86] For a device of length 50 $\mu$m the doping needs to be $10^{15}$/cm$^3$ and the device voltage would be about 15 V for a field of 3 kV/cm. If the device width is 25 $\mu$m and thickness 2 $\mu$m and the electron mobility is $5 \times 10^3$ cm$^2$/V · sec the device resistance is 12.5 k$\Omega$ and the device bias current is 1.2 mA. The power dissipation in the absence of a domain is therefore 18 mW and less (say 10 mW) in the presence of a domain. If one logic event involves transit of one domain in $0.5 \times 10^{-9}$ sec (for a 2-GHz domain repetition rate) the power involved is 5 pJ. This is appreciably larger than the best estimates for GaAs or Si FET logic gates. Furthermore, it should be noted that "normally-off" logic cannot be achieved as it can for FET logic.

The active device area considered in the example above is $25 \times 50$ $\mu$m excluding interconnection area. This is larger than areas of Si logic gates made to 5-$\mu$m design rules. There is a possibility that Si logic-gate areas may be decreased significantly as electron-beam lithography becomes more available. However, there is also room for reduction of TED-logic gate dimensions, although the upper frequency limit is likely to be about 40 GHz which corresponds to device lengths of 2 or 3 $\mu$m.

## 11.9  QUESTIONS

1. What is the resonant frequency $f_r$ (where $(2\pi f_r)^2 = 2J_0 V_s \epsilon^{-1} d\alpha/d\mathcal{E}$) and impedance $Z(j\omega)$ of an abrupt junction $p^+$-$n$-$n^+$ Ge Impatt diode of area $5 \times 10^{-4}$ cm$^2$, $n$ region thickness 15 $\mu$m, $n$-region doping $10^{16}$ cm$^{-3}$, and current density 1000 A-cm$^{-2}$.

2. Design a Si-Read diode $p^+$-$n$-$i$-$n^+$ structure to oscillate at 10 GHz and provide 0.5 W of CW power at 300°K.

3. Consider a Si Impatt diode of $p^+$-$n$-$n^+$ form and area $10^{-3}$ cm$^2$ where the $n$ region of thickness 20 $\mu$m is doped $2 \times 10^{15}$ cm$^{-3}$. What is the dc power that must be dissipated if a reverse voltage of 225 V is applied?

4. For the Si-Impatt diode of Question 2 what is the avalanche frequency at a current of 1 A? Find approximate values for the transit time in the avalanche zone, its length, and the average ionization coefficient.

5. A ring-shaped Impatt diode has a better ratio of thermal conductance to impedance than a disk-shaped diode. Consider this quantitatively for a ring with the hole diameter one-half of the outside diameter at $X$ band, 1-W output power, 15% oscillator efficiency, on a copper heat sink. (Use an approximate heat-flow treatment.)

6. Consider the power density and device area needed for a Trapatt 3-GHz oscillator having a peak RF power of 0.5 kW, a pulsewidth of 50 $\mu$s, and a 1% duty cycle.[110]

7. Show that for mature dipole-layer-mode oscillations in a Gunn diode $n_o L > \dfrac{\epsilon V_d}{q |\mu_{neg}|}$, where $n_0$ is the equilibrium doping level, $v_d$ is the dipole-drift velocity, and $|\mu_{neg}|$ is the magnitude of the electron negative mobility. Given that $n_o L$ is $2 \times 10^{12}$ cm$^{-2}$ in GaAs how long is the longest device capable of sustaining oscillations at 10 GHz? What is the minimum $n_0$ and what is the minimum operating voltage at 10 GHz?

8. Assume the transfer of electrons in a Gunn-type semiconductor is represented by $\dfrac{n_2}{n_1} = \left(\dfrac{\mathcal{E}}{\mathcal{E}_0}\right)^4$ where $\mathcal{E}_0$ is 4000 V/cm and $\mu_2 = 0.05(\mu_1)$. Find an expression for the current density as a function of device voltage if $\mu_1$ is 8000 cm$^2$/V·sec and $n$ is $10^{14}$ cm$^{-3}$.

9. Using the linear velocity versus field approximation proposed by Butcher et al.[15] estimate the field distribution and the dipole length in a GaAs Gunn diode, doped $n = 10^{15}$ cm$^{-3}$ and of length 20 $\mu$m, at an operating voltage of 30 V. (Take $\mathcal{E}_T = 3$ kV/cm, $\mathcal{E}_V = 100$ kV/cm, $\mu_1 = 7000$ cm$^2$/V·sec and $\mu_2 = 40$ cm$^2$/V·sec.)

10. Under what conditions do current limits (space-charge limited) or thermal limits determine the performance or scaling of Impatt diodes?[150]

11. Discuss the causes of premature avalanche effects in Impatt diodes and how to avoid them.

12. Discuss the use of $Q$ as a figure of merit for Impatt oscillator performance. What experimental evidence has been published to prove that it is a sound concept?

13. Compare the relative merits of $n^+$-$p$-$p^+$ and $p^+$-$n$-$n^+$ Impatt structures for Si and GaAs.[132]

14. Locate and present some typical plots of susceptance versus negative conductance for an Impatt diode and interpret them.

15. Discuss the effects of temperature on Si and GaAs Impatt diodes in the range 77-500°K.

16. Discuss quantitatively the effect of undepleted high-resistance regions on Impatt efficiency.[3]

17. Discuss the merits of double-drift Impatt diodes.[159]

18. Discuss varactor tuning of an Impatt diode oscillator.[25]

19. Describe magnetic tuning of an Impatt diode.[69]

20. Discuss parallel operation of Impatt diodes and combining efficiency.[114]

21. Review Impatt noise theory and the extent to which theory and experimental results match.

22. It has been stated that $f^{-1/2}$ FM noise is associated with upconverted $1/f$ noise in an Impatt oscillator.[199] Explain this.

23. Discuss the assumptions made in the derivation of the relationship between Trapatt-avalanche-shock-wave velocity and the punch-through factor in Eq. (11.27).

24. Discuss the design triangle concepts for Trapatts presented by Clorfeine,[26] and Gibbons.[68] Are the concepts for Impatts significantly different?

25. Discuss the merits of $n^+$-$n$-$p$-$p^+$ Trapatt diodes.
26. Explain with clear diagrams several methods by which harmonic tuning could be obtained for a Trapatt 3-GHz diode.
27. How can harmonic power be extracted from a Trapatt oscillator and what are typical efficiencies?
28. Discuss series connections of Impatt and Trapatts. [70]
29. Explore the relevance of the reflected power expression [Eq. (11.28)] to the performance of an Impatt-diode amplifier.
30. Impatt amplifiers may contain injection-locked oscillator stages. [156] Discuss injection-locking concepts in this context.
31. Under what circumstances would Impatt-amplifier gain compression [80] be acceptable or valuable?
32. Consider the design of multiple-stage Impatt amplifiers. [203]
33. Discuss the merits and limitations of Baritt diodes. (See *Electronics*, March 7, 1974, p. 38.) What are not all Impatt diodes made this way?
34. Compare the performance, efficiency, noise, cost, weight, reliability, etc. of transistor, Impatt, Gunn, and klystron oscillators with 1-W (CW) power output at 5 GHz.
35. Explain the factors that influence domain shape and the current waveform in Gunn oscillators.
36. Explain the LSA mode of oscillation in transferred-electron devices.
37. Discuss power/frequency relations in Gunn oscillators.
38. Discuss electronic tuning of Gunn oscillators.
39. Discuss FM and AM noise in Gunn oscillators.
40. Consider the effects of doping nonuniformities in Gunn and LSA modes of operation.
41. Discuss the frequency-temperature stability of GaAs Gunn domain oscillators.
42. Discuss the performance of Gunn oscillators at frequencies above 20 GHz.
43. Discuss the development of InP transferred-electron oscillators.
44. Discuss Gunn-effect amplifiers involving circulators and also traveling-wave versions.
45. Discuss radar applications of Gunn-effect oscillators.
46. Discuss several ways in which third electrode triggering of a domain may be obtained and compare their usefulness.
47. Review the most promising Schottky-barrier-gated TED logic circuits for AND, OR, NAND, NOR, INHIBIT, and Exclusive-OR.
48. Review the most promising TED logic circuits based on transverse domain spreading.
49. Discuss three-level logic-circuit possibilities with TEDs.
50. Discuss TED shift registers in some detail.
51. Consider the integration of TEDs and GaAs MESFETs for logic functions.
52. Transferred-electron devices with profiled device-channel areas have been proposed for analog-to-digital and digital-to-analog conversion. Discuss these possibilities.
53. Discuss frequency division with transverse electron oscillators.
54. Discuss the possible applications of TEDs to gigapulse communications systems.
55. Discuss the interaction of Gunn domains with light including the possible electro-optic modulation of a laser beam.
56. Discuss packing density and power-dissipation problems in TED logic circuits and compare with an equivalent Si logic circuit with a high-clock pulse rate.

## 11. REFERENCES AND FURTHER READING SUGGESTIONS

1. Ahamd, S., and J. Freyer, "High-power Pt Schottky Baritt diodes," *Electronics Lett.,* 13 May 1976, p. 238.
2. Anderson, W.A., "Experimental radiation effects in Gunn and Impatt diodes (a sum-

mary)," *Third Biennial Cornell Electrical Engineering Conference*, Ithaca, N.Y., 1971, p. 245.

3. Aono, Y., and Y. Okuto, "Effect of undepleted high-resistivity region on microwave efficiency of GaAs Impatt diodes," *Proc. IEEE*, **63**, 1975, p. 724.

4. Aspnes, D.E., "Lower conduction band structure of GaAs," *Gallium Arsenide and Related Compounds* (St. Louis) 1976, Inst. Phys. Conf. Ser. No. 33(b), Bristol, p. 110.

5. Barber, M.R., "High-power quenched Gunn oscillators," *Proc. IEEE*, **56**, 1968, p. 752.

6. Bearse, S.V., "Impatts, GaAs FET's and TWT's deemed most likely to succeed," *Microwaves*, **5**, July 1976, p. 14.

7. Berson, B.E., "Transferred-electron devices take on more roles in microwave systems," *Electronics*, **45**, Dec. 4, 1972, p. 102.

8. Blair, P.K., A. Pearson, and S. Hecks, "250 watt module using series operated Gunn diodes," *Third Biennial Cornell Electrical Engineering Conference*, Ithaca, N.Y., 1971, p. 309.

9. Bott, I.B., and Hilsum, C., "An analytic approach to the LSA mode," *IEEE Trans. Electron Devices*, **ED-14**, 1967, p. 492.

10. Bowen, J.H., et al., "Analytic and experimental techniques for evaluating transient thermal characteristics of Trapatt diodes," *IEEE Trans. Electron Devices*, **ED-21**, 1974, p. 480.

11. Bowman, L.S., and C.A. Burrus, Jr., "Pulse-driven silicon p-n junction avalanche oscillators for the 0.9 to 20 mm band," *IEEE Trans. Electron Devices*, **ED-14**, 1967, p. 411.

12. Bravman, J., and J. Frey, "High performance varactor tuned Gunn oscillators," *Third Biennial Cornell Electrical Engineering Conference*, Ithaca, N.Y., 1971, p. 335.

13. Brook, P., "The breakdown voltage of double-sided p-n junctions," *IEEE Trans. Electron Devices*, **ED-21**, 1974, p. 730.

14. Buswell, R.N., "VCO's in modern ECM systems," *Microwave Journal*, **17**, May 1975, p. 43.

15. Butcher, P.N., W. Fawcett, and C. Hilsum, "A simple analysis of stable domain propagation in the Gunn effect," *British J. Appl. Phys.*, **17**, 1966, p. 841.

16. Butcher, P.N., W. Fawcett, and N.R. Ogg, "Effect of field-dependent diffusion on stable domain propagation in the Gunn effect," *British J. Appl. Phys.*, **18**, 1967, p. 755.

17. Bulman, P.J., G.S. Hobson, and B.C. Taylor, *Transferred Electron Devices*, Academic Press, London, 1972.

18. Bybokas, J., and B. Farrell, "The Gunn flange – a building block for low-cost microwave oscillators," *Electronics*, **44**, March 1, 1971, p. 47.

19. Caldwell, J.F., and F.E. Rosgtoczy, "Gallium arsenide Gunn diodes for millimeter wave and microwave frequencies," *1972 Symp. GaAs*, and Related Compounds, p. 240.

20. Camp, W.O., Jr., J.S. Bravman, and D.W. Woodard, "The operation of very high power LSA transmitters," *Third Biennial Cornell Electrical Engineering Conference*, Ithaca, N.Y., 1971, p. 373.

21. Carroll, J.E., *Hot Electron Microwave Generators*, Elsevier, N.Y., 1970.

22. Carroll, J.E., "Oscillations covering 4 Gc/s to 31 Gc/s from a single Gunn diode," *Electronics Lett.*, **2**, 1966, p. 141.

23. Chaturvedi, P.K., and W.S. Khokle, "Thermal limitations of CW and pulsed silicon Trapatt diodes," *IEEE Trans. Electron Devices*, **ED-20**, 1973, p. 353.

24. Cho, A.Y., et al., "Preparation of GaAs microwave devices by molecular beam epitaxy," *IEEE Int. Electron Devices Meeting, Technical Digest*, 1974, p. 579. See also 1975, p. 429.

25. Claxton, D.H., and P.T. Greiling, "Broad-band varactor-tuned Impatt-diode oscillator," *IEEE Trans. Microwave Theory and Techniques*, **MTT-23**, 1975, p. 501.

26. Clorfeine, A.S., "Guidelines for the design of high-efficiency mode avalanche diode oscillators," *IEEE Trans. Electron Devices*, ED-18, 1971, p. 550.
27. Clorfeine, A.S., "Circuit considerations for Trapatt oscillators and amplifiers," *Microwave J.*, 18, June 1975, p. 46.
28. Clorfeine, A.S., R.J. Ikola, and L.S. Napoli, "Theory for the high-efficiency mode of oscillation in avalanche diodes," *RCA Rev.*, 30, 1969, p. 397.
29. Clorfeine, A.S., H.J. Prager, and R.D. Hughes, "Lumped-element high-power Trapatt circuits," *RCA Rev.*, 34, 1973, p. 580.
30. Colliver, D., and B. Prew, "Indium phosphide: is it practical for solid state microwave sources?" *Electronics*, 45, April 10, 1972, p. 110.
31. Colquhoun, A., et al., "Stationary Gunn domains created by anode doping gradient current-density reduction," *IEEE Trans. Electron Devices*, ED-21, 1974, p. 681.
32. Constant, E., et al., "Effect of transferred electron velocity modulation in high-efficiency GaAs Impatt diodes," *J. Appl. Phys.*, 46, 1975, p. 3934.
33. Copeland, J.A., "A new mode of operation for bulk negative resistance oscillators," *Proc. IEEE*, 54, 1966, p. 1479.
34. Copeland, J.A., "Computer simulation of bulk semi-conductor devices," *Second Biennial Cornell Electrical Engineering Conference*, Ithaca, N.Y., 1969, p. 36.
35. Copeland, J.A., "Doping uniformity and geometry of LSA oscillator diodes," *IEEE Trans. Electron Devices*, ED-14, 1967, p. 497.
36. Copeland, J.A., "Electrostatic domains in two-valley semiconductors," *IEEE Trans. Electron Devices*, ED-13, 1966, p. 187.
37. Copeland, J.A., "LSA oscillator-diode theory," *J. Appl. Phys.*, 38, 1967, p. 3096.
38. Copeland, J.A., "Stable space-charge layers in two-valley semiconductors," *J. Appl. Phys.*, 37, 1966, p. 3602.
39. Copeland, J.A., "The LSA oscillator: theory and applications," *First Biennial Cornell Electrical Engineering Conference*, Ithaca, N.Y., 1967, p. 4.
40. Copeland, J.A., "Theoretical study of a Gunn diode in a resonant circuit," *IEEE Trans. Electron Devices*, ED-14, 1967, p. 55.
41. Cottrell, P.E., J.M. Borrego, and R.J. Gutmann, "Impatt oscillators with enhanced leakage current," *Solid-State Electronics*, 18, 1975, p. 1.
42. Crowell, C.R., and S.M. Sze, "Temperature dependence of avalanche multiplication in semiconductors," *Appl. Phys. Lett.*, 9, 1966, p. 242.
43. Dean, M., and M.J. Hawes, "An electronically tuned Gunn oscillator circuit," *IEEE Trans. Electron Devices*, ED-20, 1973, p. 597.
44. Dean, M., and M.J. Hawes, "Electronic tuning of stable transferred electron oscillators," *IEEE Trans. Electron Devices*, ED-21, 1974, p. 563.
45. Dean, R.H., and R.J. Matarese, "The GaAs traveling-wave amplifier as a new kind of microwave transistor," *Proc. IEEE*, 60, 1972, p. 1486.
46. Decker, D.R., "Impatt diode quasi-static large-signal model," *IEEE Trans. Electron Devices*, ED-21, 1974, p. 469.
47. DeKoning, J.G., et al., "Fullband millimeter-wave Gunn-effect amplification," *1975 IEEE Int. Solid-State Circuits Conference Digest*, p. 130.
48. DeLeon, J.C. (Ed.), "InP Gunn-effect devices begin to surface in the U.S.," *Microwaves*, 16, 1977, p. 12.
49. DeLoach, B.C. Jr., and D.L. Scharfetter, "Device physics of Trapatt oscillators," *IEEE Trans. Electron Devices*, ED-17, 1970, p. 9.
50. DeLoach, B.C., and R.L. Johnston, "Avalanche transit-time microwave oscillators and amplifiers," *IEEE Trans. Electron Devices*, ED-13, 1966, p. 181.
51. DeSa, B.A.E., and G.S. Hobson, "Design criteria for C.W. Gunn oscillators with good frequency-temperature stability," *Solid-State Electronics*, 16, 1973, p. 1261.
52. DeSa, B.A.E., and G.S. Hobson, "Thermal effects in the bias circuit frequency modulation of Gunn oscillators," *IEEE Trans. Electron Devices*, ED-18, 1971, p. 557.

53. DiLorenzo, J.V., et al., "Beam-lead plated heat sink GaAs Impatt: Part I–performance," *IEEE Trans. Electron Devices*, ED-22, 1975, p. 509.
54. Doumbia, I., G. Salmer, and E. Constant, "High-frequency limitation on silicon Impatt diode: velocity modulation," *J. Appl. Phys.*, 46, 1975, p. 1831.
55. Dydyk, M. "Ferrimagnetically tuneable Gunn-effect oscillator," *Proc. IEEE*, 56, 1968, p. 1363.
56. East, J.R., H. Nguyen-ba, and G.I. Haddad, "Microwave and mm wave Baritt doppler detectors," *Microwave J.*, 19, 1976, p. 51.
57. Eastman, L.F., *Gallium Arsenide Microwave Bulk and Transit-Time Devices*, Artech House, Dedham, Mass., 1972.
58. Eastman, L., and W. Wilson, Jr., "Optimization of performance and control of LSA oscillators," *Third Biennial Cornell Electrical Engineering Conference*, Ithaca, N.Y., 1971, p. 305.
59. Edridge, A.L., and J.J. Purcell, "High power Q-band (26-40GHz) pulsed Gunn oscillators," *Third Biennial Cornell Electrical Engineering Conference*, Ithaca, N.Y., 1971, p. 305.
60. Eisenhart, R.L., "An X-band GaAs Impatt power amplifier," *1975 IEEE Int. Solid-State Circuits Conference*, Philadelphia, p. 132.
61. Fentem, P.J., and B.R. Nag, "Short pulse techniques for estimating the temperature and impurity concentration of the active layer of a Gunn diode," *Solid-State Electronics*, 16, 1973, p. 1297.
62. Fleri, D.A., and R.J. Socci, "Amplifying properties of Gunn oscillator in injection locked mode," *Proc. IEEE*, 57, 1969, p. 1205.
63. Freeman, K.R., and G.S. Hobson, "A survey of CW and pulsed Gunn oscillators by computer simulation," *IEEE Trans. Electron Devices*, ED-20, 1973, p. 891.
64. Frey, J., "Gunn-effect logic devices," *Physics Today*, 27, Aug. 1974, p. 49.
65. Furukawa, S. "Avalanche and Gunn diodes research and development results in Japan," *Third Biennial Cornell Electrical Engineering Conference*, Ithaca, N.Y., 1971, p. 273.
66. Gaylord, T.K., "Gunn effect bibliography supplement," *IEEE Trans. Electron Devices*, ED-16, 1969, p. 490.
67. Gilden, M., and M.E. Hines, "Electronic tuning effects in the Read microwave avalanche diode," *IEEE Trans. Electron Devices*, ED-13, 1966, p. 169.
68. Gibbons, G., *Avalanche-Diode Microwave Oscillators*, Clarendon Press, Oxford, 1973.
69. Glance, B., "A magnetically tunable microstrip Impatt oscillator," *IEEE Trans. Microwave Theory and Techniques*, MTT-21, 1973, p. 425.
70. Gleason, K.R., et al., "Experimental study of series connected Trapatt diodes," *IEEE Trans. Microwave Theory and Techniques*, MTT-22, 1974, p. 804.
71. Goedbloed, J.J., "Intrinsic AM noise in singly tuned Impatt diode oscillators," *IEEE Trans. Electron Devices*, ED-20, 1973, p. 752.
72. Goedbloed, J.J., and M.T. Vlaardingerbroek, "Noise in Impatt-diode oscillators at large-signal levels," *IEEE Trans. Electron Devices*, ED-21, 1974, p. 342.
73. Goldwasser, R.E., and F.E. Rosztoczy, "35-GHz transferred electron amplifiers," *Proc. IEEE*, 61, 1973, p. 1502.
74. Goto, G., et al., "Gunn-effect logic device using transverse extension of a high field domain," *IEEE Trans. Electron Devices*, ED-23, 1976, p. 21.
75. Goto, G., T. Nakamura, and T. Isobe, "Two-dimensional domain dynamics in a planer Schottky-gate Gunn-effect device," *IEEE Trans. Electron Devices*, ED-22, 1975, p. 120.
76. Gough, R.A., and B.H. Newton, "An integrated wide-band varactor-tuned Gunn oscillator," *IEEE Trans. Electron Devices*, ED-20, 1973, p. 863.
77. Grace, M.I. (Ed.), "Solid-state power amplifiers," *Microwave J.*, 18, 1975, p. 29.
78. Gummel, H.K., and J.L. Blue, "A small-signal theory of avalanche noise in Impatt diodes," *IEEE Trans. Electron Devices*, ED-14, 1967, p. 569.

79. Gummel, H.K., and D.L. Scharfetter, "Avalanche region of Impatt diodes," *Bell Syst. Tech. J.*, **45**, 1966, p. 1797.
80. Gupta, M-S, "Large signal equivalent circuit for Impatt-diode characterization and its application to amplifiers," *IEEE Trans. Microwave theory and Techniques*, MTT-21, 1973, p. 689.
81. Gutmann, R.J., J.M. Borrego, and S.K. Ghandhi, "Radiation effects in transit-time microwave diodes," *Proc. IEEE*, **62**, 1974, p. 1256.
82. Haddad, G.I. (Ed.), *Avalanche Transit Time Devices*, Artech House, Inc., Dedham, Mass. 1973.
83. Haddad, G.I., C.M. Lee, and W.E. Schroder, "An approximate comparison between $n^+pn^+$ and $p^+nn^+$ silicon Trapatt diodes," *IEEE Trans. Microwave Theory and Techniques*, MTT-21, 1973, p. 501.
84. Hardeman, L.J., "X-band amplifiers go solid-state," *Electronics*, **45**, March 27, 1972, p. 62.
85. Harth, W., and M. Claassen, "Microwave Baritt diodes," *Nachrichtentechnische Z.*, **2**, 1973, p. 87.
86. Hartnagel, H., *Semiconductor Plasma Instabilities*, Elsevier, N.Y., 1969.
87. Hartnagel, H.L., *Gunn-Effect Logic Devices*, Elsevier, N.Y., 1973.
88. Hashizume, N., and K. Tomizawa, "Requirement on carrier concentration and geometry of Schottky-electrode-triggered Gunn device," *Electronic Lett.*, **12**, 1976, p. 232.
89. Hashizume, N., et al., "GaAs 4 bit gate device of integrated Gunn elements and MESFETS," *Gallium Arsenide and Related Compounds*, (St. Louis), 1976, *Inst. Phys. Conf. Serv. No. 33 b*, UK.
90. Hashizume, N., et al., "Gunn effect high-speed carry finding device for 8-bit binary adder," *IEEE Int. Electron Devices Meeting, Technical Digest*, Washington, D.C., 1977, p. 209.
91. Hasty, T., et al., "Procedures for the design and fabrication of high power, high efficiency, C.W. Gunn devices," *Third Biennial Cornell Electrical Engineering Conference*, Ithaca, N.Y., 1971, p. 325.
92. Haydl, W.H., "Planar Gunn diodes with ideal contact geometry," *Proc. IEEE*, **61**, 1973, p. 497.
93. Heaton, J., and T.B. Ramachandran, "Measurement of Gunn diode thermal resistance," *Microwave J.*, **19**, 1976, p. 43.
94. Hilsum, C., "New developments in transferred electron effects," *Third Biennial Cornell Electrical Engineering Conference*, Ithaca, N.Y., 1971, p. 1.
95. Hines, M.E., "Noise theory for the Read type avalanche diode," *IEEE Trans. Electron Devices*, ED-13, 1966, p. 158.
96. Hisatsugu, T., et al., "A 160 mW 65 GHZ Gunn diode," *1976 IEEE Int. Electron Devices Meeting Technical Digest*, p. 94.
97. Hodowanec, G., "Microwave transistor oscillators," *Microwave J.*, June 17, 1974, p. 39.
98. Hoefflinger, B., "Recent developments on avalanche diode oscillators," *Microwave J.*, March 12, 1969, p. 101.
99. Holmstrom, R., "Small-signal behavior of Gunn diodes," *IEEE Trans. Electron Devices*, ED-14, 1967, p. 464.
100. Howes, M.J., and M.L. Jeremy, "Large signal circuit characterization of solid-state microwave oscillator devices," *IEEE Trans. Electron Devices*, ED-21, 1974, p. 488.
101. Huang, H-C, et al., "High-efficiency operation of GaAs Schottky-barrier Impatts," *Proc. IEEE*, **60**, 1972, p. 464.
102. Iglesias, D.E., J.C. Irvin, and W.C. Niehaus, "10-W and 12-W GaAs Impatts," *IEEE Trans. Electron Devices*, ED-22, 1978, p. 200.

103. Irvin, J.C., et al., "Fabrication and noise performance of high power GaAs Impatts," *Proc. IEEE*, **59**, 1971, p. 1212.

104. Ivanek, F., "Solid-state amplifiers vs. TWTA's," *Microwave J.*, March 18, 1975, p. 22.

105. Izadpanah, S.H., "Pulse generation and processing with GaAs bistable switches," *Solid-State Electronics*, **19**, 1976, p. 129.

106. Izadpanah, B., et al., "A high current drop GaAs bistable switch," *Proc. IEEE*, **62**, 1974, p. 1166.

107. Johnson, R.L., D.L. Scharfetter, and D.J. Bartelink, "High-efficiency oscillations in germanium avalanche diodes below the transit-time frequency," *Proc. IEEE*, **56**, 1968, p. 1611.

108. Joshi, J.S., "Wide-band varactor-tuned X-band Gunn oscillators in full-height waveguide cavity," *IEEE Trans. Microwave Theory and Techniques*, **MTT-21**, 1973, p. 137.

109. Kataoka, S., N., et al., "High field domain functional logic devices with multiple control electrodes," *Cornell Conf. Microwave Semiconductor Devices, Circuits and Applications*, 1973, p. 225.

110. Kawamoto, H., et al., "S-band Trapatt amplifiers with four-layer diode structures," *RCA Rev.*, **35**, 1974, p. 373.

111. Kawashima, M., K. Ohta, and S. Kataoka, "$Ga_x In_{1-x}$ Sb For High Speed Transferred Electron Devices," *IEEE Int. Electron Devices Meeting, Technical Digest*, Washington, D.C. 1977, p. 372.

112. Kawashima, M., and S. Kataoka, "Measurement of transverse spreading velocity of a high field domain in a 3-terminal Gunn device," *Electronic Lett.*, **6**, 1970, p. 781.

113. Kennedy, W.K., and W.E. Kunz, "Solid state zeros in at 8.0 to 16 GHz," *Microwave Systems News*, **2**, Sept./Oct. 1971.

114. Knerr, R.H., and J.H. Murray, "Microwave amplifier using several Impatt diodes in parallel," *IEEE Trans. Microwave Theory and Techniques*, **MTT-22**, 1974, p. 569.

115. Komizo, H., et al., "Improvement of nonlinear distortion in an Impatt stable amplifier," *IEEE Trans. Microwave Theory and Techniques*, **MTT-21**, 1973, p. 721.

116. Koning, J.G., et al., "Gunn-Effect amplifiers for microwave communication systems in X, Ku, and Ka bands," *IEEE Trans. Microwave Theory and Techniques*, **MTT-23**, 1975, p. 367.

117. Kramer, B., and A. Mircea, "Determination of saturated electron velocity in GaAs," *Appl. Phys. Lett.*, **26**, 1975, p. 623.

118. Kramer, N.B., "Impatt diodes and millimeter-wave applications grow up together," *Electronics*, **44**, Oct. 11, 1971, p. 78.

119. Kroemer, H., "Negative bulk mobility devices – what next?" *IEEE Int. Electron Devices Meeting, Technical Digest*, Washington, D.C., Dec. 1974, p. 3.

120. Kroemer, H., "Effect of a parasitic series resistance on the performance of bulk negative conductivity amplifiers," *Proc. IEEE*, **54**, 1966, p. 1980.

121. Kroemer, H., "Nonlinear space-charge domain dynamics in a semi-conductor with negative differential mobility," *IEEE Trans. Electron Devices*, **ED-13**, 1966, p. 27.

122. Kroemer, H., "Detailed theory of the negative conductance of bulk negative mobility amplifiers, in the limit of zero ion density," *IEEE Trans. Electron Devices*, **ED-14**, 1967, p. 476.

123. Kroemer, H., "Negative conductance in semiconductors," *IEEE Spectrum*, **5**, 1968, p. 47.

124. Kroemer, H., "Gunn effect – bulk instabilities," in *Topics in Solid State and Quantum Electronics*, W.D. Hershberger (Ed.), Wiley, N.Y., 1972, Chap. 2.

125. Kroemer, H., "Review of theoretical and experimental aspects of hot electrons in devices," *Int. Conf. Hot Electrons in Semiconductors, 1977, Bull. Am. Phys. Soc.*, **22**, June 1977, p. 702.

126. Kuno, H.J., "Analysis of nonlinear characteristics and transient response of Impatt

amplifiers," *IEEE Trans. Microwave Theory and Techniques*, **MTT-21**, 1973, p. 694.

127. Kurokawa, K., "Injection locking of microwave solid-state oscillators," *Proc. IEEE*, **61**, 1973, p. 1386.

128. Kuru, I., and Y. Tajima, "Domain suppression in Gunn diodes," *Proc. IEEE*, **57**, 1969, p. 1215.

129. Kurumada, T. Mizutani, and M. Fujimoto, "GaAs planar Gunn digital devices with subsidiary anode," *Electronics Lett.*, **10**, 1974, p. 161.

130. Kuvas, R.L., and W.E. Schroeder, "Premature collection mode in Impatt diodes," *IEEE Trans. Electron Devices*, **ED-22**, 1975, p. 549.

131. Laton, R.W., and R.W. Sudbury, "Effect of unequal ionization rates on GaAs Impatt device admittance," *IEEE Trans. Electron Devices*, **ED-21**, 1974, p. 729.

132. Lee, C.M., G.I. Haddad, and R.J. Lomax, "A comparison between $n^+pp^+$ and $p^+nn^+$ silicon Impatt diodes," *IEEE Trans. Electron Devices*, **ED-21**, 1974, p. 137.

133. Lee, C.A., et al., "Technological developments evolving from research on Read diodes," *IEEE Trans. Electron Devices*, **ED-13**, 1966, p. 175.

134. Liu, S.G., and F.C. Duigon, "Planar Trapatt diodes," *IEEE Int. Electron Device Meeting Abstracts*, Washington, D.C., Dec. 1974, p. 138.

135. Luther, L.C., and J.V. DiLorenzo, "Growth of epitaxial GaAs structures for high efficiency Impatts," *J. Electrochem. Soc.*, **122**, 1975, p. 760.

136. Magarshack, J., A. Robier, and R. Spitalnik, "Optimum design of transferred-electron amplifier devices in GaAs," *IEEE Trans. Electron Devices*, **ED-21**, 1974, p. 652.

137. Matthei, W.G., "State of the art of GaAs Impatt diodes," *Microwave J.*, June 16, 1973, p. 29.

138. Mause, K., "Simple integrated circuit with Gunn devices," *Electronic Lett.*, **8**, 1972, p. 62.

139. Mause, K., et al., "Gunn device gigabit rate digital microcircuits," *IEEE J. Solid-State Circuits*, **SC-10**, 1975, p. 2.

140. McCumber, D.E., and A.G. Chynoweth, "Theory of negative-conductance amplification and of Gunn instabilities in "two-valley" semi-conductors," *IEEE Trans. Electron Devices*, **ED-13**, 1966, p. 4.

141. Meyer, M., "A simple method for the measurement of Gunn diode thermal resistance," *IEEE Trans. Electron Devices*, **ED-21**, 1974, p. 175.

142. Mircea, A., E. Constant, and R. Perrichon, "FM noise of high-efficiency GaAs Impatt oscillators and amplifiers," *Appl. Phys. Lett.*, **26**, 1975, p. 245.

143. Misawa, T., "Negative resistance in pn junctions under avalanche breakdown conditions, Parts I and II," *IEEE Trans. Electron Devices*, **ED-13**, 1966, Part I, p. 137; Part II, p. 143.

144. Misawa, T., "Multiple uniform layer approximation in analysis of negative resistance in pn junction in breakdown," *IEEE Trans. Electron Devices*, **ED-14**, 1967, p. 795.

145. Mun, J., "High-efficiency and high-peak power InP transferred-electron oscillators," *Electronic Lett.*, **13**, 1977, p. 275.

146. Nagano, S., and S. Ohmaka, "Low noise 80 GHz silicon Impatt oscillator highly stabilized with a transmission cavity," *IEEE Trans. Microwave Theory and Techniques*, **MTT-22**, 1974, p. 1152.

147. Nakamura, T., et al., "Picosecond Gunn-effect carry generator for binary adders," *1975 IEEE Int. Solid-State Circuits Conf. Digest*, p. 166.

148. Nawata, K., M. Ikeda, and Y. Ishii, "Millimeter-wave GaAs Schottky barrier Impatt diodes," Proc. 4th Conf. Solid State Devices, Tokyo, 1972, *Suppl. J. Japan Soc. Appl. Phys.*, **42**, 1973, p. 58.

149. Niehaus, W.C., T.E. Seidel, and D.E. Iglesias, "Double-drift Impatt diodes near 100 GHz," *IEEE Trans. Electron Devices*, **ED-20**, 1973, p. 765.

150. O'Hara, S., and J.R. Grierson, "A study of the power handling ability of gallium

arsenide and silicon, single and double drift Impatt diodes," *Solid-State Electronics,* 17, 1974, p. 137.

151. Ohta, K., M. Kawashima, and S. Kataoka, "Electro-optic modulation of a laser beam by a traveling high-field domain in a GaAs Gunn diode," *Appl. Phys.,* 46, 1975, p. 1318.

152. Okamoto, H., "Noise characteristics of GaAs and Si Impatt diodes for 50 GHz range operation," *IEEE Trans. Electron Devices,* ED-22, 1975, p. 558.

153. Olson, H.M., "Thermal runaway of Impatt diodes," *IEEE Trans. Electron Devices,* ED-22, 1975, p. 165.

154. Omori, M., "Gunn diodes and sources," *Microwave J.,* June 16, 1973, p. 57.

155. Omori, M., "Octave electronic tuning of CW Gunn diode using a YIG sphere," *Proc. IEEE,* 57, 1969, p. 97.

156. Paik, S.F., P.J. Tanzi, and D.J. Kelley, "Impatt-diode power amplifiers for digital communications systems," *IEEE Trans. Microwave Theory and Techniques,* MTT-21, 1973, p. 716.

157. Perlman, B.S., "C.W. microwave amplification from circuit-stabilized epitaxial GaAs transferred electron devices," *IEEE J. Solid-State Circuits,* SC-5, 1970, p. 331.

158. Peterson, D.F., and D.H. Steinbrecher, "Circuit model for characterizing the nearly linear behavior of avalanche diodes in amplifier circuits," *IEEE Trans. Microwave Theory and Techniques,* MTT-21, 1973, p. 19.

159. Pfund, G., C.P. Snapp, and A. Podell," Pulsed double-drift silicon Impatt diodes and their application," *IEEE Trans. Microwave Theory and Techniques,* MTT-22, 1974, p. 1134.

160. Powell, R.S., "Solid-state amplifiers vs TWTA's," *Microwave J.,* 18, Feb. 1975, p. 14.

161. Prager, H.J., K.K.N. Chang, and S. Weisbrod, "High power, high efficiency silicon avalanche diodes at ultra-high frequencies," *Proc. IEEE,* 55, 1967, p. 586.

162. Read, W.T., "A proposed high-frequency negative resistance diode," *Bell Syst. Tech. J.,* 37, 1958, p. 401.

163. Rosen, A., et al., "Wideband class-C Trapatt amplifiers," *RCA Rev.,* 33, 1972, p. 729.

164. Ruch, J.G., and G.S. Kino, "Measurement of the velocity-field characteristic of gallium arsenide," *Appl. Phys. Lett.,* 10, 1967, p. 40.

165. Ruch, J.G., and G.S. Kino, "Transport properties of GaAs," *Phys. Rev.,* 174, 1968, p. 921.

166. Sadler, W.G., "An FM/CW Gunn powered subminiature radar altimeter," *Third Biennial Cornell Electrical Engineering Conference,* Ithaca, N.Y., 1971, p. 265.

167. Sandbank, C.P., "Synthesis of complex electronic functions by solid state bulk effects," *Solid-State Electronics,* 10, 1967, p. 369.

168. Scharfetter, D.L., "Power-frequency characteristics of the Trapatt diode mode of high efficiency power generation in germanium and silicon avalanche diodes," *Bell Syst. Tech. J.,* 49, 1970, p. 799.

169. Scharfetter, D.L., "Power-impedance-frequency limitations of Impatt oscillators calculated from a scaling approximation," *IEEE Trans. Electron Devices,* ED-18, 1971, p. 536.

170. Scharfetter, D.L., and H.K. Gummel, "Large-signal analysis of a silicon Read diode oscillator," *IEEE Trans. Electron Devices,* ED-16, 1969, p. 64.

171. Schlachetzki, A., and E. Hesse, "Current pulses in planar GaAs Gunn Devices," *Solid-State Electronics,* 17, 1974, p. 633.

172. Schneider, S.M., and W.K. Kennedy, Jr., "Cavity design for millimeter-wavelength gallium arsenide devices," *Proc. IEEE,* 57, 1969, p. 1213.

173. Sechi, F.N., and D. Zieger, "A new design technique for transferred electron oscillators," *Microwave J.,* 16, Apr. 1973, p. 47.

174. Seddik, M.M., and G.I. Haddad, "Effects of ionization rates on Impatt device admittance," *IEEE Trans. Electron Devices*, ED-20, 1973, p. 1164.
175. Seddik, M.M., and G.I. Haddad, "Properties of millimeter-wave Impatt diodes," *IEEE Trans. Electron Devices*, ED-21, 1974, p. 809.
176. Seidel, T.E., W.C. Niehaus, and D.E. Iglesias, "Double-drift silicon Impatts at X band," *IEEE Trans. Electron Devices*, ED-21, 1974, p. 523.
177. Sobol, H., and F. Sterzer, "Microwave power sources," *IEEE Spectrum*, 9, April 1972, p. 20.
178. Solomon, P.R., et al., "An experimental study of the influence of boundary conditions on the Gunn effect," *IEEE Trans. Electron Devices*, ED-22, 1975, p. 127.
179. Sterzer, F., "Information processing with transferred-electron devices," *RCA Rev.*, 34, 1973, p. 153.
180. Stillman, G.E., "Unequal electron and hole impact ionization coefficients in GaAs," *Appl. Phys. Lett.*, 24, 1974, p. 471.
181. Su, S., and S.M. Sze, "Design considerations of low-noise high-efficiency silicon Impatt diodes," *IEEE Trans. Electron Devices*, ED-20, 1973, p. 755.
182. Sugeta, T., et al., "Characteristics and applications of a Schottky-barrier-gate Gunn-effect digital device," *IEEE Trans. Electron Devices*, ED-21, 1974, p. 504.
183. Sugeta, T., H. Yanai, and K. Schido, "Schottky-gate bulk effect digital devices," *Proc. IEEE*, 59, 1971, p. 1629.
184. Sugeta, T., and H. Yanai, "Logic and memory applications of the Schottky-gate Gunn-effect digital device," *Proc. IEEE*, 60, 1972, p. 238.
185. Suguira, T., and S. Sugimoto, "FM noise reduction of Gunn-effect oscillators by injection locking," *Proc. IEEE*, 57, 1969, p. 77.
186. Swartz, G.A., et al., "Performance of p-type epitaxial silicon millimeter wave Impatt diodes," *IEEE Trans. Electron Devices*, ED-21, 1974, p. 165.
187. Sze, S.M., "Impact-avalanche transit-time diodes (Impatt diodes)," in *Physics of Semiconductor Devices*, Wiley, N.Y., 1969, Chap. 5, p. 200.
188. Takeuchi, M., A. Higashisaka, and K. Sekido, "GaAs planar Gunn diodes for DC-biased operation," *IEEE Trans. Electron Devices*, ED-19, 1972, p. 126.
189. Talwar, A.K., and W.R. Curtice, "An experimental study of stabilized transferred-electron amplifiers," *IEEE Trans. Microwave Theory and Techniques*, MTT-21, 1973, p. 477.
190. Tantraporn, W., and S.P. Yu, "Efficiencies of Schottky-barrier GaAs and both complementary structures of Si Impatt diodes," *IEEE Trans. Electron Devices*, ED-20, 1973, p. 492.
191. Tatsuguchi, I., and J.W. Gewartowski, "A 10-W 6-GHZ Impatt amplifier for microwave radio systems," *1975 IEEE Int. Solid-State Circuits Conference*, Philadelphia, p. 134.
192. Thim, H.W., and M.R. Barber, "Microwave amplification in a GaAs bulk semiconductor," *IEEE Trans. Electron Devices*, ED-13, 1966, p. 110.
193. Tomizawa, K., and S. Kataoka, "Dependance of transverse spreading velocity of a high-field domain in a GaAs bulk element on the bias electric field," *Electronic Lett.*, 8, 1972, p. 130.
194. Upadhyayula, L.C., et al., "High efficiency GaAs Impatt structures," *RCA Rev.*, 35, 1974, p. 567.
195. Upadhyayula, L.C., and S.Y. Narayan, "Microwave frequency dividers," *RCA Rev.*, 34, 1973, p. 595.
196. Wada, O., S. Yanagisawa, and H. Takanashi, "Electric-field control in planar Gunn-effect device with Schottky-barrier anode," *Electronic Lett.*, 12, 1976, p. 319.
197. Wada, O., S. Yanagisawa, and H. Takanashi, "Fabrication of planar Gunn-effect logic device with self-aligned Schottky-barrier gates," *Electronic Lett.*, 12, 1976, p. 215.

198. Watson, H.A. (Ed.), *Microwave Semiconductor Devices and Their Circuit Applications*, McGraw-Hill, N.Y., 1969.
199. Weller, K.P., "A study of millimeter-wave GaAs Impatt oscillator and amplifier noise," *IEEE Trans. Electron Devices*, ED-20, 1973, p. 517.
200. Weller, K.P., et al., "Fabrication and performance of GaAs $p^+n$ junction and Schottky barrier millimeter Impatts," *IEEE Trans. Electron Devices*, ED-21, 1974, p. 25.
201. Weller, K.P., R.S. Ying, and E.M. Nakagi, "Millimeter-wave 94 GHZ Silicon IMPATT Amplifiers," *1975 IEEE Int. Solid-State Circuits Conference*, Philadelphia, p. 138.
202. White, G., and R.F. Adams, "A 2-GHZ multiple Gunn device logic circuit," *Proc. IEEE*, 57, 1969, p. 1684.
203. Willing, H.A., "A two-stage Impatt-diode amplifier," *IEEE Trans. Microwave Theory and Techniques*, MTT-21, 1973, p. 707.
204. Wilson, W.L., Jr., "Precise frequency and phase control of LSA oscillators," *IEEE Trans. Microwave Theory and Techniques*, MTT-21, 1973, p. 146.
205. Wilson, W.E., and B.L. Gregory, "Neutron induced degradation and failure in high power LSA diode oscillators," *Third Biennial Cornell Electrical Engineering Conference*, Ithaca, N.Y., 1971, p. 247.
206. Wisseman, W.R., et al., "GaAs Schottky-Read diodes for X-band operation," *IEEE Trans. Electron Devices*, ED-21, 1974, p. 317.
207. Yanagisawa, S., O. Wada, and H. Takanashi, "Gigabit rate-Gunn-effect shift register," *IEEE Int. Electron Device Meeting, Technical Digest*, Washington, D.C., Dec. 1975, p. 317.
208. Yu, S.P., and W. Tantraporn, "Device physics of a new Trapatt oscillator," *IEEE Trans. Electron Devices*, ED-22, 1975, p. 140.
209. Zetsche, H., "Stability criterion for Gunn diodes with injection-limiting cathodes," *IEEE Trans. Electron Devices*, ED-21, 1974, p. 142.
210. Snapp, C.P., and L.F. Eastman, "Introduction to special issue on microwave semiconductor devices," *IEEE Trans. Electron Devices*, ED-25, 1978, p. 557.
211. Midford, T.A., and R.L. Bernick, "Millimeter-wave CW IMPATT diodes and oscillators," *IEEE Trans Microwave Theory and Techniques*, MTT-27, 1979, p. 483.

# 12

# Solar Cells

## CONTENTS

Solar cells are junction devices of large area in which the barrier electric field separates the photon-induced carrier pairs and produces voltage and a current flow in a resistance load. Direct conversion of solar energy to electric power is possible with efficiencies of up to 20%. The cost at present is five to ten times greater than for generation of power from coal, oil, or nuclear electric power stations. However, there is reasonable hope that in the next few decades photovoltaic costs will be reduced to a level where we will see a significant fraction, say 10-20%, of electric power needs supplied in this way.

This chapter discusses solar cells primarily for terrestrial power generation. The cells may be of Si, or of III-V semiconductors such as GaAs, or heterojunction cells such as $Cu_2S/CdS$ and others. Tandem cells with dual junctions may be as high in efficiency as 30% when fully developed. Applications of cells in solar concentration systems for cost reductions are also considered.

## 12.1 SOLAR ENERGY

The solar spectrum in outer space represents an energy of about 140 mW/cm². Since there is no absorption by the atmosphere this is termed an air-mass-zero (AM0) spectrum. At sea level with the sun at zenith, the atmospheric absorption has reduced the power level to about 100 mW/cm² (1 kW/m²) and the spectrum is termed AM1. For the sun at an angle that results in twice the pathlength through the atmosphere, the power is further reduced to about 80 mW cm⁻² and the spectrum is termed AM2.

AM0 and AM2 spectra are shown in Fig. 12.1(a). A photon of high energy, say 2.0 eV, incident on a Si junction creates a carrier pair of 1.1-eV separation energy and the remaining 0.9 eV appears as heat in the lattice.

Semiconductors of larger bandgap provide more efficient use of high-energy photons but photons of energy less than the bandgap are not absorbed. Fig. 12.1(b) shows the number of carrier pairs produced for semiconductors of bandgaps 2.25 eV (GaP), 1.45 eV (GaAs), 1.02 eV (Si), and 0.68 eV (Ge) from an AM1 spectrum. The efficiency of a solar cell in an AM1 spectrum is always greater (often by about 20%) than in an AM0 spectrum because the UV part of AM0 radiation cannot be used effectively. The higher the bandgap of the semiconductor the higher the open-circuit voltage is of the cell but the lower the short-circuit current. Figure 12.2 shows some calculated idealized current densities assuming unity spectral response. In practice $V_{OC}$ for a Si cell is typically 0.55 V and the short-circuit current about 35 mA/cm², whereas for a GaAs cell $V_{OC}$ is about 0.95 V and the current density about 25 mA/cm². The optimum bandgap for matching the solar spectrum is seen from Fig. 12.3(a) to be in the region 1.4-1.6 eV. The power-conversion efficiencies shown in Fig. 12.3(a) are unrealistically large and Fig. 12.3(b) shows more representative efficiencies after allowances have been made for losses in performance. Figure 12.4 shows a bar chart for Si cells in which various losses are displayed.

Concepts for the use of solar cells include:

- direct mounting of panels of tens of square meters in area on roofs of residences and commercial and industrial buildings;
- cells used in conjunction with concentrating reflectors, or low-cost plastic Fresnel lenses, that track the sun. Some concepts are shown in Fig. 12.5.

In such systems, the cell is provided with coolant loops to form combined photovoltaic-thermal systems. Figure 12.6 gives a comparison of a photovoltaic-thermal arrangement and a thermal system without solar cells.

Thermal systems may take the form of power towers in which hundreds of large planar mirrors (several meters square in area) are focused onto a receptor on a tower hundreds of feet in height. The heat-transfer medium, helium gas,

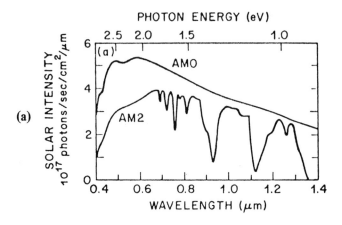

(a)

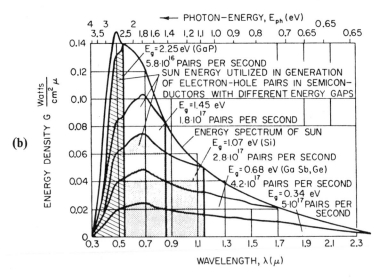

(b)

Fig. 12.1.  Solar insolation.
(a)   AM0 and AM2 spectra. (After Shay, et al. Reprinted with permission from *Eleventh Photovoltaic Specialists Conference*, 1975, p. 504.)
(b)   AM1 spectrum with the carrier pairs produced in semiconductors of various bandgaps. (After Wolf. Reprinted with permission from *Proc. IEEE*, 1960, 48, p. 1248.)

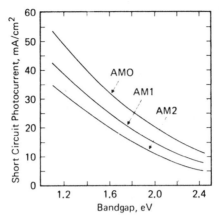

Fig. 12.2. Short-circuit current versus bandgap for unity-spectral response. (After Hovel, from *Semiconductors and Semimetals*, **11**, *Solar Cells*, © 1975; Academic Press, N.Y., p. 38.)

is raised to a temperature of 600°C or more and produces superheated steam in a heat exchanger to drive a turbine generator. Heat may be stored for use at night (or other periods when the sun is obscured) by passing part of the helium through a pressure vessel in the ground containing rocks. Calculations of the overall efficiency of power-tower systems have shown 18-20% efficiency. The predicted costs of such systems are high at present in relation to coal fired or nuclear power stations.[92]

The capital cost of a coal-fired power station is $600-$800/kW peak-generating capacity. A nuclear power station costs $1000-$1200/kW. Photovoltaic panels of $500/kW and 14% efficiency would result in a plant cost of about $2000/kW (see Fig. 12.7) when allowances are made for installation and other expenses. If photovoltaics are to be feasible economically, attention must be given to reducing cell costs below the 50¢/W level and to achieving high efficiency so that mounting and related expenses are minimized. Because of the high costs of batteries or other efficient means of storage, the problems of energy flow with photovoltaic stations as part of a larger network must be solved. In a hot region of the United States, air conditioning results in a peak demand during the day which is when a photovoltaic station could be expected to generate power. However, there are problems associated with the timing of the demand (see Fig. 12.8) and the differences between summer and winter. Most of these factors represent added effective costs.

The present electric-power capacity of the United States is of the order of $3 \times 10^{11}$ W. With cells of 10% efficiency, $10^9$ m$^2$ of panel area would be needed if it were desirable to generate one-third of this power by photoelectrics. If this area is multiplied by three to allow for support and manufacturing areas it would

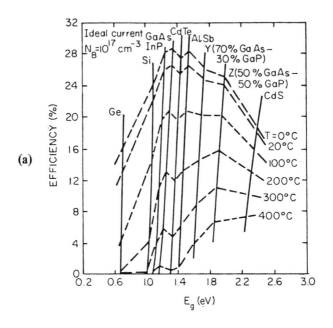

**(a)**

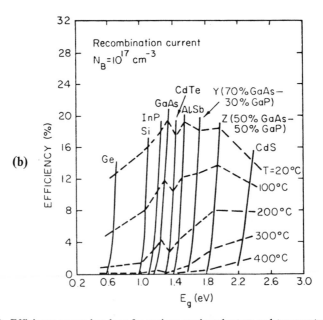

**(b)**

Fig. 12.3.  Efficiency versus bandgap for various semiconductors and temperatures.
(a)  With recombination current neglected.
(b)  With typical recombination currents. (After Wysocki and Rappaport. Reprinted with permission from *J. Appl. Phys.*, **31**, 1960, p. 574.)

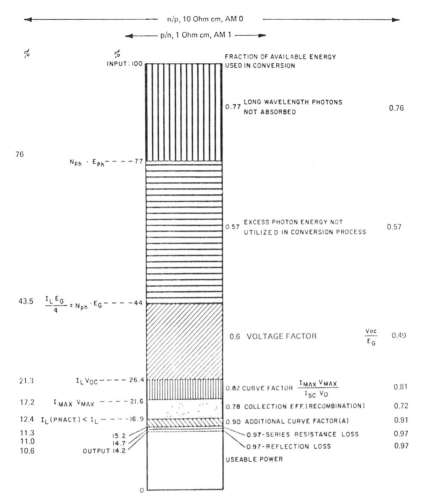

Fig. 12.4. Efficiency losses for a typical Si solar cell in bar-chart form. The values given are for *np*, 10 ohm-cm AM0 and *pn*, 1 ohm-cm AM1 cells. (After Wolf. Reprinted from *Energy Conversion*, **11**, 1971, p. 63.)

represent about 1% of the area of a state such as Arizona, or 6% of the land area in the United States presently covered by roads.

For an overall cost, of say $1/W, the capital investment needed would be $100 billion. The largest business organization in the United States is the General Motors Corporation with 750,000 employees and gross yearly revenues of about $40 billion. The United States gross national product is about $2000 billion/year (1978 dollars). Thus the enterprise, capital, and labor, needed to provide one-third of the United States electrical power by solar cells would

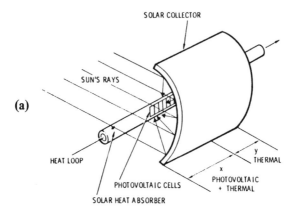

(a)

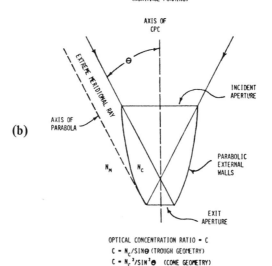

(b)

Fig. 12.5.  Various solar concentrators.

(a)  An integrated photovoltaic-thermal solar-concentrating collector. (After Schueler, et al. Reprinted with permission from *Eleventh IEEE Photovoltaic Specialists Conference*, 1975, p. 331.)

(b)  Cross-sectional view of a compound-parabolic-concentrator unit showing an extreme meridional ray reflected off the exit edge of the external wall. (After Gorski, et al. Reprinted with permission from *Twelfth IEEE Photovoltaic Specialists Conference*, 1976, p. 768.)

(c)  Sketch showing imaging details of square Fresnel-lens design. The central portion of the lens has a common focal point and the corner areas have variable focal lengths. Image on cell is shown; nonshaded areas are lens-corner projections. (After Burgess and Edenburn. Reprinted with permission from *Twelfth IEEE Photovoltaic Specialists Conference*, 1976, p. 779.)

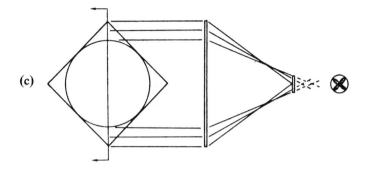

(c)

Fig. 12.5  (continued)

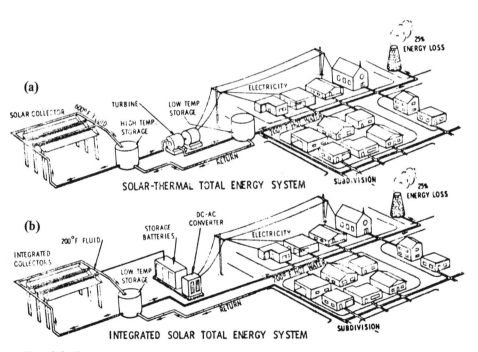

Fig. 12.6.  Concepts for solar energy systems.
(a)  A solar energy system with solar thermal conversion and thermal cascading.
(b)  A solar energy system using combined photovoltaic and thermal conversion. (After Schueler, et al. Reprinted with permission from *Eleventh IEEE Photovoltaic Specialists Conference*, 1975, p. 331.)

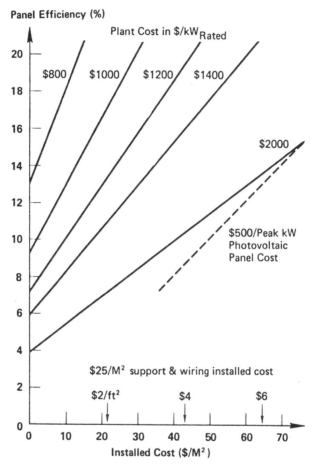

Fig. 12.7. Relationship between efficiency, solar-panel cost, and total plant cost per kilowatt of capacity. (After DeMeo, Spencer, and Bos. Reprinted with permission from *Twelfth IEEE Photovoltaic Specialists Conference*, 1976, p. 657.)

be substantial even if spread over a ten year span. The present (1978) business base is only some tens of millions of dollars and less than a 1000 employees.

## 12.2 SILICON SOLAR CELLS

### 12.2.1 Structure

A conventional Si solar cell is a 100-200-$\mu$m thick (4-8 mil) $p$-type wafer (1-10 $\Omega \cdot$cm) of 5-7-cm diameter with an $n$-type phosphorous-diffused junction about 0.2 $\mu$m deep on the front face. Contact to the diffused layer is by a finger

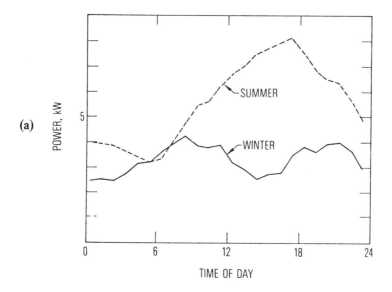

(a)

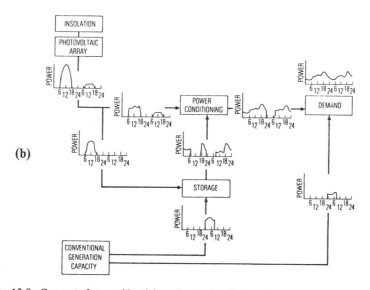

(b)

Fig. 12.8. Concepts for a residential on-site photovoltaic system.
(a)  Residential demand profiles, Arizona.
(b)  Typical energy-flow patterns. (After Leonard. Reprinted with permission from *Eleventh IEEE Photovoltaic Specialists Conference*, 1975, p. 250.)

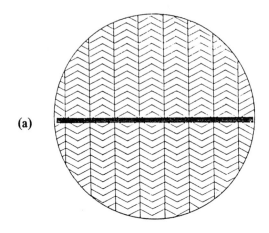

**(a)**

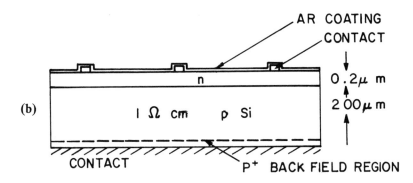

**(b)**

Fig. 12.9.  Some structural details of Si solar cells.
(a)    Metal finger pattern on front face (Solarex Cell).
(b)    Cross section of an *n* on *p* cell.
(c)    Reflectance versus wavelength. (After Hovel, from *Semiconductors and Semimetals,* **11**, *Solar Cells,* © 1975; Academic Press, N.Y., p. 204.)
(d)    Textured cell face to reduce reflectance.

structure of silver (or Ti-Pd-Ag) that provides a low-resistance contact resistance and obscures only about 5% of the front face area of the cell. A recent finger pattern is shown in Fig. 12.9(a) and a cross section is shown in Fig. 12.9(b). An antireflection coating, perhaps SiO or $Ta_2O_5$, of quarter-wave thickness (about 800 Å) may be provided. The reflection coefficient of a cell without an antireflection coating is given by $R = (n-1)^2/(n+1)^2$ where *n* is the refractive index. For Si, *n* is about 3.6 and the reflection loss could be as high as 30%. With a quarter-wave antireflection coating of index $n_{AR}$ to air of index 1, the

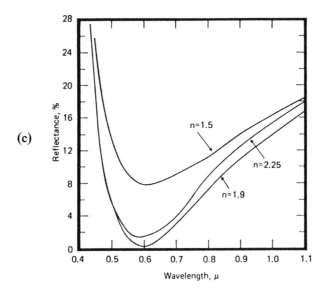

**(c)**

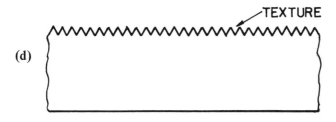

**(d)**

Fig. 12.9 (continued)

reflection is $[(n_{AR}^2-n)/(n_{AR}^2+n)]^2$ This is zero if $n_{AR} = n^{1/2}$ but the quarter-wave length condition can be satisfied at only one wavelength [see Fig. 12.9(c)] and average reflection losses of 5% or more are usual.

Some cells are etched to provide a nonreflecting velvet-textured surface (in place of or in conjunction with an AR layer) as suggested by Fig. 12.9(d). A cover glass is usually provided to protect the front face of the cell. This is important in cells for space applications in satellites to reduce the harmful effects of high-energy irradiation. In terrestrial applications it may be there as part of the hermetic seal needed to ensure a cell life of 20 years or more.

At the back face of the wafer the cell is provided with a $p^+$-diffused region and the contact metal may be any convenient metal for soldering (usually tin-based). The $p$-$p^+$ structure forms a built-in field that aids the carrier collection at the $n$-$p$ junction and improves the cell voltage and current performance.

Although Si cells are usually *n* on *p* structures, *p* on *n* structures have very similar performance characteristics. The reason for the interest in *p*-base structures is that the minority-carrier-diffusion length in the base is usually greater because the mobility of electrons exceeds that of holes. In space applications, *n* on *p* cells are found to have better radiation resistance characteristics than *p* on *n* base cells.

### 12.2.2  Theory and Equivalent Circuit Model

For an *n-p* junction under illumination the current from the simple Shockley model is

$$I = I_L - I_0 \left( \exp \left( q V / kT \right) - 1 \right) \tag{12.1}$$

where $I_0$ is the leakage current of the junction without illumination and $I_L$ is the carrier-injection current into the junction-depletion region produced by the illumination.

As shown in Fig. 12.10(a) the electric field of the junction-depletion region causes electrons induced in the *p* base to move to the *n* layer which therefore becomes negative with respect to the base. The $I/V$ characteristic with light applied is the solid line in Fig. 12.10(b). The current with illumination is similar in shape to the dashed line for the junction $I/V$ characteristic without illumination but displaced downwards by the current $I_{SC}$. This displacement may be explained by a superposition model.[80] An equivalent circuit for the solar cell is shown in Fig. 12.10(c) where $R_S$ is the series resistance of the cell, $R_{Sh}$ is a shunt loss term, and $R_L$ is the load resistance. More elaborate equivalent circuits have been suggested but need not be considered here.

If $R_S$ is negligibly small and $R_{Sh}$ is large enough to be neglected, the open-circuit voltage may be obtained from Eq. (12.1) by setting $I=0$: whence

$$V_{OC} = \frac{kT}{q} \ln \left[ \left( \frac{I_L}{I_0} \right) + 1 \right] \tag{12.2}$$

From this, the junction dark generation-recombination current $I_0$ must be small if the open-circuit voltage is to be large. The maximum open-circuit voltage that can be expected under very high-illumination conditions corresponds to near leveling of the band edges in Fig. 12.10(a) and is always less than the bandgap of the semiconductor. Increased doping in the base region decreases the quantity $\delta$ and also narrows the depletion-region width and so reduces $I_0$ and results in a larger output voltage. However, the open-circuit voltage of a Si solar cell at 300°K does not usually exceed about 630 mV.

The power output of the solar cell is

$$P = IV = \left( I_L - I_0 \left[ \exp \left( q V / kT \right) - 1 \right] \right) V \tag{12.3}$$

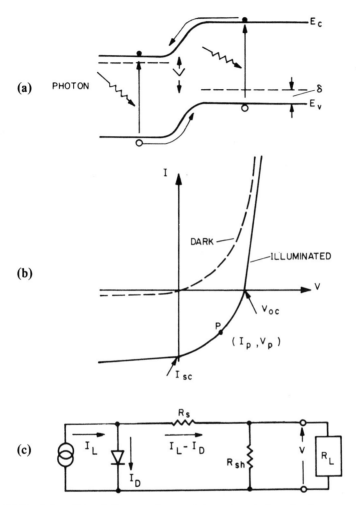

Fig. 12.10. Solar cell band diagram, characteristics and equivalent circuit.
(a)   Energy band diagram for an $n$ on $p$ cell showing carrier flow and a generated voltage $V$.
(b)   Current versus voltage characteristics.
(c)   Equivalent circuit.

This is a maximum at the point $P$ in Fig. 12.10(b) where $dP/dV$ is 0. Hence $V_P$ is given by

$$\left(\frac{qV_p}{kT} + 1\right) + \exp(qV_p/kT) = \frac{I_L}{I_0} + 1 \qquad (12.4)$$

and

$$I_P = I_L - I_0 \left[\exp(q(V_P/kT) - 1)\right] \qquad (12.5)$$

The quantity $I_P V_P/I_{SC} V_{OC}$ is the fill-factor (FF) of the cell and is a measure of the "squareness" of the $I$-$V$ curve.[81] From the equation presented

$$FF = \frac{(I_L - I_0 \left[\exp\left(q V_P/kT\right)-1\right]) V_P}{I_{SC} \dfrac{kT}{q} \ln\left(\dfrac{I_L}{I_0}+1\right)} \qquad (12.6)$$

Since $I_{SC}$ approximately equals $I_L$ this may be rearranged as

$$FF = \frac{V_P}{V_{OC}} \left[1 - \frac{\exp\left(q V_P/kT\right)-1}{\exp\left(q V_{oc}/kT\right)-1}\right] \qquad (12.7)$$

For a Si cell ($298°K$) the ratio $V_P/V_{OC}$ is about 0.86 for a $V_{OC}$ value of 0.6 V and the corresponding fill factor is 0.83.

If the junction behavior is represented by an ideality factor $n$ greater than unity so that the dark characteristic is $I = I_0$ (exp $(q V/nkT) - 1$) the fill factor expression becomes

$$FF = \frac{V_P}{V_{OC}} \left[1 - \frac{\exp\left(q V_p/nkT\right)-1}{\exp\left(q V_{OC}/nkT\right)-1}\right] \qquad (12.8)$$

For a Si cell with $V_{OC}$ value of 0.6 V and $n$ equal to 2.0, the ratio of $V_P/V_{OC}$ is 0.80 and the fill factor is about 0.72. Detailed studies may be found elsewhere.[55] The treatment above neglects the role of series resistance that can have a serious degrading effect on fill-factor and efficiency of a solar cell. One method of estimating the internal series resistance of a cell is to add external resistance and study the degeneration of the fill-factor. For a 10-cm$^2$ cell delivering say 0.35 A, the cell series resistance must be less than 0.06 $\Omega$ if the voltage drop is to be acceptably small ($\sim$20 mV).

The dark $I/V$ characteristics of solar cells are often studied as the sum of two exponentials of the form

$$I = I_{01} \left[\exp\left(\frac{q V}{A_1 kT}\right) - 1\right] + I_{02} \left[\exp\left(\frac{q V}{A_2 kT}\right) - 1\right] \qquad (12.9)$$

$I_{01}$ is assumed to represent the leakage current generated within a diffusion length of the depletion region and $A_1$ is usually taken as unity to conform with the simple model of a diode. The current $I_{02}$ is considered to be the recombination-generation current of the depletion region. The ideality factor $A_2$ is determined by fitting the data and often has a value not very different from 2.0. For a Si cell, $I_{01}$ may have a 300°K value of about $10^{-9}$ A/cm$^2$ and an activation energy about equal to the Si bandgap. The value of $I_{02}$ is much larger, say $10^{-6}$-$10^{-5}$ A/cm² at 300°K, and the activation energy is about half the bandgap of Si as would be expected for recombination-generation centers near the middle of the bandgap. However, this modeling, in many instances, is an oversimplification of a more complex situation.

### 12.2.3  Spectral Response of Silicon Solar Cells

The absorption of photons in a semiconductor is represented by

$$N(x) = N(0) \exp (-\alpha x) \qquad (12.10)$$

where $N(0)$ is the incident photon flux, $N(x)$ is the photon flux density remaining at a distance $x$ within the semiconductor, and $\alpha$ is the absorption coefficient. The variation of $\alpha$ with photon energy is shown in Fig. 12.11 for Si and several

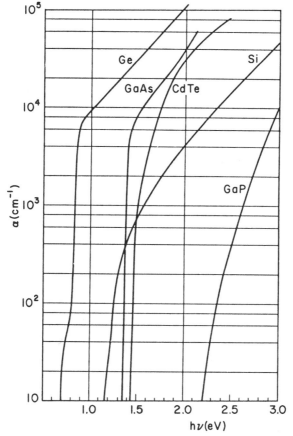

Fig. 12.11.  Absorption coefficient versus photon energy for some semiconductors of interest as solar cells. (After Loferski. Reprinted with permission from *Proc. IEEE*, **63**, 1963, p. 669.)

other semiconductors of interest. The absorption coefficient in Si is much less than in GaAs for photons of energy above 1.5 eV because Si is an indirect-gap semiconductor. Hence the main part of the collected current in a Si solar cell comes from the base region. The collection efficiency is reduced if the base is thin or if the minority-carrier-diffusion length is small in the base. Calculated current-efficiency curves in Fig. 12.12 for an $n$-$p$ structure (AM0) show that a Si base thickness in excess of 100 $\mu$m is desirable for collection efficiencies of greater than 80% and that the minority-carrier-diffusion length must not be much less than the base thickness. However recently 50 $\mu$m thickness Si cells have given over 13% AM0 efficiency.

The Si absorption curve in Fig. 12.11 shows that for photons of energy greater than 2.5 eV, carriers are produced within about 1 $\mu$m of the incident surface. Care must be taken that the junction-diffusion process does not result in a region of low-minority-carrier lifetime (dead layer), or high-surface-recombination velocity. Figure 12.13 presents the spectral responses of two cells to show the greatly improved response to high-energy photons if the dead layer is minimized. This results in a power-efficiency improvement of several percent for the violet responsive cell. A further improvement is possible by texturizing the surface of the cell for better absorption. The nonreflective cell (CNR) of Fig. 12.14 produces a power of 21 mW/cm$^2$ and a short-circuit current of 46 mA/cm$^2$ under AM0 conditions, and further improvements have been made recently.

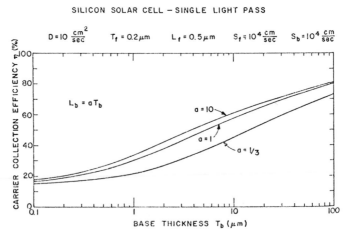

Fig. 12.12. Calculated Si $p$-$n$ junction collection efficiency versus base thickness for a single light pass. (After Esher and Redfield. Reprinted with permission from *Appl. Phys. Lett.*, 25, 1974, p. 702.)

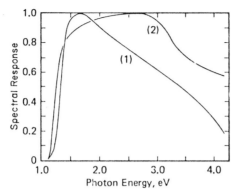

Fig. 12.13. Measured relative spectral responses of *n-p* Si solar cells.
(1)  Low-lifetime (dead layer) high-surface recombination-velocity devices ($x_j$ = 0.3-0.4 $\mu$m).
(2)  With no dead layer ($x_j$ = 0.1-0.2 $\mu$m) the response to the high-energy violet end of the spectrum is improved. (After Hovel, from *Semiconductors and Semimetals,* **11,** *Solar Cells,* 1975; Academic Press, N.Y., p. 34.)

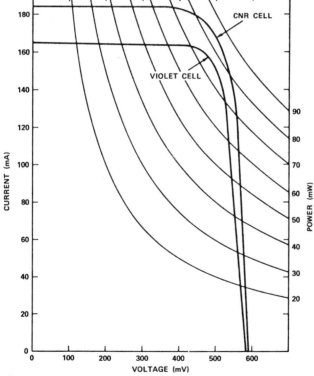

Fig. 12.14. Current-voltage characteristics under AM0 illumination for textured nonreflective cells (CNR) and violet cells (4 cm²). (After Arndt, et al. Reprinted with permission from *Eleventh Photovoltaic Specialists Conference,* 1975, p. 42.)

### 12.2.4  Back-Surface Field Effects

For an *n-p* solar cell with region widths $W_n$ and $W_p$ and with uniform doping so that diffusion effects are dominant rather than field-aided collection effects, the diode-saturation-current expression is

$$I_0 = \frac{qn_i^2 D_p}{N_D} \left\{ \frac{\dfrac{s_f L_p}{D_p} + \tanh \dfrac{W_n}{L_p}}{1 + \dfrac{s_f L_p}{D_p} \tanh \dfrac{W_n}{L_p}} \right\} \frac{1}{L_p} + \frac{qn_i^2 D_n}{N_A} \left\{ \frac{\dfrac{s_r L_n}{D_n} + \tanh \dfrac{W_p}{L_n}}{1 + \dfrac{s_r L_n}{D_n} \tanh \dfrac{W_p}{L_n}} \right\} \frac{1}{L_n} \tag{12.11}$$

where $s_f$ and $s_r$ are, respectively, the front and rear surface recombination velocities. More complicated expressions are obtained if field-aided collection is considered.

The behavior of an abrupt low-high $p$-$p^+$ transition must also be considered. The $p$-$p^+$ transition inhibits the flow of excess minority carriers (electrons) from the $p$ to the $p^+$ region, but majority carriers cross this region freely. The minority-carrier concentrations at the space-charge region edges depend on the barrier potential ($\psi$) associated with the low-high junction (LHJ):

$$n_p = n_{p^+} \exp \frac{q\psi}{kT} \tag{12.12}$$

The diffusion currents at the $p^+$ and $p$ region space-charge edges are proportional to the gradients of the minority-carrier densities there, and are given by

$$qD_p \nabla n_p = qD_{p^+} \nabla n_{p^+} \tag{12.13}$$

Equation (12.12)' represents a subregion boundary condition which differs significantly from that assumed in the drift-field model.[14] Recombination effects and thickness of the LHJ space-charge region are assumed negligible. The $I_0$ using this model is given by

$$I_{0p} = \frac{qn_i^2 D_n}{N_A L_p} \left[ \frac{S + \tanh \dfrac{W_p}{L_p}}{1 + S \tanh \dfrac{W_p}{L_p}} \right] \tag{12.14}$$

where

$$S = \frac{N_A}{N_{A^+}} \frac{D_{p^+}}{D_p} \frac{L_p}{L_{p^+}} \left[ \frac{\dfrac{s_r L_{p^+}}{D_{p^+}} + \tanh \dfrac{W_{p^+}}{L_{p^+}}}{1 + \dfrac{s_r L_{p^+}}{D_{p^+}} \tanh \dfrac{W_{p^+}}{L_{p^+}}} \right] \tag{12.15}$$

The parameter $S$ is a normalized surface-recombination velocity. It contains the LHJ barrier factor $(N_A/N_A^+)$ and the mobility and diffusion-length ratios for the $p$ and $p^+$ region. $S$ also contains a geometry factor for the $p^+$ region, the surface-recombination velocity for the metal-$p^+$ contact $(s_r)$, and the $p^+$-region thickness $(W_{p^+})$. The total $I_0$ must include the $n$-base component $I_{0_n}$ but generally $I_{0_n}$ is small compared with $I_{0_p}$.

The basic merit of the back-surface-field low-high transition is that it reduces the loss of minority carriers to the back contact and so increases the cell performance. The back field may be obtained by diffusion of boron or aluminum or by an epitaxial structure as shown in Fig. 12.15. The open-circuit voltage is improved by as much as 10%.

For an $n^+$-$p$ solar cell the efficiency with a uniformly doped base is found to peak for a base resistivity of about 0.1 $\Omega$-cm corresponding to an acceptor concentration of $7 \times 10^{17}$ cm$^{-3}$ because of increasing radiative or Auger recombination and because of a decline of electron mobility with increasing acceptor concentration. The ultimate efficiency derived from calculations that neglect other loss terms is 24.5% (AM0, 300°K) for thick cells. However, with increased doping the density of recombination centers increases and the minority-carrier lifetime decreases. The effect on efficiency of $\tau_{n_0}$ as a parameter in the calculations is shown in Fig. 12.16. The provision of a low-high back-field structure improves the calculated efficiency by a worthwhile extent as shown by the difference between the dashed solid curves. In practice most $n^+$-$p$-$p^+$ cells show peak performance in the $10^{17}$ cm$^{-3}$ 1 $\Omega$-cm range of base doping and resistivity.

### 12.2.5 Schottky and MIS Barrier Solar Cells

Schottky barrier solar cells offer low-temperature fabrication possibilities and thus in polycrystalline cells the problems of impurity diffusion down grain boundaries are minimized.[149]

The energy-band diagram for a metal $n$-type Si solar cell is shown in Fig. 12.17(a). The upper limit of efficiency possible depends on the barrier height $\phi_{B_n}$ since this determines the barrier reverse current and so affects the cell voltage. For a barrier height of 0.8 eV (Au on $n$ Si), the calculated 300°K efficiency is 10% and for 0.9 eV it is 14% (Pt on $n$ Si). Although these appear to be interesting values the overall performance is significantly lower when allowance is made for losses in the metal film [see Fig. 12.17(b)]. The use of a metal-grid-type structure instead of a continuous metal layer allows some reduction of these losses.

Schottky barrier solar cells studies, however, have led to the discovery that the cell output power and open-circuit voltage is increased by 35 or 40% if an insulating layer of 20 or 30 Å exists between the semiconductor and the metal thus converting it into an MIS structure.[104] Figure 12.18(a) shows experimental

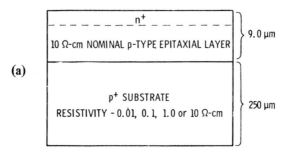

(a)

VARIATION OF EPITAXIAL CELL SHORT-CIRCUIT
CURRENT WITH SUBSTRATE RESISTIVITY

(b)

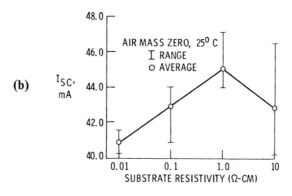

VARIATION OF EPITAXIAL CELL OPEN
CIRCUIT VOLTAGE WITH SUBSTRATE RESISTIVITY

(c)

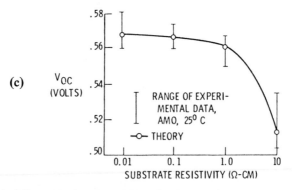

Fig. 12.15. Effect of back-surface field on Si cell performance.
(a)  Epitaxial structure that creates the $pp^+$ barrier.
(b)  Variation of short-circuit current with substrate resistivity.
(c)  Variation of open-circuit voltage with substrate resistivity. (After Brandhorst, et al. Reprinted with permission from *Tenth IEEE Photovoltaic Specialists Conference*, 1973, p. 216.)

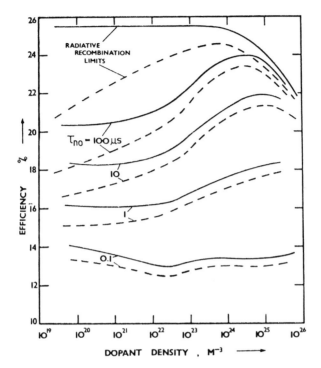

Fig. 12.16. Dependence of conversion efficiency of Si $n^+$-$p$-$p^+$ cells (solid curves) and $n^+$-$p$ cells (dashed curves) on dopant density in the $p$ region for various recombination parameters. (After Green. Reprinted with permission from *IEEE Trans. Electron Devices,* ED-23, 1976, p. 15.)

results for Au-SiO$_2$-$n$Si structures. The insulator thickness is seen to be critical. The improving effect of an oxide layer is seen also for other semiconductors such as GaAs [see Fig. 12.18(b)].

Theoretical and experimental studies of the effect show that over a certain bias range the MIS structure operates in a mode in which the tunnel current through the insulator is large enough to disturb the semiconductor from thermal equilibrium. Such nonequilibrium diodes are classified as majority-carrier, surface-state, or minority-carrier controlled, depending on whether the dominant tunnel current flow near zero bias is between the metal and the majority-carrier band, between surface states, or between the minority-carrier band. Which action dominates depends on whether the semiconductor oxide interface is accumulated, depleted, or in strong inversion at zero bias. This is strongly affected by the metal work function. On $p$-type Si the work function should be low (less than 3.4 eV) and Al-SiO$_2$-Si is a good structure for minority-carrier action that provides the desired enhancement of solar cell voltage.[153] Compared to normal Schottky diode action where the metal and semiconductor are in

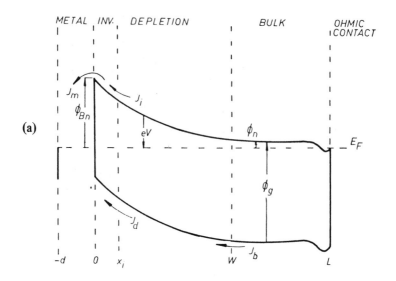

**(a)**

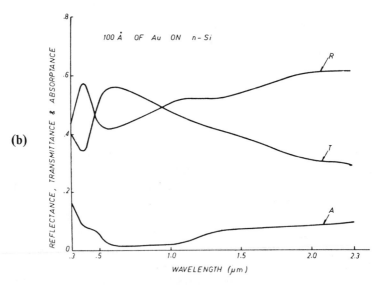

**(b)**

Fig. 12.17. Schottky barrier solar cell with thin metal front face.
(a) Energy band diagram showing barrier, inversion, and depletion regions.
(b) Optical characteristics expected of a 100-Å gold film on Si. (After McOuat and Pulfrey. Reprinted with permission from *Eleventh IEEE Photovoltaic Specialists Conference*, 1975, p. 373.)

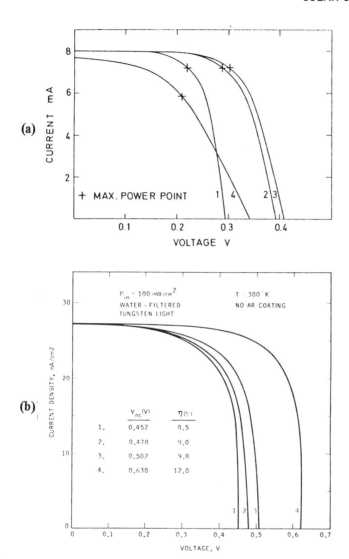

Fig. 12.18. Metal/insulator/semiconductor solar cells.
(a)  Current-voltage characteristics (100 mW-cm² tungsten illumination) for four Schottky barrier Au-*n* Si cells. Curve 1: clean surface; curves 2, 3, and 4: 15, 19, and 23 Å of oxide. (After Lillington and Townsend. Reprinted with permission from *Appl. Phys. Lett.*, **28**, 1976, p. 98.)
(b)  Au-*n* GaAs Schottky-barrier solar cells with and without oxide interfacial layer with no antireflection coating: (1) "clean" interface; (2) GaAs exposed to air at room temperature for 4 hr; (3) same as (2) but for 95 hr; and (4) GaAs exposed to air at 105°C for 70 hr. Efficiencies of greater than 15% AM1 have now been obtained for MIS GaAs cells. (After Stirn and Yeh. Reprinted with permission from *Eleventh IEEE Photovoltaic Specialists Conference*, 1975, p. 438.)

direct contact, an oxide layer of suitable thickness eliminates Fermi level pinning effects and results in an increased barrier height and therefore large open-circuit voltages. Conversion efficiencies as high as for *n-p* solar cells may be expected for MIS cells. The tight control required on insulator thickness, however, is a problem in the fabrication of large area cells that must be stable over many years of operating life.

### 12.2.6  Cost, Energy Payback Time, and Concentration Systems

Conventional Si solar cells at present (1978) cost about $10/W in small panels, exclusive of hermetic sealing for a 20-year life. Predictions are that $5 or even $2/W might be reached by the economies of large-scale production and the development of manufacturing techniques that eliminate expensive processing steps. The cost reductions postulate a government-supported market for cells so that a fall of price along a 70% slope curve as shown in Fig. 12.19 may be achieved as the volume of production increases. Research support should stimulate new cell concepts that will further accelerate the price decline.

Various process changes have been advocated:

- Cells may be produced by ion-implantation processes to eliminate many of the wet chemical steps. However, this depends on the development of large implant machines capable of many hundreds of square meters of cells per hour.

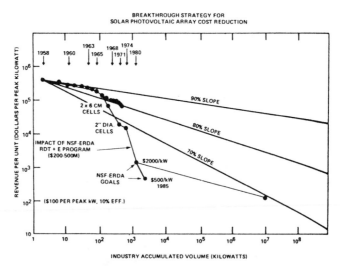

Fig. 12.19. Postulated learning curves of cost per kilowatt versus industry-accumulateᴜ production to illustrate hoped-for price reductions. (After Maycock and Wakefield. Reprinted with permission from *Eleventh IEEE Photovoltaic Specialists Conference*, 1975, p. 254.)

- Growth of Si in ribbon form is being examined to eliminate wafer sawing and polishing costs. Growth of a Si web between two dendrites is such a process [see Fig. 12.20(a)] and another is the edge-defined film obtained by capillary action from a carbon die [see Fig. 12.20(b)]. Both of these processes require the growth of three or four wide ribbons simultaneously to make the economics attractive. Pull rates of about 5 cm/min are typical. The edge-defined film-growth technique requires a die that does not produce Si carbide grains in the ribbon since these cause polycrystalline regions to develop and low-minority carrier lifetime. The possibility exists of improving polycrystalline regions by float-zone gas-laser annealing of ribbons. The EFG process apparently does not allow segregation of impurities in the melt to the same extent as the dendritic-web process.

- Cost savings might be possible with the use of a lower purity Si, to be known as *solar grade* Si (see Fig. 12.21). However, if the cells produced from this material are low in efficiency because of low-minority-carrier lifetime the economics may be offset by the costs of mounting larger area cells.

- The idea of producing Si sheet directly from silane compounds such as $SiH_4$, $SiHCl_3$, or $SiCl_4$ on low-cost substrates is potentially appealing.[27] However, polycrystalline grains of very small size (micrometer) are obtained which result in low-cell efficiencies (1 or 2%). Also the silane epitaxy machines available at present have a low throughput of material per hour.

- Amorphous Si thin films have an apparent bandgap of 1.60 V and good photon absorption and $V_{OC}$ values greater than for conventional Si cells.[18-20] However, there is a problem of minority-carrier collection which limits the efficiency at present to 6%.

- Casting of Si under certain conditions produces blocks of large grain size (millimeter). This yields cells of reasonable (10-12%) efficiency and is a promising approach.[79]

- Carbon-coated ceramic sheets dipped in a melt of Si and slowly withdrawn produce large-grain-size films and cells of greater than 8% efficiency. Satisfactory adhesion is obtained, although in other work with Si the anomalously low-expansion coefficient ($3 \times 10^{-6}/°C$) of this material is a difficulty that often causes strain and adhesion problems to develop on cooling from a high temperature.

Since Si is an indirect bandgap material, fairly thick films are needed (see Fig. 12.12). This is expensive in material, in growth time, and in energy requirements. The energy payback time in the fabrication of solar cell systems is the time that the system must operate to return the total energy expended in fabricating the system. The accounting of the energy must include the energy spent in fabricating all of the materials and the machinery employed. Such

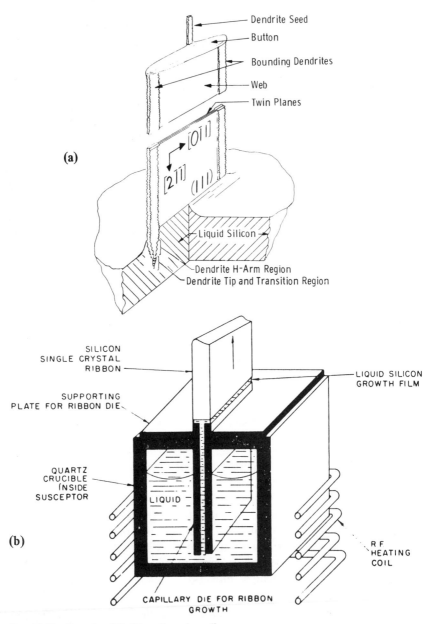

Fig. 12.20. Growth of Si ribbon for solar cells.

(a) Dedritic web. (After Seidensticker, Scudder, Brandhorst, courtesy of Westinghouse Corporation, NAS 3-19439. Reprinted with permission from *Eleventh IEEE Photovoltaic Specialists Conference*, 1975, p. 301.)

(b) Edge-defined film growth (EFG) of ribbon. (After Bates, Jewett, and White. Reprinted with permission from *Tenth IEEE Photovoltaic Specialists Conference*, 1973, p. 199.)

SOLAR SILICON DEFINITION

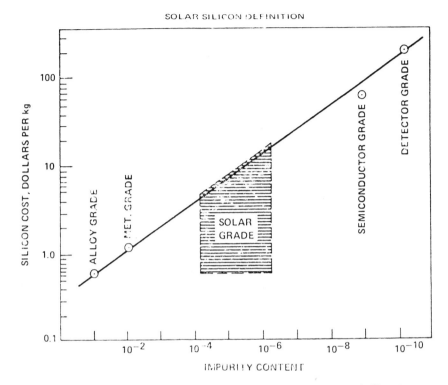

Fig. 12.21.  Cost and impurity content postulated for solar-grade Si.

studies for conventional Si solar cell manufacturing processes and system assembly suggest payback times of three or more years. This is a significant fraction of the expected 20-year operating life of a solar cell system.

One approach to obtaining more power per solar cell is the use of solar concentrator systems. If a ×20 concentrator can be constructed for a cost of say $10 for a $10 solar cell of 1 W normal AM1 rating, the power generated is 20 W for a total cost of $20. Thus a 10 to 1 reduction in effective cost per watt has been achieved.

The problems in the approach are that:

- The cell must be designed to have very-low-series resistance because of the increased current flow. The change of current and voltage with power level and temperature is shown in Fig. 12.22 for a vertical-junction cell designed to have low resistance. A vertical-junction arrangement is not the only way of minimizing series resistance. Low-series resistance may be obtained by placing $n$ and $p$ junctions and contacts on the reverse (nonilluminated) face of the cell in an interdigitated geometry. The cell

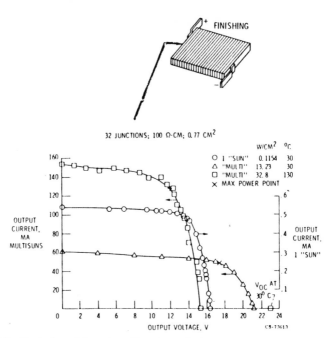

Fig. 12.22. Vertical multijunction Si solar cells for use in high-intensity concentrator systems. Structure and output characteristics for various illumination levels. (After Sater and Goradia. Reprinted with permission from *Eleventh IEEE Photovoltaic Specialists Conference*, 1975, p 362.)

must have a diffusion length for minority carriers that is large compared with the cell thickness, if carriers from the illuminated face are to reach the junction regions. For a Si cell of 100 $\mu$m thickness an electron minority-carrier lifetime of 300 $\mu$s would allow an efficiency of 24% at 300 suns intensity.[75]

- A 15% efficient cell generating 20 W at X20 concentration receives about 130 W of total energy of which 110 W must be dissipated as heat. The cell-receiving area may be 65 cm$^2$ and the concentrator-receiving area must be about 1300 cm$^2$ for an AM1 spectrum of 0.1 W/cm$^2$. Natural convection heat loss from a cooling plate dissipates about 0.1 W/cm$^2$ (front and back area) for a 100°C rise in temperature above the air-ambient temperature. Therefore, reflectors and solar cells at high-concentration ratios have to be provided with forced cooling as suggested by the combined system shown in Fig. 12.23.

- Concentrator systems of high-concentration ratio must track the sun. This adds to the cost and involves maintenance problems. Tracking systems also have the disadvantage of being ineffective in periods of diffuse daylight, whereas nonconcentrator flat-plate systems still deliver power.

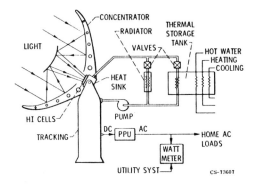

Fig. 12.23. Concept for concentrator with solar cells cooled as part of a combined photo-voltaic and thermal system. (After Sater and Goradia. Reprinted with permission from *Eleventh IEEE Photovoltaic Specialists Conference*, 1975, p. 362.)


## 12.3 SOLAR CELLS OF GaAs, InP, AND RELATED SEMICONDUCTORS

GaAs for solar cells has the advantage that it is a better match than Si to the solar spectrum (see Fig. 12.3) and therefore offers the possibility of higher efficiency. The steep-photon absorption edge of GaAs because of the direct bandgap (see Fig. 12.11) means that carrier pairs are created very close to the surface and may be lost by recombination at the surface instead of being separated by the junction field. This effect can be minimized, however, by use of a thin heteroface layer of $Al_{0.85}Ga_{.15}As$ as shown in Fig. 12.24. $Al_xGa_{1-x}As$ has an excellent lattice match to GaAs and so the recombination velocity at the interface with the GaAs is low. If the aluminum content is as high as 0.85 the material has an indirect bandgap of about 2.0 eV and the direct gap is about 2.6 eV. Photons of energy above about 2.6 eV create carriers in the $Al_xGa_{1-x}As$ that may be lost to the surface. The spectral responses of $pAl_{0.85}Ga_{0.15}As/pGaAs/nGaAs$ structures are illustrated in Fig. 12.25(a). The minority-carrier diffusion length in the base is a factor of importance in the response of the cell to low-energy photons even though only a small fraction, such as 20%, of the current collection comes from the base. Also of importance is the recombination-generation current in the junction region since this influences the open-circuit voltage. If the ingot-grown $n$-type GaAs is low in diffusion length ($L_p < 1.5$ $\mu$m), it is necessary to provide an $n$ epilayer (about $10^{17}$ cm$^{-3}$ in doping) on the substrate by liquid-phase epitaxy before growing the $pAl_{0.85}Ga_{0.15}As$. The $pGaAs$ region is formed by diffusion of zinc during the growth of the $Al_{0.85}Ga_{0.15}As$ and the field in this region assists in the collection process.

Large area (4 × 4 cm) AlGaAs/GaAs cells have power efficiencies of 20-24% AM1, and 16-20% AM0. Open-circuit voltages are 0.98-1.0 V and short-circuit

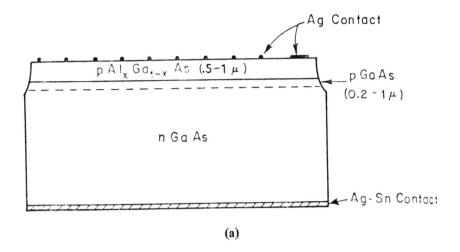

(a)

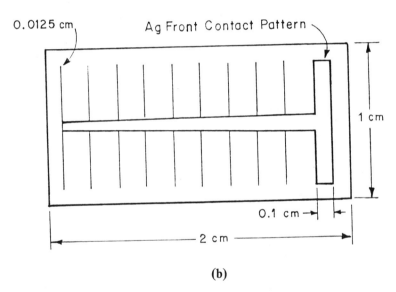

(b)

Fig. 12.24. $Al_xGa_{1-x}As/GaAs$ solar-cell structure.
(a)  Cross section.
(b)  Contact pattern on top of the cell may be this or a finer grid. (After Sekela, Milnes, and Feucht, from Proc. Int. *Symp. on Solar Energy*; This figure was originally presented at the Spring 149th Meeting of The Electrochemical Society, Inc., held in Washington, D.C., May 5-10, 1976.)

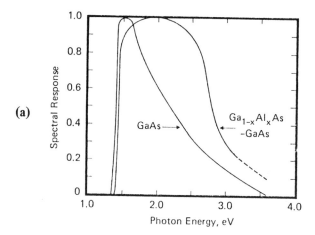

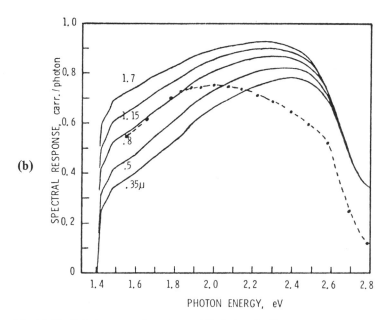

Fig. 12.25. Relative spectral responses of GaAs-based cells.

(a) Measured response for diffused $p$-$n$ GaAs solar cell and $p\mathrm{Al}_{0.85}\mathrm{Ga}_{0.15}\mathrm{As}/p$-$n\mathrm{GaAs}$ cell (heteroface layer = 0.7 μm).

(b) Spectral responses of $p\mathrm{Al}_{0.85}\mathrm{Ga}_{0.15}\mathrm{As}/\mathrm{GaAs}$ structures ($x_j \simeq 0$) calculated as a function of base-diffusion length. The dotted line is the response of an experimental device. (After Hovel. (a) Reprinted from *Semiconductors and Semimetals*, **11**, *Solar Cells*, © 1975; Academic Press, N.Y., p. 35; and (b) reprinted with permission from *Eleventh IEEE Photovoltaic Specialists Conference*, 1975, p. 435.)

current densities are 18-21 mA/cm$^2$ for AM1 sunlight of about 100 mW/cm$^2$.

The energy-band diagram of an AlGaAs/GaAs solar cell is shown in Fig. 12.26(a) where the assumption is made that all the regions are uniformly doped. However, usually the $p$ regions have doping gradients that introduce built-in fields to aid in current collection. Also in some experimental designs the regions have been made graded in composition to further aid the collection process. The importance of using only a thin (less than 1 $\mu$m) AlGaAs or AlAs layer may be seen from Fig. 12.26(b). An antireflection coating is needed. Good performance has also been obtained with cells in which the structure is $n$AlAs on $p$GaAs base regions.

Another approach to the achievement of high performance in GaAs solar cells, dispenses with the AlGaAs and uses an $np$ homostructure with the front-

(a)

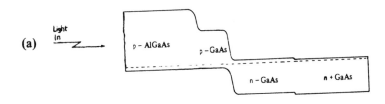

(b)

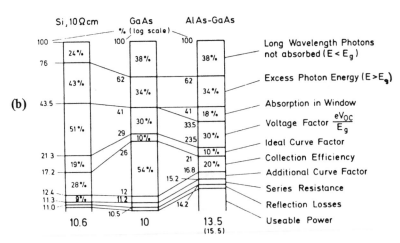

Fig. 12.26. Al$_x$Ga$_{1-x}$As/$p$-$n$GaAs heteroface solar cells.

(a)  Energy band diagram, in the absence of illumination.

(b)  Losses in Si, GaAs, and $p$AlAs-$p$-$n$GaAs cells compared for AM0 spectrum. (After Huber and Bogus 1973.) Efficiency improvements have been obtained since this work for all three classes of cells. (After Huber and Bogus. Reprinted with permission from *Tenth IEEE Photovoltaic Specialists Conference*, 1973, p. 102.)

face $n$ layer very thin (<0.3 $\mu$m) so that carriers are created in the junction region or in the $p$ type base region and are therefore not much affected by the high surface recombination velocity (150). Such cells have even been grown in thin-film single-crystal form on single-crystal Ge substrates (to reduce the amount of Ga needed) and have given efficiencies of 20%, AM1.

GaAs solar cells decrease in efficiency with temperature at only about half the rate observed for Si cells. Line (a) in Fig. 12.27 shows the falloff usually observed for Si cells, line (b) is the usual falloff in GaAs (calculated and observed), and line (c) is typical of calculated and measured efficiencies for AlGaAs/GaAs cells.

GaAs heteroface solar cells are much more costly to fabricate than Si cells because of the high-initial cost of the Ga and because the liquid-phase epi-processes involved are not well adapted to low-cost mass-production methods. Recently vapor-phase-epitaxial growth methods using trimethylgallium, trimethylaluminum and arsine have been found to make cells of excellent performance, but at present these methods are also quite expensive.

Concentrator systems, therefore, have been considered as the approach

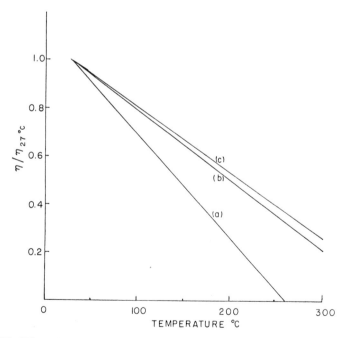

Fig. 12.27. Effect of temperature on Si and AlGaAs/GaAs solar-cell efficiencies.
(a)   Typical Si cell.
(b)   GaAs cell (experimental).
(c)   AlGaAs/GaAs cells (calculated and experimental). (After Sekela, 1976.)

needed to make GaAs-based systems cost effective. The performance of AlGaAs/GaAs cells at high-concentration levels (see Fig. 12.28) is excellent. Also under development are tandem solar cells using six or seven layers of GaAs and AlGaAs materials to form an internally connected two-junction structure. The upper cell may be a *pn* AlGaAs cell of bandgap 1.7 eV and this, via an $n^+p^+$ tunnel junction, is connected electrically and optically to an underlying *pn* cell of GaAs or InGaAs (bandgap 1.4 to 1.1 eV). Such tandem cells, although not simple to fabricate, have been calculated to have the potential for efficiencies in excess of 30% (AM1). In concentrator approaches, tandem cell action may be obtained by using beam splitters to send different parts of the solar spectrum to separate cells of different bandgaps.

Another approach to achieving cost effectiveness with GaAs is to explore the possibility of making low-cost thin-film cells. Since the photon-absorption action in GaAs is almost complete within a depth of 2 or 3 μm the thin-film cell need be no more than 5 μm thick. If the polycrystalline grain size is of

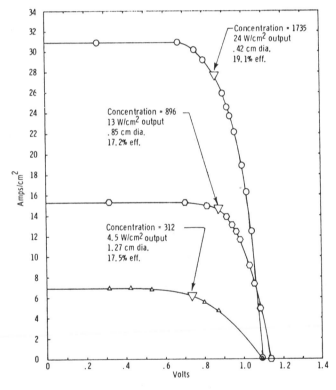

Fig. 12.28. Current versus voltage characteristics for AlGaAs/GaAs under conditions of high-solar concentration. (After James and Moon. Reprinted with permission from *Eleventh IEEE Photovoltaic Specialists Conference*, 1975, p. 408.)

comparable dimensions, minority-carrier-diffusion lengths within the grain should be sufficient to ensure good collection and efficient cells. Thin-film GaAs solar cell efficiencies in excess of 6% are difficult to achieve by conventional methods. However, a concept such as rheotaxy which involves growth of the GaAs on a thin-molten layer of some compound that buffers the growing layer from the polycrystalline or amorphous substrate may offer the solution to the problem of achieving the large-grain size needed for highly efficient thin-film cells. The rheotaxial buffer layer may be a low-melting point semiconductor such as GaSb or InSb or a glassy layer such as obtained from compounds of In, Ge, S, Se. The cell may be completed by a diffused $p$ region and a $p$ AlGaAs heteroface layer or by a Schottky barrier with an interfacial oxide [as shown in Fig. 12.18(b)].

In selecting any semiconductor other than Si for the possible generation of large amounts of power, consideration must be given to the availability of the material in sufficient quantities as well as its cost. There is no problem with the supply of elements such as As, P, Cd, and S, however, In and Ga are not so plentiful. The United States resources of economically recoverable In and Ga are estimated to be in the low and high $10^3$ metric tons, respectively (1 metric ton is $10^3$ Kg). The production, at present, is measured in tens of tons per year. Ga is $750/Kg and In is $80/Kg compared with Cd at $7/Kg and Al at less than $1/Kg.

The electric-power-production capacity in the United States is about $3 \times 10^{11}$ W. Figure 12.29 shows the power capacity that could be expected for

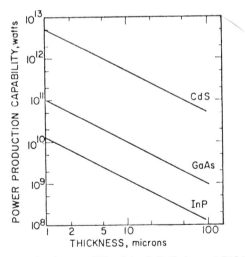

Fig. 12.29. Peak-power production capability from InP, GaAs, and CdS if all the identified United States material resources were used for solar cells. The efficiency assumed is 10% and the input power 100 mW/cm². No concentration considered. (After Hovel. Reprinted from *Semiconductors and Semimetals*, 11, *Solar Cells*, © 1975; Academic Press, N.Y., p. 221.)

InP, GaAs, and CdS thin-film solar cells from identified United States resources. Considering worldwide resources the numbers would increase by roughly a factor of ten. It is therefore important to develop thin-film structures for compound semiconductors, or use concentration systems or do both.

InP single-crystal solar cells have an efficiency potential almost matching that of GaAs. Heterojunction cells with single-crystal InP $p$-type substrates and $n$CdS polycrystalline front-face layers have given efficiencies (AM2) of 15%. The energy-band structure representing this heterojunction[31] is shown in Fig. 12.30 with some typical characteristics. Polycrystalline CdS/InP thin-film cells are at present about 6% efficient.

## 12.4  CELLS OF CdS AND RELATED SEMICONDUCTORS

Studies of thin-film solar cells of *II-VI* compounds have concentrated on CdS because of good availability compared with Se and Te compounds. The structure of interest is the heterojunction of $Cu_2S$ ($p$-type, bandgap 1.2 eV) on poly-crystalline CdS ($n$-type, bandgap 2.4 eV) grown by vacuum evaporation usually on a Zn-coated Cu sheet. The $Cu_2S$ layer is a fraction of a micrometer thick and may be obtained by dipping in hot cuprous chloride solution, followed by a heat treatment of a few minutes at 220-250°C to develop the heterojunction properties. Figure 12.31(a) gives an impression of the columnar-type grain structure in the CdS, and Fig. 12.31(b) shows the energy diagram inferred for the heterojunction.

The open-circuit voltage under illumination (AM1) is about 0.4 V. The efficiencies (AM1) of such cells have been raised in recent years from 6% to over 9%.[152] A target of 14% is considered possible for cells based on $Cu_2S/Zn_{0.1}Cd_{0.9}S$ because of the larger bandgap and more suitable hetero-junction conditions.

$Cu_2S/CdS$ structures have a history of being affected by temperature cycling, light-exposure cycles, and exposure to oxygen and humidity.[35] The crystal structure of $Cu_xS$ undergoes a change from orthochalcocite to djurleite if $x$ falls below 1.995 and this seriously reduces cell performance. The cells must be protected against any factors such as attack by $O_2$ or $H_2O$ on the $Cu_2S$ or any combination of light, temperature, and electrical stress that might affect the stoichiometry. Stability over many years of operation is now being claimed for $Cu_2S/CdS$ cells that are suitably prepared and hermetically protected.

CdS films 2 $\mu$m thick can be made by spraying a solution of $CdO_2$, thiourea, and other chemicals onto glass, hence eliminating vacuum processing. The $Cu_2S$ layer may also be achieved by spraying, by dipping, or electrolytic-ion exchange. The spraying process lends itself to large-scale continuous production in association with a float-glass plant as shown in Fig. 12.32. Cells from such a process line might cost no more than 10¢/W. Allowing for humidity protection,

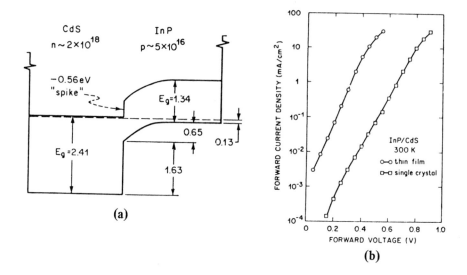

**(a)**

**(b)**

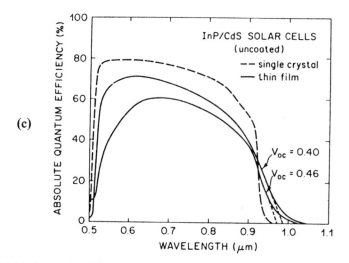

**(c)**

Fig. 12.30. Energy band diagram for an $n$CdS/$p$InP solar cell and cell characteristics.

(a)  Energy band diagram inferred from capacitance versus voltage measurements. (After Shay, Wagner, and Phillips. Reprinted with permission from *Appl. Phys. Lett.*, **28**, 1976, p. 32.)

(b)  Dark $I/V$ characteristics. The thin-film cells are VPE InP growths on carbon substrates coated with GaAs. The diffusion length $L_n$ ($\sim 0.6$ $\mu$m) limits the polycell performance.

(c)  Collection-efficiency spectrum. The single-crystal cell has a power efficiency of 15% (AM2) with a SiO antireflection coating. The polycrystalline cells have efficiencies of 5.2 and 5.7% with an SiO coating. (After Shay, et al. Reprinted with permission from *Twelfth IEEE Photovoltaic Specialists Conference*, 1976, p. 543.)

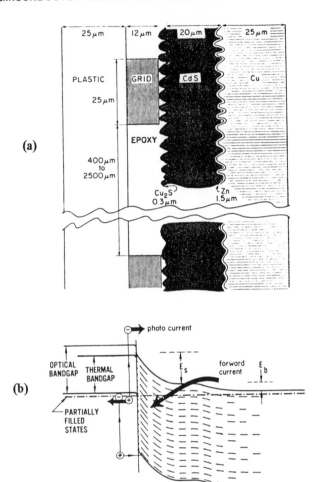

Fig. 12.31. Cu$_2$S/CdS thin-film solar cells.
(a) Impression of cell structure (After Rothwarf and Barnett. Reprinted with permission from *IEEE Trans. Electron Devices*, **ED-24**, 1977, p. 381.)
(b) Energy band diagram. The optical bandgap is 1.2 eV, the thermal bandgap is 0.95 eV, the energy $E_S$ is 0.8 eV, the $E_b$ activation energy is 0.2 eV, and the $\triangle E_v$ barrier is about 1.7 eV. (After Lindmayer and Revesz, from *Solid-State Electronics*, **14**, 1970, p. 648. Reprinted with permission from Pergamon Press, Ltd.)

installation, site preparation, and power-handling equipment, 30¢/W overall cost might be achievable.

There are several other heterojunction combinations involving *II-VI* semiconductors that might be of interest for solar cells.[36] These include *p*CdTe/*n*CdS, *p*CdTe/*n*Zn$_x$Cd$_{1-x}$S, *p*ZnTe/*n*CdTe, and *p*CdTe/*n*ZnSe. At present only a few such structures have given good performance in thin-film form.[146] Other

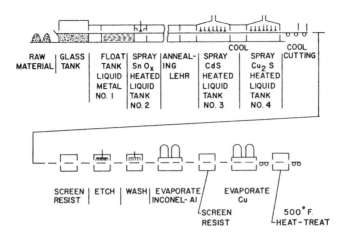

Fig. 12.32. Schematic drawing of float-glass plant with additional sections for automatic production of $Cu_2S/CdS$ solar cells. (After Jordan. Reprinted with permission from *Eleventh IEEE Photovoltaic Specialists Conference*, 1975, p. 512.)

structures that have been proposed include $CuInSe_2/CdS$ and cells involving *II-IV-V$_2$* compounds such as $ZnGeAs_2$.

## 12.5 DISCUSSION

Solar cells for space missions must have the ability to survive electron and proton irradiation, and have a favorable power-to-weight ratio in a deployable array. These matters are outside the range of this chapter which is concerned with cells for terrestrial applications. However, ideas have been developed for solar power stations in synchronous orbit with power beamed back to earth by microwaves. In such schemes both radiation resistance and power-to-weight ratio are of extreme importance.[105]

The state of the engineering art in Si solar cells has not yet reached the low-price levels (less than 30¢/W cell costs) needed to result in power systems at less than $1000/kW. Neither are the energy payback times of several years short enough in terms of the projected 20 years of life. Another problem is that efficient low-cost energy storage is not available for use with photovoltaic power generation.

Thin-film low-cost solar cells of high performance have not yet been demonstrated. GaAs- and CdS-based structures appear to offer possibilities here. Packaging cost considerations require, however, that these films be 10% or more in efficiency. Such cells may possibly be used in tandem structures (a front and a rear cell in optical series, but electrically isolated) to use the solar spectrum more efficiently.

Concentrator systems offer possibilities of cost effectiveness although they involve moving parts and cooling systems, and are unable to use diffuse sunlight.

Thermophotovoltaic conversion is also under consideration in which a hot metallic element, heated by concentrated sunlight, radiates infrared electromagnetic energy onto a solar cell.

Since the technology of solar cells is in a rapidly developing state, the Proceedings of the IEEE Photovoltaic Specialists Conferences and other International Conferences are recommended as sources of up-to-date information.

## 12. 6 QUESTIONS

1. A Schottky barrier solar cell is formed by a 50-Å layer of Au on an $n$-type 1 $\Omega$-cm Si substrate of thickness 200 $\mu$m and hole-diffusion length of 50 $\mu$m. If illuminated through the Au by AM1 solar energy, what is the $I_{SC}$, $V_{OC}$ and efficiency (negligible series-resistance loss) (a) neglecting reflection from and absorption in the Au? (b) allowing for losses associated with the Au layer? (c) For negligible Au loss what is the probable effect on performance of raising the temperature from 300°K to 350°K?

2. A Si $p$-$n$-junction solar cell of diffused structure and area 1 cm² is characterized by a dark $I$ versus $V$ equation (300°K)

$$I = 10^{-11} \left( \exp \left( \frac{qV}{kT} \right) -1 \right) + 10^{-8} \left( \exp \left( \frac{qV}{2kT} \right) -1 \right)$$

Solar illumination is applied that produces a short-circuit current of 40 mA. What is likely to be the open-circuit voltage under these conditions? What is the maximum-power-point fill-factor if the diode-series resistance is negligible?

3. Consider a hypothetical semiconductor structure which contains a plane at which penetrating light is generating electron-hole pairs as free carriers at a rate $G$ pairs per cm²/sec. The top side of this plane consists of a $p$-type $10^{14}$ cm⁻³ doped semiconductor material of thickness $W$ and the upper surface is of infinite recombination velocity. The underside of the plane consists of a thickness $2W$ of the same $p$-type semiconductor with the lower surface also of infinite recombination velocity. If the semiconductor is free of recombination (i.e., infinite bulk lifetime) sketch the electron- and hole-excess concentration and currents throughout the structure produced by the generation effect of the illumination.

4. A certain Si semiconductor structure can be approximately represented as a $p$-type $10^{16}$ cm⁻³ doped face region of thickness 0.005 cm of infinite-minority-carrier lifetime (and zero-surface recombination) on a thick $p$-region substrate, $10^{16}$ cm⁻³, of zero-minority-carrier lifetime. The structure face region is illuminated by monochromatic light that is absorbed in a depth of $10^{-5}$ cm. If the electron-current density produced is 40 mA/cm², what must be the light flux in photons per cm²/sec? What is the change in the majority-carrier concentration at the surface and throughout the bulk during illumination? If the minority-carrier lifetime in the $p$-face region became $10^{-6}$ sec (the substrate remaining a sink for electrons) what then would be the electron density across the $p$ region for the same illumination? If the face region of bulk lifetime $10^{-6}$ sec, has in addition, a surface recombination of $10^5$ cm/sec what would be the effect?

5. A $p^+$-$n$ GaAs solar cell is made on $n$-type $10^{17}$ cm⁻³ GaAs of 2-$\mu$m hole diffusion length $L_p$. The $p^+$ region is obtained by a Zn diffusion 0.3 $\mu$m deep. The front-finger contact is Ag and obscures 5% of the cell area. The surface-recombination velocity is high, $10^5$ cm/sec, but assume the built-in field ensures that all the carriers created in the $p^+$

region reach the junction. The series resistance of the 4-cm² area cell is 0.2 Ω. Calculate the spectral response of the cell (collected current versus photon energy) in air-mass unity (AM1) illumination allowing for the loss with no antireflection coating present. What is the total current collected per square centimeter? What is the open-circuit voltage? What is the optimum-load resistance, the fill-factor, and the power output at the maximum-power point? What is the cell efficiency? What would these numbers become if the series resistance of the cell was made much smaller? What is the reflection loss?

6. The cell of Problem 5 is coated with an $SiO_2$ antireflection coating 950 Å thick. What then is the change in the reflection loss? The change in $I_{SC}$, $V_{OC}$, and efficiency?

7. The cell of Problem 5 has the $p^+$ GaAs region covered with a 1 μm thick layer of $Al_{0.8} Ga_{0.2}$ As. What then is the change of spectral response, $I_{SC}$, $V_{OC}$, and efficiency?

8. A p-n GaAs solar cell has a 3 μm thick $10^{17}$ cm⁻³ uniformly doped (Ge) p region with an electron-diffusion length of 6 μm. The p region is coated with a 0.5-μm layer of pAlAs that has a surface-recombination velocity of $10^5$ cm/sec but reduces the interface-recombination velocity between the pAlAs and the p GaAs to $10^3$ cm/sec. The n-GaAs region is doped $10^{17}$ cm⁻³ and has a hole-diffusion length of 1 μm. The cell-series resistance is negligible. There is no antireflection coating and the contact covers 5% of the front face. If the illumination is air-mass two (AM2) what is expected to be $I_{SC}$, $V_{OC}$, and the cell efficiency? If the J versus V dark-forward characteristic is represented approximately by:

$$J = J_{01} \exp \left( \frac{q V}{kT} \right) + J_{02} \exp \left( \frac{q V}{2kT} \right)$$

what are the expected values of $J_{01}$ and $J_{02}$ (make any assumption necessary)?

9. An $n^+$-p-$p^+$ Si solar cell of 10 Ω-cm p-region resistance has an air-mass zero (AM0) efficiency of 14% at 300°K. What is the power generated per square centimeter? It is then exposed to an air-mass unity (AM1) spectrum. What is now the efficiency and the power generated? The same cell in the AM1 spectrum becomes heated to 350 and 400°K. What then are the efficiencies obtained?

10. A 4-cm² Si $n^+$-p solar cell under AM1 illumination conditions has $V_{OC} = 0.6$ V, a ratio $V_p/V_{OC}$ of 0.80, and a fill-factor of about 0.72. What upper limit value can be ascribed to the cell-series resistance? Can the cell efficiency be estimated? If the cell develops an additional series resistance of 0.2 Ω due to contact-resistance deterioration, what then is the exact shape of the I versus V curve under illumination?

11. Discuss the design of antireflection coatings for solar cells.

12. Discuss front-face grid-pattern designs for Si solar cells and the resulting series resistance.

13. Si solar cells in an AM2 spectrum are more efficient than in AM1 insolation. For a typical cell calculate the efficiences and power developed for the two conditions.

14. Several methods exist for measuring the series resistance of solar cells. Describe two methods.

15. Discuss the factors determining the fill-factor of Si solar cells.

16. Discuss surface-recombination effects and low-performance dead layers at the front face of Si solar cells.

17. Explain the concept of back-surface fields in Si cells and the efficiency improvements produced.

18. What is the optimum-base-doping level (or resistivity) for Si solar cells and the probable fundamental limits of efficiency?

19. Consider a single-crystal Si solar cell of thickness 20 × $10^{-3}$ cm and 14% AM1 power efficiency. If this is redesigned to be successively 6 × $10^{-3}$ and 2 × $10^{-3}$ cm thick what

efficiencies may be expected? Would a reflecting contact on the rear face provide multiple-pass action that would improve the efficiency?

20. Calculate the decrease in $V_{OC}$ and power efficiency in Si and GaAs solar cells with increase in temperature for the range 300-400°K.

21. Examine the concept of representing the dark current of a solar cell by:

$$I = I_{01} \left[ \exp \frac{qV}{A_1 kT} - 1 \right] + I_{02} \left[ \exp \frac{qV}{A_2 kT} - 1 \right]$$

How successful has this approach been with Si and GaAs cells?

22. Discuss vertical multijunction Si solar cells as a means of obtaining low-series resistance for cells operated in high-intensity illumination.

23. Review the spectral-response characteristics of Si solar cells.

24. Discuss the base-diffusion lengths needed for efficient Si and GaAs single-crystal solar cells.

25. Discuss the probable effects of grain size on the efficiencies of polycrystalline Si and GaAs solar cells.

26. Many solar cells would be improved by the use of transparent metal or semiconductor contacts. Discuss transmission losses in thin metal films and the availability of highly conducting transparent semiconductors as contacts.

27. Review the role of a thin insulating layer in metal/insulator/Si solar cells or $In_2O_3$/$SiO_2$/Si heterojunction cells.

28. Discuss ohmic contact and interconnection arrangements for Si solar cells.

29. Discuss energy payback times for the fabrication of Si single-crystal solar cells.

30. Discuss the probable economics of Si ribbon for solar cells grown by either dendritic-web or edge-defined film (EFG) processes.

31. Si solar cells may be made by ion-implantation methods. Discuss the economic promise.

32. Does the concept of solar grade Si make economic and technical sense?

33. Discuss the probable future economics of the growth of 10% efficient thin-films Si solar cells from $SiHCl_3$ or $SiH_4$. Assume that the grain-boundary recombination problems are unimportant because of large grain sizes achieved by recrystallization or rheotaxy processes.

34. Discuss concentrator designs for photovoltaic cells for the range 10 to 500 suns. What accuracy is required for sun tracking for each concentration?

35. Consider the merits of operating two solar cells of bandgap 2.0 and 1.1 eV in optical series to utilize more effectively an AM1 spectrum.

36. Consider the causes of current, voltage, and power loss in a typically proportioned AlGaAs/GaAs solar cell.

37. AlGaAs/GaAs solar cells may be fabricated by liquid-phase or vapor-phase epitaxy. Discuss the relative advantages and economics.

38. What is the role of built-in electric fields produced by doping or graded composition in the performance of AlGaAs/GaAs cells?

39. Review progress in the design, fabrication, and performance of InP-based solar cells.

40. Review the fabrication and properties of $n$CdS/$p$InP solar cells.

41. Discuss current-transport processes in $Cu_2S$/CdS thin-film polycrystalline solar cells.

42. What differences in performance are to be expected if a $Cu_2S$/CdS cell is operated in a back-wall mode, with illumination through a transparent substrate, rather than in a normal front-wall mode with light incident on the $Cu_2S$?

43. Explain the degradation effects that have been reported for $Cu_2S$/CdS solar cells.

44. The solar cell system $p$CdTe/$n$CdS has been estimated to have a maximum theoretical solar efficiency of 17%. Show how this figure is developed.

45. Review the problem of obtaining high efficiency from CdTe-based solar cells.

46. Summarize the properties of $II$-$IV$-$V_2$ semiconductors such as $ZnGeAs_2$ that might be

of interest in solar cells and discuss possible growth and fabrication processes.
47. Discuss the solar cell potential of $I$-$III$-$VI_2$ semiconductors, such as $CuInSe_2$.
48. Discuss rheotaxial-layer concepts for fabricating thin-film large-grain-size solar cells of various materials.
49. Discuss electron and proton-irradiation effects on solar cells for space applications.
50. Review power-to-weight ratios in various solar cell designs for space applications.

## 12.7 REFERENCES AND FURTHER READING SUGGESTIONS

1. Allison, J.F., R.A. Arndt, and A. Meulenberg, "A comparison of the COMSAT violet and non-reflective solar cells," *COMSAT Tech. Rev.,* 5, 1975, p. 211.
2. Anderson, R.L., "Photocurrent suppression in heterojunction solar cells," *Appl. Phys. Lett.,* 27, 1975, p. 691.
3. Anderson, W.A., and R.A. Milano, "I-V characteristics for silicon Schottky solar cells," *Proc. IEEE,* 63, 1975, p. 206.
4. Arndt, R.A., et al., "Optical properties of the Comsat non-reflective cell," *Conference Record of the Eleventh IEEE Photovoltaic Specialists Conference,* 1975, p. 40.
5. Backus, C.E., *Solar Cells,* IEEE Press, Piscataway, N.J., 1976.
6. Bates, H.E., D.N. Jewett, and V.E. White, "Growth of silicon ribbon by edge-defined, film-fed growth," *Conference Record of the Tenth IEEE Photovoltaic Specialists Conference,* 1973, p. 197.
7. Barnett, A.M., and A. Rothwarf, "Advances in the development of efficient thin film $CdS/Cu_2S$ solar cell," *Conference Record of the Twelfth IEEE Photovoltaic Specialists Conference,* 1976, p. 544.
8. Bell, A.E., "Thermal analysis of single-crystal silicon ribbon growth processes," *RCA Rev.,* 38, 1977, p. 109.
9. Berkowitz, J.B., and I.A. Lesk, *Proc. Int. Symp. on Solar Energy,* Electrochemical Society, N.J., 1976.
10. Bettini, M., K.J. Bachmann, and J.L. Shay, "CdS/InP and CdS/GaAs heterojunctions by chemical-vapor deposition of CdS," *J. Appl. Phys.,* 49, 1978, p. 865.
11. Bell, R.O., H.B. Serreze, and F.V. Wald, "A new look at CdTe solar cells," *Conference Record of the Eleventh IEEE Photovoltaic Specialists Conference,* 1975, p. 497.
12. Böer, K.W., "The $CdS/Cu_2S$ heterojunction," *Proc. Int. Symposium on Solar Energy,* Electrochemical Society, N.Y., 1976.
13. Böer, K.W., and J. Olson, "The photovoltaic CdS converter on the Delaware solar house," *Conference Record of the Tenth IEEE Photovoltaic Specialists Conference,* 1973, p. 254.
14. Brandhorst, H.W., Jr., C.R. Baraona, and C.K. Swartz, "Performance of epitaxial back surface field cells," *Conference Record of the Tenth IEEE Photovoltaic Specialists Conference,* 1973, p. 212.
15. Burgess, E.L., and M.W. Edenburn, "One kilowatt photovoltaic subsystem using Fresnel lens concentrators," *Conference Record of the Twelfth IEEE Photovoltaic Specialists Conference,* 1976, p. 774.
16. Card, H.C., and E.S. Yang, "MIS–Schottky theory under conditions of optical carrier generation in solar cells," *Appl. Phys. Lett.,* 29, 1976, p. 51.
17. Card, H.C., and E.S. Yang, "Opto-electronic processes at grain boundaries in polycrystalline semiconductors," *IEEE Electron Devices Meeting Abstracts,* Washington, D.C., 1976.
18. Carlson, D.E., "The effects of impurities and temperature on amorphous silicon solar cells," *IEEE Int. Electron Devices Meeting, Technical Digest,* Washington, D.C., 1977.
19. Carlson, D.E., and C.R. Wronski, "Amorphous silicon solar cell," *Appl. Phys. Lett.,* 28, 1976, p. 671.

20. Carlson, D.E., et al., "Properties of amorphous silicon and a-silicon solar cells," *RCA Rev.*, 38, 1977, p. 211.
21. Chai, Y.G., and W.W. Anderson, "Semiconductor–electrolyte photovoltaic cell energy conversion efficiency," *Appl. Phys. Lett.*, 27, 1975, p. 183.
22. Chappell, T.I., and R.M. White, "Characteristics of a water absorber in front of a silicon solar cell," *Appl. Phys. Lett.*, 28, 1976, p. 422.
23. Charlson, E.J., and J.C. Lien, "An Al p-silicon MOS photovoltaic cell," *J. Appl. Phys.*, 46, 1975, p. 3982.
24. Chopra, K.N., J.N. Maggo, and G.S. Bhatnager, "Performance analysis of photodiodes and phototransistors operating in photon integration mode," *Int. J. Electronics*, 35, 1973, p. 713.
25. Christensen, O., and H.L. Bay, "Production of solar cells by recoil implantation," *Appl. Phys. Lett.*, 28, 1976, p. 491.
26. Chu, T.L., H.C. Mollenkopf, and S.S. Chu, "Polycrystalline silicon on coated steel substrates," *J. Electrochem. Soc.*, 122, 1975, p. 1681.
27. Chu, T.L., H.C. Mollenkopf, and S.S.C. Chu, "Deposition and properties of silicon on graphite substrates," *J. Electrochem. Soc.*, 123, 1976, p. 106.
28. D'Aiello, R.V., P.H. Robinson, and H. Kressel, "Epitaxial silicon solar cells," *Appl. Phys. Lett.*, 28, 1976, p. 231.
29. Dalal, V.L., H. Kressel, and P.H. Robinson, "Epitaxial silicon solar cell," *J. Appl. Phys.*, 46, 1975, p. 1283.
30. DeMeo, E.A., D.F. Spencer, and P.B. Box, "Nominal cost and performance objectives for photovoltaic panels in nonconcentrating central station applications," *Conference Record of the Twelfth IEEE Photovoltaic Specialists Conference*, 1976, p. 653.
31. Dunbar, P.M., and J.R. Hauser, "A study of efficiency in low resistivity silicon solar cells," *Solid-State Electronics*, 19, 1976, p. 95.
32. Escher, J.S., and D. Redfield, "Analysis of carrier collection efficiencies of thin-film silicon solar cells," *Appl. Phys. Lett.*, 25, 1974, p. 702.
33. Ewon, J., G.S. Kamath, and R.C. Knechtli, "Large area GaAlAs/GaAs solar cell development," *Conference Record of the Eleventh IEEE Photovoltaic Specialists Conference*, 1975, p. 409.
34. Fabre, E., M. Mautref, and A. Mircea, "Trap saturation In silicon solar cells," *Appl. Phys. Lett.*, 27, 1975, p. 239.
35. Fahrenbruch, A.L., and R.H. Bube, "Thermally-restorable optical degradation and the mechanism of current transport in $Cu_2$ S-CdS photovoltaic cells," *Conference Record of the Tenth IEEE Photovoltaic Specialists Conference*, 1973, p. 85.
36. Fahrenbruch, A.L., F. Buch, K. Mitchell, and R.H. Bube, "II-VI photovoltaic heterojunctions for solar energy conversion," *Conference Record of the Eleventh IEEE Photovoltaic Specialists Conference*, 1975, p. 490.
37. Fang, P.H., "Analysis of conversion efficiency of organic-semiconductor solar cells," *J. Appl. Phys.*, 45, 1974, p. 4672.
38. Fischer, H., and W. Pschunder, "Impact of material and function properties on silicon solar cell efficiency," *Conference Record of the Eleventh IEEE Photovoltaic Specialists Conference*, 1975, p. 25.
39. Fonash, S.J., "The role of the interfacial layer in metal-semiconductor solar cells," *J. Appl. Phys.*, 45, 1975, p. 1286.
40. Fonash, S.J., "Metal-thin film insulator-semiconductor solar cells," *Conference Record of the Eleventh IEEE Photovoltaic Specialists Conference*, 1975, p. 376.
41. Fossum, J.G., and D.G. Schueler, "Design optimization of silicon solar cells for concentrated-sunlight, high-temperature applications," *IEEE Electron Devices Meeting Abstracts*, Washington, D.C., 1976.
42. Fossum, J.G., F.A. Lindholm, and C.T. Sah, "Physics underlying improved efficiency of high-low-junction emitter silicon solar cells," *IEEE Int. Electron Devices Meeting, Technical Digest*, Washington, D.C., 1977.

43. Fraas, L.M., "Basic grain-boundary effects in polycrystalline heterostructure solar cells," *J. Appl. Phys.*, **49**, 1978, p. 871.
44. Godlewski, M.P., H.N. Brandhorst, Jr., and C.R. Baraona, "Effects of high doping levels on silicon solar cell performance," *Conference Record of the Eleventh IEEE Photovoltaic Specialists Conference*, 1975, p. 32.
45. Godlewski, M.P., C.R. Baraona, and H.W. Brandhorst, Jr., "Low-high junction theory applied to solar cells," *Conference Record of the Tenth IEEE Photovoltaic Specialists Conference*, 1973. p. 40.
46. Goldhammer, L.J., "Particulate irradiation of an advanced silicon solar cell," *Conference Record of the Eleventh IEEE Photovoltaic Specialists Conference*, 1975, p. 172.
47. Gorski, A., et al., "Novel versions of the compound parabolic concentrators for photovoltaic power generation," *Conference Record of the Twelfth IEEE Photovoltaic Specialists Conference*, 1976, p. 764.
48. Gray, P.E., "The saturated photovoltage of a p-n junction," *IEEE Trans. Electron Devices*, **ED-16**, 1969, p. 424.
49. Green, M.A., "Enhancement of Schottky solar cell efficiency above its semiempirical limit," *Appl. Phys. Lett.*, **27**, 1975, p. 287.
50. Green, M.A., "Resistivity dependence of silicon solar cell efficiency and its enhancement using a heavily doped back contact region," *IEEE Trans. Electron Devices*, **ED-23**, 1976, p. 11.
51. Green, M.A., F.D. King, and J. Shewchun, "Minority carrier MIS tunnel diodes and their application to electron- and photovoltaic energy conversion, I and II," *Solid-State Electronics*, **17**, 1974, pp. 551, 563.
52. Haas, G.M., and S. Bloom, "Mitre terrestrial photovoltaic energy system," *Conference Record of the Eleventh IEEE Photovoltaic Specialists Conference*, 1975, p. 256.
53. Heller, A., (Ed.), "Semiconductor liquid-junction solar cells," The Electrochemical Society, Proc. Volume 77-3, Princeton, N.J.
54. Hovel, H.J., "The effect of depletion region recombination currents on the efficiencies of Si and GaAs solar cells," *Conference Record of the Tenth IEEE Photovoltaic Specialists Conference*, 1973, p. 34.
55. Hovel, H.J., *Solar Cells*, Vol. 11 of *Semiconductors and Semimetals*, Academic Press, N.Y., 1975. See also "Novel materials and devices for sunlight concentrating systems," *IBM J. Res. Develop.*, **22**, 1978, p. 112.
56. Hovel, H.J., "Diffusion length measurement by a simple photoresponse technique," *Conference Record of the Twelfth IEEE Photovoltaic Specialists Conference*, 1976, p. 913.
57. Hovel, H.J., "Transparency of thin metal films on semiconductor substrates," *J. Appl. Phys.*, **47**, 1976, p. 4968.
58. Hovel, H.J., and J.M. Woodall, "Theoretical and experimental evaluations of $Ga_{1-x}Al_xAs$-GaAs solar cells," *Conference Record of the Tenth IEEE Photovoltaic Specialists Conference*, 1973, p. 25.
59. Hovel, H.J., and J.M. Woodall, "Diffusion length improvements in GaAs associated with Zn diffusion during $Ga_{1-x}Al_xAs$ growth," *Conference Record of the Eleventh Photovoltaic Specialists Conference*, 1975, p. 433.
60. Hovel, H.J., and J.M. Woodall, "Improved GaAs solar cells with very thin junctions," *Conference Record of the Twelfth IEEE Photovolaic Specialists Conference*, 1976, p. 945.
61. Howell, P., "Generating electricity from solar energy," *Electronics and Power*, **21**, 1975, p. 625.
62. Huber, D., and K. Bogus, "GaAs solar cells with AlAs windows," *Conference Record of the Tenth IEEE Photovoltaic Specialists Conference*, 1973, p. 100.
63. Hutchby, J.A., "High-efficiency graded band-gap $Al_xGa_{1-x}As$-GaAs solar cell," *Appl. Phys. Lett.*, **26**, 1975, p. 457.
64. Iles, P.A., and S.I. Soclof, "Effect of impurity doping concentration on solar cell out-

put," *Conference Record of the Eleventh IEEE Photovoltaic Specialists Conference*, 1975, p. 19.

65. Iles, P.A., and D.K. Zemmrich, "Improved performance from thin silicon solar cells," *Conference Record of the Tenth IEEE Photovoltaic Specialists Conference*, 1973, p. 200.

66. James, L.W., and R.L. Moon, "GaAs concentrator solar cell," *Appl. Phys. Lett.*, **26**, 1975, p. 467.

67. James, L.W., and R.L. Moon, "GaAs concentrator solar cells," *Conference Record of the Eleventh IEEE Photovoltaic Specialists Conference*, 1975, p. 402.

68. Johnston, W.D., and W.M. Callahan, "Vapor-phase-epitaxial growth, processing and performance of AlAs-GaAs heterojunction solar cells," *Conference Record of the Twelfth IEEE Photovoltaic Specialists Conference*, 1976, p. 934.

69. Jordan, J.F., "Low cost CdS-Cu$_2$S solar cells by the chemical spray method," *Conference Record of the Eleventh IEEE Photovoltaic Specialists Conference*, 1975, p. 508.

70. Kalibjian, R., and K. Mayeda, "Photovoltaic and electron-voltaic properties of diffused and Schottky barrier GaAs diodes," *Solid-State Electronics*, **14**, 1971, p. 529.

71. Konagai, M., and K. Takahashi, "Graded-band-gap p Ga$_{1-x}$Al$_x$As-n GaAs heterojunction solar cells," *J. Appl. Phys.*, **46**, 1975, p. 3542.

72. Konagai, M., and K. Takahashi, "Thin film GaAlAs-GaAs solar cells by peeled film technology," *Proc. Int. Symp. on Solar Energy*, Electrochemical Society, N.J., 1976.

73. Konagai, M., and K. Takahashi, "Theoretical analysis of graded-band-gap gallium-alluminum arsenide/gallium arsenide p Ga$_{1-x}$Al$_x$As/p-GaAs/n-GaAs solar cells," *Solid-State Electronics*, **19**, 1976, p. 259.

74. Kruse, P.W., L.D. McGlauchlin, and R.B. McQuistan, *Elements of Infrared Technology: Generation, Transmission and Detection*, Wiley, N.Y., 1962.

75. Lammert, M.D., and R.J. Schwartz, "The interdigitated back contact solar cell for use in concentrated sunlight," *IEEE Trans. Electron Devices*, **ED-24**, 1977, p. 337.

76. Landsberg, P.T., and J. Mallinson, "Determination of larger-than-silicon band gaps for optimal conversion of the diffuse component," *Conference Record of the Eleventh IEEE Photovoltaic Specialists Conference*, 1975, p. 241.

77. Leonard, S.L., "Mission analysis of photovoltaic conversion of solar energy for terrestrial applications," *Conference Record of the Eleventh IEEE Photovoltaic Specialists Conference*, 1975, p. 245.

78. Lillington, D.R., and W.G. Townsend, "Effects of interfacial oxide layers on the performance of Schottky-barrier solar cells," *Appl. Phys. Lett.*, **28**, 1976, p. 97.

79. Lillington, D.R., and W.G. Townsend, "Cast polycrystalline silicon Schottky-barrier solar cells," *Appl. Phys. Lett.*, **31**, 1977, p. 471.

80. Lindholm, F.A., J.G. Fossum, and E.L. Burgess, "Basic corrections to predictions of solar cell performance required by nonlinearities," *Conference Record of the Twelfth IEEE Photovoltaic Specialists Conference*, 1976, p. 33.

81. Lindmayer, J., "Theoretical and practical fill factors in solar cells," *COMSAT Tech. Rev.*, **2**, 1972, p. 105.

82. Lindmayer, J., and A.G. Revesz, "Electronic processes in Cu$_x$S-CdS photovoltaic cells," *Solid-State Electronics*, **14**, 1971, p. 647.

83. Lloyd, WW., "Fabrication of an improved vertical multijunction solar cell," *Conference Record of the Eleventh IEEE Photovoltaic Specialists Conference*, 1975, p. 349.

84. Loferski, J.J., "Recent research on photovoltaic solar energy converters," *Proc. IEEE*, **51**, 1973, p. 6697.

85. Loferski, J.J., "Theoretical and experimental investigation of "grating" type photovoltaic cells," *Conference Record of the Eleventh IEEE Photovoltaic Specialist Conference*, 1975, p. 58.

86. Luft, W., and R.E. Patterson, "Lightweight rigid solar array development," *Conference*

*Record of the Eleventh IEEE Photovoltaic Specialists Conference*, 1975, p. 94.

87. Mandelkorn, J., and J.H. Lamneck, "Advances in the theory and application of BSF cells," *Conference Record of the Eleventh IEEE Photovoltaic Specialists Conference*, 1975, p. 36.

88. Mandelkorn, J., J.H. Lamneck and L.R. Scudder, "Design, fabrication and characteristics of new types of back surface field cells," *Conference Record of the Tenth IEEE Photovoltaic Specialists Conference*, 1973, p. 207.

89. Manifacier, J.C., and L. Szepessy, "Efficient sprayed $In_2O_3$:Sn n type silicon heterojunction solar cell," *Appl. Phys. Lett.*, **31**, 1977, p. 459.

90. Maycock, P.D., and G.F. Wakefield, "Business analysis of solar photovoltaic energy conversion," *Conference Record of the Eleventh IEEE Photovoltaic Specialists Conference*, 1975, p. 252.

91. McOuat, R.F., and D.L. Pulfrey, "Analysis of silicon Schottky barrier solar cells," *Conference Record of the Eleventh IEEE Photovoltaic Specialists Conference*, 1975, p. 371.

92. Metz, W.D., "Solar thermal electricity: power tower dominates research," *Science*, **197**, 1977, p. 353.

93. Moore, A.R. "Short-circuit capacitance of illuminated solar cells," *Appl. Phys. Lett.*, **27**, 1975, p. 26.

94. Moore, W.C., et al., "Terrestrial applications of solar cell powered systems." *Conference Record of the Tenth IEEE Photovoltaic Specialists Conference*, 1973, p. 227.

95. Nakayama, N., "Ceramic CdS solar cell," *Jap. J. Appl. Phys.*, **8**, 1969, p. 450.

96. Nakayama, N., A. Gyobu, and N. Morimoto, "Electrochemical synthesis and photovoltaic effect of copper sulfides on CdS single crystals," *Jap. J. Appl. Phys.*, **10**, 1971, p. 1415.

97. Napoli, L.S., et al., "High level concentration of sunlight on silicon solar cells," *RCA Rev.*, **38**, 1977, p. 76.

98. Nozik, A.J., "p-n photoelectrolysis cells," *Appl. Phys. Lett.*, **29**, 1976, p. 150.

99. Pai, Y.P., H.C. Lin, M. Peckerar, and R.L. Kocher, "Ion-implanted Schottky barrier solar cell," *IEEE Electron Devices Meeting Abstracts*, Washington, D.C., 1976.

100. Palz, W., et al., "Review of CdS solar cell activities," *Conference Record of the Tenth IEEE Photovoltaic Specialists Conference*, 1973, p. 69.

101. Parrott, J.E., "The saturation photovoltage of a pn junction," *IEEE Trans. Electron Devices*, **ED-21**, 1974, p. 84.

102. Penner, S.S., and L. Icerman, *Energy, Vol. II Non-Nuclear Technologies*, Addison-Wesley, Reading, Mass., 1975.

103. Pelegrini, B., and G. Salardi, "A model of ohmic contacts to semiconductors," *Solid-State Electronics*, **18**, 1975, p. 791.

104. Pulfrey, D.L., "MIS solar cells: a review," *IEEE Int. Electron Devices Meeting, Technical Digest*, Washington, D.C., 1977.

105. Pulfrey, D.L., *Photovoltaic Power Generation*, Van Nostrand-Reinhold, N.Y., 1978.

106. Pulfrey, D.L., and R.F. McOuat, "Schottky-barrier solar-cell calculations," *Appl. Phys. Lett.*, **24**, 1974, p. 167.

107. Redfield, D., "Multiple-pass thin-film silicon solar cells," *Appl. Phys. Lett.*, **25**, 1974, p. 647.

108. Rittner, E.S., "An improved theory of the silicon p-n junction solar cell," *IEEE Electron Devices Meeting Abstracts*, Washington, D.C., 1976. See also *J. of Energy*, **1**, 1977, p. 9.

109. Rothwarf, A., L.C. Burton, H.C. Hadley, Jr., and G.M. Storti, "Reflection mode of operation of the $Cu_2S$-CdS solar cell," *Conference Record of the Eleventh IEEE Photovoltaic Specialists Conference*, 1975, p. 476.

110. Rothwarf, A., and A.M. Barnett, "Design analysis of the thin-film CdS-$Cu_2S$ solar cell," *IEEE Trans. Electron Devices*, **ED-24**, 1977, p. 381.

111. Rothwarf, A., and K.W. Böer, "Direct conversion of solar energy through photovoltaic cells," *Progress in Solid-State Chemistry,* **10**, 1975, p. 71.

112. Salter, G.C., and R.E. Thomas, "Induced junction silicon solar cells," *Conference Record of the Eleventh IEEE Photovoltaic Specialists Conference,* 1975, p. 364.

113. Sater, B.L., and C. Goradia, "The high intensity solar cell-key to low cost photovoltaic power," *Conference Record of the Eleventh IEEE Photovoltaic Specialists Conference,* 1975, p. 356.

114. Sayed, M., and L. Partain, "Effect of shading on CdS/CuS solar cells and optimal solar array design," *Energy Conversion,* **14**, 1975, p. 61.

115. Schueler, D.G., J.G. Fossum, E.L. Burgess, and F.L. Vook, "Integration of photovoltaic and solar-thermal energy conversion systems," *Conference Record of the Eleventh IEEE Photovoltaic Specialists Conference,* 1975, p. 327.

116. Seidensticker, R.G., "Dendritic, web silicon for solar cell application," *Crystal Growth,* **39**, 1977, p. 17.

117. Sekela, A.M., Jr., "Aluminum-gallium arsenide heteroface solar cells," Ph.D. Thesis, E.E. Dept. Carnegie Mellon University, Pittsburgh, Pa., 1976.

118. Shah, P., "Analysis of vertical multijunction solar cell, using a distributed circuit model," *Solid-State Electronics,* **18**, 1975, p. 1099.

119. Shay, J.L., S. Wagner, K. Bachman, E. Buehler, and H.M. Kasper, "Preparation and properties of InP/CdS and CuInSe$_2$/CdS solar cells," *Conference Record of the Eleventh IEEE Photovoltaic Specialists Conference,* 1975, p. 503.

120. Shay, J.L., S. Wagner, and H.M. Kasper, "Efficient CuInSe$_2$/CdS solar cells," *Appl. Phys. Lett.,* **27**, 1975, p. 89.

121. Shay, J.L., M. Bettini, S. Wagner, K.J. Bachmann, and E. Buehler, "InP/CdS solar cells," *Conference Record of the Twelfth IEEE Photovoltaic Specialists Conference,* 1976. p. 540.

122. Shay, J.L., S. Wagner, and J.C. Phillips, "Heterojunction band discontinuities," *Appl. Phys. Lett.,* **28**, 1976, p. 31.

123. Shewchun, J., "The operation of the semiconductor-insulator-semiconductor (SIS) solar cell: theory," *J. Appl. Phys.,* **49**, 1978, p. 855.

124. Shewchun, J., M.A. Green, and F.D. King, "Minority carrier MIS tunnel diodes and their application to electron-and photovoltaic energy conversion–II. Experiment," *Solid-State Electronics,* **17**, 1974, p. 563.

125. Shewchun, J., R. Singh, and M.A. Green, "Theory of metal-insulator-semiconductor solar cells," *J. Appl. Phys.,* **48**, 1977, p. 765.

126. Smeltzer, R.K., D.L. Kendall, and G.L. Varnell, "Vertical multijunction solar cell fabrication," *Conference Record of the Tenth IEEE Photovoltaic Specialists Conference,* 1973, p. 194.

127. Stanley, A.G., "Review of CdS-Cu$_2$S research through 1973," in *Applied Solid State Sciences,* Vol. 5, Academic Press, N.Y., 1975, pp. 251-366.

128. Stirn, R.J., and Y.C.M. Yeh, "Solar and laser energy conversion with Schottky barrier solar cells," *Conference Record of the Tenth IEEE Photovoltaic Specialists Conference,* 1973. p. 15.

129. Stirn, R.J., and Y.C.M. Yeh, "The AMOS cell-an improved metal-semiconductor solar cell," *Conference Record of the Eleventh IEEE Photovoltaic Specialists Conference,* 1975, p. 437.

130. Stirn, R.J., and Y.C.M. Yeh, "Technology of GaAs-metal-oxide-semiconductor solar cells," *IEEE Trans. Electron Devices,* ED-24, 1977, p. 476.

131. Stokes, E.D., and T.L. Chu, "A novel method for the measurement of diffusion lengths in solar cells," *IEEE Electron Devices Meeting Abstracts,* Washington, D.C., 1976.

132. Swartz, G.A., L.S. Napoli, and N.Klein, "Silicon solar cells for use at high solar concentration," *IEEE Int. Electron Devices Meeting, Technical Digest,* Washington, D.C., 1977.

133. Tomlinson, R.D., E. Elliott, J. Parkes, and M.J. Hampshire, "Homojunction fabrication in CuInSe$_2$ by copper diffusion," *Appl. Phys. Lett.*, **26**, 1975, p. 383.
134. Usami, A., and M. Yamaguchi, "Lithium-doped drift-field radiation-resistant p/n-type silicon solar cells," *Conference Record of the Eleventh IEEE Photovoltaic Specialists Conference*, 1975, p. 227.
135. Wagner, S., J.L. Shay, P. Migliorato, and H.M. Kasper, "CuInSe$_2$/CdS heterojunction photovoltaic detectors," *Appl. Phys. Lett.*, **25**, 1974, p. 434.
136. Wagner, S., J.L. Shay, K.J. Bachmann, and E. Buehler, "p-InP/n-CdS solar cells and photovoltaic detectors," *Appl. Phys. Lett.*, **26**, 1975, p. 229.
137. Wang, E.Y., F.T.S. Yu, and V.L. Simms, "Optimum design of antireflection coating for silicon solar cells," *Conference Record of the Tenth IEEE Photovoltaic Specialists Conference*, 1973, p. 168.
138. Wang, E.Y., and R.N. Legge, "Semi-empirical calculation of depletion region width in n$^+$p silicon solar cells," *J. Electrochem Soc.*, **122**, 1975, p. 1562.
139. Weisberg, L.R., C.R. Grain, and R.R. Addiss, "Particulate semiconductor solar cells," *Appl. Phys. Lett.*, **27**, 1975, p. 440.
140. Wolf, M., "Limitations and possibilities for improvement of photovoltaic solar energy converters," *Proc. IRE*, **48**, 1960, p. 1246.
141. Wolf, M., "A new look at silicon solar cell performance," *Energy Conversion*, **11**, 1971, p. 63.
142. Woodall, J.M., and H.J. Hovel., "High-efficiency Ga$_{1-x}$Al$_x$As-GaAs solar cells," *Appl. Phys. Lett.*, **21**, 1972, 1972, p. 379.
143. Wysocki, J.J., and P. Rappaport, "Effect of temperature on photovoltaic solar energy conversion," *J. Appl. Phys.*, **31**, 1960, p. 571.
144. Yeh, Y-C. M., and R.J. Stirn, "Improved Schottky barrier solar cells," *Conference Record of the Eleventh IEEE Photovoltaic Specialists Conference*, 1975, p. 391.
145. Yu, P.W., S.P. Faile, and Y.S. Park, "Cadmium-diffused CuInSe$_2$ junction diode and photovoltaic detection," *Appl. Phys., Lett.*, **26**, 1975, p. 384.
146. Bucher, E., "Solar cell materials and their basic parameters," *Appl. Phys.* **17**, 1978, p. 1.
147. Cohen, M.J. and J.S. Harris, Jr., "Polymer semiconductor solar cells," *IEEE International Electron Devices Meeting, Technical Digest*, Washington, D.C., Dec. 1978, p. 247.
148. Bedair, S.M. et al., "Growth and characterization of a two-junction, stacked solar cell," *IEEE International Electron Devices Meeting, Technical Digest*, Washington, D.C., Dec. 1978, p. 250.
149. Gupta, D., "The effect of solutes in grain boundary diffusion and related phenomena and its origin," *Thin Films Phenomena - Interfaces and Interactions* Ed. J.EE Baglin and J.M. Poate, *The Electrochemical Society, Proceedings.* Volume 78-2, p. 498.
150. Fan, J.C.C. and C.O. Bozler, "High efficiency GaAs shallow-homojunction solar cells," *Conference Record of the Thirteenth IEEE Photovoltaic Specialists Conference-*1978, p. 953.
151. Dapkus, P.D. et al., "The properties of polycrystalline GaAs materials and devices for terrestrial photovoltaic energy conversion," *Conference Record of the Thirteenth IEEE Photovoltaic Specialists Conference-*1978, p. 960.
152. Meakin, J.D., "The status of CdS/Cu$_2$S solar cell development," *IEEE International Electron Devices Meeting, Technical Digest,* Washington, D.C., Dec. 1978, p. 235.
153. Tarr, N.G. and D.L. Pulfrey, New experimental evidence for minority carrier MIS diodes, *Abstracts Photovoltaic Solar Energy Conference*, Berlin, 1979, p. 3.

# 13

# Light Detecting Semiconductor Devices

## CONTENTS

Photodetectors and light emitting sources (LEDs or lasers) form the basis of many optical control or information processing systems at visible and infrared wavelengths. The photodetector may involve conductivity changes in bulk semiconductors by excitation of carriers from doping levels at low temperatures for infrared detection (as in Zn- or Au-doped Ge detectors) or by excitation across the bandgap in an intrinsic semiconductor detector. In some intrinsic detectors, for example CdS and CdSe, the sensitivity is increased by trapping of one carrier type. The increase in sensitivity is at the expense of speed of response.

Other light sensing structures depend on separation of photon-induced carrier pairs at *p-n* junctions or Schottky barriers. A valuable feature of such structures is that current gain may be obtained by the use of avalanche multiplication or transistor action. Tradeoffs are involved in quantum efficiency, wavelength sensitivity, internal gain, temperature dependence, and noise.

Electron emission into vacuum is possible from widegap semiconductors such as $p^+$GaAs and $p^+$GaInAs when excited with light if they are treated with very thin films of a low-work-function material Cs or $Cs_2O$. These structures are known as negative electron affinity-emitting photocathodes. Their output response may be raised by subsequent electron multiplication at a series of dynodes based on GaP.

Large area sensing is achieved by semiconductor-coated vidicon surfaces or Si targets composed of large arrays of diodes. Diode array Si vidicons are sensitive to lower light intensities and longer wavelengths than conventional $As_2S_3$ or $PbO_2$ vidicon surfaces.

This chapter concludes with a brief discussion of semiconductor surfaces used for electrophotographic copying purposes.

## 13.1 PHOTODIODES

Some fundamentals of photon absorption in junctions have been discussed in Chapter 12 on solar cells. Photodiodes, unlike solar cells, are normally operated at reverse bias voltage, say $V_R$ as shown in Fig. 13.1, or at a higher voltage $V_M$ if use is to be made of avalanche multiplication to increase the sensitivity.

Large response in the absence of multiplication requires collection of the photon-induced carriers with few recombination losses. The quantum efficiency at an optical power level $P_{opt}$ and photon energy $h\nu$ is

$$\eta = \frac{I_{ph}}{q} \left( \frac{h\nu}{P_{opt}} \right) \qquad (13.1)$$

where $I_{ph}/q$ is the average number per second of electron-hole pairs collected across the junction and $P_{opt}/h\nu$ is the optical flux in photons per second.

A photodetector diode is normally followed by a low-noise amplifier. The transient risetime $T_r$ of the optical system is limited by carrier transport or multiplication times or by circuit $RC$ time constants. If $C_I$ is the photodiode capacitance, $C_A$ the amplifier capacitance, and $R_A$ the amplifier input resistance, then

$$T_r \geq 2.3\, R_A (C_I + C_A) \qquad (13.2)$$

and the bandwidth $B$ is

$$B \leq 1/2\pi R_A (C_I + C_A) \qquad (13.3)$$

In an ideal optical detector the sensitivity to very weak light signals is limited

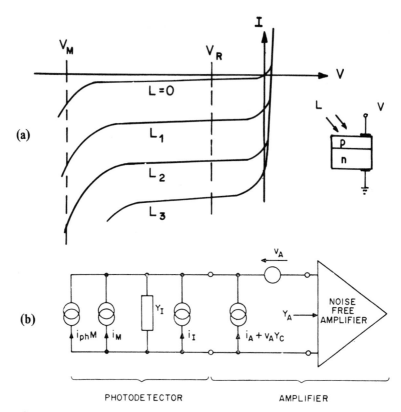

Fig. 13.1. Diode photodetector concepts:
(a) characteristics in reverse bias; and
(b) equivalent circuit of a generalized optical receiver showing the principal signal and noise sources of a photodetector with internal current gain and a noisy output amplifier. (After Melchior. Reprinted from *J. Lumin.*, 7, 1973, p. 390.)

by fluctuations in the photoexcitation of carriers. For a Poisson distribution the mean square noise current is

$$\overline{i_{ph}^2} = 2qI_{ph}\,\delta f \tag{13.4}$$

where $\delta f$ is the amplifier bandwidth. Often, however, the sensitivity level is limited by noise in the load or in the amplifier.

The power signal-to-noise ratio at the output of an optical receiver which contains a photodetector with internal current gain $M$ and an amplifier, that is not noise-free, is given by

$$\frac{S}{N} = \frac{\overline{i_{ph}^2}M^2\,\dfrac{|A(\omega)|^2}{|Y_I + Y_A|^2}}{\displaystyle\int_0^\infty \left\{2q(I_{ph}+I_B+I_D)M^2F(M) + \frac{\overline{i_I^2}}{\delta f} + \frac{\overline{i_A^2}}{\delta f} + \frac{\overline{v_A^2}}{\delta f}|Y_I+Y_C|^2\right\}\dfrac{|A(\omega)|^2}{|Y_1+Y_A|^2}\,df} \tag{13.5}$$

The form of this expression can be understood from the equivalent circuit of Fig. 13.1(b). The noise generator $i_M$ is given by

$$i_M{}^2 = 2q(I_{ph}+I_B+I_D)M^2 F(M)df \qquad (13.6)$$

where $I_B$ is the background-radiation-induced photocurrent and $I_D$ is the part of the dark current that is multiplied within the photodetector. $F(M)$ is a noise factor 1.5-3, corresponding to the current gain processes. The current $i_I$ represents the noise in any shunt or load conductance associated with the detector or the nonmultiplied part of the dark current. The output amplifier is represented by a noise-free amplifier with input admittance $Y_A = G_A + j\omega C_A$ and voltage gain $A(\omega)$ and by a noise voltage source $v_A{}^2$ in series with its input, an uncorrelated noise current source $i_A^2$ in parallel with the input, and a noise current source $v_A{}^2 |Y_C|^2$ that is fully correlated to the noise voltage source $v_A{}^2$ by the complex correlation admittance $Y_C = \mathrm{RE}Y_C + j\mathrm{IM}Y_C$.

Preferably the internal current gain in the detector should be made high enough to raise the signal level above the amplifier noise level.

Structures for various photodiodes are shown in Fig. 13.2. The junctions are usually small in diameter, 50 to 1000 $\mu$m, so that the leakage currents and diode capacitances are small. Point contact diodes provide the ultimate in low area but require accurate focusing of the light. The series resistance of diodes may be kept low by the use of fully depleted intrinsic or lightly doped regions as in p-i-n or metal-i-n diodes. For a depletion region width of 200 $\mu$m at a voltage of 30 V the electron transit time in Si is about 9 ns, as seen from Fig. 13.3. Table 13.1 lists the performance characteristics of a selection of photodiodes, including several with avalanche action. The spectral response characteristics of some photodiodes are shown in Fig. 13.4.

Avalanche photodiodes are frequently designed with guard-ring-type structures to provide uniform field conditions over the active area and to limit edge-leakage current effects (see Fig. 13.5). Typical multiplication gains are $10^2$-$10^3$ and current gain $\times$ bandwidth products of 100 GHz are possible for diodes of area about $10^{-5}$ cm$^2$. The response of an avalanche diode varies strongly with applied voltage and with temperature as shown in Fig. 13.6.

The avalanche buildup time, with the ionization coefficients of electrons ($\alpha$) and holes ($\beta$) assumed equal, is given by

$$t = \frac{1}{2} t_p M \qquad (13.7)$$

where $t_p$ is the depletion layer transit time. More generally the avalanche buildup time[75] can be written as

$$t = \tau_1 M \qquad (13.8)$$

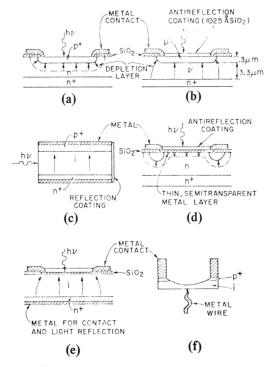

Fig. 13.2. Typical photodiode structures:
(a)    *p-n* diode;
(b)    *p-i-n* diode (Si optimized for 0.63 μm);
(c)    *p-i-n* diode with illumination parallel to junction;
(d)    metal semiconductor diode;
(e)    metal-*i-n* diode; and
(f)    semiconductor point contact diode. (After Melchior, Fisher, and Arams. Reprinted with permission from *Proc. IEEE,* 58, 1970, p. 1471.)

where $\tau_1$ is an intrinsic response time. This time can be determined experimentally by measurement of the shot-noise power $P$ versus the multiplication factor, since

$$P_n = \frac{2qI_pBM^xR_{eq}}{1 + (\omega t)^2} \tag{13.9}$$

where $I_p$ is the photocurrent before multiplication, $B$ the bandwidth, $R_{eq}$ the equivalent resistance, $\omega$ the angular frequency, and $t = \tau_1M$.

From measurements on Si avalanche photodiodes $x$ is 2.7 and the relationship obtained is

$$t = 5 \times 10^{-13}M \tag{13.10}$$

hence $\tau_1$ is $5 \times 10^{-13}$ sec.

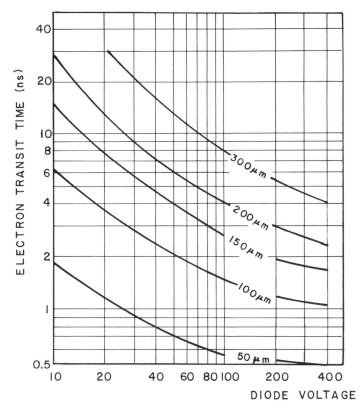

Fig. 13.3. Electron transit times for Si *p-i-n* diodes as a function of depletion width and voltage, assuming uniform field strength and the 300°K drift velocity vs. field strength data of Canali et al., *J. Phys. Chem. Solids*, **32**, 1971, p. 1707.

The intrinsic response time $\tau_1$ is considered to be closely related to the mean free time between ionizing collisions, $\tau$. If the maximum electric field is $5 \times 10^5$ V/cm, the ionization coefficients give mean-free times $\alpha V_S$ and $\beta V_S$ in the range $1\text{-}4 \times 10^{-12}$ sec, for carrier saturation velocities of $10^7$ cm/sec. The value of $\tau_1$ is smaller than $\tau$, because the carriers are multiplied exponentially in the avalanche process.

## 13.2 DETECTIVITY

The signal-to-noise ratio per watt of incident power is termed the detectivity

$$D = \frac{S/N}{EA} \tag{13.11}$$

where $E$ is the irradiance rms power density and $A$ is the detector area. Com-

Table 13.1. Performance Characteristics of Photodiodes

| Diode | Wavelength Range (μm) | Peak Efficiency (%) or Responsivity | Sensitive Area (cm²) | Capacitance (pF) | Series Resistance (Ω) | Response Time (seconds) | Dark Current | Operating Temperature (°K) | Comments |
|---|---|---|---|---|---|---|---|---|---|
| Si $n^{+}-p$ | 0.4-1 | 40 | $2 \times 10^{-5}$ | 0.8 at −23 V | 6 | 130 ps with 50-Ω load | 50 pA at −10 V | 300 | avalanche photodiode |
| Si $p-i-n$ | 0.6328 | > 90 | $2 \times 10^{-5}$ | < 1 | ~ 1 | 100 ps with 50-Ω load | $< 10^{-9}$ A at −40 V | 300 | optimized for 0.6328 Å |
| Si $p-i-n$ | 0.4-1.2 | > 90 at 0.9 μm  > 70 at 1.06 μm | $5 \times 10^{-2}$ | 3 at −200 V  3 at −200 V | < 1  < 1 | 7 ns  7 ns | 0.2 μA at −30 V | 300 | |
| Metal–$i$–$n$Si | 0.38-0.8 | > 70 | $3 \times 10^{-2}$ | 15 at −100 V | | 10 ns with 50-Ω load | $2 \times 10^{-2}$ A at −6 V | 300 | |
| Au–$n$Si | 0.6328 | 70 | 2 | | | < 500 ps | | 300 | Schottky barrier, antireflection coating |

760

| Material | λ (μm) | Responsivity / η | NEP (W) | Bias | Gain $M$ | Response time | | $T$ (K) | Remarks |
|---|---|---|---|---|---|---|---|---|---|
| PtSi-$n$Si | 0.35-0.6 | ~40 | $2 \times 10^{-5}$ | | < 1 | 120 ps | | 300 | Schottky barrier avalanche photodiode |
| Ge $n^+-p$ | 0.4-1.55 | 50 uncoated | $2 \times 10^{-5}$ | 0.8 at $-16$ V | < 10 | 120 ps | $2 \times 10^{-8}$ | 300 | Germanium avalanche photodiode |
| Ge $p-i-n$ | 1-1.65 | 60 | $2.5 \times 10^{-5}$ | 3 | | 25 ns at 500 V | | 77 | illumination entering from side |
| InAs $p-n$ | 0.5-3.5 | > 25 | $3.2 \times 10^{-4}$ | 3 at $-5$ V | 12 | $< 10^{-6}$ | | 77 | |
| InSb $p-n$ | 0.4-5.5 | > 25 | $5 \times 10^{-4}$ | 7.1 at $-0.2$ V | 18 | $5 \times 10^{-6}$ | | 77 | |
| $Pb_{1-x}Sn_xTe$ $x = 0.16$ | 9.5 μm | 45 V/W $\eta = 60$ | $4 \times 10^{-3}$ | | | $\sim 10^{-9}$ | | 77 | shunt resistance $R_i = 10\ \Omega$ |
| $Pb_{1-x}Sn_xSe$ $x = 0.064$ | 11.4 μm | 3.5 V/W $\eta = 15$ | $7.8 \times 10^{-3}$ | | | $\sim 10^{-9}$ | | 77 | shunt resistance $R_i = 2.5\ \Omega$ |
| $Hg_{1-x}Cd_xTe$ $x = 0.17$ | 15 μm | $\eta \sim 10$-$30$ | $4 \times 10^{-4}$ | | 8 | $< 3 \times 10^{-9}$ | | 77 | shunt resistance $R_i > 100\ \Omega$ |

After Melchior, Fisher, Arams, Reprinted with permission from *Proc. IEEE*, **58**, 1970, p. 1473.

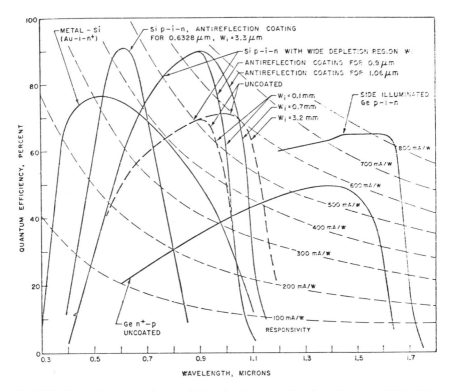

Fig. 13.4. Spectral response characteristics of various junction photodetectors. (After Melchior. Reprinted from *J. Lumin.*, 7, 1973, p. 390.)

parison of detectors is simplified by introducing a detectivity $D^*$ for which the detector areas are normalized to 1 cm² and the amplifier bandwidth to 1 Hz. Then assuming the detector noise varies as $A^{1/2}$ and $(\Delta f)^{1/2}$, the normalized detectivity is given by

$$D^* = \frac{(S/N)(\Delta f)^{1/2}}{E(A)^{1/2}} \tag{13.12}$$

Measurement of detectivity is made by determining the signal-to-noise ratio when the detector is exposed to a standard blackbody radiator, at a temperature such as 500°K for an infrared detector.

If thermal noise is dominant, a photovoltaic detector has a detectivity

$$D^*_\lambda = \frac{\eta q \, R^{1/2} A^{1/2}}{2E_\lambda \, (kT)^{1/2}} \tag{13.13}$$

where $\eta$ is the quantum efficiency at the wavelength $\lambda$ and $R$ is equivalent to

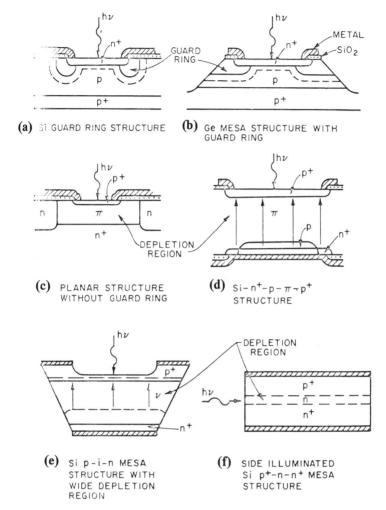

Fig. 13.5. Construction of avalanche photodiodes. (After Melchior. Reprinted from *J. Lumin.*, 7, 1973, p. 390.)

the diode resistance. For an ideal diode, $R$ is given by $kT/qJ_SA$ where $J_S$ is the reverse saturation current density. Since the generation recombination current $J_S$ may be determined from the diode dark characteristics, or estimated from minority carrier lifetime and mobility information, the detectivity may be calculated as a function of wavelength and temperature. The detectivity of a $Pb_{1-x}Sn_xTe$ photodiode for 10 $\mu$m use at 77°K may be expected to have an upper value of about $10^{12}$ cm $Hz^{1/2}/W$ as shown in Fig. 13.7.

The intrinsic detectivity of a detector is not achieved unless background

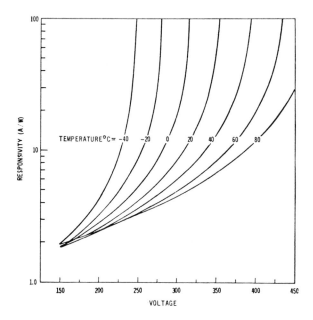

Fig. 13.6. Typical responsivity at $\lambda = 1.06$ $\mu$m as a function of voltage and temperature for a reach-through Si avalanche diode of 100-$\mu$m depletion width. (After Webb, McIntyre, and Conradi. Reprinted from *RCA Review*, 35, 1974, p. 264.)

radiation impinging on the device is made small by reducing the field of view with a cooled shroud or filter. Photon detectors for which further cooling brings about no continued improvement in signal-to-noise ratio are termed background limited in operation (BLIP). Figure 13.8 shows the background-limited $D^*_\lambda$ as a function of field of view and cutoff wavelength, calculated for a 300°K thermal background. Actual detector performances are usually factors of 5 or 10 less than these background-limited values. Typical $D^*_\lambda$ behavior is shown in Fig. 13.9 for some small-bandgap semiconductors.

## 13.3 PHOTOCONDUCTIVE DETECTORS

Now consider detectors that depend on conductivity modulation by the received light rather than junction separation of induced carrier pairs. Assume the detector is a bar of $n$-type semiconductor of the dimensions shown in Fig. 13.10, supplied with a constant current $I$. In the absence of light the detector resistance is

$$R_D = \frac{\ell}{\sigma d w} \tag{13.14}$$

where the electric conductivity $\sigma$ is given by $q(n\mu_n + p\mu_p)$. When light is applied

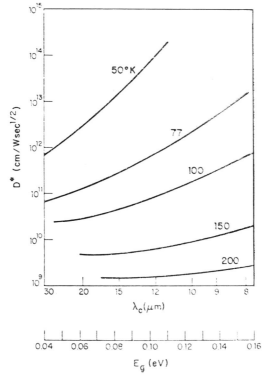

Fig. 13.7. Calculated detectivity $D^*$ vs. bandgap of $Pb_{1-x}Sn_xTe$ photodiodes for various temperatures, assuming a minority carrier concentration of $5 \times 10^6 \, cm^{-3}$, a mobility of $2 \times 10^4 \, cm^2/V \cdot sec$, and a lifetime of $10^{-8}$ sec. (After Melngailis. Reprinted from *J. Lumin.*, 7, 1973, p. 503.)

the electron concentration increases from $n$ to $n + \Delta n$ and the hole concentration from $p$ to $p + \Delta p$. Neglecting the possibility of carrier trapping, $\Delta n = \Delta p$ since charge neutrality must be maintained in the bar by the free carriers. The conductivity change is therefore

$$\Delta \sigma = q \, \Delta n \, (\mu_n + \mu_p) \tag{13.15}$$

and this produces a conductance change

$$\Delta G_D = \frac{dw}{\ell} q \, \Delta n \, (\mu_n + \mu_p) \tag{13.16}$$

At constant current $I$ this produces an output voltage change proportional to $I/G_D - I/(G_D + \Delta G_D)$.

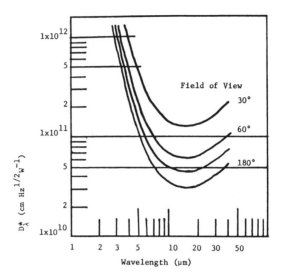

Fig. 13.8. Background-limited detectivity vs. wavelength for various fields of view (calculated for 300°K). (After Levinstein and Mudar. Reprinted with permission from *Proc. IEEE,* **63**, 1975, p. 9.)

In the circuit shown in Fig. 13.10(b) the photodetector $R_D$ is subjected to light sinusoidally modulated at a frequency $\omega$, in series with a load resistor $R_L$ and supplied by a voltage $E$. The sinusoidal output voltage signal $V_S$ obtained is

$$V_S = \frac{E R_L R_D^2 \Delta G_D}{(R_D + R_L)^2 (1 + \omega^2 \tau^2)^{1/2}} \tag{13.17}$$

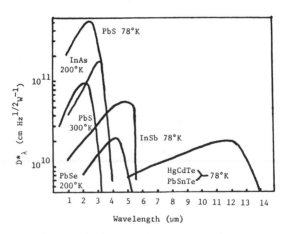

Fig. 13.9. Infrared detectivity for various intrinsic semiconductor detectors. (After Levinstein and Mudar. Reprinted with permission from *Proc. IEEE,* **63**, 1975, p. 9.)

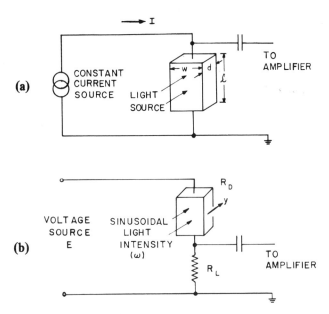

Fig. 13.10. Photoconductive detector circuits:
(a)  with constant current source; and
(b)  with voltage source and load $R_L$.

where $\tau$ is the minority carrier recombination time constant. For a given conductivity change $\Delta G_D$ the maximum responsivity $V_S$ occurs for $R_D = R_L$. Then

$$V_{S_{max}} = \frac{E R_D \Delta G_D}{4(1 + \omega^2\tau^2)^{1/2}} \tag{13.18}$$

Inspection of this equation shows that the sensitivity is improved if $R_D$ is large and so a lightly doped bar should be used. For a given photon flux the carrier increase $\Delta n$, reflected in $\Delta G_D$, is proportional to the carrier lifetime $\tau$ and to the photon quantum efficiency. Hence the lifetime $\tau$ should be large provided the condition $(\omega\tau)^2 \ll 1$ can be met.

### 13.3.1  Intrinsic Photodetectors

A more elaborate treatment commences with the equation[123]

$$D \frac{\partial^2}{\partial y^2} (\Delta p) - \frac{\Delta p}{\tau} = \alpha N_\phi e^{-\alpha y} \tag{13.19}$$

where $D$ is the ambipolar diffusion constant, $N_\phi$ is the photon flux per unit area per second, and $\alpha$ is the absorption coefficient. Then the photocarrier density integrated across the specimen thickness $d$ is given by

$$\Delta p = \frac{\eta N_\phi \tau}{(\alpha^2 L^2 - 1)} \left[ \frac{KL \left[ \left( \frac{sL}{D} - \alpha L \right) e^{-\alpha d} + \frac{sL}{D} + \alpha L \right]}{1 + \frac{sL}{D} \coth \left( \frac{d}{2L} \right)} - \left( 1 - e^{-\alpha d} \right) \right] \quad (13.20)$$

where $s$ is the recombination velocity at the front and back surfaces. Then from Eq. (13.18)

$$V_s = \frac{E(b+1) \, \Delta p}{4d(bn+p)(1+\omega^2\tau^2)^{1/2}} \quad (13.21)$$

where $b = \mu_n/\mu_p$ and the condition $R_L = R_D$ is satisfied.

For highest sensitivity the surface recombination should be low, so assume $s = 0$. Equations (13.20) and (13.21) then give

$$V_s = \frac{E(b+1) \, \eta \, N_\phi \, \tau \, (1 - e^{-\alpha d})}{4d(bn+p)(1+\omega^2\tau^2)^{1/2}} \quad (13.22)$$

If $n$ is written as $n_i^2/p$ and the equation is differentiated with respect to $p$ the maximum value of $V_s$ is found to occur when $p = bn$. Since $b$ is greater than unity for semiconductors, this implies the use of $p$-type material (lightly doped) for optimum responsivity. The equation also suggests that the detector, assuming zero surface recombination, should be thin, since although fewer photons are absorbed they are concentrated in a smaller volume.

If the system performance is limited by thermal Johnson noise the detectivity is a maximum for $p = bn$ and has the value

$$D_\lambda^* = \frac{E\eta\tau\lambda(1-e^{-\alpha d})(q\mu_p)^{1/2}(b+1)}{4hc(2pkT)^{1/2}N_\phi d^{1/2}(1+\omega^2\tau^2)^{1/2}} \quad (13.23a)$$

Hence the driving voltage $E$ should be large and the material should have a high value of the product $\mu_p(b+1)$ and a high lifetime.

However if generation-recombination noise within the semiconductor is dominant the detectivity expression is

$$D_\lambda^* = \frac{\eta\lambda\tau^{1/2}(1-e^{-\alpha d})(n+p)^{1/2}}{2d^{1/2}hcn_i} \quad (13.23b)$$

The detectivity is now independent of $b$ and less dependent on lifetime. The doping should be such that either $n$ or $p$ is large. This differs from the light doping desirable for good detectivity under thermal-noise-limited conditions. In practice, conditions may be such that both noise mechanisms are important and some compromise in doping is needed.

The detectivity expressions given suggest maximum values for thicknesses where $\alpha d \simeq 1.25$. Thicker values may be desirable however if significant surface recombination is present.

### 13.3.2  Trapping Effects to Provide Quantum Gain

A photon of energy $h\nu > E_g$, absorbed in an $n$-type semiconductor bar produces one electron-hole pair. This pair may recombine for a loss or the carriers may be swept to the ohmic contacts for a current flow of one electron in the external circuit. However, if the semiconductor contains a level in the bandgap that traps the minority carrier (hole) for a time $\tau_T$, the semiconductor bar has an extra free carrier electron during the entire trapping time. If the drift velocity of this electron is $v_d$ (depending on the electric field and mobility) the transit time through the bar is $\ell/v_d$. The number of traversals of the bar in the trapping time $\tau_T$ is therefore $\tau_T v_d/\ell$ and this is the quantum gain of the photoconductor produced by the minority carrier trapping.

In photoconductors such as CdS or CdSe the semiconductors are specially provided with trapping levels (typically by processing with Cu and Cl levels) to achieve large quantum efficiencies. If the trapping time is $10^{-4}$ sec and if $v_d$ is $10^6$ cm sec$^{-1}$ a detector of length $10^{-1}$ cm will have a quantum gain of the order of $10^3$. This gain, however, is achieved at the expense of the transient response of the photoconductor to illumination changes.

### 13.3.3  Impurity Photodetectors

For low-energy long-wavelength detectors, semiconductors of very small bandgaps such as InSb, $Pb_x Sn_{1-x} Te$ and $Hg_x Cd_{1-x} Te$ are available. However, another approach is the use of a deep or shallow impurity in a larger bandgap semiconductor. Cooling is necessary so that optical excitation of carriers from the impurity to the nearest bandedge controls the conductivity. An example is the use in Si of In which has an acceptor level at $E_v + 0.155$ eV. The response of the Si:In system at $77°K$ or less is good in the infrared atmospheric window between 3 and 5 $\mu$m (see Fig. 13.11). The spectral absorption of Si:In is given in Fig. 13.12. This absorption curve may be fitted by the Lucovsky absorption model which shows that:

$$\sigma(h\nu) = \frac{1}{n} \left[ \frac{\mathscr{E}_{eff}}{\mathscr{E}_0} \right]^3 \left[ \frac{8e^2 h}{3m^* c} \right] E_{In}^{1/2} \frac{(h\nu - E_{In})^{3/2}}{(h\nu)^3} \tag{13.24}$$

The use of Si for extrinsic photoconductors allows good photolithographic processes to be applied and permits integration of photoconductive sensors with

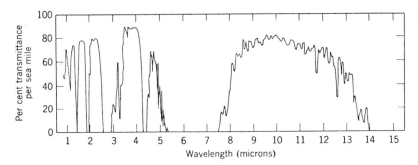

Fig. 13.11. Atmospheric transmission spectrum, showing windows from 2-2.6 μm, 3.4-4.2 μm, 4.5-5.0 μm, and 8-13 μm. (After Kruse, McGlauchlin, and McQuistan. Reprinted from *Elements of Infrared Technology: Generation, Transmission, and Detection,* © John Wiley & Sons, N.Y., 1962, p. 164.)

other Si devices such as charge-coupled photosensing arrays. Impurities that have been studied in Si for extrinsic detectors include relatively deep ones such as Au, Zn, S, In and shallow ones such as B, Al, P, As, Sb. These cover wavelength ranges from a few microns to 30 μm. For detectors in the wavelength range from 3 to 14 μm cooling is required to dry ice or liquid nitrogen temperatures or even lower. Desirable impurity properties are high solubilities in Si and large optical capture cross sections. A relatively low-diffusion coefficient may also be needed in integrated device applications so that the impurity may be restricted to the desired part of the structure.

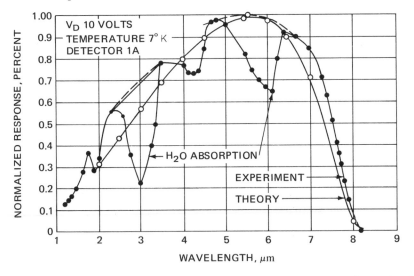

Fig. 13.12. Spectral response of In-doped Si for infrared detection in the 3-5-μm band. (After Pines and Baron. Reprinted with permission from *IEEE Int. Electron Devices Meeting, Technical Digest*, 1974, p. 449.)

Ge has also been extensively employed as a host lattice. Ge as an intrinsic semiconductor has an activation energy of 0.7 eV and the corresponding long-wavelength threshold is 1.7 $\mu$m. Ge:Au provides a response to 10 $\mu$m and Hg is an effective impurity for the 8-14-$\mu$m range at a temperature such as $30°$K. Ge:Cu at $10\text{-}18°$K provides response to 30 $\mu$m, and Ge:Zn to 40 $\mu$m, with detectivity values $D^*$ in the mid $10^{10}$ cm Hz$^{1/2}$ W$^{-1}$ range (see Fig. 13.13). Response times of such detectors range from $10^{-6}\text{-}10^{-9}$ sec.

The equation governing extrinsic photoconductivity in an $n$-type semiconductor is

$$\Delta n = \eta N_\phi S(N_D - N_A - n)d\tau_0 \qquad (13.25)$$

where $\eta$ is the quantum efficiency, $N_\phi$ is the photon flux, $S$ is the optical absorption cross section, $N_D$ is the density of the dopant level being optically ionized, $N_A$ is the compensating acceptor level density present, $n$ is the dark carrier concentration, $d$ is the bar thickness, and $\tau_0$ is the effective lifetime given by

$$\tau_0 = 1/v_{th}\, \sigma\,(2n + N_A) \qquad (13.26)$$

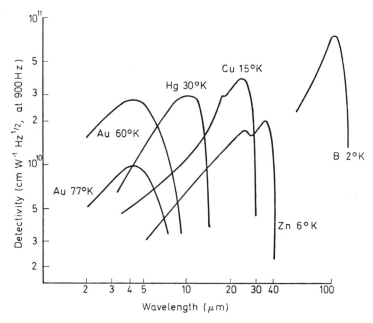

Fig. 13.13. Spectral response of Ge extrinsic detectors with various dopants. The field of view is 60° (10° for B). (After Levinstein. Reprinted from *Appl. Optics,* 4, 1965, p. 644.)

In the absence of $N_A$ the response time in Eq. (13.26) is inversely proportional to the free electron concentration $n$ and is strongly temperature dependent. The responsivity, voltage per unit power, expression for a detector of length $\ell$ and width $w$, for the $R_D = R_L$ condition is

$$\left(\frac{V_s}{P}\right)_\lambda = \frac{E\,\eta\lambda\,S\,(N_D - N_A - n)}{4\,hc(2n + N_A)\,v_{th}\,\sigma n\ell w\,(1 + \omega^2\tau^2)^{1/2}} \qquad (13.27)$$

For high responsivity the impurity concentration should be large and the degree of compensation small. If the temperature is so low that the carrier concentration is largely determined by the light signal, the responsivity decreases with increasing signal [$n$ is in the denominator of Eq. (13.27)] and from Eq. (13.26) the response time increases.

Extrinsic detectors are operated at low temperatures and the dominant noise mechanism is usually generation-recombination noise and not thermal Johnson noise. The detectivity expression with generation-recombination noise dominant is

$$D^*(\lambda) = \frac{\eta\lambda S d^{1/2}\,(N_D - N_A - n)}{2hc\,(v_{th}\sigma)^{1/2}\,n^{1/2}(2n + N_A)^{1/2}}$$

Hence $(N_D - N_A)$ should be large, the sample thick, and the temperature low.

## 13.4 PHOTOTRANSISTORS

Bipolar-junction and field-effect transistors may be used as light detectors. In the bipolar device the transistor is designed so that illumination creates carriers in the base region. The supply voltage is applied between the collector and the emitter, and the base floats in potential to suit the current flow and photoinduced carrier conditions as shown in Fig. 13.14. There is a buildup of excess electrons in the base and development of a voltage $V_{EB}$ that allows a small electron current $I_e$ related to the photon flux to flow into the emitter. However, the voltage $V_{EB}$ causes a much larger current of holes $I_h$ to flow between emitter and collector. Thus the photon-induced carriers are multiplied by the injection gain of the transistor, provided the base is free to find its own potential.

Figure 13.15 shows the use of phototransistors as part of an optoelectronic keyboard, in which the key strokes code the patterns of light falling onto a phototransistor array so that electrical contacts are not needed.

Bipolar transistors used in imaging arrays operate in a charge-storage mode. The ratio of collector capacitance $C_C$ to emitter capacitance $C_E$ should be large from consideration of dynamic output signal range and crosstalk problems. This may be achieved by having the $n^+$ emitter much smaller in area than the

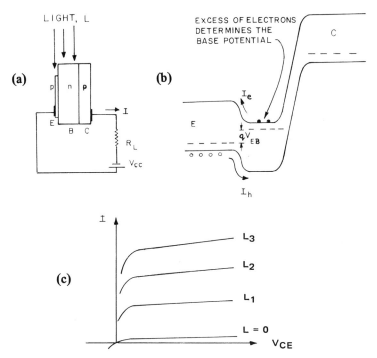

Fig. 13.14. Bipolar transistor (p-n-p) operated as a light detector:
(a)  bias diagram with the base floating in potential;
(b)  energy diagram under illumination; and
(c)  characteristics for various light levels.

n collector [see Fig. 13.16(a)]. However, a better ratio of $C_C/C_E$ is obtained by the use of an $n^+$ collector surface diffusion that extends over the base diffusion and is thin enough to allow photon penetration [see Fig. 13.16(b)].

Now consider photodetection with field-effect transistor structures. Figure 13.17(a) illustrates a field-effect structure with light input through a $p$Si gate and with top and bottom gates electrically connected to enclose the channel. With the top gate thin in relation to the electron minority carrier diffusion length $L_n$, the carriers created by the light are collected by the junction of the top gate and the channel. This acts as a photodiode with a narrow base region that contains a voltage gradient as suggested by Fig. 13.17(b). Analysis using the Shockley gradual channel model, assuming the photocurrent is supplied uniformly along the channel, gives

$$-\frac{I_\phi}{2} = G_0 \left\{ V_\phi - \frac{2}{3}\sqrt{\frac{8\epsilon}{qN_Dd^2}} \left[ (-V_\phi + \phi_B - V_G)^{3/2} - (\phi_B - V_G)^{3/2} \right] \right\} \quad (13.28)$$

$$G_0 = \frac{q\mu_n N_D dw}{\ell} \quad (13.29)$$

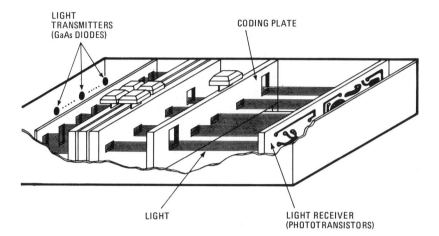

Fig. 13.15. Optoelectronic keyboard using phototransistors. Depressing a key in the optoelectronic keyboard alters the pattern of light falling on the phototransistors. Thus the keys can be coded into a digital form in the absence of any electrical contacts. (Reprinted from *Electronics*, July 24, 1975, © McGraw-Hill, Inc., N.Y, 1975.)

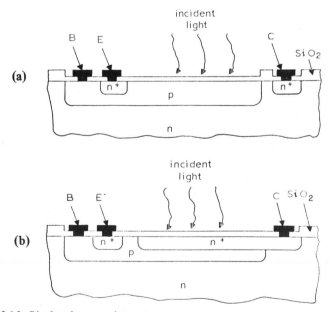

Fig. 13.16. Bipolar phototransistor structures:
(a)  conventional base illumination; and
(b)  modified collector overdiffusion to increase the ratio $C_C/C_E$. (After Holmes and Salama. Reprinted with permission from IEE, from *Electronics Letters,* 8, 1972, p. 23.)

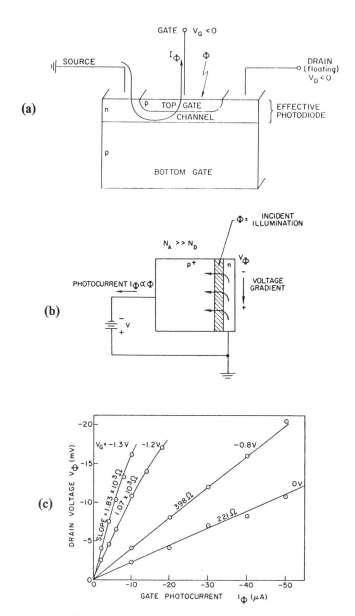

Fig. 13.17. Junction FET photodetector concepts:
(a)   channel structure where $I_\phi$ is the photocurrent induced by the photon flux $\phi$;
(b)   representation as a photodiode with a narrow base containing a voltage gradient;
(c)   photodetecting characteristic for a commercial JFET; and
(d)   electrical characteristics showing resemblance to (c). (After Bandy and Linvill. Reprinted with permission from *IEEE Trans. Electron Devices*, ED-20, 1973, pp. 793-795.)

where $\phi_B$ is the junction built-in voltage, $\epsilon$ is the Si dielectric constant, $\mu_n$ is the channel mobility, $N_D$ is the channel doping (much less than the gate doping), $d$ is the channel depth (undepleted), $w$ is the channel width, and $\ell$ is the channel length.

These are identical to conventional results except that the drain current $I_D$ is replaced by $-I_\phi/2$ and $V_D$ is relabeled as $V_\phi$ the photodetector voltage.

Commercial Si junction FETs may be made to exhibit gate photoconductivity action [see Fig. 13.17(c)], although not well designed for the purpose. For comparison Fig. 13.17(d) shows the electrical characteristics with the factor of two in channel resistance apparent. The operation is in the linear portion of the drain characteristic in the third quadrant with $V_D$ and $V_G$ having the same sign, because the channel current is reversed with a corresponding reversal in the sign of the drain voltage.

Responsivity of the device, defined as the photodetector output voltage per unity power flux $P$, is

$$\frac{V_\phi}{P} = \frac{V_\phi}{I} S = \frac{1}{2} A_R R_0 S \qquad (13.30)$$

where $A_R$ is the multiplicative increase in the channel impedance from its value $R_0 = G_0^{-1}$ at $V_G = 0$ and $S$ is proportional to the device area and to the quantum efficiency. Since the channel resistance $R_0$ is also approximately proportional to the device area, the responsivity is proportional to the square of the device area.

Now consider a thin-film field-effect modulation structure. The detectivity of a photoconductor such as PbS increases as the carrier concentration decreases. With the thin-film transistor structure shown in Fig. 13.18 it is possible, by applying a negative gate potential, to deplete the PbS film of carriers and thus enhance the photoconductive response by a factor of $10^2$ or more.

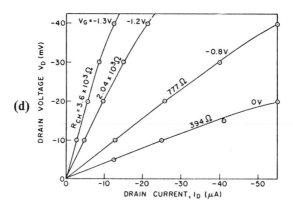

Fig. 13.17 (continued)

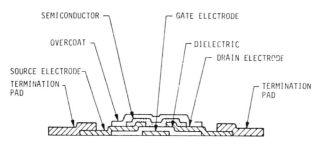

SEMICONDUCTOR
OVERCOAT
SOURCE ELECTRODE
TERMINATION
PAD
GATE ELECTRODE
DIELECTRIC
DRAIN ELECTRODE
TERMINATION
PAD

Fig. 13.18. Cross section of a PbS photoconductive thin-film transistor, in which the insulated gate is used to increase sensitivity by depleting the channel of carriers. (After Kramer and Levine. Reprinted with permission from *Appl. Phys. Lett.*, 28, 1976, p. 101.)

Detectivities of $2 \times 10^{10}$ cm $Hz^{1/2}/W$ may be obtained at 173°K (500 Hz) for a 180° field of view and 300°K background.

Infrared detectors may also be made by doping the channel of a JFET with a deep impurity as shown by the structure and characteristics of Fig. 13.19. The quantum efficiency is low, $3 \times 10^{-5}$, but the gain (ratio of number of electrons flowing in the external circuit to the number of incident photons) is high (300) because the trapping time is long (20 sec at 100°K). The resulting responsivity (output power/input energy) is 4 mW/$\mu$J.

One method of obtaining high sensitivity from light detecting devices is to operate in an integration mode in which a diode discharges a capacitance by a leakage current that is proportional to the light intensity falling on the diode. This is well suited to a Si integrated-circuit structure with MOSFET amplification. Such a chip is shown in Fig. 13.20, where $D_1$ is the integrating diode and $D_2$ is a similar diode that compensates for dark current. $M_3$ and $M_5$ constitute the differential amplifier with $M_4$ and $M_6$ as their corresponding drain loads. $M_2$ operates as a constant current source while $M_1$ and $M_7$ act as analog switches. When the clock pulse $\phi$ is more negative than the threshold voltage $V_T$, $M_1$ and $M_7$ conduct and diodes $D_1$ and $D_2$ charge up to $-(\phi - V_T)$. Then $\phi$ returns to zero and $M_1$ and $M_7$ are cut off. The voltage across $D_2$ starts decaying at a rate corresponding to its dark leakage current, and to the charging of the gate capacitance of $M_5$. The voltage across $D_1$ decays at a rate that is proportional to the light intensity falling on $D_1$. The difference of the voltage on the gate of $M_3$ and the voltage on the gate of $M_5$ is amplified and the output is obtained across the drain of $M_3$. The sensitivity is about $10^{-12}$ W·cm$^2$ at 0.9 $\mu$m with an integration time of 1 sec at 300°K or $10^{-14}$ W·cm$^2$ at 200°K. This allows detection of objects hundreds of feet away under night-sky conditions.

Position-sensitive photodetectors are also possible. These are area type devices with four sensing electrodes on the perimeter that provide electrical signals from which the location of a light spot on the surface may be determined. The sensing action depends on detecting lateral current flow effects in a resistive sheet

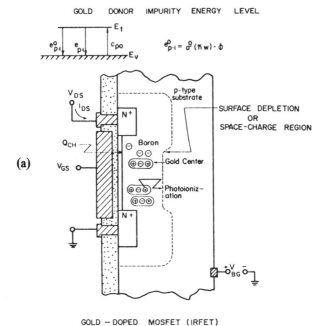

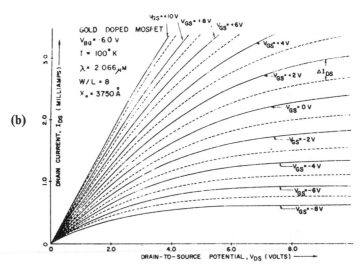

Fig. 13.19. MOSFET infrared sensor with Au-doped Si:
(a) schematic of the Au center photoionized in the depletion region and $Q_{CH}$ the charge contributed by electrons in the surface channel; and
(b) *I-V* characteristics of the IRFET before illumination (solid line) and after illumination (dashed line). (After Parker and Forbes. Reprinted with permission from *IEEE Trans. Electron Devices*, **ED-22**, 1975, pp. 917, 922.)

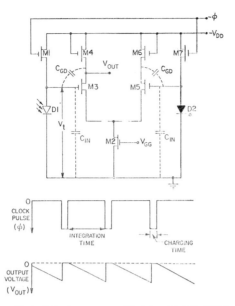

Fig. 13.20. Integrating diode light detector on IC chip with MOSFET amplification. (After Chamberlain and Aggarwal. Reprinted with permission from *IEEE J. Solid-State Circuits*, SC-7, 1972, p. 202.)

associated with the large area junction as shown in Fig. 13.21. For the dual-axis detector [Fig. 13.21(b)] with extended contacts at all four sides—except for small gaps at the vertices—an approximately linear relation is found between the light spot position along an axis and the log ratio of the corresponding output currents. Figure 13.21(c) shows a structure with resistive sheets on the front and back sides (produced by ion implantation of boron and phosphorous, respectively) into a lightly doped *n*-type wafer. With electrodes of the shape shown, this structure has better linearity than the structure shown in Fig. 13.21(b). Position sensitivity is possible to a few microns in laser-pulse position sensing. Delays in the system require consideration of the charge collection process with transmission-line-type equations.

## 13.5 PHOTOCATHODES AND NEGATIVE-ELECTRON-AFFINITY-EMIT-TING DEVICES

Alkali metals have low work functions (Cs, 1.9 eV; Na, 2.35 eV, and K, 2.2 eV) and provide surfaces that emit electrons when subjected to photons of visible light. Typical photocathode surface properties are summarized in Table 13.2 and spectral response curves are given in Fig. 13.22.

The Ag-O-Cs cathode, known as the *S*1 surface, is the only material suitable for use in the wavelength range out to 1 $\mu$m (1.2 eV). However, even at the

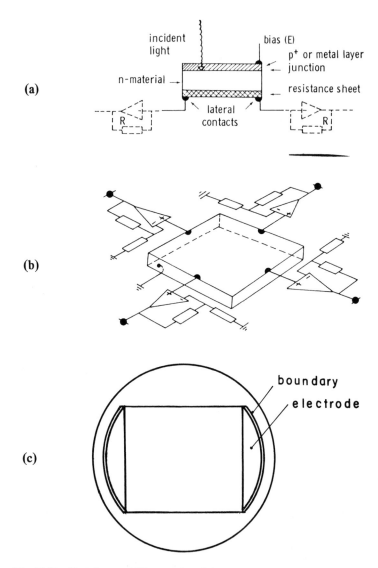

Fig. 13.21. Photobeam position sensing plates:
(a)  cross section of a single-axis lateral photodiode. The difference in the two lateral con-
     tact currents indicates the position of the light spot; and
(b)  dual-axis square Wallmark diode with floating junction (bottom view); (After Woltring.
     Reprinted with permission from *IEEE Trans. Electron Devices*, **ED-22**, 1975, pp. 582-
     584.)
(c)  one side of a double-sided two-dimensional detector with the active area boundary
     outlines. The layout of the reverse side is identical apart from being rotated 90°.
     (After Lindholm and Petersson. *IEEE Int. Electron Devices Meeting, Technical Digest*,
     1976).

**Table 13.2.  Composition and Characteristics of Various Photocathodes**

| Nominal Composition | Response Designation | Type of Photocathode[a] | Conversion Factor at λmax (lm/W) | Luminous Sensitivity (μA/lm) | Wavelength of Maximum Response max (nm) | Sensitivity at λmax (mA/W) | Quantum Efficiency at λmax (%) | Dark Emission at 25°C A × 10^-15/cm² |
|---|---|---|---|---|---|---|---|---|
| Ag-O-Cs | S-1 | O | 92.7 | 25 | 800 | 2.3 | 0.36 | 900 |
| Ag-O-Rb | S-3 | O | 285 | 6.5 | 420 | 1.8 | 0.55 | — |
| $Cs_2Sb$ | S-19 | O | 1603 | 40 | 330 | 64 | 24 | 0.3 |
| $Cs_3Sb$ | S-4 | O | 1044 | 40 | 400 | 42 | 13 | 0.2 |
| $Ca_3Sb$ | S-5 | O | 1262 | 40 | 340 | 50 | 18 | 0.3 |
| $Cs_3Bi$ | S-8 | O | 757 | 3 | 365 | 2.3 | 0.77 | 0.13 |
| Ag-Bi-O-Cs | S-10 | S | 509 | 40 | 450 | 20 | 5.6 | 70 |
| $Cs_3Sb$ | S-13 | S | 799 | 60 | 440 | 48 | 14 | 4 |
| $Cs_3Sb$ | S-9 | S | 683 | 30 | 480 | 20 | 5.3 | — |
| $Cs_3Sb$ | S-11 | S | 808 | 60 | 440 | 48 | 14 | 3 |
| $Cs_3Sb$ | S-21 | S | 783 | 30 | 440 | 23 | 6.7 | — |
| $Cs_3Sb$ | S-17 | O | 667 | 125 | 490 | 83 | 21 | 1.2 |
| $Na_2KSb$ | S-24 | S | 1505 | 32 | 380 | 64 | 23 | 0.0003 |
| K-Cs-Sb | — | S | 1117 | 80 | 400 | 89 | 28 | 0.02 |
| $(Ca)Na_2KSb$ | — | S | 429 | 150 | 420 | 64 | 18.9 | 0.4 |
| $(Cs)Na_2KSb$ | S-20 | S | 428 | 150 | 420 | 64 | 19 | 0.3 |
| $(Cs)Na_2KSb$ | S-25 | S | 276 | 160 | 420 | 44 | 13 | — |
| $(Cs)Na_2KSb$ | ERMA | S | 169 | 265 | 575 | 45 | 10 | 1 |

[a] O = Opaque; S = Semitransparent. (Courtesy RCA Corporation.)

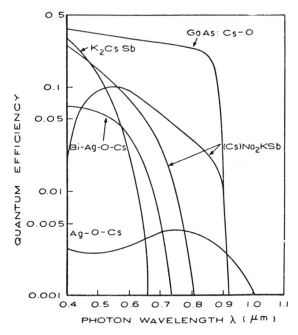

Fig. 13.22. Photoemission spectral response curves of several conventional cathodes compared with that of a NEA GaAs:Cs-O cathode, showing that the NEA cathode has higher quantum efficiency throughout the visible and near IR. (After Sommer. Reprinted with permission from *Phys. Colloque C-6*, Nov-Dec. 1973, supplement, **34**, p. C6-57.)

wavelength for peak sensitivity, the quantum efficiency is only 0.36% and the sensitivity only about 25 $\mu A\ lm^{-1}$. These values decline rapidly for longer wavelengths. Typically, sensitivity values are given relative to incident rather than absorbed radiation. At 5550 Å (2.24 eV), 1 W of radiant energy is equivalent to 660 lm. Hence 1 lm at this wavelength represents $4.2 \times 10^{15}$ photons $sec^{-1}$. If the quantum efficiency were 100% (i.e., 1 electron emitted per incident photon), the current yield would be 680 $\mu A\ lm^{-1}$. As the sensitivity of the eye decreases for longer wavelengths, more photons are required per lumen and current yields in excess of 680 $\mu A\ lm^{-1}$ become possible without exceeding a quantum yield of 1 electron per photon.

Electronic processes involved in Ag-O-Cs cathodes have remained somewhat obscure to this day. Cs has a work function of 1.9 eV and creates lower effective work functions when in monatomic layers on other metals (for instance, 1.54 eV on W). However, the Ag also plays a role in the performance of Ag-O-Cs surfaces. The performance of the $Cs_3Sb$ photocathode is attributed to the material being a $p$-type semiconductor with a bandgap of 1.6 eV and an electron affinity of ~0.4 eV. Hence the cutoff edge expected is about 2.0 eV: this corresponds to a 620-nm cutoff wavelength.

Studies of $Cs_2O$ show this to be an $n$-type semiconductor with a bandgap of 2.0 eV and an electron affinity of 0.7-1.0 eV. However, such studies suggested no improvements for Ag-O-Cs surfaces and did not lead to the development of other long-wavelength photocathodes. For a considerable number of years therefore, the performance of long-wavelength photocathodes remained almost unchanged. Recently, however, the concept of zero or negative work function photocathodes based on III-V semiconductors with Cs or O-Cs coated surfaces has provided a breakthrough to new performance levels. The sensitivity in the region of 1.4 eV has been increased by about two orders of magnitude over that of Ag-O-Cs surfaces (see the GaAs:Cs-O curve of Fig. 13.22) and useful performance has been obtained at lower energies by the use of ternary III-V compounds such as (In,Ga)As:Cs-O.

### 13.5.1 Negative-Electron-Affinity Emitters

For negative-electron-affinity action a very thin (monolayer) of Cs or Cs:O is applied to the surface of a III-V semiconductor such as $In_xGa_{1-x}As$. The work function energy $\phi$ (see Fig. 13.23) should be as small as the choice of the coating material will allow, and the processing should be such that the band bending is as large as possible. Maximum possible band bending has been obtained if the bottom of the notch approximately lines up with the valence bandedge in the bulk. Furthermore, the bending should take place in the shortest possible distance so that most of the radiation penetrates to the flat region of the conduction band and the photoinduced electrons have the highest energy. The shorter the band-bending distance the smaller the loss of conduction-band electrons thermalizing into the notch by scattering processes instead of being emitted. The bulk semiconductor must be heavily doped ($N_A \sim 10^{19}$ cm$^{-3}$) if the band bending is to occur over a suitably short distance.

Reflection-mode NEA photocathodes have the light incident on the cathode-vacuum surface as in photodiode multipliers, whereas transmission-mode photocathodes are thin-film structures with the light incident from the rear as for image tubes.

In reflection-mode operation the photoemissive quantum efficiency $Y_R$ is given by

$$Y_R(\lambda) = \frac{B(1-R_{cv})}{1 + 1/(\alpha L)} \qquad (13.31)$$

where $B$ is the electron escape probability, $R_{cv}$ is the reflectivity of the cathode/vacuum interface, $L$ is the electron diffusion length, and $\alpha$ is the optical absorption coefficient. If the acceptor doping is low, the band notch region is wide and the escape probability $B$ is low. If the doping is very high, however, the diffusion length (or $\alpha L$) becomes low and so the quantum efficiency declines for this reason. Optimum doping conditions must be found by experiment.

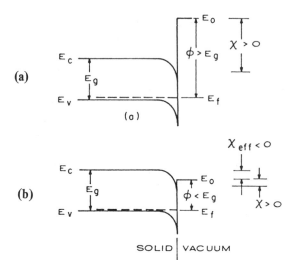

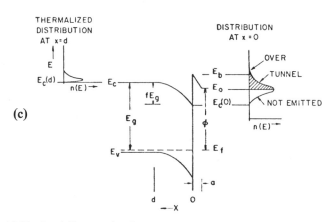

Fig. 13.23.  Band diagrams showing:
(a)  positive-electron affinity:
(b)  negative-electron-affinity conditions in relation to bandgap energy $E_g$ and work function $\phi$; and
(c)  thermalized distributions of electron energy in the bulk and at the surface with those
that tunnel through the barrier indicated. (After Martinelli and Fisher. Reprinted with
permission from *Proc. IEEE*, **62**, 1974, pp. 1339, 1343.)

The spectral photoresponse curves of (In,Ga)As:Cs-P photocathodes of various
bandgaps are shown in Fig. 13.24 to illustrate the doping effect. Figure 13.25
shows the decline in quantum yield for InP and InAsP as the bandgap is
progressively decreased.

Negative-electron-affinity action can also be achieved in $p^+$Si with a photo-

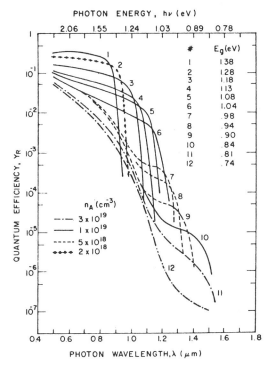

PHOTON ENERGY, $h\nu$ (eV)

| # | $E_g$(eV) |
|---|---|
| I | I 38 |
| 2 | 1.28 |
| 3 | I.18 |
| 4 | 1 13 |
| 5 | 1.08 |
| 6 | 1.04 |
| 7 | .98 |
| 8 | .94 |
| 9 | .90 |
| 10 | .84 |
| II | .81 |
| 12 | .74 |

$n_A$ (cm$^{-3}$)

—·— $3 \times 10^{19}$
——— $1 \times 10^{19}$
----- $5 \times 10^{18}$
••••• $2 \times 10^{18}$

PHOTON WAVELENGTH,$\lambda$ ($\mu$m)

Fig. 13.24. Reflection-mode spectral-photoresponse curves for (In,Ga)As:Cs-O photocathodes. Doping densities and bandgap energies are shown. (After Martinelli and Fisher. Reprinted with permission from *Proc. IEEE*, 62, 1974, p. 1345.)

response to about 1.1 $\mu$m, but the dark current density is $10^{-10}$ A/cm$^2$ at room temperature, compared with $10^{-14}$ A/cm$^2$ for In(AsP) and other ternary III-V compounds of comparable bandgap. The difference is associated with interface and Cs-O layer generation effects.

For GaAs, white-light sensitivity values have been reported of 2060 $\mu$A/lm for (111B) LPE GaAs and 1700 $\mu$A/lm for (100) VPE GaAs, both doped in the mid $10^{18}$ cm$^{-3}$ range. These cathodes have quantum efficiencies in the range 0.2 to 0.4 for $\lambda \leq 0.85$ $\mu$m. At room temperature, the dark current from GaAs:Cs-O is $10^{-16}$ A/cm$^2$ and, at $-20°$C, it is reduced by about three orders of magnitude. The dark current originates in the Cs-O layer and contributions to the dark current arising from the bulk are negligible.

At 1.06 $\mu$m, the Nd:YAG laser wavelength, the ternary or quaternary III-V compounds (In,Ga,As,P) give quantum efficiencies of 3 to 5%.

For longer wavelength response, studies are underway with multiple heterojunction layer structures and field-aided collection. For instance, the light may be absorbed in a $p^+$GaSb layer and the electrons passed through a thin

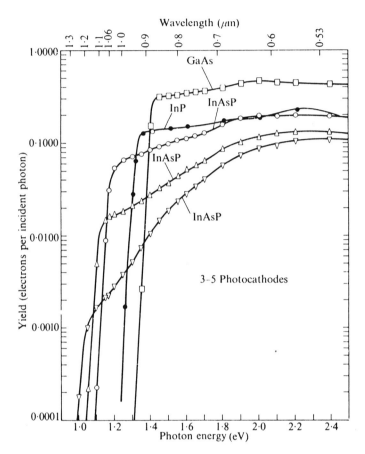

Fig. 13.25. Yield curves of negative-affinity 3-5 photocathodes activated with (Cs,O), show-
ing the effect of progressively decreasing the 3-5 bandgap. (By courtesy of G.A. Antypas,
R.L. Moon, and L.W. James, from R.L. Bell, *Negative Electron Affinity Devices*, Oxford
University Press, 1973.)

wide-bandgap layer GaAlAs or GaAlSb into an emitting layer $p^+$GaAs with a
Cs-O surface. The role of the wide-bandgap intermediate layer is to provide
a barrier in the valence band to excessive hole current flow when the GaSb
absorber is biased negatively with respect to the $p^+$GaAs emitter.

Another approach for long-wavelength response is to use the transferred
electron effect in a semiconductor such as InP with an upper conduction band
valley that can be populated by subjecting electrons to moderate field strengths
$\sim 10^4$ V/cm. A typical structure of this kind is shown in Fig. 13.26(a) and
consists of a $p$ InGaAs substrate with a thin coating of lightly doped $p$ InP,
followed by a 200 Å film of Ag and a negative-electron-affinity coating of CsO.
The thin Ag film is biased positively to form a reverse-biased Schottky barrier

($\phi_B \geq 0.6$ eV) with fields in the InP depletion region of the order of $10^4$ V/cm over a distance approaching $10^{-4}$ cm. A fraction of the electrons generated by infrared photons in the $p$ InGaAs or $p$ InP [processes 1 and 2 in Fig. 13.26(b)] transfer to the higher $L$ conduction band under these field conditions and are emitted over the CsO barrier. The thin Ag film is permeable to electrons but controls the hole current at 300°K to a low value ($10^{-11}$-$10^{-12}$ A/cm$^2$ dark current) for reverse voltages of 3 to 5 V. At higher reverse voltages the current rises by the impact ionization of hot holes [process 3 in Fig. 13.26(b)].

The spectral response obtained is shown in Fig. 13.26(c) for bias voltages of 2 to 5 V at 300°K and compared with the behavior without bias. In$_{0.53}$Ga$_{0.47}$As has a bandgap of 0.75 eV and matches InP in lattice constant. For efficient photoelectron transfer from the InGaAs absorber into the emitter InP, the emitter doping level and thickness must allow the depletion field to reach the interface at modest bias voltages and the interface must be graded and have few recombination centers. At present transferred electron fractions of 0.1 to 5% and quantum yields (emitted electrons/photon) of $10^{-3}$ to $10^{-2}$ have been achieved.

Direct emission is also possible from InGaAs without the presence of the InP heterojunction since the conduction-band conditions are also favorable for transferred electrons to achieve a negative-electron affinity, but the quantum yield seen at present is not quite as high as that for heterojunction structures.

### 13.5.2 Secondary Emitters for Photomultipliers

The low-work-function conditions that have been discussed for photoemission are also of interest with respect to the emission of electrons excited by exposure of the surface to bombardment by high-energy electrons. The secondary-emission yield in a simple model may be written as

$$\delta = \int (-\epsilon^{-1} dE/dx) f(x) dx \qquad (13.32)$$

where $E$ is the energy remaining in a primary electron at a depth $x$ from the surface and $\epsilon$ is the energy required to produce a secondary excited electron in the material. The function $f(x)$ is the probability that an excited electron produced at $x$ reaches the surface and is emitted. This probability may be written as

$$f(x) = B_1 B_2 \exp(-x/L) \qquad (13.33)$$

where the coefficient $B_1$ is the fraction of the excited electrons that diffuse towards the surface, $B_2$ is the probability that escape occurs when an electron reaches the surface, and $L$ is the mean-free path (escape depth) for recombination of the excited electrons while diffusing towards the surface. If the energy loss per unit pathlength is constant for a primary electron with a penetration

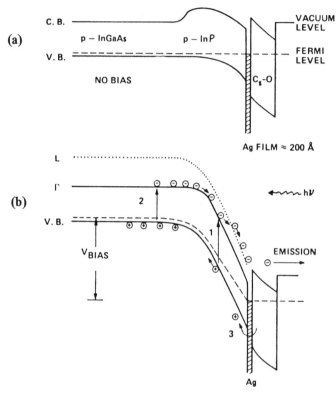

Fig. 13.26. Infrared sensitive negative-electron-affinity, heterojunction bias-assisted transferred-electron electron emitter:
(a)  structure with $p$InGaAs the absorber, $p$InP the transferred-electron emitter, and Ag the Schottky barrier for field creation;
(b)  energy diagram under biased condition showing the optical generation of electrons and the transferred-electron effect to an upper conduction band prior to emission over the electron-affinity barrier; and
(c)  reflection mode quantum yield curves. (After Escher, et al. Reprinted with permission from *IEEE Int. Electron Devices Meeting, Technical Digest*, 1977, p. 464.

depth $R$, then

$$\delta = (B_1 B_2 / \epsilon)(E_0 L / R) [1 - \exp(-R/L)] \qquad (13.34)$$

where $E_0$ is the initial energy of the primary electron.

This model results in a yield curve that increases with primary energy up to a certain value of $E_0$ and then declines somewhat, since the secondary electrons created by very deeply penetrating primaries are unable to reach the surface.

For $p$-GaP(Cs) secondary-emission yields of over 100 are possible (see Fig. 13.27) since good diffusion lengths are obtainable in heavily doped material.

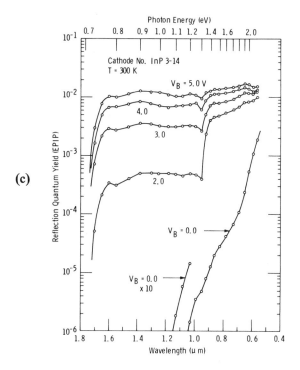

Fig. 13.26 (continued)

These yields are much better than the values for MgO, the highest gain secondary emitter previously known.

Typical photomultiplier arrangements are shown in Fig. 13.28. In the circular electrostatic multiplier tube shown in Fig. 13.28(a), the light falls on the front face of the photocathode and the electrons emitted are multiplied by nine dynode stages. In the previously widely used 1P21 phototube, the photocathode is of the S4 class and the sensitivity is about 40 $\mu$A/lm$^{-1}$. The overall gain for the nine dynode stages is $2 \times 10^6$ corresponding to an average gain of about five per stage. Conventional dynode materials are $Cs_3Sb$, Mg-Ag, or Be-Cu although now being replaced by GaP(Cs). Figure 13.28(b) shows a partition-type electron multiplier in which the photocathode is transparent and mounted in the end of the tube. It is screened from the secondary-emission dynode electrodes by a partition and aperture that provides convenient separation during activation.

Since the gain per stage is only of the order of five for conventional secondary emitters, the statistical fluctuation in the number of secondary electrons emitted by the first dynode limits the tube performance. To provide discrimination between signals representing the emission of one or of two photoelectrons it is necessary that the first dynode have a gain greater than 15 to 20.

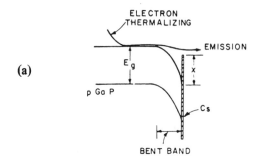

(a)

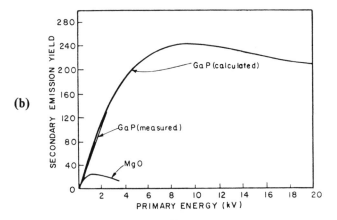

(b)

Fig. 13.27. Secondary emission from $p$-GaP(Cs):
(a)   band diagram, showing negative-effective-electron affinity; and
(b)   yield as a function of primary-electron energy. The GaP (calculated) curve is derived
      from Eq. (13.34) and $R = 1.15/p \times 10^{-6}E_0{}^{1.35}$, where $B_1 B_2 = 5$, $\epsilon = 8.7$ eV, $L =$
      2000 Å, and $\rho = 5.35$ g·cm$^{-3}$. (After Simon and Williams. Reprinted with permission
      from *IEEE Trans. Nuclear Science*, **NS-15**, 1968, p. 169.)

Even higher gains are needed to distinguish between $n$ and $(n+1)$ photoelectrons
when $n$ is greater than 1. For a GaP(Cs) first dynode at 600 V, gains of between
20 and 40 are obtainable. With such tubes excellent pulse-height resolution
capabilities are available and it is possible to distinguish the emission of one,
two, and up to five photoelectrons. Such technology has applications in the
detection and measurement of very low-light-level scintillations (for example
in tritium counting, carbon counting, and Cerenkov-type counters).

### 13.5.3  Cold-Cathode Electron Emitters

An efficient room temperature (or near room temperature) cathode, capable
of emitting electrons for electron beam applications on the application of a

(a)

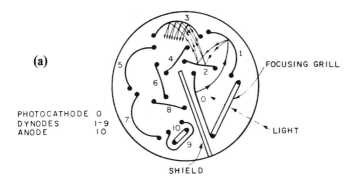

PHOTOCATHODE  0
DYNODES      1-9
ANODE        10

FOCUSING GRILL

LIGHT

SHIELD

(b)

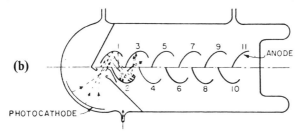

ANODE

PHOTOCATHODE

Fig. 13.28.  Typical photomultiplier electrode arrangements:
(a)  circular configuration of electrodes and a nontransparent photocathode; and
(b)  linear configuration of the dynode section, with a partitioned transparent photocath-
     ode. (After Simon and Williams. Reprinted with permission from *IEEE Trans. Nuclear
     Science,* **NS-15,** 1968, p. 170.)

voltage to a junction, is a device objective that has been under study for many
years with only limited success until recently.

One approach that has been examined is the emission of hot electrons from
reverse-biased shallow *p-n* junctions in Si.[11] The *n*-type region through which
photoexcited hot electrons must pass is ∿1000 Å in thickness. However, the
mean-free path for optical-photon emission in Si is only about 60 Å and for
impact ionization about 190 Å. Therefore, the efficiency of emission is very
low (∿$10^{-6}$ electrons per photon).

However, with the recent availability of negative-electron-affinity processing,
good emission has been obtained from forward-biased Si *n-p* junctions with
NEA activation of the 2 μm thick *p*-surface (see Fig. 13.29). Emitted electron
current to bias current ratios as high as 10% have been observed. Emitted current
densities as large as 225 A/cm$^2$ and currents as large as 7 mA have been obtained
under pulsed conditions, while continuous current densities of 2 A/cm$^2$ and
currents of tens of microamperes have been drawn for many hours. Such
cathodes have been used experimentally in applications such as vidicons, where

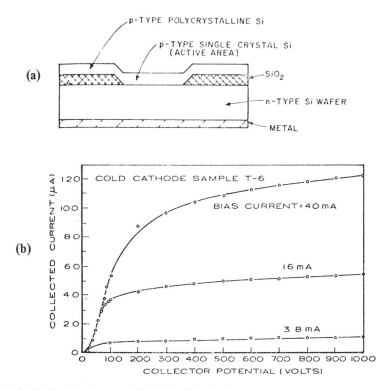

Fig. 13.29. Negative-electron-affinity Si cold cathode:
(a) junction structure (the CsO activation layer on the p active area is not shown); and
(b) collected electron current vs. collector voltage for three values of bias current. (After Kohn. Reprinted with permission from *Appl. Phys. Lett.*, **18**, 1971, pp. 272-273.)

some aspects of performance are presently limited by the thermionic cathode, and where power consumption and heat generation must be minimized.

GaAs cold-cathode structures have also been studied. One approach is to couple high-yield electron emitters of GaAs with efficient light-emitting diodes of (Al,Ga)As. Efficiencies of about 2% for the ratio of current emitted into vacuum to the input current have been reported for the optoelectronic structure shown in Fig. 13.30(a) and 4% is observed for the direct injection structure shown in Fig. 13.30(b). Operational lifetime at present is limited to a few hours because of electron-stimulated desorption of Cs-O from the surface or bombardment of the cathode by impurities in the tube.

In another (completely different) approach efficient photoemission has been obtained from uniform large-area arrays of field emitter points fabricated at densities of $10^5$ cm$^{-2}$ by microetching techniques in $p$-type Si. Such arrays were shown to emit electrons into vacuum when biased by a closely spaced plane anode (125 to 500 $\mu$m spacing) at a few kilovolts. No high-vacuum cesiation or

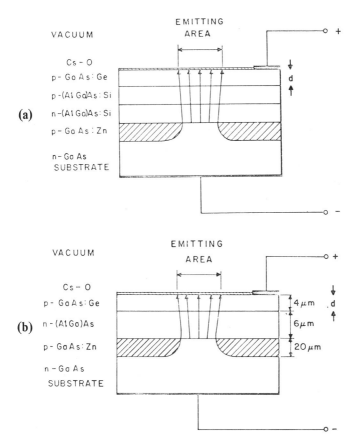

Fig. 13.30. GaAs cold-cathode structures incorporating an aperture for lateral electron confinement:
(a)  optoelectronic device; and
(b)  direct injection device. (After Kressel, Schade, and Nelson. Reprinted from *J. Lumin.*, 7, 1973, p. 150.)

high-temperature cleaning is required to observe this emission. Stable reflective photosensitivities exceeding 1500 $\mu$A/lm and quantum efficiencies of 2% at 1.06 $\mu$m and 28% at 0.90 $\mu$m have been demonstrated in thick *p*-type Si arrays.[170]

## 13.6  VIDICON CAMERA TUBES AND SILICON DIODE ARRAY TARGETS

Conventional vidicon camera tubes have an image-illuminated transparent faceplate coated with a thin (1-2 $\mu$m) polycrystalline layer of a photoconductive semiconductor such as antimony trisulfide ($Sb_2S_3$). A fixed potential of 10-30 V positive, relative to the thermionic cathode, is applied to the $Sb_2S_3$ layer [see

Fig. 13.31(a)]. A light image falling on the film increases the conductivity from the dark level of $10^{-12}$ mho and the illuminated areas of the film become positive by a volt or two with respect to the cathode during the frame time (1/30 sec) between successive scans. The beam accessing each picture element deposits sufficient electrons to neutralize the charge accumulated during the frame time and in doing so generates the video signal in the signal plate lead. Photosensitivities of 50-300 $\mu A/lm$ are typical for white-light illumination.

$Sb_2S_3$ vidicon surfaces have troublesome after-image effects associated with photoconductive lags. Lower dark currents are also desirable in image surfaces so that low-light levels may be used. These problems are fewer in lead oxide *p-i-n* blocking-type targets, as in the Philip's Plumbicon tube. This target is fabricated with a highly insulating layer between *p-* and *n*-type surface layers [see Fig. 13.31(b)]. Such targets have white-light sensitivities typically in the range 100-450 $\mu A/lm$ and resolution modulations of 35-55% under TV conditions (625-400 lines).

Other tube surfaces are shown in Fig. 13.31(c) and (d). The CdSe-based structure has good sensitivity and is capable of operation with a low-faceplate illumination (0.1-1 lx) and a lag of 10% for a 200-nA signal current. The structure shown in Fig. 13.31(d) is a multilayer Se(As,Te)-based structure. As doping is used to prevent Se crystallization. Te doping is used near the signal electrode to increase red sensitivity. Along with the hole-blocking contact at the signal electrode, a blocking structure is formed on the beam-scanned surface of the target to prevent injection of the beam electrons. The tube with this target has a sensitivity of 300-450 $\mu A/lm$, the gamma is nearly unity, and the spectral response covers the entire visible region. A 2/3-in tube shows a modulation transfer curve as high as 30% at 400 TV lines. Its lag is 3% at a signal current of 200 nA.

Camera tubes operating on other principles are known. In the image orthicon, photoelectrons emitted from a photocathode are accelerated onto a storage target by an electrostatically or magnetically focused image section. This creates secondary electrons and a positive charge on the target and a return beam scan is used with electron multiplication to obtain a high-signal input to the video amplifier.

### 13.6.1 Silicon Array Vidicon Targets

Si cannot readily be used as a direct substitute for photoconductive film vidicon surfaces ($Sb_2S_3$, $Se,As_2S_3$, PbO) since in crystalline form it is too conductive because of the low bandgap, and in amorphous form it does not have mobilities and electrooptic properties that are suitable. However, Si can be used in single-crystal form if the surface is formed into a mosaic of *p-n* junctions, as shown in Fig. 13.32, to limit the dark current of each picture element. The optical-sensing mechanism of the target is the discharge of the elements of the array by light incident on the $n^+$ surface that is held at a potential $V_T$. The target face

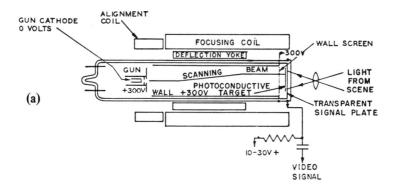

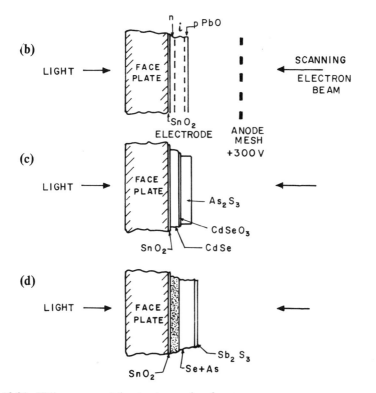

Fig. 13.31. Vidicon camera tube structure and surfaces:
(a)  structure. (After Weimer, Forgue, and Goodrich. Reprinted from *Electronics*, May 1950 and *RCA Review*, **12**, 1951, p. 309.)
(b)  *p-i-n* PbO surface, as in the Plumbicon tube;
(c)  CdSe/As$_2$S$_3$ surface; and
(d)  Saticon, Se + As surface. (After Goto, et al. Reprinted with permission from *IEEE Trans. Electron Devices*, **ED-21**, 1974, pp. 662-663.)

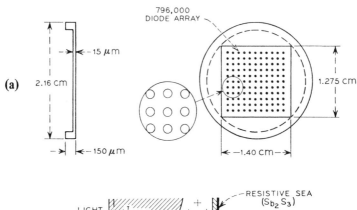

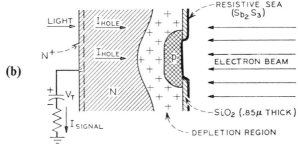

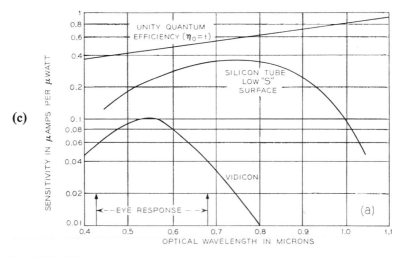

Fig. 13.32. Si diode array camera-tube target:
(a) structure for "Picturephone";
(b) picture element in scanning mode; and
(c) sensitivity relative to a conventional vidicon surface. (After Crowell and Labuda, and
Beadle and Schoor. Reprinted with permission from *Bell Syst. Tech. J.,* © 1969, 1970,
The American Telephone and Telegraph Co.)

potential is periodically reduced to ground potential by the scanning electron beam. The target unit cell is reverse biased by an amount $V_T$ immediately after scanning. During the frame time, $\tau_F$, when the electron beam is scanning the remainder of the target the diode cell is discharged by the photon-induced holes that diffuse to the depletion region of the junction.

The individual diode steady-state signal current can be related to the diode discharge by

$$I = \frac{1}{\tau_F} \int_{V_F}^{V_T} C(V) dV \qquad (13.35)$$

where $C(V)$ is the diode cell capacitance, $V_T$ is the voltage across the diode cell at the start of a frame, and $V_F$ is the voltage across the diode cell at the completion of a frame.

A resistive film of $Sb_2S_3$ is provided to prevent accumulation of negative charge on the $SiO_2$ surface that would repel the electron beam and prevent it from impinging on the $p$ regions.

The advantages of a Si diode array tube are the following:

- It cannot be readily overheated and burned by accidental exposure to intense light images or prolonged exposure to fixed images of normal intensity.
- It is chemically very stable and so can be given a high-temperature vacuum bake (400°C) to de-gas the tube envelope for long life.
- There is no undesirable image persistence caused by photoconductive lag effects.
- The response extends further into the red and infrared than a conventional vidicon surface and so a lower level of signal illumination may be used. The spectral response is almost uniform over the wavelength range from 0.45 to 0.90 $\mu$m with an effective quantum yield greater than 50%. The sensitivity relative to that of a conventional vidicon is shown in Fig. 13.31(c).
- Electronic zoom can be achieved by varying the size of the raster on the mosaic of diodes.

### 13.6.2  Variations of Silicon Array Target Tubes

Si diode target arrays may be constructed in which the light-illuminated side is provided with regions of different $n^+$-diffusion depths to produce variation in spectral response.[129] This can be the basis of a color TV system.

Methods of enhancing the signal so that greater usable sensitivity is obtained have also been explored. One approach is to increase the gain by secondary-electron multiplication as in a Si return-beam vidicon. An intermediate image-gain stage can be added to the tube, or gain prior to electron-beam readout can be obtained by using phototransistors instead of diodes in the Si vidicon array. Figure 13.33 shows the phototransistor picture elements and the equivalent circuit that must be considered.

Phototransistor signal gain is possible during both the read and write portions of the operating cycle. Since the base is floating in potential, the transistor is open circuited during the write (store) interval and supplied with emitter current by the electron beam during the read (sample) interval. In the equivalent circuit, the junction capacitances are shown exterior to the transistor and the transistor is considered simply as a current amplifier. Here $C_{CB}$ is the collector-base depletion capacitance, $C_{EB}$ is the emitter-base depletion capacitance plus the parasitics such as would result from a metal field plate contacting the emitter and extending over the oxide that is over the base region, and $C_{CE}$ is the direct capacitance between the emitter and the collector. $i_p$ is the elemental photo-current generated by a constant source illumination. The illumination generates a charge $\Delta q_p$ in a storage time $\tau$, so that

$$\Delta q_p = \tau i_p \tag{13.36}$$

Near the end of the read interval the emitter-base junction is slightly forward biased. When the write period is started with the removal of the scanning electron beam, the emitter current decreases greatly. At this time the emitter-collector capacitance is charged to a slightly greater voltage than the base-collector capacitance. If there is no appreciable direct leakage between the emitter and collector, the discharge of the collector-base capacitance by the illumination will maintain a forward bias on the emitter-base junction. Thus the collected photogenerated charge causes charge to flow from $C_{CE}$ and be injected through the transistor. The emitter conductance will control the situation so that over the write interval,

$$\Delta V_{CB} \simeq \Delta V_{CE} \tag{13.37}$$

Thus

$$\frac{\Delta q_{CB}}{C_{CB}} \simeq \frac{\Delta q_{CE}}{C_{CE}} \tag{13.38}$$

The ratio of charge injected from the emitter into the base to charge recombining in the base is $1/(1 - \alpha)$. If $\alpha_W$ is $\alpha$ during the write interval and is assumed to be independent of current for small signals, then

$$\Delta q_{CB} = \Delta q_P - \Delta q_{CE}(1 - \alpha_W) = \Delta q_P[1 + (C_{CE}/C_{CB})(1 - \alpha_W)]^{-1} \tag{13.39}$$

The total incremental charge which the electron beam must supply, $\Delta q_G$, is

$$\Delta q_G = \Delta q_{CE} + \frac{\Delta q_{CB}}{1 - \alpha_R} = \frac{\Delta q_P}{1 - \alpha_R} \left[ \frac{1 + (C_{CE}/C_{CB})(1 - \alpha_R)}{1 + (C_{CE}/C_{CB})(1 - \alpha_W)} \right] \qquad (13.40)$$

where $\alpha_R$ is $\alpha$ during the read interval.

The effective capacitance of a transistor from the point of view of the electron gun is

$$C_{\text{eff}} \equiv \frac{\Delta q_G}{\Delta V_{CB}} = C_{CE} + \frac{C_{CB}}{1 - \alpha_R} \qquad (13.41)$$

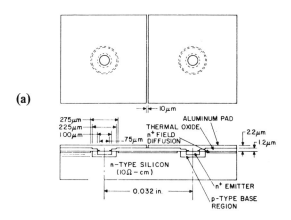

(a)

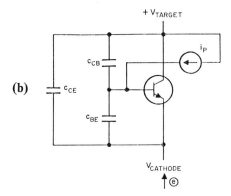

(b)

Fig. 13.33. Experimental vidicon phototransistor array:
(a)  adjacent picture element transistors; and
(b)  equivalent circuit of a picture element. (After Madden, Kiewit, and Crowell. Reprinted with permission from *IEEE Trans. Electron Devices*, **ED-18**, 1971, pp. 1044, 1046.)

and the effective gain is

$$G_{\text{eff}} = \frac{1}{1 - \alpha_R} \left[ \frac{1 + (C_{CE}/C_{CB})(1 - \alpha_R)}{1 + (C_{CE}/C_{CB})(1 - \alpha_W)} \right] \tag{13.42}$$

Signal gains of 20-70 have been obtained.

Image-intensified Si scan converter tubes represent another approach that gives much higher electron gain, for instance, a 10-kV electron creates about 2800 hole-electron pairs. In operation an image is focused onto a planocurved fiber-optic faceplate. The radiation is guided through the glass fibers and strikes a photocathode, usually of the S-20 type, that has been deposited on the curved face of the fiber faceplate. The emitted photoelectrons are then accelerated and focused by a high-quality electron optical system onto a fairly conventional Si diode target. With 10-kV acceleration, the gain is at least 2000 after collection and the tube may be operated at a light level as low as $10^{-6}$ fc faceplate illumination.[158]

Electronic image storage is possible with vidicon-type tubes in which the photoconducting faceplate or conventional Si diode target is replaced by a storage target fabricated by selective etching of thermally grown $SiO_2$ films on Si substrates. Targets may have a storage density of 600,000 elements/cm$^2$ and be designed with continuous readout times in the range of 3 to 30 min, corresponding to erase times of 2 to 60 TV frames, respectively. Images may be stored for as long as 45 days without loss of resolution. The present limiting resolution is approximately 500 TV lines, making such tubes capable of storing excellent TV quality images with selective erase, write, and read functions.[155]

Flat-plate display panels will not be discussed. A general impression of the state of the art may be obtained by reading the September 1975 and July 1977 issues of the IEEE Transactions on Electron Devices.

## 13.7 ELECTROPHOTOGRAPHIC COPYING

Numerous copying systems have been developed in the last 20 or 30 years that depend on charged-pigment interaction with latent electrostatic images on high-resistance amorphous semiconductor films. Xerography is one of the commonly used methods and the process steps involved in this method are shown in Fig. 13.34. The photoreceptor is a 50 $\mu$m thick vacuum-deposited layer of Se. The outer surface is sensitized by coating with a layer of positive charge from a corona discharge. Electric fields of $10^5$ V/cm are involved in the layer. An optical image is formed on this surface and in the illuminated areas photoconductivity occurs and the plate is discharged by the movement of holes through the film to the substrate. Thus an electrostatic potential distribution is obtained that matches the optical image. This image is developed by dusting on a powder (10 $\mu$m diameter) of black toner particles. The powder is attracted

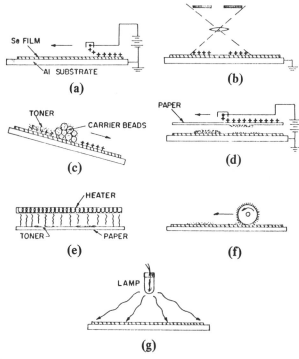

Fig. 13.34. Schematic of the process steps in xerography:
(a)   sensitization;
(b)   imaging;
(c)   development;
(d)   image transfer;
(e)   fixing;
(f)   cleaning; and
(g)   image erasure. (After Tabak, Ing, and Scharfe. Reprinted with permission from *IEEE Trans. Electron Devices*, **ED-20**, 1973, p. 132.)

by the surface fringe field at the light-dark boundaries, or in systems using a development electrode by the absolute potential in the dark area. Electrostatically charged paper is then used to pick up the image toner particles and the image is made permanent by fusing the plastic-based toner into the surface of the paper. The photoreceptor surface is then cleaned by mechanical or electrostatic means and an intense light is used to erase any residual image.

   The photoconductive semiconductor used should have a large carrier range (proportional to $\mu\tau$) so that each photon absorbed results in transport of one electronic charge completely through the photoreceptor layer with minimum trapping effects. To minimize lateral surface conductivity that would wash out the electrostatic image, the semiconductor should have almost no mobile carriers that are not photogenerated. The thermal generation rate, $g_{th}$, and the dark carrier concentration, $p_0$, should be low. The dark conductivity is proportional to $p_0\mu$ which is equal to $g_{th}\mu\tau$, since $g_{th} = p_0/\tau$. Hence a very low dark conductivity by itself, even though the dielectric relaxation time (resistivity $\times$ permitivity) exceeds the development time, is not an adequate criterion for judging the semiconductor layer since $\mu\tau$ must be large. Another important factor is that the interface to the metal back-electrode must be a good blocking contact that does not allow charge injection into the photoreceptor or the accumulation of an interfacial potential drop.

The mobility of holes in amorphous Se at room temperature is about 0.13 $cm^2/V\cdot sec$ and the hole lifetime varies from 1-100 $\mu s$ depending on the layer-processing details. Hence at the electric fields involved, $10^5$ V/cm, the range is good. On the other hand $As_2Se_3$ has mobilities of the order of $10^{-4}$ $cm^2/V\cdot sec$ and is less useful as a photoreceptor. The performance of $As_2S_3$ is severely limited by bulk-trapping effects. ZnO powder in a resin binder is found to give a usable receptor surface and some organic surfaces also are in use.

## 13.8 QUESTIONS

1. In a silicon $p^+$-$n^-$-$n^+$ planar photodiode, the lightly doped region is $10^{15}$ cm$^{-3}$ and the $n^+$ region is $10^{18}$ cm$^{-3}$ and the $n^-$ thickness is 20 $\mu m$. What reverse voltage $V_R$ applied to the diode just causes the avalanche multiplication factor $M(=(1 - \int\alpha dx)^{-1})$ to be 1.1? What are the transit times of holes and electrons through the depletion region (300°K)? What voltage causes $M$ the avalanche multiplication factor to be 20? What is the avalanche buildup time?

2. Evaluate the spectral response from 1.0 eV to 3.0 eV for the diode of Problem 1 if the reverse voltage applied is 50 V. The minority carrier hole lifetime in the $n^-$ region is to be taken as $10^{-6}$ sec and in the $n^+$ region as $10^{-8}$ sec. The $p^+$ region may be assumed to be 0.5 $\mu m$ thick and carriers generated in this region are not collected. How will the spectral response change if the voltage is raised to 150 V?

3. A Schottky barrier Au/$n$ Si photodiode has a 100 Å thick gold layer as the barrier. The $n$ region is doped $10^{14}$ cm$^{-3}$, is of thickness 100 $\mu m$, and has a transparent indium-tin-oxide ohmic contact on the back face. The hole lifetime in the $n$ region is $10^{-6}$ sec. The diode is intended to operate in the spectral range 0.8 to 3.0 eV. It may be illuminated through the partially transparent Au layer (frontwall) or through the fully transparent back-side contact.
   (a) If zero voltage bias is applied, what is the spectral response for front-wall illumination? How could this be improved by an antireflection coating?
   (b) If back-wall illumination is used, what is the change in the spectral response?
   (c) If 100-V reverse voltage is applied, what changes occur in the front-wall and back-wall response?

4. Figure 13.6 shows the responsivity at 1.06 $\mu m$ for a reach-through Si avalanche diode of 100-$\mu m$ depletion width. Attempt to explain quantitatively the temperature dependence shown.

5. The detectivity D* of a $Pb_{1-x}Sn_xTe$ photodiode of bandgap 0.10 eV is seen from Fig. 13.7 to vary from $5 \times 10^{13}$ at 50°K to $8 \times 10^{10}$ at 100°K. Show quantitatively that this would be expected.

6. Calculate the background-limited detectivity for 3, 5, 10, 30 $\mu m$ for a 60° field of view detector at 300°K to confirm the findings of Fig. 13.8.

7. A photoconductive bar of $n$-type Si is of resistivity 50 $\Omega$-cm (dark) and is 1 cm long, 0.5 cm wide, and 100 $\mu$m thick. The hole minority carrier lifetime is $10^{-6}$ sec and the surface recombination is negligible. If illuminated on a wide face with $10^{17}$ photons per $cm^{-2}$ per second of energy 1.5 eV what is the change of resistance of the bar? What would be the change of resistance in air-mass-unity (ground level) sunlight? If the bar is now assumed to contain a hole trap such that one-half of the holes are trapped for a period of $10^{-3}$ sec what would then be the current flow with 10 V applied and the $10^{17}$ $cm^{-2}$ $sec^{-1}$ 1.5-eV photon illumination? Can the generation-recombination noise of the photodetector be estimated from the information given?

8. Square-wave chopped-light is applied to a photoconductor of dark resistance 15 k$\Omega$. The light intensity is such that the resistance becomes 3 k$\Omega$ under constant illumination. The bar is in series with a resistor $R_L$ = 1 k$\Omega$ as shown in Fig. 13.10 and the dc voltage source is 2 V. The minority carrier lifetime in the photoconductor is $10^{-5}$ sec. What is the waveform of the voltage across $R_L$ if the square wave is of period $10^{-3}$ sec (0.5 ms on and 0.5 ms off)?

9. An $n$CdSe photoconductive cell is 0.3 cm long by 0.1 cm wide and $10^{-2}$ cm thick. The bandgap is 1.7 eV and the electron mobility is 400 $cm^2$/V·sec. Some induced holes are trapped at levels that have an average reemission time to the valence band of $10^{-1}$ sec at $300°$K. Excess electrons recombine through recombination centers that result in pulses decaying with a time constant of $10^{-2}$ sec. The cell is illuminated with 1 mW/$cm^2$ of light of wavelength 0.4 $\mu$m and it is assumed that all the photons create carrier pairs. What is the increase in the electron density in the cell under illumination? What is the photocurrent caused by the light if 10 V is applied between strip contacts on the width of the cell? What is the gain factor associated with the cell?

10. A $p^+$-$n$-type Si junction depletion region ($N_D$ = 2 × $10^{16}$ $cm^{-3}$) contains a deep impurity with a concentration $N_T$ = $10^{15}$ $cm^{-3}$. The depletion region is biased at 10 V reverse and monochromatic illumination of constant intensity (photons $cm^{-2}$ $sec^{-1}$) is applied at $77°$K and the current observed as a function of wavelength. The spectral response is

| eV | 0.58 | 0.60 | 0.62 | 0.64 | 0.68 |
|---|---|---|---|---|---|
| I(A) | 4 × $10^{-12}$ | 8 × $10^{-12}$ | 1.1 × $10^{-11}$ | 2.3 × $10^{-11}$ | 4.5 × $10^{-11}$ |

| eV | 0.72 | 0.76 | 0.80 | 0.85 | 0.90 |
|---|---|---|---|---|---|
| I | 7 7 × $10^{-11}$ | 9 × $10^{-11}$ | 1.4 × $10^{-10}$ | 2.2 × $10^{-10}$ | 2.6 × $10^{-10}$ |

| eV | 0.95 | 1.0 | 1.05 | 1.1 | 1.2 |
|---|---|---|---|---|---|
| I | 8.3 × $10^{-10}$ | 4 × $10^{-10}$ | 4.7 × $10^{-10}$ | 8 × $10^{-10}$ | 6 × $10^{-9}$ |

What can be inferred about the impurity in Si by use of the Lucovsky absorption model?

11. The detectivity of Au in a Ge extrinsic photodetector is shown for $77°$K and $60°$K in Fig. 13.13. Account for the difference in performance at 4 $\mu$m and predict the detectivity expected at $30°$K.

12. A Si MESFET has a 100 Å thick Au gate 20 $\mu$m long and 200 $\mu$m wide. The Si is a 5 $\mu$m thick, single-crystal epilayer doped $N_D$ = $10^{14}$ $cm^{-3}$ grown on clear sapphire. How would you bias this to make a sensitive photodetector of light from a GaAs light-emitting diode? What is the responsivity expected as output voltage change per milliwatt of light input?

13. Verify the calculated yield curve for GaP secondary emission shown in Fig. 13.27.

14. Secondary electrons are produced inside a secondary-emission material and arrive at the surface in a random direction. Assume the conduction band lowest edge is 5 eV below the vacuum level and consider a secondary electron reaching the surface with an energy 7 eV but random direction. What is the probability of emission?

15. A layer of $p$-type GaAs (electron affinity 4.07 eV) in vacuum is illuminated with ultra-violet photons of energy 10 eV. Electrons are emitted into vacuum and collected on an aluminum plate 2 cm away. What negative voltage must be applied to the aluminum plate to just cut off the collection of the electrons emitted from the GaAs?

16. A Si phototransistor in a vidicon array such as shown in Fig. 13.33 has a capacitance ratio $C_{CE}/C_{CB}$ = 15. The steady-state photoresponse to a 1-$\mu$m source for a dc bias of 5 V is

| Illumination power | $10^{-7}$ | $10^{-6}$ | $10^{-5}$ | Wcm$^{-2}$ |
|---|---|---|---|---|
| Emitter current | $10^{-10}$ | $2 \times 10^{-9}$ | $3 \times 10^{-8}$ | A |

100 transistors are being scanned during the read interval. What is the effective gain as defined by Eq. (13.42)?

17. Discuss guard-ring structures applied to avalanche photodiodes.

18. Examine the relative merits of Ge, Si, and GaAs avalanche photodiodes.

19. Discuss the merits and performance of Schottky barrier-type relative to junction-type avalanche photodiodes.

20. Discuss how photodetectors may be chosen and optimized for use with particular gas lasers such as He-Ne, YAG, and $CO_2$ lasers.

21. Discuss InSb and InAs photodetector performance. Are both available in the photoconductive and photovoltaic modes of operation?

22. Are there conditions under which the $\Delta\sigma/\sigma$ of a photoconductor is proportional to the light intensity or to the square root of the light intensity?

23. Examine the tradeoff between responsivity, detectivity, and bandwidth in an intrinsic photoconductor such as CdSe containing traps.

24. Examine the relative merits of In $(E_v + 0.16$ eV$)$ and S $(E_c - 0.16$ eV$)$ as dopants for extrinsic Si photoconductive detectors (perhaps to be used in conjunction with a charge-coupled (CCD) array).

25. If a junction detector of Ge, without avalanche gain, is compared to an intrinsic Ge photoconductor for detection of 1.06-$\mu$m laser light what are the relative advantages?

26. Compare the merits of $Pb_x Sn_{1-x} Te$ and $Hg_x Cd_{1-x} Te$ photodetecting structures.

27. Infrared atmospheric windows exist in certain sections of the spectrum to 30 $\mu$m. Discuss the kinds of detectors appropriate for each of these main windows.

28. Certain detectors are available that use a photoelectromagnetic (PEM) effect. Describe them and discuss their performances relative to photoconductive detectors.

29. Insulated-gate field-effect electrodes may be used to decrease the conductivity of a channel and so to increase the photosensitivity. Discuss the magnitude of the improvement expected for some typical photoconductors.

30. MOS gated diodes have been proposed as light sensors. Explain the principle of operation and discuss the performance and limitations.

31. Discuss the desirable proportions of the junction field-effect photodetector shown in Fig. 13.17.

32. Discuss the tradeoffs between responsivity, signal-to-noise detectivity, temperature, integrating time, and bandgap for an integrating diode photodetector.

33. In bipolar phototransistor imaging arrays explain why the ratio of collector-to-emitter capacitance must be large.

34. Present the analysis of a single-axis lateral photodetector with illumination.

35. Conceive two uses for a position-sensitive photodetector involving an area-type lateral-photoeffect device. Will the pulse-response delay effect matter in these applications?

36. Discuss the physical actions that may take place in a junction field-effect transistor in grounded-gate connection illuminated by a strip of light at a distance $l$ from the source.

37. Discuss the use of photodetecting diodes or transistors as part of a keyboard without moving electrical contacts.

38. Discuss photosensitive readers for computer punched cards.
39. Discuss various types of infrared image-conversion systems for night vision uses.
40. Discuss the role of the minority carrier (electron) diffusion length in negative-electron affinity $p^+$GaAs/Cs photocathodes.
41. What success has been had with $p^+$Si/Cs negative-electron-affinity photocathodes?
42. Heterojunction structures have been proposed to extend the wavelength range of negative-electron-affinity photocathodes. Review the possibilities and requirements.
43. Discuss dynode electron multiplication systems involving semiconductors.
44. Review progress in cold-cathode design and performance.
45. Certain semiconductor display devices consist of a layer of photoconductor sandwiched with a light-modulating medium between two electrodes. In the absence of an input signal illumination, the field between the two electrodes is more or less equally impressed across the photoconductor and the light-modulating layer. If the photoconductor is illuminated, the field across it is reduced and the field across the light-modulating layer is increased sufficiently to cause significant modulation. Discuss with a numerical example how much light is required to shift the field from the photoconductor to the light-modulating layer.
46. Many PC-EL display devices use photoconductive sheets in conjunction with electroluminescent layers. Trace the performance developments and features of this class of display.
47. Discuss the design of Si diode vidicon arrays.
48. Discuss the action of the $p\text{-}i\text{-}n$ structure in a PbO image-tube surface.
49. Discuss return-beam image tubes involving Si targets.
50. Review image-storage tube systems with Si/SiO$_2$ targets.
51. Describe the performance merits of field-emission large-area arrays.
52. Describe two of the most promising thin-film flat-plate display image TV-type systems.
53. Describe the toner coating process in Xerox type electrocopier processing.
54. Describe a photoelectrocopier process based on ZnO.

## 13.9 REFERENCES AND FURTHER READING SUGGESTIONS

1. Amano, Y., "A flat-panel TV display system in monochrome and color," *IEEE Trans. Electron Devices,* **ED-22**, 1975, p. 1.
2. Andrushka, A.I., D.N. Nasledov, and S.V. Slobodchikov, "Investigations of the kinetics of the photo-EMF of InAs p-n junctions," *Soviet Physics-Semiconductors,* **6**, 1972, p. 712.
3. Antypas, G.A., and J. Edgecumbe, "Glass-sealed GaAs-AlGaAs transmission photocathode," *Appl. Phys. Lett.,* **26**, 1975, p. 371.
4. Arai, H., "Color display system using liquid crystals," *Japan Electronic Engr.,* **72**, Nov. 1973, p. 40.
5. Arai, H., T. Yoshizama, K. Awazu, K. Kurahashi, and S. Ibuki, "EL panel display," *IEEE Conference Record of 1970 IEEE Conference on Display Devices*, Dec. 1970, p. 52.
6. Arams, F.R., *Infrared-to-Millimeter Wavelength Detectors*, Artech House, Dedham, Mass., 1973.
7. Armstrong, L., "Fiber optic systems advance," *Electronics*, Aug. 7, 1975, p. 73.
8. Astlo, B., "Liquid Crystals—a viable new medium," *Optical Spectra*, July 1973, p. 35.
9. Balkarshi, M., "Band structure and optical properties of small gap semiconductors and alloys," *J. Lumin.,* **7**, 1973, p. 451.
10. Bandy, S.G., and J.G. Linvill, "The design, fabrication and evaluation of a silicon junction field effect photodetector," *IEEE Trans. Electron Devices,* **ED-20**, 1973, p. 793.

11. Bartelink, D.J., J.L. Moll, and N.I. Meyer, "Hot electron emission from shallow p-n junctions in silicon," *Phys. Rev.,* **130,** 1963, p. 972.
12. Beadle, W.E., and A.J. Schorr, "Picturephone silicon target signal analysis," *Bell Syst. Tech. J.,* **49,** 1970, p. 921.
13. Bell, R.L., *Negative Electron Affinity Devices,* Oxford Clarendon Press, England, 1973.
14. Berman, L.V., A.G. Zhukov, K.K. Rychkov, and V.I. Smirnov, "Low temperature detectors for long-wavelength infrared spectral instruments," *Instruments and Experimental Techniques,* **6,** 1969, p. 1630.
15. Bertolini, G., and A. Coche, Eds., *Semiconductor Detectors,* Wiley Interscience, N.Y., 1968.
16. Birenbaum, L., "Bivicon,–a new double vidicon," *Proc. IEEE,* **60,** 1972, p. 1236.
17. Blount, G.H., M.K. Preis, and R.L. Yamada, "Photoconductive properties of chemically deposited PbS with dielectric overcoatings," *J. Appl. Phys.,* **46,** 1975, p. 3489.
18. Blumenfeld, S.M., G.W. Ellis, R.W. Redington, and R.H. Wilson, "The epicon camera tube: an epitaxial diode array vidicon," *IEEE Trans. Electron Devices,* **ED-18,** 1971, p. 1036.
19. Brandinger, J.J., "Will matrix displays replace the cathode-ray tube?" *The 1970 IDEA Symposium. Information, Display, Evaluation and Advances,* Digest of Papers, Society for Information Display, May 1970, p. 48.
20. Brody, T.P., et al., "A 6 x 6-in. 20-lpi electroluminescent display panel," *IEEE Trans. Electron Devices,* **ED-22,** 1975, p. 739.
21. Buzanava, L.K., A.Ya Gliberman, A.S. Lisin, and Yu. V. Prichko, "Photocurrent kinetics in thin-base silicon photodiodes," *Radio Eng. Electron. Phys.,* **16,** 1971, p. 644.
22. Carleton, H.R., "Acoustically read pulsed image converter," *Appl. Phys. Lett.,* **27,** 1975, p. 105.
23. Castellano, J.A., "Liquid crystals for electro-optical application," *RCA Review,* **33,** 1972, p. 296.
24. Celler, G.K., S. Mishra, and R. Bray, "Saturation of impurity photoconductivity in n-GaAs with intense YAG laser light," *Appl. Phys. Lett.,* **27,** 1975, p. 297.
25. Chambauleyron, I., J.M. Besson, and M. Balkanski, "Photovoltaic effect in lead selenide p-n junctions," *J. Appl Phys.,* **44,** 1973, p. 3222.
26. Chamberlain, S.G., and V.K. Aggarwal, "Photosensitivity and characterization of a solid-state integrating photodetector," *IEEE J. Solid-State Circuits,* **SC-7,** 1972, p. 202.
27. Chamberlain, S.G., and M. Kuhn, Eds., "Issue on optoelectronic devices and circuits," *IEEE Trans. Electron Devices,* **ED-25,** 1978, p. 273.
28. Chang, I.F., and W.E. Howard, "Performance characteristics of electrochromic displays," *IEEE Trans. Electron Devices,* **ED-22,** 1975, 749.
29. Chang, I.F., B.L. Gilbert, and T.I. Sun, "Electrochemichromic systems for display applications," *J. Electrochem. Soc.,* **122,** 1975, p. 955.
30. Christensen, C.P., R. Joiner, S.T.K. Nieh, and W.H. Steier, "Investigation of infrared loss mechanisms in high-resistivity GaAs," *J. Appl. Phys.,* **45,** 1974, p. 4957.
31. Ciarlo, D.R., "A positive-reading silicon vidicon for x-ray imaging," *IEEE Trans. Electron Devices,* **ED-20,** 1973, p. 362.
32. Clarke, D., "Experiments with an integrated circuit photo-transistor," *Observatory,* **90,** #979, 1970, p. 249.
33. Clément, G., and C. Loty, "The use of channel plate electron multipliers in cathode-ray tubes," *ACTA Electron.,* **16,** 1973, p. 101.
34. Crowell, M.H., and E.F. Labuda, "The silicon diode array camera tube," *Bell Syst. Tech. J.,* **48,** 1969, 1481.
35. Crowell, C.R., and R.M. Madden, "Large-signal transient analysis and effects of lateral charge spreading in television camera tubes," *IEEE Trans. Electron Devices,* **ED-18,** 1971, p. 1049.

36. deGennes, P.G., *The Physics of Liquid Crystals*, Oxford University Press, N.Y., 1974.
37. Dekker, A.J., "Secondary electron emission," *Solid-State Phys.*, **6**, 1958, or "Solid State Physics," Chap. 17, Prentice Hall, N.J., 1957.
38. Dimmock, J.O., "Capabilities and limitations of infrared imaging systems," *Proc. of the Society of Photo-Optical Instrumentation Engineers*, **32**, San Mateo, Calif., Oct. 16-17, 1972, p. 9.
39. Dyck, R.H., "Self-scanned photosensor array," *Wescon Technical Papers*, **14**, 1970, 15/4, p. 1.
40. Eklund, J.K., and D. Baeu, "Looking at the world through an infrared eye," *Industrial Research*, **17**, April 1975, p. 52.
41. Elliot, J.K., R.L. Gunshor, and R.F. Pierret, "Zinc oxide-silicon monolithic acoustic surface wave optical image scanner," *Appl. Phys. Lett.*, **27**, 1975, p. 179.
42. Engeler, W.E., M. Blumenfeld, and E.A. Taft, "The 'Epicon' array: a new semiconductor array-type camera tube structure," *Appl. Phys. Lett.*, **16**, 1970, p. 202.
43. Engeler, W.E., J.J. Tiemann, and R.D. Baertsch, "Surface charge, collection, storage and transport for visible image sensing arrays," *J. Lumin.*, **6-7**, 1973, p. 415.
44. Engstrom, O., and H.G. Grimmeiss, "Optical properties of sulfur-doped silicon," *J. Appl. Phys.*, **47**, 1976, p. 4090.
45. Engstrom, R.W., and J.H. Sternberg, "The silicon return–beam vidicon–a high-resolution camera tube," *RCA Review*, **33**, 1972, p. 501.
46. Escher, J.S., and R. Sankaran, "Transferred-electron photomission to 1.4 μm," *Appl. Phys. Lett.*, **29**, 1976, p. 87.
47. Escher, J.S., et al., "Schottky-barrier height of Au/p InGaAsP alloys lattice matched to InP," *J. Vac. Sci. Technol.*, **13**, 1976, p. 874.
48. Escher, J.S., et al., "Bias-assisted photoemission in the 1-2 micron range," *IEEE Int. Electron Devices Meeting, Technical Digest*, 1977, p. 460.
49. Fields, S.W., "Solid-state camera uses photodiodes," *Electronics*, Feb. 1, 1973, p. 121.
50. Fischer, A.G., "Infrared imaging devices," *J. Lumin.*, **6-7**, 1973, p. 427.
51. Fisher, D.G., R.E. Enstrom, J.S. Escher, and B.F. Williams, "Photoelectron surface escape probability of (Ga,In)As:Cs-O in the 0.9 to ∿ 1.6 μm range," *J. Appl. Phys.*, **43**, 1972, p. 3815.
52. Fisher, D.G., and R.V. Martinelli, "Negative electron affinity materials for imaging devices," *Advances in Image Pick up and Display*, **1**, 1974, Academic Press, N.Y.
53. Fisher, D.G., R.E. Enstrom, J.S. Escher, H.F. Gossenberger, and J.R. Appert, "Photoemission characteristics of transmission-mode negative electron affinity GaAs and (In,Ga) as vapor-grown structures," *IEEE Trans. Electron Devices*, **ED-21**, 1974, p. 641.
54. Flynn, J.B., J.M. Epstein, D.R. Palmer, and J.V. Egan, "Total active area silicon photodiode array," *IEEE Trans. Electron Devices*, **ED-16**, 1969, p. 877.
55. Forbes, L., and L.L. Wittmer, "Experimental verification of operation of the indium-doped infrared-sensing MOSFET," *IEEE Trans. Electron Devices*, **ED-22**, 1975, p. 1100.
56. Geiger, D.F., "Increase phototransistor bandwidth without sacrificing output voltage," *Electronic Design 8*, **21**, April 12, 1973, p. 102.
57. Gibbons, D.J. "Choosing a vidicon," *Wireless World*, **77**, 1971, pp. 89 and 135.
58. Goodman, L.A., "Liquid-crystal displays—Electro-optic effects and addressing techniques," *RCA Review*, **35**, 1974, p. 613.
59. Goto, N., Y. Isozaki, K. Shidara, E. Maruyama, T. Hirai, and T. Fugjita, "SATICON: A new photoconductive camera tube with Se-As-Te target," *IEEE Trans. Electron Devices*, **ED-21**, 1974, p. 662.
60. Gree, M., and P.R. Collings, "The burn resistant SEC camera tube," *Proc. Technical Program, Electro-Optical Systems Design Conference*, N.Y., Sept. 1970, p. 574.
61. Guldberg, J., H.C. Nathanson, D.L. Balthis, and A.S. Jensen, "An aluminum/SiO$_2$/sili-

con-on-sapphire light valve matrix for projection displays," *Appl. Phys. Lett.*, **26**, 1975, p. 391.

62. deHaan, E.F., A. van der Drift, and P.P.M. Schampers, "The 'Plumbicon,' a new television camera tube," *Philips Technical Rev.*, **25**, 1963, p. 133.

63. Hardeman, L., "How optoelectronic components fit in the optical spectrum," *Electronics*, Oct. 11, 1973, p. 111.

64. Harmon, W.J., Jr., "Dot matrix display features inherent scanning ability," *Electronics*, March 2, 1970, p. 121.

65. Hayes, R., R.G. Culter, and K.W. Hawken, "Storage tube with silicon target captures very fast transients," *Electronics*, Aug. 30, 1973, p. 97.

66. Hiraki, H., S.⸱ Hasegawa, H. Shirakusa, S. Noguchi, and M. Hirashima, "Still picture TV receiver using a silicon target storage tube," *Nat. Tech. Rep.*, **19**, 1973, p. 113.

67. Holmes, F.E., and C.A.T. Salama, "Modified bipolar-phototransistor structure," *Electronics Letters*, **8**, 2, 1972, p. 23.

68. Hudson, R.D., and J.W. Hudson, *Infrared Detectors*, Halsted Press, N.Y., 1975.

69. Hurwitz, C.E., and J.P. Donnelly, "Planar InSb photodiodes fabricated by Be and Mg ion implantation," *Solid-State Electronics*, **18**, 1975, p. 753.

70. Hyder, S.B., et al., "Liquid-phase-epitaxial growth of lattice matched $In_{.53}Ga_{.47}As$ on (100)-oriented InP," *Appl. Phys. Letters*, **31**, 1977, p. 551.

71. Jacobson, A.D., et al., "A liquid crystal light valve—a new display device," *IEEE Int. Electron Devices Meeting, Technical Digest*, Dec. 1976, p. 624.

72. James, L. W., et al., "Band structure and high-field transport properties of InP," *Phys. Rev. B*, **1**, 1970, p. 3998.

73. Jamieson, J.A., R.H. McFee, G.N. Ploss, R.H. Grube, and R.G. Richards, *Infrared Physics and Engineering*, McGraw-Hill, N.Y., 1963.

74. Jurgen, R.K., "Competing display technologies struggle for superiority," *IEEE Spectrum*, **11**, 1974, p. 90.

75. Kaneda, T., and H. Takanashi, "Avalanche buildup time of silicon avalanche diodes," *Appl. Phys. Lett.*, **26**, 1975, p. 642.

76. Kazan, B., *Advances in Image Pickup and Display*, **1**, Academic Press, N.Y., 1974.

77. Kazan, B., and M. Knoll, *Electronic Image Storage*, Academic Press, N.Y., 1968.

78. Kiess, H., "The physics of electrical charging and discharging of semiconductors," *RCA Review*, **36**, 1975, p. 667.

79. Kim, C.W., and W.E. Davern, "InAs charge–storage photodiode infrared vidicon targets," *IEEE Trans. Electron Devices*, ED-18, 1971, p. 1062.

80. Kohn, E.S., "Cold-cathode electron emission from silicon," *Appl. Phys. Lett.*, **18**, 1971, p. 272.

81. Kohn, E.S., "The silicon cold cathode," *IEEE Trans. Electron Devices*, **ED-20**, 1973, p. 321.

82. Korotin, V.G., S.N. Krivonogov, D.N. Nasledov, and Yu.S. Smetannikova, "Photoconductivity of n-type InSb under optical and electric carrier heating conditions," *Soviet Physics-Semiconductors*, **7**, 1973, p. 447.

83. Kortlandt, J., "Light sensitive CdS thin films with temperature resistant contacts," *Microelectronics and Reliability*, **10**, 1971, p. 261.

84. Kovac, M.G., "Charge-transfer readout and white-video-defect suppression in x-y image sensors," *IEEE Trans. Electron Devices*, **ED-22**, 1975, p. 168.

85. Kramer, G., "Thin-film transistor switching matrix for flat panel displays," *IEEE Trans. Electron Devices*, **ED-22**, 1975, p. 733.

86. Kramer, G., and M.A. Levine, "Field modulation of detectivity in PbS photoconductors," *Appl. Phys. Lett.*, **28**, 1976, p. 101.

87  Kressel, H., H. Schade, and H. Nelson, "Heterojunction cold-cathode electron emitters of (AlGa)As-GaAs," *J. Lumin.*, 6-7, 1973, p. 146.

88. Kruse, P.W., L.D. McGlauchlin, and R.B. McQuistan, *Elements of Infrared Technology: Generation Transmission and Detection*, Wiley, N.Y., 1962.

89. Landsman, A.P., and D.S. Strebkov, "Some electrical and optical characteristics of photoelectric generators with thin active surface perpendicular to the plane of the p-n junctions," *Radio Engr. Electron. Phys.*, 15, 1970, p. 2337.

90. Levinstein, H., Ed., *Photoconductivity*, Pergamon Press, Oxford, England 1962.

91. Levinstein, H., "Extrinsic detectors," *Appl. Opt.*, 4, 1965, p. 639.

92. Levinstein, H., "Infrared detectors," *Physics Today*, Nov. 1977, p. 23.

93. Levinstein, H., and J. Mudar, "Infrared detectors in remote sensing," *Proc. IEEE*, 63, 1975, p. 6.

94. Li, S.S., F.A. Lindholm, and C.T. Wang, "Quantum yield of metal-semiconductor photodiodes," *J. Appl. Phys.*, 43, 1972, p. 4123.

95. Lindholm, L.E., and G. Petersson, "Position sensitive photodetectors with high linearity," *IEEE Int. Electron Devices Meeting, Technical Digest*, Dec. 1976, p. 408.

96. List, W.F., "Characteristics and limitations of light sensing arrays based on LSI technologies," *IEEE Int. Convention Digest*, March 1971, p. 254.

97. Madden, R.M., D.A. Kiewit, and C.R. Crowell, "Silicon phototransistor vidicon," *IEEE Trans. Electron Devices*, ED-18, 1971, p. 1043.

98. Marlowe, F., F. Wendt, and C. Wine, "A display system using the alphechon storage tube," *The 1970 IDEA Symposium. Information Display, Evaluation and Advances*, Digest of Papers, Society for Information Display, May 1970, p. 112.

99. Martinelli, R.U., and D.G. Fisher, "The application of semiconductors with negative electron affinity surfaces to electron emission devices," *Proc. IEEE*, 62, 1974, p. 1339.

100. Maruyama, E., et al., "Graded-composition chalcogenide-glass," *Proc. 6th Conf. Solid-State Devices*, Tokyo, 1974; supplement to the *J. Jap. Soc. Appl. Phys.*, 44, 1975, p. 97.

101. McGee, J.D., D. McMullan, and E. Kahan, Eds., *Photo-Electronic Image Devices*, Vol. 33A, Academic Press, N.Y., 1972.

102. McIntyre, R., "Light detection using P-I-N photodiodes," Application Note No. 1, RCA Ltd., Ste-Anne-de-Bellevue, Quebec, Canada.

103. Meitzler, A.H., J.R. Maldonado, and D.B. Fraser, "Image storage and display devices using fine-grain ferroelectric ceramics," *Bell Syst. Tech. J.*, 49, 1970, p. 953.

104. Melchior, H., "Sensitive high speed photodetectors for the demodulation of visible and near infrared light," *J. Lumin.*, 7, 1973, p. 390.

105. Melchior, H., "Detectors for lightwave communication," *Physics Today*, Nov. 1977, p. 32.

106. Melchior, H., M.B. Fisher, and F.R. Arams, "Photodetectors for optical communication systems," *Proc. IEEE*, 58, 1970, 1466.

107. Melchior, H., and A.R. Hartman, "Epitaxial silicon $n^+p\pi p^+$ avalanche photodiodes for optical fiber communication at 800 to 900 nanometers," *IEEE Int. Electron Devices Meeting, Technical Digest*, Dec. 1976, p. 412.

108. Melngailis, I., "Small bandgap semiconductor infrared detectors," *J. Lumin.*, 7, 1973, p. 501.

109. Melz, P.J., "Gain in electrophotography-I: Phototransistor configuration," *IEEE Trans. Electron Devices*, ED-19, 1972, p. 433.

110. Melz, P.J., and U. Yahtra, "Gain in electrophotography-II: Charge transfer configuration," *IEEE Trans. Electron Devices*, ED-19, 1972, p. 437.

111. Mende, S.B., "Single photoelectron recording by an image intensifier TV camera system," *Appl. Opt.*, 10, 1971, p. 829.

112. Migliorato, P., A.W. Vere, and C.T. Elliott, "Sulphur-doped silicon background–limited infrared photodetectors near 77K," *Appl. Phys.,* 11, 1976, p. 295.
113. Milnes, A.G., *Deep Impurities in Semiconductors*, Wiley, N.Y., 1973.
114. Milnes, A.G., and D.L. Feucht, "Heterojunction photocathode concepts," *Appl. Phys. Lett.,* 19, 1971, p. 383.
115. Milnes, A.G., and D.L. Feucht, *Heterojunctions and Metal-Semiconductor Junctions*, Academic Press, N.Y., 1972, Chap. 8.
116. Mito, S., et al., "TV imaging system using electroluminescent panels," *Digest of Technical Papers*, 1974 Int. Symp. Society for Information Display, San Diego, Calif., May 1974.
117. Moore, T.H., and J.D. Pace, "Design of an electrooptic light valve projection display." *IEEE Trans. Electron Devices,* ED-17, 1970, p. 423.
118. Morelière, R., P. Vignes, "On the analysis of the photocurrent of a directly polarized p-n junction," *Comptes Rendus Hebd. Seances Acad. Sci., B* (France), 275, 1972, p. 951.
119. Morris, F.J., "Sheet resistivity measurements of thin-film resistance overlays on silicon diode array camera tube targets,"*IEEE Trans. Electron Devices,* ED-19, 1972, p. 1139.
120. Mort, J., and D.M. Pai, Eds., *Photoconductivity and Related Phenomena*, Elsevier, Amsterdam, 1976.
121. Morten, J.D., and M.H. Jervis, "Infra-red detectors using mercury cadmium telluride at 5 $\mu$m at ambient temperatures," *Proc. Technical Program Electro-Optical Systems Design Conference*, New York, Sept. 1970, p. 166; available from Industrial and Scientific Conference Management, Inc., 222 West Adams Street, Chicago, Ill. 60606.
122. Moss, T.S., *Optical Properties of Semi-Conductors*, Butterworths Scientific Publications, London, 1959.
123. Moss. T.S., G.J. Burrell, and B. Ellis, *Semiconductor Opto-Electronics*, Wiley, N.Y., 1973.
124. Müller, J., "Ultrafast multireflection–and transparent thin film silicon photodiodes," *IEEE Int. Electron Devices Meeting, Technical Digest*, Dec. 1976, p. 420.
125. Müller, J., and A. Ataman, "Double-mesa thin-film reach-through silicon avalanche photodiodes with large gain-bandwidth product," *IEEE Int. Electron Devices Meeting, Technical Digest*, Dec. 1976, p. 416.
126. Needham, M.J., "Microchannel plates advance night-viewing technology," *Electronics*, Sept. 27, 1973, p. 117.
127. Nevin, J.H., and H.T. Henderson, "Thallium-doped silicon ionization and excitation levels by infrared absorption," *J. Appl. Phys.,* 46, 1975, p. 2130.
128. Nobutoki, S., S. Nagahara, and T. Takazi, "A color separating filter integrated vidicon for frequency multiplex system single pickup tube color television camera," *IEEE Trans. Electron Devices,* ED-18, 1971, p. 1094.
129. Noda, T., A. Kouno, and T. Ando, "An economical color-television camera utilizing a silicon vidicon for electronic color separation," *IEEE Trans. Electron Devices,* ED-22, 1975, p. 248.
130. Niu, H., et al., "Application of lateral photovoltaic effect to the measurement of the physical quantities of p-n junctions–sheet resistivity and junction conductance of $N_2^+$ implanted Si," *Japan J. Appl. Phys.,* 15, 1976, p. 601.
131. Olsen, G.H., et al., "High performance GaAs photocathodes," *J. Appl. Phys.,* 48, 1977, p. 1007.
132. Osborne, W.E., "Long-range infrared intruder alarm resists fault triggering," *Electronics*, Nov. 20, 1972, p. 111.
133. Pankove, J.I., "Phenomena useful for displays," *IEEE Int. Electron Devices Meeting, Technical Digest*, Dec. 1976, p. 621.
134. Pankove, J.I., and J.E. Berheyheiser, "Properties of Zn-doped GaN II Photoconductivity," *J. Appl. Phys.,* 45, 1974, p. 3892.

135. Parker, W.C., and L. Forbes, "Experimental characterization of gold-doped infrared-sensing MOSFET's," *IEEE Trans. Electron Devices*, **ED-22**, 1975, p. 916.
136. Pines, M.Y., and R.Baron, "Characteristics of indium doped silicon infrared detectors," *IEEE Int. Electron Devices Meeting, Technical Digest*, 1974, p. 446.
137. Piqueras, J., and E. Muñoz, "A novel GaAs avalanche photodiode," *Appl. Phys. Lett.*, **25**, 1974, p. 214.
138. Piqueras, J., and E. Muñoz, "Photo-response of the GaAs controlled field avalanche photodiode (CFAPD)," *J. Appl. Phys.*, **46**, 1975, p. 3532.
139. Riezenman, M.J., "Special report: The new displays complement the old," *Electronics*, April 12, 1973, p. 91.
140. Robinson, L.C., *Physical Principles of Far-Infrared Radiation, Methods of Experimental Physics*, Vol. 10, Academic Press, N.Y., 1973.
141. Roehrig, H., et al., "High-resolution low-light-level video systems for diagnostic radiology," *Proc. Society of Photo-optical Instrumentation Engineers, Low Light Level Devices*, 78, 1976, p. 102.
142. Ryzhii, U.I., "Photoconductivity characteristics in thin films subjected to crossed electric and magnetic fields," *Soviet Phys.–Sol. State*, **11**, 1970, p. 2078.
143. Schaffert, R.M., *Electrophotography*, rev. ed., Focal Press, London, 1965; Halstead Press, N.Y., 1975.
144. Schmidlin, F.W., "Electrophotography," Chap. 11 in *Photoconductivity and Related Phenomena*, J. Mort and D.M. Pai, Eds., Elsevier, Amsterdam, 1976.
145. Schroder, D.K., R.N. Thomas, J. Vine, and H.C. Nathanson, "The semiconductor field-emission photocathode," *IEEE Trans. Electron Devices*, **ED-21**, 1974, p. 785.
146. Schulze, R.G., and P.E. Petersen, "Photoconductivity in solution-grown copper-doped GaP," *J. Appl. Phys.*, **45**, 1974, p. 5307.
147. Schut, Th. G., and W.P. Weijland, "30 mm Plumbicon camera tubes with fibre-optic faceplate, anti-comet-tail gun, and light pipe," *Mullard Tech. Commun.*, **11**, #109, 1971, p. 186.
148. Schut, Th. G., and W.P. Weyland, "Plumbicon tubes with fibre optic faceplate, anti-comet-tail gun and light pipe," *Proc. Tech. Program, Electro-Optical Syst. Des. Conf.*, New York, Sept. 1970, p. 594.
149. Seck, A., "Survey of photosensitive materials and devices," *Wescon Technical Papers*, **14**, 1970, 15/2, p. 1.
150. Sedgwick, T.O., M.E. Cowher, I.F. Chang, and J.F. O'Hanlon, "A field effect controlled storage display device using polycrystalline silicon film," *J. Electronic Materials*, **2**, 1973, p. 309.
151. Seki, H., and I.P. Batra, "Photocurrents due to pulse illumination in the presence of trapping II," *J. Appl. Phys.*, **42**, 1971, p. 2407.
152. Shallcross, F.V., W.S. Pike, P.K. Weimer, and G.M. Fryszman, "A Photoconductive sensor for card readers," *IEEE Trans. Electron Devices*, **ED-17**, 1970, p. 1086.
153. Shaunfield, W.R., D.W. Boone, and D.G. Deppe, "Broadband avalanche photodetector modules for 1.06 μm and 1.54 μm," *Proc. Technical Program. Electro-Optical Systems Design Conference*, New York, Sept. 1970, p. 491; available from Industrial and Scientific Conference Management, Inc., 222 West Adams St., Chicago, Ill. 60606.
154. Shimizu, K., O. Yoshida, S. Aihara, and Y. Kiuchi, "Characteristics of experimental CdSe vidicons," *IEEE Trans. Electron Devices*, **ED-18**, 1971, p. 1058.
155. Silver, R.S., and E. Luedicke, "Electronic image storage utilizing a silicon dioxide target," *IEEE Trans. Electron Devices*, **ED-18**, 1971, p. 229.
156. Simon, R.E., "A solid-state boost for electron-emission devices," *IEEE Spectrum*, **9-12**, 1972, p. 74.
157. Simon, R.E., and B.F. Williams, "Secondary electron emission," *IEEE Trans. Nuclear Sci.*, **NS-15**, 1968, p. 167.
158. Singer, B., "Analysis and performance characteristics of an intensified silicon vidicon tube," *IEEE Trans. Electron Devices*, **ED-18**, 1971, p. 1016.

159. Sommer, A.H., *Photoemissive Materials*, Wiley, N.Y., 1968.
160. Sommer, A.H., "Conventional and negative electron affinity photoemitters," *J. Phys. (Paris)*, 34, 1974, p. C6-51.
161. Spalding, R.L., S.A. Ochs, and E. Luedicke, "The Binicon camera tube–a new double vidicon, *RCA Review*, 34, 1973, p. 121.
162. Steckl, A.J., et al., "Application of charge-coupled devices to infrared detection and imaging," *Proc. IEEE*, 63, 1975, p. 67.
163. Stelzer, E.L., J.L. Schmit, and M.W. Scott, "Mercury cadmium telluride as a short wavelength (2-7 μm) detector," *Proc. Technical Program. Electro-Optical Systems Design Conference*, New York, Sept. 1970, p. 159; available from Industrial and Scientific Conference Management, Inc., 222 West Adams St., Chicago, Ill. 60606.
164. Swartz, G.A., A. Gonzalez, and A. Dreeben, "A field-effect phototransistor," *Int. Solid-State Circuits Conference, Digest of Technical Papers*, Feb. 17-19, 1971, p. 134.
165. Tabak, M.D., S.W. Ing, and M.E. Scharfe, "Operation and performance of amorphous selenium-based photoreceptors," *IEEE Trans. Electron Devices*, ED-20, 1973, p. 132.
166. Takeda, M., Y. Kahihara, M. Yoshida, Y. Nahata, M. Kamaguchi, H. Kisheshita, Y. Ya-maulki, T. Inoguthi, and S. Mito, "Inherent memory effects in ZnS:Mn thin-film EL devices," *Proc. 6th Conf. Solid-State Devices*, Tokyo, 1974; supplement to the *J. Jap. Soc. Appl. Phys.*, 44, 1975, p. 103.
167. Tani, Z., et al., "GaAs light coupled device," *Proc. 4th Conf. Solid-State Devices*, Tokyo, 1972; supplement to the *J. Jap. Soc. Appl. Phys.*, 42, 1973, p. 258.
168. Tanigawa, H., and T. Ando, "Influence of the charge pumping effect on an integrated solid-state image sensor using MOST switches," *Proc. IEEE*, 61, 1973, p. 491.
169. Thomas, C.M., "Photocathodes," *Proc. 17th Annual Technical Meeting of Soc. of Photo-optical Instrumentation Engineers*, Image Intensifiers: Technology, Perform-ance, Requirements and Applications, 42, 1973, p. 71.
170. Thomas, R.N., and H.C. Nathanson, "Photosensitive field emission from silicon point arrays," *Appl. Phys. Lett.*, 21, 1972, p. 384.
171. Thomas, R.N., and H.C. Nathanson, "Transmissive-mode silicon field emission array photoemitter," *Appl. Phys. Lett.*, 21, 1972, p. 387.
172. Thomas, R.N., R.A. Wickstrom, D.K. Schroder, and H.C. Nathanson, "Fabrication and some applications of large-area silicon field emission arrays," *Solid-State Electron-ics*, 17, 1974, p. 155.
173. Trishenkov, M.A., "An investigation of the lag properties of a photodiode in the presence of trapping," *Radio Engineering and Electronic Phys.*, 15, 1970, p. 659.
174. Tseng, C.-C., and S. Wang, "Integrated grating-type Schottky-barrier photodetector with optical channel waveguide," *Appl. Phys. Lett.*, 26, 1975, p. 632.
175. Twu, B-l., and S.E. Schwarz, "Mechanism and properties of point-contact metal-insu-lator-metal diode detectors at 10.6 μ," *Appl. Phys. Lett.*, 25, 1974, p. 595.
176. VanRoosmalen, J.H.T., "New possibilities for the design of plumbicon tubes," *IEEE Trans. Electron Devices*, ED-18, 1971, p. 1087.
177. van der Ziel, A., "Classical infrared mixing with solid state diodes," *Solid-State Elec-tronics*, 18, 1975, p. 355.
178. Volkov, A.S., A.A. Gutkin, O.V. Kosogov, and S.E. Kumekov, "Photosensitivity spectra of InSb p-n junctions in the photon energy range up to 3.3 eV," *Sov. Phys. Semicon-ductors*, 6, 1973, p. 1928.
179. Wang, C.C., and S.R. Hampton, "Lead telluride-lead tin telluride heterojunction diode array," *Solid-State Electronics*, 18, 1975, p. 121.
180. Watton, R., G. Smith, and B. Harper, "Infrared television: performance of pyroelectric vidicon at 8 to 14 μm," *Electronics Lett.*, 9, 1973, p. 534.
181. Watton, R., C. Smith, B. Harper, and W.M. Wreathall, "Performance of the pyroelectric vidicon for thermal imaging in the 8-14 micron band," *IEEE Trans. Electron Devices*, ED-21, 1974, p. 462.

182. Waxman, A., and F. Heiman, "A new storage tube terminal utilizing the Lithocon™ silicon storage tube," *The 1970 IDEA Symposium. Information Display, Evolution and Advances*, Digest of Papers, Society for Information Display, May 1970, p. 62.

183. Webb, P.P., R.J. McIntyre, and J. Conradi, "Properties of avalanche photodiodes," *RCA Review*, 35, 1974, p. 234.

184. Weimer, P.K., "Systems and technologies for solid state image sensor," *RCA Review*, 32, 1971, p. 251.

185. Weimer, P.K., J.V. Forgue, and R.R. Goodrich, "The vidicon photoconductive camera tube," *RCA Review*, 12, 1951, p. 306.

186. Weimer, P.K., W.S. Pike, G. Sadasiv, F.V. Shallcross, and L. Meray-Horvath, "Multi-element self-scanned mosaic sensors," *IEEE Spectrum*, 6, March 1969, p. 52.

187. Weimer, P.K., F.V. Shallcross, and V.L. Fraritz, "Phototransistor arrays of simplified design," *IEEE J. Solid-State Circuits*, SC-6, 1971, p. 135.

188. Whelan, M.V., "Resistive MOS-gated diode light sensor," *Solid-State Electronics*, 16, 1973, p. 161.

189. Wieder, H.H., and D.A. Collins, "Minority carrier lifetime in InAs epilayers," *Appl. Phys. Lett.*, 25, 1974, p. 742.

190. Woltring, H.J., "New possibilities for human motion studies by real-time light spot position measurement," *Biotelemetry*, 1, 1974, p. 132.

191. Woltring, H.J., "Single and dual-axis lateral photodetectors of rectangular shape," *IEEE Trans. Electron Devices*, ED-22, 1975, p. 581.

192. Woody, W.R., "The image Isocon for low light level television applications," *Proc. Tech. Program, Electro-Optical Syst. Design Conf.*, New York, Sept. 1970, p. 585.

193. Wright, H.C., and J.E. Slawek, Jr., "Lead-tin-telluride 10 micron photovoltaic detectors," *Proc. Technical Program, Electro-Optical Systems Design Conf.*, New York, Sept. 1970, p. 288; available from Industrial and Scientific Conference Management, Inc., 222 West Adams Street, Chicago, Ill. 60606.

194. Wronski, C.R., B. Abeles, and A. Rose, "Granular metal-semiconductor vidicon," *Appl. Phys. Lett.*, 27, 1975, p. 91.

195. Yamamoto, H., et al., "A contact type linear photosensor array using an amorphous thin film," *Proc. 9th Conf. Solid-State Devices*, Tokyo, 1977.

196. Yamamoto, J., H. Yoshinaga, and S. Kon, "Far infrared resonant photoconductive properties in n-type InSb," *Japanese J. Appl. Phys.*, 8, 1969, p. 242.

197. Yamato, T., S. Yoshikawa, K. Kabayashi, J. Matsuzaki, and I. Tagoshima, "A new silicon vidicon with a CdTe-Si target," *IEEE Trans. Electron Devices*, ED-19, 1972, p. 385.

198. Yamauchi, Y., H. Kishishita, M. Takeda, T. Inoguchi, and S. Mito, "Red electro-luminescence from ZnS:Mn-F thin film," *IEEE Int. Electron Devices Meeting, Technical Digest*, Dec. 1974, p. 352.

199. Yamauchi, Y., M. Takeda, Y. Kahihara, Y. Yoshida, J. Kawaguchi, H. Kishishita, Y. Nakata, T. Inoguchi, and S. Mito, "Inherent memory effects in ZnS:Mn thin film EL devices," *IEEE Int. Electron Devices Meeting, Technical Digest*, Dec. 1974, p. 348.

200. Yonezu, H., and A. Kawaji, "Computer-aided design of a Si avalanche photodiode," *IEEE Trans. Electron Devices*, 16, 1969, p. 923.

201. Yoshifumi, A., "New flat panel TV display system," *IEEE Int. Electron Devices Meeting, Technical Digest*, Dec. 1973, p. 196.

202. Yoshikawa, M., R. Watanabe, K. Yamamoto, and M. Kikuchi, "Characteristics of ultra-high sensitivity vidicons with a photoconductor of antimony trisulfide," *Hitachi Rev.*, 20, 1971, p. 23.

203. Yoshikawa, S., and J-i. Chikawa, "Electrical effect of growth striations in the silicon vidicon-type camera tubes," *Appl. Phys. Lett.*, 23, 1973, p. 636.

204. Zamfir, G.N., and V.F. Zolotarev, "Photoelectric image converter based on high-field domains," *Sov. Phys. Semiconductors*, 4, 1971, p. 1470.

# 14

# Light-Emitting Diodes and Injection Lasers

## CONTENTS

## 14.1 LIGHT EMISSION FROM DIRECT-GAP GaAs$_{1-x}$P$_x$

Since GaAs has a bandgap of 1.45 eV the recombination radiation that is emitted is at about 850 nm for the full bandgap transition; therefore, it is in the infrared region and not visible to the human eye. The bandgap of GaAs$_{0.6}$P$_{0.4}$, however, is about 1.9 eV which corresponds to emission at 650 nm in the red region of the spectrum. Hence GaAsP is a widely used material for light-emitting diodes.

The relative response of the human eye at various wavelengths (for constant energy input) is shown in Fig. 14.1. The wavelengths corresponding to selected materials for light-emitting diodes (LEDs) are indicated. Since the eye response peaks strongly in the green, the more nearly the emitted light approaches the green wavelength the greater the perceived brightness for a given power level. Thus a green GaP LED of efficiency 0.1% may be as acceptable as a red GaP LED of efficiency 3%. Further details of the eye response and optical units of interest to electronics engineers for photometry and radiometry may be found in a review by Zaha.[224]

For the ternary compound GaAs$_{1-x}$P$_x$, increase of $x$ up to the value of

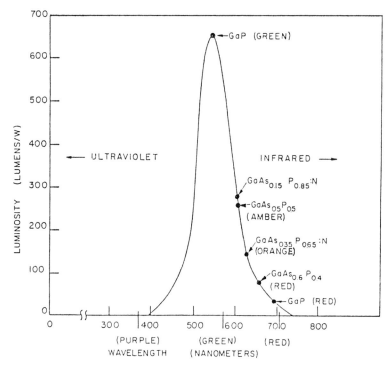

Fig. 14.1. Standard eyeball. The average luminosity of four major types of LEDs is shown by their wavelength location on the CIE standard photopic luminosity curve. The eye's response to light energy, in lumens per watt of radiant energy, peaks at green, and falls off rapidly at the edges of the visible-light spectrum. GaP is less luminous than GaAsP when emitting red light. (After Smith. Reprinted from *Electronics*, Oct. 25, 1971; Copyright © McGraw-Hill, Inc., N.Y., 1971.)

0.4 results in an increasing direct bandgap that is favorable to recombination radiation emission. Further increase of the phosphorus content causes the bandgap to become indirect (as shown in Fig. 14.2) and then recombination requires phonon cooperation and light-emission recombination becomes much less. In Fig. 14.3 curves are given of efficiency and brightness as a function of phosphorus content.

Initial studies of GaP showed low ($\sim 10^{-5}$) emission efficiencies, however, by control of Zn,O doping good efficiency for red light emission has been achieved and by N doping satisfactory green emission is achieved. These matters are discussed in detail a little later.

The brightness perceived is a function of the spectrum of the radiation emitted in relation to the eye-luminosity curve. A "red" emitting GaP diode may have a much wider spectrum than a GaAsP diode which may appear as bright even though less efficient and lower in total radiated energy (see Fig. 14.4).

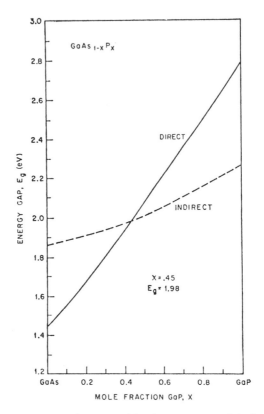

Fig. 14.2. The dependence on the compositional parameter $x$ of the lowest energy direct $(\Gamma_8 \to \Gamma_1)$ and indirect $(\Gamma_8 \to x_1)$ energy gaps in $GaAs_{1-x}P_x$. (After Bergh and Dean. Reprinted with permission from *Proc. IEEE,* **60**, 1972, p. 190.)

Other ternary semiconductors that are light emitters include GaInP, GaAlAs, GaSbP, and InAlAs. The bandgap structures for these are roughly illustrated in Fig. 14.5. Another approach to visible light emission has been to use the infrared emission of GaAs to excite phosphors that up-convert in energy to give visible output. Such up-conversion is low in efficiency and in overall performance as seen from Fig. 14.6.

Light generated in a *p-n* junction may not emerge because of internal reflections at the refractive index discontinuity at the semiconductor-air or semiconductor-resin encapsulated surface. For a refractive index ratio of 3.5:1 at a plane interface, only the radiation in a cone of half-angle 17 deg $[\sin^{-1}(1/3.5)]$ emerges. The performance can be improved by designing the diode geometry so that more of the light arrives at the surface within the critical angle and by coating the diode with a medium that has a refractive index between that of the semiconductor and air to increase the critical angle. Some

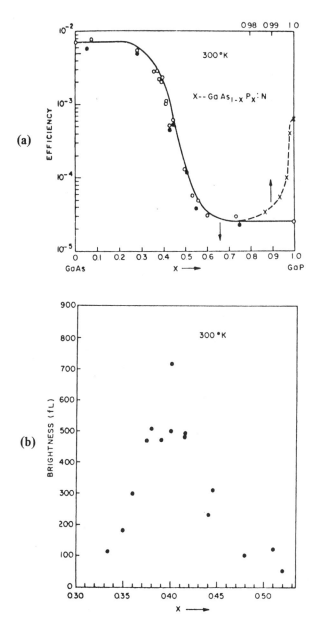

Fig. 14.3. Light from GaAs$_{1-x}$P$_x$ LEDs:

(a) External quantum efficiencies of GaAs$_{1-x}$P$_x$ LEDs at 300°K as a function of the compositional parameter $x$ without (full curve) and with (dashed portion near $x = 1$) deliberate doping with the isoelectronic trap $N$ in the high $10^{18}$-cm$^{-3}$ range. Note the change in scale for the dashed points, and

(b) The brightness of GaAs$_{1-x}$P$_x$ LEDs as a function of the compositional parameter $x$. The experimental points are for a diode current of 4.4 Acm$^{-2}$. (Adapted from Henzog, Groves, and Craford, *Appl. Phys.*, **40**, p. 1830. Reprinted with permission from *Proc. IEEE*, **60**, 1972, p. 191.

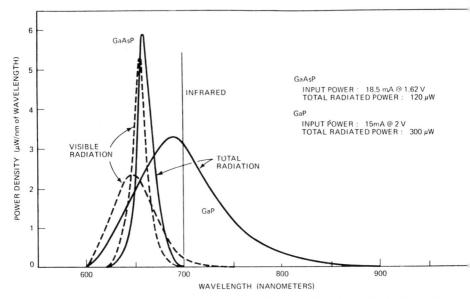

Fig. 14.4. Visible shift. Emission spectra of typical red GaAsP and GaP diodes show why GaAsP appears as bright despite its lower quantum efficiency. GaAsP emits a narrow band of wavelengths and the shift between luminous flux visible radiation and the total radiated output is small. The GaP spectrum is broad, much of the energy is invisible, and the power density at any given wavelength is lower. (After Smith. Reprinted from *Electronics*, Oct. 25, 1971; Copyright © McGraw-Hill, Inc., N.Y., 1971.)

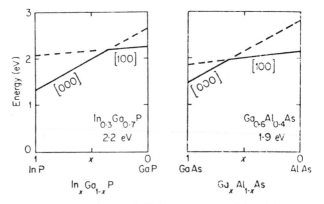

Fig. 14.5. Band-structure parameters of III-V ternary materials, giving composition and bandgap of direct-indirect transition materials at 300°K. (After Gooch, from *Injection Electroluminescent Devices*; Copyright © 1973 John Wiley and Sons Ltd., N.Y.)

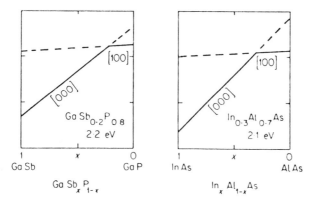

Fig. 14.5. (continued)

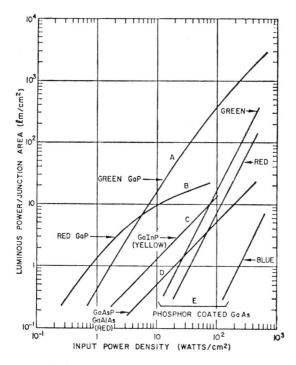

Fig. 14.6. Best reported values of luminous power conversion for LEDs of technological importance. For direct-gap semiconductors the (luminous power)/(junction area) is approximately equal to brightness in lamberts (L). (After Bergh and Dean. Reprinted with permission from *Proc. IEEE*, **60**, 1972, p. 159.)

geometries that have been considered are shown in Fig. 14.7. Their figures of merit (uncoated and in air) are given in Table 14.1. A truncated cone can be coupled to a flat detector with a high transfer efficiency (about 20% compared with 10% for a hemisphere and 1.3% for a flat diode). However, specially shaped semiconductor structures are expensive to make.

GaAsP for LEDs is usually obtained by growth of an epitaxial layer on a GaAs substrate and light that enters the substrate is absorbed and lost. However, with the recent improved availability of GaP, the epitaxial GaAsP may be grown on GaP and a reflecting back contact may be used to obtain increased light output as shown in Fig. 14.8.

Multiplex pulsing (time-shared excitation) of LEDs is sometimes used since this reduces the amount of decoders, drivers, and peripheral circuitry required. The effective light output seen by the viewer is related to the peak absolute value of light output during the pulse and to the duty factor. For some diodes there is a gain in the average light output under pulse conditions, where an average pulse current is matched to a direct current, because of the slope of the light output-versus-current characteristic of the LED. The relative light output per unit current for several diodes is shown in Fig. 14.9. If the light for a peak current of 40 mA at 25% duty factor is compared with the light for a direct current of 10 mA the gain factors are 0.68, 1.03, and 1.35 for the SLA-1, MAN-1, and MAN-4, respectively. However, the SLA-1 has a higher absolute

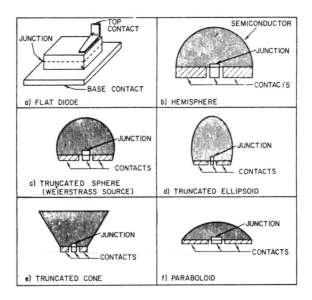

Fig. 14.7. LED geometries to increase the light extraction or optical efficiency. The effectiveness of the various geometries is listed in Table 14.1. (After Carr. Reprinted with permission from *Infrared Physics*, 6, 1966; © 1966 Pergamon Press.)

Table 14.1.  **Figures of Merit for Various LED Geometries per Unit Internal Light Flux Generation ($n = 3.6$)**

| Geometry | Radiant Flux $P$ | Maximum Radiant Intensity $J(0)$ $\theta = 0$ | Average Radiant Intensity $\langle J(0) \rangle$ $\theta = 26°$ |
|---|---|---|---|
| Flat plane diode area emission | 0.013 | 0.0042 | 0.0039 |
| Hemisphere | 0.34 | 0.054 | 0.054 |
| Weierstrasse sphere | 0.34 | 1.4 | 0.52 |
| Truncated ellipsoid | 0.25 | 9.8 | 0.39 |
| Truncated cone | 0.20 | 0.063 | 0.059 |
| Paraboloid source | | | |
| $R_j/F_p = 0.1$ | 0.34 | 0.84 | 0.52 |
| $R_j/F_p = 0.05$ | 0.34 | 3.3 | 0.52 |

(After Carr, reprinted from Infrared Physics, 1966, V. 6)

output at 40 mA (not apparent from the plot because of the normalized form) and gives the brightest output under either pulsed or dc conditions.

## 14.2 RADIATIVE AND NONRADIATIVE RECOMBINATION IN GaAs DIODES

The internal quantum efficiency of an LED is given by

$$\eta_i = \frac{\tau}{\tau_r} = 1 - \frac{\tau}{\tau_{non}} \tag{14.1}$$

where $\tau$, $\tau_r$, and $\tau_{non}$ are the minority carrier, the radiative, and the nonradiative lifetimes and

$$\tau = \left( \frac{1}{\tau_r} + \frac{1}{\tau_{non}} \right)^{-1} \tag{14.2}$$

The nonradiative recombination processes must be made as low as possible to increase $\tau_{non}$. This is achieved by selecting the fabrication processes to be as clean as possible and low in crystal imperfections.

Consider a single deep-level acceptor of density $N_{TA}$, at an energy level $E_i - E_{TA}$ below the center of the bandgap, dominating the nonradiative recombination process in the linearly graded region of an LED. In the $p$ side of the junction the recombination center is either in the state $N_{TA}^0$ or $N_{TA}^-$ and recombination of excess injected electrons takes place by the process

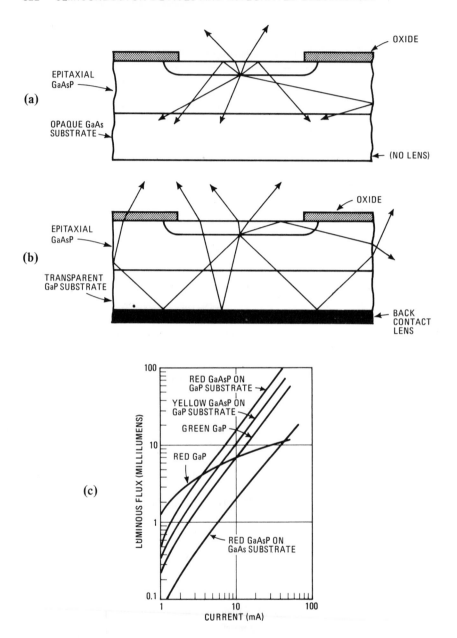

Fig. 14.8. GaAsP LEDs grown on GaP instead of GaAs provide more light:
(a)    On GaAs the rearward directed light is lost,
(b)    On GaP the rearward light can be redirected, and
(c)    Superior brightness results. (After Kniss. Reprinted from *Electronics*, May 2, 1974; Copyright © McGraw-Hill, Inc., N.Y., 1974.)

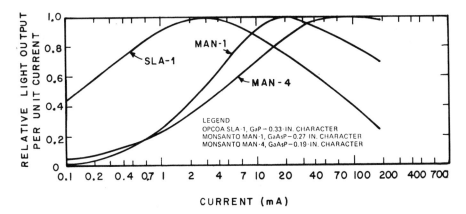

Fig. 14.9. Curves of light output (normalized) for three commercially available seven-segment displays. (After Ahrons. Reprinted from *Electronics*, Nov. 22, 1971; Copyright © McGraw-Hill, Inc., N.Y., 1971.)

$N_{TA}^0 + e \rightarrow N_{TA}^-$ followed by $N_{TA}^- + h \rightarrow N_{TA}^0$. The recombination statistics then give the familiar expression

$$\tau_{\text{non}} = \frac{\tau_{n0}(p_0 + p_1) + \tau_{p0}(n_0 + n_1)}{n_0 + p_0} \tag{14.3}$$

which if the electron capture process is rate limiting reduces to

$$\tau_{\text{non}} = \tau_{no} = \frac{1}{N_{TA}^0 \sigma_n v_{th}} \tag{14.4}$$

where $\sigma_n$ is the capture cross section for electrons and $v_{th}$ is about $\sqrt{3\ kT/m^*}$. The neutral trap density $N_{TA}^0$ equals $N_{TA} - N_{TA}^-$ where

$$\frac{N_{TA}^-}{N_{TA}} = \frac{\sigma_n n + \sigma_p p_1}{\sigma_n(n + n_1) + \sigma_p(p + p_1)} \tag{14.5}$$

where

$$p = n_i \exp\ [(E_i - E_{Fp})/kT]$$

$$p_1 = n_i \exp\ [(E_i - E_{TA})/kT]$$

$$n = n_i \exp\ [(E_{Fn} - E_i)/kT]$$

$$n_1 = n_i \exp\ [(E_{TA} - E_i)/kT]$$

To characterize the dominant nonradiative recombination process one therefore needs to know the recombination center density, its energy level, and its capture cross section(s). Characterization by density and energy is valuable in the study of the effects of fabrication process variations. Transient capacitance techniques, particularly deep-level transient spectroscopy DLTS,[124] are now the standard methods of making such studies and much work is in progress although full understanding is not available yet.

Prior to this development the determination of the recombination center density in linearly graded junctions presented some difficulty, but estimates could be made from second-order effects in the capacitance-voltage relationship. The analytical $C$-$V$ equation for a linearly graded junction with a deep-center $N_{TA}$ present is

$$V = \frac{qa\epsilon^2 A^3}{12C^3} - \frac{q\epsilon N_{TA} V_2 A^2}{2VC^2} + \frac{q\epsilon N_{TA} V_2^2 A^2}{2V^2 C^2} \qquad (14.6)$$

where $a$ is the shallow impurity grading, $A$ is the diode area, and $V_2$ is $(E_{TA} - E_{Fp})/q$ where $E_{TA} - E_{Fp}$ is the difference of the trap level and the hole quasi-Fermi level. The voltage $V$ is the total voltage obtained from the applied bias voltage $V_A$ and the built-in voltage $V_B$, by the relationship

$$V = V_B - V_A - 0.125 \qquad (14.7)$$

where

$$V_B = \frac{kT}{q} \ln \left( \frac{aW}{n_i} \right)^2$$

and

$$W = \frac{\epsilon A}{2C}$$

Equation (14.6) may be rearranged as

$$C^3 V = qa\epsilon^2 A^3 - q \frac{\epsilon A^2 V_2 N_{TA}}{2} \left( \frac{C(V - V_2)}{V^2} \right) \qquad (14.8)$$

Hence $N_{TA}$ may be determined from the slope of the $C^3 V$ versus $C(V - V_2)/V^2$ graph provided $V_2$ is known. Fortunately, $V_2$ may be determined from the recombination current versus temperature measurements at a small bias voltage (for which nonradiative recombination is dominant). Such current versus temperature curves are shown in Fig. 14.10(a) for Zn-diffused GaAs diodes on $2 \times 10^{18}$ cm$^3$ doped GaAs(Sn) substrates. The values inferred for $V_2$

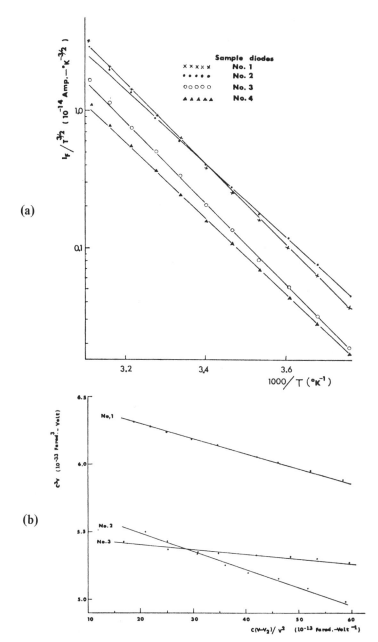

Fig. 14.10. The characteristics of GaAs LEDs studied in these ways reveal information on nonradiative recombination processes:

(a) $I_F/T^{3/2}$ vs $1000/T$ obtained from current-temperature measurements at a small forward bias, and

(b) Plots of $C^3 V$ vs $C(V - V_2)/V^2$ indicate from their slopes the densities of nonradiative recombination centers. (After Lo and Yang. Reprinted with permission from *IEEE Trans. Electron Devices*, **ED-20**, 1973, p. 690.)

ranged from 0.592 V for diode 2 to 0.497 V for diode 4. The plots of $C^3V$ versus $C(V - V_2)/V^2$ are shown in Fig. 14.10(b) for three diodes. Diodes 1 and 2 were found from these lines to have a recombination level at 0.1 eV from the center of the bandgap. Diodes 3 and 4 (fabricated under slightly different conditions) had a recombination level at about 0.2 eV from the bandgap center. The density of recombination centers was about $10^{16}$ cm$^{-3}$. Unfortunately, the symmetry of the situation does not allow determination of whether the centers are acceptors or donors, or above or below midgap.

The minority carrier lifetime and external quantum efficiency of a GaAs LED may be measured by the circuit of Fig. 14.11(a) in which the diode is pulsed to give the storage time. Typically the GaAs lifetime decreases from $5 \times 10^{-9}$ sec to $4 \times 10^{-9}$ sec over the temperature range 200 to 400°K. The external quantum efficiency decreases from 1.5% to 0.5% in the same temperature range.

The external efficiency of an LED may be expressed as

$$\eta_{ext} = \eta_i \left( \frac{1}{1 + \alpha V_p/ATr} \right) \tag{14.9}$$

where $\alpha$ is the absorption coefficient of the light, $V_p$ is the volume of the $p$ region of the diode, and $Tr$ is the transmission coefficient of the light emerging from the surface $A$ of the diode. Since $\eta_i$ is of the order of 60% and $\eta_{ext}$ is about 1%

$$\eta_i \simeq \eta_{ext}\, \alpha\, \frac{V_p}{ATr} \tag{14.10}$$

From Eqs. (14.1), (14.2), (14.4), and (14.10)

$$1 - \eta_{ext}\frac{\alpha V_p}{ATr} = \tau T^{1/2}(\sigma_n N_{TA}^0 \sqrt{3}\; k/m^*) \tag{14.11}$$

This gives

$$\tau T^{1/2} = \left( \frac{V_p/ATr}{\sigma_n N_{TA}^0 \sqrt{3}\, k/m^*} \right) \alpha\eta_{ext} + \left( \frac{1}{\sigma_n N_{TA}^0 \sqrt{3}\, k/m^*} \right) \tag{14.12}$$

Hence if $\tau T^{1/2}$ is plotted against $\alpha\eta_{ext}$ the intercept allows the product $\sigma_n N_{TA}^0$ to be inferred. Such plots are shown in Fig. 14.11(b) and capture cross-section values of about $1 \times 10^{-16}$ cm$^2$ are obtained for the known trap densities 1-3 $\times 10^{16}$ cm$^{-3}$.

Although densities, energy levels, and capture cross sections may be determined in the ways described, little can be said about the physical nature of the recombination center; however, the values determined would be compatible with those of copper in GaAs.

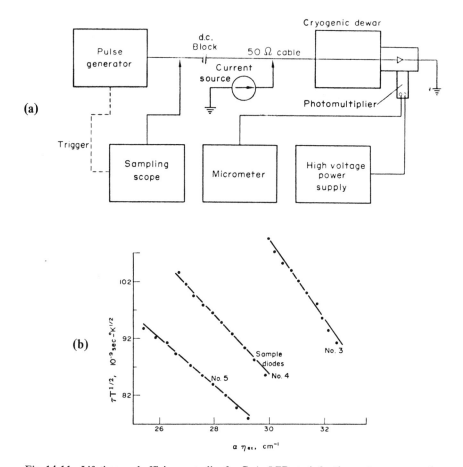

Fig. 14.11. Lifetime and efficiency studies for GaAs LEDs to infer the capture cross section of the dominant recombination center:

(a) Experimental system for measuring the minority carrier lifetime and the external quantum efficiency simultaneously, and

(b) Plots of $\tau T^{1/2}$ vs $a\eta_{ex}$ for diodes allow the product $\sigma_n N_T$ to be determined. (After Lo and Yang. Reprinted from *Solid-State Electronics*, 17, 1974, pp. 115-116; copyright © 1974 Pergamon Press.)

Other studies have been made of recombination levels in GaAs and GaAsP by transient capacitance methods.[74] In general a simple localized center is not sufficient to account for the effects seen and nonlocalized, multiple and coupled level defect centers must be considered. Deep-level junction transient spectroscopy (DLTS) is then a good measurement technique. In LPE, $n$ GaAs energy levels are commonly seen at $E_v + 0.40$ eV and $E_v + 0.71$ eV. Both are presently nonidentified defects and the shallower one seems to be the controlling factor in recombination events. In VPE GaAs, other defects are present, and the one that is frequently the dominant recombination level is at $E_c - 0.39$ eV and appears to be indirectly related to the presence of Ni in the crystal.

## 14.3 GaP LIGHT-EMITTING DIODES

GaP is a semiconductor with an indirect bandgap of 2.26 eV at room temperature and therefore is semitransparent to light and exhibits normally a yellow-brown honey color. The band structure at $0°$K is shown in Fig. 14.12, from which it will be noted that the direct gap $\Gamma_1 - \Gamma_8$ is about 0.5 eV larger than the indirect gap. N-type dopants are group VI elements such as S, Se, and Te which substitute on the phosphorus site. P-type dopants are group II elements such as Zn, Cd, Be, Mg that are one electron short on a Ga site. Si may be either an n- or p-type dopant depending on its site. Sn is n type on a Ga site and Ge is p type since it usually substitutes for phosphorus. Oxygen is a deep donor at $E_C - 0.9$ eV. Some precise energy levels are given in Table 14.2.

In diodes created by Zn or Cd diffusions into n-type GaP the dominant emission is red light with a peak energy of about 1.8 eV and a rather broad spectrum. This is associated with a transition from a Zn-O (or Cd-O) near-

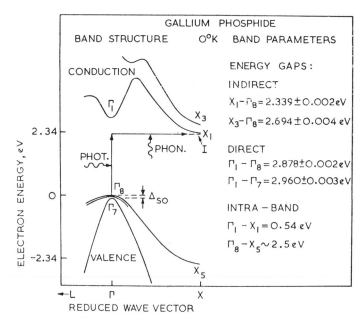

Fig. 14.12. The energy band structure of GaP. The arrows indicate a typical phonon-assisted absorption-edge transition. (After Lo and Yang. Reprinted from *J. Lumin.*, 7, 1973, p. 52.)

**Table 14.2. Ionization Energies of Impurities in GaP**

| Impurity | Donor (eV) | Acceptor (eV) | Ref. |
|---|---|---|---|
| Sn | $E_c -$ 0.065 | | Dean et al., 1970 |
| Si | $E_c -$ 0.082 | | Casey and Trumbore, 1970 |
| Te | $E_c -$ 0.0895 | | Casey and Trumbore, 1970 |
| Se | $E_c -$ 0.102 | | Casey and Trumbore, 1970 |
| S | $E_c -$ 0.104 | | Casey and Brumbore, 1970 |
| Unknown | $E_c - \leq$ 0.165 | | Gershenzon et al., 1965 |
| Unknown | $E_c -$ 0.24 | | Gloriozova et al., 1969 |
| Unknown | $E_c -$ 0.6 | Electron trap | Goldstein and Perlman, 1966 |
| O | $E_c -$ 0.896 | | Dean and Henry, 1968 |
| Cu | | $E_v +$ 0.68 | Bowman, 1967 |
| Co | | $E_v +$ 0.41 | Loescher et al., 1966 |
| Ge | | $E_v +$ 0.30 | Dean 1970 |
| Si | | $E_v +$ 0.203 | Dean 1970 |
| Cd | | $E_v +$ 0.097 | Dean 1970 |
| Zn | | $E_v +$ 0.064 | Dean 1970 |
| Be | | $E_v +$ 0.056 | Dierschke and Pearson, 1970 |
| Mg | | $E_v +$ 0.054 | Dean et al., 1970 |
| C | | $E_v +$ 0.048 | Dean et al., 1970 |
| C | | $E_v +$ 0.041 | Bortfield et al., 1972 |
| N | Isoelectronic trap | $E_c -$ 0.008 | Dean 1970 |
| Bi | Isoelectronic trap | $E_v +$ 0.038 | Dean et al., 1969 |
| Zn–O | Isoelectronic trap | $E_c -$ 0.30 | Cuthbert et al., 1968 |
| Cd–O | Isoelectronic trap | $E_c -$ 0.40 | Cuthbert et al., 1968 |

After, Milnes in *Deep Impurities* in Semiconductors, © 1973, John Wiley & Sons, Inc.

neighbor pair of energy 0.3 or 0.4 eV below the conduction band edge. Oxygen tends to be present in the crystal as grown and its concentration can be increased if necessary by the addition of $Ga_2O_3$. The spectral emission of a Cd-doped *p-n* junction is shown in Fig. 14.13. The red light appears from the *p* side of the junction since this is where the Cd (or Zn) is concentrated. But a small amount of green light is emitted as a result of near bandgap transitions.

The amount of light emitted from early GaP *p-n* junctions was small, but improvements were rapidly made by increasing the fraction of the current carried by injection of electrons into the *p* region and by increasing the (Zn-O) complex concentration to a few times $10^{16}$ cm$^{-3}$. Then followed growth-procedure studies to determine the conditions producing minimum nonradiative recombination processes. In general liquid-phase epitaxy growth has been found to give superior results to vapor-phase epitaxy and to diffused diodes created on crystal-pulled ingot GaP bulk material. Study has been made of nonradiative recombination processes such as auger recombination, where energy is given to another carrier, to the role of excitons, and to the nature of the Zn-O

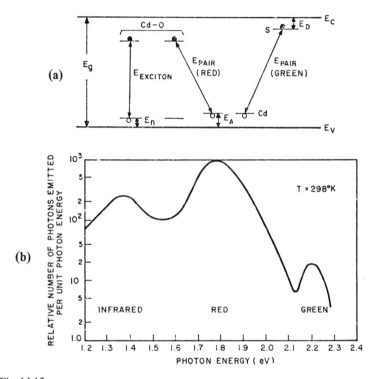

Fig. 14.13.
(a) Energy band diagram for a Cd-doped GaP *p-n* junction. Cd-O is the cadmium-oxygen complex. Transitions between the exciton level of the Cd-O complex to the acceptor level of Cd give rise to the red light emission. Transitions between the donor level (S) and acceptor level (Cd) give rise to the green light emission. (After Henry, Dean, and Cuthbert. Reprinted from *Phys. Rev.,* **116**, 1968, p. 455.)
(b) Measured emission spectrum from a GaP diode. (After Gershenzon. Reprinted from *Bell Syst. Tech. J.,* **45**, 1966, p. 1600; Copyright © 1966, The American Telephone and Telegraph Company.)

complex.[18,19] Figure 14.14 illustrates some recombination processes considered and Fig. 14.15 shows some photoexcited luminescent measurements.

By now much has been learned and red GaP diodes with external quantum efficiencies of 7-12% have been reported.[150,184] Usual external efficiencies, however, are only a few percent. The nonradiative deep centers that are limiting the internal quantum efficiency have yet to be identified with any assurance.

In parallel with the improvement of red-emitting GaP diodes, considerable effort has been applied to the development of green-emitting diodes. Green luminescence is associated with an isoelectronic trap formed by N substituting for P in the *p* side of the junction. The term isoelectronic relates to the fact that N is in the same column of the periodic table as P and on casual consideration

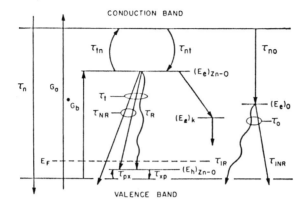

Fig. 14.14. A schematic recombination scheme for minority electrons in $p$-type GaP:Zn,O. The diagram includes generation processes $G_a$ and $G_b$, auger processes at Zn, O traps ($\tau_{NR}$) and deep O donors ($\tau_{INR}$) as well as thermal ionization of the electron ($\tau_{in}$) and hole ($\tau_{xp}$) from the Zn, O traps and the converse processes. The $\tau$'s are all partial lifetimes; the Fermi level is $E_F$. Note the shunt path $\tau_n$ of unknown origin. (After Bergh and Dean. Reprinted with permission from *Proc. IEEE*, **60**, 1972, p. 182.)

might not be expected to be an interesting dopant or trap center. However, since the cores of the N and P atoms are different a short-range perturbation exists which acts as a shallow electron trap (at $E_C - 0.008$ eV). (The band structure of the GaP favors such action because the direct and indirect bandgaps are not too different in value.) Some of the electrons injected into the $p$ side of the junction are trapped by the isoelectronic N and a hole recombines with the trapped electron to give green-emission peaking at about 2.2 eV. Details of the process are summarized in recent reviews.[18, 19, 58] External efficiencies of 0.6% (300°K) have been reported for N concentrations of $10^{19}$ cm$^{-3}$ in $p$-$n$ junctions with Zn, 2 to $5 \times 10^{17}$ cm$^{-3}$, and S, 3 to $6 \times 10^{16}$ cm$^{-3}$. As the eye is about 30 times more sensitive to the green band than to the red (Zn-O) band, the luminous effect of a 0.1% efficient diode is very satisfactory. Such diodes require current densities of about 5 A cm$^{-2}$.

By increasing the N doping to greater than $1 \times 10^{19}$ cm$^{-3}$ and by epilayer growth with a vapor-phase process involving $PH_3$ and GaCl, yellow light emission from GaP is possible with an external efficiency of 0.15% (at 20 A cm$^{-2}$). Si-P complexes or Ga vacancy complexes possibly play a role in such diodes.[216] Yellow light may also be obtained from $GaAs_{0.15}P_{0.85}$:N with good efficiency. GaP that contains Mg-O complexes emits yellow but not efficiently.

A dual-diode structure that emits either red or green depending on the junction that is injecting is shown in Fig. 14.16.

In general the quantum efficiency of a GaP LED increases with increasing current density but peaks at a few tens of A-cm$^{-2}$ as the radiative centers

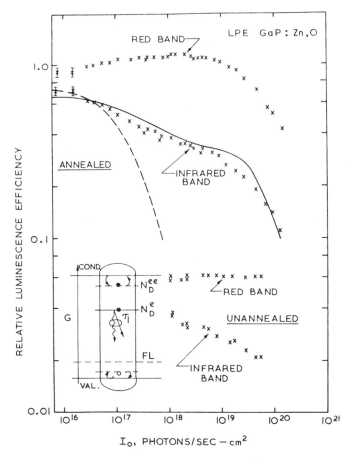

Fig. 14.15. Photoinduced luminescence studies of liquid-phase epitaxial GaP:Zn,O versus excitation intensity (300°K) provide information on luminescence mechanisms. Red and infrared excitation is used. The points are experimental. The dashed curve is the theoretical saturation curve assuming the deep (O) donor can bind but one electron as $N_D^e$ (inset), while the solid theoretical curve assumes that a second electron can be tightly bound to $(O)_p$, as $N_D^{ee}$, by an energy $> 0.4$ eV. The saturation of the infrared luminescence, when the red luminescence stays constant (unannealed) or increases (efficient annealed material) is attributed to auger nonradiative recombinations of the $N_D^{ee}$ state with either a bound or a free hole. These recombinations set an intrinsic limit to the red photoluminescence efficiency in optimally doped and annealed material, estimated to be about 40% compared to $\sim$17% observed experimentally from the best available l.p.e. material. The superlinear dependence for the red luminescence below $10^{18}$ photons sec$^{-1}$ is attributed to saturation of an unidentified recombination center. (After Dean. Reprinted from *J. Lumin.*, 7, 1973, p. 58.)

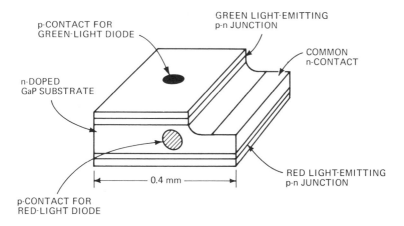

Fig. 14.16. Glowing results. This LED consists of two layers on an *n*-doped gallium-phosphide substrate: one is nitrogen-doped for green light, the other zinc-doped for red. (Reprinted from *Electronics*, 1974; Copyright © McGraw-Hill, Inc., N.Y., 1974.)

become saturated. As the diode becomes hot the efficiency decreases, say from 2% to 0.5% for a red diode over the temperature range 20° to 100°C. The minority carrier lifetime in GaP LEDs varies from a few nanoseconds to some hundreds of nanoseconds depending on the recombination centers present.[12, 55] The turn-on and turn-off times of the light therefore vary over a comparable range.

Degradation of brightness of LEDs with operating time or with current density is not fully understood but is presumably caused by accumulation of recombination (killer) centers in the junction region. Passivation with an oxygen treatment reduces the degradation rate as shown in Fig. 14.17(a). The presence of a fast-diffusing deep-level impurity such as Cu adds to the degradation rate as shown in Fig. 14.17(b). Degradation in well prepared, relatively Cu-free, diodes may have other (yet to be defined) causes. Useful lifetimes of many thousands, or even tens of thousands, of hours are obtainable in normal operating conditions for well-fabricated diodes.

## 14.4 OTHER LIGHT-EMITTING MATERIALS

$Al_xGa_{1-x}As$ becomes indirect gap when $x$ is about 0.35 and the bandgap is then 1.8 eV (see Fig. 14.5). This allows light emission at the red end of the spectrum with a performance comparable to that of GaAsP (see Fig. 14.6). However, GaAsP diodes have proven satisfactory and there has been little

**(a)**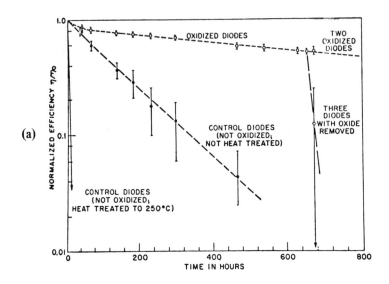

**(b)**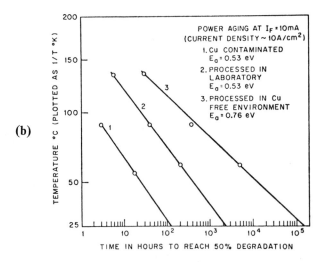

Fig. 14.17.  Degradation effects in red LPE GaP diodes:
(a)  The time change in external quantum efficiency at 200°C and 8 A cm$^{-2}$ forward current. Previous heat treatment to 250°C for 2 hours dramatically accelerates the degradation rate for the control (unoxidized) diodes. The degradation rate is greatly reduced
in diodes which are initially oxidized (surface passivation) at 20°C, then heat treated
at 250°C. Removal of the oxide layer during the degradation study produces a pre-
cipitous increase in the degradation rate. (After Hartman, Schwartz, and Kuhn. Reprinted with permission from *Appl. Phys. Lett.,* 18, 1971, p. 305.)
(b)  Activation energy plot of degradation with and without Cu ions present. (After Bergh.
Reprinted with permission from *IEEE Trans. Electron Devices,* ED-18, 1971, p. 168.)

commercial interest in LEDs of AlGaAs (although there has been a great deal of interest in heterojunction lasers involving GaAs-AlGaAs). Materials studies are in progress to assess recombination actions in AlGaAs and to consider isoelectric trap possibilities in the material when of indirect gap composition.

External quantum efficiencies well over 15% have been attained in the infrared (8100 Å) with double epitaxial $n$-$p$ junctions of AlGaAs on GaAs substrates.[8] Such diodes are useful with photodiodes as fast photon-coupled devices. The switching times are less than 10 ns. The current densities are large ($10^3$ A cm$^{-2}$) and a typical diode current is 0.2 A. However, no degradation occurs in times of tens of thousands of hours for room temperature operation at this current density.

SiC light-emitting junctions are difficult to fabricate because of the high temperatures involved (1650-2500°C). N is a typical $n$ dopant and Al and B are $p$ dopants. The bandgaps of typical polytypes are 2.3-3.1 eV and therefore junctions in high bandgap material can, in principle, emit blue light. However, the observed luminescence is generally yellow or green since the dopants available are not very shallow. The quantum efficiencies are in the $10^{-5}$ range. Operation at high temperatures such as 400°C is possible. Since SiC is an intractable material to work with, interest in such diodes at present is very small.[19]

Light emission studies with II-VI compounds have also been relatively unrewarding. The main problem here is that of achieving satisfactory doping conditions. Only CdTe can be made $p$ and $n$ type with any ease and this is too small in bandgap ($\sim$1.5 eV) to emit visible radiation. ZnTe (2.3 eV) can only be made $p$ type and the other II-VIs such as ZnSe (2.67 eV) and ZnS (3.58 eV) can only be made $n$ type because of compensating effects of native defects (or very lightly $p$ type by special processes). Thus homojunction formation is difficult or impossible. Heterojunctions of II-VI compounds tend to give almost no light emission because of lattice-mismatch-induced interface states. Graded junctions of ZnSe and ZnTe, or $p$-$n$ junctions in the mixed crystal ZnSe$_{0.36}$Te$_{0.64}$ give $10^{-5}$ photons per electron at room temperature.

Diodes formed by Au or Ag rectifying contacts on a thin high-resistance layer (an MIS structure) of ZnS (on a low resistivity $n$-ZnS substrate) give blue light emission for forward bias voltages of 5 V or more. The quantum efficiency (external) is about $5 \times 10^{-4}$ at 300°K. The possibility of further improvement does not seem great. In MIS structures the high fields in the insulating region produce hot carriers and avalanche effects that result in the injection and recombination effect.

Blue light has also been observed from GaN (bandgap $\sim$ 3.5 eV) with external quantum efficiencies of $2 \times 10^{-4}$. The material tends to be heavily $n$ type and presents junction fabrication problems.

Blue luminescence is possible from rare-earth phosphors coated on the surface of infrared emitting GaAs (or Al$_x$Ga$_{1-x}$As) diodes. Red and green phosphors (such as $NaYF_4$:Yb,Er) require two absorption pumping steps by 1.4-eV photons to allow the required transitions and are not very efficient on this account.[182] But blue emission requires three pumping steps and so is even less efficient (see Fig. 14.6).

Visible light emission is obtainable from ac excitation of thin powder layers of polycrystalline ZnS that has been treated with activators such as Cu, Cl for blue and green luminescence and Cu, Cl, and Mn for yellow luminescence. The power efficiencies tend to be in the range $10^{-2}$ to $10^{-3}$ and the cells degrade in times of some hundreds or thousands of hours even when hermetically sealed. The presence of the Cu creates narrow conducting $Cu_2S$ decorated imperfection lines that cause high field concentrations as in ② in Fig. 14.18. The field concentration then creates field emission of carriers which become trapped. Decay of the field is associated with emergence from the traps and recombination with the emission of radiation. The brightness of the luminescence for ac excitation is found to vary with voltage $V$ as

$$B = B_1 e^{-(V_1/V)^{1/2}} \qquad (14.13)$$

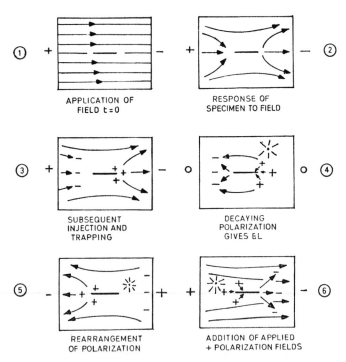

Fig. 14.18. Probable sequence of events occurring during ac excitation of ZnS powder suspensions. (After Thornton, from *Physics of Electroluminescent Devices*; Copyright © 1967, E. & F.N. Spon Ltd.)

and the light increases with increasing ac frequency. However, the degradation rate also increases. It is as though there is only a finite amount of light available from a powder electroluminescence panel within its operating life and that this may be taken either at low intensity for a long period of time or at higher intensity for a shorter period of time. Perhaps the lattice-coupled energy of a fraction of the nonradioactive recombination events creates ionic movement or native defects that form nonradiative recombination sites.

DC excited powder layer cells in which minority carrier injection must occur by some mechanism, also have been studied.[214] Good brightnesses can be obtained and power efficiencies of up to $5 \times 10^{-3}$. Degradation in 1000 hours may be by a factor of two from a level of 1000 fL at 100 V. The voltage level required is much higher than for LEDs of GaAsP or GaP and this limits the potential uses to large area displays.

Infrared to visible light up-conversion is possible by multilayer heterojunction structures of GaAs and GaAlAs as shown in Fig. 14.19. A light-detecting diode of $p$-$n$ GaAs receives incoming infrared radiation of photon energy greater than the GaAs bandgap. The electrons created by the infrared are multiplied by avalanche action, raised in energy and pass into the larger bandgap GaAlAs $n$-$p$ junction where recombination takes place with the emission of red light. The $p\text{Ga}_x\text{Al}_{1-x}\text{As}$ layer is slightly increased in bandgap ($E_{G2}' > E_{G2}$) to allow easy extraction of the light. The $p$GaAs layer on the extreme right is to provide ohmic contact to the $\text{Ga}_x\text{Al}_{1-x}\text{As}$ and is etched very thin so that it absorbs very little light.

The conversion efficiency (photons out/photons in) is

$$\eta_C = \eta_D m \eta_{LED} \tag{14.14}$$

where $\eta_D$ is the number of photocarriers primarily generated in the detector diode divided by the number of incident photons, $m$ is the avalanche multiplication factor, and $\eta_{LED}$ is the number of photons emitted by the LED divided by the number of electrons flowing into the LED. Typically $\eta_D$ may be 0.3, $m$ is 6, and $\eta_{LED}$ is 1% so the overall photon conversion efficiency is 2% at 300°K.

## 14.5 APPLICATIONS OF LIGHT-EMITTING DIODES

Some steps in the fabrication of light-emitting GaP diodes are shown in Fig. 14-20. The ohmic contact forms only a small part of the back surface, which is otherwise $SiO_2$ coated to allow internal reflections of the radiation from the rear to enhance the useful light output. The small diode may be mounted in a metallized reflector (see Fig. 14.21), the facets of which produce an uneven distribution that increases visibility. Sometimes filters must be used over the LED to improve the contrast between emitted and reflected light

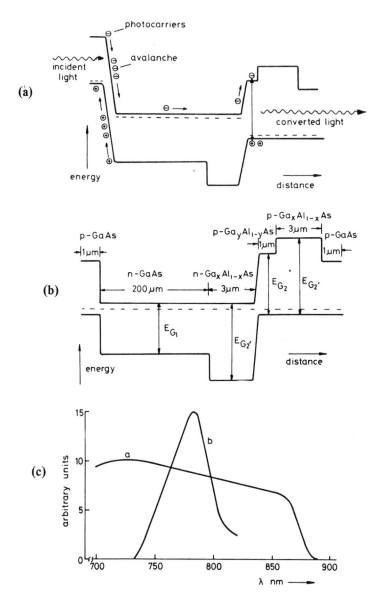

Fig. 14.19. Up-conversion of light in energy with a GaAs detector injecting electrons into an AlGaAs LED:
(a)  Energy band diagram and concept,
(b)  Structure, and
(c)  Wavelength dependence of detector photocurrent and emission spectrum of convertor. Curve a is the detector photocurrent and curve b is the emission spectrum. (After Beneking, Schul, and Mischel. Reprinted with permission from *IEEE Int. Electron Devices Meeting, Technical Digest*, 1974, p. 69.)

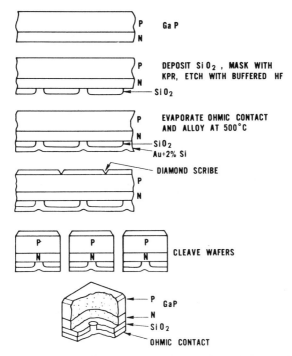

Fig. 14.20. Schematic diagram showing preparation of GaP EL diode with a composite contact. (After Bergh and Strain, from *Ohmic Contacts to Semiconductors*; Copyright © 1969, Electrochemical Society, p. 127.) This figure was originally presented at the 134th Fall 1968 Meeting of The Electrochemical Society, Inc., held in Montreal, Canada.

so that the distinction between "on" and "off" in reference to other diodes becomes more obvious.

The seven-bar pattern as shown in Fig. 14.22(a) is one of the most common numeric displays. It may consist of one or two diodes per bar, or of a mesa etched bar structure as shown in Fig. 14.22(b). An assembly of diodes on a circuit board with a metallized reflector cover plate is shown in Fig. 14.22(c). Recent studies of fabrication methods have resulted in much improved surface brightness.[226,227]

As an illustration of drive circuits, Fig. 14.23 shows part of the circuit for an LED watch display. The market for LEDs in the United States in recent years has been over $100 million in a total display market of $300 million/year. The CMOS integrated circuit chip performs timekeeping and multiplex drive functions. More discussion of watch circuits has been given in Chapter 9.

Optical couplers are composite devices in which a light-emitting diode, often an infrared GaAs device, couples into a Si-detecting device to provide signal transmission with complete electrical isolation between the two parts of the

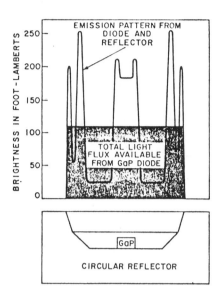

Fig. 14.21. Schematic diagram of a GaP-reflector LED structure along with the emission pattern of the reflector. For a 1% diode at 10 mA, the shaded area represents the total emitted light flux. The facets of the reflector produce an uneven distribution of light with much higher peak intensities than the average level of emission. (After Bergh and Johnson. Reprinted with permission from *Bell Lab Record,* 47, 1969, p. 323; Copyright © 1969 American Telephone & Telegraph Co.)

circuit. The output device may be a diode but is more usually a phototransistor or a photo-Darlington as shown in Fig. 14.24(a) and (b) or dual SCRs or a bidirectional SCR as shown in Fig. 14.24(c) and (d).

Some impressions of the uses of photocouplers may be gathered by inspecting the circuits and the associated text of Figs. 14.25 and 14.26. Achievement of circuit isolation without the use of transformers, that may be expensive, heavy, and frequency-limiting, is a considerable boon.

## 14.6  HETEROJUNCTION $Al_xGa_{1-x}As$-GaAs INJECTION LASERS

Laser action was first observed in Cr doped $Al_2O_3$ in 1960. This was followed by the development of gas lasers in 1961 and of GaAs diode lasers in 1962. The evolution of GaAs injection lasers up to about 1968 is reviewed in the book *Gallium Arsenide Lasers*, edited by Gooch.[80] A recent book by Kressel[120] brings the account up to 1977.

Injection diode lasers at first were slow to find useful applications in competition with other laser systems, primarily because of low power capability and the need for operation below room temperature. With the development of heterojunction confinement lasers, continuous-wave (CW) operation has been

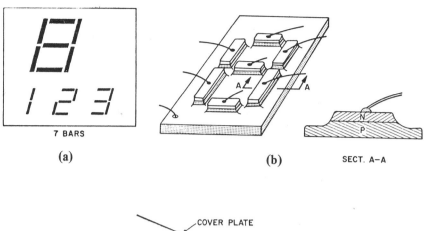

(a)          (b)          SECT. A-A

7 BARS

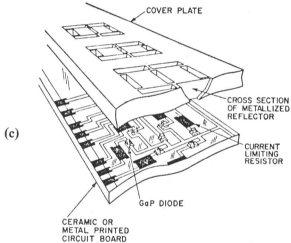

(c)

COVER PLATE

CROSS SECTION OF METALLIZED REFLECTOR

CURRENT LIMITING RESISTOR

GaP DIODE

CERAMIC OR METAL PRINTED CIRCUIT BOARD

Fig. 14.22. Numeric display assembly of the seven-bar format. The GaP diodes and the Si integrated circuits are mounted on a ceramic or metal printed circuit board. A plastic or ceramic cover plate distributes the light via the metallized reflectors. (After Bergh and Dean. Reprinted with permission from *Proc. IEEE,* **60,** 1972, pp. 215-216.)

achieved at room temperature and above. The threshold current density for lasing action has been lowered by a factor of at least 20 from 30 A for 1 W of output to 1.5 A, because of the greater ease of achieving high carrier densities and high photon flux in the recombination region.

The advantages of injection lasers include small size and ease of modulation by signals superimposed on the injected current. Factors, however, that continue to limit the application of injection lasers are that they do not have the narrow spectral line width, large exit aperture size, narrow beam angle, and coherence of competing lasers.

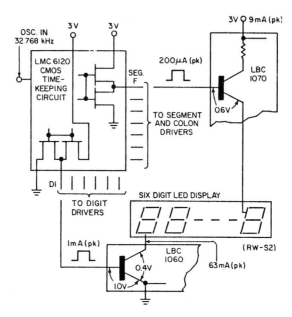

Fig. 14.23. Drive circuits for a six digit 7 bar LED display in a watch. (After Laws & Ady. Reprinted with permission from *Electronic Design*, 1974, **v.22** #23, p. 89.)

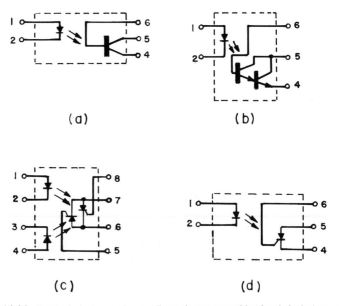

Fig. 14.24. Typical photocoupler configurations to provide circuit isolation.

Heterojunction diode laser concepts began to develop in 1963 with the concept that injected carrier pile-up, or confinement, in the narrow active region would make the achievement of population inversion possible at lower current densities. It was also recognized that dielectric confinement or waveguiding of the photon flux within the semiconductor optical cavity region would be advantageous. Some consideration was also given to the possibility that heterojunctions might open the way to the achievement of laser action in indirect-gap semiconductors. For instance, it was speculated that a GaAs-Ge heterojunction might be used to achieve direct injection into the $k = (000)$ valley of Ge. However, studies show that the residence time in this valley is extremely short before scattering to the indirect-gap valley, and laser action from indirect-gap materials such as Ge and Si is not likely to be achieved in this way.

Physical realization of the heterojunction laser concept in relation to GaAs depended on finding a wider gap semiconductor with an excellent lattice match (to minimize interface state recombination) and with suitable barriers and refractive index conditions to provide both carrier and photon confinement. Such a material exists, $Al_x Ga_{1-x} As$, but some years of work were necessary with this alloy system before striking results were obtained. The bandgap of AlGaAs is shown in Fig. 14.5 from which it is seen that an Al content of 0.2-0.4 is adequate to give a bandgap difference of sufficient size to represent a good heterojunction barrier against GaAs. As seen from Fig. 14.27 the lattice constant match between GaAs and AlAs is excellent and so the heterojunction interface state density should be low and not form a surface of large recombination velocity. The refractive index changes also are favorable for waveguiding action since the 20% Al alloy has an index of about 3.27 compared with 3.43 for GaAs.[103]

From barrier studies that have been made it appears that the energy discontinuity in the valence band is small for AlGaAs-GaAs junctions ($\Delta E_v \sim 0.15 \ \Delta E_g$).[67] (This is compatible with the concept illustrated in Fig. 2.12 that the anion (As) determines the valence band energy.) It follows that in the conduction band of an abrupt heterojunction the barrier $\Delta E_c$ is nearly equal to the difference in energy gap of the heterojunction components.

For $Al_{0.3}Ga_{0.7}As$-GaAs, the value of $\Delta E_c$ is about 0.4 eV and the bandgap of the $Al_{0.3}Ga_{0.7}As$ is 1.85 eV. The energy band diagrams that may be expected therefore are approximately as shown in Fig. 14.28. For an $n$-$p$ $Al_x Ga_{1-x} As$-GaAs junction the band diagram with zero voltage applied is Fig. 14.28(a). The bending of the bands shown is caused by electrons moving from the $n$-$Al_x Ga_{1-x} As$ into the $p$-GaAs to provide alignment of the Fermi levels on the two sides of the junction. Interface state effects are assumed negligible in this diagram, which is probably permissible because of the excellent lattice-match conditions. Application of a forward voltage, the $Al_x Ga_{1-x} As$ being made negative, results in the band diagram of Fig. 14.28(b). Here the Fermi levels are separated by an energy corresponding to an applied voltage $V_a$, and a

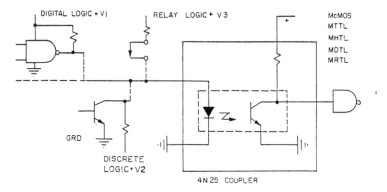

### 1. Logic-to-Logic Interfacing

Optical couplers afford total flexibility in logic family interfacing. Logic supplies can float with respect to each other, ground loops and intricate interface techniques involving voltage translators are eliminated.

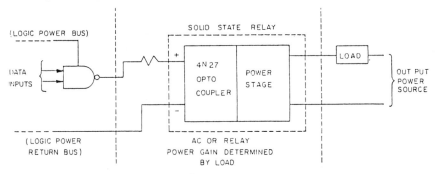

### 2. Logic-to-High Power Interfacing

One unidirectional optical coupler minimizes transient feedback from high power loads conventionally isolated through bidirectional transformers or RC coupling.

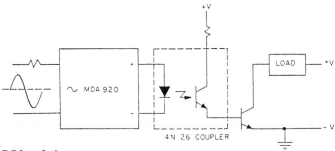

### 3. AC-to-DC Interfacing

AC signals actuate the optical coupler which controls the current through the DC load. Highly economical, the coupler replaces step-down transformers and obviates transient feedback.

Fig. 14.25. Some applications of photocouplers. (Courtesy Motorola Corp. Reprinted from *Electronics*, Nov. 8, 1973; Copyright © McGraw-Hill, Inc., N.Y. 1973.)

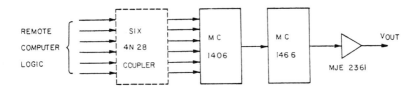

### 4. Remote Control Of Digitally-Programmed Power Supply
100-billion-ohm isolation affords control of a remote floating power source from the computer/peripheral without intricate biasing networks.

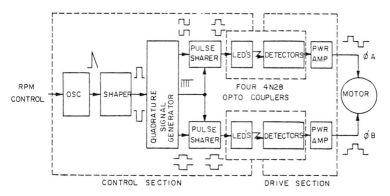

### 5. Logic-to-AC Control/Motor
Easy, economical control of high-level AC power without electromechanical relays and transformers is possible using low-level, 500-2500 V optical isolation.

Fig. 14.25. (continued)

quasi-Fermi level is shown in the $p$-GaAs to represent the density of injected electrons before they recombine with holes farther from the interface. Since in Fig. 14.28(b) the Fermi level spacing $\delta_2$ is less than $\delta_1$, it may be inferred that the density of injected electrons in the GaAs is greater than the electron density in the $n$-$Al_xGa_{1-x}As$, that is providing the injection (the densities of states in the $Al_xGa_{1-x}As$ and in the GaAs are assumed to be not very different). This effect is known as "superinjection" and is a special feature of hetero-junctions.[144]

The energy band diagram for $p$-$n$ rather than $n$-$p$ junctions of $Al_xGa_{1-x}As$-GaAs is shown in Fig. 14.28(c). From the barriers that exist, it is apparent that current flow in this junction is predominantly by injection of holes into the $n$-GaAs. The energy diagram for a $p$-GaAs/$p$-$Al_xGa_{1-x}As$ is shown in Fig. 14.28(d) and $\Delta E_c$ is seen to be the main barrier in the conduction band.

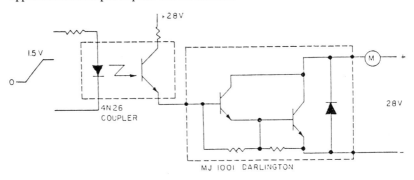

### 6. Zero-Crossing Detection

For applications requiring use of line voltage for synchronization purposes, an economical approach uses a coupler in place of a transformer.

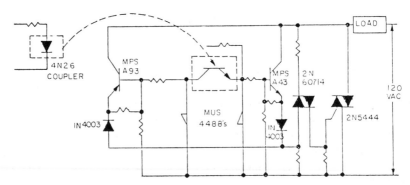

### 7. Logic-to-DC Motors

Traditional, long-term solid-state operation without arcing, bounce or wear-out is ensured with a no-moving-parts optical coupler and power Darlington transistor.

### 8. Logic-to-AC Relay

Control of AC loads from logic is easily implemented using optical isolation. Speed for zero-crossing actuation unavailable through E-M means, is provided by the total solid-state approach.

Fig. 14.26. Further circuit applications of photocouplers. (Courtesy Motorola Corp. Reprinted from *Electronics*, Nov. 8, 1973; Copyright © McGraw-Hill, Inc., N.Y. 1973.)

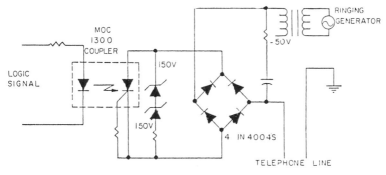

### 9. Telephone Bell Ringing Actuator
Limited contact life, high maintenance and EMI problems are eliminated with an opti-cally-coupled bell actuator.

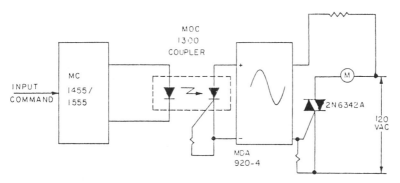

### 10. Long Time Delay Relays
Processing requirements needing precise mixing/metering through time delay can be met using the MC1555 timer and a coupler to isolate the logic/AC power source.

Fig. 14.26. (continued)

If a double heterojunction structure $n\text{-Al}_x\text{Ga}_{1-x}\text{As}/p\text{-GaAs}/p\text{-Al}_x\text{Ga}_{1-x}\text{As}$ is considered, the resulting energy band structure may be envisaged by imagining Figs. 14.28(b) and (d) placed side by side. Hence $\Delta E_c$ in Fig. 14.28(d) will act as a confinement barrier for injected electrons in the GaAs and the valence band barrier of Fig. 14.28(b) will provide confinement of holes.

In a heterojunction laser structure, two confinement actions have to be considered. There is the confinement of the injected carriers by the energy barriers in the conduction and valence bands. Also there is the waveguide confinement of the photons caused by the refractive index changes at the GaAs-$\text{Al}_x\text{Ga}_{1-x}\text{As}$ interfaces. Both confinement effects contribute to the lowering of the threshold current density for laser action. The carrier confine-ment predominantly controls the population inversion and therefore the gain $\beta$

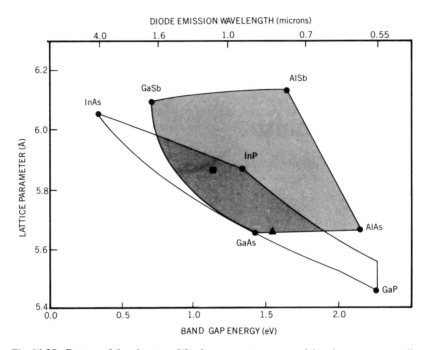

Fig. 14.27. Extent of bandgaps and lattice parameters covered by the quaternary alloys $Al_xGa_{1-x}As_ySb_{1-y}$ and $In_xGa_{1-x}As_yP_{1-y}$. The boundaries of the areas represent ternary, and the vertices binary, alloys. Two particularly useful alloys are indicated: The square (■) locates $In_{0.80}Ga_{0.20}As_{0.35}P_{0.65}$, the lattice parameter of which matches InP substrates and makes diodes that emit at 1.1 microns; the triangle (▲) shows the position of $Al_{0.1}Ga_{0.9}As$ which matches the GaAs lattice and emits at about 0.82 microns. (After Kressel, et al. Reprinted with permission from *Physics Today*, May 1976, p. 40.)

of the laser cavity, and allows the width of the GaAs active region to be made narrower. The refractive index confinement reduces the photon losses from the laser cavity and relates therefore mainly to the loss term $\alpha$ in laser analysis.

A comparison of homojunction, single-heterojunction, and double-heterojunction structures is shown in Fig. 14.29. The confinement barriers in a homojunction structure are seen to be quite small and are primarily a consequence of the doping differences. The spread of the light on either side of the active region is seen to be quite large. In the single-heterostructure laser there is an electron confinement barrier of about 0.4 eV and roughly a 5% decrease in refractive index in passing from the $p$-GaAs to the $p$-$Al_xGa_{1-x}As$. This confines the spreading of the light at this interface. However, at the $n$-$p$ GaAs interface there is only a small change of refractive index and considerable photon loss occurs into the $n$-GaAs. At high bias conditions, with thin widths of the $p$-GaAs region, injection of holes may occur into the $n$-GaAs and laser action may be affected.

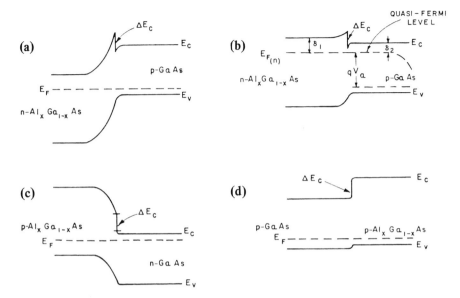

Fig. 14.28. Energy band diagrams of $Al_xGa_{1-x}As$-GaAs heterojunctions assuming $\Delta E_c \sim$ 0.4 eV and $\Delta E_v$ is negligibly small:
(a)   For an $n$-$p$ $Al_xGa_{1-x}As$-GaAs junction, with no voltage applied. Grading of the Al content over a few hundred angstroms at the interface would reduce the effect of $\Delta E_c$,
(b)   Forward voltage $V_a$ applied to junction (a). The quasi-Fermi level in the $p$-GaAs represents the injected electron density, which may be greater than the electron density of the emitter,
(c)   Energy band diagram for a $p$-$n$ $Al_xGa_{1-x}As$-GaAs junction, with no voltage applied, and
(d)   Energy band diagram for $p$-$p$ heterojunction, neglecting interface state effects. (After Milnes and Feucht, from *Heterojunction and Metal-Semiconductor Junctions*, 1972, p. 145; Copyright © 1972, Academic Press, N.Y.)

For double-heterojunction structures, as shown in Fig. 14.29, carrier and photon confinement can be expected for both sides of the active region.

Some typical double-heterojunction structures are shown in Fig. 14.30, where in Fig. 14.30(a) the active region is $p$-GaAs and in Fig. 14.30(b) the active region is $Al_{0.1}Ga_{0.9}As$ and thus provides light of shorter wavelength (although still in the infrared). The lasers have cleaved faces that form a reflecting Fabry-Pérot cavity for buildup of the photon flux. The contact stripe $S$ provides a convenient way of obtaining a small device area with some lateral heat flow that allows a high CW operating temperature. The optimum stripe width is between 10 and 15 $\mu$m for typical laser structures. The lasing region of the junction may be somewhat wider than the stripe width because of current spreading. Typically

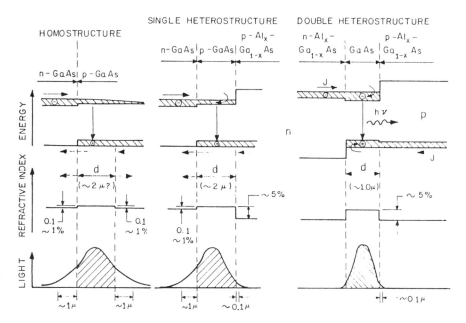

Fig. 14. 29. Physical structure, bandshapes under operating voltage, refractive index steps, and optical power distribution in homostructure, single-heterostructure, and double-heterostructure laser diodes. (After Hayashi, Panish, and Reinhart. Reprinted with permission from *J. Appl. Phys.*, 42, 1971, p. 1920.)

the laser is mounted with the stripe side down on a metallized diamond heat sink having five times the thermal conductivity of copper.

In order to achieve optical gain in semiconductors, the stimulated emission rate must exceed the absorption rate. The common laser condition of "population inversion" between upper and lower energy levels can be conveniently expressed for semiconductors by using Fermi statistics for the injecting nonequilibrium condition, which implies the assumption of separate quasi-Fermi levels, $E_{F_u}$ and $E_{F_\ell}$ for electrons and holes. The condition for stimulated emission is then

$$E_{F_u} - E_{F_\ell} > E_u - E_\ell \qquad (14.15)$$

For direct band-to-band transitions the laser energy is $\hbar\omega = E_u - E_\ell$ and degeneracy of at least one band is necessary to satisfy Eq. (14.15).[168] If the laser transition occurs between impurity levels, no degeneracy is needed even for direct transitions and the laser condition is

$$np > n_i^2 e^{\hbar\omega/kT} \qquad (14.16)$$

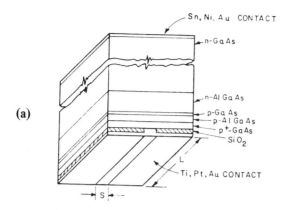

**(a)**

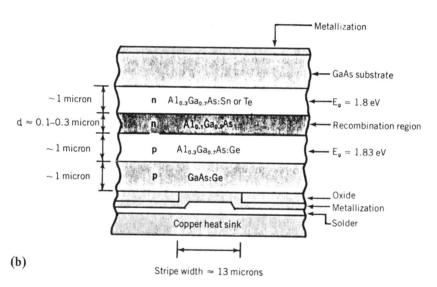

**(b)**

Stripe width ≈ 13 microns

Fig. 14.30. Stripe geometry double-heterojunction injection lasers:
(a)  GaAs active region. (After Dyment and D'Asaro. Reprinted with permission from *Appl. Phys. Lett.,* **11**, 1967, p. 292.)
(b)  $Al_{0.1}Ga_{0.9}As$ active region. (After Kressel. Reprinted with permission from *Physics Today*, May 1976, p. 42.)

The dependence of the threshold current density upon the cavity length $L$ may be expected to follow the relationship:

$$J_{th} = (1/L\beta) [\alpha L + \ln (1/R)] \qquad (14.17)$$

where $\alpha$ is the internal loss per unit length, $\beta$ is the gain factor per unit length and per unit current density, and $R$ is the reflectivity coefficient.[169] Observed variations of $J_{th}$ with $1/L$ for heterojunction AlGaAs-GaAs lasers at 300°K are shown in Fig. 14.31 and from such curves the gain and loss terms $\alpha$ and $\beta$ may be estimated.

For general use, laser lengths are typically in the range 200-500 $\mu$m. From the $\alpha$ and $\beta$ values given, it is seen that the successive degrees of confinement decrease the loss from 60 to 10 cm$^{-1}$ and increase the gain factor from 1.7 to 20 cm$^{-1}$kA$^{-1}$. Further study of double-heterojunction lasers suggest that the data may be fitted by a straight line corresponding to $J_{th}^2$ proportional to $1/L$. The gain is found to be an increasing function of the operating current density according to a power law $J^m$, where $m$ is two or somewhat higher.

The threshold current density in a double heterostructure is almost inversely proportional to the thickness, $d$, of the active region, as shown in Fig. 14.32. The smaller the active thickness that has to sustain population inversion, the smaller the current and the current density needed to provide it. For a single-heterostructure laser however, loss of holes occurs at 300°K if the active thickness $d$ is made smaller than about 2 $\mu$m because of imperfect confinement at the $n$-$p$ GaAs interface, and therefore the threshold current density passes through a minimum as shown by the upper curve in Fig. 14.32(a).

A spectral response curve for a stripe laser structure [see Fig. 14.30(a)] with a GaAs active region is shown in Fig. 14.33 at very high resolution. Many

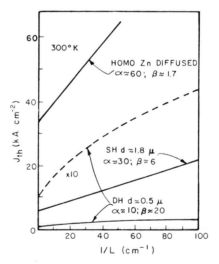

Fig. 14.31. Threshold current density $J_{th}$ versus $1/L$ for homostructure, SH, and DH GaAs lasers. The units for the loss term $a$ are cm$^{-1}$ and for the gain term $\beta$ are cm$^{-1}$kA$^{-1}$. (After Panish and Hayashi, from *Proc. International Conf. of Semiconductors, Heterojunctions, Layer Structures*, Budapest, 2, 1970, Hungarian Academy of Sciences.)

longitudinal and spatial modes are seen: the separations $(\Delta\lambda)_q$ correspond to longitudinal modes and $(\Delta\lambda)_n$ are between transverse modes. These separations can be calculated from Fabry-Pérot cavity resonance equations.[56]

A more recent treatment of the power spectrum of an injection laser begins with:[197]

rate equations in which the gains and losses are approximated as prorated averages over one round trip. The gain contains an explicit nonlinearity described by a critical power at which the gain begins to saturate. The steady-state solution at low level is the linear theory. At high level, the excited power in each cavity mode is an explicit function of its particular gain and loss coefficients. To express the characteristics of the radiation in terms of observables, the dependence of gain and loss coefficients on mode number are deduced by modeling. The unsaturated gain is taken as parabolic in frequency, while the loss coefficients are deduced by decomposing each cavity mode into three independent sets of waveguide modes. For a cavity with sawed side walls, this gives a scattering loss proportional to the square of the lateral mode number, which determines the shape of the lateral profile of the beam. The summation of the modal powers over all cavity modes expresses the lateral profile, the frequency spectrum, and the total power in terms of an overdrive parameter $X$ associated with the dynamical state of the laser, which is an implicit function of the population inversion. Relating $X$ to the spontaneous emission at short wavelength gives a set of equations, involving only observables, which determines all the parameters in the summation.

Such equations provide relations between power, spontaneous emission, beam width, spectral width, and polarization. The critical power may be expressed from first principles in terms of phenomenological optical constants.

The far-field pattern of the beam of light from an injection laser depends on the resonant modes and diffraction limits imposed by the rectangular symmetry of the active region. However, under certain conditions, the distribution can be Gaussian which is convenient for subsequent optical processing. For a homostructure or a single-heterostructure laser the beam spread perpendicular to the junction plane may correspond to a half-angle of 10-15 deg. For effective collimation, an optical system with an aperture of about $f/2$ must be used.[80] Single-heterostructure lasers, with spectral line widths $\sim$20 Å, have been shown to be capable of retrieving holographic information with about $200 \times 200$ resolvable lines.[73] In double-confinement structures with very narrow active-region thicknesses the half-angle for the beam spread is 40 deg or more and the collimation problem is correspondingly more difficult.

The lasing light output from a double-heterojunction diode is generally polarized with the optical electric field vector of the radiation parallel with the plane of the heterojunction. This corresponds to TE modes within the laser

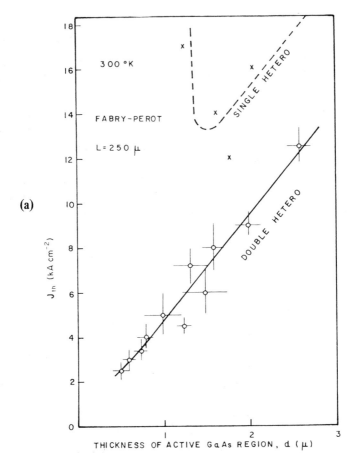

**(a)**

Fig. 14.32. Variation of threshold current density with active region width for heterojunction lasers:

(a)   For GaAs active region. (After Panish and Hayashi. Reprinted with permission from *J. Appl. Phys.*, **42**, 1974, p. 1932.)

(b)   For $Al_{0.1}Ga_{0.9}As$ active region. (After Kressel, et al. Reprinted with permission from *Physics Today*, May 1976, p. 42.)
      $\Delta x$ of 0.20 (triangles) and 0.65 (circles). The theoretical curves are for the discontinuities in the refractive index $\Delta n$ shown where the relation $\Delta n = 0.62\Delta x$ has been assumed to apply.

cavity and is in contrast to single-heterojunction or homostructure lasers in which the field distribution may be approximated by TEM modes and the emission generally does not have a well-defined polarization.

In an injection laser there is a rapid increase in light output as the current is increased above threshold (see Fig. 14.34). At rated current the 300°K overall quantum efficiency (photons per electron) may be 1 or 2%. As the temperature

**(b)**

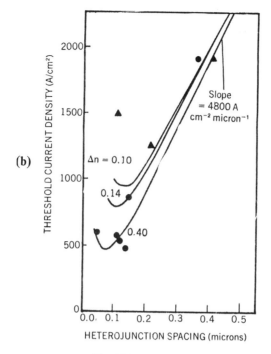

Fig. 14.32. (continued)

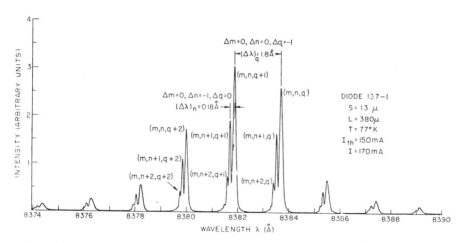

Fig. 14.33. GaAs injection laser (Fabry-Pérot). Frequency spectrum with excitation of many longitudinal modes $(q)$ and transverse modes along the junction plane $(n)$. All the modes have the same transverse mode number perpendicular to the junction plane $(m)$. (After Zachos and Ripper. Reprinted with permission of Bell Laboratories from *IEEE J. Quantum Electronics*, QE-5, 1969, p. 34.)

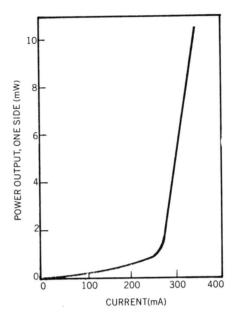

Fig. 14.34. Optical power output from one facet of a typical AlGaAs CW laser as a function of the drive current, measured at room temperature. (After Kressel, et al. Reprinted with permission from *Physics Today*, May 1976, p. 45.)

is lowered the efficiency increases and the threshold current at 77°K may be a tenth of that at 300°K.

If the active region of a confinement laser is $Al_x Ga_{1-x} As$ instead of GaAs, the wavelength of the emitted light is shortened and enters the visible range of the spectrum. For the spectral range 9000-8000 Å the quantum efficiency is relatively constant and the threshold current densities are about 1000 A cm$^{-2}$, but for shorter wavelengths the efficiency begins to drop rapidly and is down by a factor of 10 at 7000 Å which is obtained for $x$ about 0.30.[116] Typical devices have stripe widths of 10-20 $\mu$m with 300°K emission levels of 5-15 mW at currents of 350 mA or less.

Two basic degradation modes are seen in injection lasers. These are the internal formation of isolated or clustered nonradiative centers and the development of facet damage on the reflecting faces. Clustered recombination regions begin on dislocation networks and grow by the migration of vacancies or interstitials and produce dark lines seen against a radiating background. The inclusion of a small fraction of Al in the active region, so that it is AlGaAs instead of GaAs, helps reduce the dislocation density. The dislocation density is also reduced by care in the growth and structure fabrication process. Facet damage formation is related to the optical flux density and the ambient on the cleaved face. The use of dielectric coatings on the faces improves matters considerably. The result of all these measures is that lifetimes in excess of 10,000 hours with very little degradation have been logged for 10-mW injection lasers.

Many special laser structures have become possible as greater control is being achieved in liquid-phase epitaxial growth and in various masking and ion-milling techniques. Much of this development is aimed at achieving low threshold current densities in conjunction with lasering-mode pattern control and less beam divergence to match more closely the performance of gas lasers.

One variation of the stripe geometry that has been investigated is the mesa stripe structure—versions of which are shown in Fig. 14.35(a) and (b). These give nearly single-mode operation and threshold currents less than 50 mA were seen in structures of width 6 $\mu$m, length 100 $\mu$m, and active region thickness $\sim$0.5 $\mu$m. However, the threshold current density was above 6 kA/cm$^2$ compared with 1 kA/cm$^2$ for a broader width structure. Technology has also been developed to completely bury the active region as in Fig. 14.36(a) and (b). The buried GaAs lasering region may be as small as 0.5 $\mu$m $\times$ 0.5 $\mu$m and the threshold current as low as 15 mA for a 390 $\mu$m length (this represents a fairly high threshold current density). Such diodes with active regions 1 $\mu$m square or smaller lase mostly in the lowest order transverse mode (TE$_{00}$).

The cleaved faces of GaAs lasers that form Fabry-Pérot cavities have good reflection coefficients ($R$ = 0.31 from $[(n-1)/(n+1)]^2$ for a refractive index of

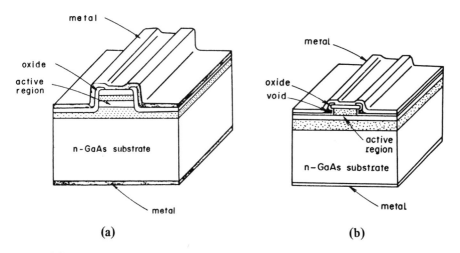

**(a)**                                    **(b)**

Fig. 14.35. Mesa-stripe-geometry double-heterostructure lasers:
(a) High-mesa-type (HMS) lasers in which mesa etching was effected up to the first epitaxial layer ($n$-GaAlAs) of the DH crystal, and
(b) Low-mesa-type (LMS) lasers in which mesa etching was stopped just above the active layer. The fourth layer ($p$-GaAs) overhangs the third layer ($p$-GaAlAs) due to lateral etching of the third layer. (After Tsukada. Reprinted with permission from *Proc. of the 4th Conf. on Solid-State Devices*, Tokyo, 1972. Suppl. to the *Trans. Japan Society of Appl. Phys.*, **42**, 1973, p. 251.)

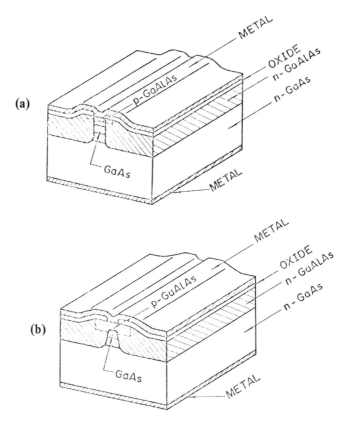

Fig. 14.36. Buried-heterostructure injection lasers. In both (a) and (b) the GaAs active regions are completely surrounded by GaAlAs. The lasers are typically 400 $\mu$m long, 300 $\mu$m wide, and 100 $\mu$m thick. The height of the raised area near the stripe region is about 1 $\mu$m or less. The drawings are not to scale in order to show the various regions clearly. (After Tsukada. Reprinted with permission from *J. Appl. Phys.*, 45, 1974, p. 4900.)

3.5 relative to air). However, the cleaving of the complete wafer to achieve these faces makes it difficult to envisage the integration of such lasers with other devices such as modulator, film waveguide, and beam-steering structures on a single GaAs slice. This difficulty may be resolved by the use of selective etches that cut mirrorlike faces with reflection coefficients of about 0.2-0.25. These give differential quantum efficiencies (photons per increment of current) of 10 to 18% and room temperature threshold current densities of as low as 4.2 kA/cm². This must be compared with 25-30% and about 3.4 kA/cm² for cleaved-face diodes.

However, a more interesting approach is to obtain the required optical feedback by backward Bragg scattering from a periodic perturbation of refractive index and/or gain in the laser waveguide. The perturbation may be achieved by etching a periodic corrugation at the interface of the $p$-GaAs active layer and the $p$-GaAlAs layer in a single-heterojunction structure as shown in Fig. 14.37(a). The grating structure is achieved by interferrometric exposure of photoresist on the $p$-GaAs surface and ion-milling through the resist, followed by the GaAlAs layer growth.

The period ($\Lambda$) of the perturbation is selected from the Bragg condition

$$\Lambda = \frac{m\pi}{\beta} \approx \frac{m\lambda_0}{2n} \tag{14.18}$$

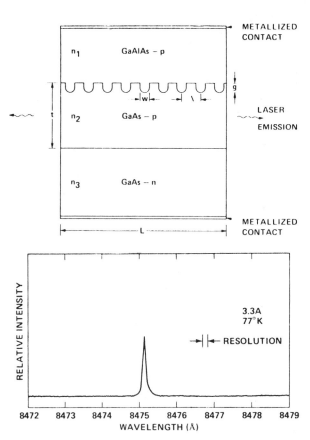

Fig. 14.37. A single-heterojunction AsAs/GaAlAs laser diode with distributed feedback:
(a)   The ion-milled corrugation produces Bragg feedback, and
(b)   Narrow linewidth single-longitudinal-mode laser spectrum for third-order grating feedback (one Fabry-Pérot face has been given an antireflection coating). (After Scifres, Burnham, and Streifer. Reprinted with permission from *IEEE Trans. Electron Devices*, **ED-22**, 1975, p. 610.)

where $\lambda_0$ is the free-space laser wavelength, $\beta$ is the propagation constant of the wave in the guide, $m$ is an integer ($m = 1, 2, 3, \ldots$), and $n$ is the refractive index of the waveguide material.

If third-order Bragg feedback is required $\Lambda$ is about 3600 Å and for fourth-order about 4800 Å. If the diode ends are cleaved, laser action occurs from the cooperation of the distributed and discrete feedback. However, if one face is given an antireflection coating the laser action depends on the distributed feedback and becomes much more a single-mode action of narrow line width as shown for the third-order ($\Lambda = 3600$ Å) SH DFB-laser of Fig. 14.37(b).

The emission wavelength of distributed feedback GaAs/GaAlAs lasers varies about 0.5 Å/°K because of refractive index change with temperature. For Fabry-Pérot cavity lasers the variation is 4 Å/°K and is mainly caused by bandgap change with temperature. Other features of DFB structures include wavelength selectivity (for instance, with gratings of different period six wavelengths have been obtained on the same GaAs wafer), longitudinal mode control, and well-collimated output beams. The threshold current densities tend, at present, to be about twice that of conventional stripe geometry lasers.

The third-order grating control is a special condition and consideration must be given again to the general condition in which cleaved faces are acting in concert with a grating. Figure 14.38(a) shows a magnified view of a corrugated grating of period $\Lambda$. From geometric optics the rays scattered from successive teeth are all in phase if, for example, ray 2 scattered from tooth 2 is in phase with ray 1. This will be true if the additional distance traveled is an integral multiple of a wavelength in the material, $\lambda_0/n$, so that

$$b + \Lambda = m(\lambda_0/n) \tag{14.19}$$

where $n$ is the refractive index, $\lambda_0$ is the free-space wavelength, $m$ is an integer, and $b$ is the distance shown in Fig. 14.38(a). Since $b = \Lambda \sin\theta$,

$$\sin\theta = m\,\lambda_0/n\Lambda - 1 \tag{14.20}$$

If $\lambda_0/n = \Lambda$, $\theta$ is zero for the solution $m = 1$ and the wavefront is parallel to the junction and the rays are orthogonal.

If $m$ is zero the solution describes light scattered in the forward direction and the $m = 2$ solution corresponds to light scattered backwards—the distributed feedback condition already discussed. Both forward and reverse traveling waves are scattered, propagating above and below the grating plane. The exit angles in air after refractions are given by

$$\sin\phi = m\lambda_0/\Lambda - n \tag{14.21}$$

where $\phi$ is the angle of the wavefront measured from the normal. Since $\sin\phi$ is

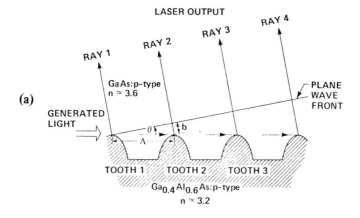

**(a)**

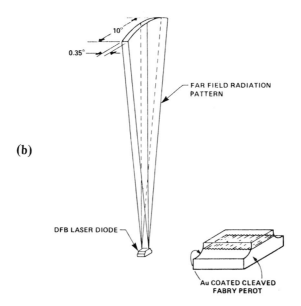

**(b)**

Fig. 14.38. Distributed feedback grating acting in concert with Fabry-Pérot faces in a laser:

(a) Rays scattered from the waveguide by the periodic corrugations. Note that $\theta$ is the angle that the wavefront makes with the plane of the junction. This wavefront is in phase if $\Lambda + b = \lambda_0/n$. (After Scifres, Burnham, and Streifer. Reprinted with permission from *Appl. Phys. Lett.*, **26**, 1975, p. 49.)

(b) Representation of the radiation pattern of a DFB injection laser radiating perpendicular to the plane of the *p-n* junction. (After Scifres, Burnham, and Streifer. Reprinted with permission from *IEEE Trans. Electron Devices*, **ED-22**, 1975, p. 611.)

also limited in magnitude by unity and $n \simeq 3.6$ for GaAs or AlGaAs, only beams within a $\pm$ 16-deg angle $\theta$ will be visible externally. The optimum grating spacing for coupling orthogonal to the junction is

$$\Lambda(\lambda_0/n)^{-1} = p, \qquad p = 1, 2, 3, \ldots \qquad (14.22)$$

corresponding to Bragg scattering of order $2p$ in the DFB structure. For $p = 1$ only the radiation mode normal to the plane of the grating exists. This is not true for larger values of $p$.

For a device with a grating of $\sim 4700$ Å corresponding to fourth-order Bragg scattering ($p=2$) the radiation pattern observed is shown in Fig. 14.38(b). The pulsed threshold current at 77°K was 1.2 kA/cm$^2$ with the cleaved faces gold coated to reduce reflection losses.

The output beam divergence along the corrugation direction is 0.35 deg and is determined by the lasing spectral bandwidth $\Delta\lambda$ and by the finite radiating length. Differentiation of Eq. (14.21) gives

$$d\phi = \frac{1}{\cos \phi} \left[ \frac{m\lambda_0}{n\Lambda} - \frac{\lambda_0}{n} \frac{dn}{d\lambda} \right] \frac{\Delta\lambda}{\lambda_0} n \qquad (14.23)$$

Hence if the bandwidth $\Delta\lambda$ is 6 Å, the angular spread is 0.21 deg. To this must be added the 0.1 deg of divergence expected for a 500 $\mu$m long radiator, to give a total spread close to that observed. The filamentary character of the laser oscillation produces the 10-deg divergence in the orthogonal direction, which is quite comparable to the far-field radiation divergence of a conventional laser in this direction.

The output is nearly completely polarized with the &-vector parallel to the groves. The power output (77°K) in the orthogonal beam at 7-A pulse current (1.7 times the threshold level) is only 10 $\mu$W. The threshold current is large because of damage created by the ion milling.

Interface nonradiative recombination tends to be high even if the grating is made by chemical etching.[6] A solution to the problem is to separate the optical and carrier confinement regions. In the separate confinement heterojunction structure (SCH) in Fig. 14.39(a) the GaAs active layer is separated from the corrugated interface. With a third-order grating (3770 Å) SCH diodes operate at 340°K with pulsed threshold current densities as low as 3 kA/cm$^2$ for light output levels of a few milliwatts.

Distributed feedback laser structures may be taper coupled into an integral single-heterojunction waveguide. The layers and their compositions and dimensions are shown in Fig. 14.39(b) and (c). The distributed Bragg reflector is a third-order grating ion-milled on a passive single-heterostructure waveguide section with taper coupling to the active double-heterojunction section. The threshold current density is 5 kA/cm$^2$ at 300°K and the half-power spectral bandwidth is about 1 Å.

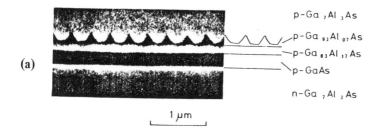

(a)

p-Ga$_{.7}$Al$_{.3}$As
p-Ga$_{.93}$Al$_{.07}$As
p-Ga$_{.83}$Al$_{.17}$As
p-GaAs

n-Ga$_{.7}$Al$_{.3}$As

1 μm

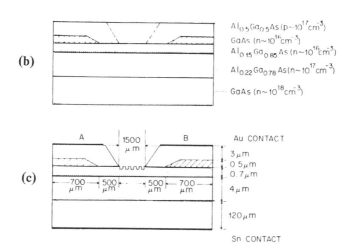

(b)

Al$_{0.5}$Ga$_{0.5}$As (p~$10^{17}$cm$^{-3}$)
GaAs (n~$10^{16}$cm$^{-3}$)
Al$_{0.15}$Ga$_{0.85}$As (n~$10^{16}$cm$^{-3}$)
Al$_{0.22}$Ga$_{0.78}$As(n~$10^{17}$cm$^{-3}$)
GaAs (n~$10^{18}$cm$^{-3}$)

(c)

A          1500
           μm          B          Au CONTACT

                                   3 μm
                                   0.5 μm
                                   0.7 μm

-700-  500     500    700
 μm    μm      μm     μm            4 μm

                                   120 μm

                                   Sn CONTACT

Fig. 14.39. Some variations of distributed feedback injection lasers:
(a)  The grating may be displaced from the active *p*-GaAs region to separate the optical and carrier confinement regions. Recombination at the grating is thus reduced. (After Aiki, et al. Reprinted with permission from *Appl. Phys. Lett.*, **27**, 1975, p. 145.)
(b,c) Distributed feedback laser structures may be taper coupled into an integral single-heterojunction waveguide. (After Reinhart, Logan, and Shank. Reprinted with permission from *Appl. Phys. Lett.*, **27**, 1975, p. 46.)

Another method of integrating a laser into a waveguide is shown in Fig. 14.40, where the active and waveguide regions are separated. Although optical pumping is shown the concept is the same for current injection. A block diagram showing a typical optical pumping arrangement for test studies is given in Fig. 14.41.

In gas-laser studies the concept of a ring laser is well known. Figure 14.42 shows an electrically pumped GaAs/GaAlAs ring laser with a fourth-order distributed feedback grating for output coupling. The holes etched through the *p*-GaAs active region at each corner are to eliminate Fabry-Pérot lasering that would compete with ring modes.

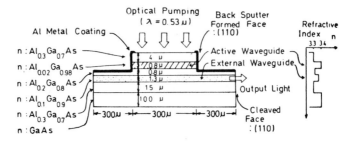

Fig. 14.40. Cross section of an integrated twin-guide AlGaAs laser. The active waveguide is $Al_{0.02}Ga_{0.98}As$ layer and the external waveguide is $Al_{0.1}Ga_{0.9}As$ layer. The end mirrors of the active waveguide are fabricated by backsputtering. (After Suematsu, Yamada, and Hayashi. Reprinted with permission from *Proc. IEEE,* 63, 1975, p. 208.)

## 14.7 OTHER INJECTION LASERS

Laser action occurs in many semiconductor materials with electron beam or optical excitation where it has proven impossible to obtain efficient *p-n* junction injection because of doping difficulties. Table 14.3 (a) and (b) lists some of the semiconductors found to lase and the methods of excitation. The photon energy for stimulated emission is slightly less than bandgap energy if shallow impurity states are involved in the transition.

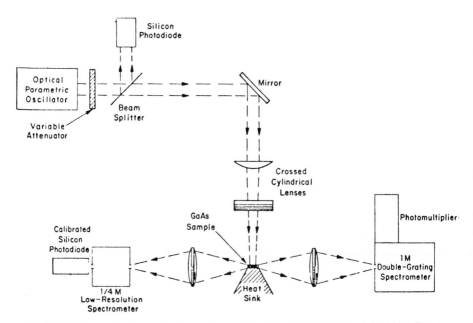

Fig. 14.41. Test setup for optical pumping experiments of lasering materials. (After Rossi, et al. Reprinted with permission from *J. Appl. Phys.,* 45, 1974. p. 5383.)

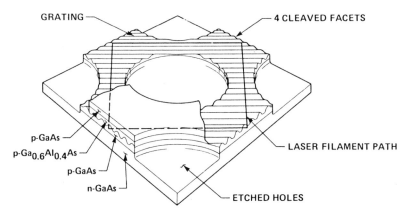

Fig. 14.42. Ring injection laser with grating. (After Scifres, Burnham, and Streifer. Reprinted with permission from *Appl. Phys. Lett.*, **28**, 1976, p. 681.)

The study of such lasers may be categorized in three major areas: (a) visible-light lasers; (b) lasers that are in the near infrared to match the low-loss transmission regions of optical fibers; and (c) lasers in the middle or far infrared that have uses in spectroscopy and meet infrared military needs.

Since double-heterojunction structures are essential for reduction of threshold current density, interest in the visible-light light range is in the

**Table 14.3 (a). Semiconductor Laser Materials — *p-n* Junction Lasers**

| Material | | Photon Energy (eV) | Wavelength (μm) |
|---|---|---|---|
| Gallium arsenide-phosphide | $GaAs_xP_{1-x}$ | 1.9-1.4 | 0.65-0.9 |
| Gallium-aluminium arsenide | $Ga_xAl_{1-x}As$ | 1.9-1.4 | 0.65-0.9 |
| Gallium arsenide | GaAs | 1.4 | 0.9 |
| Indium phosphide | InP | 1.35 | 0.91 |
| Gallium antimonide | GaSb | 0.83 | 1.5 |
| Indium arsenide-phosphide | $InAs_xP_{1-x}$ | 0.8 | 1.6 |
| Gallium-indium arsenide | $Ga_xIn_{1-x}As$ | 0.7-0.6 | 1.8-2.1 |
| Gallium-indium phosphide | $Ga_xIn_{1-x}P$ | 1.6 | 0.76 |
| Indium arsenide | InAs | 0.4 | 3.1 |
| Indium arsenide-antimonide | $InAs_xSb_{1-x}$ | 0.39 | 3.2 |
| Indium antimonide | InSb | 0.23 | 5.4 |
| Lead sulphide | PbS | 0.29 | 4.3 |
| Lead telluride | PbTe | 0.19 | 6.5 |
| Lead selenide | PbSe | 0.14 | 8.5 |
| Lead-tin telluride | $Pb_xSn_{1-x}Te$ | 0.2-0.05 | 6.5-30 |
| Lead-tin selenide | $Pb_xSn_{1-x}Se$ | 0.1 | 10-12 |

**Table 14.3 (b).   Semiconductor Laser Materials — Electron-Beam or Optically Pumped Lasers**

| Material | | Photon Energy (eV) | Wavelength ($\mu$m) |
|---|---|---|---|
| Zinc sulphide | ZnS | 3.8 | 0.33 |
| Zinc oxide | ZnO | 3.3 | 0.38 |
| Zinc selenide | ZnSe | 2.7 | 0.46 |
| Zinc telluride | ZnTe | 2.3 | 0.53 |
| Cadmium sulphide | CdS | 2.5 | 0.50 |
| Cadmium sulphide-selenide | $CdS_xSe_{1-x}$ | 2.5-1.8 | 0.5-0.7 |
| Cadmium selenide | CdSe | 1.8 | 0.69 |
| Cadmium telluride | CdTe | 1.6 | 0.78 |
| Gallium nitride | GaN | 3.45 | 0.36 |
| Gallium selenide | GaSe | 2.1 | 0.59 |
| Gallium arsenide-phosphide | $GaAs_xP_{1-x}$ | 1.8 | 0.7 |
| Gallium arsenide | GaAs | 1.4 | 0.85 |
| Gallium antimonide | GaSb | 0.83 | 1.5 |
| Indium arsenide | InAs | 0.4 | 3.1 |
| Indium antimonide | InSb | 0.23 | 5.4 |
| Cadmium silicon arsenide | $CdSiAs_2$ | 1.6 | 0.77 |
| Cadmium tin phosphide | $CdSnP_2$ | 1.24 | 1.0 |
| Cadmium phosphide | $Cd_3P_2$ | 0.6 | 2.1 |
| Tellurium | Te | 0.33 | 3.7 |
| Mercury-cadmium telluride | $Hg_xCd_{1-x}Te$ | 0.3 | 3.8-4.1 |
| Lead sulphide | PbS | 0.29 | 4.3 |
| Lead telluride | PbTe | 0.19 | 6.5 |
| Lead selenide | PbSe | 0.14 | 8.5 |
| Lead-tin telluride | $Pb_xSn_{1-x}Te$ | 0.08 | 15 |

(Courtesy C. H. Gooch, Injection Electroluminescent Devices, Wiley 1973)

quaternaries of the three-five compounds such as $In_{1-x}Ga_xP_{1-z}As_z$ [see Fig. 14.43(a)]. The use of quaternaries opens up the possibility of achieving the required bandgap conditions in two alloys while still retaining lattice-match conditions. A quaternary diagram of energy gaps and lattice constants for $In_{1-x}Ga_xP_{1-z}As_z$ is shown in Fig. 14.44. A yellow laser DH structure in this alloy system has the spectral response and threshold characteristics shown in Fig. 14.45.

For near-infrared lasers interest is strong in the 1-$\mu$m wavelength region and slightly beyond, where optical fiber losses are low (see Fig. 14.46). Dispersion, namely the propagation characteristics as a function of wavelength, is also of interest in optical fibers (see Fig. 14.47). This is a reason why lasers are preferable in optical communications systems to the use of light-emitting diodes of broader spectral bandwidths unless at the wavelength 1.27 $\mu$m where the dispersion is low.

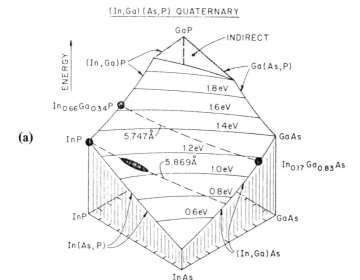

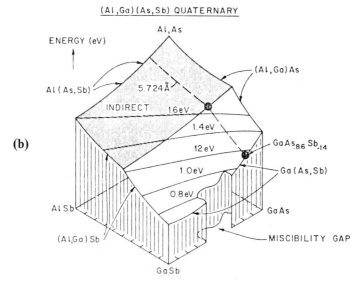

Fig. 14.43. Energy bandgap versus composition for quaternary compounds. (InGa) (AsP) and (AlGa) (AsSb). A few lines of matched lattice constant are shown. (After Nuese. Reprinted with permission from *IEEE Int. Electron Devices Meeting, Technical Digest,* 1976, p. 127.)

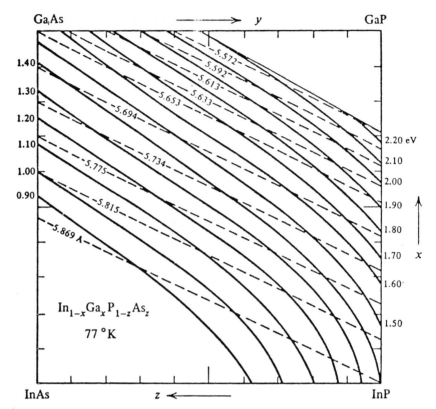

Fig. 14.44. Lattice constant and energy gap (77°K) as a function of composition for the quaternary alloy $In_{1-x}Ga_xP_{1-z}As_z$. The solid lines are constant energy gap curves; the dashed lines are lattice constant curves obtained by application of Vegard's law. The intersection of the energy gap curves with the lattice constant curves makes it apparent that the energy gap may be varied while a fixed lattice constant is maintained, indicating a variety of heterojunction possibilities from the infrared to the yellow for the quaternary alloy. (After Coleman, et al. Reprinted with permission from *J. Appl. Phys.*, 47, 1976, p. 2016.)

Fig. 14.45. Characteristics of a double-heterojunction InGaPAs injection laser:

(a) Emission spectra (300°K) of a relatively large-heterobarrier ($\Delta E \sim 137$ meV) $In_{1-x}Ga_xP_{1-z}As_z$ double-heterojunction laser diode. Curve (a) shows the low-level spontaneous emission spectrum. At higher levels, the line narrows and laser operation occurs at $\lambda \sim 6470$ Å and a threshold current density of $J_{th} \sim 2.0 \times 10^4$ A/cm². Curve (c) shows the shift in energy of the laser peak for a decrease in the temperature from 300° to 77°K, and

(b) Laser threshold current density as a function of temperature for $In_{1-x}Ga_xP_{1-z}As_z$ double-heterojunctions of various heterobarrier heights ($\Delta E$). The reference curve (a) represents the behavior of an $In_{1-x}Ga_xP$ homojunction laser diode ($\Delta E=0$) operating in the yellow portion of the spectrum. Curves (b)-(d) are for heterobarriers of 63,90, and 137 meV, respectively. In the linear regions, the curves fit $J_{th} \sim J_0 \exp(T/T_0)$. As the heterobarrier height increases, the magnitude of the threshold current density decreases at all temperatures, and the threshold current slope decreases (larger $T_0$). Room temperature laser operation is achieved, for a heterobarrier of 137 meV, at a current density of $\leq 2 \times 10^4$ A/cm². (After Coleman, et al. Reprinted with permission from *Appl. Phys. Lett.*, 29, 1976, pp. 167, 168.)

**(a)**

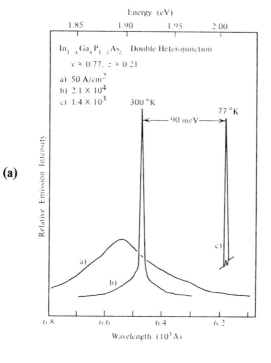

**(b)**

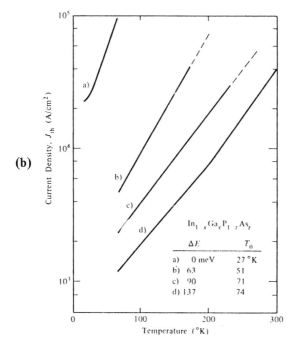

Some double-heterojunction lasers that have been studied for the $\sim 1$-$\mu$m wavelength region are

- $Ga_{0.84}In_{0.16}As(active)/Ga_{0.32}In_{0.68}P$
- $Ga_{0.12}In_{0.88}As_{0.23}P_{0.77}(active)/InP$
- $GaAs_{0.88}Sb_{0.12}(active)/Ga_{0.6}Al_{0.4}As_{0.88}Sb_{0.12}$

Threshold current densities for the first structure were 15 kA/cm$^2$ at room temperature and 1 kA/cm$^2$ at 77°K, but were as low as 2.8, 2.1 kA/cm$^2$ at 300°K for the second and third structures. The third structure is grown on a GaAs substrate with several layers of step-graded $GaAs_{1-z}Sb_z$ to provide stress relief before the actual DH laser is grown. Differential power efficiencies tend to be of the order of 10% and the external quantum efficiencies are below those for GaAs/AlGaAs lasers in the present state of the art. However, with further development there seems good potential for such lasers used in optical fiber communications systems in the low loss 1-$\mu$m wavelength region.

The middle- and far-infrared wavelength region is dominated by lead-based alloy lasers such as Pb-Sn-Te and others shown in Fig. 14.48. The diagram also

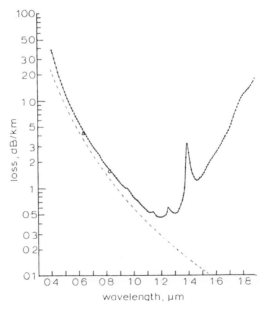

Fig. 14.46. Loss spectrum of optical fiber consisting of borosilicate cladding and phosphosilicate core. Triangular and circular points represent the total losses measured by an He-Ne laser and a GaAs laser, respectively. Broken line shows the Rayleigh scattering loss. The large loss near 1.4 $\mu$m is an OH vibrational mode. (After Horiguchi. Reprinted from *Electronics Lett.*, **12**, 1976, p. 310, Institution of Electrical Engineers.)

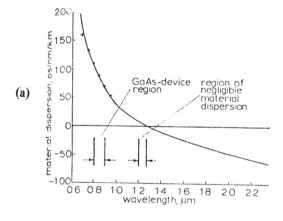

(a)

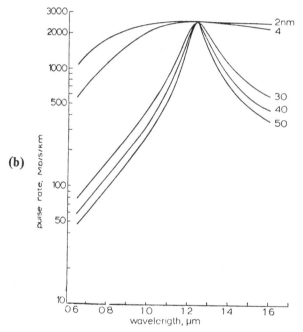

(b)

Fig. 14.47. Dispersion effects in optical fibers:
(a) Material dispersion $d\lambda = (-\lambda/c)d^2 n/d\lambda^2$ as function of wavelength. The solid curve is calculated for silica from the data of Mallitson and the points have been measured in a fiber having a phosphosilicate-glass core, and
(b) Pulse rate characteristics for fibers having the material dispersion in (a) and with source linewidth of injection laser type (2-4 nm) or of light-emitting diode type (30-50 nm). Note the wavelength region at 1.27 $\mu$m where the dispersion is very low and the LED performance matches that of an injection laser. (Assumes waveguide dispersion of 0.2 ns/km.) (After Payne and Gambling. Reprinted from *Electronics Lett.*, **11**, 1975, p. 177, Institution of Electrical Engineers.)

shows the strongly absorbing (vibration-rotation resonance) regions of a number of gases. These lasers in conjunction with similar junction detectors therefore are appropriate for pollution monitoring of various gases. In an automobile exhaust the spectral range of the major pollutants is 4.2 to 10 $\mu$m (CO is at 4.2 $\mu$m, $CO_2$ at 5 $\mu$m, and unburned hydrocarbons at longer wavelengths). However, for low bandgap materials temperatures of liquid nitrogen and below are needed for good lasing action and this is somewhat inconvenient.

In the lead-salts $Pb_{1-x}Sn_xTe$, $PbS_{1-x}Se_x$, and $Pb_{1-x}Cd_xS$ the ability to control the composition has made it possible to produce diode lasers with any desired wavelength in the range 2.5 to 34 $\mu$m. Continuous-wave operation generally requires 4.2°K operation. Single-mode output powers of $10^{-5}$ to $10^{-3}$ W are readily obtained and threshold current densities are typically 50-100 A/cm$^2$.[140] Some tuning of such laser-emission wavelengths is possible by temperature change, variation of the diode current, and magnetic field tuning. Fully resolved absorption spectra of NO, $NO_2$, $CH_4$, and CO have been obtained via magnetic field tuning of $PbS_{1-x}Se_x$ and PbTe lasers operating in the 4-$\mu$m to 7-$\mu$m region. The rate of mode-frequency tuning with magnetic field is proportional to the rate of change of the index of refraction with field. In $PbS_{0.82}Se_{0.18}$ at 10°K the change of frequency is of the order of 1000 MHz for a kilogauss change of field in the range of 10 to 20 kG.

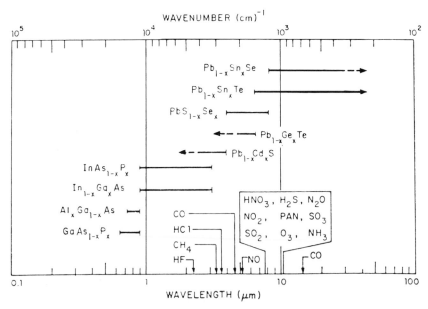

Fig. 14.48. Wavelength ranges for semiconductor lasers made from different alloys and compositions. Also shown are some strongly absorbing regions for several common gases. (After Calawa. Reprinted from *J. Lumin.*, 7, 1973, p. 478.)

Hydrostatic pressure changes the bandgap significantly in lead-salt lasers. A PbSe diode laser operating pulsed at $77°K$ has been pressure tuned from 7.5-23 $\mu$m with a hydrostatic pressure of 0-14 kBar. Continuous-wave operation has been reported at $77°K$ for double-heterostructure $p$-PbTe/Pb$_{0.88}$ Sn$_{0.12}$Te(active)/$n$-PbTe lasers at 8.2 $\mu$m. Lowering the temperature to $12°K$ increases the wavelength to 10.5 $\mu$m so the temperature tuning range is substantial. A power output of 10 mW at $12°K$, with a threshold current density of 1.6 kA/cm$^2$, falls to 1.2 mW at $77°K$ and the threshold current density is then 4.2 kA/cm$^2$.[83]

## 14.8 INJECTION LASERS AND LEDs AS LIGHT SOURCES FOR OPTICAL COMMUNICATIONS SYSTEMS

In optical communications systems data rates of as high as 1 Gbit/sec may be expected in the future. Figure 14.49 shows a schematic of a 100 Mbit/sec experimental system with two subchannels multiplexed. This operates well over an optical loss range of 10 dB; corresponding to a fiber length of greater than 1 km and up to 5 km for 2 dB/km loss fibers. In 1978, a 32 Mbit/sec pcm signal system has operated, without intermediate repeaters, over a distance of 53 km with a 1.27 $\mu$m InGaAs/InP double heterojunction laser source. The graded index fiber had a loss of 0.66 dB/km and the transmission bandwidth was 17 MHz. For comparison a 1 cm (0.375 in.) coaxial cable has an attenuation of 5 dB/km at about 10 MHz and a simple twisted-wire pair attenuates 5 dB/km at about 10 KHz. At about $1 per meter high-grade optical fibers are price-competitive with coaxial cable, and price reductions to less than 10¢ per meter can be expected in mass production. Thus optical fiber communication systems using diode lasers or light-emitting diodes can provide improved wide bandwidth transmission for long distance telephone communication, for shorter intracity routes, or for interconnection within a building. Fiber optics transmission is free from electromagnetic interference from motors, fluorescent lights, and control

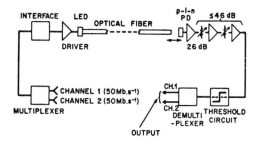

Fig. 14.49. Schema of fiber-optic communication system. (After White and Chin. Reprinted with permission from *Proc. IEEE*, **61**, 1973, p. 683.)

equipment. Other advantages are freedom from signal leakage between adjacent fibers, lightweight (< 1% of coaxial cable), and small size. Fiber optics and closely associated devices are expected to constitute a $200 million market by 1982.

The coupling of an LED to an optical fiber is shown in Fig. 14.50. There is presently much study of methods to integrate injection lasers to modulators, multiplexers, and fibers. Light coupling to thin-film waveguides may be made by prisms or gratings as illustrated in Fig. 14.51. Modulation or beam steering for multiplexing can be achieved by electrooptic diffraction, as shown in Fig. 14.52, or by Bragg refraction from surface waves. An impression of a GaAs integrated optical transmitter is shown in Fig. 14.53. Regeneration of light pulses may be possible with diodes operating as amplifiers in the lasering condition.[175]

Many applications exist in which high power output is required for line-of-sight uses. The assembly steps in the fabrication of a stacked array of series-connected laser diodes is shown in Fig. 14.54. Power levels of 1.5 kW (30 W average) have been achieved at 77°K for arrays of about 100 diodes.[64]

Wide-band signal recording on films is possible with 710-nm AlGaAs CW injection lasers of 5-10-mW power at 77°K.[179] An f/2 collection lens collimates the output of the laser diode. A polygon rotating mirror, located within the

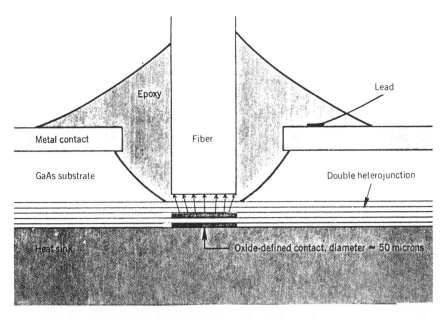

Fig. 14.50. An etched-well surface-emitting LED designed for fiber optics, in a schematic cross section. The emission from this type of diode is essentially isotropic. (After Kressel, et al. Reprinted with permission from *Physics Today*, May 1976, p. 46.)

focus of the imaging lens, deflects the modulated spot in an arc at a rate of 63,500 cm/sec. Silver halide film is transported past the scanning laser beam and is exposed by it along transverse tracks in a format similar to that of rotary-head videotape recording. After the film is developed, it is moved past the scanning beam, which once again becomes intensity modulated, this time by the transmittance variations of the film. The light is collected using appropriate optics and is photodetected, yielding a reproduction of the original signal. Excellent performance is obtained with the AlGaAs laser for recordings of signal frequencies up to 100 MHz (160 cycles/mm).

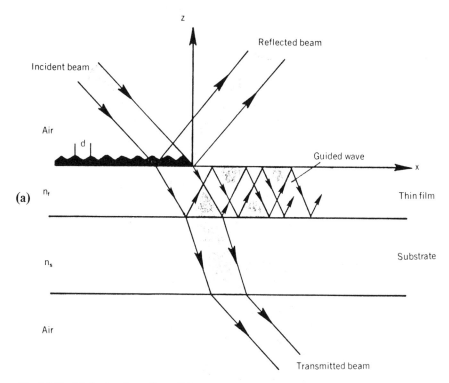

Fig. 14.51. Methods of coupling a light beam into a thin waveguide:
(a) Grating coupler. The grating is usually made of photoresist that has been exposed to an optical interference pattern and then developed; it need not be as rounded and symmetrical as shown here. It is not easy to obtain an efficiency as high as with prism couplers, although theory predicts efficiencies approaching 100% with appropriate grating design, and
(b) Prism coupler (above) and the corresponding "potential barrier" (below). Leakage of light across the gap can be expressed in quantum-mechanical terms as tunneling through the potential barrier. (After Conwell. Reprinted with permission from *Physics Today*, 29, 1976, pp. 53-54.)

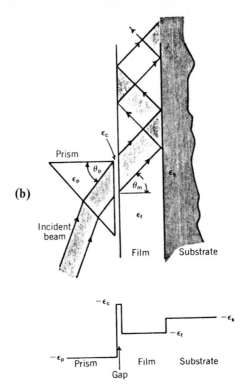

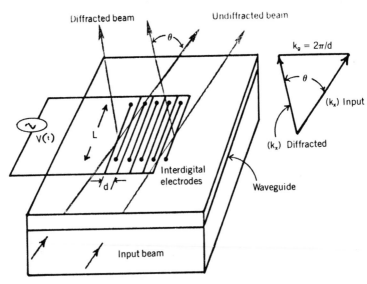

Fig. 14.51. (continued)

Fig. 14.52. Electrooptic diffraction modulation in a thin-film waveguide. Voltage applied to the interdigital electrodes causes a periodic variation in refractive index, which deflects part of the guided beam through an angle $\theta$. (After Conwell. Reprinted with permission from *Physics Today*, **29**, 1976, p. 55.)

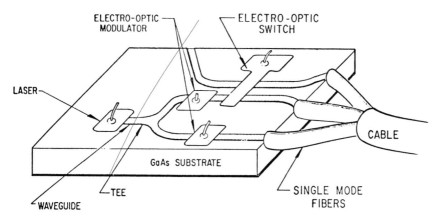

Fig. 14.53. Artist's impression of a monolithic GaAs integrated optical transmitter with its output coupled to single mode-optical fibers forming a transmission cable. (After Blum, Lawley, and Holton. Reprinted from *Naval Res. Rev.,* **18**, 1975, p. 1.)

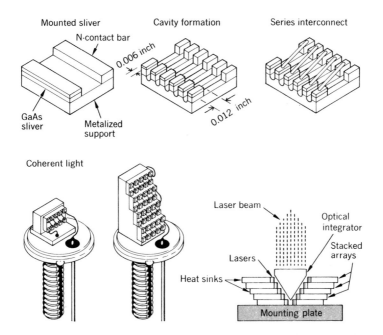

Fig. 14.54. Assembly steps in the fabrication of series-connected laser diodes designed for high-power pulsed operation at room temperature. (After Kressel, Lockwood, and Ettenberg. Reprinted with permission from *IEEE Spectrum*, May 10, 1973, p. 59.)

Injection lasers are presently under study for TV video players and for optical memory readout.[189] Pulse-position modulation techniques for a voice communication system have been described for use with low-duty cycle laser diodes.[171]

In summary, the injection laser field of study is thriving and numerous applications are developing. The military laser market for radar, communications, navigation, and guidance systems was about $250 million in 1976 and is expected to be about double this in 1981.

International conferences that serve these device areas in the United States include the IEEE International Semiconductor Laser Conference (Sixth Conference was held October 1978) and the IEEE Specialists Conference on the Technology of Electroluminescent Diodes (Third Conference was held November 1978).

## 14.9 QUESTIONS

1. A planar $p$-$n$ junction diode is made from a material of indirect bandgap 2.2 eV and relative dielectric constant 9.0. The radiative recombination occurs from the conduction band edge to an acceptor 0.2 eV above the valence band edge and only 1% of the recombinations are radiative. If the diode current is 10 mA and the chip area is $10^{-2}$ cm$^2$ with a junction area of 20% of this, what is the intensity of the light seen by an observer 1-m distant at 90 deg to the plane of the junction and at 60 deg to the plane of the junction? How visible would this diode be in bright sunlight (100 mW cm$^{-2}$)? How visible in ordinary room light in comparison with a 1-W tungsten lamp of bulb diameter 1 cm? How do you express the brightness in optical units?

2. A GaP green light diode of external efficiency 0.1% is being considered as a replacement for a red Ga As$_{0.6}$P$_{0.4}$ diode of external efficiency 0.3% that draws the same current, 10mA. What are the relative luminosities seen at a distance of 1 m? What are approximately the powers (watts) collected by the human eye observing the two lamps? How does this compare with moonlight ($\sim$5.10$^{-4}$W/cm$^2$)?

3. Some optical measurement terms are defined in the article by Zaha, Electronics, Nov. 6, 1972. Use as many of these as possible to characterize: (a) a GaP green LED (0.2% QE, 20mA); (b) a 40-W fluorescent lamp: both observed at a distance of 1 m.

4. The red GaP and GaAsP diodes of Fig. 14.4 are mounted in reflectors that collimate the radiated powers (300 $\mu$W and 120 $\mu$W, respectively) in beams of diameter 0.5 cm. The beams are observed at a distance of 1 m by the human eye. Which LED will appear brighter?

5. If band-to-band transitions are assumed radiative, what color light may be expected from LEDs of the following ten materials?

Ga As$_{0.5}$P$_{0.5}$         In As$_{0.5}$P$_{0.5}$         In$_{0.5}$Ga$_{0.5}$P

Ga$_{0.5}$Al$_{0.5}$As         Ga$_{0.5}$Al$_{0.5}$P         Ga Sb$_{0.5}$P$_{0.5}$

In$_{0.5}$Al$_{0.5}$P          Al As$_{0.5}$P$_{0.5}$        Al Sb$_{0.5}$P$_{0.5}$

Al$_{0.5}$Ga$_{0.5}$As$_{0.5}$P$_{0.5}$

Add comments on which may be expected to be bright emitters and explain.

6. Assume the forward current voltage relationship of an LED is given by $I = 10^{-18}$ exp $\left(\dfrac{qV}{1.6kT}\right)$. If the light output power is proportional to $I^2$ with an efficiency of a 1% at 10 mA, consider the light output (average) for a peak current of 40 mA at 25% duty factor compared with the output for a direct current (100% duty factor) of 10 mA. Also what would be the average power dissipated in the diode for these two conditions? If the power dissipated is to remain the same as that for 10-mA dc and the duty factor remains 25%, what is the permitted peak current and the effect on the light power?

7. In Fig. 14.8(c) it is shown that red GaAsP on a GaAs substrate has a luminous flux of 2 millilumens for a current of 10 mA, but that red GaAsP on a GaP substrate gives 20 millilumens at 10 mA. Infer the diode area and show that the 2 millilumens at 10 mA is reasonable. Then consider the 20 millilumens for a GaP substrate with the rearward light redirected. Does this number seem reasonable and how would you design to achieve it?

8. From the GaAs LED curves of Figs. 14.10 and 14.11 infer the depth of the dominant impurity, its density and its capture cross section.

9. Suppose a red-emitting GaP diode with a current density of 1 A/cm$^2$ has an internal light-emitting efficiency of 2% at 300°K. The efficiency is assumed limited by a single dominant recombination level. Make some reasonable assumptions about this level such as its position in the bandgap and its density and capture cross section and then attempt to predict the internal light-emitting efficiency of the diode at 400°K.

10. Discuss in detail the problem of achieving high light emission through the surface of an LED.[17] What can be the role of back-surface reflectors?

11. Review temperature dependences of GaAsP and GaP LEDs.

12. Discuss the state of the art in yellow-emitting LEDs and injection lasers.

13. Discuss delay times in the emission and turnoff of light from an LED and from a laser. What is the tradeoff between brightness and delay time in an LED?

14. Discuss the frequency response of GaAs, GaAsP, and GaP light-emitting diodes.[153]

15. Describe the use of capacitance transients to determine the density and level of non-radiative recombination centers in LEDs.[74]

16. What are the limitations of steady-state capacitance studies in determining the density and level of nonradiative recombination centers in LEDs.[131,132] For instance can only centers that are near the middle of the bandgap be found? Compare in potential with DLTS study methods.[124]

17. The reactive properties of GaP light-emitting junctions have not received much examination. What can be learned from such studies?[2]

18. Review the evidence that Cu may be responsible for degradation of light output from LEDs. What other causes should be considered?[17]

19. Describe the role of excitons in light-emitting diodes.

20. How important are auger recombination processes in light-emitting diodes?[18,19]

21. List all the isoelectronic traps that have been reported and comment on the results.

22. Review the present state of the art for 1-$\mu$m wavelength LED and injection laser emission.

23. Although the well-known seven-bar system produces all ten digits, it does not allow letters to be displayed. Discuss systems that are fully alphanumeric.

24. Discuss the performance of phosphor-coated diodes excited by infrared from GaAs LEDs.[19]

25. Review progress with SiC devices including rectifiers, LEDs, transistors, and thyristors.

26. A company needs a blue light-emitting diode developed within two years by two engineers and a technician. What program would you initiate?

27. Review the present state of the art in electroluminescent II-VI light-emitting diodes and powder ac and dc electroluminescent panels.

28. Discuss various kinds of antistokes light upconverters. What new materials possibilities seem promising for investigation using quaternary three-five alloys?

29. Estimate the power requirements of LED ten-digit seven-bar displays for 0.5-in. size digits in normal room light. Compare with the requirements of a liquid crystal display.

30. Discuss factors such as nonlinearity, isolation mode rejection, and frequency response for optical coupling devices and compare with isolation transformer performance.

31. Discuss *p-n-p-n* light-emitting structures that have current-controlled negative resistance characteristics. How could these be used in light-logic circuits?[221]

32. Consider light-coupled diode arrays for logic circuits.[205]

33. Review liquid-phase epitaxy processes for the fabrication of injection lasers and light-emitting diodes. Discuss doping matters for the III-V compounds including solubility and segregation problems.

34. Injection laser reflecting faces are usually formed by cleaving. Discuss instead the growth or etching of optical facets.[23]

35. What advantages does the molecular-beam-epitaxy process have over liquid-phase epitaxy for heterojunction laser growth?

36. How does the threshold current density of a DH injection laser vary with laser dimensions?

37. What factors limit the power output from a double-heterostructure GaAs-GaAlAs injection laser at room temperature, and what can be done to raise the output limits?

38. Review what is known of filament formation in semiconductor lasers.

39. Discuss emission line width and stability with temperature and drive current in a semiconductor laser intended for the $0.8$-$1.1$-$\mu$m wavelength region. Compare with similar properties for a gas laser.

40. What progress has been made towards achieving highly collimated beams from injection lasers? If a double-heterojunction laser $Al_{0.3}Ga_{0.7}As/GaAs/Al_{0.3}Ga_{0.7}As$ of length 500 $\mu$m is emitting from a face of width 20 $\mu$m what is the beam spread expected? How effectively could this be collimated by a lens of aperture $f/2$?[80]

41. Discuss typical polarization effects in the beam of a GaAs laser and the dependence on the structure.

42. Discuss the performance of injection lasers with buried active regions.

43. Discuss arrangements for separate optical and carrier confinement in diode lasers.

44. In an injection laser there is a difference in the threshold for a driving current pulse with and without another pulse preceding it. To what extent can charge storage explain this?[127]

45. The room temperature lattice mismatch of GaAs and AlGaAs may be reduced by the use of an AlGaAsP alloy where the P content is very small. How successful is this technique?[4]

46. Discuss Fabry-Pérot injection laser mode outputs.[56,197]

47. Describe grating methods of third-order Bragg distributed feedback in injection lasers.

48. Discuss measurements that have been reported of the laser loss coefficient term $\alpha$ and the gain term $\beta$ for Fabry-Pérot type semiconductor lasers [see Eq. (14.17)].

49. Review the progress in injection lasers capable of continuous-wave $77°$K operation in the 8-14-$\mu$m atmospheric window.

50. Discuss temperature tuning of a lead-salt injection laser. Why is this technique more appropriate to the lead salts than to GaAs?[140]

51. Discuss magnetic tuning of lead-salt lasers.[140]

52. What progress has been made in the use of 4-10-$\mu$m injection lasers for gas spectroscopy?

53. To what extent have the spikes $\Delta E_c$ and $\Delta E_v$ been confirmed as present and playing any role in double-heterojunction confinement lasers?

54. With what degree of success is it possible to use injection lasers as light regenerators or amplifiers?

55. Metal-insulator-semiconductor sandwich structures have been proposed for use as pulsed lasers. What are the problems and how likely are they to be resolved?[109]

56. Review methods or ideas for integrating injection lasers and LEDs into optical waveguides.

57. What progress has been made in ring-injection laser structures?

58. Discuss pulse-position modulation as a technique for transmitting information on a pulsed injection laser.[172]

59. Review techniques for electrical or accoustical beam steering of an injection laser beam (C. S. Tsai, *Proc. IEEE*, May 1974).

60. Review the present state of the art in fiber-optics digital communications systems.

61. Describe uses of injection laser diodes in holography; also in beam-addressable memories, such as EuO films.[118]

62. How may an injection laser be used as a range finder?

## 14.10 REFERENCES AND FURTHER READING SUGGESTIONS

1. Abagyan, S.A., et al., "Yellow electroluminescence of GaP," *Sov. Phys. Semicond.,* 7, 1973, p. 412.

2. Abdullaev, G.B., et al., "Reactive properties of GaP light-emitting diodes," *Sov. Phys. Semicond.,* 6, 1973, p. 1822.

3. Adams, M.J., and P.T. Landsberg, *Gallium Arsenide Lasers* (edited by C.H. Gooch), Wiley-Interscience, N.Y., (1969), Chap. 2.

4. Afromowitz, M.A., and D.L. Rode, "Limitations on stress compensation in $Al_xGa_{1-x}As_{1-y}P_y$-GaAs LPE layers," *J. Appl. Phys.,* 45, 1974, p. 4738.

5. Ahrons, R.W., "In strobed LED displays how bright is bright?" *Electronics*, Nov. 22, 1971, p. 78.

6. Aiki, K., et al., "GaAs-GaAlAs distributed-feedback diode lasers with separate optical and carrier confinement," *Appl. Phys. Lett.,* 27, 1975, p. 145.

7. Alferov, Zh.I., et al., "Electroluminescence of GaP:N diodes at high excitation levels," *Sov. Phys. Semicond.,* 7, 1973, p. 433.

8. Alferov. Zh.I., et al., "100% internal quantum efficiency of radiative recombination in three-layer AlAs-GaAs heterjunction light-emitting diodes," *Sov. Phys. Semicond.,* 9, 1975, p. 305.

9. Aven, M., and J.Z. Devie, "Advances in injection luminescence of II-VI compounds," *J. Lumin.,* 7, 1973, p. 195.

10. Bachrach, R.Z., R.W. Dixon, and O.G. Lorimor, "Hemispherical GaP:N green electroluninescent diodes," *Solid-State Electronics*, 16, 1973, p. 1037.

11. Bachrach, R.Z., W.B. Joyce, and R.W. Dixon, "Optical-coupling efficiency of GaP:N green-light-emitting diodes," *J. Appl. Phys.,* 44, 1973, p. 5458.

12. Bachrach, R.Z., P.D. Dapkus, and O.G. Lorimor, "Room-temperature deep-state emission spectra, radiative efficiency, and lifetime of some GaP:Te, N crystals," *J.Appl. Phys.,* 45, 1974, p. 4971.

13. Barnoski, M.K. (Ed.), *Fundamentals of Optical Fiber Communications*, Academic Press, N.Y., 1976.

14. Bass, J.C., "Options in solid state data displays," *New Electronics, G.B.,* 6(17), 1973, p. 45.

15. Beneking, H., G. Schul, and P. Mischel, "Efficient semiconductors antistokes light converters," *IEEE Electron Devices Meeting Abstracts*, Washington, D.C., Dec. 1974, p. 69.

16. Beppu, T., "Luminescent properties variation toward growth direction in nitrogen doped GaP n-LPE layer," *Jap. J. Appl. Phys.,* 14, 1975, p. 761.

17. Bergh, A.A., "Bulk degradation of GaP red LED's," *IEEE Trans. Electron Devices,* **ED-18,** 1971, p. 166.
18. Bergh, A.A., and P.J. Dean, "Light-emitting diodes," *Proc. IEEE,* **60,** 1972, p. 156.
19. Bergh, A.A., and P.J. Dean, *Light Emitting Diodes,* Clarendon Press, Oxford, 1976.
20. Bergh, A.A., and B.H. Johnson, "A solid future for telphone lamps," *Bell Lab. Rec.,* **47,** 1969, p. 320.
21. Bergh, A.A., and R.J. Strain, "A contact for gallium phosphide electroluminescent diodes," in *Ohmic Contacts to Semiconductors,* B. Schwartz, Ed., Electrochem. Soc., N.Y., 1969, p. 115.
22. Blackwell, G.R., "Materials for 1. e. d. s," *New Electronics, G.B.,* **6**(17), 1973, p. 64.
23. Blum, F.A., K.L. Lawley, and W.C. Holton, "Monolithic $Ga_{1-x}In_xAs$ mesa lasers with grown optical facets," *J.Appl. Phys.,* **46,** 1975, p. 2605.
24. Blum, F.A., K.L. Lawley, and W.C. Holton, "Integrated optical circuits," *Naval Research Reviews,* **18,** 1975, p. 1.
25. Bortfield, D.P., B.J. Curtis, and H. Meier, "Electrical properties of carbon doped gallium phosphide," *J. Appl. Phys.,* **43,** 1972, p. 1293.
26. Bouley, J.C., "Luminescence in highly conductive n-type ZnSe," *J.Appl. Phys.,* **46,** 1975, p. 3549.
27. Bowman, D.L., "Photoconductivity and photo-Hall measurements on high resistivity GaP," *J.Appl. Phys.,* **38,** 1967, p. 568.
28. Brantley, W.A., "Effect of dislocations on green electroluminescent efficiency in GaP grown by liquid phase epitaxy," *J.Appl. Phys.,* **46,** 1975, p. 2629.
29. Brown, R.L., and R.G. Sobers, "Stress compensation in $GaAs\text{-}Al_{0.24}Ga_{0.76}As_{1-y}P_y$ LPE binary layers," *J. Appl. Phys.,* **45,** 1974, p. 4735.
30. Buchsbaum, S.J., "Lightwave communications: an overview," *Physics Today,* May 1976, p. 23.
31. Burnham, R.D., D.R. Scrifes, and W. Streifer, "Low-divergence beams from grating-coupled composite guide heterostructure GaAlAs diode lasers," *Appl. Phys. Lett.,* **26,** 1975, p. 645.
32. Calawa, A.R., "Small bandgap lasers and their uses in spectroscopy," *J. Lumin.,* **7,** 1973, p. 477.
33. Camassel, J., "Temperature dependence of the band gap and comparison with the threshold frequency of pure GaAs lasers," *J. Appl. Phys.,* **46,** 1975, p. 2683.
34. Campbell, J.C., et al., "Band structure enhancement and optimization of radiative recombination in $GaAs_{1-x}P_x$:N (and $In_{1-x}Ga_xP$:N)," *J. Appl. Phys.,* **45,** 1974, p. 4543.
35. Carr, W.N., "Photometric figures of merit for semiconductor luminescent sources operating in spontaneous mode," *Infrared Phys.,* **6,** March 1966, p. 1.
36. Casey, H.C., and F.A. Trumbore, "Single crystal electroluminescent materials," *Mater. Sci. Eng.,* **6,** 1970, p. 69.
37. Casey, H.C., and M.B. Panish, "Influence of $Al_xGa_{1-x}As$ layer thickness on threshold current density and differential quantum efficiency for $GaAs\text{-}Al_xGa_{1-x}As$ DH lasers," *J. Appl. Phys.,* **46,** 1975, p. 1393.
38. Casey, H.C., Jr., S. Somekh, and M. Illegems, "Room temperature operation of low-threshold separate-confinement heterostructure injection laser with distributed feedback," *Appl. Phys. Lett.,* **27,** 1975, p. 144.
39. Chamberlain, S.G., and M. Kuhn (Eds.), "Issue on Optoelectronic Devices and Circuits," *IEEE Trans. Electron Devices,* **ED-25,** 1978, p. 273.
40. Chase, B.D., and D.B. Holt, "Scanning electron microscope studies of electroluminescent diodes of GaAs and GaP," *Phys. Stat. Sol.,* **(a)19,** 1973, p. 467.
41. Chiang, S.Y., and G.L. Pearson, "Properties of vacancy defects in GaAs single crystals," *J. Appl. Phys.,* **46,** 1975, p. 2986.

42. Chiao, S.H., B.L. Mattes, and R.H. Bube, "Photoelectronic properties of LPE GaAs:Cu," *J. Appl. Phys.*, **49**, 1978, p. 261.

43. Chiao, S.H., and G.A. Antypas, "Photocapacitance effects in deep traps in n type InP," *J. Appl. Phys.*, **49**, 1978, p. 466.

44. Cho, A.Y., and H.C. Casey, Jr., "GaAs-Al$_x$Ga$_{1-x}$As double-heterostructure lasers prepared by molecular-beam epitaxy," *Appl. Phys. Lett.*, **25**, 1974, p. 288.

45. Christensen, C.P., et al., "Investigation of infrared loss mechanisms in high-resistivity GaAs," *J. Appl. Phys.*, **45**, 1974, p. 4957.

46. Chynoweth, A.G., "The fiber lightguide," *Physics Today*, May 1976, p. 28.

47. Coleman, J.J., "Yellow In$_{1-x}$Ga$_x$P$_{1-z}$As$_z$ double-heterojunction lasers," *J. Appl. Phys.*, **47**, 1976, p. 2015.

48. Coleman, J.J., et al., "Liquid phase epitaxial In$_{1-x}$Ga$_x$P$_{1-z}$As$_z$/GaAs$_{1-y}$P$_y$ quaternary (LPE)-ternary (VPE) heterojunction lasers ($x\sim0.70$, $z\sim0.01$, $y\sim0.40$; $\lambda < 6300$ Å, 77° K)," *Appl. Phys. Lett.*, **25**, 1974, p. 725.

49. Coleman, J.J., et al., "Heterojunction laser operation of N-free and N-doped GaAs$_{1-y}$P$_y$ ($y = 0.42-0.43$, $\lambda\sim6200$ Å, 77°K) near the direct-indirect transition ($y\sim y_c$ 0.46)," *J. Appl. Phys.*, **46**, 1975, p. 3556.

50. Coleman, J.J., et al., "Pulsed room temperature operation of In$_{1-x}$Ga$_x$P$_{1-z}$As$_z$ double heterojunction lasers at high energy (6470Å, 1.916 eV)," *Appl. Phys. Lett*, **29**, 1976, p. 167.

51. Collier, R.J., C.B. Burkhardt, and L.H. Lin, *Optical Holography* (Student Edition), Academic Press, N.Y., 1977.

52. Conwell, E.M., "Integrated optics," *Physics Today*, May 1976, p. 48.

53. Cook, D.C., and F.R. Nash, "Gain-induced guiding and astigmatic output beam of GaAs lasers," *J. Appl. Phys.*, **46**, 1975, p. 1660.

54. Cuthbert, J.D., "Luminescence and free carrier decay times in semiconductors containing isoelectronic traps," *J. Appl. Phys.*, **42**, 1971, p. 739.

55. Dapkus, P.D., et al., "Kinetics of recombination in nitrogen-doped GaP," *J. Appl. Phys.*, **45**, 1974, p. 4920.

56. D'Asaro, L.A., "Advances in GaAs junction lasers with stripe geometry," *J. Lumin.*, **7**, 1973, p. 310.

57. Dean, P.J., "Recombination processes associated with 'deep states' in gallium phosphide," *J. Lumin.*, **1, 2**, 1970, p. 398.

58. Dean, P.J., "Isolectronic traps in semiconductors (experimental)," *J. Lumin.*, **7**, 1973, p. 51.

59. Dean, P.J., J.D. Cuthbert, and R.T. Lynch, "Interimpurity recombinations involving the isoelectronic trap bismuth in gallium phosphide," *Phys. Rev.*, **179**, 1969, p. 754.

60. Dean, P.J., and C.H. Henry, "Electron-capture ("internal") luminescence from the oxygen donor in gallium phosphide," *Phys. Rev.*, **176**, 1968, p. 928.

61. Dean, P.J., E.A. Schonherr, and R.B. Zettlerstrom, "Pair spectra involving shallow acceptor Mg in GaP," *J. Appl. Phys.*, **41**, 1970, p. 3475.

62. Dean, P.J., et al., "The isoelectronic trap antimony in gallium phosphide," *Bull. Amer. Phys. Soc.*, **14**, 1969, p. 395.

63. Dentai, A.G., T.P. Lee, and C.A. Burrus, "Small-area high-radiance C.W. InGaAsP L.E.D.S. emitting at 1.2 to 1.3 $\mu$m," *Electronics Lett.*, **13** (16), Aug. 4, 1977, p. 484.

64. DeVilbiss, W.F., "Using gallium arsenide laser arrays as pulsed infrared sources," Proc. of the Technical Program, Electro-optical Systems Design Conf., New York, Sept. 1970, p. 111.

65. Dickey, R.K., "Isolator circuit permits scope to check underground voltages," *Electronics*, March 21, 1974, p. 121.

66. Dierschke, E.G., and G.L. Pearson, "Effect of the donor concentration on the green electroluminescence from gallium phosphide diodes," *J. Appl. Phys.*, **41**, 1970, p. 321.

67. Dingle, R., W. Wiegmann, and C.H. Henry, "Quantum states of confined carriers in very thin $Al_xGa_{1-x}As$-GaAs-$Al_xGa_{1-x}As$ heterostructures," *Phys. Rev. Lett.,* **33,** 1974, p. 827.

68. Dmitriev, A.G., and B.V. Tsarenkov, "Emission kinetics of electroluminescent diodes," *J. Appl. Phys.,* **46,** 1975, p. 1739.

69. Dyment, J.C., Y.C. Cheng, and A.J. SpringThorpe, "Temperature dependence of spontaneous peak wavelength in GaAs and $Ga_{1-x}Al_xAs$ electroluminescent layers," *J. Appl. Phys.,* **46,** 1975, p. 1739.

70. Ettenberg, M., and H. Kressel, "Heterojunction diodes of (AlGa)As-GaAs with improved degradation resistance," *Appl. Phys. Lett.,* **26,** 1975, p. 478.

71. Ettenberg, M., and C.J. Nuese, "Comparison of Zn-doped GaAs layers prepared by liquid-phase and vapor-phase techniques, including diffusion lengths and photoluminescence," *J. Appl. Phys.,* **46,** 1975, p. 3500.

72. Everest, F.G., "Gallium arsenide laser rangefinder," Proc. of the Technical Program, Electro-optical Systems Design Conf., New York, Sept. 1970, p. 528.

73. Firester, A.H., and M.E. Heller, "Use of diode lasers to recover holographically stored information," *IEEE J. Quantum Electronics,* QE-6, 1970, p. 572.

74. Forbes, L., "Non-radiative recombination centers in $GaAs_{0.6}P_{0.4}$ red light-emitting diodes," *Solid-State Electronics,* **18,** 1975, p. 635.

75. Gershenzon, M.F., "State of the art in GaP electroluminescent junctions," *Bell Syst. Tech. J.,* **45,** 1966, p. 1599.

76. Gershenzon, M.F., et al., "Radiative recombination between deep-donor-acceptor pairs in GaP," *J. Appl. Phys.,* **36,** 1965, p. 1528.

77. Glicksman, R., "Recent progress in injection lasers," Proc. of the Technical Program, Electro-optical Systems Design Conf., New York, Sept. 1970, p. 93.

78. Gloriozova, R.I., M.I. Iglitsyn, and L.I. Kolesnik, "Energy spectrum of deep centers in high resistivity gallium phosphide," *Sov. Phys. Semicond.,* 3, 1969, p. 790.

79. Goldstein, B., and S.S. Perlman, "Electrical and optical properties of high-resistivity gallium phosphide," *Phys. Rev.,* 148, 1966, p. 715.

80. Gooch, C.H. (ed.), *Gallium Arsenide Lasers,* Wiley-Interscience, N.Y., 1969.

81. Gooch, C.H., *Injection Electroluminescent Devices,* Wiley, N.Y., 1973.

82. Goodwin, A.R., et al., "Threshold temperature characteristics of double heterostructure $Ga_{1-x}Al_xAs$ lasers," *J. Appl. Phys.,* **46,** 1975, p. 3126.

83. Groves, S.H., K.W. Nill, and A.J. Strauss, "Double heterostructure $Pb_{1-x}Sn_xTe$-PbTe lasers with cw operation at 77 K," *Appl Phys. Lett.,* 25, 1974, p. 331.

84. Hakki, B.W., "GaAs double heterostructure lasing behavior along the junction plane," *J. Appl. Phys.,* **46,** 1975, p. 292.

85. Hakki, B.W., "Striped GaAs lasers: mode size and efficiency," *J. Appl. Phys.,* **46,** 1975, p. 2723.

86. Hakki, B.W., and F.R. Nash, "Catastrophic failure in GaAs double-heterostructure injection lasers," *J. Appl. Phys.,* **45,** 1974, p. 3907.

87. Harikoshi, Y., "(Al,Ga)As, GaAs double heterostructure lasers prepared by new liquid phase epitaxial growth technique," *IEEE Int. Electron Devices Meeting, Technical Digest,* Washington, D.C., 1976, p. 121.

88. Hartman, R.L., and R.W. Dixon, "Reliability of DH GaAs lasers at elevated temperatures," *Appl. Phys. Lett.,* **26,** 1975, p. 239.

89. Hartman, R.L., B. Schwartz, and M. Kuhn, "Degradation and passivation of GaP light emitting diodes," *Appl. Phys. Lett.,* 18, 1971, p. 304.

90. Hass, G. (Ed.), *Physics of Thin Films: Advances in Research and Development,* Vol. 8, Academic Press, N.Y., 1975.

91. Hayashi, I., and M.B. Panish, "GaAs-$Ga_xAl_{1-x}As$ heterostructure injection lasers which exhibit low thresholds at room temperature," *J. Appl. Phys.,* 41, 1970, p. 150.

92. Hayashi, I., M.B. Panish, and F.K. Reinhart, "GaAs-$Al_xGa_{1-x}As$ double heterostructure injection lasers," *J. Appl. Phys.,* 42, 1971, p. 1929.

93. Henry, C.H., P.J. Dean, and J.D. Cuthbert, "New red pair luminescence from GaP," *Phys. Rev.,* **166**, 1968, p. 754.
94. Herzog, A.H., W.O. Groves, and M.G. Craford, "Electroluminescence of diffused GaAs$_{1-x}$P$_x$ diodes with low donor concentrations," *J. Appl. Phys.,* **40**, 1969, p. 1830.
95. Hitchens, W.R., et al., "Liquid phase epitaxial (LPE) grown junction In$_{1-x}$Ga$_x$P (x∿0.63) laser of wavelength λ∿ 5900 Å (2.10 eV, 77°K)," *Appl. Phys. Lett.,* **25**, 1974, p. 352.
96. Hitchens, W.R., et al., "Low threshold LPE In$_{1-x}$Ga$_x$P$_{1-z}$As$_z$ yellow double-heterojunction laser diodes ($J$<10$^4$ A/cm$^2$, λ 5850 Å, 77°K)," *Appl. Phys. Lett.,* **27**, 1975, p. 245.
97. Horiguchi, M., "Spectral losses of low-OH-content optical fibres," *Electron. Lett.,* **12**, 1976, p. 310.
98. Hsieh, J.J., and J.A. Rossi, "GaAs:Si double heterostructure LED's," *J. Appl. Phys.,* **45**, 1974, p. 1834.
99. Hurwitz, C.E., "Integrated GaAs-AlGaAs double-heterostructure lasers," *Appl. Phys. Lett.,* **27**, p. 241.
100. Hutchinson, P.W., and P.S. Dobson, "Defect structure of degraded heterojunction GaAlAs-GaAs lasers," *Appl. Phys. Lett.,* **26**, 1975, p. 250.
101. Hwang, C.J., et al., "Threshold behavior of (GaAl)As-GaAs lasers at low temperatures," *J. Appl. Phys.,* **49**, 1978, p. 29.
102. Ito, R., H. Nakashima, and O. Nakada, "Growth of dark lines from crystal defects in GaAs-GaAlAs double heterostructure crystals," *Jap. J. Appl. Phys.,* **13**, 1974, p. 1321.
103. Jensen, S.M., et al., "Low-loss optical waveguides in single layers of Ga$_{1-x}$Al$_x$As," *J. Appl. Phys.,* **46**, 1975, p. 3547.
104. Jordan, A.S., et al., "Solid composition and gallium and phosphorous vacancy concentration isobars for GaP," *J. Appl. Phys.,* **45**, 1974, p. 3472.
105. Joyce, W.B., and R.W. Dixon, "Thermal resistance of heterostructure lasers," *J. Appl. Phys.,* **46**, 1975, p. 855.
106. Kameda, S., and W.N. Carr, "Analysis of proposed MIS laser structures," *IEEE J. Quantum Electronics,* QE-9, 1973, p. 374.
107. Kaminow, I.P., and A.E. Siegman, *Laser Devices and Applications,* IEEE Press, N.Y., 1973.
108. Kan, H., et al., "Continuous operation over 10000 h of GaAs/GaAlAs double-heterostructure laser without lattice mismatch compensation," *Appl. Phys. Lett.,* **27**, p. 138.
109. Kawabe, M., et al., "Heterostructure CdS$_{1-x}$Se$_x$-CdS surface lasers for integrated optics," *Appl. Phys. Lett.,* **26**, 1975, p. 46.
110. Kazan, B. (Ed.), *Advances in Image Pickup and Display,* Vol. 3, Academic Press, N.Y., 1977.
111. Kim, C.K., "Total oxygen content of gallium phosphide grown by the Czochralski technique using liquid encapsulation," *J. Appl. Phys.,* **45**, 1974, p. 243.
112. Kishino, S., et al., "X-ray topographic study of dark-spot defects in GaAs-Ga$_{1-x}$Al$_x$As double heterostructure wafers," *Appl. Phys. Lett.,* **27**, 1975, p. 207.
113. Kniss, F., "GaP under GaAsP brightens LED colors," *Electronics,* May 2, 1974, p. 34.
114. Koechner, W., "Extremely small and simple pulse generator for injection lasers," *Rev. Scientific Instruments,* **38**, 1967, p. 17.
115. Kressel, H., "Gallium arsenide and aluminum gallium arsenide devices prepared by liquid phase epitaxy," *J. Electron. Matls.,* **3**, 1974, p. 747.
116. Kressel, H., H.F. Lockwood, and H. Nelson, "Low threshold Al$_x$Ga$_{1-x}$As visible and IR light-emitting diode lasers," *IEEE J. Quantum Electronics,* QE-6, 1970, p. 278.
117. Kressel, H., H.F. Lockwood, and M. Ettenberg, "Progress in laser diodes," *IEEE Spectrum,* May 1973, p. 59.
118. Kressel, H., et al., "Light sources," *Physics Today,* May 1976, p. 38.
119. Kressel, H., H. Schade, and H. Nelson, "Heterojunction cold-cathode electron emitters of (AlGa)As-GaAs," *J. Lumin.,* **6-7**, 1973, p. 146.

120. Kressel, H., *Lasers* (Edited by Levine and DeMaria), Vol. 2, Dekker, N.Y., 1971, p. 2. *Laser Handbook* (Edited by F.T. Arecchi and E.O. Schultz-DuBois), Vol. 1, North Holland, Amsterdam, 1972, p. 441. See also: Kressel, H., and J.K. Butler, *Semiconductor Lasers and Heterojunction LED's*, Academic Press, N.Y., 1977.

121. Ladany, I., and H. Kressel, "An experimental study of high-efficiency GaP:N green-light-emitting diodes," *RCA Rev.,* 33, 1972, p. 517.

122. Ladany, I., and H. Kressel, "Influence of device fabrication parameters on gradual degradation of (AlGa)As cw laser diodes," *Appl. Phys. Lett.,* 25, 1974, p. 708.

123. Ladany, I., and H. Kressel, "Visible CW (AlGa)As heterojunction laser diodes," *IEEE Int. Electron Devices Meeting, Technical Digest*, Washington D.C., 1976, p. 129.

124. Lang, D.V., "Deep level transient spectroscopy: A new method to characterize traps in semiconductors," *J. Appl. Phys.,* 45, 1974, p. 3023.

125. Lauer, R.B., and J. Zucker, "Narrow-bandwidth (Al, Ga)As: Si double-heterostructure LED's," *J. Appl. Phys.,* 46, 1975, p. 1837.

126. Laws, D.A., and R.R. Ady, "Should you use LCD or LED displays for portable equipment? The choice hinges more on drive and interface requirements than on display appearance," *Electronics Design,* 23, 1974, p. 88.

127. Lee, T.P., and R.M. Derosier, "Charge storage in injection lasers and its effect on high-speed pulse modulation of laser diode," *Proc. IEEE,* 62, 1974, p. 1176.

128. Lefévre, H., and M. Schulz, "DDLTS—the extension of DLTS to double correlation, to resolve the deep level profile and to exclude the field dependence of the capture cross-section and the contact effects," *Appl. Phys.,* 12, 1977, p. 45.

129. Levinstein, H., and J. Mudar, "Infrared detectors in remote sensing," *Proc. IEEE,* 63, 1975, p. 6.

130. Li, S.S., J. R. Anderson, and J. K. Kennedy, "Electrical properties of epitaxially grown $InAs_{0.61}P_{0.39}$ films," *J. Appl. Phys.,* 46, 1975, p. 1223.

131. Lo, W., and E.S. Yang, "A technique for the investigation of deep-level states in diffused p-n junction devices: application to GaAs electroluminescent diodes," *IEEE Trans. Electron Devices,* ED-20, 1973, p. 684.

132. Lo, W., and E.S. Yang, "A method for the determination of electron capture cross-section at imperfection centers in gallium arsenide of electroluminescent diodes," *Solid-State Electronics,* 17, 1974, p. 113.

133. Loescher, D.H., J.W. Allen, and G.L. Pearson, "The application of crystal field theory to the electrical properties of Co impurities in GaP," Proceedings of the International Conference on the Physics of Semiconductors, Kyoto, 1966, *J. Phys. Soc. Japan*, Suppl. 21, 1966, p. 239.

134. Lopez, C., A. Garcïa, and E. Munoz, "Deep-level changes associated with the degradation of $GaAs_{0.6}P_{0.4}$ L.E.D.'s," *Electronic Lett.,* 13 (16), Aug. 4, 1977, p. 460.

135. Lorimor, O.G., W.H. Hackett, Jr., and R.Z. Bachrach, "Reproducible high-efficiency GaP green-emitting diodes grown by overcompensation," *J. Electrochem. Soc.,* 120, 1973, p. 1424.

136. Lorimor, O.G., et al., "High capacity liquid phase epitaxy apparatus utilizing thin melts," *Solid-State Electronics,* 16, 1973, p. 1289.

137. Lorimor, O.G., P.D. Dapkus, and W.H. Hackett, Jr., "Very high efficiency GaP green light emitting diodes," *J. Electrochem. Soc.,* 122, 1975, p. 407.

138. Mataré, H.F., "Light emitting devices, Part I: Methods," in *Advances in Electronics and Electron Physics*, Vol. 42, Academic Press, N.Y., 1976.

139. Maurer, R.D., "Glass fibers for optical communications," *Proc. IEEE,* 61, 1973, p. 452.

140. Melngailis, I., "Narrow-gap semiconductor lasers and detectors," Proc. 4th Conf. on Solid State Devices, Tokyo, 1972: *Suppl. J. Japan Soc. Appl. Phys.,* 42, 1973, p. 3.

141. Migliorato, P., G. Margaritondo, and P. Perfetti, "Subquadratic I-V dependence and double injection in GaP," *Solid-State Communication,* 13, 1973, p. 499.

142. Miller, B.I., and W.D. Johnston, Jr., "Successful liquid phase epitaxial growth and optically pumped laser operation of $In_{0.5}$ $P$-$Ga_{0.4}$ $Al_{0.6}$ As double-heterostructure material," *Appl. Phys. Lett.,* **25**, 1974, p. 216.

143. Miller, R.C, et al., "Optically pumped taper-coupled $GaAs$-$Al_xGa_{1-x}As$ laser with a second-order Bragg reflector," *J. Appl. Phys.,* **49**, 1978, p. 539.

144. Milnes, A.G., and D.L. Feucht, *Heterojunctions and Metal-Semiconductor Junctions,* Academic Press, N.Y., 1972.

145. Minden, H.T., and R.Premo, "High-temperature GaAs single heterojunction laser diodes," *J. Appl. Phys,* **45**, 1974, p. 4520.

146. Moorehead, F.F., Jr., "Light-emitting diodes," *Scientific American,* **216**, 1967, p. 108.

147. Nagai, H., and Y. Noguchi, "A new grading layer for liquid epitaxial growth of $Ga_xIn_{1-x}As$ on GaAs substrate," *Appl. Phys. Lett.,* **26**, 1975, p. 108.

148. Nagi, H., and Y. Noguchi, "Ge-doped $Ga_xIn_{1-x}As$ LED's in 1 $\mu$m wavelength region," *J. Appl. Phys.,* **49**, 1978, p. 450.

149. Nahory, R.E., M.A. Pollack, and J.C. DeWinter, "Growth and characterization of liquid-phase epitaxial $In_xGa_{1-x}As$," *J. Appl. Phys.,* **46**, 1975, p. 775.

150. Naito, M., and A. Kasami, "GaP red-emitting diodes with an external quantum efficiency of 12.6%," Presented at the 1974 IEEE Specialist Conference on the Technology of Electroluminescent Diodes, Atlanta, Ga., Nov. 1974.

151. Nakada, O., et al., "Continuous operation over 2500 h of double heterostructure laser diodes with output powers more than 80 mW," *Jap. J. Appl. Phys.,* **13**, 1974, p. 1485.

152. Nakamura, M., et al., "GaAs-$Ga_{1-x}Al_xAs$ double-heterostructure distributed-feedback diode lasers," *Appl. Phys. Lett.,* **25**, 1974, p. 487.

153. Namizaki, H., M. Nagano, and S. Nakahara, "Frequency response of $Ga_{1-x}Al_xAs$ light-emitting diodes," *IEEE Trans. Electron Devices,* **ED-21**, 1974, p. 688.

154. Nash, F.R., "Laser-excited photoluminescence of three-layer GaAs double-heterostructure laser materials," *Appl. Phys. Lett.,* **27**, 1975, p. 234.

155. North, D.O., and H.S. Sommers, Jr., "Saturation of the junction voltage in stripe-geometry (AlGa)As double-heterostructure junction lasers: a comment," *Appl. Phys. Lett.,* **30**, Jan. 1977, p. 116.

156. Nuese, C.J., "Diode sources for 1.0 to 1.2 $\mu$m emission," *IEEE Int. Electron Devices Meeting, Technical Digest*, Washington, D.C., 1976, p. 125.

157. Nuese, C.J., and G.H. Olsen, "Room-temperature heterojunction laser diodes of $In_xGa_{1-x}As/In_yGa_{1-y}P$ with emission wavelength between 0.9 and 1.15 $\mu$m," *Appl. Phys. Lett.,* **26**, 1975, p. 528.

158. Nyul, P., and A. Limm, "High average power GaAs injection laser arrays," Proc. of the Technical Program, Electro-optical Systems Design Conf., New York, Sept. 1970, p. 473.

159. Olsen, G.H., et al., "Reduction of dislocation densities in hetero-epitaxial III-V VPE semiconductors," *J. Appl. Phys.,* **46**, 1975, p. 1643.

160. Osamura, K., S. Naka, and Y. Murakami, "Preparation and optical properties of $Ga_{1-x}In_xN$ thin films," *J. Appl. Phys.,* **46**, 1975, p. 3432.

161. Ota, T., K. Kobayashi, and K. Takahashi, "Light-emitting mechanism of ZnTe-CdS heterojunction diodes," *J. Appl. Phys.,* **45**, 1974, p. 1750.

162. Ozsan, M.E., and J. Woods, "Electroluminescence in zinc selenide," *Solid-State Electronics,* **18**, 1975, p. 519.

163. Panish, M.B., and I. Hayashi, "Low threshold injection lasers utilizing a p-n heterojunction and a p-p heterojunction between GaAs and $Al_xGa_{1-x}As$," *Proc. Int. Conf. Phys. Chem. Semicond. Heterojunctions Layer Structures*, Budapest 1970, **2**, p. 419, Hung. Acad. Sciences, Budapest, Hungary.

164. Pankove, J.I., "Blue anti-stokes electroluminescence in GaN," *Phys. Rev. Lett.,* **34**, 1975, p. 809.

165. Payne, D.N., and W.A. Gambling, "Zero material dispersion in optical fibres," *Electronics Lett.*, 11, 1975, p. 176.
166. Petroff, P., and R.L. Hartman, "Rapid degradation phenomenon in heterojunction GaAlAs-GaAs lasers," *J. Appl. Phys.*, 45, 1974, p. 3899.
167. Philipp-Rutz, E.M., "Spatially coherent radiation from an array of GaAs lasers," *Appl. Phys. Lett.*, 26, 1975, p. 475.
168. Pilkuhn, M.H., "Fundamentals of stimulated emission in semiconductors," *J. Lumin.*, 7, 1973, p. 269.
169. Pilkuhn, M.H., and H. Rupprecht, "Spontaneous and stimulated emission from GaAs diodes with three-layer structures," *J. Appl. Phys.*, 37, 1966, p. 3621.
170. Pilkuhn, M. H., and H. Rupprecht, "Optical and electrical properties of epitaxial and diffused GaAs injection lasers," *J. Appl. Phys.*, 38, 1967, p. 5.
171. Ralston, R.W., et al., "Double heterostructure $Pb_{1-x}Sn_xTe$ waveguides at 10.6 $\mu$m," *Appl. Phys. Lett.*, 26, 1975, p. 64.
172. Rao, B.S.S., A. Subrahmanyam, and P. Swarup, "A technique of modulating pulsed semiconductor lasers," *IEEE Trans. Comm.*, COM-21, 1973, p. 284.
173. Rediker, R.H., "Semiconductor lasers," *Physics Today*, Feb. 1965, p. 42.
174. Rees, C., "Display economics," *New Electronics, G.B.*, 6(17), 1973, p. 56.
175. Reinhart, F.K., R.A. Logan, and C.V. Shank, "GaAs-$Al_xGa_{1-x}As$ injection lasers with distributed Bragg reflectors," *Appl. Phys. Lett.*, 27, 1975, p. 45.
176. Rezek, E.A., et al., "Single and multiple thin-layer ($L_z \leq 400$ Å) $In_{1-x}Ga_xP_{1-z}As_z$-InP heterostructure light emitters and lasers ($\lambda \sim 1.1$ $\mu$m, 77° K)," *J. Appl. Phys.*, 49, 1978, p. 69.
177. Robinson, R.J., and Z.K. Kun, "p-n Junction zinc sulfo-selenide and zinc selenide light-emitting diodes," *Appl. Phys. Lett.*, 27, 1975, p. 74.
178. Rode, D.L., "How much Al in the AlGaAs-GaAs laser?" *J. Appl. Phys.*, 45, 1974, p. 3887.
179. Roddy, J.E., "Wideband signal recording on film using (AlGa) As CW injection lasers," *RCA Rev.*, 36, 1975, p. 746.
180. Ross, M. (Ed.), *Laser Applications*, Vols. 2,3 Academic Press, N.Y., 1974, 1977.
181. Rossi, J.A., et al., "Comparison of optical to injection excitation in GaAs heterostructure lasers," *J. Appl. Phys.*, 45, 1974, p. 5383.
182. Saitoh, T., and S. Minagawa, "Multicolor light-emitting diodes with double junction structure," *IEEE Trans. Electron Devices*, ED-22, 1975, p. 29.
183. Sangster, R.C., *Compound Semiconductors*, (Edited by R.K.W. Hardson and H.Z. Guering), Vol. 1, Reinhold, N.Y., 1962, p. 241.
184. Saul, R.H., J. Armstrong, and W.H. Hackett, Jr., "GaP red electroluminescent diodes with an external quantum efficiency of 7%," *Appl. Phys. Lett.*, 15, 1969, p. 229.
185. Schicketany, D., J. Wiltmann, and G. Zeidler, "Effects of degradation-induced absorption in GaAs-d.h.s. laser diodes," *Electronics Lett.*, 10, 1974, p. 252.
186. Schul, G., and P. Mischel, "$Ga_xIn_{1-x}P$-$Ga_yAl_{1-y}$ as heterojunction close-confinement injection laser," *Appl. Phys. Lett.*, 27, 1975, p. 394.
187. Scifres, D.R., R.D. Burnham, and W. Streifer, "Highly collimated laser beams from electrically pumped SH GaAs/GaAlAs distributed-feedback lasers," *Appl. Phys. Lett.*, 26, 1975, p. 48.
188. Scifres, D.R., R.D. Burnham, and W. Streifer, "Grating coupled GaAs/GaAlAs ring laser," *Appl. Phys. Lett.*, 28, 1976, p. 681.
189. Seko, A., et al., "Self-quenching in semiconductor lasers and its applications in optical memory readout," *Appl. Phys. Lett.*, 27, 1975, p. 140.
190. Shaklee, K. L., R.E. Nahory, and R.F. Leheny, "Optical gain in semiconductors," *J. Lumin.*, 7, 1973, p. 284.

191. Skobel'tsyn, D.V., "Radiative recombination in semiconducting crystals," Consultants Bureau, N.Y., 1975.
192. Smith, B.L., "A deep center associated with the presence of nitrogen in GaP," *Appl. Phys. Lett.*, **26**, 1975, p. 122.
193. Smith, G.E., "The display job calls the shots in contest between GaP and GaAsP," *Electronics*, Oct. 1971, p. 74.
194. Smith, W.V., "Solid state light sources: lasers," in *Topics in Solid State Quantum Electronics*, W.D. Hershberger, Ed., John Wiley and Sons, Inc., N.Y., 1972, p. 241.
195. Sommers, H.S., Jr., "Performance of injection lasers with external gratings," *RCA Rev.*, **38**, 1977, p. 33.
196. Sommers, H.S., "On the internal quantum efficiency of injection lasers," *Appl. Phys. Lett.*, **32**, 1978, p. 547.
197. Sommers, H.S., and D.O. North, "The power spectrum of injection lasers: the theory and experiment on a non-linear model of lasing," *Solid-State Electronics*, **19**, 1976, p. 675.
198. Sonomura, H., et al., "Composition determination of $Al_xGa_{1-x}P$ alloys using diffracted x-ray intensities," *J. Appl. Phys.*, **45**, 1974, p. 5109.
199. Sonomura, H., et al., "Homogeneity of solution grown $Al_xGa_{1-x}P$ alloys," *J. Appl. Phys.*, **46**, 1975, p. 3693.
200. Stern, F., *Laser Handbook* (Edited by F.T. Arecchi and E.O. Schulz-DuBois), Vol. 1, North Holland, Amsterdam, 1972, p. 425.
201. Streifer, W., D.R. Scifres, and R.D. Burnham, "Status of distributed feedback lasers," *IEEE Int. Electron Devices Meeting, Technical Digest*, Washington, D.C., 1976, p. 117.
202. Suematsu, Y., M. Yamada, and K. Hayashi, "A multi-hetero-AlGaAs laser with integrated twin guide," *Proc. IEEE*, **63**, 1975, p. 208.
203. Suzuki, T., and Y. Matsumoto, "Effects of dislocations on photoluminescent properties in liquid phase epitaxial GaP," *Appl. Phys. Lett.*, **26**, 1975, p. 437.
204. Tada, K., and K. Hirose, "A new light modulator using perturbation of synchronism between two coupled guides," *Appl. Phys. Lett.*, **25**, 1974, p. 651.
205. Tani, Z., et al., "GaAs light coupled device," Proc. 4th Conf. on Solid-State Devices, Tokyo, 1972. Suppl. to the *J. Japan. Soc. Appl. Phys.*, **42**, 1973, p. 258.
206. Thornton, P.R., *The Physics of Electroluminescent Devices*, E. & F.N. Spon Limited, London, 1967.
207. Tomasetta, L.R., and C.G. Fonstad, "Threshold reduction in $Pb_{1-x}Sn_xTe$ laser diodes through the use of double heterojunction geometries," *Appl. Phys. Lett.*, **25**, 1974, p. 440.
208. Tsarenkov, B.V., et al., "Delay of electroluminescence emitted by GaP p-n structures after application of a forward pulse," *Sov. Phys. Semicond.*, 7, 1973, p. 146.
209. Tsukada, T., "Optical characteristics of mesa-stripe-geometry double-heterostructure injection lasers," Proc. 4th Conf. on Solid-State Devices, Tokyo, 1972. Suppl. to the *J. Japan. Soc. Appl. Phys.*, **42**, 1973, p. 251.
210. Tsukada, T., "$GaAs-Ga_{1-x}Al_xAs$ buried-heterostructure injection lasers," *J. Appl. Phys.*, **45**, 1974, p. 4899.
211. Tsukada, T., "Buried-heterostructure injection lasers," Proc. 6th Conf. on Solid-State Devices, Tokyo, 1974. Suppl. to the *J. Japan. Soc. Appl. Phys.*, **44**, 1975, p. 33.
212. Vander Sande, J.B., and E.T. Peters, "On precipitatelike zones in as-grown GaAs," *J. Appl. Phys.*, **46**, 1975, p. 3689.
213. Vecht, A., "Electroluminescent displays," *J. Vac. Sci. Tech.*, **10**, 1973, p. 789.
214. Vecht, A., and N.J. Werring, "Direct current electroluminescence in ZnS," *J. Phys. D: Appl. Phys.*, **3**, 1970, p. 105.

215. Vecht, A., et al., "Materials control and dc electroluminescence in ZnS: Mn, Cu, Cl powder phosphors," *J. Phys. D: Appl. Phys.*, 2, 1969, p. 953.
216. Wessels, B.W., "Vapor deposition of GaP for high-efficiency yellow solid-state lamps," *J. Electrochem. Soc.*, 122, 1975, p. 402.
217. White, G., and G.M. Chin, "A 100-Mb.s$^{-1}$ fiber optic communication channel," *Proc. IEEE*, 61, 1973, p. 683.
218. Wieder, H.H., and D.A. Collins, "Minority carrier lifetime in InAs epilayers," *Appl. Phys. Lett.*, 25, 1974, p. 742.
219. Wright, P.D., et al., "In$_{1-x}$Ga$_x$P$_{1-z}$As$_z$ double heterojunction laser operation (77°K, yellow) in an external grating cavity," *J. Appl. Phys.*, 47, 1976, p. 3580.
220. Wu, T.Y., "Thermochemical analysis and optimum conditions for vapor epitaxy of GaAs$_{1-x}$P$_x$ ($0.7 < x < 0.9$)," *J. Electrochem. Soc.*, 122, 1975, p. 778.
221. Yano, S., T. Sakurai and T. Inoguchi, "(Ga.Al)As P-N-P-N diode," Proc. 2nd Conf. on Solid-State Devices, Tokyo, 1970. Suppl. to *J. Japan. Soc. Appl. Phys.*, 40, 1971, p. 166.
222. Yariv, A., A. Katzir, and H.W. Yen, "GaAs-GaAlAs distributed-feedback diode lasers with separate optical and carrier confinement," *Appl. Phys. Lett.*, 27, 1975, p. 145.
223. Zachos, T.H. and J.E. Ripper, "Resonant modes of GaAs junction lasers," *IEEE J. Quantum Electronics*, QE-5, 1969, p. 29.
224. Zaha, M.A., "Shedding some needed light on optical measurements," *Electronics*, Nov. 6, 1972, p. 91.
225. "Current variation changes LED from red to green," *Electronics*, Jan. 24, 1974, p. 25.
226. Kahn, F.J. et al., "Introduction, Special Issues on Displays and LED's," *IEEE Trans. on Electron Devices*, ED-24, 1977, p. 785.
227. Craford, M.G., "Recent developments in light-emitting-diode technology," *IEEE Trans. on Electron Devices*, ED-24, 1977, p. 935.
228. Tannas, L.E. and W.F. Goede, "Flatpanel displays: a critique," *IEEE Spectrum*, July 1978, p. 26.

# 15

# Semiconductor Sensors and Transducers

## CONTENTS

Semiconductor effects make possible a wide range of sensors responding to physical effects such as light, magnetic fields, position, acceleration, strain, temperature, high energy particles, gamma rays, and ambient gases. Light detection has been discussed earlier (Chapters 12 and 13).

This chapter begins with a section on sensors involving magnetics. Semiconductor sensors involving strain effects and temperature are then discussed. A discussion of oxide semiconductor layers that are the basis for detection of gases such as $H_2$ and $CH_4$ is given and semiconductor detectors of particles and gamma rays in high energy and radiation physics are reviewed briefly.

## 15.1 SEMICONDUCTOR SENSORS INVOLVING MAGNETICS

### 15.1.1 Hall Effect Devices

Magnetic sensors fall into three categories, namely Hall effect sensors, magnetoresistive devices, and junction structures in which minority carrier injection streams are deflected by or modified by a magnetic field.

The Hall effect is illustrated in Fig. 15.1 for a moderately heavily doped semi-conductor. The applied current $I$ corresponds to an electron flow $n\mu_n \mathscr{E}_x$ where $n$ is the electron (majority carrier) density and $\mathscr{E}_x$ is the field strength in the direction of flow. This electron flow interacts with the magnetic field $B$ applied in the $z$ direction perpendicular to the plane of the bar. This deflects electrons on the $y$ direction and lead 3 becomes negative with respect to 4. This corresponds to a field $\mathscr{E}_y$ which in equilibrium produces a force $q\mathscr{E}_y$ that balances the $v \times B$ force. Hence

$$q\mathscr{E}_y = q\mu_n \mathscr{E}_x \times B \qquad (15.1)$$

or

$$\frac{-V_H}{y} = \mu_n \frac{V_x}{L} \times B$$

where

$$V_H = -\frac{1}{qn}\left(\frac{I}{t}\right) \cdot B$$

$$= \frac{R_H}{t}\left(I \cdot B\right) \qquad (15.2)$$

where $R_H$ is the Hall coefficient. If $B$ has the units of $V \cdot sec/cm^2$, $I$ is in amperes, and $t$ is in centimeters, then the units of $R_H$ are $cm^3/A \cdot sec$. The magnetic field may also be expressed in kilogauss where 1 kG is $10^{-5}$ $V \cdot sec/cm^2$. More exact analysis for a lightly doped semiconductor gives

$$R_H = r\frac{1}{q}\left(\frac{p - b^2 n}{(p + bn)^2}\right) \qquad (15.3)$$

where $b$ is $\mu_n/\mu_p$ and $r$ is $3\pi/8$ (= 1.18) for phonon scattering and $315\pi/512$ (= 1.93) for ionized impurity scattering.

One application of the Hall effect is the measurement of magnetic field strength. Figure 15.1(b) shows the variation of output voltage with field strength for an InAs Hall plate of typical doping and thickness and with the applied current as the parameter. From the relationship that $R_H \simeq -1/qn$ it seems that any semiconductor may be used and that $R_H$ will be large provided $n$ is small. However the sensitivity of the device is

$$\frac{V_H}{IB} = \frac{R_H}{t} \qquad (15.4)$$

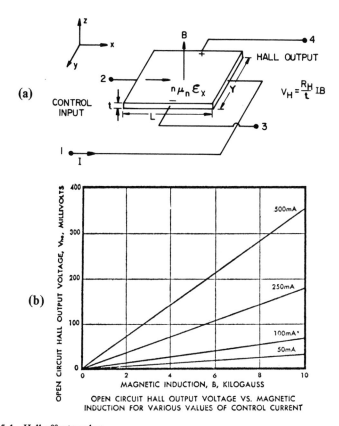

Fig. 15.1. Hall effect probe:
(a)  Structure, and
(b)  InAs Hall probe characteristics. (After Bulman. Reprinted with permission from *Solid-State Electronics,* 9, 1966, p. 361, Pergamon Press.)

where $I$ is limited by the input power $P_{in}$ permitted by the allowed heat dissipation of the Hall slice. The input power is given by

$$P_{in} = I^2 R \qquad (15.5)$$

hence

$$\frac{V_H}{P_{in}^{1/2}B} = \left(\frac{R_H \mu_n}{t}\right)^{1/2}\left(\frac{y}{L}\right)^{1/2} \qquad (15.6)$$

So for a given input power and magnetic field, the Hall voltage increases as $(R_H \mu_n)^{1/2}$. Thus the greater the ratio $\mu_n/n$ the better. However, the ratio $\mu_n/n$ is not a sufficient merit figure if the Hall device must deliver power to a matched load. For a large Hall power, $n$ must be large to reduce the device internal resistance and therefore $\mu_n$ must be large. The Hall power output $P_H$ can be shown to be proportional to $P_{in} \, (\mu_n B)^2$. This is one reason for choosing materials of high mobility such as InSb ($\mu_n = 70{,}000$ cm$^2$/V · sec at 300°K), InAs($\mu_n = 20{,}000$ cm$^2$/V · sec), and GaAs ($\mu_n = 5000$ cm$^2$/V · sec) for Hall devices instead of Si or Ge. This is necessary only if the Hall device must deliver power under matched load conditions. If the Hall unit is to be followed by a high-impedance amplifier, Si or Ge devices may be used.

Hall devices are the basis of magnetic field sensors. For intrinsic InSb sensors of thickness 100 $\mu$m or 5 $\mu$m and input power 100 mW at 300°K the $V_H/B$ sensitivity is 1.1 or 5 V per V · sec/m$^2$ for allowed currents of 310 and 70 mA and internal resistances of 1 and 20 $\Omega$ respectively.[91] The temperature sensitivity depends upon the variation of $n$ with temperature and from this viewpoint higher bandgap materials are preferable. Commercial gaussmeters are available with ranges from a fraction of a gauss to tens of kilogauss, and can be used in reference feedback systems to obtain magnetic field regulation to better than 1 part in $10^6$.

One feature of Hall effect sensors is that they provide a response that is proportional to the instantaneous field, whereas coil-type field sensors that depend upon a $d\phi/dt$ flux cutting effect must be controlled in speed to give correct results. Thus Hall effect devices are suitable, for instance, for reading magnetically coded labels with handheld wands in point-of-sales systems. Some problems in magnetic sensing head design for magnetic tape systems are illustrated in Fig. 15.2.

Since Si is a suitable Hall device material in a high-impedance circuit it is possible to combine a sensor and a Si integrated circuit. A typical Si Hall generator may have a sensitivity of 30 mV/kG but an output current only in the low microampere range. In some applications a logic type 1 or 0 output is desired in response to a magnetic field change. The integrated circuit chip of Fig. 15.3 switches from a 1 level (several volts) to a 0 level when the Hall generator experiences a field level of 500 G. The Schmitt trigger section prevents turn on of the 1 level until the field falls back below 200 G. These levels were chosen since 500 G can easily be provided by small permanent magnets and because 200 G is larger than stray magnetic fields (30-100 G) likely to be produced by inductors or transformers that may be close by. Such chips may be used as part of Hall effect keyboards to avoid the contact bounce effects of mechanical switches.

Hall IC sensors in digital tape and disk memories are improved by the absence of significant inductance between the device and the amplifier in the integrated circuit form. If a linear output is required for reading analog tapes the Schmitt trigger may be replaced by a differential amplifier. Such Hall pick-ups are relatively insensitive to the effects of tape speed fluctuations.

In a Hall plate the two Hall sense electrodes cannot usually be aligned perfectly because of fabrication tolerances and thus the output voltage $V_H$ is not exactly zero at zero magnetic field strength [see Fig. 15.4(a)]. One solution to this is to

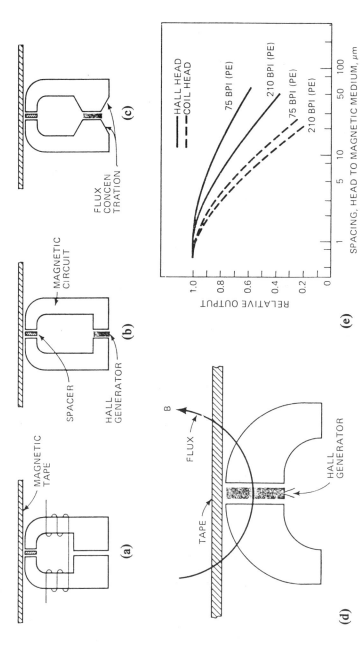

Fig. 15.2. Magnetic sensing head designs. A coil-wound head usually has a wide-faced gap in the rear (a), which seems to be a good place to put the Hall generator (b). However, the Hall generator requires both a low reluctance, and thus wide faces, and a concentration of flux, and therefore narrow, tapered faces (c). This dilemma cannot be resolved with conventional head shapes, and requires a new approach, the front-gap design of (d). Here the InSb Hall generator is brought close to the tape, while the horseshoe-shaped pole pieces act as an open magnetic circuit and concentrate the magnetic flux in the generator and perpendicular to it, if the gap is narrow. But the design leaves no room for an electrode at the top of the generator for which therefore must be recessed into the gap to protect it from wear and to permit front electrode connections. This Hall head tolerates irregularities in the surface being read, as well as gaps, and allows considerable head wear before sensitivity drops appreciably. (BPI stands for bits per inch.) (After Murai. Reprinted from *Electronics*, Feb. 1, 1973; Copyright © McGraw-Hill, Inc., N.Y., 1973.)

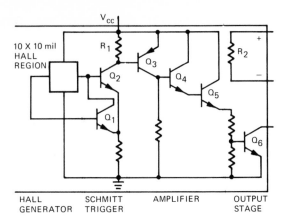

Fig. 15.3. An IC Hall effect sensor. The Hall generator is a square section of the Si. Resistor $R_2$ helps to tailor the output. When it connects $Q_6$'s collector to a 5-V supply, IC directly drives TTL and DTL. With 15-V supply, circuit operates MOS devices. (After Oppenheimer. Reprinted from *Electronics*, Aug. 2, 1971; Copyright © McGraw-Hill, Inc., N.Y., 1971.)

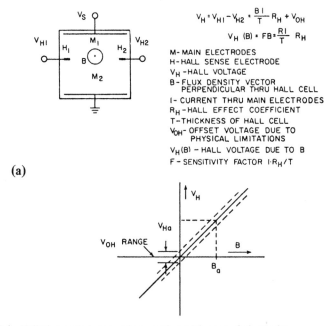

$$V_H = V_{H1} - V_{H2} = \frac{B\,I}{T} R_H + V_{OH}$$

$$V_H(B) = FB = \frac{R\,I}{T} R_H$$

M- MAIN ELECTRODES
H- HALL SENSE ELECTRODE
$V_H$ - HALL VOLTAGE
B- FLUX DENSITY VECTOR PERPENDICULAR THRU HALL CELL
I- CURRENT THRU MAIN ELECTRODES
$R_H$ - HALL EFFECT COEFFICIENT
T- THICKNESS OF HALL CELL
$V_{OH}$ - OFFSET VOLTAGE DUE TO PHYSICAL LIMITATIONS
$V_H(B)$ - HALL VOLTAGE DUE TO B
F- SENSITIVITY FACTOR $I \cdot R_H / T$

(a)

Fig. 15.4. Hall plate unbalance and correction with control electrode:

(a) A conventional Hall cell does not have $V_H$ exactly zero with zero magnetic field, because of physical inaccuracies and nonuniformities. This finite offset voltage is shown as the $V_{OH}$ range in the graph. The total Hall voltage is the sum of the $V_{OH}$ and $V_H(B)$, and

(b) Current $I_c$, injected into electrode $C$ causes a linear change $V_H(I_c)$ in the Hall voltage. The change can be used to compensate for undesirable offsets, to introduce bias, or to simulate a magnetic field. (After Braun. Reprinted by permission from *Electronic Design*, May 24, 1974, pp. 88,89; Copyright © Hayden Publishing Company, 1974.)

896

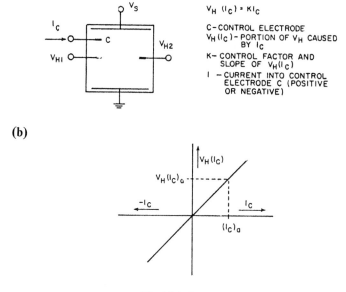

**(b)**

Fig. 15.4. (continued)

add an extra electrode with a control current applied that compensates for the undesirable offset effect. This solution is shown in Fig. 15.4(b). The addition of more than one control electrode allows circuit feedback effects that can produce switching response with any switching point offset that may be desired (see Fig. 15.5).

A miniswitch with a Si Hall sensor that involves a total push-button travel of only 2.5 mm is shown in Fig. 15.6(a). In the fully released position shown the Hall cell sees maximum flux in one direction (about 1000 G) and full movement results in reversal of this flux. Since the Hall transducer circuit switches at ± 60 G the miniswitch actually operates at about 0.02 mm (0.8 mil) on either side of the push-button midposition.

Figure 15.6(b) shows the application of a similar sensor to detect step rotation or the rotational speed of a wheel with magnets on the perimeter.

Hall sensors of displacement, compared with inductance-type sensors such as differentially wound transformers, have the advantage of practically no reaction forces of magnetic attraction. Often the system is simpler than for capacitance displacement sensors although there is the problem of supplying four wires to the moving part if this is the Hall plate (although not if the magnet is small enough to be moved). However, generally the magnet is large for high-sensitivity systems having a measurement range of 1 or 2 mm. Typical magnet systems are illustrated in Fig. 15.7. The arrangement Fig. 15.7(c) has a field gradient of 10 kG/mm and an output of about 100 mV for 10-kG change. The power sensitivity for an InAs Hall plate in this system may be 2 mW/mm and about a tenth of this for a Ge

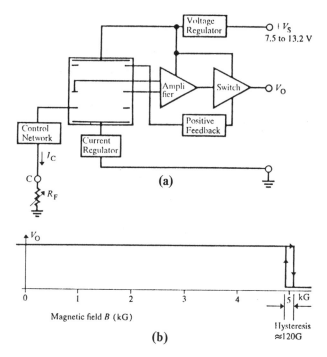

Fig. 15.5. Hall effect magnetic sensing switching circuit with current regulation and switching point offset, capability:
(a)   Circuit, and
(b)   Switching diagram with a 5000-G switching point offset. (After Braun. *IBM J. Res. Develop.,* **19,** July 1975, p. 346; Copyright © 1975 by International Business Machines Corporation, reprinted with permission.)

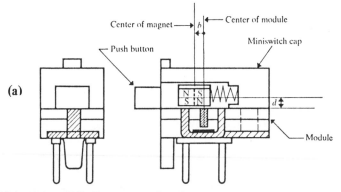

Fig. 15.6. Various Hall offset sensors of motion:
(a)   Miniswitch built with magnet detector module, and
(b)   Magnet detector used to detect rotation or to measure the rotational speed of an emitter wheel. (After Braun. *IBM. J. Res. Develop.,* **19,** July 1975, pp. 349, 351; Copyright © 1975 by International Business Machines Corporation, reprinted with permission.)

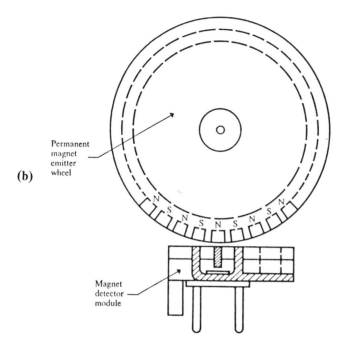

**(b)**

Permanent
magnet
emitter
wheel

Magnet
detector
module

Fig. 15.6. (continued)

sensor. Some applications of Hall sensors that are open-loop nonfeedback systems are shown in Fig. 15.8. The accuracy, linearity, and insensitivity to input voltage changes and to load changes are much improved by the use of feedback. A typical closed-loop system performance is shown in Fig. 15.9.

Angular position $\theta$ may be determined by having two Hall elements arranged at right angles on a shaft in a magnetic field and resolution to a few minutes of arc is possible (see Fig. 15.10).

Application to a syncho (a servo component device used to signal angular position or to couple two shafts together electrically) is shown in Fig. 15.11. Such devices are as small as 0.5 in. in diameter. The use of the permanent magnet rotor and the Hall plates eliminates windings, laminations, brushes, and slip rings.

A Hall effect sensor may be used as a compass or as a magnetometer to determine field strength.

An important class of applications is based on the use of the multiplication property $I \cdot B$ in Eq. (15.2). For a wattmeter the product of line current and line voltage is required. If in Fig. 15.12(a) the line current is $I_L \cos(\omega t)$ in the coil in the magnetic circuit, the magnetic induction is $B \cos(\omega t + \delta_1)$ where $\delta_1$ is a phase shift caused by the hysteresis of the core. The voltage transformer across the line

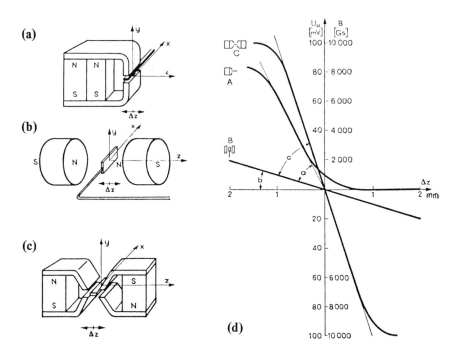

**(a)**

**(b)**

**(c)**

**(d)**

Fig. 15.7. Hall effect motion sensors require an appropriate magnet geometry for high sensitivity. (After Nalecz and Warsza. Reprinted with permission from *Solid-State Electronics*, **9**, 1966, p. 485, Pergamon Press.)

voltage $V_L$ cos $\omega t$ gives a Hall sensor control current $I_c$ cos $(\omega t + \theta + \delta_2)$. The Hall output voltage is then

$$V_H = \frac{R_H}{t} I_c \cos(\omega t + \theta + \delta_2) B \cos(\omega t + \delta_1) \tag{15.7}$$

and since $I$ and $B$ are very closely proportional to $V_L$ and $I_L$ the Hall voltage has a dc component

$$V_H = K V_L I_L \cos(\theta + \delta_2 - \delta_1) \tag{15.8}$$

With careful design the phase shift error $\delta_2 - \delta_1$ may be made small and the instrument indicates the true power $V_L I_L$ cos $\theta$. InAs is often the Hall material used and this has a temperature coefficient of $-0.1\%/°C$. If accurate measurement is required over a wide temperature range, compensation of the temperature effect error is needed and may be provided by a thermistor network. The accu-

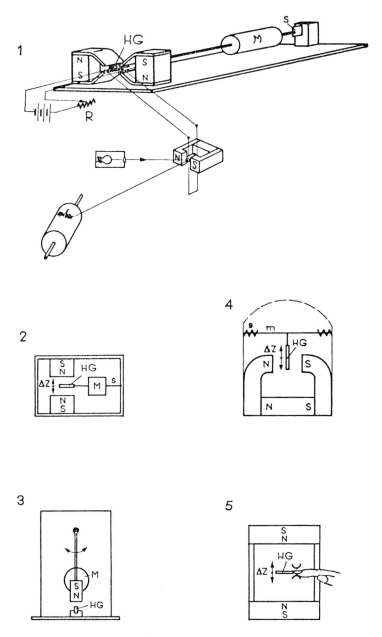

Fig. 15.8. Some applications of Hall generators as the displacement and motion transducers operating in feedback-free systems: 1. seismograph; 2. accelerometer; 3. pendulum; 4. microphone; 5. quivering meter for measuring vibrations (for instance of the hands for medical purposes). (After Nalecz and Warszu. Reprinted from *Solid-State Electronics*, 9, 1966, p. 485, Pergamon Press.)

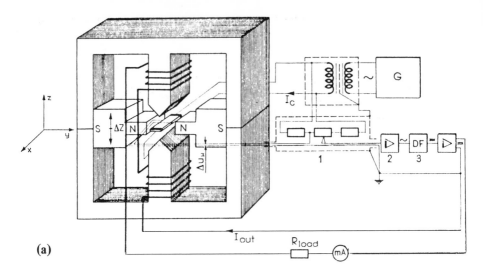

(a)

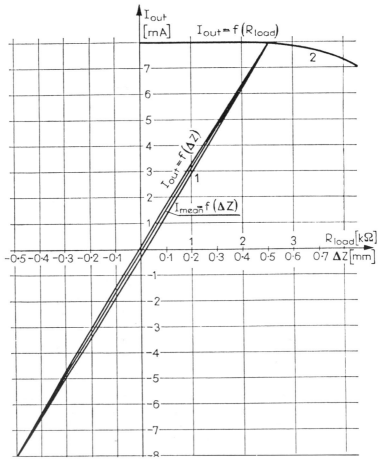

(b)

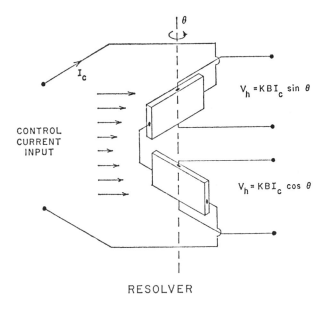

$$V_h = KBI_c \sin \theta$$

$$V_h = KBI_c \cos \theta$$

CONTROL
CURRENT
INPUT

$I_c$

$\theta$

RESOLVER

Fig. 15.10. Schematic of two Hall plates at 90 deg for angle sensing as in resolvers. (After Bulman. Reprinted from *Solid-State Electronics,* 9, 1966, p. 361, Pergamon Press.)

racy attainable in a wattmeter rated at 500 W, 50 to 2500 Hz, with a power factor range 0.1 lag to 0.1 lead, is ± 0.25% of full scale for a temperature range of 0° to 50°C.

Since the transducer output is a product function it is capable of measuring reactive volt amps (vars) if the input phase relationship is right. There are various methods of obtaining the proper phase relationship to measure vars. The simplest technique is shown in Fig. 15.12(b) where the voltage transformer is replaced by capacitors that provide the required 90-deg phase shift. If the output must be isolated from the line then the magnetic circuit is driven from the line voltage and provides the 90-deg phase shift and a current transformer supplies drive for the Hall element that is proportional to the line current.

The Hall effect may also be used to measure power in waveguides at microwave frequencies. Figure 15.13 shows an early application in which a rectangular

---

Fig. 15.9. Example of a closed-loop Hall effect position sensor ($\Delta Z$ displacement):
(a)   Circuit where (1) balances the zero-field residual voltage, (2) is an ac amplifier, and (3) is a phase discriminator, and
(b)   Characteristics of the transducer. (1) output current versus displacement of the Hall generator. (2) Output current versus load resistance. (After Nalecz and Watsza. Reprinted from *Solid-State Electronics,* 9, 1966, p. 485, Pergamon Press.)

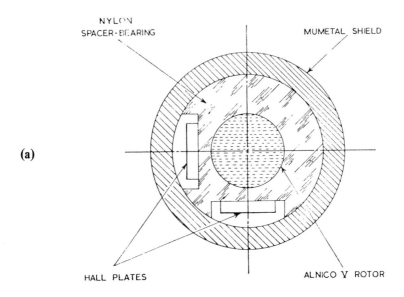

**(a)**

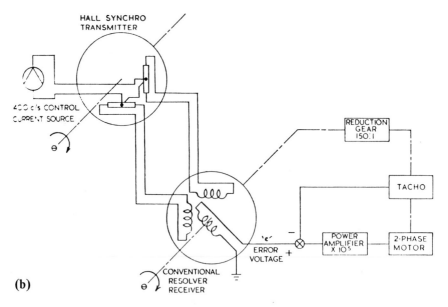

**(b)**

Fig. 15.11. Hall effect syncho control system:
(a)  Sectional end elevation of a tangential synchro (diameter about 0.8 cm), and
(b)  Schematic of the position control system used for testing synchros. (After Inglis and Donaldson. Reprinted from *Solid-State Electronics*, 9, 1966, p. 541, Pergamon Press.)

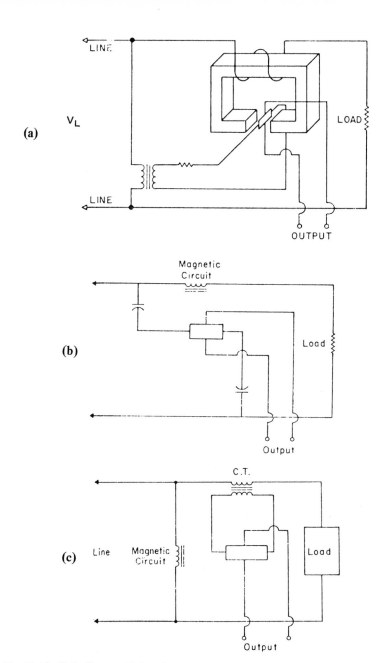

**(a)**

**(b)**

**(c)**

Fig. 15.12. Hall effect multipliers for wattmeters:
(a) For a wattmeter the product $IB$ provides an output that is proportional to $V_L I_L \cos\theta$,
(b) For measurement of reactive VAR the capacitors provide the required 90 deg phase shift, and
(c) VAR meter in which the magnetic circuit provides the phase shift and the current transformer provides isolation of the output from the line. (After Crawford. Reprinted from *Solid-State Electronics, 9,* 1966, pp. 529, Pergamon Press.)

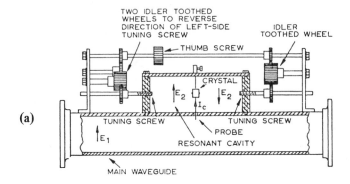

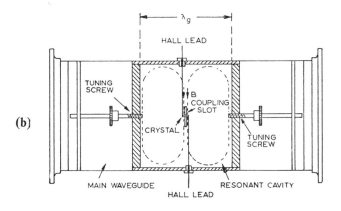

Fig. 15.13.  Hall effect device for power measurement in a 4-GHz rectangular cavity:
(a) Elevation, and
(b) Plan showing the coupling slot for the current probe. (After Stephenson and Barlow. Reprinted from *Proc. IEE* (London), **106**, pt. B, 1959, p. 27.

$H_{01}$ resonant cavity about one wavelength long contains the Ge Hall sensor and is magnetically coupled by means of a narrow slot to the main waveguide carrying the power to be measured. A probe passes through the slot and induces in the Hall sensor a displacement current proportional (in time-quadrature) to the electric field in the main waveguide. At the sensor position the magnetic field in the cavity at resonance is proportional to the magnetic field in the main waveguide and also in time-quadrature to it. The voltage on the Hall leads then gives the required power measurement. In a 4-GHz system of this kind a Hall voltage of about 2 $\mu$V is obtained for a power level of 100 mW. The accuracy is ± 3% for a power range of 30 mW to 20 W and a standing-wave ratio between unity and 0.1.

A Hall plate has gyrator properties that are useful in some microwave circuits. Considering Fig. 15.14(a) as a four-pole device, the application of a voltage $V_1$

GYRATOR

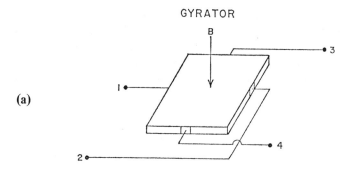

**(a)**

APPLY $V_I$ BETWEEN I & 2, GET $kV_I$ BETWEEN 3 & 4
APPLY $V_I$ BETWEEN 3 & 4, GET $-kV_I$ BETWEEN I & 2

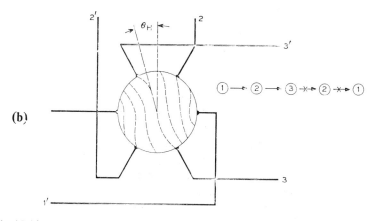

**(b)**

Fig. 15.14.  Hall gyrators and circulators:
(a)   A gyrator plate is an antireciprocal four-pole device, and
(b)   A three-port circulator with the equipotential lines shown for an input at port 1.
Notice that there is an output at port 2 but no voltage output between ports 3 and
3'. The plate may be Ge and the field strength about 14 kOe for a load impedance of
100-1000 $\Omega$. (After Grubbs. Reprinted with permission from *Bell Syst. Tech.*
**38,** 1959, p. 853; Copyright © American Telephone and Telegraph Company.)

between terminals 1 and 2 gives an output voltage $kV_1$ between terminals 3 and
4 but application of $V_1$ between terminals 3 and 4 gives $-kV_1$ between terminals
1 and 2. Hence the device is an antireciprocal four-pole structure. The minimum
power loss in a Hall effect gyrator is 7.6 dB.[93]

If external shorting resistances are connected between terminals 1 and 3 and
between terminals 2 and 4 the gyrator can be made to function as an isolator that
transmits signals in only one direction. For an InSb device the reverse isolation
may be 100 dB, and the forward loss about 7.6 dB.

If more than two ports are provided [see Fig. 15.14(b)], a device known as a circulator is obtained in which signals are transmitted between adjacent ports in a particular (say clockwise) direction. Hence in a three-port structure a transmitter can be connected to an antenna, and the transmission does not enter the receiver. However, a signal received on the antenna can be circulated to a receiver coupled to port 3.

Hall effect multipliers may be used for various signal processing applications, although these are usually limited to frequencies below 1 MHz. Applications have been described of the generation of single sideband and phase-modulated signals and the use of Hall effect devices for the measurement of noise-power spectra by correlation techniques.[22]

Hall effect devices may also be used to detect magnetic bubble domains in orthoferrite memory planes with good sensitivity.[39]

### 15.1.2 Magnetoresistive Devices

In some magnetic sensor applications a magnetoresistive effect is used instead of the Hall effect because the impedance levels are more suitable and because only two leads are needed.

The increase of resistance of a semiconductor in a magnetic field is associated with the curvature of the lines of current flow caused by the $v \times B$ effect. This curvature results in a long flow path for a disk structure [see Fig. 15.15(a)] in which the current enters from a central electrode and flows to the perimeter electrode along a curve in the presence of a magnetic field perpendicular to the plane of the disk. The deflection is by an angle $v_H$, the Hall angle. The structure, known as a Corbino disk, has few practical uses. Instead a zigzag-shaped structure to give a narrow width to length is used or a raster plate with short-circuiting metal stripes which lie transverse to the longitudinal direction [see Fig. 15.15(b)]. Such structures of InSb provide a resistance increase of 15 for a field of 10 kG (1 V · sec/m²). Without the stripes the resistance increase is less than 10% for a typical rectangular geometry.

A zigzag-shaped sensor is shown in Fig. 15.16 with a magnetic field introduced in a longitudinal mode. The relationship between change of resistance and the overlap $x$ is

$$\frac{\Delta R}{R_0} = \frac{x}{L} \left( \frac{\rho(B) - 1}{\rho_0} \right) \tag{15.9}$$

where $\rho(B)$ is the resistivity in the magnetic field $B$, $\rho_0$ is the resistivity out of the field, and $R_0$ is the resistance of the sensor ($\rho_0 L/WT$) when out of the field. The InSb magnetoresistance sensor has four sections so that it may be connected as the four arms of a Wheatstone bridge. The full-scale output is 5 V dc at 500-$\mu$m (20 mil) displacement for a magnetic field of 12 kG and an excitation of 15 V dc at an impedance level of about 550 $\Omega$. A voltage output of 1 mV is observable for a

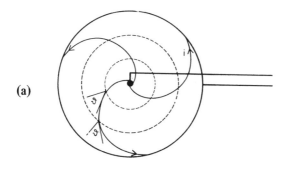

(a)

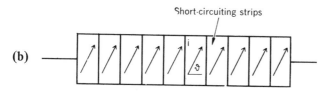

(b)

Fig. 15.15.  Magnetoresistive plates:
(a)  Corbino disk with center and perimeter contacts and the magnetic field perpendicular to the surface, and
(b)  Rectangular field plate with short-circuiting metal strips to provide suitable resistance increase. (After Weiss, 1968. Reprinted with permission from *IEEE Spectrum*, 5, 1968, p. 78.)

displacement of 0.1 $\mu$m. The combined nonlinearity and hysteresis effects are less than 1% of full scale. With moving permanent magnet systems, magnetoresistive sensors may be used as contactless potentiometers.

A magnetoresistive element may also be used for the measurement of power in a waveguide. The principle is shown in Fig. 15.17. A small magnetoresistance element is erected at the center of the waveguide supporting a dominant mode $H_{01}$, so that a small current $I$ proportional to and in phase with the electric field $E$ of the wave is induced through the element and a transverse component of the magnetic field $H$ of the wave penetrates the element at right angles. A constant magnetic bias $B_0$ is applied to the element transversely. The time-average voltage across the element is

$$\bar{V} = \left(\frac{dR}{dB}\right)_{B_0} \mathrm{Re}\ [I \cdot B]$$

$$\propto \mathrm{Re}\ [E \cdot H] \tag{15.10}$$

and is proportional to the total power. For InSb at 10 GHz a power of about 1 W gives an output of the order of 10 $\mu$V.

Magnetoresistive elements may also be used as microwave mixer devices.

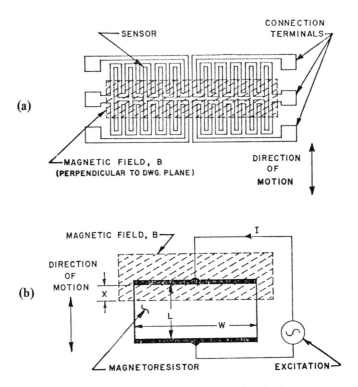

Fig. 15.16. InSb magnetoresistive sensor with four sections in zigzag:
(a)  Plate, overall dimensions $\sim$ 0.5 × 1.3 cm, and
(b)  Direction of motion introduces the long edge into the field. (After Yuan. Reprinted
     from *Solid-State Electronics,* 9, 1966, p. 497, Pergamon Press.)

### 15.1.3 Hall Effect in a MOSFET Structure

By the addition of Hall contacts to an enhancement MOSFET structure of large
channel dimensions, operated either in the pinch-off or triode region, the sensitivity to magnetic fields is made larger than for standard bulk or diffused Hall
plates. A typical Si MAGFET may have a sensitivity of 40 mV/mA kG at $I_D$ = 0.1
mA (between five and ten times larger than for a diffused Si Hall plate).

The structure is shown in Fig. 15.18 together with theoretical and experimental characteristics. For a channel width $W$ and length $L$, and Hall contacts
located at a distance $x$ from the source, the Hall voltage for the pinch-off condition assuming constant mobility is

$$V_H = -\frac{W\mu_H \, B \, V_{DS}}{2L\sqrt{1-x/L}} \qquad (15.11)$$

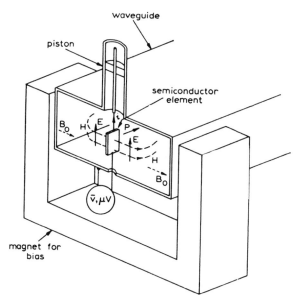

Fig. 15.17.  Principle of a magnetoresistive microwave watt meter. (After Kataoka. Reprinted with permission from *Proc. IEE* (London), **III**, 1964, p. 1742.)

In the triode region of operation the equation becomes

$$V_H = \frac{W \, \mu_H \, BI_0}{\beta \sqrt{\left(\dfrac{I_D L}{V_{DS}\beta} + \dfrac{V_{DS}}{2}\right)^2 - \dfrac{2x I_D}{\beta}}} \tag{15.12}$$

where $\beta = \mu_p \epsilon_{ox} W / t_{ox}$.

More complete analysis takes into account the variation of channel hole mobility with effective gate voltage and the shunting effect of the source and drain ohmic contacts. Even so the theoretical results tend to overestimate the sensitivity actually observed as seen in Fig. 15.18(b).

Cascading of MAGFETs is possible to give output Hall voltages in excess of 1 V.

### 15.1.4  Hall Effect Applied to Control an Injected Electron-Hole Current Filament

In minority carrier injection devices a magnetic field may be used to deflect the injection stream lines between two different collector regions. If the semiconductor has high mobility the effect can be reasonably large. Such structures have been given the name Madistor.[54] No use has been found for them as logic switch-

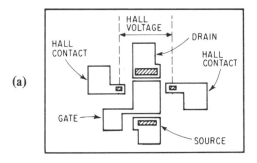

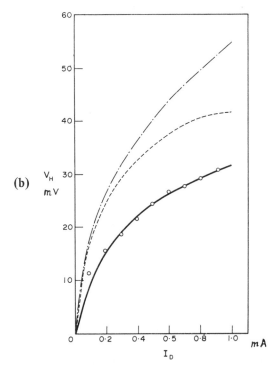

Fig. 15.18. A MAGFET is an MOS transistor with Hall effect contacts at either side of the channel:

(a) A p-enhancement Si MOSFET of channel dimensions 180 × 116 μm (7 x 4.6 mil) with side Hall contacts, and

(b) Hall voltage versus drain current, operation at pinch-off for a [111] oriented substrate. The chain-dot line is the theoretical characteristic for constant mobility and the dashed line is for a field-dependent mobility model. The magnetic field is 2.84 kG. (After Mohan, Rao, and Carr. Reprinted from *Solid-State Electronics,* **14**, 1971, p. 1000, Pergamon Press.)

ing devices because the need to create a magnetic field as the control parameter involves magnetic cores and the inductance makes the frequency response and switching speed uninteresting.

A magnetic field detector that operates on a somewhat similar principle is shown in Fig. 15.19. This is a distributed Si planar *p-n-i-p-n* structure that gives an output up to 30 times that of a comparable Hall detector. Its sensitivity stems from the lateral displacement of the injected electron-hole current filament. The device also is free of the offset voltage problem that exists in conventional Hall plates where control of alignment of Hall contacts is hard to achieve. Details of the device operation are left for independent study.

## 15.2  STRAIN SENSORS AND RELATED TRANSDUCERS

The application of strain to almost all semiconductors causes change of resistance and the size of the effect may be 100 times greater than for the effect in metals. Therefore, metal strain gages have been replaced by Si semiconductor strain sensors in the last decade or so.

### 15.2.1  Piezoresistive Coefficients

The piezoresistive effect in a semiconductor is a function of strain interacting with the band structure. Consider for example *n*-type Si for which there are six ellipsoidal conduction band minima along principal crystal axes on the momentum diagram. Four of these are shown in Fig. 15.20—the orthogonal *z*-axis pair being

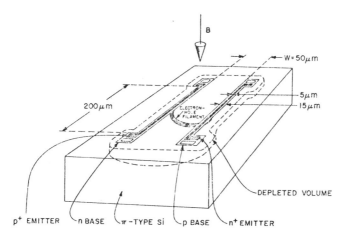

Fig. 15.19. Schematic view of a distributed planar *p-n-i-p-n* structure with filament that is deflected by a magnetic field. (After Bartelnik and Persky. Reprinted with permission from *Appl. Phys. Lett.*, **25**, 1974, p. 590.)

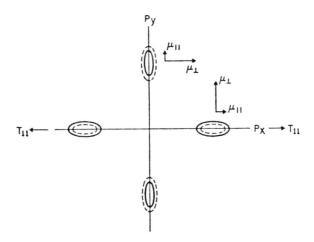

Fig. 15.20. Position of energy valleys with reference to crystallographic axes for *n*-type Si. (After Mason, Forst, and Tornillo. Reprinted Courtesy of Bell Telephone Laboratories, from *Semiconductor and Conventional Strain Gages*, M. Dean and R.D. Douglas, Eds., Academic Press, N.Y., © 1962.)

omitted for simplicity in drawing the diagram. The corresponding mobilities $\mu_\perp$ and $\mu_\parallel$ perpendicular and along the direction of an applied field are determined by the shape of the ellipsoid and in Si $\mu_\parallel$ is one fifth as large as $\mu_\perp$. The effect of a tension $T_{11}$ along the x axis is to raise the energy level for those valleys lying along x and to lower the energy levels for those valleys perpendicular to x. Hence the same energy is present as the smaller dotted-line ellipse in the x direction and the large dotted-line ellipse in the y direction. The relative numbers of electrons in the two types of valleys, by Boltzmann's principle, will be

$$\frac{N_{[010]}}{N_{[100]}} = \frac{e^{[V_0 + \alpha_1 T_{11}]/RT}}{e^{[V_0 - \alpha_2 T_{11}]/RT}} = e^{(\alpha_1 + \alpha_2)T_{11}/RT} \tag{15.13}$$

where $V_0$ is the energy level for no stress, $R$ the gas constant equal to 2 cal/mol/°K, $T$ is the absolute temperature, $\alpha_1 T_{11}$ is the energy rise of the [100] direction valleys, and $-\alpha_2 T_{11}$ is the energy fall in the [010] y direction valleys. The total number of electrons $N_0$ is unchanged and is given by

$$N_0 = 2N_{[100]} + 4N_{[010]} \tag{15.14}$$

and the conductivity in the x direction is

$$\sigma = q\left[2N_{[100]}\,\mu_\parallel + 4N_{[010]}\,\mu_\perp\right] \tag{15.15}$$

From Eqs. (15.13)-(15.15) the resistivity in the absence of stress is

$$= \frac{3}{N_0 q \, [\mu_{\parallel} + 2\mu_{\perp}]} \qquad (15.16)$$

and in the presence of stress is

$$\rho = \frac{1}{\sigma} = \frac{1 + 2e^{(\alpha_1 + \alpha_2)T_{11}/RT}}{N_0 e \, [\mu_{\parallel} + 2\mu_{\perp} e^{(\alpha_1 + \alpha_2)T_{11}/RT}]} \qquad (15.17)$$

Hence the fractional change of resistivity is

$$\frac{\Delta\rho}{\rho_0} = \frac{\rho - \rho_0}{\rho_0} = \frac{2}{3} \frac{(\mu_{\perp} - \mu_{\parallel})(1 - e^{(\alpha_1 + \alpha_2)T_{11}/RT}}{\mu_{\parallel} + 2\mu_{\perp} e^{(\alpha_1 + \alpha_2)T_{11}/RT}}. \qquad (15.18)$$

By expansion in powers of the applied stress $T_{11}$,

$$\frac{\Delta\rho}{\rho_0} = -\frac{2}{3} \frac{(\mu_{\perp} - \mu_{\parallel})}{(\mu_{\parallel} + 2\mu_{\perp})} \frac{(\alpha_1 + \alpha_2)T_{11}}{RT}$$

$$\left[ 1 - \left( \frac{2\mu_{\perp}}{\mu_{\parallel} + 2\mu_{\perp}} - \frac{1}{2} \right) \frac{(\alpha_1 + \alpha_2)T_{11}}{RT} + \cdots \right] \qquad (15.19)$$

For $n$ Si, $\mu_{\perp} = 5\mu_{\parallel}$ and $\Delta\rho/\rho_0 T_{11} = -0.95 \times 10^{-10}$ cm$^2$/dyn from measurements for small stresses. Hence $(\alpha_1 + \alpha_2)/RT$ is $3.92 \times 10^{-10}$ at room temperature. The numerical expression becomes

$$\frac{\Delta\rho}{\rho_0} = -0.95 \times 10^{-10} T_{11} [1 - 1.6 \times 10^{-10} T_{11}] \qquad (15.20)$$

Using the value of $1.3 \times 10^{12}$ dyn/cm$^2$ for the value of Young's modulus for $n$-type Si measured along a crystallographic axis, one obtains

$$\frac{\Delta\rho}{\rho_0} = 125 S_1 - 26,000 S_1^2$$

Hence, the gage factor for small stresses is 125, and for a strain of say $10^{-4}$ the resistivity change is about 1.22%.

For the more general case of three orthogonal tensions along principal crystal axes, as shown in Fig. 15.21, where $E_{1,2,3}$ are applied electric fields, and $i_{1,2,3}$ are corresponding current densities

$$E_1/\rho_0 = i_1[1 + \pi_{11}T_{11} + \pi_{12}(T_{22} + T_{33})] + \pi_{44}(i_2T_{12} + i_3T_{13}) \qquad (15.21\text{a})$$

$$E_2/\rho_0 = i_2[1 + \pi_{11}T_{22} + \pi_{12}(T_{11} + T_{33}) + \pi_{44}(i_1T_{12} + i_3T_{23}) \qquad (15.21\text{b})$$

$$E_3/\rho_0 = i_3[1 + \pi_{11}T_{33} + \pi_{12}(T_{11} + T_{22})] + \pi_{44}(i_1T_{13} + i_2T_{23}) \qquad (15.21\text{c})$$

Some values for the piezoresistive coefficients of Si and Ge are given in Table 15.1.

The longitudinal piezoresistive coefficient along an axis other than the crystallographic axis is

$$\pi_l = \pi_{11} + 2(\pi_{44} + \pi_{12} - \pi_{11})(l^2m^2 + l^2n^2 + m^2n^2) \qquad (15.22)$$

where $l, m, n$ are the direction cosines of the direction associated with $\mu_l$ referred to the crystallographic axes. The term $(l^2m^2 + l^2n^2 + m^2n^2)$ has a maximum value for $l = m = n = 1/\sqrt{3}$ which is the 111 axis of the crystal and has a minimum value of zero along a crystallographic axis. It follows from Eq. (15.22) that $\pi_l$ is a maximum along the 111 axis for $(\pi_{44} + \pi_{12} - \pi_{11} > 0)$. If $(\pi_{44} + \pi_{12} - \pi_{11} < 0)$ the maximum piezoresistance coefficient occurs along a crystallographic axis. The maximum value of $\pi_l$ for both types of Ge and for $p$-type Si occurs along the 111 axis and for $n$-type Si occurs along a 100 crystal axis.

The transverse coefficient $\pi_t$ is given by

$$\pi_t = \pi_{12} + (\pi_{11} - \pi_{12} - \pi_{44})(l_1^2l_2^2 + m_1^2m_2^2 + n_1^2n_2^2) \qquad (15.23)$$

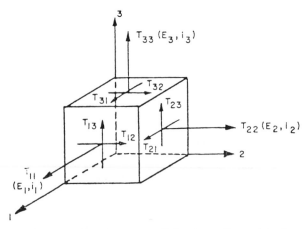

Fig. 15.21. An element of a semiconductor crystal showing orthogonal tensions along principal crystal axes. (After O'Regan from *Semiconductor and Conventional Strain Gages*, M. Dean and R.D. Douglas, Eds., Academic Press, N.Y., © 1962.)

Table 15.1.  Adiabatic Piezoresistance Coefficients at Room Temperature

| Material | $\rho$ $\Omega$-cm | $\pi_{11}$ | $\pi_{12}$ | $\pi_{44}$ | $(\pi_l)_{111}$ | Young's Modulus $10^{12}$ dyn/cm$^2$ | Gage Factor |
|---|---|---|---|---|---|---|---|
| | | | $10^{-12}$cm$^2$/dyn | | | | |
| Si | | | | | | 1.87 | |
| ($n$-type) | 7.8 | +6.6 | −1.1 | +138.1 | +93.6 | | +175 |
| ($p$-type) | 11.7 | −102.2 | +53.4 | −13.6 | −81 | | −142 |
| Ge | 1.5 | −2.3 | −3.2 | −138.1 | −94.9 | 1.55 | −147 |
| ($n$-type) | 5.7 | −2.3 | −3.9 | −136.8 | −94.7 | | −147 |
| | 9.9 | −4.7 | −5.0 | −137.9 | −96.9 | | −150 |
| | 16.6 | −5.2 | −5.5 | −138.7 | −101.2 | | −157 |
| ($p$-type) | 1.1 | −3.7 | +3.2 | +96.7 | +65.4 | | +101.5 |
| | 15.0 | −10.6 | +5.0 | +46.5 | +31.4 | | +48.7 |

## 15.2.2 Strain Gage Concepts

Consider a thin-wire strain gage of resistance $R = \rho L/A$, where $L$ is the unstrained length and $A$ is the cross-sectional area. With tension applied the length increases by $\Delta L$, the area decreases by $\Delta A$, and the resistivity changes by $\Delta \rho$. Hence

$$\frac{\Delta R}{R} = \frac{\Delta L}{L} - \frac{\Delta A}{A} + \frac{\Delta \rho}{\rho}$$

$$= (1 + 2\sigma + Y\pi_e)S \qquad (15.24)$$

where $\sigma$ is Poisson's ratio (usually about 0.3), $Y$ is Young's Modulus $(T/S)$, $\pi_e$ is the effective piezoresistive coefficient, and $S$ is the longitudinal strain. For a metal $\pi_e$ is quite small and the gage factor $(1 + 2\sigma + Y\pi_e)$ is typically about 2 and therefore much less than the 125 or so for a semiconductor. However, the semiconductor gage tends to be more temperature sensitive as may be seen from Table 15.2.

If the Si used in a strain gage is lightly doped the resistance varies considerably with temperature as shown in Fig. 15.22. Therefore, it is usually better to use heavily doped Si even though this results in a lower gage factor. The $\Delta R/R_0$ output versus stress sensitivity at two temperatures is illustrated in Fig. 15.23 for $p$-type Si. The $n$-type (100) Si of about the same resistivity has less $\Delta R/R$ for the same stress and the relationship is less linear; however, the sensitivity to temperature is less. The better linearity of $p$-type Si in tension is a consequence of the difference of band structure for holes and electrons that results in sign differences of the piezoresistive $\pi$ parameters (see Table 15.1). The responses to compression and tension of high-resistivity Si therefore are of the form shown in Fig. 15.24. The response to tension for $n$-type Si is seen to be nonlinear.

Table 15.2.  Comparison of Several Properties of a Semiconductor
Strain Gage to Properties of a Strain Gage Wire

| Property | Material | |
| --- | --- | --- |
| | 0.1 $\Omega$-cm $p$-type Si, 111 | Karma Strain Gage Wire |
| Strain gage factor | 125 (nom) | 2.0 |
| Thermal coeff. res., $\dfrac{dR}{R_0} \cdot \dfrac{1}{\Delta T} \left(\dfrac{\%}{°C}\right)$ | $12 \times 10^{-4}$ | $0.2 \times 10^{-4}$ |
| Thermal coeff. G.F., $\dfrac{dG}{G_0} \cdot \dfrac{1}{\Delta T} \left(\dfrac{\%}{°C}\right)$ | $10 \times 10^{-4}$ | $5 \times 10^{-4}$ |
| Seebeck coeff., alpha $(uV/°C)$ | 600 | 40 |
| Linear expansion coeff., $K$ $\left(\dfrac{in./in.}{°C}\right)$ | $4 \times 10^{-6}$ | $10 \times 10^{-6}$ |

(After Padgett and Wright from *Semiconductor and Conventional Strain Gages*, M. Dean and R.D. Douglas, ed., Academic Press, © 1962)

### 15.2.3 Gage Circuits with Temperature Compensation

The voltage sensitivity $S_v$ in microvolts per microstrain $S$ for a single gage of resistance $R_g$ and gage factor $G$ $\left( \dfrac{(\Delta R_g)}{R_g} S \right)$ is

$$S_v = R_g G I_g / S \tag{15.25}$$
$$= \Delta R_g I_g$$

where $I_g$ is the gage current.

If $I_g$ is provided by a current source and is independent of temperature, the sensitivity is seen to vary with temperature as $R_g$ and not as $\Delta R_g / R_g$.

As with wire and metal-foil strain gages care must be taken to mount the sensing element so that the mounting material (typically an epoxy glue) and mounting stress (associated with shrinkage) have the lowest temperature coefficients possible.

A strain gage may be used in a Wheatstone bridge-type-circuit, as shown in Fig. 15.25, where $R_1$ is the active gage and $R_3$ may be either a fixed resistance or a nonstrained gage to provide temperature compensation. If the bridge is initially balanced and $R_1$ changes by $\Delta R$ the output is

$$E_0 = E_{in} \frac{R_4}{(R_1 + R_2)(R_3 + R_4)} \times \frac{\Delta R}{1 + [\Delta R/(R_1 + R_2)]} \qquad (15.26)$$

If $\Delta R$ is small the relationship with $E_0$ is almost linear, but if $\Delta R$ is 20% a non-linearity of 10% exists for a bridge with equal arms. This can be reduced by choosing $R_2$ and $R_4$ to be large. Fortunately, also for $p$-type Si the $\Delta R$ versus strain relationship tends to have a nonlinearity that can be made to reduce or correct the nonlinearity of the bridge.

Other compensation methods include the use of thermistors in shunt or series with bridge arms. Also $p$-type (111) Si strain elements with positive gage factors and positive temperature coefficients may be used in conjugate bridge arms with $n$-type (100) Si elements of negative gage factors but positive temperature coefficients.

### 15.2.4 Diaphragm and Other Integrated Strain Sensors

For a pressure-measuring sensor, a natural form to use is a diaphragm that elastically deforms with pressure and undergoes strains that can be read by piezoresistance changes.[103]. There are then obvious advantages in making the disphragm itself of $n$-type Si and using $p$-diffused stripes for the sensing arms with the junctions formed providing the required isolation. The problems of bonding gages to a metal diaphragm are avoided in this approach.

The theoretical stress pattern for a diaphragm subjected to a uniform pressure over its surface is shown in Fig. 15.26. At the center of the diaphragm there are two equal components of the stress, while near the edge the two components become quite different.

The equations relating the stress with radial distance for the clamped edge condition are

$$\sigma_r = \frac{3}{8} (q/h^2)[a^2(1 + \sigma) - r^2(3 + \sigma)] \qquad (15.27a)$$

and

$$\sigma_t = \frac{3}{8} (q/h^2)[a^2(1 + \sigma) - r^2(1 + 3\sigma)] \qquad (15.27b)$$

and for the unclamped condition

$$\sigma_r = \frac{3}{8} (q/h^2)(3 + \sigma)(a^2 - r^2) \qquad (15.28a)$$

and

$$\sigma_t = \frac{3}{8} (q/h^2)[a^2(3 + \sigma) - r^2(1 + 3\sigma)] \qquad (15.28b)$$

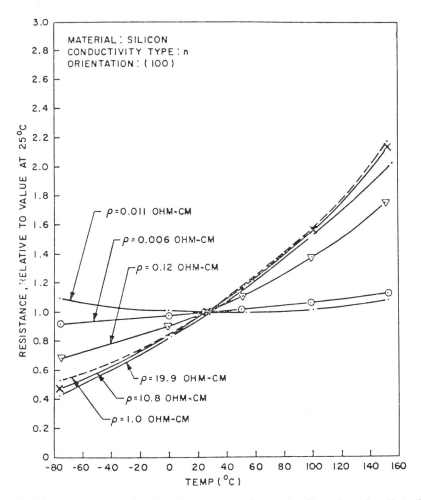

Fig. 15.22. Resistance as a function of temperature for $n$ and $p$ Si, to show the effect of resistivity in selecting for strain gage use:
(a)  $n$ type, and
(b)  $p$ type. (After Padgett and Wright from *Semiconductor and Conventional Strain Gages*, M. Dean, and R.D. Douglas, Eds., Academic Press, N.Y., © 1962.)

where $q$ is the pressure, $h$ the diaphragm thickness, $a$ the radius, and $\sigma$ Poisson's ratio.

Since the stresses vary with position on the diaphragm, average stress values $\bar{\sigma}_l$ and $\bar{\sigma}_t$ must be used for piezoresistive strips. Then

$$\frac{\Delta R}{R_0} = \pi_l \bar{\sigma}_l + \pi_t \bar{\sigma}_t \qquad (15.29)$$

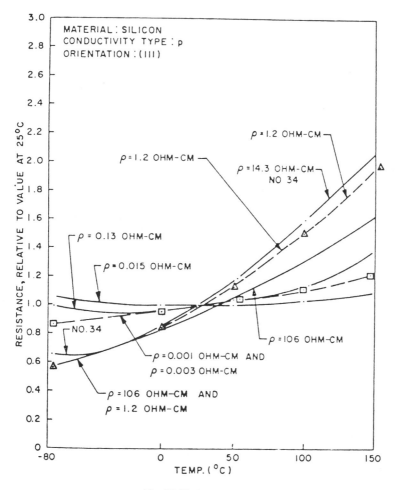

Fig. 15.22. (continued)

The values of $\pi_l$ and $\pi_t$ vary widely with crystallographic direction. Near the center of the diaphragm where $\bar{\sigma} \simeq \bar{\sigma}_t$, the maximum sensitivity $(\Delta R/R_0)$ obtainable from a $p$-type strip is $(\pi_{44}/2)\bar{\sigma}_l$ and occurs when the longitudinal direction is $[1\bar{1}0]$ and the transverse direction is $[001]$. For an $n$-type strip near the center of the diaphragm the maximum sensitivity is $(\pi_{11}/2)\bar{\sigma}_l$ and occurs when the longitudinal direction is $[001]$ and the transverse direction is $[010]$ or $[110]$. For an $n$-type strip near the center of the diaphragm having a $[1\bar{1}0]$ longitudinal direction and a $[001]$ transverse direction, the sensitivity is $-(\pi_{11}/4)\bar{\sigma}_l$ so that identical $n$-type strips near the center of the diaphragm can either show an increase or decrease of resistance with increasing pressure, depending on the orientation of the strip. If the diaphragm is a $[110]$ crystallographic plane, both

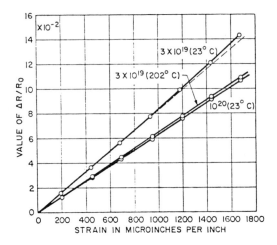

Fig. 15.23. Change in resistance as a function of the strain for a *p*-type semiconductor strain gage measured at room temperatures. (After Mason, Forst, and Tornillo from *Semiconductor and Conventional Strain Gages*, M. Dean and R.D. Douglas, Eds., Academic Press, N.Y., © 1962.)

of these strip orientations may be placed on the same diaphragm and a four-active arm bridge arrangement used. If the edge of the diaphragm is clamped, the stress near the edge is large and opposite in sign to the stress at the center as shown in Fig. 15.26. Since the stress near the edge varies rapidly with distance from the edge, the sensitivity depends on the geometry and position of the strip. However, utilizing the edge stress, a sensitive four-arm bridge of identical strips each having the same orientation may be fabricated on a diaphragm.

In all cases, the sensitivity attainable using both longitudinal and transverse effects is less than the sensitivity attainable using only the longitudinal effect (as for a normal bonded strain gage). However, this disadvantage is offset by the freedom from hysteresis creep inherent in the diffusion technique. Forming a four-active arm bridge consisting of identical diffused strips on a Si diaphragm also allows freedom from drift with temperature. Figure 15.27 shows the variation of resistance with pressure for a 2.25-cm diameter Si diaphragm of thickness 0.063 cm (25 mil) with a diffused strip 0.95 cm long centrally placed on the diagram.

Typical pressure transducers on Si diaphragms have four arm-active bridges and cover the differential pressure range 5-2000 psi. The bridge input voltage may be 5-15 V dc and the full-scale output voltage is about 100 mV. The accuracy is usually about ± 1% of full scale and the operating range is typically −20°C to 120°C with compensation to ± 1% provided over 50°C.

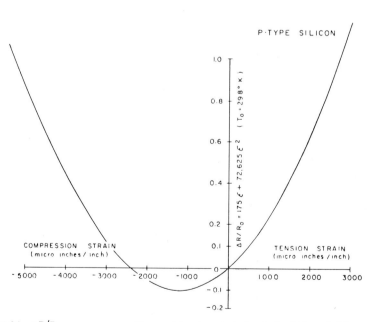

Fig. 15.24. $\Delta R/R_0$ versus composition and tension strain levels for high-resistivity $n$ and $p$ Si. (After Mason, Forst, and Tornillo. Reprinted courtesy Bell Telephone Laboratories, from *ISA Report #15-NY60, Instrument-Automation Conference*, 1960.)

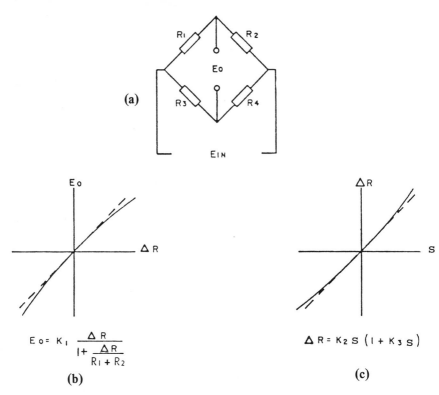

**(a)**

**(b)**

$$E_o = K_1 \frac{\Delta R}{1 + \dfrac{\Delta R}{R_1 + R_2}}$$

**(c)**

$$\Delta R = K_2 S \left( 1 + K_3 S \right)$$

Fig. 15.25. Wheatstone bridge circuit; b, bridge nonlinearity; c, Strain gage nonlinearity. (After Xavier and Vogt from *Semiconductor and Conventional Strain Gages*, M. Dean and R.D. Douglas, Eds., Academic Press, N.Y., © 1962.)

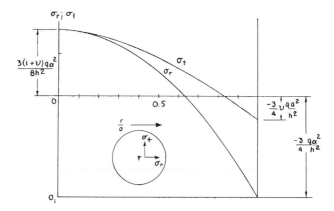

Fig. 15.26. Theoretical stress pattern in a diaphragm. The diaphragm thickness is assumed here to be much greater than its pressure-induced deflection. ($\nu$ here is Poisson's ratio.) (After Tufte, Chapman, and Long. Reprinted with permission from *J. Appl. Phys.*, **33**, 1962, p. 3324.)

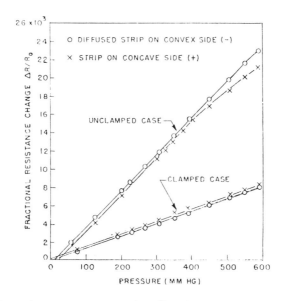

Fig. 15.27. Piezoresistance versus pressure for a Si strain gage diaphragm (diameter 2.25 cm, thickness 0.063 cm.) (After Tufte, Chapman, and Long. Reprinted with permission from *J. Appl. Phys.*, 33, 1962, p. 3325.)

Not all piezoresistive pressure sensors are of diaphragm form. In many load cells the load plate transfers strain to a Si cube, one face of which carries four diffused resistors.

Figure 15.28 shows a silicon *n*-type cantilever bar with four *p*-type diffused resistors along the [111] axis that change with the longitudinal piezoresistive effect. The nonlinearity of the device can be made less than 0.01% of full scale by arranging the four bridge resistances asymmetrically to the middle of the cantilever beam. The distance of the resistances under tension strain $(R_1, R_3)$ from the middle of the cantilever must be smaller than the distance of the resistances under compression strain $(R_2, R_4)$.

Essentially use is made of the relationship

$$\frac{\Delta R}{R} = c_1 \epsilon + c_2 \epsilon^2 \tag{15.30}$$

where $c_1$ and $c_2$ are different for tension and compression. With different strain values $\epsilon$ for the tension and compression elements there can be some compensation of the nonlinear components.

In a typical design, with 5 V dc bridge excitation, a force of 10 g applied to a 0.008-cm-thick cantilever beam will generate an output of about 160 mV; 25-$\mu$m deflection of the unclamped end will generate about 230 mV. The maximum load is 40 g. The change of sensitivity with temperature is about 0.05%/°C.

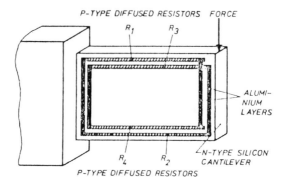

Fig. 15.28. Si cantilever with integrated Wheatstone bridge strain gage. (After Bretschi. Reprinted with permission from *IEEE Trans. Electron Devices,* **ED-23**, 1976, p. 60.)

For the measurement of accelerations an inertial mass may be used on the end of a cantilever strain gage beam. Acceleration ranges from 10 $g$ full scale to 2000 $g$ full scale are typical, with sensitivities of 15 mV/$g$ to 0.15 mV/$g$, respectively, where $g$ is the gravitational acceleration constant.

Integration of a Si pressure diaphragm on an IC chip is possible as shown in Fig. 15.29 if a vacuum cell is incorporated by etching and bonding under vacuum. A typical diaphragm of 0.23 × 0.16 cm has a pressure range up to 1 atm and the buffer and operational amplifiers provide a full-scale signal output of 7.5 V. The maximum error is about 2% of full-scale output.

Strain and pressure transducers may also be made of thin polycrystalline films with high gage factors and reasonable reproducibility.[53] Tunnel diode currents are also pressure sensitive and such junctions have been considered as strain sensors but tend to be rather too fragile for easy use.[40,84] Heterojunction structures have also been proposed for strain sensing by use of the piezoelectric properties of such junctions on barrier heights, but such devices have not come into general use.[57]

The use of a piezoelectric insulator in an IGFET structure however is of interest. The structure shown in Fig. 15.30 consists of a Si insulated-gate field-effect transistor having a layered insulator over the channel region. Strain sensitivity is initiated in an oriented ZnO film that is sputtered over thermally grown $SiO_2$. Such structures can respond to time-varying surface strains at frequencies that are ultimately determined by the response of the FET. Under optimized conditions this can approach a radian frequency equal to the reciprocal of the channel transit time. This frequency can be in the megahertz to gigahertz range if a short channel DMOS structure is used. Strains of $10^{-8}$ have been detected at 25 MHz, and gage factors ($\Delta I_D/I_{DO}$) of $10^4$-$10^5$ have been observed. Such a device may be used to respond to surface acoustic waves.

Some observations may be of interest on the size and nature of the transducer market. There are, for example, in the United States, 90 makers of pressure trans-

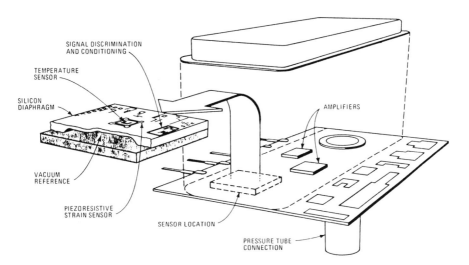

Fig. 15.29. An integrated hybrid Si pressure transducer. The Si chip is mounted on a ceramic substrate along with array of thickfilm signal-processing resistors, mechanical enclosures, and two op amps (a 747 and a 741) that serve as buffer amplifier and output amplifier, respectively. (After Zias and Hare. Reprinted from *Electronics*, Dec. 4, 1972, Copyright © McGraw-Hill, Inc., N.Y., 1972.)

ducers based on various physical principles (semiconductor, electromagnetic, etc.) and these compete for less than half of the annual $320 million transducer market. Thus this is not a high volume market and the cost per unit is further raised by the need for individual calibration and tests. For comparison some 75 manufacturers shared a $3.5 billion IC market in 1977.

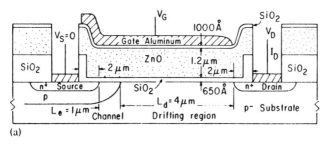

(a)

Fig. 15.30. Cross section of an $n$-channel PI-DMOS transistor for piezoelectric strain sensing. (After Yeh and Muller. Reprinted with permission from *Appl. Phys. Lett.*, 29, 1976, p. 521.)

## 15.3 TEMPERATURE SENSORS

The forward voltage across a *p-n* junction diode is related to the current by the expression

$$I = I_0 \left[ \exp \left( \frac{qV}{nkT} \right) - 1 \right] \tag{15.31}$$

or some variation of this form. Therefore, at constant current $V$ is proportional to $T$ and the sensitivity should be of the order 2-4 mV/°K. Figure 15.31 shows the characteristics for a diffused GaAs *p-n* junction as a temperature sensor. Provided the forward current is reasonably high (0.1-1 mA), the sensitivity is good between 4-400°K. At low forward currents, drift and noise become problems. Calibration of temperature-sensing diodes over the required temperature range is, of course, desirable before and after use. Some slight sensitivity to high (kilogauss) magnetic fields may have to be considered in certain applications.

The semiconductors used are typically GaAs and Si, since these have low bulk and surface leakage currents and so may be operated at low forward current densities where diode self-heating is not a problem. The dopants used should have shallow levels so that the temperatures at which nonlinearities set in due to deionization are low.

Transistor junction structures also are excellent as temperature sensors. If $V_{BE}$ is observed for a constant $I_C$ the linearity is better than for $V_{BE}$ at fixed $I_E$, since there are components of generation-recombination and surface leakage current that are drained away as base current. The circuit diagram for a differential transistor temperature sensor using Si 2N1893 transistors is given in Fig. 15.32, together with an indication of the accuracy that may be achieved.

If the two transistors are a matched pair the difference in emitter-base voltage appearing across $R_b$ is

$$\Delta V_{BE} = \frac{kT}{q} \ln \left( \frac{I_{C1}}{I_{C2}} \right) \tag{15.32}$$

Another approach to semiconductor temperature measurement, particularly appropriate at low temperatures, is to use the change of bulk resistance to indicate temperature. The variation of resistivity with temperature for a compensated semiconductor is shown schematically in Fig. 15.33(a). In region I the resistivity rises as the temperature falls because the carrier concentration is decreased with less thermal activation across the bandgap. The activation energy slope of this region corresponds to $E_g/2$. In region II the carrier concentration is fairly constant at the value corresponding to the net shallow doping concentration ($N_d$-$N_a$) but the resistivity decreases with decreasing temperature because the mobility increases

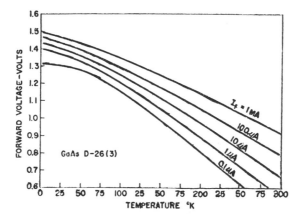

Fig. 15.31. The forward voltage drop of a GaAs *p-n* junction at constant current provides a good temperature-sensing effect. (After Cohen, Shaw, and Tretala from *Rev. Scientific Instruments,* 34 (10), 1963, p. 1091.)

with reduced lattice vibrations. If the bar is doped with donors and partially compensated by shallow acceptors the electron concentration is

$$n \simeq \frac{N_d - N_a}{1 + AN_a e^{B/T}} \qquad (15.33)$$

where $A$ and $B$ depend on the donor. At temperatures below about 100°K the term $AN_a e^{B/T}$ becomes larger than unity and the resistivity becomes proportional to $(N_a/N_d\text{-}N_a)e^{B/T}$ and increases as shown in region III where log $\rho$ tends to be proportional to $1/T$. This region extends to about 10°K after which impurity hopping of electrons (assuming $N_d$ is large and the compensation is significant) remains as the residual conduction mode (region IV). The transition from region III to region IV for a Ge thermometer doped with $3.75 \times 10^{18} \text{cm}^{-3}$Ga and compensated with $3 \times 10^{17} \text{cm}^{-3}$ Sb is shown in Fig. 15.33(b). This is very suitable for temperature measurements in the range below 4°K. Ge is used because the dopants tend to be shallower than in Si and purer material is available. For higher temperature ranges either Ge or Si may be used.

Other semiconductors, particularly those based on oxides, may be used in sintered powder form as beads or rods in which the resistance decreases considerably with increasing temperature. Figure 15.34 shows some typical temperature characteristics. The materials used tend to be mixtures of the oxides of copper, manganese, zinc, nickel, iron, vanadium, chromium, titanium, tungsten, and cobalt. Mixed ferrites may also be used such as triferric tetroxide with traces of lithium oxide. Thermistors are useful in various temperature-compensating circuit applications; however, they are not very precise in characteristics and skill is needed for reproducible design uses of these devices.

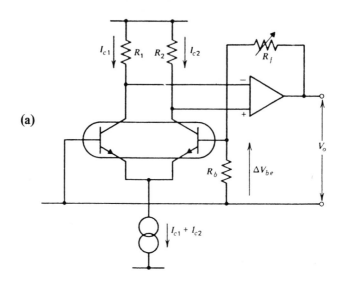

(a)

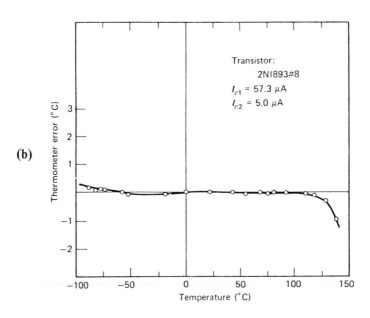

(b)

Fig. 15.32. Transistor parameters may be used as a means of temperature sensing:
(a)  Circuit diagram for dual-transistor thermometer, and
(b)  Measured error curve for a practical thermometer operating on the collector step prin-
     ciple. (After Verster from *Temperature, Its Measurement and Control in Science and
     Industry,* **4,** 1973, p. 2, p. 1117, Instrument Society of America.)

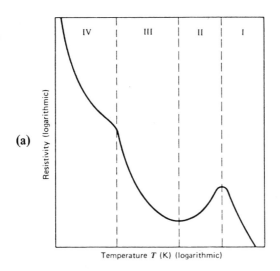

(a)

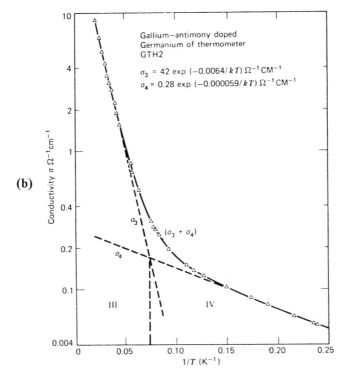

(b)

Fig. 15.33. Semiconductor bulk-resistance thermometry:
(a) Electrical resistivity as a function of the absolute temperature, plotted in a log-log scale, for a semiconductor similar to Ge, and
(b) Sharp transition knee between conductivity regions for Ge doped with $3.75 \times 10^{18}$ cm$^3$ Ga compensated with $3 \times 10^{17}$ cm$^3$ Sb. (After Blakemore from *Rev. Scientific Instruments*, **33**, 1962, p. 106.)

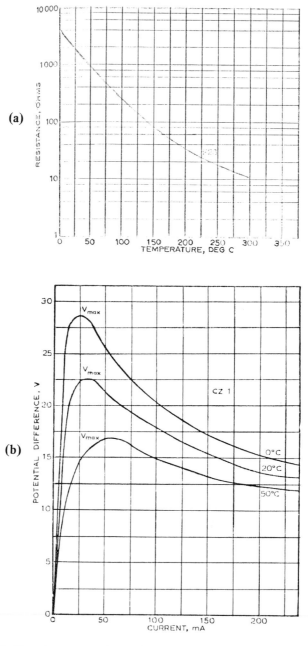

Fig. 15.34. Thermistors are polycrystalline structures having nonlinear resistance characteristics with temperature and current:

(a)  Resistance/temperature characteristic of a bead thermistor, and

(b)  Voltage/current characteristics of a rod thermistor at various ambient temperatures. (After Scarr and Setterington. Reprinted from *Proc. IEE*, **107**, pts. B-C, 1960, pp. 397-398.)

Other temperature-sensing methods include use of the positive temperature coefficient of a platinum coil and use of metal thermocouples. Platinum-resistance thermometers are very linear but are low in sensitivity ($\sim$0.4%/°K). Thermo-couples are somewhat nonlinear, and need cold-junction compensation. Typical sensitivities are 30-60 $\mu$V/°K.

Trigger action when a predetermined temperature is received can be obtained from any sensor by incorporating it in suitable circuitry; however, such action may also be obtained directly from n-p-n-p thyristor-type structures that are designed to have a temperature-sensitive gate trigger action. Figure 15.35 shows the structure and characteristics of such a device.

## 15.4 GAS SENSING SEMICONDUCTOR STRUCTURES

Many polycrystalline semiconductor films are found to be sensitive to the adsorp-tion and absorption of gas molecules. If the gas molecule is chemisorbed, a charge transfer between the absorbing species and the semiconductor can occur because of the difference in electron energy between the semiconductor surface and the molecule. The gas species may also react with the bulk material by diffusion along grain boundaries introducing scattering centers at intergranular contacts. The effect is a change in the semiconductor resistivity. The materials used tend to be metal-oxides such as sintered $SnO_2$ with trace impurities for doping purposes. For $SnO_2$, an n-type wide band gap semiconductor, the expression for the film resistance is

$$R = g/<n>q\,\mu_e \qquad (15.34)$$

where $g$ is a geometric factor, $<n>$ is the average number of carriers in the film, and $\mu_e$ is the effective mobility. It is generally observed in the polycrystalline metal-oxides that $<n>$ and $\mu_e$ are strong functions of ambient and account for the observed changes in resistance. Materials such as these, that are the basis of detectors which rely on changes in resistance as the transducer, are usually pro-vided with a heater that raises the film temperature to several hundred degrees C depending on the application. Response time tends to be some tens of seconds with recovery occuring in a similar time frame[19,98,100]

While the exact mechanism of the observed gas sensitivity in these films is not clear at this moment, several recent investigations point to the role of oxygen as the key element in the operation of these devices.[105,106] In particular, one such study of a CO detector points out the role of CO adsorption through chemi-sorbed $O_2^-$ sites by the process

physisorption → diffusion → chemical reaction of CO and $O_2^-$ → charge transfer

of an electron into the conduction band of the semiconductor and subsequent thermal desorption of neutral $CO_2$.[108]

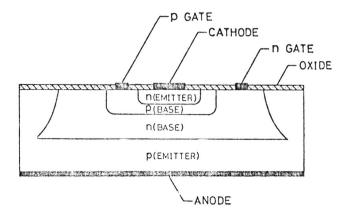

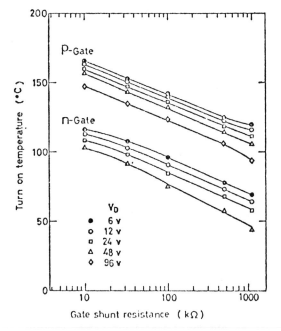

Fig. 15.35. *n-p-n-p* structures similar to thyristors may act as trigger-type temperature sensing devices:
(a) Cross section of a thermosensor, and
(b) Relationship between value of the gate shunt resistance and turn-on temperature for each gate connection as a function of off-state voltage. (After Nakata, et al. Reprinted with permission from *IEEE Int. Electron Devices Meeting, Technical Digest,* 1976, pp. 277-278.)

Fig. 15.36(a) shows the response of a surface that is particularly sensitive to carbon monoxide, hydrogen and ethanol. Fig. 15.36(b) shows another surface for which the sensitivity to carbon monoxide and butane is reversed. Such changes tend to be produced by variations of sintering temperature, grain size and impurity content of the films.

Semiconductor junction devices that involve Pd barriers are found to be sensi-tive to hydrogen and other gases. Fig. 15.37 shows an n channel Si MOS transistor with a Pd gate (10 nm thick) and the device response to hydrogen. The action is that hydrogen dissolves in the metal and diffuses to the metal-oxide interface where it gives rise to a dipole layer. The dipole layer changes the work-function difference between the metal and the semiconductor and thereby the threshold voltage of the MOS transistor. This threshold voltage change is easily measured electrically.

Ion sensitive field-effect-transistor and diode structures depend for their action on monitoring polarization changes at solution-oxide interfaces.[4,100,102,104]

The sensing of $H_2S$ with a PbSe film is similar to the response to atomic hydrogen and suggests that $H_2S$ may be dissociating on the surface.[98]

Oxygen sensing is possible with a semiconductor film of zinc oxide.[19] $Na^+$ ion sensing is a possibility with $SiO_2/Si$ junction structures.[4,100]

## 15.5 HIGH-ENERGY PARTICLE AND GAMMA RAY SENSORS

Semiconductor sensors are widely used for the detection of high-energy particles such as alpha particles, protons, neutrons, fission fragments, and fast electrons. Also high-energy gamma rays may be detected. The semiconductor used is gene-rally Ge or Si in a *p-i-n* junction form although heavier materials such as CdTe have also proven useful. Specialized conferences are held each year on the subject and the proceedings of these Scintillation and Semiconductor Counter Symposia may be found in the IEEE Transactions on Nuclear Science. The books of Berto-lini and Coche; Dearnaley and Northrop; Taylor; and Deme should be consulted for overviews on the subject. Those readers who are curious about the Ge(Li) detected neutron-activated spectrum of a lima bean will find the evidence in Bertolini and Coche.

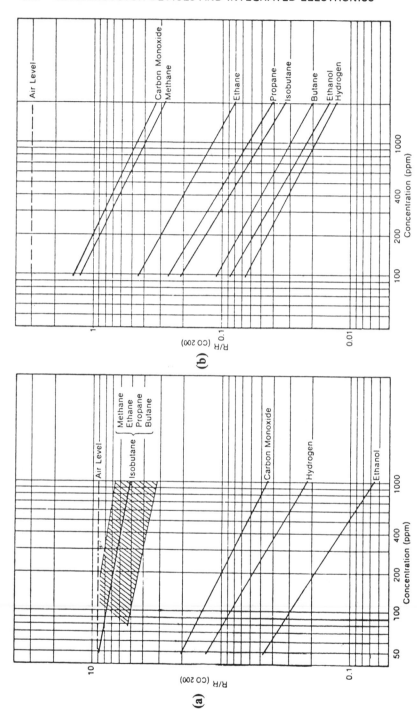

Fig. 15.36. Semiconductor gas sensing devices. Different surfaces exhibit selective sensitivity to different gases. (Courtesy Figaro Engineering, Inc.)

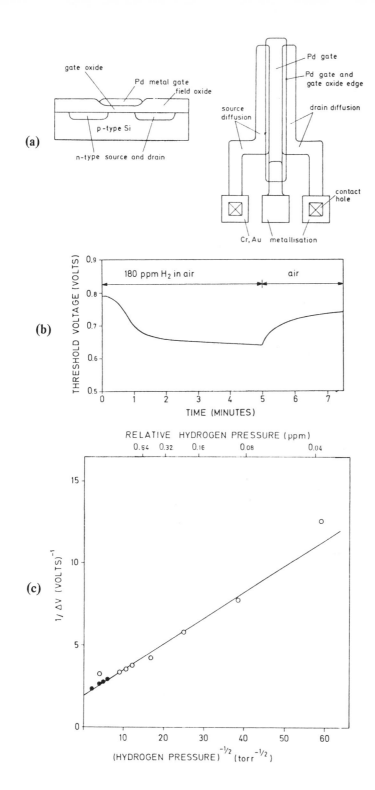

(a)

(b)

(c)

937

## 15.6 QUESTIONS

1. Design a Si Hall effect wattmeter for 0 to 2 kW on a 115-V single-phase line supply to a load of 0.8 power factor. What accuracy do you expect?

2. Design a Hall probe to measure magnetic fields in a superconducting magnet coil (10-40 kG).

3. Discuss the problem of zero signal balance of a Hall generator.

4. Discuss the design of a Hall generator sensor for profiling the flatness of a surface to 0.1 $\mu$m.

5. What are likely to be the effects of temperature ($-40$ to $100°C$) on a Hall effect displacement sensor?

6. Discuss the use of a Si Hall voltage to drive a differential transistor in IC form.

7. Discuss the problem of designing a Hall effect sensing head to read the magnetic numbers on a document such as a bank check.

8. Discuss Hall effect reading heads for magnetic tape.

9. Design a Hall effect magnetic resolver to provide synchro action with accuracy better than 0.5 deg. The size should not exceed 5 cm diameter.

10. Design a GaAs Hall effect MOSFET and discuss the performance expected.

11. Describe the problems in designing a Hall effect device to multiply or correlate two signals at 1 MHz.

12. Design a Hall effect array on a disk of about 2 cm diameter in which the individual elements are very small in size, say $20 \times 20$ $\mu$m, to be able to read out the uniformity of field distribution in a magnet region or the presence of bubble domains.

13. Design an InAs raster plate with two sections, as in Fig. 15.16, for Wheatstone bridge use in which a resistance of about 100 $\Omega$ per arm increases by a factor of four for a field of 10 kG (1 V · sec/m²) over an area of 4 cm². How will the plate resistance change with temperature in the range $77°K-400°K$? What is the power dissipation when suitably biased and what is the minimum displacement that can be sensed with normal equipment when moved in this field?

14. Design an InAs Hall effect gyrator to provide more then 20 dB of reverse isolation at 1 GHz.

15. A semiconductor Hall effect gyrator is shown in Fig. 15.14(a). It is basically a two-port circuit that inverts a load impedance. If the plate is GaAs and the magnetic field is 10 kOe over an area of 1 cm² choose a doping and thickness for the plate and state what magnitude of load resistance is suitable for use with your gyrator. With your gyrator if a capacitance $C$ is connected across terminals 3 and 4, would the input between 1 and 2 act like an inductance? If the frequency is $10^3$ Hz what capacitance would produce what apparent inductance value? How would the apparent inductance differ from a true inductance? What would be its $Q$ value?

16. The need for a magnetic field in a semiconductor gyrator is a disadvantage. Circuit gyrators that depend on operational amplifiers can be designed. Consider a vertical

---

Fig. 15.37. MOS transistor gas sensor structure and characteristics:

(a) Design of the structure. The gate oxide is 10 nm thick and field oxide about 200 nm thick. The channel is 20 $\mu$m wide and 1000 $\mu$m long,

(b) Transistor threshold voltage as function of time when hydrogen is fed to the transistor and removed again. The temperature is $150°C$, and

(c) $1/\Delta V$ versus (hydrogen pressure)$^{-1/2}$ for hydrogen in nitrogen at $150°C$ (filled circles) and at $125°C$ (open circles). The ppm values are ten times larger than shown. (After Lundstrom, Shivaraman, and Svensson. Reprinted with permission from *J. Appl. Phys.*, 46, 1975, pp. 3876, 3877, 3879.)

series chain of five impedances, $R_1$, $R_2$, $R_3$, $C$, and $R_5$. Connect the positive input terminal of amplifier 1 to the input at the top of $R_1$. The negative input terminal is connected between $R_2$ and $R_3$, and the output of the amplifier is taken to the node between $R_3$ and $C$. The positive input terminal of amplifier 2 is connected to the node between $C$ and $R_5$; the negative input terminal to the $R_2$-$R_3$ node and the output to the $R_1$-$R_2$ node. Show that the circuit input appears to be inductive. Specify some typical values for the impedance chain, assume the amplifiers to be of 40 dB gain, and calculate the $Q$ obtained for the apparent inductance.

17. Present the theory of a Hall effect circulator and discuss design problems.

18. Design a load cell based on a Si cube to weigh a truck.

19. Design a loaded cantilever-type strain gage to observe accelerations of 100 times gravity.

20. Describe torque measurement in an electric motor shaft with a semiconductor strain gage.

21. Design a 3 cm diameter Si diaphragm pressure gage to respond to a range 0-10 kg/cm² (142 psi).

22. Discuss the design of semiconductor pressure sensors for biomedical instrumentation.

23. A Si DMOS-type strain transducer is shown in Fig. 15.30. The gate insulator is 650 A thickness and the relative dielectric constant is assumed to be 2. The piezoelectric material is of thickness 1.2 $\mu$m and we assume for our calculations the relative dielectric constant to be 8 and the piezoelectric coefficient to be $2 \times 10^{-5}$ cm$^{-2}$ (where induced surface charge per unit area equals the product of the piezoelectric coefficient and the strain).

    If the strain applied is $10^{-4}$ estimate the shift in threshold voltage of the transistor. Can you convert this into an estimated decrease of the $I_{DS}$ vs $V_{DS}$ curve of the transistor at a constant gate voltage bias?

24. Discuss the limits of performance expected of a ZnO piezoelectric DMOS strain gage. What other piezoelectric semiconductor might be used in place of ZnO?

25. A Wheatstone bridge has three arms of constant resistance 10 k$\Omega$ and one arm is a Ge bar of $10^{14}$ cm$^{-3}$ boron doping of resistance 10 k$\Omega$ at 300°K. The bridge voltage applied to opposing terminals of the bridge is a sinewave of frequency $10^3$ Hz and peak voltage 1 V. The detector across the other opposing terminals is of infinite impedance. What is the output voltage detected for the temperature range 77-400°K? Would there be any advantage in using $n$-type Ge or either conductivity type of Si? Could a junction diode be used successfully as one arm of the bridge?

26. The variation of forward voltage drop of a GaAs $p$-$n$ junction for constant current is shown in Fig. 15.31. Design a Si $p$-$n$ junction diode as a temperature sensor (if possible, of comparable sensitivity) over the temperature range 50-300°K. Specify the dopants used, the assumptions made, and the curves expected. Would a Schottky barrier diode be better?

27. Design a temperature sensor block that could be integrated with a Si power transistor to protect the IC chip in the event the dissipation causes the chip temperature to exceed 180°C.

28. Describe the performance expected from an Au-doped Si bar as a temperature sensor.

29. Discuss the relative merits of GaAs, Si, and Ge diodes as forward voltage versus temperature sensors for the range 4-400°K.

30. The thermistor of Fig. 15.34(b) is assumed to be of surface area 1 cm² and is to be used as one arm of a Wheatstone resistor bridge in an attempt to control accurately the temperature of a 1 m × 1 m × 1 m hot box between 10 and 50°C and in a 0°C ambient. Design the bridge and calculate the output characteristic: assume a high-input impedance amplifier as the sensor that controls the heater inside the box.

31. Review some circuit applications of thermistors. In which applications could a Si device be used as a substitute for a thermistor?

32. Explain the turn-on performance of a temperature-sensitive thyristor.

33. Design a temperature sensor for mouth use in which the temperature shift of the light-absorption edge of the semiconductor is the basis of the temperature sensing and estimate the sensitivity expected.
34. Discuss various gas detectors in which hydrogen adsorption plays a role.
35. Discuss the effects of temperature on gas detectors.
36. How can the leakage current of a junction or Schottky barrier device be used as a gas detector?
37. Describe and discuss junction structures that might sense $Na^+$ ions.
38. Discuss the energy for pair production in Ge, Si, and GaAs and the range of various high-energy particles.
39. Discuss pulseshape in a *p-i-n* Ge (Li) detector of a 1-MeV alpha particle.
40. Discuss the relative merits of Ge, Si, and CdTe for alpha particle detection.
41. Discuss a semiconductor particle detector that uses a $dE/dx$ sensing mode.
42. Discuss gamma ray detection with Ge and Si semiconductors.

## 15.7 REFERENCES AND FURTHER READING SUGGESTIONS

1. Barlow, H.E.M., and J.C. Beal, "An experimental impedance relay using the Hall effect in a semiconductor," *Proc. IEE* (London), **107**, 1960, p. 48.
2. Bartelink, D.J., and G. Persky, "Magnetic sensitivity of a distributed Si planar pnpn structure supporting a controlled current filament," *Appl. Phys. Lett.*, **25**, 1974, p. 590.
3. Becker, J.A., C.B. Green, and G.L. Pearson, "Properties and uses of thermistors—thermally sensitive resistors," *AIEE Trans.*, **65**, 1946, p. 711.
4. Bergveld, P., "Development, operation and application of the ion-sensitive field-effect-transistor as a tool for electrophysiology," *IEEE Trans. Biomed. Eng.*, **BME-19**, 1972, p. 342.
5. Bertolini, G., and A. Coche, *Semiconductor Detectors*, Wiley-Interscience, N.Y., 1968.
6. Blakemore, J.S., "Design of germanium for thermometric applications," *Rev. Sci. Instr.*, **33**, 1962, p. 106.
7. Braun, R.J., "Give the Hall transducer flexibility by adding a control electrode," *Electronic Design*, **22** (11) May 24, 1974, p. 88.
8. Braun, R.J., "Modular Hall masterslice transducer," *IBM J. Res. Develop.*, **19**, 1975, p. 344.
9. Bretschi, J., "A silicon integrated strain-gage transducer with high linearity," *IEEE Trans. Electron Devices*, **ED-23**, 1976, p. 59.
10. Bulman, W.E., "Applications of the Hall effect," *Solid-State Electronics*, **9**, 1966, p. 361.
11. Bursky, D., "Sensors in 5 areas are getting tinier, cheaper and more precise," *Electronic Design*, **15**, July 19, 1974, p. 34.
12. Caldren, L.A., "Zinc oxide on silicon memory cells scanned by acoustic surface waves," *Appl. Phys. Lett.*, **26**, 1975, p. 137.
13. Christensen, D.A., "Light monitors tissue temperature" Int. Microwave Symposium, San Diego, June 1977; see also *Science News*, July 2, 1977.
14. Cohen, B.G., W.B. Snow, and A.R. Tretola, "GaAs pn junction diodes for wide range themometry", *Rev. Sci. Instr.*, **34**, 1963, p. 1091.
15. Crawford, R.W., "Industrial measurements utilizing watt transducers," *Solid-State Electronics*, **9**, 1966, p. 527.
16. Davidson, R.S., and R.D. Gourlay, "Applying the Hall effect to angular transducers," *Solid-State Electronics*, **9**, 1966, p. 471.
17. Dean, M., III, and R.D. Douglas, *Semiconductor and Conventional Strain Gages*, Academic Press, N.Y., 1962.

18. Dearnaley, G., and D.C. Northrop, *Semiconductor Counters for Nuclear Radiations*, John Wiley, Inc., N.Y., 1963.
19. Degn, H., and M. McK. Nobbs, "Solid-state oxygen meter using zinc oxide," *Appl. Phys. Lett.*, **26**, 1975, p. 526.
20. Deme, S., *Semiconductor Detectors for Nuclear Radiation Measurement*, Wiley-Interscience, N.Y., 1971.
21. Dorey, A.P., "A high sensitivity [MOS] semiconductor strain sensitive device," *Solid-State Electronics*, **18**, 1975, p. 295.
22. Epstein, M., and J.J. Brophy, "Applications of Hall-effect multipliers," *Solid-State Electronics*, **9**, 1966, p. 507.
23. Fischler, A.S., and J.A. Collins, "Self-compensating silicon load cell with an electronic converter," *IEEE Trans. Electron Devices*, **ED-16**, 1969, p. 861.
24. Fleming, W.J., "Physical principles governing non-ideal behavior of the zirconia oxygen sensor," *J. Electrochem. Soc.*, **124**, 1977, p. 21.
25. Frobenius, W.D., A.C. Sanderson, and H.C. Nathanson, "A microminiature solid-state capacitive blood pressure transducer with improved sensitivity," *IEEE Trans. Biomed. Eng.*, **BME-20**, 1973, p. 312.
26. Fry, P.W., and S.J. Hoey, "A silicon MOS magnetic field transducer of high sensitivity," *IEEE Trans. Electron Devices*, **ED-16**, 1969, p. 35.
27. Fulkerson, D.E., "A silicon integrated circuit force sensor," *IEEE Trans. Electron Devices*, **ED-16**, 1969, p. 867.
28. Gallagher, R.C., and W.S. Corak, "A metal-oxide-semiconductor (MOS) Hall element," *Solid-State Electronics*, **9**, 1966, p. 571.
29. Greeneich, E.W., and R.S. Muller, "Theoretical transducer properties of piezoelectric insulator FET transducers," *J. Appl. Phys.*, **46**, 1975, p. 4632.
30. Grover, T.P., "Precise voltage to frequency convertor for telemetry applications of strain gage pressure transducers," *IEEE Trans. Biomed. Eng.*, **BME-22**, 1975, p. 441.
31. Grubbs, W.J., "Hall effect devices," *Bell Syst. Tech. J.*, **38**, 1959, p. 853.
32. Hemment, R.S., "Invariance of the Hall effect MOSFET to gate geometry," *Solid-State Electronics*, **17**, 1974, p. 1039.
33. Iltis, R., "Solid state sensors: strain gages," *Control Engineering*, **17**, Jan. 1970, p. 71.
34. Inglis, B.D., and G.W. Donaldson, "A new Hall-effect synchro," *Solid-State Electronics*, **9**, 1966, p. 541.
35. Johnson, H.H., and D. Midgley, "The Corbino disc and its selfmagnetic field," *Proc. IEEE*, **109**, pt.B, 1962, p. 283.
36. Kamarinos, G., and P. Viktorovitch, "Schottky diode magnetic sensor of high sensitiveness," *IEEE Int. Electron Devices Meeting, Technical Digest*, 1976, p. 271.
37. Kataoka, S., "Multiplying action of the magnetoresistive effect in semiconductors and its application to power measurements," *Proc. IEE* (London), **III**, 1964, p. 1937.
38. Kataoka, S., and H. Naito, "Magnetoresistive microwave devices," *Solid-State Electronics*, **9**, 1966, p. 459.
39. Kataoka, S., H. Yamada, and Y. Sugiyama, "New semiconductor bubble domain detector employing the Hall effect associated with a locally inverted magnetic field," *Proc. 4th Conf. on Solid-State Devices*, Tokyo, 1972. Suppl. to the *J. Japan. Soc. Appl. Phys.*, **42**, 1973, p. 91.
40. Kiggins, T.R., and A.G. Milnes, "A solid state tunnel-diode strain gage," *Proc. Instrument Society of America Annual Conference* **18**(pt. 2), Sept. 1963.
41. Klein, C.A., and R.W. Bierig, "Pulse-response characteristics of position-sensitive photodetectors," *IEEE Trans. Electron Devices*, **ED-21**, 1974, p. 532.
42. Kuhrt, F., "New Hall generator applications," *Solid-State Electronics*, **9**, 1966, p. 567.
43. Levine, J.D., "Theory of varistor electronic properties," *CRC Critical Reviews in Solid-State Sciences*, Nov. 1975, p. 597.

44. Levinson, L.M., and H.R. Philipp, "The physics of metal oxide varistors," *J. Appl. Phys.*, **46**, 1975, p. 1332.
45. Levinson, L.M., and H.R. Philipp, "AC properties of metal-oxide varistors," *J. Appl. Phys.*, **47**, 1976, p. 1117.
46. Lundström, K.I., M.S. Shivaraman, and C.M. Svensson, "A hydrogen-sensitive Pd-gate MOS transistor," *J. Appl. Phys.*, **46**, 1975, p. 3876.
47. Luo, F.C., and M. Epstein, "Strain sensitivity of thin-film InSb transistor," *Proc. IEEE*, **61**, 1973, p. 129.
48. Lynch, T.H., "The right gyrator trims the fat off active filters," *Electronics*, July 21, 1977, p. 115.
49. Mallard, T.M., et al., "Portable gas chromatograph for the acetylene reduction assay for nitrogenase," *Analyt. Chem.*, **49**, 1977, p. 1275.
50. Malm, H.L., C. Canali, J.W. Mayer, M.A. Nicolet, K.R. Zanio, and W. Ahutagawa, "Gamma-ray spectroscopy with single-carrier collection in high-resistivity semiconductors," *Appl. Phys. Lett.*, **26**, 1975, p. 344.
51. Mason, W.P., and R.N. Thurston, "Use of piezoresistive materials in the measurement of displacement, force and torque," *J. Acoust. Soc. Amer.*, **29**, 1957, p. 1096.
52. Mason, W.P., J.J. Forst, and L.M. Tornillo, "Recent developments in semiconductor strain transducers," in *Semiconductor and Conventional Strain Gages*, M. Dean and R.D. Douglas, Eds., Academic Press, N.Y., 1962, p. 109.
53. A.H., Meiksin, "On the strain coefficient of resistance of very thin continuous metal films," *Thin Films*, **1**, 1968, p. 211.
54. Melngailis, I., and R.H. Rediker, "Madistor—magnetically controlled semiconductor plasma device," *Proc. IRE*, **50**, 1962, p. 2428.
55. Miller, G.L., "Recent advances in compound semiconductors for radiation detectors," *IEEE Trans. Nuclear Science*, NS-19, 1972, p. 251.
56. Mohan Rao, G.R., and W.N. Carr, Magnetic sensitivity of a MAGFET of uniform channel current density," *Solid-State Electronics*, **14**, 1971, p. 995.
57. Moore, R.M., and C.J. Busanovich, "The heterode strain sensor: an evaporated heterojunction device," *IEEE Trans. Electron Devices*, ED-16, 1969, p. 850.
58. Murai, M., "Hall-effect magnetic sensor reads data at any speed," *Electronics*, February 1, 1973, p. 91.
59. Nadkarni, G.S., A. Simani, and J.G. Simmon, "Fabrication of high sensitivity thin-film indium antimonide magnetoresistors," *Solid-State Electronics*, **18**, 1975, p. 393.
60. Nakata, J., et al., "Thermosensor—a new temperature-sensitive switching device," *IEEE Int. Electron Devices Meeting, Technical Digest*, 1976, p. 275.
61. Nalecz, M., and Z.L. Warsa, "Hall effect transducers for measurement of mechanical displacements," *Solid-State Electronics*, **9**, 1966, p. 485.
62. Nathanson, H.C., and J. Goldberg, "Topologically structured thin films in semiconductor device operation," *Physics of Thin Films*, **8**, p. 251.
63. Nishizawa, J.I., and T. Nonaka, "Solid-state frequency indicator," *IEEE Trans. Electron Devices*, ED-21, 1974, p. 391.
64. Oppenheimer, M., "In IC form, Hall-effect devices can take on many new applications," *Electronics*, August 2, 1971, p. 46.
65. O'Reagan, R., "Development of the semiconductor strain gage and some of its applications," in *Semiconductor and Conventional Strain Gages*, M. Dean and R.D. Douglas, Eds., Academic Press, N.Y., 1962, p. 245.
66. Ott, W.E., "Monolithic converter augments ac-measurement capabilities," *Electronics*, January 23, 1975, p. 79.
67. Padgett, E.D., and W.V. Wright, "Silicon piezoresistive devices," in *Semiconductor and Conventional Strain Gages*, M. Dean, III, and R.D. Douglas, Eds., Academic Press, N.Y., 1962, p. 1.

68. Pavese, F., and S. Limbarinu, "Accuracy of gallium arsenide diode thermometers in the range 4-300°K," in *Temperature, Its Measurement and Control in Science and Industry*, Vol. 4, H.H. Plumb, Ed., Instrument Society of America, Pittsburgh, Pa., 1972, p. 1103.

69. Pehl, R.H., "Germanium gamma-ray detectors," *Physics Today*, Nov. 1977, p. 50.

70. Putley, E.H., *The Hall Effect and Related Phenomena*, Butterworths, London, 1960.

71. Riezenman, M.J., "Integrated temperature transducers," *Electronics*, November 14, 1974, p. 130.

72. Rudberg, H., "Detectors tout versatility, low cost, simple operation," *Chem. Eng.*, February 17, 1975, p. 56.

73. Ruehle, R.A., "Solid-state temperature sensor out performs previous transducers," *Electronics*, March 20, 1975, p. 127.

74. Sachse, H.B., *Semiconducting Temperature Sensors and Their Applications*, Wiley, Inc., N.Y., 1975.

75. Samaun, K.D. Wise, and J.B. Angell, "An IC piezoresistive pressure sensor for biomedical instrumentation," *IEEE Trans. Biomed. Eng.*, **BME-20**, 1973, p. 101.

76. Sanchez, J.C., and W.V. Wright, "Recent developments in flexible silicon strain gages," in *Semiconductor and Conventional Strain Gages*, M. Dean and R.D. Douglas, Eds., Academic Press, N.Y., 1962, p. 307.

77. Scarr, R.W.A., and R.A. Setterington, "Thermistors, their theory, manufacture and application," *Proc. IEEE*, **107**, pts. B-C, 1960, p. 395.

78. Stalinski, E.J., and Z.H. Meiksin, "Thin-film distributed RC parameter strain gauges," *IEEE Trans. Electron Devices*, **ED-22**, 1975, p. 102.

79. Steele, M.C., J.W. Hile, and B.A. MacIver, "Hydrogen-sensitive palladium gate MOS capacitors," *J. Appl. Phys.*, **47**, 1976, p. 2537.

80. Steele, M.C., and B.A. MacIver, "Palladium/cadmium-sulfide Schottky diodes for hydrogen detection," *Appl. Phys. Lett.*, **28**, 1976, p. 687.

81. Stephenson, L.M., and H.E.M. Barlow, "Power measurement at 4 Gc/s by the application of the Hall effect in a semiconductor," *Proc. IEE* (London), **106**, 1959, p. 27.

82. Stockert, J., and E.R. Nave, "Operational amplifier circuit for linearizing temperature readings from thermistors," *IEEE Trans. Biomed. Eng.*, **BME-21**, 1974, p. 164.

83. Swartz, J.M., and J.R. Gaines, "Wide range thermometry using GaAs sensors," *Temperature, Its Measurement and Control in Science and Industry*, Vol 4, H.H. Plumb, Ed., Instrument Society of America, Pittsburgh, Pa., 1972, p. 1117.

84. Sze, S.M., *Physics of Semiconductor Devices*, Wiley-Interscience, N.Y., 1969.

85. Takamiya, S., and K. Fujikawa, "Differential amplification magnetic sensor," *IEEE Trans. Electron Devices*, **ED-19**, 1972, p. 1085.

86. Taylor, J.M., *Semiconductor Particle Detectors*, Butterworths, Inc., Washington, D.C., 1963.

87. Tufte, O.N., P.N. Chapman, and D. Long, "Silicon diffused - element piezoresistive diaphragms," *J. Appl. Phys.*, **33**, 1962, p. 3322.

88. Verster, T.C., "The silicon transistor as a temperature sensor," in *Temperature, Its Measurement and Control in Science and Industry*, Vol. 4, H.H. Plumb, Ed., Instrument Society of America, Pittsburgh, Pa., 1972, p. 1125.

89. Wagner, J.P., A Fookson, and M.May, "Performance characteristics of semiconductor sensors under pyrolytic, flaming and smoldering combustion conditions," *J. Fire and Flammability*, 7, 1976, p. 71.

90. Watson, J., and D. Tanner, "Applications of the Taguchi gas sensor to alarms for inflammable gases," *The Radio and Electronic Engineer*, **44**, 1974, p. 85.

91. Weiss, H., "Galvanomagnetic devices," *IEEE Spectrum*, **5**, 1968, p. 75.

92. Weiss, H., *Structure and Applications of Galvanomagnetic Devices*, Pergamon Press, Oxford, England, 1969.

93. Wick, R.F., "Solution of the field problem of the germanium gyrator," *J. Appl. Phys.,* **25**, 1954, p. 741.
94. Wieder, H.H., *Intermetallic Semiconducting Films*, Pergamon Press, Oxford, England, 1970.
95. Woltjen, J.A., et al., "Bladder motility detection using the Hall effect," *IEEE Trans. Biomed. Eng.,* **BME-20**, 1973, p. 295.
96. Xavier, M.A., and C.O. Vogt, "Characteristics and applications of a semiconductor strain gage," *Semiconductor and Conventional Strain Gages*, M. Dean and R.D. Douglas, Eds., Academic Press, N.Y., 1962, p. 169.
97. Yeh, K.W., and R.S. Muller, "Piezoelectric DMOS strain transducers," *Appl. Phys. Lett.,* **29**, 1976, p. 521.
98. Young, J.J., and J.N. Zemel, "Photolytic detection of $H_2 S$ with a PbSe sensor," *Appl. Phys. Lett.,* **27**, 1975, p. 455.
99. Yuan, L.T., "Magnetoresistive transducer," *Solid-State Electronics,* **9**, 1966, p. 497.
100. Zemel, J.N., "Ion-sensitive field effect transistors and related devices," *Analyt. Chem.,* **47**, 1975, p. 255A.
101. Zias, A.R., and W.F.J. Hare, "Integration brings a generation of low-cost transducers," *Electronics*, December 4, 1972, p. 83.
102. Zemed, J.N., "Chemical sensitive semiconductor devices," *Research/Development,* **28**, April 1977, p. 38.
103. Wise, K.D. and S.K. Clark, "Diaphragm formation and pressure sensitivity in batch-fabricated silicon pressure sensors," *IEEE International Electron Devices Meeting, Technical Digest*, Washington, D.C., Dec. 1978, p. 96.
104. Wen, C.C., T.C. Chen and J.N. Zemel, "Ion controlled diodes (ICD), *IEEE International Electron Devices Meeting, Technical Digest*, Washington, D.C., Dec. 1978, p. 108.
105. Advani, G.N. et al., "Studies of $SnO_2$ as a material for gas detection, *153rd Electrochemical Society* Meeting Seattle, Washington, May 1978. Late News Paper.
106. Leary, D.J. et al., "Studies of ZnO as a material for gas detection," *153rd Electrochemical Society Meeting*, Seattle, Washington, May 1978. Late News Paper.
107. Zemel, J.N., "Surface Physics of Phosphors and Semiconductors," C.G. Scott and C.E. Reed, eds, Academic Press, London, 1975.
108. Windischmann, H. and P. Mark, "A model for the operation of a thin-film $SnO_x$ conductance modulation carbon monoxide sensor" *153rd Electrochemical Society Meeting* Seattle, Washington, 78, May 1978, Extended Abstract No. 91 p. 217.

# References

Abeles, F. (Ed.), *Optical Properties of Solids*, Elsevier, New York, 1972.

Adler, D., *Amorphous Semiconductors* (reprinted from CRC Critical Reviews in Solid State Sciences), CRC Press, Cleveland, Ohio, 1971.

Adler, R.B., A.C. Smith, and R.L. Longini, *Introduction to Semiconductor Physics*, Vol. 1, Semiconductor Electronics Education Committee, Wiley & Sons, New York, 1964.

Agajanian, A.H., *Semiconducting Devices, A Bibliography of Fabrication Technology, Properties, and Applications*, Plenum, New York, 1976.

Aigrain, P. and M. Balkanski (Eds.), *Tables of Constants and Numerical Data Affiliated to the International Union of Pure and Applied Chemistry*, 1. "Selected Constants Relative to Semiconductors," Pergamon Press, Oxford, England, 1961.

Alder, B., and S. Fernbach, *Methods in Computational Physics; Advances in Research and Applications*, Vol. 8, Energy Bands of Solids, Academic Press, New York, 1968.

Allen, B.M., *Soldering Handbook*, Drake Publishers, Inc., New York, 1972.

Alley, C.L., and K.W. Atwood, *Semiconductor Devices and Circuits*, Wiley, New York, 1971.

Allison, J., *Electronic Integrated Circuits: Their Technology and Design*, McGraw-Hill, New York, 1975.

Almazov, A.B., *Electronic Properties of Semiconducting Solid Solutions*, Plenum, New York, 1968.

Altman, L. (Ed.), *Microprocessors*, Electronics Book Series, McGraw-Hill, New York, 1975.

Altman, L. (Ed.), *Large Scale Integration*, McGraw-Hill, New York 1976.

Altman, L. and S.E. Scrupski (Eds.), *Applying Microprocessors*, McGraw Hill, New York, 1977.

Altman, L., (Ed.), *Memory Design: Microcomputers to Mainframes*, McGraw-Hill, New York, 1978.

Ambroziak, A., *Semiconductor Photoelectric Devices*, Iliffe, London, 1968.

*American Committee for Crystal Growth*, Conference on Crystal Growth, Gaithersburg, August 1969, Office of International Relations, National Bureau of Standards, Washington, D.C. 20234.

American Micro-Systems, Inc. (Staff), "MOS integrated circuits: theory, fabrication, design and systems applications of MOS LSI," Van Nostrand Reinhold, New York, 1972.

Anderson, C.A., *Microprobe analysis*, Wiley-Interscience, New York, 1973.

Anderson, J.C. (Ed.), *The Use of Thin Films in Physical Investigations*, Academic Press, New York, 1965.

Anderson, J.B., *Gasdynamic Lasers; An Introduction*, Academic Press, New York, 1976.

Angelo, Jr., E.J., *Electronics: BJT's, FET's and Microcircuits*, McGraw-Hill, New York, 1969.

Arams, F.R., *Infrared-to-Millimeter Wavelength Detectors*, Artech House, Dedham, Mass., 1973.

Arecchi, F.T. and V. Degiorgio (Eds.), *Coherent Optical Engineering*, North Holland, New York, 1977.

Arnaud, J.A., *Beam and Fiber Optics*, Academic Press, New York, 1976.

Ash, E.A. (Ed.), *Solid State Devices 1977*, Conference Series 40, Adam Hilger Ltd., Bristol, England, 1978.

Atkinson, P., *Thyristors and Their Applications*, Crane, Russak Co., New York, 1971.

Auleytner, J., *X-Ray Methods in the Study of Defects in Single Crystals*, Pergamon Press, New York, 1966.

Aven, M., and J.S. Prener, *Physics and Chemistry of II-VI Compounds*, North-Holland Publ., Amsterdam, and Wiley, New York, 1967.

Azaroff, L., *X-Ray Spectroscopy*, McGraw-Hill, New York, 1974.

Backus, C.E., *Solar Cells*, IEEE Press, Piscataway, New Jersey, 1976.

Bailar, J.C., H.J. Emeleus, R. Nyholm, and A.F. Trotman-Dickenson, (Eds.), *Comprehensive Inorganic Chemistry*, Vol. 1-5, Pergamon Press, Oxford and New York, 1973.

Bailey, F.J., *Introduction to Semiconductor Devices*, George Allen & Unwin Ltd., London, England, 1972.

Baker, A.D., and D. Belteridge, *Photoelectron Spectroscopy*, Pergamon Press, Oxford and New York, 1972.

Bakish, R. (Ed.), *International Conference on Electron and Ion Beam Science and Technology*, Vol. 1-6, The Electrochemical Soc., Inc., Princeton, New Jersey, 1964, 1966, 1968, 1970, 1972, 1974.

Bardsley, W., *Progress in Semiconductors,* Vol. 4, The Electrical Effects of Dislocations in Semiconductors, Wiley, New York, 1960.

Bar-Lev, A., *Semiconductors and Electronic Devices*, Prentice Hall, Englewood Cliffs, New Jersey, 1979.

Barna, A., and D.I. Porat, *Integrated Circuits in Digital Electronics*, Wiley, New York, 1973.

Barna, A., and D.I. Porat, *Introduction to Microcomputers and Microprocessors*, Wiley Interscience, New York, 1976.

Barnard, R.D., *Thermoelectricity in Metals and Alloys*, Halsted, New York, 1973.

Baron, R., *Effects of Diffusion on Double Injection in Semiconductors*, Academic Press, New York, 1970.

Barrer, P.M., *Diffusion in and Through Solids*, Cambridge University Press, London, England 1951.

Barrett, C.S., and T.B. Massalski, *Structure of Metals*, McGraw-Hill, New York, 1963.

Basov, N.G. (Ed.), *Optical Properties of Semiconductors*, Lebedev Trudy, Vol. 75, Plenum, New York, 1976.

Basov, N.G. (Ed.), *Electrical and Optical Properties of III-IV Semiconductors*, P.N. Lebedev Physics Institute Series, Vol. 89, Plenum, New York, 1978.

Beam, W.R., *Electronics of Solids*, McGraw-Hill, New York, 1965.

Becke, H., et al., *Gate Turn-Off Silicon Controlled Rectifiers—A User's Guide*, RCA Solid State Division, Application Note AN-6357, 1974.

Bedford, B.D., and R.G. Holt, *Principles of Inverter Circuits*, Wiley-Interscience, New York, 1964.

Beeforth, T.H., and H.J. Goldsmid, *Physics of Solid State Devices*, Academic Press, New York and London, 1970.

Beer, A.C., *Galvanometric Effects in Semiconductors*, Academic, New York, 1963.

Beeston, B.E.P., Horne, R.W., and Markham, R., *Electron Diffraction and Optical Diffraction Techniques*, North-Holland Publ., Amsterdam, Netherlands, 1972.

Bell, R.L., *Negative Electron Affinity Devices*, Clarendon Press, Oxford, England 1973.

Berger, M., *Neutron Radiography*, American Elsevier, New York, 1965.

Bergh, A.A. and P.J. Dean, *Light Emitting Diodes*, Oxford Press, London, England 1976.

Berker, P.W., and F. Jensen, *Design of Systems and Circuits for Maximum Reliability or Maximum Production Yield*, McGraw-Hill, New York, 1977.

Bertolini, G., and A. Coche, (Eds.), *Semiconductor Detectors*, Wiley Interscience, New York, 1968

Bibbero, R.J., *Microprocessors in Instruments and Control*, Halsted Press, Wiley-Interscience, New York, 1977.

Biondi, F.J. (Ed.), *Transistor Technology*, Vol. 3, Van Nostrand Reinhold, Princeton, New Jersey, 1958.

Bir, G.L., and G.E. Pikus, *Symmetry and Strain-Induced Effects in Semiconductors*, translated from Russian by P. Shelnitz, Wiley, New York, 1974.

Birks, L.S., *Electron Probe Microanalysis*, Wiley-Interscience, New York, 1971.

Blackwell, L.A., and K.L. Kotzebue, *Semiconductor-Diode Parametric Amplifiers*, Prentice-Hall, Englewood Cliffs, New Jersey, 1961.

Blakemore, J.S., *Semiconductor Statistics*, Vol. 3, International Series of Monographs on Semiconductors, Pergamon Press, New York, 1962.

Blakeslee, T.R., *Digital Design with Standard MSI and LSI*, Wiley-Interscience, New York, 1975.

Blatt, F.J. et al., *Thermoelectric Power of Metals*, Plenum, New York 1976.

Blicher, A. *Thyristor Physics*, Applied Physics and Engineering, Vol. 12, Springer-Verlag, New York, 1976.

Blom, G.M., S.L. Blank, J.M. Woodall, (Eds.), *Liquid Phase Epitaxy*, North-Holland Publ., Amsterdam, 1974.

Blum, W., and G.B. Hogaboom, *Principles of Electroplating and Electroforming*, McGraw-Hill, New York, 1949.

Bockris, J.O'M., and S. Srinivasan, *Fuel Cells: Their Electro-chemistry*, McGraw-Hill, New York, 1969.

Bode, H.W., *Network Analysis and Feedback Amplifier Design*, Van Nostrand, Princeton, New Jersey, 1945.

Boltaks, B.I., *Diffusion in Semiconductors*, Academic Press, New York, 1963.

Bonch-Bruyevich, V.L., *Electronic Theory of Highly Doped Semiconductors*, American Elsevier, New York, 1966.

Bond, W.I., *Crystal Technology*, Wiley-Interscience, New York, 1975.

Bosch, B.G., and R.W.H. Engelmann, *Gunn Effect Electronics*, Halsted Press (Division of John Wiley), New York, 1975.

Boylestad, R., and L. Nashelsky, *Electronic Devices and Circuit Theory*, 2nd ed, Prentice Hall, Englewood Cliffs, New Jersey, 1978.

Boynham, A.C., and A.D. Boardman, *Plasma Effects in Semiconductors: Helicon and Alfvén Waves*, Taylor and Francis, London, England and Barnes and Noble, New York, 1971.

Bremer, J.W., *Superconductive Devices*, McGraw-Hill, New York, 1962.

Brice, D.K., *Ion Implantation Range and Energy Deposition Distributions: High Incident Ion Energies*, Vol. I, IFI Plenum, New York, 1975.

Brice, J.C., *The Growth of Crystals from Liquids*, North-Holland/American Elsevier, New York, 1973.

Bridgers, H.E., J.H. Scoff, and J.R. Shive (Eds.), *Transistor Technology*, Vol. 1, Van Nostrand-Reinhold, Princeton, New Jersey, 1958.

Brignell, J. and G. Rhodes, *Laboratory on Line Computing: An Introduction for Engineers and Physicists*, Halsted Press, New York, 1975.

Brodsky, M.H., S. Kirkpatrick, and D. Weaire, (Eds.), *International Conference on Tetrahedrally Bonded Amorphous Semiconductors*, Yorktown Heights, New York, 1974, American Institute of Physics, New York, 1974.

Brooks, H., *Advances in Electronics and Electron Physics*, Vol. 7, Electrical Properties of Germanium and Silicon, Academic Press, New York, 1955.

Brophy, J.J., *Semiconductor Devices*, McGraw-Hill, New York, 1964.

Brown, H.E., *Solution of Large Networks by Matrix Methods*, Wiley, New York, 1975.

Brugler, J.S., *Low-Light-Level Limitations of Silicon Junction Photodetectors*, Stanford University Thesis, 1968. Available from University Microfilms, Ann Arbor, Michigan, #69-201.

Bruins, P.F. (Ed.), *Packaging with Plastics*, Gordon and Breach, New York, 1975.

Bube, R.H., *Photoconductivity of Solids*, Wiley, New York, 1960.

Bullough, R., and R.C. Newman, *The Interaction of Impurities with Dislocations in Silicon and Germanium*, Vol. 7, Progress in Semiconductors, Wiley, New York, 1963.

Bulman, P.J., G.S. Hobson, and B.C. Taylor, *Transferred Electron Devices*, Academic Press, New York and London, 1972.

Burford, W.B., and H.G. Verner, *Semiconductor Junctions and Devices*, McGraw-Hill, New York, 1965.

Burger, R.M., and R.P. Donovan (Eds.), *Fundamentals of Silicon Integrated Device Technology*, Vol. 1, Oxidation, Diffusion and Epitaxy, Prentice-Hall, Englewood Cliffs, New Jersey, 1967.

Burger, R.M., and R.P. Donovan (Eds.), *Fundamentals of Silicon Integrated Device Technology*, Vol. II, Bipolar and Unipolar Transistors, Prentice-Hall, Englewood Cliffs, New Jersey, 1968.

Burgess, R.E. (Ed.), *Fluctuation Phenomena in Solids*, Academic Press, New York and London, 1965.

Burstein, E., and S. Lundquist (Eds.), *Tunneling Phenomena in Solids*, Plenum Press, New York, 1969.

Cali, J.P. (Ed.), *Trace Analysis of Semiconductor Materials*, Pergamon, New York, 1963.

Calvert, J.M., and M.A.H. McCausland, *Electronics*, Wiley-Interscience, New York, 1978.

Camenzind, H.R., *Electronic Integrated Systems Design*, Van Nostrand Reinhold, New York, 1972.

Campbell, R.W. and F.W. Mims, *Semiconductor Diode Lasers*, Foulsham, Slough, England, 1972.

Candano, R. and J. Verbist, (Eds.), *Electron Spectroscopy*, Elsevier Scientific Publishing Co., New York, 1975.

Carr, W.N., and J.P. Mize, *MOS/LSI Design and Application*, McGraw-Hill, New York, 1972.

Carroll, J.E., *Hot Electron Microwave Generators*, American Elsevier Publishing Co., New York, 1970.

Carroll, J.E., *Physical Models for Semiconductor Devices*, Crane-Russak Co., New York, 1974.

Carson, R.S., *High-Frequency Amplifiers*, Wiley-Interscience, New York, 1975.

Carter, D.L., and R.T. Bate (Eds.), *Physics of Semimetals and Narrow-Gap Semiconductors*, Pergamon, New York, 1971.

Carter, G., and D. Mrazek, *The Systems Approach to Character Generators*, National Semiconductor Application Note AN-40, June 1970.

Carter, G., and W.A. Grant, *Ion Implantation of Semiconductors*, Halsted Press, New York, 1976.

Casasent, D. (Ed.), *Optical Data Processing*, Springer-Verlag, New York, 1978.

Casasent, D., *Electronic Circuits*, Quantum Pubs., New York, 1973.

Casey, H.C., Jr., and M.B. Panish, *Heterostructure Lasers, Part A: Fundamental Principles*, Academic Press, New York, 1978.

Casey, H.C., Jr., and M.B. Panish, *Heterostructure Lasers, Part B: Materials and Operating Characteristics*, Academic Press, New York, 1978.

Chadwick, G.A., and D.A. Smith, *Grain Boundary Structure and Properties*, Academic Press, New York, 1976.

Chaffin, R.J., *Microwave Semiconductor Devices: Fundamentals and Radiation Effects*, Wiley-Interscience, New York, 1973.

Chance, B., et al., *Waveforms*, Radiation Laboratory Series, McGraw-Hill, New York, 1949.

Chang, H., *Magnetic-Bubble Memory Technology*, Marcel Dekker, New York, 1978.

Change, K.K.N., *Parametric and Tunnel Diodes*, Prentice Hall, Englewood Cliffs, New Jersey, 1964.

Cheng, Y.C., *Electronic States at the Silicon-Silicon Dioxide Interface*, Pergamon Press, Fairview Park, New York, 1978.

Cherry, E.M., and D.E. Hooper, *Amplifying Devices and Low-Pass Amplifier Design*, Wiley, New York, 1968.

Chirlian, P.M., *Electronic Circuits: Physical Principles, Analysis and Design*, McGraw-Hill, New York, 1971.

Chow, W.F., *Principles of Tunnel Diode Circuits*, Wiley & Sons, New York 1964.

Chua, L.O., and P.M. Lin, *Computer-Aided Analysis of Electronics Circuits: Algorithms and Computational Techniques*, Prentice-Hall, Englewood Cliffs, New Jersey, 1975.

Clarricoats, P.J.B., *Optical Fibre Waveguides*, Peregrinus (for) the Institution of Electrical Engineers, Stevenage, 1975.

Cobbold, R.S.C., *Theory and Applications of Field-Effect Transistors*, Wiley-Interscience, New York, 1970.

Coekin, J.A., *High Speed Pulse Techniques*, Pergamon, Elmsford, New York, 1975.

Cohen, M.M., *Introduction to the Quantum Theory of Semiconductors*, Gordon and Breach, New York, 1972.

Collins, D.H. (Ed.), *Power Sources 5; Research and Development in Non-Mechanical Electrical Power Sources*, (Proceedings of the 9th International Symposium held at Brighton, Sept., 1974), Academic Press, New York, 1975.

Colliver, D., *The Technology of Compound Semiconductor Materials and Devices*, Artech House, Dedham, Mass., 1976.

Comer, D.J., *Semiconductor Circuits Laboratory Manual*, Prentice-Hall, Englewood Cliffs, New Jersey, 1969.

Connelly, J.A. (Ed.), *Analog Integrated Circuits*, Wiley, New York, 1975.

Connolly, T.F., and D.T. Hawkins (Eds.), *Ferroelectrics Literature Index*, IFI/Plenum, New York, 1974.

Conwell, E.M., *High Field Transport in Semiconductors*, Academic Press, New York, 1967.

Cooper, W.D., *Electronic Instrumentation and Measurement*, Prentice-Hall, Englewood Cliffs, New Jersey, 1970.

Corbett, J.W., *Electron Radiation Damage in Semiconductors and Metals*, Academic Press, New York, 1966.

Corbett, J.W., and G.D. Watkins, *Radiation Effects in Semiconductors*, Gordon and Breach, London, New York and Paris, 1971.

1st Cornell Biennial Conference on Engineering Applications of Electronic Phenomena, 1967 Topic: *High Frequency Generation and Amplification*, Ithaca, New York, 1967.

2nd Biennial Cornell Electrical Engineering Conference. 1969 Topic: *Computerized Electronics*, Ithaca, New York, 1969.

3rd Biennial Cornell Electrical Engineering Conference, 1971 Topic: *High Frequency Generation and Amplification: Devices and Applications*, Ithaca, New York, 1971.

4th Cornell Electrical Engineering Conference, *Microwave Semiconductor Devices, Circuits and Applications*, Ithaca, New York, 1973.

Cornetet, W.H., Jr., and F.E. Battocletti, *Electronic Circuits by Systems and Computer Analysis*, McGraw-Hill, New York, 1975.

Corning, J.J., *Transistor Circuit Analysis and Design*, Prentice-Hall, Englewood Cliffs, New Jersey, 1965.

Cottrell, A.H., *Dislocations and Plastic Flow in Crystals*, Oxford Univ. Press (Clarendon), London and New York, 1953.

Coughlin, R.F., *Semiconductor Fundamentals*, Prentice-Hall, Englewood Cliffs, New Jersey, 1976.

Coutts, T.J., *Electrical Conduction in Thin Metal Films*, Elsevier Scientific Book Co., New York, 1974.

Cowles, L.G., *Transistor Circuits and Applications*, Prentice-Hall, Englewood Cliffs, New Jersey, 1974.

Cowles, L.G., *Transistor Circuit Design*, Prentice-Hall, Englewood Cliffs, New Jersey, 1972.

Crawford, J.H. (Ed.), *Semiconductor and Molecular Crystals: Point Defects in Solids*, Vol. 2, Plenum, New York, 1975.

Crawford, J.H., Jr., and Slifkin, L.M., *Point Defects in Solids, General and Ionic Crystals*, Vol. 1, Plenum Press, New York, 1972.

Crawford, R.H., *MOSFETs in Circuit Design*, McGraw-Hill, New York, 1967.

Crowder, B.L. (Ed.), *Ion Implantation in Semiconductors and Other Materials*, Plenum, New York, 1973.

Crowhurst, N.H. (Ed.), *Understanding Solid State Electronics*, TAB Books, Blue Ridge Summit, Pa., 1970.

Cullen, G.W., and C.C. Wang (Eds.), *Heteroepitaxial Semiconductors for Electronic Devices*, Springer-Verlag, New York, 1978.

Cullen, G.W., et al. (Eds.), *Vapour Growth & Epitaxy* (Conference Proceedings Second International Conference on Vapour Growth and Epitaxy, May 21-25, 1972, Jerusalem, Israel), North-Holland, Amsterdam, 1972.

Cullity, B.D., *Introduction to Magnetic Materials*, Addison Wesley, Reading, Mass., 1972.

*Custom Microcircuit Design Handbook*, Fairchild Semiconductor, Mountain View, Calif., 1963.

Daglish, H.N., J.G. Armstrong, J.C. Alling, and C.A.P. Foxell, *Low Noise Microwave Amplifiers*, Cambridge University Press, Cambridge, U.K., 1968.

Dainty, J.C., and R. Shaw, *Image Science; Principles, Analysis and Evaluation of Photographic-Type Imaging Processes*, Academic Press, New York, 1974.

Darr, J., *Transistor Audio Amplifiers*, Foulsham, Slough, 1971.

Dascalu, D., *Electronic Processes in Unipolar Semiconductor Devices*, International Scholarly Book Services, Inc., Forest Grove, Oregon, 1976

Dascalu, D., *Transit-Time Effects in Unipolar Solid State Devices*, International Scholarly Book Services, Inc., Forest Grove, Oregon, 1976.

Davies, D.W., and D.L.A. Barber, *Communication Networks for Computers*, Wiley-Interscience, New York, 1973.

Davis, R.M., *Power Diode and Thyristor Circuits*, Cambridge University Press, Cambridge, U.K., 1971.

D'Azzo, J.J., and C.H. Houpis, *Linear Control System Analysis and Design: Conventional and Modern*, McGraw-Hill, New York, 1975.

Dean, M., III, and R.D. Douglas (Eds.), *Semiconductor and Conventional Strain Gages*, Academic Press, New York, 1962.

Dearnaley, G., *Ion Implantation*, North-Holland, Amsterdam, 1973.

Dearnaley, G., J.H. Freeman, R.S. Nelson, and J. Stephen, *Ion Implantation*, Series: Defects in Crystalline Solids, Elsevier, New York, 1973.

Dearnaley, G., and D.C. Northrop, *Semiconductor Counters for Nuclear Radiations*, Wiley, New York, 1963.

Dearnaley, G., and D.C. Northrop, *Semiconductor Counters for Nuclear Radiations*, 2nd ed., Wiley, New York, 1966.

Deboo, G.J., and C.N. Burrous, *Integrated Circuits and Semiconductor Devices, Theory and Application*, McGraw-Hill, New York, 1971.

Deboo, G.J., and C.N. Burrous, *Integrated Circuits and Semiconductor Devices: Theory and Application*, Gregg Division, McGraw-Hill, New York, 1977. 2nd. Ed.

DeForest, W.S., *Photoresist: Materials and Processes*, McGraw-Hill Book Co., New York, 1975.

de Gennes, P.G., *The Physics of Liquid Crystals*, Oxford Univ. Press, New York, 1974.

Dekker, A.J., *Solid State Physics*, Prentice-Hall, Englewood Cliffs, New Jersey, 1958.

Delhom, L.A., *Design and Application of Transistor Switching Circuits*, McGraw-Hill, New York, 1968.

Deme, S., *Semiconductor Detectors for Nuclear Radiation Measurement*, Hilger, London, England 1972.

Dempsey, J.A., *Basic Digital Electronics with MSI Applications*, Addison-Wesley Publishing Co., Reading, Mass., 1977.

DeRenzo, D., *Polymers in Lithography*, Noyes Data Corp., Park Ridge, New Jersey, 1971.

Désirant, M., and J.L. Michiels (Eds.), *Solid State Physics in Electronics and Telecommunications*, Vol. 1, Academic Press, New York, 1960.

DeSoete, D. et al., *Neutron Activation Analysis*, Wiley-Interscience, New York, 1972.

Dewan, S.B., and A. Straughen, *Power Semiconductor Circuits*, Wiley, New York, 1975.

Diggle, J.W., *Oxides and Oxide Films*, Vol. 1 and 2, Marcel Dekker, New York, 1972.

Diggle, J.W., and A.K. Vijh (Eds.), *Oxides and Oxide Films*, Vol. 4, Marcel Dekker, New York, 1976.

Dinaburg, M.S., *Photosensitive Diazo Compounds*, Focal Press, New York, 1965.

Director, S.W., (Ed.), *Computer-Aided Circuit Design*, Dowden, Hutchinson and Ross, Inc., Stroudsburg, Pa., 1974.

Dixon, R.C., *Spread Spectrum Systems*, Wiley-Interscience, New York, 1976.

Dobkin, R.C., *Linear Brief 8*, National Semiconductor Corp., Aug. 1969.

Dorsey, J., *Semiconductor Strain Gage Handbook*, Section 1 "Theory", Baldwin-Lima-Hamilton Corp., Electronics Division, Waltham, Mass., May 1963.

Dosse, J., *The Transistor*, 4th ed., Van Nostrand, Princeton, New Jersey, 1964.

Douglas, A., and S. Astley (Eds.), *Transistor Electric Organs for the Amateur*, Soccer, London, 1969.

Drabble, J.R., and H.J. Goldsmid, *Thermal Conduction in Semiconductors*, Vol. 4, International Series of Monographs on Semiconductors, Pergamon Press, Oxford, England, 1961.

Duke, C.B., *Tunneling in Solids*, Academic Press, New York, 1969.

Dummer, G.W.A. (Ed.), *Microelectronics and Reliability* (bimonthly) Pergamon, Elmsford, New York, 1962.

Dummer, G.W., and N.B. Griffin (Eds.), *Electronic Reliability*, Pergamon, Oxford and New York, 1966.

Dunlap, W.C., Jr., *Impurities in Germanium, Progress in Semiconductors*, Vol. 2, Wiley, New York, 1956.

Eastman, L.F. (Ed.), *Gallium Arsenide Microwave Bulk and Transit-Time Devices*, Artech, Dedham, Mass., 1972.

Eckertová, L., *Physics of Thin Films*, Plenum, New York, 1977.

Ehrenreich, H., F. Seitz, and D. Turnbull (Eds.), *Solid State Physics, Advances in Research and Applications*, Vol. 1, Academic Press, New York, 1955.

Ehrenreich, H., F. Seitz and D. Turnbull (Eds.), *Solid State Physics*, Vol. 25, Academic Press, New York, 1970.

Ehrenreich, H., F. Seitz, and D. Turnbull (Eds.), *Solid State Physics, Advances in Research and Applications*, Supplement 14: *Liquid Crystals*. Ed. L. Liebert, Academic Press, New York, 1978.

Ehrsam, W., *Audio Power Generation Using IC Operational Amplifiers*, Motorola Semiconductor Products, Inc., Application Note AN-275.

Eimbinder, J., *Designing with Linear Integrated Circuits*, Wiley-Interscience, New York, 1969.

Eimbinder, J. (Ed.), *Semiconductor Memories*, Wiley-Interscience, New York, 1971.

Eisen, F.H., and L.T. Chadderton (Eds.), *Ion Implantation*, Gordon and Breach, London, England 1971.

Eisenman, W.L. (Ed.), *Utilization of Infrared Detectors,* SPIE Vol. 132, SPIE Meeting Jan. 16-18, 1978, Los Angeles,, Calif., SPIE, Bellingham, Washington, 1978.

Electrochemical Society Extended Abstracts, 134th National Meeting, Symposium on Ohmic Contacts, Montreal, Fall 1968.

Electrochemical Society Meeting, 1st International Symposium on Silicon Materials Science and Technology, New York, May 1969.

Electrochemical Society, Proceedings of Conferences on *Chemical Vapor Deposition*, 1973, 1975, 1977.

Electro-Optical Systems Design Conference, *Proceedings of the Technical Program*, New York, Sept. 1970. Available from Industrial and Scientific Conference Management, Inc., 222 W. Adams St., Chicago, Ill. 60606.

Elliott, R.J., and A.F. Gibson, *An Introduction to Solid State Physics and its Applications*, Barnes & Noble, New York, 1974.

Elmer, W.B., *The Optical Design of Reflectors*, Elmer, Mass., 1974.

Elphick, M.S., *Microprocessor Basics*, Hayden Book Co., Rochelle Park, New Jersey, 1977.

Emsley, J.W., J. Feeney, and L.H. Sutcliffe, *High Resolution Nuclear Magnetic Resonance Spectroscopy*, Pergamon Press, New York, 1965.

Ertl, G., and J. Kupper, *Low Energy Electrons and Surface Chemistry*, Verlag Chemie, Weinheim, Germany, 1974.

Eshbach, O.W. (Ed.), *Handbook of Engineering Fundamentals*, 2nd ed. Wiley and Sons, New York, 1952.

European Microwave Conference, 4th, Montreux, Switzerland, Sept. 10-13, 1974.

European Conference on Ion Implantation, Reading, Berks, England, Sept. 1970, Proceedings were published by: Peter Perregrinus, Stevenage, Herts, England, 1970.

1st *European Semiconductor Device Research Conference on Solid-State Devices*, Proceedings 1971, Institute of Physics, London, 1971.

Faulkenberry, L.M., *An Introduction to Operational Amplifiers*, Wiley, New York, 1977.

Feldman, J.M., *The Physics and Circuit Properties of Transistors*, Wiley, New York, 1972.

Ferrari, R.L., and A.K. Jonscher (Eds.), *Problems in Physical Electronics*, Academic Press, New York, 1973.

Fink, D.G. (Ed.), *Standard Handbook for Electronic Engineers* McGraw-Hill, New York, 1974

Finkel, J., *Computer-Aided Experimentation*, Wiley, New York, 1975.

Fistul, V.I., *Heavily Doped Semiconductors*, Plenum, New York, 1969.

Fitchen, F.C., *Electronic Integrated Circuits and Systems*, Van Nostrand Reinhold, New York, 1970.

Flugge, S. (Ed.), *Handbuch der Physik*, Vol. 17, Springer, Berlin, 1956.

Flynn, C.P., *Point Defects and Diffusion*, Clarendon Press, Oxford, England, 1972.

Flynn, G., *MOS and CMOS Logic IC's*, H.W. Sams, Indianapolis, Ind., 1973.

Flynn, G., *Transistor-Transistor Logic*, H.W. Sams, Indianapolis, Ind., 1973.

Fogiel, M., *Modern Microelectronics; Basic Principles, Circuit Design, Fabrication Technology*, Research and Education Association, New York, 1972.

Fogiel, M., *Microelectronics: Principles, Design Techniques, Fabrication Processes*, Research and Education Association, New York, 1968.

Foner, S., and B.B. Schwartz (Eds.), *Superconducting Machines and Devices*, Plenum, New York, 1974.

Fox, J. (Ed.), *Symposium on Optical and Acoustical Microelectronics*, Proceedings, Polytechnic Press, New York, 1975.

Francombe, M.H., and H. Sato (Eds.), *Single Crystal Films*, Mac Millan, New York, 1964.

Frank, F.C., R. Kern, R.A. Laudise, M. Schieber, and R.L. Parker (Eds.), *Journal of Crystal Growth*, Vol. 27, North-Holland Publishing Co., Amsterdam, 1974.

Frankl, D.R., *Electrical Properties of Semiconductor Surfaces*, Vol. 7, International Series of Semiconductors, Pergamon Press, New York, 1967.

Frantsevich, I.N. (Ed.), *Silicon Carbide; Structure, Properties, and Uses*, Plenum Press, New York, 1970.

Freeman, R.L. *Telecommunication Transmission Handbook*, Wiley, New York, 1975.

Frey, J., *Microwave Integrated Circuits*, Artech, Dedham, Mass., 1974.

Friedland, B., O. Wing, and R. Ash, *Principles of Linear Networks*, McGraw-Hill, New York, 1961.

Friedman, A.D., and P.R. Menon, *Theory and Design of Switching Circuits*, Computer Science Press, Fall River Lane, Potomac, Md., USA, 1975.

Frova, A. (Ed.), *The Physics and Technology of Semiconductor Light Emitters and Detectors*, Proc. of the International Symposium at Pugnochiuso, Italy, Sept. 4-10, 1972, North Holland Publishing, Co., Amsterdam, 1973.

Frumkin, A.N. (Ed.), *Surface Properties of Semiconductors*, Consultants Bureau, New York, 1966.

Früngel, F.B.A., *Capacitor Discharge Engineering*, Academic Press, New York, 1976.

Gaertner, W.W., *Adaptive Electronics*, Artech House, Dedham, Mass., 1973.

Gagliardi, R.M., and S. Karp, *Optical Communications*, Wiley-Interscience, New York, 1976.

Gardner, F.M., *Phaselock Techniques*, Wiley-Interscience, New York, 1966.

Garland, H., and R. Melen, *Understanding CMOS Integrated Circuits*, Sams, Indianapolis, Ind., 1975.

Garrett, P.H., *Analog Systems for Microprocessors and Minicomputers*, Prentice-Hall, Englewood Cliffs, New Jersy, 1978.

Garver, R.V., *Microwave Diode Control Devices*, Artech, Dedham, Mass., 1976.

Gatos, H.C. (Ed.), Metallurgical Society Conferences, *Properties of Elemental and Compound Semiconductors*, Vol. 5, Interscience, New York and London, 1960.

Gentry, F.E., et al., *Semiconductor Controlled Rectifiers*, Prentice-Hall, Englewood Cliffs, New Jersey, 1964.

Geschwind, S., *Electron Paramagnetic Resonance*, Plenum Publ., New York, 1972.

Getreu, I., *Modeling the Bipolar Transistor*, Tektronix, Inc., Beaverton, Oreg., 1976, Part No. 062-2841-00.

Ghandhi, S.K., *The Theory and Practice of Microelectronics*, Wiley-Interscience, New York, 1968.

Ghandhi, S.K., *Semiconductor Power Devices*, Wiley-Interscience, New York, 1977.

Ghausi, M.S., *Electronic Circuits Devices, Models, Functions, Analysis and Design*, text ed., Van Nostrand Reinhold, New York, 1971.

Ghausi, M.S., *Principles and Design of Linear Active Circuits*, McGraw-Hill, New York, 1965.

Giacoletto, L.J. (Ed.), *Electronics Designers' Handbook*, 2nd ed., McGraw-Hill, New York, 1978.

Giacoletto, L.J., *Differential Amplifiers*, Wiley-Interscience, New York, 1970.

Gibbons, G., *Avalanche Diode Microwave Oscillators*, Clarendon Press, Oxford, England 1973.

Gibbons, J.F., *Semiconductor Electronics*, McGraw-Hill, New York, 1966.

Gibbons, J.F., W.S. Johnson, and S.W. Mylroie, *Projected Range Statistics, Semiconductor and Related Materials*, 2nd ed., Dowden, Hutchinson & Ross, Stroudsburg, Pa., 1975.

Gibson, A.F., R.E. Burgess, and P. Aigrain (Eds.), *Progress in Semiconductors*, Vol. 1-8, John Wiley & Sons, New York, 1956-1964.

Glaser, A.B., and G.E. Subak-Sharpe, *Integrated Circuit Engineering, Design, Fabrication and Applications*, Addison Wesley, New York, 1977.

Glazov, V.M., *Liquid Semiconductors*, Plenum, New York, 1969.

Glazov, V.M., and V.S. Zemskov, *Physicochemical Principles of Semiconductor Doping*, Israel Program for Scientific Translation, Jerusalem, Israel, 1968.

Gloge, D. (Ed.), *Optical Fiber Technology*, IEEE Press, NY, 1976.

Glorioso, R.M., *Engineering Cybernetics*, Prentice-Hall, Englewood Cliffs, New Jersey, 1975.

Glotin, P. (Ed.), *Int. Conference on Applications of Ions Beams to Semiconductor Technology*, Editions Ophrys, France, 1967.

Goetzberger, A. et al., *Interface States on Semiconductor Insulator Surfaces*, Vol. 6, (in CRC Critical Reviews in Solid State Sciences edited by D.E. Schuele and R.W. Hoffman), CRC Press, Cleveland, Ohio, 1976.

Goetzberger, A., and S.M. Sze, *Metal Insulator Semiconductor MIS Physics*, Vol. 1, Applied Solid State Science, R. Wolfe, Ed., Academic Press, New York, 1969.

Goldsmith, H.J. (Ed.), *Diffusion in Semiconductors*, Academic Press, New York, 1963.

Goldstein, J.I., and H. Yakowitz (Eds.), *Practical Scanning Electron Microscopy: Electron and Ion Microprobe Analysis*, Plenum Press, New York, 1975.

Gooch, C.H. (Ed.), *Gallium Arsenide Layers*, Wiley Interscience, New York, 1969.

Gooch, C.H., *Injection Electroluminescent Devices*, Wiley-Interscience, New York, 1973.

Goodman, C.H.L. (Ed.), *Crystal Growth*, Theory and Techniques, Vol. 1, Plenum, New York, 1974.

Goodman, C.H.L., *Crystal Growth*, Plenum, New York, 1978.

Gosling, W., *Introduction to Microelectronic Systems*, McGraw-Hill, New York, 1968.

Gossick, B.R., *Potential Barriers in Semiconductors*, Academic, New York, 1964.

Graeme, J.G., *Designing With Operational Amplifiers*, McGraw-Hill, New York, 1978.

Graeme, J.G., *Designing with Operational Amplifiers: Applications Alternatives*, McGraw-Hill, New York, 1977.

Graeme, J.G., *Applications of Operational Amplifiers: Third Generation Techniques*, McGraw-Hill, New York, 1973.

Graeme, J.G., C.E. Tobey, and L.P. Huelsman, *Operational Amplifiers, Design and Applications*, McGraw-Hill, New York, 1971.

Graham, F.D., Jr., and C.W. Gwyn, *Microwave Transistors*, Artech House, Dedham, Mass., 1975.

Gray, D. (Ed.), *American Institute of Physics Handbook*, 2nd ed., McGraw-Hill, New York, 1963.

Gray, P.E., D. Dewitt, A.R. Boothroyd, and J.F. Gibbons, *Physical Electronics and Circuit Models of Transistors*, Vol. 2, Semiconductor Electronics Education Committee, Wiley, New York, 1964.

Gray, P.E., and C.L. Searle, *Electronic Principles, Physics, Models and Circuits*, Wiley, New York, 1969.

Gray, P.R., and R.G. Meyer, *Analysis and Design of Analog Integrated Circuits*, John Wiley & Sons, New York, 1977.

Grebene, A.B. (Ed.), *Analog Integrated Circuits*, IEEE, New York, 1978.

Grebene, A.B., *Analog Integrated Circuit Design*, Van Nostrand, New York, 1972.

Green, M. (Ed.), *Solid State Surface Science*, Vol. 1, Marcel Dekker, New York, 1969.

Green, W. (Ed.), *Practical Test Instruments You Can Build*, TAB Books, Blue Ridge, Summit, Pa., 1974.

Greenaway, D.L., and G. Harbeke, *Optical Properties and Band Structures of Semiconductors*, Pergamon, New York, 1968.

Greiner, R.A., *Semiconductor Devices and Applications*, McGraw-Hill, New York, 1961.

Grinich, V.H., and H.G. Jackson, *Introduction to Integrated Circuits*, McGraw-Hill, New York, 1975.

Grivet, P., *Microwave Circuits and Amplifiers*, (translated from the French by P. Hawkes), Academic Press, New York, 1976.

Gronner, A.D., *Outline of Transistor Circuit Analysis*, Simon & Schuster, New York, 1966.

Grove, A.S., *Physics and Technology of Semiconductor Devices*, John Wiley and Sons, New York, 1967.

Grubel, R.O. (Ed.), *Metallurgical Society Conferences*, Vol. 12, Conference on Metallurgy of Elemental and Compound Semiconductors, Interscience Publishers, New York and London, 1961.

Gupta, K.C., and A. Singh, *Microwave Integrated Circuits,* John Wiley and Sons, New York, 1974.

Gutmann, F., and L.E. Lyons, *Organic Semiconductors* (Science and Technology of Materials), Wiley, New York, 1967.

Gyugyi, L., and B.R. Pelly, *Static Power Frequency Changers*, Wiley-Interscience, New York, 1976.

Haberecht, R.R., and E.L. Kern (Eds.), *Semiconductor Silicon*, Electro Chemical Society, New York, 1969.

Hackforth, H.L., *Infrared Radiation*, McGraw-Hill, New York, 1960.

Haddad, G.I. (Ed.), *Avalanche Transit-Time Devices*, Artech, Dedham, Mass., 1973.

Hakim, S.S., *Junction Transistor Circuit Analysis*, Wiley, New York, 1962.

Hamer, D.W., and Biggers, J.V., *Thick Film Hybrid Microcircuit Technology*, Wiley-Interscience, New York, 1973.

Hamilton, D.J., C.S. Meyer, and D.K. Lynn, *Analysis and Design of Integrated Circuits*, McGraw-Hill, New York, 1967.

Hamilton, D.J., and W.G. Howard, Jr., *Basic Integrated Circuit Engineering*, McGraw-Hill, New York, 1975.

Hamilton, D.J., F.A. Lindholm, and A.H. Marsahk, *Principles and Applications of Semiconductor Device Modeling*, Holt, Rinehart & Winston, New York, 1971.

Hamilton, T.D.S., *Handbook of Linear Integrated Electronics for Research*, McGraw-Hill, New York, 1978.

*Handbook of Electronic Materials*, ed. by A.J. Moses, Vol. 1, Optical Materials Properties, IFI Plenum, New York, 1971.

*Handbook of Electronic Materials*, ed. by M.S. Neuberger, Vol. 2, III-V Semiconducting Compounds, IFI Plenum, New York, 1971.

*Handbook of Electronic Materials*, ed. by J.T. Milek, Vol. 3, Silicon Nitride for Microelectronic Applications, Part I: Preparation and Properties, IFI Plenum, New York, 1971.

*Handbook of Electronic Materials*, ed. by M.S. Neuberger, Vol. 5, Group IV Semiconducting Materials, IFI Plenum, New York, 1971.

*Handbook of Electronic Materials*, ed. by J.T. Milek, Vol. 6, Pt. 1, Silicon Nitride for Microelectronic Applications, Part II: Applications and Devices, IFI Plenum, New York, 1972.

Hannay, N.B., (Ed.), *Treatise on Solid State Chemistry, Defects in Solids*, Vol. 2, Plenum Press, New York, 1975.

Hannay, N.B. (Ed.), *Semiconductors*, Van Nostrand-Reinhold, Princeton, New Jersey, 1959.

Hannay, N. (Ed.), *Treatise on Solid State Chemistry*, Vol. 6A: Surfaces I and Vol. 6B: Surfaces II. Plenum, New York, 1976.

Hannay, N.B., and U. Colombo (Eds.), *Electronic Materials*, Plenum Press, New York and London, 1973.

Hansen, M., and K. Anderko, *Constitution of Binary Alloys*, 2nd ed., McGraw-Hill, New York, 1958.

Harnden, J.D., and F.B. Golden, *Power Semiconductor Applications*, Vol. 1, General Considerations, Vol. 2, Equipment and Systems, IEEE Press, New York, 1972.

Harper, C.A., *Handbook of Electronic Packaging*, McGraw-Hill, New York, 1969.

Harper, C.A., *Handbook of Thick Film Hybrid Microelectronics; a Practical Sourcebook for Designers, Fabricators, and Users*, McGraw-Hill, New York, 1974.

Harrick, N.J., *Internal Reflection Spectroscopy*, Interscience Publishers, New York, 1967.

Hartman, P., *Crystal Growth: an Introduction*, American Elsevier, New York, 1973.

Hartnagel, H., *Semiconductor Plasma Instabilities, For Microwave Generation and Amplification*, American Elsevier, New York, 1969.

Hasiguti, R.R., *Lattice Defects in Semiconductors*, Pennsylvania State University Press, University Park, Pa., 1968.

Hass, G. (Ed.), *Physics of Thin Films*, Advances in Research and Development, Vol. 1, Academic Press, New York and London, 1963.

Hass, G., and R.E. Thun, *Physics of Thin Films*, Vol. 2, Academic Press, New York, 1964.

Head, A.K., P. Humble, L.M. Clearbrough, and A.J. Morton, *Computed Electron Micrographs and Defect Identification*, North-Holland, Amsterdam, Netherlands, 1973.

Heavens, O.S., *Thin Film Physics*, Barnes and Noble, New York, 1970.

Heck, C., *Magnetic Materials and Their Application*, Newnes-Butterworths, London, England 1974.

Heinlein, W.E., and W.H. Holmes, *Active Filters for Integrated Circuits*, Springer-Verlag, New York, 1974.

Helszajn, J., *Nonreciprocal Microwave Junctions and Circulators*, Wiley, New York, 1975.

Henderson, B., *The Structures and Properties of Solids 1*, Defects in Crystalline Solids, Crane, Russak & Co., New York, 1972.

Henisch, H.K., *Rectifying Semiconductor Contacts*, Oxford University Press (Clarendon), London and New York, 1957.

Henisch, H.K., *Electroluminescence*, Vol. 5 International Series of Monographs on Semiconductors, Pergamon Press, New York, 1962.

Herring, C., and R.G. Breckenridge, (Eds.), *Photoconductivity Conference at Atlantic City, 1954*, Wiley, New York, 1956.

Hershberger, W.D. (Ed.), *Topics in Solid State and Quantum Electronics*, Wiley, New York, 1970, 1971.

Herskowitz, G.J., and R.B. Schilling (Eds.), *Semiconductor Device Modeling for Computer-Aided Design*, McGraw-Hill, New York, 1972.

Hetterscheid, W.T., *Transistor Bandpass Amplifiers*, Springer-Verlag, Berlin, 1966.

Hewlett-Packard, *Applications of PIN Diodes*, Hewlett-Packard Application Note 922.

Hewtt-Packard, *High Performance PIN Attenuator for Low Cost AGC Applications*, Hewlett-Packard Application Note 936.

Hewlett-Packard, *Hot Carrier Diode Video Detectors*, Hewlett-Packard Application Note 923.

Hewlett-Packard, *Ku-Band Step Recovery Multipliers*, Hewlett-Packard Application Note 928.

Hewlett-Packard, *S-Parameter Design*, Hewlett-Packard Application Note 154, 1972.

Hewlett-Packard, *S-Parameters . . . Circuit Analysis and Design*, Hewlett-Packard Application Note 95, 1968.

Hewlett-Packard Optoelectronics Division, Applications Engineering Staff, *Optoelectronics Applications Manual*, McGraw-Hill, New York, 1978.

Hibberd, R.G., *Integrated Circuits: A Basic Course for Engineers and Technicians*, McGraw-Hill, New York, 1969.

Hibberd, R.G., *Integrated Circuits-Questions and Answers*, Newnes-Butterworth, London, England 1974.

Hilburn, J.L., and Julich, *Microcomputers/Microprocessors*, Hardware, Software and Applications, Prentice-Hall, Englewood Cliffs, New Jersey, 1976.

Hilburn, J.L., and D.E. Johnson, *Manual of Active Filter Design*, McGraw-Hill, New York, 1973.

Hilsum, C., and A.C. Rose-Innes, *Semiconducting III-V Compounds*, Pergamon Press, New York, 1961.

Hirsch, P.B., A. Howie, R.B. Nicholson, D.W. Pashley, and M.J. Whelan, *Electron Microscopy of Thin Crystals*, Plenum Press, New York, 1965.

Hnatek, E.R., *Applications of Linear Integrated Circuits*, Wiley, New York, 1974.

Hnatek, E.R., *A User's Handbook of Integrated Circuits*, Wiley, New York, 1973.

Hobson, G.S., *The Gunn Effect*, Oxford University Press, New York, 1974.

Hodges, D.A. (Ed.), *Semiconductor Memories*, Wiley-Interscience, New York, 1972.

Hoeschele, D.F., Jr., *Analog-To-Digital/Digital-To-Analog Conversion Techniques*, Wiley-Interscience, New York, 1968.

Hoffman, A., *Physikolische Methoden zur Storstellenanalyse, beim Silizium, Halbeiterprobleme*, Vol. 5, Vieweg and Sohn, Braunschweig, 1961.

Hollahan, J.R., and A.T. Bell, *Techniques and Applications of Plasma Chemistry*, John Wiley & Sons, New York, 1974.

Holland, L., *Vacuum Deposition of Thin Films*, Halsted Press, New York, 1956.

Holmes, P.J., *The Electrochemistry of Semiconductors*, Academic Press, New York, 1962.

Holonyak, N., Jr., *Integrated Electronic Systems*, Prentice-Hall International, New York, 1970.

Holt, C.A., *Electronic Circuits, Digital and Analog*, John Wiley & Sons, New York, 1978.

Holter, M.R., S. Nudelman, G.H. Suits, W.L. Wolfe, and G.J. Zissis, *Fundamentals of Infrared Technology*, Macmillan, New York, 1962.

Hovel, H., *Solar Cells*, Vol. 11, Semiconductors and Semimetals, Academic Press, New York, 1975.

Howe, H. Jr., *Stripline Circuit Design*, Artech House, Dedham, Mass., 1975.

Howe, H.S., *Electronic Music Synthesis*, Norton, New York, 1975.

Howes, M.J. and D.V. Morgan (Eds.), *Microwave Devices: Device Circuit Interactions*, Wiley-Interscience, New York, 1976.

Hudson, R.D. Jr., and J.W. Hudson (Eds.), *Infrared Detectors*, Dowden, Hutchinson & Ross, Stroudsburg, Pa., 1975.

Huelsman, L.P., *Active Filters: Lumped, Distributed, Integrated Digital and Parametric*, McGraw-Hill, New York, 1970.

Huff, H.R. and R.R. Burgess (Eds.), *Semiconductor Silicon 1973*, Electro-chemical Society Symposium Series at 143rd Spring Meeting, Chicago, Ill., May 1973.

Hughes, A.L. and L.A. DuBridge, *Photoelectric Phenomena*, McGraw-Hill, New York, 1932.

Hulin, M. (Ed.), *Physics of Semiconductors*, Proceedings of the 7th International Conference, Paris, Academic Press, New York and London, 1964.

Hunter, L.P. *Handbook of Semiconductor Electronics*, McGraw-Hill, New York, 1970.

Hunter, L.P., *Introduction to Semiconductor Phenomena and Devices*, Addison-Wesley, Reading, Mass., 1966.

Huntley, F.A. (Ed.), *Lattice Defects in Semiconductors*, Conference Proceedings, 1974, Institute of Physics, London, 1975.

IEE International Conference on Low Light and Thermal Imaging Systems, IEE Conference Publication 124, May 1975, IEE, Herts., England, 1975.

IEE International Conference on Power Electronics, Power Semiconductors, and their Applications, Institute of Electrical Engineers, London, 1974.

IEEE 9th Annual Reliability Physics, Symposium, Proceedings, Las Vegas, Nev., 1971.

IEEE Conference Record of 1970, Conference on Display Devices, New York, Dec. 1970.

IEEE Conference Record, 5th Annual Meeting, Industry and General Applications Group, Chicago, Ill. Oct. 5-8, 1970 (New York, 1970a).

IEEE Electron Devices Group and American Vacuum Society, Thirteenth Symposium on Electron, Ion, and Photon Beam Technology, Colorado Springs, Colo. May 21-23, 1975.

IEEE 20th Electronics Components Conference, Proceedings, Washington, D.C., May 1970.

IEEE 1968 G-MTT International Microwave Symposium. Digest and Technical Program, Detroit, Mich., May 1968.

IEEE International Electron Devices Meeting, Technical Digest, Washington, D.C., Dec. 1974-1977.

IEEE International Solid State Circuits Conferences, Digest of Technical Papers, Philadelphia, Pa., Feb. 1971-1976.

IEEE Photovoltaic Specialists Conference, 9th, Silver Spring, Md., IEEE, New York, 1972; 10th, Conference Record, Palo Alto, Calif., Nov. 13-15, 1973; 11th, Conference Record, Scottsdale, Ariz. May 6-8, 1975, IEEE, New York, 1975; 12th, Conference Record, Batton Rouge, La., 1976.

Institute of Physics, Conference Series 22, *Metal-Semiconductor Contacts*, Proceedings of a conference organized by the Solid State Physics Subcommittee in association with the Thin Films and Surfaces Group of the Institute of Physics, Manchester, April 1974, Institute of Physics, London and Bristol, 1974.

Institute of Physics Conference Series, Vol. 32, *Solid State Devices 1976*, Bristol, England.

Institute of Physics, International Symposium on Gallium Arsenide and Related Compounds 6th Edinburgh and St. Louis 1976, Institute of Physics, Bristol and London, England, 1977.

Institute of Physics, Radiation Effects in Semiconductors, 9th Int. Conf. Dubrovnik 1976, Bristol and London, England, 1977.

Institute of Physics, Conference Series 16, *Radiation Damage and Defects in Semiconductors*, Proceedings of the International Conference organized by the Institute of Physics and sponsored by the International Union of Pure and Applied Physics and the U.S. Air Force, University of Reading, July 19-21, 1972, Institute of Physics, London and Bristol, England, 1973.

Institute of Physics, Conference Series 26, *Temperature Measurement 1975* Institute of Physics, London, England, American Institute of Physics, New York, 1975.

Institute of Physics, Conference Series 27, *Static Electrification 1975*, Institute of Physics, London, England, American Institute of Physics, New York, 1975.

Instrument Society of America, ISA Transducer Compendium, 2nd ed., Part 1: 3,000 Transducers—338 Model Series, Instrument Society of America, Pittsburgh, Pa., 1969.

Instrument Society of America, ISA Transducer Compendium, 2nd ed., Part 2: 8,000 Transducers—520 Model Series, Instrument Society of America, Pittsburgh, Pa., 1970.

Instrument Society of America, ISA Transducer Compendium, 2nd ed., Part 3: 2,000 Transducers—237 Model Series, Instrument Society of America, Pittsburgh, Pa., 1972.

International Conference on Technology and Applications of Charge Coupled Devices; Held at University of Edinburgh, 25-27 September 1974; Published by the University of Edinburgh, Scotland, 1974.

International Conference on the Physics and Chemistry of Semiconductor Heterojunctions and Layer Structures, Budapest, 1970, Hungarian Academy of Sciences, Budapest, Hungary, ed. by G. Szigeti, 1970.

8th International Conference on the Physics of Semiconductors, Kyoto, Japan, 1966. *J. Phys. Soc. Japan, Suppl. 21*, 1966.

9th International Conference on the Physics of Semiconductors, Moscow, Nauka, Leningrad, 1968.

10th International Conference on the Physics of Semiconductors, M.I.T. Cambridge, Mass., 1970, S.P. Keller (Ed.), Publ. CONF 700801 USAEC, National Technical Information Service, Springfield, Va. (14th Conference was in Edinburgh 1978.)

International Symposium on Gallium Arsenide, Institute of Physics and Physical Society, London, Reading, U.K., 1966, Dallas, Texas, 1968.

International Symposium on Gallium Arsenide and Related Compounds, 3rd, Aachen, Germany, October 1970; 4th , Boulder, Colorado 1972; 5th, Deauville, France, 1974; 6th, Edinburgh and St. Louis, 1976.

Ioffe, A.F., *Semiconductor Thermoelements and Thermoelectric Cooling*, Pion, London, England 1958.

Ivey, H.F., *Electroluminescence and Related Effects* (Supplement 1 to Advances in Electronics and Electron Physics), Academic Press, New York and London, 1963.

Jaffe, B., W.R. Cook, Jr., and H. Jaffe, *Piezoelectric Ceramics* (Nonmetallic Solids Series), Academic Press, New York and London, 1971.

James, R.W., *The Optical Principles of the Diffraction of X-Rays*, Bell, London, England 1958.

Jamison, J.A., R.H. McFee, G.N. Plass, R.H. Grube, and R.G. Richards, *Infrared Physics and Engineering*, McGraw-Hill, New York, 1963.

Japanese Journal of Applied Physics, *Proceedings of the 1st Conference on Solid-State Devices*, Tokyo, 1969. Supplement of the Journal, Vol. 39, 1970.

Japanese Journal of Applied Physics, *Proceedings of the 2nd Conference on Solid-State Devices*, Tokyo, 1970. Supplement to the Journal, Vol. 40, 1971.

Japanese Journal of Applied Physics, *Proceedings of the 3rd Conference on Solid State Devices*, Tokyo, 1971, Supplement to the Journal, 41, 1972.

Japanese Journal of Applied Physics, *Proceedings of the 4th Conference on Solid State Devices*, Tokyo, 1972. Supplement to the Journal, Vol. 42, 1973.

Japanese Journal of Applied Physics, *Proceedings of the 5th Conference (1973) International) on Solid State Devices,* Tokyo, 1973, Supplement to the Journal, Vol. 43, 1974.

Japanese Journal of Applied Physics, *Proceedings of the 6th Conference on Solid State Devices*, Tokyo, 1974, Supplement to the Journal, Vol. 44, 1975.

Japanese Journal of Applied Physics, *Proceedings of the Seventh Conference on Solid State Devices*, Tokyo, 1976. Supplement to the Journal, Vol. 45, 1976.

Jarzebski, Z.M., *Oxide Semiconductors* (Science of Solid State Monographs), Pergamon, New York, 1974.

Jespers, P.G., F. VandeWiele, and M.H. White (Eds.), *Solid State Imaging*, Noordhoff, Leyden, 1976.

Johnson, D.E., and J.L. Hilburn, *Rapid Practical Designs of Active Filters*, Wiley, New York, 1975.

Jones, D.V., and R.F. Shea, *Transistor Audio Amplifiers*, Wiley, New York, 1968.

Jowett, C.E., *Electronic Engineering Processes*, Business Books, London, England, 1972.

Jowett, C.E., *Semiconductor Devices: Testing and Evaluation*, Business Books, London, England, 1974.

Kaldis, E., *Current Topics in Materials Science*, Vol. 1, North-Holland, New York, 1977.

Kaldis, E., and H.J. Scheel (Eds.) *Crystal Growth and Materials*, North-Holland, New York, 1977.

Kaminow, I.P., *An Introduction to Electrooptic Devices*, Academic Press, New York, 1974.

Kaminow, I.P. and A.E. Siegman (Eds.), *Laser Devices and Applications*, IEEE Press, New York, 1973.

Kane, P.F., and G.B. Larrabee, *Characterization of Semiconductor Materials*, McGraw-Hill, New York, 1970.

Kane, P.F., and G.B. Larrabee, *Characterization of Solid Surfaces*, Plenum Press, New York, 1974.

Kanzig, W., *Ferroelectrics and Antiferroelectrics*, (Vol. 4, Solid State Physics, 1957), Academic Press, New York and London, 1964.

Kazan, B. (Ed.), *Advances in Image Pickup and Display*, Academic Press, New York, Vol. 1, 1974; Vol. 2, 1976; Vol. 3, 1977.

Kazan, B., and M. Knoll, *Electronic Image Storage*, Academic Press, New York, 1968.

Kendall, D.L., *Diffusion, in Semiconductors and Semimetals*, Vol. 4, Academic Press, New York, 1968.

Kendall, E.J. (Ed.), *Transistors*, (Selected Readings in Physics), Pergamon, New York, 1969.

Keonjian, E., *Microelectronics, Theory, Design and Fabrication*, McGraw-Hill, New York, 1963.

Keyes, R.J. (Ed.), *Optical and Infrared Detectors*, (Topics in Applied Physics, Vol. 19) Springer-Verlag, Heidelberg, 1977.

Khambata, A.J., *Introduction to Large-scale Integration*, Wiley, New York, 1969.

Kingston, R.H. (Ed.), *Semiconductor Surface Physics*, University of Pennsylvania Press, Philadelphia, Pa., 1957.

Kiver, M.S., *Transistor and Integrated Electronics*, 4th ed., McGraw-Hill, New York, 1972.

Klingman, E.E., *Microprocessor Systems Design*, Prentice-Hall, Englewood Cliffs, New Jersey, 1977.

Kmetz, A.R., and F.K. vonWillisen (Eds.) *Nonemissive Electrooptic Displays*, Plenum, New York, 1976.

Kocsis, M., *High-speed Silicon Planar-Epitaxial Switching Diodes*, Wiley, New York, 1976.

Kodak Seminars on Microminiaturization and Photoresists, Presented by Professional, Commercial, Industrial and Marketing Divisions, Eastman Kodak Co., Rochester, N.Y., 1965-1974.

Kooi, E., *Surface Properties of Oxidized Silicon*, Springer-Verlag, New York, 1967.

Korn, G.A., *Microprocessor and Small Digital Computer Systems for Engineers and Scientists*, McGraw-Hill, New York, 1977.

Korn, G.A. and T.M. Korn, *Electronic Analog and Hybrid Computers*, McGraw-Hill, New York, 1964.

Kosar, J., *Light-sensitive Systems*, John Wiley & Sons, New York, 1965.

Kressel, H., and J.K. Butler, *Semiconductor Lasers and Heterojunction LED's*, Academic Press, New York, 1977.

Kruse, P.W., L.D. McGlaughlin and R.D. McQuistan, *Elements of Infrared Technology*, Wiley, New York, 1962.

Krutz, R.L., *Introduction to Microprocessors and Digital Systems*, Wiley, New York, 1979.

Kuo, F.F. and W.G. Magnuson (Eds.), *Computer Oriented Circuit Design*, Prentice Hall, New Jersey, 1969.

Kurokawa, K., *An Introduction to the Theory of Microwave Circuits*, Academic Press, New York, 1969.

Lampert, M.A., and P. Mark, *Current Injection in Solids*, Academic Press, New York, 1970.

Lancaster, G., *Electron Spin Resonance in Semiconductors*, Plenum Press, New York, 1967.

Landee, R.W., and L.J. Giacoletto (Eds.), *Electronic Designers' Handbook*, 2nd ed., McGraw-Hill, New York, 1975.

Larach, S. (Ed.), *Photoelectronic Materials and Devices*, Van Nostrand-Reinhold, Princeton, New Jersey, 1965.

*Large Scale Integration*, ed. by L. Altman, (Electronics Book Series), McGraw-Hill, New York, 1976.

Larin, F., *Radiation Effects in Semiconductor Devices*, Wiley & Sons, New York, 1968.

Leahy, W.F., *Microprocessor Architecture and Programming*, Wiley, New York, 1977.

Leaver, K.D., and Chapman, B.N., *Thin Films*, Springer-Verlag, New York, 1971.

Leck, J.H., *Theory of Semiconductor Junction Devices: (A Textbook for Electrical and Electronic Engineers)*, Pergamon, New York, 1967.

LeComber, P.G., and J. Mort (Eds.), *Electronic and Structural Properties of Amorphous Semiconductors*, Academic Press, New York and London, 1973.

Lee, T.H., *Physics and Engineering of High Power Switching Devices*, M.I.T. Press, Cambridge, Mass., 1975.

Lenert, *Semiconductor Physics, Devices, and Circuits*, Merrill, Chicago, Ill., 1968.

Lesea, A., and R. Zaks, *Microprocessor Interfacing Techniques*, Sybex, Inc., 2161 Shattuck Ave., Berkeley, Calif.

Levin, A., *Solid State Quantum Chemistry: the Chemical Bond and Energy Bands in Tetrahedral Semiconductors*, McGraw-Hill, New York, 1976.

Levine, S.N. and R.R. Kurzrok (Eds.), *Selected Papers on Semiconductor Microwave Electronics*, Dover, New York, 1964.

Levinstein, H. (Ed.), *Photoconductivity*, Pergamon Press, Oxford, England, 1962.

Lewin, D., *Logical Design of Switching Circuits*, American Elsevier, New York, 1974.

Lin, H.C., *Integrated Electronics*, Holden-Day, San Francisco, Calif., 1967.

Lin, W.C. (Ed.), *Microprocessors: Fundamentals and Applications*, IEEE Press, Hoes Lane, Piscataway, New Jersey, 1977.

Lindmayer, J., and C.Y. Wrigley, *Fundamentals of Semiconductor Devices*, Van Nostrand Reinhold, New York, 1965.

Lines, M.E., and A.M. Glass, *Principles and Applications of Ferroelectrics and Related Materials*, Clarendon (Oxford U.P.), Oxford, England, 1977.

Linvill, J.G., *Models of Transistors and Diodes*, McGraw-Hill, New York, 1963.

Linvill, J.G., and J.F. Gibbons, *Transistors and Active Circuits*, McGraw-Hill, New York, 1961.

Lion, K.S., *Elements of Electrical and Electronic Instrumentation*, McGraw-Hill, New York, 1975.

Liu, C.L., and J.W.S. Liu, *Linear Systems Analysis*, McGraw-Hill, New York, 1975.

Lo, A.W., et al., *Transistor Electronics*, Prentice-Hall, Englewood Cliffs, New Jersey, 1956.

Long, D., *Energy Bands in Semiconductors*, Wiley-Interscience, New York, 1968.

Loretto, M.H., and R.E. Smallman, *Defect Analysis in Electron Microscopy*, Halsted Press, New York, 1976.

Lucovsky, G., and F.L. Galeener, *Structure and Excitations of Amorphous Solids*, American Institute of Physics Conf. Proc. No. 31, 1976.

Luecke, J., J.P. Mize, and W.N. Carr, *Semiconductor Memory Design and Application*, (Texas Instruments Electronics Series), McGraw-Hill, New York, 1973.

Luxenberg, H.R., and R.L. Kuehn, *Display Systems Engineering*, McGraw-Hill, New York, 1968.

Lytel, A., *Solid-State Power Supplies and Convertors*, H.W. Sams, Indianapolis, Indiana, 1971.

Madelung, O., *Physics of III-V Compounds*, John Wiley & Sons, New York, 1964.

Maissel, L. and R. Glang, *Handbook for Thin Film Technology*, McGraw-Hill, New York, 1970.

Malmstadt, H.V., et al., *Electronic Measurements for Scientists*, W.A. Benjamin, Reading, Mass., 1974.

Malvino, A.P., *Transistor Circuit Approximations*, 2nd ed., McGraw-Hill, New York, 1973.

Manasse, F.K., *Semiconductor Electronics Design*, Prentice-Hall, Englewood Cliffs, New Jersey, 1976.

Manera, A.S., *Solid State Electronic Circuits for Engineering Technology*, McGraw-Hill, New York, 1973.

Many, A., Y. Goldstien, and N.B. Grover, *Semiconductor Surfaces*, North Holland Publ., Amsterdam, 1965.

Markus, J., *Electronic Circuits Manual*, McGraw-Hill, New York, 1971.

Markus, J., *Guidebook of Electronic Circuits*, McGraw-Hill, New York, 1974.

Markus, J., *Sourcebook of Electronic Circuits*, McGraw-Hill, New York, 1968.

Marsden, C.P. (Ed.), National Bureau of Standards Special Publication 337, *Silicon Device Processing*, United States Department of Commerce, National Bureau of Standards, Proceedings of a Symposium held at Gaithersburg, Md., June 2-3, 1970, U.S. Govt Printing Office, Wash., D.C., 1970.

Martin, R.E., *Avoidance of Electrical Interference in Electronic Systems*, Research Studies Press, Forest Grove, Oregon, 1979.

Marton, L. (Ed.), *Advances in Electronics and Electron Physics*, Vol. 46, Academic Press, New York, 1978.

Marton L., *Advances in Electronics and Electron Physics*, Vol. 44 (Microwave Power Semiconductor Devices: Electron Bombarded Semiconductor Devices). Academic Press, New York, 1978.

Marton, L. (Ed.), *Advances in Electronics and Electron Physics*, Vols. 1-38, Academic Press, New York and London, 1948-1975.

Marton, L. (Ed.), Advances in Electronics and Electron Physics, Vol. 40, Parts A and B, *Sixth Symposium on Photo-Electronic Image Devices*, ed. by B.L. Morgan, Academic Press, New York, 1976.

Matare, H.F., *Defect Electronics in Semiconductors*, Wiley-Interscience, New York, 1971.

Matthews, J.W., *Epitaxial Growth*, Academic Press, New York, 1975.

Mattson, R.H., *Electronics*, Krieger, Huntington, New York, 1966.

Mavor, J., *M.O.S.T. Integrated Circuit Engineering*, Peter Peregrinus, Herts., England, 1973.

Mayer, J.W., L. Eriksson, and J.A. Davies, *Ion Implantation in Semiconductors, Silicon and Germanium*, Academic Press, New York, 1970.

Mazda, F.F., *Integrated Circuits: Technology and Applications*, Cambridge University Press, New York, 1978.

Mazda, F.F., *Thyristor Control*, Halsted Press, New York, 1973.

McCluskey, E.J., Jr., *Introduction to the Theory of Switching Circuits*, McGraw-Hill, New York, 1965.

McGee, J.D., D. McMullan, E. Kahan, and B.L. Morgan (Eds.), *Photo-Electronic Image Devices*, Vol. 28B, Proceedings of the Fourth Symposium held at Imperial College, London, Sept. 16-20, 1968, Academic Press, New York and London, 1970.

McGee, J.D., D. McMullan, and E. Kahan (Eds.), *Photo-Electronic Image Devices*, Vol. 33A, Proceedings of the Fifth Symposium held at Imperial College, London, Sept. 13-17, 1971, Academic Press, New York and London, 1972.

McKelvey, J.P., *Solid State and Semiconductor Physics*, Harper and Row, New York, 1966.

Mead, C.A., and L.A. Conway, *Introduction to VLSI Systems*, California Institute of Technology, 1978. Addison Wesley, Reading, Mass., 1979.

Meaden, G.T., *Electrical Resistance of Metals*, Plenum Press, New York, 1965.

Meindl, J.D., *Micropower Circuits*, Wiley, New York, 1969.

Melen, R., and H. Garland, *Understanding CMOS Integrated Circuits*, H.W. Sams, Indianapolis, Ind., 1975.

Melen, R., and H. Garland, *Understanding IC Operational Amplifiers*, H.W. Sam, Indianapolis, Ind., 1971.

Melen, R., and D. Buss (Eds.), *Charge-Coupled Devices: Technology and Applications*, IEEE Press, New York, 1977.

*Metallurgy of Advanced Electronic Materials*, ed. by G.E. Brock Interscience, New York, 1963.

Meyer, C.S., D.K. Lynn, and D.J. Hamilton, (Eds.), *Analysis and Design of Integrated Circuits*, McGraw-Hill, New York, 1968.

Meyer, R.G. (Ed.), *Integrated-Circuit Operational Amplifiers*, IEEE, New York, 1978.

Middlebrook, R.D., *An Introduction to Junction Transistor Theory*, Wiley, New York, 1957.

Middlehoek, S., et al., *Physics of Computer Memory Devices*, Academic Press, New York, 1977.

Milek, J.T., *Silicon Nitride for Microelectronic Applications, part 1, Preparation and Properties*, IFI Plenum, New York, 1971.

Milek, J.T., *Silicon Nitride for Microelectronics Applications*, IEI/Plenum, New York, 1972.

Miller, J.R. (Ed.), *Solid-State Communications*, Texas Instruments Electronic Series, McGraw-Hill, New York, 1966.

Miller, L.E. (Ed.), *Microwave Semiconductor Devices and Their Circuit Applications*, McGraw-Hill, New York, 1969.

Miller, L.F., *Thick Film Technology and Chip Joining*, Gordon and Beach, New York, 1972.

Millman, J., *Microelectronics: Digital and Analog Circuits and Systems*, McGraw-Hill, New York, 1978.

Millman, J., *Vacuum-Tube and Semiconductor Electronics*, McGraw-Hill, New York, 1958.

Millman, J., and C.C. Halkias, *Electron Devices and Circuits*, McGraw-Hill, New York, 1967.

Millman, J., and C.C. Halkias, *Integrated Electronics: Analog and Digital Circuits and Systems*, McGraw-Hill, New York, 1972.

Millman, J., and S. Seely, *Electronics*, 2nd ed., McGraw-Hill, New York, 1951.

Millman, J., and H. Taub, *Pulse, Digital, and Switching Waveforms*, McGraw-Hill, 1965.

Milnes, A.G., *Deep Impurities in Semiconductors*, Wiley-Interscience, New York, 1973.

Milnes, A.G., and D.L. Feucht, *Heterojunctions and Metal Semiconductor Junctions*, Academic Press, New York, 1972.

Mims, F.M., III, *Optoelectronics*, H.W. Sams, Indianapolis, Inc., 1975.

Mitra, S.K. (Ed.), *Active Inductorless Filters*, IEEE Press, New York, 1971.

Mitra, S.K. (Ed.), *Analysis and Synthesis of Linear Active Networks*, Wiley, New York, 1969.

Mlynar, P., Westinghouse High-Voltage Silicon Rectifier Designer's Handbook, Westinghouse Semiconductor Division, Youngwood, Pa.

*Modern Applications of Linear IC's*, by Editorial Staff, United Technical Publications, TAB Books, Blue Ridge Summit, Pa., 1976.

Moll, J., *Physics of Semiconductors*, McGraw-Hill, New York, 1964.

Möltgen, G., *Line Commutated Thyristor Converters*, Heyden & Son, London, England, 1972.

Moore, A.D. (Ed.), *Electrostatics and Its Applications*, Wiley-Interscience, New York, 1973.

Morgan, B.L., R.W. Airey, and D. McMullan (Eds.), *Photo-Electronic Image Devices: Proceedings of the Sixth Symposium*, Imperial College, London, Sept. 1974, Academic Press, New York, 1976.

Morgan, D.V. (Ed.), *Channeling Theory, Observation and Applications*, Wiley-Interscience, New York, 1973.

Morris, R.L., and J.R. Miller, *Designing with TTL Integrated Circuits*, Texas Instruments Electronics Series, McGraw-Hill, New York, 1971.

Mort, J., and D.M. Pai (Ed.), *Photoconductivity and Related Phenomena*, Elsvier, Amsterdam, 1976.

Mortenson, K.E., *Variable Capacitance Diodes*, Artech, Dedham, Mass., 1975.

Mortenson, K.E., and J.M. Borrego, *Design, Performance and Applications of Microwave Semiconductor Control Components*, Artech House, Dedham, Mass., 1972.

Moschytz, G.S., *Linear Integrated Networks; Design*, Van Nostrand Reinhold, New York, 1975.

Moschytz, G.S., *Linear Integrated Networks; Fundamentals*, Van Nostrand Reinhold, New York, 1974.

Moss, T.S., *Optical Properties of Semi-Conductors*, Butterworths Scientific Publications, London, England 1959.

Moss, T.S., G.J. Burrell, and B. Ellis, *Semiconductor Opto-electronics*, Halsted Press Division, Wiley, New York, 1972.

Motorola, Inc. [(R.M. Warner, Jr., and J.N. Fardemwalt (Eds.)] *Integrated Circuits*, McGraw-Hill, New York, 1965.

*Motorola Power Transistor Handbook*, Phoenix, Arizona, 1961.

Mott, N., *Metal-Insulator Transitions*, Harper, Row, Barnes and Noble, New York, 1977.

Mott, N.F., and R.W. Gurney, *Electronic Processes in Ionic Crystals*, Oxford University Press, London and New York, 1940.

Mottershead, A., *Electronic Devices and Circuits: An Introduction*, Goodyear, Pacific Palisades, Calif., 1974.

Motto, J.W., Jr. (Ed.), *Introduction to Solid State Power Electronics*, Westinghouse Electric Corp., Youngwood, Pa., 1977.

Mullard Electronics Staff, *Transistor Audio and Radio Circuits*, 2nd ed., Scholium International, Flushing, New York, 1973.

Müller, R., and E. Lange (Eds.), *Solid State Devices 1976*, Proc. of 6th European Solid-State Device Research Conf., Munich, Inst. of Physics, London, England, 1977.

Muller, R.S., and T.I. Kamins, *Device Electronics for Integrated Circuits*, John Wiley & Sons, New York, 1977.

Murphy, J.M.D., *Thyristor Control of AC Motors*, Pergamon, Oxford and New York, 1973.

Murr, L.E., *Solid State Electronics*, Electrical Engineering and Electronics Series, Vol. 4, Marcel Dekker, Inc., New York, 1978.

Murr, L.E., *Electron Optical Applications in Materials Science*, McGraw-Hill, New York, 1970.

Murt, E.M., and Guldner, W.G. (Eds.), *Physical Measurement and Analysis of Thin Films*, Plenum Press, New York, 1969.

Myamlin, V.A., and Y.V. Pleskov, *Electrochemistry of Semiconductors*, Plenum Publ., New York, 1967.

Nag, B.R., *Theory of Electrical Transport in Semiconductors*, Pergamon, New York, 1972.

Namba, S. (Ed.), *Ion Implantation in Semiconductors: Science and Technology*, (Proceedings of the Fourth International Conference), Plenum Press, New York, 1975.

Nanavati, R.P., *Semiconductor Devices: BJTS, JFETS, MOSFETS and integrated circuits*, Intext Educational Publ., New York, 1975.

Nanavati, R.P., *An Introduction to Semiconductor Electronics*, McGraw-Hill, New York, 1963.

Nashelsky, L., and R. Boylstead, *Electronic Devices and Circuit Theory*, Prentice-Hall, Englewood Cliffs, New Jersey, 1972.

National Semiconductor, *Pressure Transducer Handbook 1977*, Santa Clara, Calif.

Navon, D.H., *Electronic Devices and Materials*, Houghton Mifflin, Boston, Mass., 1975.

Nergaard, L.S., and M. Glicksman (Eds.), *Microwave Solid-State Engineering*, Van Nostrand, New York, 1964.

Neuberger, M.S., *Handbook of Electronic Materials*, Vol. 5, Group IV Semiconducting Compounds, IFI Plenum, New York, 1971.

Neuberger, M., *Handbook of Electronic Materials*, Vol. 7, III-V Ternary Semiconducting Compounds-Data Tables, IFI/Plenum, New York, 1972.

Neville, R.C., *Solar Energy Conversion: The Solar Cell*, Studies in Electrical and Electronic Engineering, Vol. 1, Elsevier Scientific Publishing Co., Amsterdam, 1978.

Newhouse, V.L. (Ed.), *Applied Superconductivity*, Academic Press, New York, 1974.

Newkirk, J.B., and J.H. Wernick (Eds.), *Direct Observation of Imperfections in Crystals*, Wiley-Interscience, New York, 1962.

Newman, R.C., *Infrared Studies of Crystal Defects*, Barnes & Noble, New York, 1973.

Nichols, K.G. and E.V. Vernon, *Transistor Physics*, Halsted Press, New York, 1973.

Norris, B., *Semiconductor Circuit Design*, Vols. 1 and 2, Texas Instruments, Dallas, Texas, Marton Lane, Bedford, England, 1972-1973.

Norris, B., *Power Transistor and TTL Integrated-Circuit Applications*, Texas Instruments, McGraw-Hill, New York, 1977.

Norris, B. (Ed.), *Digital Integrated Circuits and Operational-Amplifier and Optoelectronic Circuit Design*, McGraw-Hill, New York, 1978.

Norris, B. (Ed.), *Electronic Power Control and Digital Techniques*, McGraw-Hill, New York, 1978.

Norris, B. (Ed.), *MOS and Special-Purpose Bipolar Integrated Circuits and R-F Power Transistor Circuit Design*, McGraw-Hill, New York, 1978.

Nosov, Yu. R., *Switching in Semiconductors*, Plenum, New York, 1969.

Nowick, A.S., and Burton, J.J., *Diffusion in Solids: Recent Development*, Academic Press, New York, 1975.

Oberman, R.M., *Electronic Counters*, Barnes and Noble, New York, 1974.

O'Dwyer, J.J., *The Theory of Electrical Conduction and Breakdown in Solid Dielectrics*, Clarendon Press, Oxford, England, 1973.

Oldham, W.G., and S.E. Schwarz, *An Introduction to Electronics*, Holt, Rinehart, Winston, New York, 1972.

Olesen, H.L., *Radiation Effects on Electronic Systems*, Plenum Press, New York, 1966.

Oliner, A.A. (Ed.), *Acoustic Surface Waves*, Springer-Verlag, New York, 1978.

Oppenheimer, S., *Semiconductor Logic and Switching Circuits*, Merrill, Chicago, Ill., 1967.

Osborne, A., *An Introduction to Microcomputers*, Vols. 0-11, Adam Osborne Associates, Box 2036, Berkeley, Calif., 1976-1977

Pamplin, B.R. (Ed.), *Crystal Growth*, Pergamon Press, Oxford, England and Elmsford, New York, 1975.

Pankove, J.I. (Ed.), *Electroluminescence*, Vol. 17, Springer-Verlag, New York, 1977.

Pankove, J.I., *Optical Processes in Semiconductors*, Prentice Hall, Englewood Cliffs, New Jersey, 1971.

Parker, R.L., et al. (Ed.), *Crystal Growth 1977*, North-Holland, New York, 1977.

1975 Particle Accelerator Conference, held at Washington, D.C., March 1975; Proceedings published in: *IEEE Trans. Nucl. Sci.*, NS-22, June 1975.

Paul, R., *Field-effect Transistors: Physical Principles and Properties*, VEB Verlag Technik, Berlin, Germany, 1972, (in German).

Pauling, L., *The Nature of the Chemical Bond*, Cornell University Press, Ithaca, New York, 1960.

Paushkin, Ya M., et al., *Organic Polymeric Semiconductors*, Wiley & Sons, 1974.

Peatman, J.B., *The Design of Digital Systems*, McGraw-Hill, New York, 1971.

Peatman, J.B., *Microcomputer-Based Design*, McGraw-Hill, New York, 1978.

Peiser, H.S., ed., *Proceedings of International Conference on Crystal Growth*, Boston, Mass., Pergamon Press, Oxford, England, 1967.

Pell, E.M., *Proceedings of the 3rd International Conference on Photoconductivity*, Stanford University, 1969, Pergamon, New York, 1971.

Pelly, B.R., *Thyristor Phase-Controlled Converters and Cycloconverters*, Wiley, New York, 1971.

Pendry, J.B., *Low energy electron diffraction*, Academic Press, New York, 1974.

Penfield, P. Jr. and R.P. Rafuse, *Varactor Applications*, M.I.T. Press, Cambridge, 1962.

Penner, S.S. and L. Icerman, *Energy*, Addison-Wesley, Reading, Mass. 1974.

Penney, W.M., and L. Yau, *MOS Integrated Circuits*, Van Nostrand Reinhold, New York, 1972.

Phillips, A.B., *Transistor Engineering and Introduction to Integrated Semiconductor Circuits*, McGraw-Hill, New York, 1962.

Phillips, J.C., *Bonds and Bands in Semiconductors*, Academic Press, New York, 1973.

Phillips, V.A., *Modern Metallographic Techniques and Their Applications*, Wiley-Interscience, New York, 1971.

Picraux, S.T., E.P. EerNisse, and F.L. Vook, (Eds.), *Applications of Ion Beams to Metals*, Plenum Press, New York, 1974.

Pincherle, L., *Electronic Energy Bands in Solids*, Beckman Pubs., New York, 1973.

Planer, G.V., and L.S. Phillips, *Thick Film Circuits*, Crane Russak, New York, 1973.

Platzman, P.M., and P.A. Wolff, *Waves and Interactions in Solid State Plasmas*, Academic Press, New York, 1973.

Poate, J.M., K.N. Tu, and J.W. Mayer (Eds.), *Thin Films—Interdiffusion and Reactions*, Wiley-Interscience, Somerset, New Jersey, 1978.

Powell, C.F., J.H. Oxley, and J.M. Blocher (Eds.), *Vapor Deposition*, Wiley, New York, 1966.

Pridham, G.J., *Solid State Circuits*, Pergamon, New York, 1973.

Pritchard, R.L. *Electrical Characteristics of Transistors*, McGraw-Hill, New York, 1967.

Proceedings of the First National Conference on Crystal Growth and Epitaxy from the Vapor Phase, Held at: Zurich, Switzerland, September 1970, Published in: *J. Crystal Growth*, Vol. 9, May 1971.

Proceedings of the Second International Conference on Vapor Growth and Epitaxy, Held at: Jerusalem, Israel, May 1972, Published in: *J. Crystal Growth*, Vol. 17, December 1972.

Proceedings of the Third International Conference on Vapor Growth and Epitaxy, Held at: Amsterdam, the Netherlands, August 1975, Published in: *J. Crystal Growth*, Vol. 31, December 1975.

*Proceedings of the SPIE National Seminar on Efficient Transmission of Pictorial Information*, Vol. 66, San Diego, Calif., August 21-22, 1975.

*Proceedings of the SPIE National Seminar on Guided Optical Communications*, Vol. 63, edited by F.L. Thiel, San Diego, Calif., August 19-20, 1975.

*Proceedings of the SPIE National Seminar on Long Wavelength*, Vol. 67, San Diego, Calif., August 21-22, 1975.

*Proceedings of the SPIE National Seminar on Modern Utilization of Infrared Technology, Civilian and Military*, Vol. 62, edited by I.J. Spiro, San Diego, Calif., August 19-20, 1975.

Proebster, W.E., ed., *Digital Memory and Storage*, Heyden, Philadelphia, 1978.

Prywes, N.S. (Ed.), *Amplifier and Memory Devices: with Films and Diodes*, McGraw Hill, NY, 1965.

Pulfrey, D.L., *Photovoltaic Power Generation*, Van Nostrand-Reinhold, NY, 1978.

Putley, E.H., *The Hall Effect and Related Phenomena*, Butterworths, London, 1960.

Queisser, H.J. ed., *Festkörper Probleme XV Advances in Solid State Physics*, Pergamon Press, NY, 1975.

Ranney, M.W., *"Microencapsulation technology"*, Noyes Development Corporation, Park Ridge, N.J., 1969.

Ravich, Yu I., B.A. Efimova and I.A. Smirnov, *Semiconducting Lead Chalcogenides*, Plenum, NY, 1970.

*RCA Linear Integrated Circuits*, Radio Corporation of America, Harrison, New Jersey, 1967.

*RCA Photomultiplier Manual*, RCA Technical Series, PT-61, Radio Corporation of America, Harrison, New Jersey, 1970.

*RCA Solid State Power Circuits*, Radio Corporation of America, Harrison, New Jersey, 1971.

Ready, J.F. ed., *Industrial Applications of Lasers*, Academic Press, New York, 1978.

Rebane, K.K., *Impurity Spectra of Solids*, Plenum Press, New York, 1970.

Reed, S.J.B., *Electron Microprobe Analysis*, Cambridge University Press, New York, 1975.

Renwick, W., and A.J. Cole, *Digital Storage Systems*, 2nd ed., Chapman and Hall, London, England 1971.

Rhoderick, E.H., *Metal Semiconductor Contacts*, Oxford University Press, New Jersey, 1978.

Rhodes, R.G., *Imperfections and Active Centers in Semiconductors*, Vol. 6, International Series of Monographs on Semiconductors, MacMillan, New York, 1964.

Richards, C.J., *Electronic Display and Data Systems: Constructional Practice*, McGraw-Hill, New York, 1973.

Richman, P., *Characteristics and Operation of MOS Field-Effect Devices*, McGraw-Hill, New York, 1967.

Richman, P., *MOS Field-Effect Transistors and Integrated Circuits*, Wiley, New York, 1973.

Ricketts, L.W., *Fundamentals of Nuclear Hardening of Electronic Equipment*, Wiley-Interscience, New York, 1972.

Rieck, H., *Semiconductor Lasers*, Beckman Pubs, New York, 1970.

Rikoski, R.A., *Hybrid Microelectronic Circuits*, Wiley-Interscience, New York, 1973.

Riley, W.B., *Electronic Computer Memory Technology*, McGraw-Hill, New York, 1975.

Robbins, M.S., *Electronic Clocks and Watches*, H.W. Sams, Indianapolis, Ind., 1975.

Roberge, J.K., *Operational Amplifiers; Theory and Practice*, Wiley, New York, 1975.

Robinson, F.N.H., *Noise and Fluctuations in Electronic Devices and Circuits*, Oxford University Press, Oxford, 1974.

Robinson, L.C., *Physical Principles of Far-Infrared Radiation*, Vol. 10, Methods of Experimental Physics, Academic Press, New York, 1973.

Robson, P.N. (Ed.), *Solid State Devices, 1972*, (Conference Proceedings of the 2nd European Conference, Sept. 12-15, 1972), Institute of Physics, London, England, 1973.

Roddy, D., *Introduction to Microelectronics (2nd Ed.)*, Pergamon Press, Oxford, England, 1977.

Rosenthal, M.P., *Understanding Integrated Circuits*, Hayden, Rochelle Park, New Jersey, 1975.

Roy, D.K., *Tunnelling and Negative Resistance Phenomena in Semiconductors*, Pergamon, London, England, 1977.

Ruddell, R.L. (Ed.), *Developments in Semiconductor Microlithography III*, SPIE Vol. 135, SPEI Meeting April 10-11, 1978, San Jose, Calif., SPIE, Bellingham, Washington, 1978.

Ruge, I., and J. Graul, (Eds.), *Ion Implantation in Semiconductors*, Springer-Verlag, New York, 1971. (Proceedings of the Second International Conference).

Runyan, W.R., *Silicon Semiconductor Technology*, Texas Instruments Electronic Series, McGraw-Hill, New York, 1965.

Runyan, W.R., *Semiconductor Measurements and Instrumentation*, McGraw-Hill, New York, 1975.

Rutkowski, G.B., *Handbook of Integrated-Circuit Operational Amplifiers*, Prentice-Hall, Englewood Cliffs, New Jersey, 1975.

Rymer, T.B., *Electron Diffraction*, Halsted Press, London, England, 1974.

Ryvkin, S.M., *Photoelectric Effects in Semiconductors* (translated from the Russian by A. Tybulewicz), Consultants Bureau, New York, 1964.

Sachse, H.B., *Semiconducting Temperature Sensors and Their Applications*, Wiley-Interscience, New York, 1975.

Sargent, M. III, M.O. Scully, and W.E. Lamb, Jr., *Laser Physics*, Addison-Wesley, Reading, Mass., 1974.

Sawin, D.H. III, *Microprocessors and Microcomputer Systems*, D.C. Heath, Lexington, Mass., 1977.

Scarlett, J.A., *Transistor:Transistor Logic and Its Interconnections*, Van Nostrand Reinhold, New York, 1972.

Schilling, D.C., *Electronic Circuits: Discrete and Integrated*, McGraw-Hill, New York, 1968.

Schmid, H., *Electronic Analog/Digital Conversions*, Van Nostrand Reinhold, New York, 1970.

Schneider, H.G., V. Ruth, and T. Kormany, *Advances in Epitaxy and Endotaxy, Selected Chemical Problems*, Elsevier Scientific Publ., New York, 1976.

Schroeder, J.B. (Ed.), *Metallurgy of Semiconductor Materials*, Metallurgical Society Conference, Vol. 15, Wiley (Interscience), New York, 1962.

Schwartz, B (Ed.), *Ohmic Contacts to Semiconductors*, Electrochemical Society, New York, 1969.

Schwartz, M. and L. Shaw, *Signal Processing: Discrete Spectral Analysis, Detection and Estimation*, McGraw-Hill, New York, 1975.

Schwartz, S., *Integrated Circuit Technology, Instrumentation, and Techniques for Measurement, Process, and Failure Analysis*, McGraw-Hill, New York, 1967.

Schwartz, S. (Ed.), *Selected Semiconductors Circuits Handbook*, Wiley, New York, 1960.

Schuele, D.E., and R.W. Hoffman (Eds.), *CRC Critical Review in Solid State and Materials Science*, Vols. 1-7, CRC Press, Cleveland, Ohio, 1977-1978.

Scoles, G.J.S., *Handbook of Electronic Circuits; Design, Operation, Applications,* Halsted Press, Wiley, New York, 1975.

Scott, A.W., *Cooling of Electronic Equipment*, Wiley-Interscience, New York, 1974.

Scott, C.G. and C.E. Reed (Eds.), *Surface Physics of Phosphors and Semiconductors*, Academic Press, New York, 1975.

Searle, C.L. ed., *Semiconductor Electronic Education Committee Monographs*, 7V, Wiley, New York, 1967, *Vol. 1 Introduction to Semiconductor Physics, Vol. 2 Physical Electronics and Circuit Models of Transistors, Vol. 3 Elementary Circuit Properties of Transistors, Vol. 4 Characteristics and Limitations of Transistors, Vol. 5 Multistage Transistor Circuits, Vol. 6 Digital Transistor Circuits, Vol. 7 Handbook of Basic Transistor Circuits and Measurements*

Seeger, K., *Semiconductor Physics*, Springer-Verlag, New York, 1974.

Seely, J.H. and Chu, R.C., *"Heat transfer in microelectronic equipment"*, Marcel Dekker, *Inc.*, New York, 1972.

Seippel, R.G. and R.L. Nelson, *Designing Circuits with IC Operational Amplifiers*, American Technical Society, Chicago, 1975.

Seitz, F. and D. Turnbull, eds., *Solid State Physics*, V.8, Academic Press, New York, 1959.

Selvage, C., R. Winston, and R. Schmidt, eds., *Optics in Solar Energy Utilization II*, V. 85, Proceeding of the SPIE meeting August 24-25, 1977, San Diego, CA, Bellingham, Washington, 1978.

Séquin, C.H., and M.F. Tompsett, *Charge Transfer Devices* (Supplement 8 to Advances in Electronics and Electron Physics), Academic Press, New York, 1975.

Seraphin, B.O. (Ed.), *Optical Properties of Solids: New Developments*, North-Holland Publ., Amsterdam, 1976.

Seraphin, B.O. (Ed.), *Solar Energy Materials*, North-Holland Publ., Amsterdam, 1978.

Sessions, K.W. (Ed.), *Master Handbook of 1001 Practical Electronic Circuits*, Tab Books, Blue Ridge Summit, Pa., 1975.

Sessions, K.W., *Discrete/Transistor Circuit Sourcemaster*, John Wiley & Sons, Inc. New York, 1978.

Sessions, K.W., *IC Schematic Sourcemaster*, John Wiley & Sons, Inc., New York, 1978.

Sevin, L.J., *Field-effect Transistors*, McGraw-Hill, New York, 1965.

Seymour, J. (Ed.), *Semiconductor Devices in Power Engineering*, Pitman, New York, 1968.

Sharma, B.L., *Diffusion in Semiconductors*, Trans Tech Publication, Clausthal-Zellerfeld, Germany, 1970.

Sharma, B.L., and R.K. Purohit, *Semiconductor Heterojunctions*, Pergamon Press, Oxford and New York, 1974.

Shaskov, Y.M., *Metallurgy of Semiconductors*, Pittman, New York, 1960.

Shaw, D. (Ed.), *Atomic Diffusion in Semiconductors*, Plenum, London, Pitman, New York, 1960.

Shaw, M.P., H.L. Grubin and P.R. Solomon, *"The Gunn-Hilsum Effect"*, Academic Press, New York, 1979.

Shay, J.L., and J.H. Wernick (Eds.), *Ternary Chalcopyrite Semiconductors, Growth, Electronic Properties, and Applications*, Pergamon Press, Oxford and New York, 1974.

Shea, R.F., *Transistor Applications*, Wiley, New York, 1964.

Sheflol, N.N. (Ed.), *Growth of Crystals*, Vol. 8, Plenum Press, New York, 1969.

Shive, J.N., *Semiconductor Devices*, Van Nostrand, Princeton, New Jersey, 1959.

Shockley, W., *Electrons and Holes in Semiconductors*, Van Nostrand-Reinhold, New York, 1950.

Shurmer, H.V., *Microwave Semiconductor Devices*, Electrical Engineering Series, Halsted Press, New York, 1971.

Shwop, J.E., and H.J. Sullivan, *Semiconductor Reliability*, Engineering Publishers, Elizabeth, New Jersey (Reinhold), 1961.

Sibberlen, T.P., and V. Vartanian, *Digital Electronics with Engineering Applications*, Prentice-Hall, Englewood Cliffs, New Jersey, 1970.

Siddall, G., and J.N. Zemel (Eds.), *Thin Solid Films*, Vol. 1, Elsevier Sequoia S.A., Lausanne, Switzerland 1967.

Sideris, G., *Microelectronic Packaging Interconnection and Assembly of Integrated Circuits*, McGraw-Hill, New York, 1968.

Siliconix, Inc., *An Introduction to FETs*, Application Note AN73-7, Dec. 1973. Siliconix, Inc., 2201 Laurelwood Rd., Santa Clara, Calif., 95054.

Sittig, M., *Semiconductor Crystal Manufacture*, Noyes Development Corporation, New Jersey, 1969.

Smith, B., *Ion Implantation Range Data for Silicon and Germanium Device Technologies*, Research Studies Press, Forest Grove, Oregon, 1978.

Smith, P.H., *Electronic Applications of the Smith Chart: In Waveguide, Circuit, and Component Analysis*, McGraw-Hill, New York, 1969.

Smith, R.A., *Semiconductors*, Cambridge University Press, London and New York, 1959.

Smith, R.A., *Wave Mechanics of Crystalline Solids*, Chapman & Hall, London, England, 1961.

Smith, R.J., *Electronics: Circuits and Devices*, Wiley, New York, 1973.

Smith, R.E., *Circuits, Devices and Systems*, Wiley, New York, 1976.

Society for Information Display, 16th International Symposium and Seminar, Digest of Technical Papers, Washington, D.C., April 1975.

Society Photo-Optical Instrumentation Engineers, *Semiconductor Microlithography II*, Vol. 80, SPIE, Box 1146, Palos Verdes Estates, Calif. 90274.

Society Photo-Optical Instrumentation Engineers, *Technological Advances in Micro and Submicro Photo Fabrication Imagery*, Vol. 55, SPIE, Box 1146, Palos Verdes Estates, Calif. 90274.

Society of Photo-Optical Instrumentation Engineers, *Solid State Imaging Devices*, Vol.

116, Proceedings of the SPIE National Seminar meeting, August 23-24, 1977, San Diego, Calif., Bellingham, Washington, 1978.

*Solid State Physics* (Springer tracts in modern physics), Springer-Verlag, Berlin and New York, 1974.

Sommer, A.H., *Photoemissive Materials*, Wiley, New York, 1968.

Sorkin, R.B., *Integrated Electronics*, McGraw-Hill, 1970.

Soucek, B., *Microprocessors and Microcomputers*, Wiley-Interscience, New York, 1976.

Spangenberg, K.R., *Fundamentals of Electron Devices* (Electrical and Electronic Engineering Series), McGraw-Hill, New York, 1957.

Sparkes, J.J., *Transistor Switching and Sequential Circuits*, Pergamon, New York, 1969.

SPE Symposium on Photopolymers: Principles, Processes and Materials, Society of Plastics Engineers, Mid-Hudson Section, 1967, 1970, 1973.

Spenke, E., *Electronic Semiconductors*, McGraw-Hill, New York, 1958.

Spiegel, M.R., *Laplace Transforms: Including 225 Solved Problems* (Schaum's Outline Series), McGraw-Hill, New York, 1967.

Spilker, J.J., Jr., *Digital Communications by Satellite*, Prentice-Hall, Englewood Cliffs, New Jersey, 1975.

Stern, F., *Evidence for a Mobility Edge in Inversion Layers*, Joint Solid State Seminar, IBM Thomas J. Watson Research Center, Yorktown Heights, New York.

Stern, L., *Fundamentals of Integrated Circuits*, Hayden, New York, 1968.

Stockman, H.E., *Transistor & Diode Laboratory Course*, Hayden, New York, 1969.

Stoneham, A.M., *Theory of Defects in Solids: Electronic Structure of Defects in Insulators and Semiconductors*, Oxford University Press, London, 1975.

Stout, D.F., *Handbook of Operational Amplifier Circuit Design* edited by M. Kaufman, McGraw-Hill, New York, 1978.

Streetman, B.G., *Solid State Electronic Devices*, Prentice-Hall, Englewood Cliffs, N.J., 1972.

Strutt, M.J.O., *Semiconductor Devices*, Academic Press, New York, 1966.

Stuke, J., and W. Brenig (Eds.), *Amorphous and Liquid Semiconductors*, Conference Proceedings, Halsted Press (Wiley), New York, 1974.

Svoboda, A., and D.E. White, *Advanced Logical Circuit Design Techniques*, Garland STPM Press, New York, 1978.

13th Symposium on Electron, Ion, and Photo Beam Technology, Proceedings were published in: *J. Vac. Sci. Technol.*, Vol. 12(6), December 1975.

Symposium on Ion Sources and Formation of Ion Beams, Conference held at Upton, New York, Oct. 1971; Proceedings published by: AIP and BNL, New York, 1971.

Sze, S.M., *Physics of Semiconductors*, Wiley-Interscience, New York, 1969.

Szilard, J., *Sealing and Potting Compounds*, Noyes Press, Park Ridge, N.J., 1972.

Tallan, N.M. (Ed.), *Electrical Conductivity on Ceramics and Glass*, Pt. B, Marcel Dekker, New York, 1974.

Talley, H.E., and D.G. Daugherty, *Physical Principles of Semiconductor Devices*, Iowa State University Press, Ames, Iowa, 1976.

Taub, H., and D. Schilling, *Digital Integrated Electronics*, McGraw-Hill, New York, 1977.

Tauc, J. (Ed.), *Amorphous and Liquid Semiconductors*, Plenum, New York, 1974.

Tauc, J., *Photo and Thermoelectric Effects Semiconductors*, Pergamon Press, Oxford, England, 1962.

Taylor, J.M., *Semiconductor Particle Detectors*, Butterworths, Washington, D.C., 1963.

Tegart, W.J., McG., *The Electrolytic and Chemical Polishing of Metals in Research and Industry*, Pergamon Press, New York, 1959.

Terman, F.E., *Electronic and Radio Engineering*, 4th ed., McGraw-Hill, New York, 1955.

Terman, F.E., and J.M. Pettit, *Electronic Measurements*, 2nd ed., McGraw-Hill, New York, 1952.

Texas Instruments, Inc., *Understanding Solid-State Electronics*, Texas Instruments Learning Center, Dallas, Texas, 1972.

Texas Instruments, Inc., *DC-DC Germanium Power Converters*, Application Notes, August 1961.

Texas Instruments, Inc., *Transistor Circuit Design*, ed. by J. Miller, McGraw-Hill, New York, 1963.

Texas Instruments, Inc., *Solid-State Communications*, McGraw-Hill, New York, 1966.

Texas Instruments, Inc., *Circuit Design for Audio, AM/FM and TV*, ed. by W.A. Stover, McGraw-Hill, New York, 1967.

Texas Instruments, Inc., *Designing with TTL Integrated Circuits*, McGraw-Hill, New York, 1971.

Texas Instruments, Inc., *The Integrated Circuits Catalog for Design Engineers*, Dallas, Texas, 1971.

Tharma, P., *Transistor Audio Amplifiers*, Van Nostrand Reinhold, New York, 1971.

Thomas, D.G. (Ed.), *II-VI Semiconducting Compounds*, 1967 International Conference, Brown University, W.A. Benjamin, New York and Amsterdam, 1967.

Thomas, H.E., *Handbook of Integrated Circuits*, Prentice Hall, Englewood Cliffs, New Jersey, 1971.

Thorik, Yu. A., *Transients in Pulsed Semiconductor Diodes*, International Scholarly Book Services, Inc., Forest Grove, Oregon.

Thorton, P.R., *The Physics of Electroluminescent Devices*, Halsted Press (Wiley), New York, 1967.

Thornton, R.D., et al., *Characteristics and Limitations of the Transistors*, Vol. 4, Semiconductor Electronics Education Committee, Wiley, New York, 1966.

Thornton, R.D., C.L. Searle, D.O. Pederson, R.B. Adler, and E.J. Angelo, Jr., *Multistage Transistor Circuits*, SEEC Series, Vol. 5, Wiley, New York, 1965.

Tickle, A.C., *Thin-Film Transistors, a New Approach to Microelectronics*, Wiley & Sons, 1969.

Tobey, G.E., L.P. Huelsman, and G.G. Graeme, *Operational Amplifiers: Design and Application*, McGraw-Hill, New York, 1971.

Todd, C.D., *Junction Field-Effect Transistors*, Wiley & Sons, New York, 1968.

Townsend, P.D., and J.C. Kelly, *Colour Centres and Imperfections in Insulators and Semiconductors*, Crane, Russak, New York, 1973.

Tremaine, H.M., *Audio Cyclopedia*, 2nd ed., H.W. Sams, Indianapolis, Ind., 1969.

Treusch, J. (Ed.), *Advances in Solid State Physics*, Vol. XVI, Heyden, Philadelphia, Pa. 1976.

Tsidel'kovskii, I.M., *Thermomagnetic Effects in Semiconductors*, Academic Press, New York, 1962.

Tuck, B., *Introduction to Diffusion in Semiconductors*, IEE Monograph Series 16, Peter Peregrinus Ltd., Stevenage, England, 1974.

Tucker, D.G., *Circuits with Periodically-Varying Parameters*, Van Nostrand, New York, 1964.

Turner, R.F., *Solar Cells and Photocells*, H.W. Sams, Indianapolis, Ind., 1975.

Ueda, R., and J.B. Mullin (Eds.), *Crystal Growth and Characterization: Proceedings of the ISSGG2*, Japan 1974, North-Holland Publ., Amsterdam, 1976.

Uman, M.F., *Introduction to the Physics of Electronics*, Prentice-Hall, Englewood Cliffs, New Jersey, 1974.

U.S.-Japan Seminar on Ion Implantation in Semiconductors, Conference held at Kyoto, Japan, Aug. 1971; Proceedings published by: Japan Society for Promotion of Science, Tokyo, Japan, 1972.

Uzunoglu, V., *Semiconductor Network Analysis and Design*, McGraw-Hill, New York, 1964.

Valdes, L.B., *The Physical Theory of Transistors*, McGraw-Hill, New York, 1961.

VanCleemput, W.M., *Computer Aided Design of Digital Systems: a Bibliography*, Computer Science Press, Fall River Lane, Potomac, Maryland, 1976-1977.

Van der Ziel, A., *Electronics*, Allyn and Bacon, Boston, Mass., 1966.

Van der Ziel, A., *Noise*, Prentice-Hall, Englewood Cliffs, New Jersey, 1954.

Van der Ziel, A., *Solid State Electronics*, 2nd ed., Prentice-Hall, Englewood Cliffs, New Jersey, 1968.

Vavilov, V.S., *Effects of Radiation on Semiconductors*, Plenum Publ., New York, 1965.

Vavilov, V.S. and N.A. Ukuim, *Radiation Effects in Semiconductors and Semiconductor Devices*, Plenum, New York, 1976.

Vasicek, A., *Optics of Thin Films*, North-Holland Publ., Amsterdam, 1960.

Veinott, C.G., *Computer-Aided Design of Electronic Machinery*, M.I.T. Press, Cambridge, Mass., 1972.

Veronis, A.M., *Microprocessors: Design and Applications*, Prentice-Hall, Englewood Cliffs, New Jersey, 1978.

Vigdorovich, V.N., *Purification of Metals and Semiconductors by Crystallization*, Freund Publishing House. TelAviv, Israel, 1978.

Vijh, A.K., *Electrochemistry of Metals and Semiconductors*, The Application of Solid State Science to Electrochemical Phenomena, Marcel Dekker, New York, 1973.

Vishnyakov, B.A. and K.A. Osipov, *Production of Thin Films of Chemical Compounds by Electron Beam Bombardment*, Nauka, Moscow, U.S.S.R., 1970.

Vol'kershtein, F.F., *The Electronic Theory of Catalysis on Semiconductors*, Pergamon Press, London, England, 1963.

Vratny, F. (Ed.), *Thin Film Dielectrics*, The Electrochemical Society, Inc., New York, 1969.

Wait, J.V., L.P. Huelsman, and G.A. Korn, *Introduction to Operational Amplifier Theory and Applications*, McGraw-Hill, New York, 1975.

Wakerly, J., *Logic Design Projects Using Standard Integrated Circuits*, John Wiley & Sons, New York, 1976.

Wallmark, J.T., and H. Johnson, *Field Effect Transistors*, Prentice-Hall, Englewood Cliffs, New Jersey, 1966.

Wallmark, J.T., and L.G. Carlstedt, *Field-Effect Transistors in Integrated Circuits*, John Wiley & Sons, New York, 1974.

Wang, S. *Solid-State Electronics*, McGraw-Hill, New York, 1966.

Warner, R.M., and J.N. Fordemwalt, *Integrated Circuits-Design Principles and Fabrication*, McGraw-Hill, New York, 1965.

Watson, B., *Audio-frequency Noise Characteristics of Junction FETs*, Application Note AN74-4, Siliconix Inc., 2201 Laurelwood Rd., Santa Clara, Calif., 95054.

Watson, H.A. (Ed.), *Microwave Semiconductor Devices and Their Circuit Applications*, McGraw-Hill, New York, 1969.

Watson, J., *Semiconductor Circuit Design 3rd Ed. for a.c. and d.c. Amplification and Switching*, Halsted Press, Wiley-Interscience, New York.

Watson, J., *Semiconductor Circuit Design for a.f. and d.c. Amplification and Switching*, Adam Hilger Ltd., Bristol, England, 1978.

Weber, S. (Ed.), *Circuits for Electronics Engineers*, McGraw-Hill, New York, 1978.

Weber, S. *Large and Medium Scale Integration*, McGraw-Hill, New York, 1974.

Weiss, H., *Structure and Application of Galvanomagnetic Devices*, Pergamon Press, Oxford, England, 1967.

Welford, W.T., and R. Winston, *The Optics of Nonimaging Concentrators, Light and Solar Energy*, Academic Press, New York, 1978.

Wells, O.C., *Scanning Electron Microscopy*, McGraw-Hill, New York, 1974.

Wertheim, G.K., A. Hausmann, and W. Sander, *The Electronic Structure of Point Defects* (As Determined by Mossbauer Spectroscopy and By Spin Resonance), North-Holland Publ., Amsterdam, 1971.

Wetterau, L.C., Jr., *Packaging Opto-Electronic Devices*, Semiconductor Integrated Circuit Processing and Production Conference, Anaheim, Calif., Feb. 9-11, 1971, Industrial and Scientific Conference Management, 222W. Adams St., Chicago, Ill. 60606.

Weyrick, R.C., *Fundamentals of Automatic Control*, McGraw-Hill, New York, 1975.

White, J., *Semiconductor Control*, Artech House, Dedham, Mass., 1977.

Whitehouse, J.E. (Ed.), *Radiation Damage and Defects in Semiconductors*, Institute of Physics, London, 1973.

Wickes, W.E., *Logic Design with Integrated Circuits*, Wiley-Interscience, New York, 1968.

Wieder, H.H., *Intermetallic Semiconducting Films*, Pergamon Press, Oxford, England, 1970.

Wieder, H.H., *Hall Generators and Magnetoresistors*, Academic Press, New York and London, 1971.

Willardson, R.K., and A.C. Beer (Eds.), *Semiconductors and Semimetals*, Vol 1, *Physics of III-V Compounds*, 1966; Vol. 2, *Physics of III-V Compounds*, 1966; Vol. 3, *Optical Properties of III-V Compounds*, 1967; Vol. 4, *Physics of III-V Compounds*, 1968; Vol. 5, *Infrared Detectors*, 1970; Vol. 6, *Injection Phenomena*, 1970; Vol. 7, *Applications and Devices*, 1971; Vol. 8, *Transport and Optical Phenomena*, 1972; Vol. 9, *Modulation Techniques*, 1972; Vol. 10, *Transport Phenomena*, 1975; Vol. 11, *Solar Cells*, 1975; Vol. 12, *Infrared Detectors II*, 1977; Vol. 13, *Cadmium Telluride*, 1978; Academic Press, New York.

Williams, E.L., *Liquid Crystals for Electronic Devices*, Noyes Data Corp Park Ridge, New Jersey, 1975.

Williams, E.W. (Ed.), *Solar Cells*, IEE Special Publication, IEE, Herts., England, 1978.

Williams, E.W., and R. Hall, *Luminescence and the Light Emitting Diode*, International Series in the Science of the Solid State, Vol. 13, Pergamon Press, Elmsford, England, 1978.

Wilson, R.G., *Ion Mass Spectra*, Wiley-Interscience, New York, 1974.

Wilson, R.G., and G.R. Brewer, *Ion Beams with Applications to Ion Implantation*, Wiley & Sons, New York, 1973.

Winston, R., and A.L. Mlavsky (Eds.), *Optics Applied to Solar Energy Conversion*, Vol. 114, Proceedings of the SPIE meeting Aug. 23-24, 1977, San Diego, Calif., SPIE, Bellingham, Washington, 1978.

Wissernan, L., and J.J. Robertson, *High Performance Integrated Operational Amplifiers*, Motorola Semiconductor Products, Inc., Application Note AN-204.

Wöhlbier, F.H. (Ed.), *Diffusion and Defect Data*, (Materials Reference Series 1), Vol. 10, Nos. 1-4, Trans Tech House, Bay Village, Ohio, 1975.

Wolf, H.F., *Silicon Semiconductor Data*, Pergamon, New York, 1969.

Wolf, H.F., *Semiconductors*, Interscience, New York, 1971.

Wolfe, R. (Ed.), *Applied Solid State Science: Advances in Materials and Device Research*, Academic Press, New York, 1975.

Wolkenstein, T., and K. Hauffe (Eds.), *Symposium on Electronic Phenomena in Chemisorption and Catalysis on Semiconductors*, De Gruyter, New York, 1969.

Wooten, F., *Optical Properties of Solids*, Academic, New York, 1972.

Wright, H.C., *Infrared Techniques*, Clarendon Press, Oxford, England, 1973.

Young, L., and H. Sobol (Eds.), *Advances in Microwaves*, Vol. 8, Academic Press, New York, 1974.

Zaks, R., *Microprocessors: from Chips to Systems*, Sybex, Inc., 2161 Shattuck Ave., Berkeley, Calif., 94704.

Zeines, B., *Transistor Circuit Analysis and Application*, Reston, Reston, Va., 1976.

Zingaro, R.A., and W.C. Cooper (Eds.), *Selenium*, Van Nostrand Reinhold, New York, 1974.

Zworykin, V.K., and E.G. Ramberg, *Photoelectricity and Its Applications*, Wiley, New York, 1949.

# SUBJECT INDEX

Abrupt junctions, 8, 43, 79, 116, 139, 172, 205, 264, 371, 508, 563, 722
Absorption, photons, 719, 729, 738, 767, 826
ac drift field, 647
ac motors, 330, 964
ac parameters, 520
ac to dc rectifiers, 58
Accelerometer, 451
Acceptor, 49, 51, 80, 382, 411, 723, 769, 771, 783, 826, 829
Access time, memory, 554, 559, 566, 568, 572, 617, 620
Accumulation, carrier, 205, 631, 679, 725
Acoustic waves, 81, 132, 610, 807, 940, 965
Acquisition time, 526, 535
Activation energy, 440, 566, 718, 771, 791
Activators, 836
Active filters, 526, 544, 546, 583
Active loads, 261
Actuators, 553
a/d conversion, 527, 532-537, 575, 586, 612, 957
Adaptive electronics, 953
Adders, 573, 689, 698, 700
Address functions, 561, 570, 807
Adhesion, 729
Admittance, 110, 133, 700, 702, 757
Ag (silver), 133, 782, 786, 787
Ag-Bi-O-Cs, 782
Ag/GaP, 163
Ag-O-Rb, 782
Ag/GaS, 136
Ag-O-Cs, 779, 782, 783

AGC, (automatic gain control), 537
Agfa-Gevaert camera, 552
Air-mass-one (AM1), 705, 732, 740
Air-mass-two (AM2), spectrum, 705
Air-mass-zero (AM0), spectrum, 705
Al (aluminum), 2, 82, 91, 98, 120, 132, 136, 330, 387, 405, 412, 425, 733, 739, 770, 807
AlAs, 736, 749
AlGaAs, 83, 131, 270, 374, 733, 736, 809, 833, 843, 845, 862, 885, 887
AlGaP, 889
$Al_2O_3$ (aluminum oxide), 416, 464, 535, 840
Alfvén waves, 947
Alignment, 516, 557
Alkali metals, 779
Alpha, 286, 331, 487, 935
Aluminum isopropoxide, 416
AM rejection, 537
Ambipolar diffusion constant, 767
Amorphous semiconductors, 945, 948, 961, 970
Amplification, 69, 74, 159, 166, 227, 277, 334, 343, 359, 365, 372, 423, 459, 521, 524, 529, 535, 551, 586, 615, 668, 674, 696, 701, 755-757, 941-950, 955-959, 964, 971
  broadband, 359, 373
  chopper type, 354
  computer aided design, 589
  differential, 355, 376, 540, 613, 777, 953
  electron beam, 82

975